Ecological Studies, Vol. 204

Analysis and Synthesis

Ecological Studies

Further volumes can be found at springer.com

Gil Rilov • Jeffrey A. Crooks
Editors

Biological Invasions in Marine Ecosystems

Ecological, Management, and Geographic Perspectives

Springer

Editors
Dr. Gil Rilov
Department of Zoology
Cordley 3029
Oregon State University
Corvallis, OR 97331-2914,
USA

Dr. Jeffrey A. Crooks
Tijuana River National Estuarine Research
Reserve and the Southwest Wetlands
Interpretive Association
301 Caspian Way
Imperial Beach
CA 91932,
USA

ISBN: 978-3-540-79235-2 e-ISBN: 978-3-540-79236-9

Ecological Studies ISSN 0070-8356

Library of Congress Control Number: 2008929900

Cover illustration: Vectors that have transported marine species around the globe have been operating for millennia, but the diversity and intensity of vector activity has increased dramatically in recent years. The drawing on the book cover depicts the change in the types of one particularly important vector, oceangoing vessels, over this time frame. In the first millennium, large canoes were used by the Polynesian people to travel around the Pacific and occupy new lands, and the Vikings crossed the Atlantic in longships. These were replaced by large sailing ships that traversed the oceans during most of the second millennium, only to be replaced by huge freighters and tankers that dominate the seas today. With this change, the vectors changed in size and mode of transfer of marine invaders from fouling only, to fouling and dry ballast, to fouling and ballast waters. With the change in the size of vessels and the number of routes and voyages as well as the proliferation of other invasion vectors such as canals, fisheries and trade in pets, bait and seafood, the number of invasive species has dramatically increased. The graph represents this increase, from just a few species such as woodborrowing shipworms (*Teredo spp*) to a multitude of species that take advantage of the many ways in which marine species now move around the world (Design Gil Rilov).

Cover design: WMXDesign GmbH, Heidelberg, Germany

Printed on acid-free paper

9 8 7 6 5 4 3 2 1

springer.com

Preface

The timing of the publication of this book couldn't be better as we celebrate the 50th anniversary of Charles Elton's seminal book, *The Ecology of Invasions by Animals and Plants*. Since this influential book was published in 1958, the study of bioinvasions has developed exponentially, alongside the exponential growth in the magnitude of the invasion problem itself. Today, bioinvasion, a highly complex ecological process and environmental concern, has become a specific branch in ecology and environmental studies, with many disciplines developing within it.

In early 2004 one of us (G.R.) received a letter from Springer Publishers asking to consider writing or editing a book on marine bioinvasions for their Ecological Studies Series. Springer's editors have been thinking of such a book for quite a while — and rightfully so. Over the past several decades, many books have been written on management, evolutionary or ecological perspectives of bioinvasions, but those books have focused mostly on terrestrial invasions with only a few marine examples. Research on marine bioinvasions has been mounting in the last two decades; a special biannual international symposium on the topic was established in 1999, signifying the fact that it has become a discipline in its own right, but no comprehensive marine bioinvasions book existed at that time. Today, four years later, this is still the first of its kind.

The challenge was exciting and, because the book's purpose was to reflect the full breadth of this fascinating topic, it warranted the inclusion of knowledge of many expert contributors. Dr. Jeff Crooks joined in as a co-editor, and we launched on this long journey together. It was a perfect match: both of us have worked on invasive mussels and their interactions with the native communities, one (G.R.) in one of the hottest hotspots of marine bioinvasions on earth, the eastern Mediterranean, and the other (J.C.) in one of the hotspots on the West Coast of North America, Southern California.

Together, we envisioned a book that will focus mainly on the ecological aspects of the invasion process in the marine environment, which in many ways is quite different to the terrestrial one. With the help of Prof. Jim Carlton, we crafted an outline for a book that would depict the different stages of the invasion process, discuss management issues, and present illustrative case studies from different regions of the world. Identifying prospective contributors was the easy part of the task: many excellent experts could address these different processes. The big question was would these very prolific, busy, and active researchers be willing and able to contribute to this effort? The response was heartening: when we presented the idea to our prospective contributors, the excitement and willingness to commit were overwhelming. Within ten days, we had a list of chapter authors that would cover most of the topics we envisioned. This rapid and positive response reemphasized the great thirst for such a book.

One of the challenges in producing this book is that the topic is so "hot" and rapidly evolving that it is hard to keep up with the new science. This is for two main reasons. The first is that new invasions are being discovered weekly in many regions, and the second is that new basic research is continuously being produced on new, as well as older, invasions. With every iteration of the book's production, new material seemed appropriate for inclusion and we were bound to draw the line somewhere. Yet this is what made this project so exciting. It became clear that it is impossible to include everything we might have wanted in this one book. Luckily, there are other excellent books on the general theory of bioinvasions that cover many of the topics of which we could only scratch the surface. Many are mentioned in the introductory chapter.

The dedicated contributing authors of this book made great efforts to keep the content as updated and relevant as possible. This enables our book, we believe, to serve as an up-to-date learning tool for students and a valuable asset for professionals in the field. Editing such an encompassing book wasn't always an easy assignment, and quite challenging at times; still, we have thoroughly enjoyed this creative work. We are grateful to Springer Publishers for their support during this process, and mostly we thank our devoted chapter authors without which this book would not exist. We also thank our patient families, Iris, Shai and Shir Rilov, and Emma, Kai, and Aiden Crooks, for allowing us many hours of seclusion working on the book. We hope that *Biological Invasions in Marine Ecosystems: Ecological, Management, and Geographic Perspectives* will serve its purpose and broaden the knowledge and understanding of this important topic.

June 2008

Gil Rilov, USA
Jeff Crooks, USA

Contents

Section II Invader Arrival

Section IV Invader Integration into Ecosystems

Section V Management Perspectives

Contributors

Boris Alexandrov
Institute of Biology of the Southern Seas, National Academy of Sciences of Ukraine, Odessa Branch, Odessa, Ukraine
obibss@paco.net

Matthew E. S. Bracken
Northeastern University, Marine Science Center, 430 Nahant Road, Nahant, MA 01908, USA
m.bracken@neu.edu

James E. Byers
Department of Zoology, University of New Hampshire, 46 College Rd., Durham, NH 03824, USA
jeb.byers@unh.edu

Jarrett Byrnes
UC Davis Bodega Marine Laboratory, PO Box 247, Bodega Bay, CA 94923, USA
jebyrnes@ucdavis.edu

Marnie Campbell
Department of Conservation and Ecology, National Centre for Marine Conservation & Resource Sustainability, Australian Maritime College, Locked Bag 1370, Newnham Tasmania 7250, Australia
m.campbell@amc.edu.au

James T. Carlton
Maritime Studies Program, Williams College-Mystic Seaport, PO Box 6000, Mystic, CT 06355, USA
James.T.Carlton@williams.edu

Juan Carlos Castilla
Departamento de Ecología y Centro de Estudios Avanzados en Ecología y Biodiversidad. Faltad de Ciencias Biológicas. Pontificia Universidad Católica de Chile. Santiago, Chile
jcastilla@bio.puc.cl

Graeme F Clark
Centre for Evolution and Ecology, School of Biological, Earth and Environmental Sciences, University of New South Wales, Sydney 2052, NSW, Australia

Andrew N. Cohen
San Francisco Estuary Institute, 7770 Pardee Lane, Oakland, CA 94621, USA
acohen@sfei.org

Jeffrey A. Crooks
Tijuana River National Estuarine Research Reserve and the Southwest Wetlands Interpretive Association, 301 Caspian Way, Imperial Beach, CA 91932, USA
jcrooks@tijuanaestuary.org

Megan N. Dethier
Friday Harbor Laboratories, 620 University Road, University of Washington, Friday Harbor, WA 98250, USA
mdethier@u.washington.edu

Carlos Eduardo Leite Ferreira
Departamento de Biologia Marinha, Universidade Federal Fluminense, Niterói-RJ, 24001-970, Brazil
cadu@vm.uff.br

Paul W. Fofonoff
Smithsonian Environmental Research Center, PO Box 28, Edgewater, MD 21037, USA
fofonoffp@si.edu

Bella Galil
National Institute of Oceanography, Israel Oceanographic and Limnological Research, P.O.B. 8030, Haifa 31081Israel
bella@ocean.org.il

Stephan Gollasch
GoConsult, Grosse Brunnenstr. 61, 22763 Hamburg, Germany
sgollasch@aol.com

Charles Griffiths
Zoology Department, University of Cape Town, Private Bag X3, Rondebosch 7701, South Africa
Charles.Griffiths@uct.ac.za

Edwin D. Grosholz
Department of Environmental Science and Policy, One Shields Avenue, University of California Davis, Davis, CA 95616, USA
tedgrosholz@ucdavis.edu

Sally D. Hacker
Department of Zoology, 3029 Cordley Hall, Oregon State University, Corvallis, OR 97331, USA
hackers@science.oregonstate.edu

Deniz Haydar
University of Groningen, Department of Marine Benthic Ecology and Evolution, PO Box 14, 9750 AA Haren, The Netherlands
d.haydar@rug.nl

Barbara Hayden
National Institute of Water & Atmospheric Research Ltd (NIWA), PO Box 8602, Christchurch 8440, New Zealand
b.hayden@niwa.co.nz

Keith R. Hayes
CSIRO Division of Mathematical and Information Sciences, GPO Box 1538, Hobart Tasmania, 7001, Australia
keith.hayes@csiro.au

Chad L. Hewitt
Australian Maritime College, Locked Bag 1399, Launceston, Tasmania 7250, Australia
chewitt@amc.edu.au

Anson H. Hines
Smithsonian Environmental Research Center, PO Box 28, Edgewater, MD 21037, USA
hinesa@si.edu

Randall Hughes
UC Davis Bodega Marine Laboratory, PO Box 247, Bodega Bay, CA 94923, USA
arhughes@ucdavis.edu

Kristin Hultgren
Smithsonian Tropical Research Institute, Unit 0948, APO AA 34002-0948, USA
hultgrenk@si.edu

Graeme Inglis
National Institute of Water & Atmospheric Research Ltd (NIWA), PO Box 8602, Christchurch 8440, New Zealand
g.inglis@niwa.co.nz

Emma L Johnston
Centre for Evolution and Ecology, School of Biological, Earth and Environmental Sciences, University of New South Wales, Sydney 2052, NSW Australia,
e.johnston@unsw.edu.au

Andrea de Oliveira Ribeiro Junqueira
Instituto de Biologia, Universidade Federal do Rio de Janeiro, Rio de Janeiro-RJ, 21949-900, Brazil
ajunq@biologia.ufrj.br

Kevin D. Lafferty
U.S. Geological Survey, Western Ecological Research Center, c/o Marine Science Institute, University of California, Santa Barbara, CA 93106, USA
Lafferty@lifesci.ucsb.edu

Yoon Lee
Environment Research Department, National Fisheries Research and Development Institute, Ministry for Food, Agriculture, Forestry and Fisheries, 152-1, Haean-ro, Gijanag-eup, Gujang-gun
yoonlee@nfrdi.go.kr

Erkki Leppäkoski
Åbo Akdademi University, Environmental and Marine Biology, 20500 Turku, Finland
eleppako@abo.fi

Mark A. Lewis
Centre for Mathematical Biology, Department of Biological Sciences and Department of Mathematical & Statistical Sciences, University of Alberta, Edmonton, Alberta, Canada T6G 2G1
lewis@math.ualberta.ca

Steve I. Lonhart
Monterey Bay National Marine Sanctuary, NOAA and Institute of Marine Sciences, University of California 110 Shaffer Road, Santz Cruz, CA 95060, USA
Steve.Lonhart@noaa.gov

Rubens Mendes Lopes
Instituto Oceanográfico, Universidade de São Paulo, São Paulo-SP, 05508-900, Brazil
rubens@usp.br

A. Whitman Miller
Smithsonian Environmental Research Center, PO Box 28, Edgewater, MD 21037, USA
millerw@si.edu

Dan Minchin
Marine Organism Investigations, 3 Marina Village, Ballina, Killaloe, Co Clare, Ireland
moiireland@yahoo.ie

Paula Emily Neill
Universidad Católica de la Santísima Concepción, Facultad de Ciencias, Departamento de Ecología Costera, Casilla 297, Concepción, Chile
pneill@ucsc.cl

Sergej Olenin
Coastal Research and Planning Institute, Klaipeda University, H. Manto 84, 92294 Klaipeda, Lithuania
sergej@corpi.ku.lt

Suzanne Olyarnik
UC Davis Bodega Marine Laboratory, PO Box 247, Bodega Bay, CA 94923, USA
svolyarnik@ucdavis.edu

Richard F Piola
Centre for Evolution and Ecology, School of Biological, Earth and Environmental Sciences, University of New South Wales, Sydney 2052, NSW Australia
richard.piola@cawthron.org.nz

Rikke Kvist Preisler
University of California, Santa Cruz, Center for Ocean Health, 100 Shaffer Rd, Santa Cruz, CA 95060, USA
preisler@biology.ucsc.edu

Karsten Reise
Alfred Wegener Institute for Polar and Marine Research, Wadden Sea Station Sylt, 25992 List, Germany
kreise@awi-bremerhaven.de

Gil Rilov
Department of Zoology, Cordley 3029, Oregon State University Corvallis, OR 97331-2914, USA
rilovg@science.oregonstate.edu

Tammy B. Robinson
Zoology Department, University of Cape Town, Private Bag X3; Rondebosch 7701, South Africa

Gregory M. Ruiz
Smithsonian Environmental Research Center, PO Box 28, Edgewater, MD 21037, USA
ruizg@si.edu

David Schiel
School of Biological Sciences, Canterbury University, Private Bag 4800, Christchurch 8140, New Zealand
david.schiel@canterbury.ac.nz

Kyung Suk Seo
Marine Resources Team, Korea Institute of Marine Science and Technology Promotion 275-6 Yangjae-Dong, Seocho-ku, Seoul 137-941, Republic of Korea
dino_seo@hanmail.net

Tamara Shiganova
P.P. Shirshov Institute of Oceanology, Russian Academy of Sciences, Moscow, Russian Federation
shiganov@ocean.ru

Cathryn Sliwa
Centre for Research on Introduced Marine Pests, CSIRO Division of Marine Research, GPO Box 1538, Hobart, Tasmania 7001, Australia
Cath.Sliwa@deh.gov.au

L. David Smith
Biological Sciences, Director, Environmental Science & Policy Program, 235 Sabin-Reed Hall, Smith College, Northampton, MA 01063 , USA
ldsmith@email.smith.edu

John J. Stachowicz
Department of Evolution and Ecology, University of California, Davis, 1 Shields Avenue, Davis, CA 95616, USA
jjstachowicz@ucdavis.edu

Brian D. Steves
Smithsonian Environmental Research Center, PO Box 28, Edgewater, MD 21037, USA
stevesb@si.edu

Mark Erik Torchin
Smithsonian Tropical Research Institute, Box 0843-03092, Balboa, Ancon, Republic of Panama
torchinm@si.edu

Megan C. Tyrrell
Wells National Estuarine Research Reserve, 342 Laudholm Farm Rd, Wells, ME 04090, USA
mtyrrell@alumni.macalester.edu

Maria Célia Villac
Departamento de Biologia, Universidade de Taubaté, Taubaté-SP, 12030-010, Brazil
celiavillac@gmail.com

Kerstin Wasson
Elkhorn Slough National Estuarine Research Reserve 1700 Elkhorn Rd., Watsonville, CA 95076, USA
kerstin.wasson@gmail.com

Wim J. Wolff
University of Groningen, Department of Marine Benthic Ecology and Evolution, PO Box 14, 750 AA Haren, The Netherlands
W.J.Wolff@rug.nl

Marjorie J. Wonham
Centre for Mathematical Biology, Department of Biological Sciences and Department of Mathematical & Statistical Sciences, University of Alberta, Edmonton, Alberta, Canada T6G 2G1
mwonham@ualberta.ca

Section I
Perspectives on Marine Invasions

Chapter 1
Marine Bioinvasions: Conservation Hazards and Vehicles for Ecological Understanding

Gil Rilov and Jeffrey A. Crooks

1.1 Introduction – The Problem of Biological Invasions

The old ocean is gone. Life in today's seas is changing in an alarming rate. While many species are dwindling due to overfishing and habitat destruction (Roberts 2007), others invade new regions using anthropogenic vectors (Carlton 1996). These changes are rooted in human activities more than a thousand years old, but have accelerated dramatically in the past few decades due to new technology and increased connectivity (Carlton 1989; Carlton et al. 1999; Crooks and Suarez 2006). Huge fleets extract fish and shellfish with deadly efficiency, and at the same time the growth of trade facilitates the dispersal of organisms attached to the hulls of ships and, more recently, carried within ballast water. Aquaculture, live marine seafood and bait, and the aquarium trade have also become important vectors for the invasion of exotic marine species.

Invasions, in and of themselves, are rated high as a cause of native biodiversity loss and economic damage (Primack 2004; Mooney et al. 2005). But invasions also interact with other factors compromising the integrity of marine ecosystems, including habitat destruction, pollution and climate change. Habitat destruction causes disturbance which opens up space for newcomers such as invaders. Pollution can make environmental conditions less tolerable for native species, and perhaps provide opportunities for opportunists, among them exotic species. Climate change will also play a large role in the invasion process (Mooney and Hobbs 2000). Modifications to ocean temperature, biogeochemistry, salinity, sea level, and current circulation patterns have all been detected within the last few decades, and are expected to continue (IPCC 2007). The ecological 'footprint' of these changes has been observed both in terrestrial and marine ecosystems worldwide (Walther et al. 2002, 2005). Documented ecological changes in the oceans include modifications to the phenology of pelagic organisms resulting in trophic "mismatches" between predators and prey (Edwards and Richardson 2004), severe events of coral bleaching that negatively influence the structure of coral reef communities (Hughes et al. 2003), and a mostly northward shift in fish distributions in the North Sea presumably in response to warming temperatures (Perry et al. 2005). Evidence has also started to show shifts

G. Rilov, J.A. Crooks (eds.) *Biological Invasions in Marine Ecosystems.*
Ecological Studies 204,

in the distributional limits of benthic organisms in temperate coastal environments (Sagarin et al. 1999; Helmuth et al. 2006). Apart from range extension of native species due to climate change, increasing temperatures at medium and high latitudes have the potential to facilitate the establishment of species invading from warmer waters, thus affecting community structure and potentially function.

The invasion of non-native species has increased exponentially in the past 200 years, and does not show signs of slowing down or leveling off. For example, Dr. James Carlton, one of the world experts on bioinvasions and a contributor to this book, estimates that at any one time, over 7000 species might be moving around in ballast tanks in ships on the world's oceans. This translates into invasion rates that far exceed background rates of natural invasion (Crooks and Suarez 2006). For example, in San Francisco Bay alone, an average of one new species was introduced every 14 weeks between 1961 and 1995 (Cohen and Carlton 1995, 1998). In the Mediterranean, a new invader was discovered every week in the past 5 years. In some cases we are witnessing an "invasional meltdown," in which on or a group of exotic species facilitate new invaders in various ways, increasing the likelihood of survival and possibly even the magnitude of invasion impact (Simberloff and Von Holle 1999).

Although we have been much slower to realize the extent and impacts of invasions in the sea compared to those on land, our experience with problematic aquatic invasions continues to mount. The alga *Caulerpa taxifolia*, a popular aquarium species, now carpets many square kilometers of sea floor in the Mediterranean, a comb jelly native to the western Atlantic caused the collapse of fisheries when introduced into the Caspian Sea, and invasive marsh grasses and mangroves are transforming wetlands around the world. Non-native species are also agents of disease, such as the introduced protozoan MSX, which contributed to the collapse of Chesapeake Bay oyster populations. Also, escaped farmed species can hybridize with natives, leading to concern about aquaculture practices with species such as salmon. The truly troubling thing about marine invasions, however, is that although we are sure that we are doing a better job documenting their impacts, we are also sure that most invader impacts have gone, and continue to go, unnoticed.

Marine bioinvasions can be enormously costly in terms of economic and ecological damages as well as costs associated with management. Thus, especially when coupled with others changes in the sea, invasions are of growing concern to those charged with managing and protecting marine resources. In order to be most effective at addressing invasions, the conventional wisdom is that it is better to try to prevent an invasion rather than try to manage it after the fact. In order to help stem the flow of marine invasions, we are beginning to implement ballast water management practices such as open ocean exchange, but realize this is only a partial solution that must be employed until technological advances can be made to help better prevent invasions. When species do invade, drastic management action is sometimes necessary. For example, in order to stop the incipient invasion of an exotic mussel in its tracks, an entire marina in Darwin, Australia, was poisoned with a lethal cocktail of bleach and copper. Despite such efforts, though, the science and art of marine invasion management is still in its infancy.

1.2 Invasions as a Tool to Study Nature

As ecologically disastrous as they can be, however, biological invasions offer a unique opportunity to study fundamental processes in population, community, ecosystem, and evolutionary ecology across many taxonomic groups. It thus attracts the interests of a wide variety of scientists, and can be used as a vehicle for understanding some basic ecological and evolutionary questions (see Sax et al. 2005; Cadotte et al. 2006). We agree with the others who state that the relatively few studies that treat invasion in this way are "but the tip of a large iceberg" of scientific investigation that will grow in the coming decades (Sax et al. 2005).

One of the principal goals of ecology is to understand the abundance and distribution patterns of organisms in their environment. There is general agreement on some factors related to species distribution patterns (for reviews see, Brown 1984; Guisan and Thuiller 2005), but other ideas remain controversial (for example, the 'abundance center' assumption; Sagarin and Gaines 2002). It has long been recognized that abundances and distribution patters of species is highly dynamic both in time and space. These changes relate to seasonal, annual and decadal cycles, but can also be stochastic. They also relate to species interactions and to disturbance. A natural world that was once thought to be controlled by a stable equilibrium is now known to be largely governed by dynamic non-equilibrium processes (Rohde 2006).

Throughout the history of life, species have appeared and gone extinct. Species shifted their geographical ranges as they crossed barriers, on land or in the sea, or as the climate changed. As species arrived at new places, they had to cope with a new physical environment, as well as with the other species that they encountered. They could compete with them for resources, eat them or be eaten by them, parasitize them, or be indifferent to them. Many of them evolved in order to adapt better to the new environment. It is thought that most of these changes of geographical ranges occurred at a relatively slow pace. However, this is not the case today. Nowadays, species can catch a ride on a plane or a boat and cross oceans in hours or days, often in great numbers (hundreds of potentially invasive species can be found in the ballast water of a single ship arriving at a port; Carlton and Geller 1993). Invasions are fast and easy. The world is "hyperconnected," and the biota on earth is changing and homogenizing at unprecedented rates (Crooks and Suarez 2006).

Biological invasion touches on many facets of ecological disciplines. It allows population ecologists to follow the rise (and fall) of populations from their first arrival or establishment. It lets community ecologists investigate the interactions of novel species with long-time residents, including the effects of invasions on the diversity of communities and vice versa – the effects of communities and ecosystems on invaders. These have important implications for topics such as invasion resistance and ecosystem resilience (see a few recent examples: van Ruijven et al. 2003; Meiners et al. 2004; Herben 2005; Eppstein and Molofsky 2007; Shurin et al. 2007). Invasions are also important for applied disciplines such as conservation biology, restoration ecology, and pest management.

Biological invasions also are central to evolutionary considerations (Baker and Stebbins 1965; Cox 2004). Adaptations to new physical and biological environments are fundamental to invasions, which can be accomplished via changes to genotypes and/or phenotypes. By examining invaders, scientists can witness the arms race between invaders and natives at a very fast pace, as one changes to accommodate to the existence of the other. Molecular ecologists can follow hybridization processes among invaders and natives, or watch genetic drift in real time. There is evidence, for example, that invaders adapt rapidly to their new environment (see Lee 2002 for a review) and going through a genetic bottleneck, as many invaders presumably do due to a founder effect, may even eventually lead to speciation. There is already evidence that recent, anthropogenic invasions have prompted the evolution of new species (Zimmerman 1960; Filchak et al. 2000).

As well as being used to answer evolutionary questions, genetics has become invaluable in addressing a variety of other issues related to biological invasions. Molecular techniques can be used to identify hard-to-identify or cryptogenic species, including larval forms so common in marine environments, or can allow the identification of coevolved enemies for use as possible biocontrol agents. Genetics can also be used to identify geographic sources of invasive populations and invasion pathways, thus aiding invasion management (Holland 2000). Genetics is also central to other invasion-related issues, such as the development of genetically modified organisms (GMOs) and hybridization between foreign and domestic fish stocks. For a fuller treatment of genetic and evolutionary issues related to invasions, we would refer the reader to a steadily growing body of work (e.g., Baker and Stebbins 1965; Mooney and Cleland 2001; Lee 2002; Cox 2004), including some related specifically to a marine invasions (e.g., Geller 1996; Holland 2000 Chapman et al. 2008; Cunningham 2008; McGlashan et al. 2008).

1.3 Invasion Biology – The Discipline and its Application in the Marine Environment

Despite the interaction of invasion with basic disciplines, invasion biology as a recognizable entity in its own right is relatively new (see Davis 2006 for a history of invasion biology). Although naturalists have been noticing and thinking about the problems caused by the establishment of naturalized and unwanted species in new environments for centuries, the seminal book of Charles Elton (Elton 1958) drew serious attention to the phenomenon of biological invasions and their impacts on ecological communities. Since Elton's work, a growing wealth of studies on biological invasions have begun to spring up. These appear in scientific books on the topic (e.g., Drake et al. 1989; Williamson 1996; Cox 1999, 2004; Mooney and Hobbs 2000; Mooney et al. 2005; Sax et al. 2005; Lockwood et al. 2007; Nentwig 2007), dedicated journals (Biological Invasions and Aquatic Invasions), and scientific conferences (e.g., Pederson 2000). In addition, there have been an increasing number of treatments of biological invasions in the popular media, including in magazines, newspapers, web sites, blogs, television shows, and books (e.g., Lesinski 1996; Matsumoto 1996; Bright 1998; Carlton 2001; Todd 2001; Burdick 2005). This latter type of communication is particularly important, as it reflects, and in turn helps shape, public interest in the topic.

Scientific studies on the cause and effects of biological invasions, as well as the formulation of theoretical frameworks and models, have focused in the past primarily on islands or on mainland terrestrial and freshwater environments. These efforts offer some guidance for the study of marine invasions, but will not be universally applicable. One fundamental life history difference between organisms in marine and terrestrial environments that has major implications for invasions is that, in the sea, many benthic animals have developed a sessile life style (e.g. barnacles, mussels, tubeworms, bryozoans). Therefore they disperse, like most terrestrial plants, only via their early life stages, the propagules (or larvae). The three-dimensional and wet nature of the medium, water, also allows fertilization to occur externally during spawning events, and therefore this mode of reproduction is prevalent in many species of marine animals and plants. Because of these and other such differences, a discrete body of knowledge that deals explicitly with invasions in the marine environment inevitably began to develop. These papers appeared in a variety of applied and basic journals and, more recently, in journals devoted to the issue of biological invasions in general. Also, many invasion books have some treatment of marine invasions, but these tend to be relatively limited (e.g. Mooney and Hobbs 2000) or geographically restricted in scope (e.g., Europe; Leppäkoski et al. 2002).

What was lacking, therefore, was a comprehensive book that focused on biological invasions in the marine environment – one that overviews both the progress of, and the gaps in, the ecological understanding of the processes determining invasion success and impacts in this ecosystem. Management issues also need to be addressed with the ecological perspective in mind. Better ecological understanding improves predictions of potential future invasions, enhances risk assessment and advances the development of control measures. Although this is now a tall order and a full treatment is beyond the scope of any one volume, our book highlights these issues. In this volume we have assembled top experts in the fields of marine ecology and conservation biology to present to readers the most recent knowledge and new challenges in the research and management of marine bioinvasions.

1.4 Structure of the Book

The book is structured largely along the theme of process and pattern. In terms of considering invasion as a process, a successful invader has to go through several 'filters' before it becomes successful. It needs to be taken by the vector, survive the journey, survive the initial inoculation at the point of entry, be able to reproduce and increase in number in the receiving environment while competing for resources and avoiding getting eaten to oblivion, have the ability to spread, and to integrate itself into the environment through forging of ecological interactions. These can be boiled down to arrival, establishment, and integration phases (Vermeij 1996). A final phase of invasion could also be considered in terms of our response to these events. This process will largely play itself out wherever invasions occur, but with potentially interesting idiosyncrasies that will provide further insight into invasions.

In the book, we first provide an introduction to the book and field of study (this chapter), as well as general perspectives on and approaches to considering marine invasions (Section I: Perspectives on Marine Invasions). The next section broadly treats how invaders get from where they are native to where they are introduced, including association with vectors, transport, and release of translocated individuals (Section II: Invader Arrival). We then address the conditions necessary, both for the invader and the receiving environment, to make an invasion successful (Section III: Invader Establishment). Next, we discuss the consequences of successful invasions into the local natural communities (Section IV: Invader Integration into Ecosystems). It should be noted that an invasion is really a continuum of inter-related steps, and that many of the chapters in a section devoted to one phase of invasion address topics applicable to other phases of invasion as well.

After the treatment of the phases of invasion from the invader's standpoint, we have a short section on management, risk assessment and restoration related to marine bioinvasions (Section VI: Management Perspectives). We finish with a series of important regional perspectives, which provide local information and experience that touch on many of the issues addressed earlier in the book (Section VI: Geographic Perspectives), and a Concluding Thoughts (Section VII) which briefly highlights some considerations as the science of marine invasion biology moves forward. Each section is introduced by a short chapter, which highlights the material in the chapters and connects them to other sections of the book. Where appropriate, we also direct the reader to other sources of material.

1.5 Note on Terminology

We would likely be remiss if we did not wade (or at least dip our toes) into the troubled waters of invasion-related terminology. Invasion biology has accumulated the flotsam and jetsam of other popular and scientific terms, such as alien, exotic and invasive (Occhipinti-Ambrogi and Galil 2004; Lockwood et al. 2007), as well as words new to the world thanks to invasion biologists (e.g., pseudoindigenes; see Chap. 2, Carlton). Part of the often-discussed semantic tumult surrounding invasions arises because of inconsistent usage and potential for confusion, but it also reflects the rich social dimensions associated with the movement of species around the globe (Sagoff 2005; Coates 2006; Larson 2007), including accusations of inherent xenophobia among invasion biologists (Theodoropoulos 2003; Singer and Grismaijer 2005).

There have been a number of attempts to standardize usage, including sets of terms that are hierarchical and nested, and classified by scale and impact or mode of entry (e.g., Occhipinti-Ambrogi and Galil 2004). Also, as noted above, there has been the useful realization that an invasion is a process along a continuum, making it resistant to imposition of strict definitions (Lockwood et al. 2007). Because there is still no formal agreement within the scientific community, and after much consideration, we have decided that the authors are free to use invasion-related terms as they will. Therefore, words like invasive, which can mean species having demonstrable impacts, do not necessarily convey anything more than terms such as exotic, alien, introduced, non-indigenous, non-native, and so on. This liberal usage of invasive, in particular, largely reflects the fact that we know relatively little about impacts in marine habitats. (Perhaps we should adopt the precautionary approach for labeling these species and assume that they all have some degree of impact and thus are invasive, sensu stricto, unless we can prove otherwise). We believe, however, that the meaning behind the words used throughout the book is clear from context.

1.6 Concluding Thoughts

Producing this book has been a long, but rewarding, process, and the lively tempo of both invasions (species lists get longer by the month) and invasion science creates a moving target that is difficult to hit with any account of the field at any one point in time. However, we believe that many of the fundamental concepts and questions that are relevant to this field of research are presented in the following chapters, which were written by many experts in the field. We intended this book to be a useful textbook for students interested in this topic, and a tool for researches and practitioners studying and managing marine bioinvasions. We hope we have succeeded in this mission.

References

Baker HG, Stebbins GL (1965) The genetics of colonizing species. Academic Press, New York

Bright C (1998) Life out of bounds. W.W. Norton and Company, New York

Brown JH (1984) On the relationship between abundance and distribution of species. Am Nat 124:225–279

Burdick A (2005) Out of eden. An odyssey of ecological invasion. Farrar, Straus, and Giroux, New York

Cadotte MW, McMahon SM, Fukami T (2006) Conceptual ecology and invasion biology; reciprocal approaches to nature. Springer, Dordrecht, The Netherlands

Carlton JT (1989) Mans role in changing the face of the ocean – biological invasions and implications for conservation of near-shore environments. Conserv Biol 3:265–273

Carlton JT (1996) Pattern, process, and prediction in marine invasion ecology. Biol Conserv 78:97–106

Carlton JT (2001) Introduced species in US coastal waters: environmental impacts and management priorities. Pew Oceans Commission, Arlington, Virginia

Carlton JT, Geller JB (1993) Ecological roulette – the global transport of nonindigenous marine organisms. Science 261:78–82

Carlton JT, Geller JB, Reaka-Kudla ML, Norse EA (1999) Historical extinctions in the sea. Annu Rev Ecol Syst 30:515–538

Chapman JW, Blakeslee AMH, Carlton JT, Bellinger MR (2008) Parsimony dictates a human introduction: on the use of genetic and other data to distinguish between the natural and human-mediated invasion of the European snail *Littorina littorea* in North America. Biol Invas 10:131–133

Coates P (2006) American perceptions of immigrant and invasive species: strangers on the land. University of California Press, Berkeley, California

Cohen AN, Carlton JT (1995) U.S. Fish and Wildlife Service and National Sea Grant College Program (Connecticut Sea Grant), p 285

Cohen AN, Carlton JT (1998) Accelerating invasion rate in a highly invaded estuary. Science 279:555–558

Cox GW (1999) Alien species in North America and Hawaii. Island Press, Washington

Cox GW (2004) Alien species and evolution. Island Press, Washington

Crooks JA, Suarez A (2006) Hyperconnectivity, invasive species, and the breakdown of natural barriers to dispersal. In: Crooks KR, Sanjayan MA (eds) Connectivity conservation: maintaining connections for nature. Cambridge, New York, pp 451–478

Cunningham CW (2008) How to use genetic data to distinguish between natural and human-mediated introduction of *Littorina littorea* to North America. Biol Invas 10:1–6

Davis MA (2006) Invasion biology 1958–2005: the pursuit of science and conservation. In: Cadotte MW, McMahon SM, Fukami T (eds) Conceptual ecology and invasion biology; reciprocal approaches to nature. Springer, Dordrecht, The Netherlands, pp 35–64

Drake JA, Mooney HA, DiCastri F, Groves RH, Kruger M, Rejmanek M, Williamson M (1989) Biological invasions: a global perspective. Wiley, New York

Edwards M, Richardson AJ (2004) Impact of climate change on marine pelagic phenology and trophic mismatch. Nature 430:881–884

Elton CS (1958) The ecology of invasions by plants and animals. Chapman and Hall, New York

Eppstein MJ, Molofsky J (2007) Invasiveness in plant communities with feedbacks. Ecol Lett 10:253–263

Filchak KE, Roethele JB, Feder JL (2000) Natural selection and sympatric divergence in the apple maggot *Rhagoletis pomonella*. Nature 407:739–742

Geller JB (1996) Molecular approaches to the study of marine biological invasions. In: Ferraris JD, Palumbi SR (eds) Molecular zoology: advances, strategies, and protocols. Wiley-Liss, New York, pp 119–132

Guisan A, Thuiller W (2005) Predicting species distribution: offering more than simple habitat models. Ecol Lett 8:993–1009

Helmuth B, Mieszkowska N, Moore P, Hawkins SJ (2006) Living on the edge of two changing worlds: forecasting the responses of rocky intertidal ecosystems to climate change. Annu Rev Ecol Evol Syst 37:373–404

Herben T (2005) Species pool size and invasibility of island communities: a null model of sampling effects. Ecol Lett 8:909–917

Holland BD (2000) Genetics of marine bioinvasions. Hydrobiologia 420:63–71

Hughes TP, Baird AH, Bellwood DR, Card M, Connolly SR, Folke C, Grosberg R, Hoegh-Guldberg O, Jackson JBC, Kleypas J, Lough JM, Marshall P, Nystrom M, Palumbi SR, Pandolfi JM, Rosen B, Roughgarden J (2003) Climate change, human impacts, and the resilience of coral reefs. Science 301:929–933

IPCC (2007) Intergovernmental Panel on Climate Change Synthesis Report

Larson BMH (2007) An alien approach to invasive species: objectivity and society in invasion biology. Biol Invas 9:947–956

Lee CE (2002) Evolutionary genetics of invasive species. Trend Ecol Evol 17:386–391

Leppäkoski E, Gollasch S, Olenin S (2002) Invasive aquatic species of Europe: distribution, impacts and management. Springer, The Netherlands

Lesinski JM (1996) Exotic invaders: killer bees, fire ants, and other alien species are infesting America! Walker, New York
Lockwood JL, Hoopes MF, Marchetti MP (2007) Invasion ecology. Blackwell Publishing, Malden, Massachusetts
Matsumoto L (1996) Beyond Ohia Valley: adventures in a Hawaiian rainforest. Lehua, Honolulu, Hawaii
McGlashan DJ, Ponniah M, Cassey P, Viard F (2008) Clarifying marine invasions with molecular markers: an illustration based on mtDNA from mistaken calyptraeid gastropod identifications. Biol Invas 10:51–57
Meiners SJ, Cadenasso ML, Pickett STA (2004) Beyond biodiversity: individualistic controls of invasion in a self-assembled community. Ecol Lett 7:121–126
Mooney HA, Cleland EE (2001) The evolutionary impact of invasive species. Proc Natl Acad Sci 98:5446–5451
Mooney HA, Hobbs RJ (2000) Invasive species in a changing world. Island Press, Washington
Mooney HA, Mack RN, McNeeley JA, Neville LE, Schei PJ, Waage JK (2005) Invasive alien species. A new synthesis. Island Press, Washington
Nentwig W (2007) Biological invasions. Springer, The Netherlands
Occhipinti-Ambrogi A, Galil B (2004) A uniform terminology on bioinvasions: a chimera or an operative tool. Mar Pollut Bull 49:688–694
Pederson J (2000) Marine bioinvasions: proceedings of the first national conference. MIT Sea Grant College Program, Cambridge, Massachusetts
Perry AL, Low PJ, Ellis JR, Reynolds JD (2005) Climate change and distribution shifts in marine fishes. Science 308:1912–1915
Primack RB (2004) A primer of conservation biology. Sinauer, Sunderland, Massachusetts
Roberts C (2007) The unnatural history of the sea. Island Press, Washington
Rohde K (2006) Nonequilibrium ecology. Cambridge University Press
Sagarin RD, Gaines SD (2002) The 'abundant centre' distribution: to what extent is it a biogeographical rule? Ecol Lett 5:137–147
Sagarin RD, Barry JP, Gilman SE, Baxter CH (1999) Climate-related change in an intertidal community over short and long time scales. Ecol Monogr 69:465–490
Sagoff M (2005) Do non-native species threaten the natural environment. J Agric Environ Ethics 18:215–236
Sax DF, Stachowicz JJ, Gaines SD (2005) Species invasions: insights into ecology, evolution, and biogeography. Sinauer Associates
Shurin JB, Arnott SE, Hillebrand H, Longmuir A, Pinel-Alloul B, Winder M, Yan ND (2007) Diversity-stability relationship varies with latitude in zooplankton. Ecol Lett 10:127–134
Simberloff D, Von Holle B (1999) Positive interactions of non-indigenous species: invasional meltdown? Biol Invas 1:21–32
Singer SR, Grismaijer S (2005) Panic in paradise: invasive species hysteria and the Hawaiian coqui frog war. ISCD Press, Pahoa, Hawaii
Theodoropoulos DI (2003) Invasion biology: critique of a pseudoscience. Avvar Press, Blythe, California
Todd K (2001) Tinkering with Eden: a natural history of exotics in America. WW Norton, New York
van Ruijven J, De Deyn GB, Berendse F (2003) Diversity reduces invasibility in experimental plant communities: the role of plant species. Ecol Lett 6:910–918
Vermeij GJ (1996) An agenda for invasion biology. Biol Conserv 78:3–9
Walther GR, Post E, Convey P, Menzel A, Parmesan C, Beebee TJC, Fromentin JM, Hoegh-Guldberg O, Bairlein F (2002) Ecological responses to recent climate change. Nature 416:389–395
Walther GR, Berger S, Sykes MT (2005) An ecological 'footprint' of climate change. Proc R Soc B Biol Sci 272:1427–1432
Williamson M (1996) Biological invasions. Springer, Berlin Heidelberg New York
Zimmerman EC (1960) Possible evidence of rapid evolution in Hawaiian moths. Evolution 14:137–138

Chapter 2
Deep Invasion Ecology and the Assembly of Communities in Historical Time

James T. Carlton

"With dim light and tangled circumstance ..."

– George Elliott, *Middlemarch: A Study of Provincial Life (1871)*

2.1 Introduction

A critical component of – and a limitation on – interpreting community structure is a detailed understanding of the ecological and evolutionary history of the assemblage of species in question. There are thus compelling reasons to understand, and seek to measure, how communities have changed over both evolutionary (geological) and ecological (historical) time. Vast waves of change have swept across the Earth in the past one to two millennia as waves of humans invaded across the planet in sequential episodes of exploration, colonization, and urbanization. As an expected and inexorable result of human activity, alterations in biodiversity have impacted terrestrial, freshwater, and marine communities. These alterations include the addition of species (invasions), the deletion of species (extinctions), and altered population dynamics (such as decreasing or increasing the abundance of a species, or altering genetic structure). In even seemingly "pristine" areas – such as wave-exposed high-energy rocky intertidal shores – it is no longer tenable to assume that communities and ecosystems have remained unaltered, in part because of *supply-side impacts* – impacts that are the indirect cascades of human activity originating outside of the area in question (e.g., Butman et al. 1995; Chap. 7, Johnston et al.).

Three (among a number of) reasons drive the interest to understand the first of these alterations – the role of invasions in historical time:

1. An academic desire to understand whether community-level processes, such as predation, competition, and disturbance (Chap. 14, Byers; Chap. 16, Crooks; Chap. 17, Grosholz and Ruiz; Chap. 15, Rilov) derive in part from species interactions on an evolutionary-time scale, or from interactions on an ecological-historical time

G. Rilov, J.A. Crooks (eds.) *Biological Invasions in Marine Ecosystems.* Ecological Studies 204,

scale, such as might be due to the presence of recently-arrived species (Mooney and Cleland, 2001; Grosholz 2002, 2005; Sax et al. 2005; Strauss et al. 2006; Freeman and Byers 2006; Cadotte et al. 2006).
2. A desire that merges academic interests with management concerns to predict what phenomena and processes characterize invaders and invasible habitats (Ricciardi and Rasmussen 1998; Bax et al. 2001; Kolar and Lodge 2001; Chap. 7, Johnston et al.; Chap. 8, Miller and Ruiz; Chap. 10, Smith; Chap. 11, Torchin and Lafferty; Chap. 12, Olyarnik et al.)
3. An interest in establishing the scale of community alteration, in order to undertake environmental management if not actual restoration (Byers et al. 2002; Lotze et al. 2006; Chap. 21, Hacker and Dethier).

The foundation of all three rationales relies not only on the ability to recognize which species are introduced (Chapman and Carlton 1991, 1994) – and thus to make an adequate estimate of the number of non-native species – but also, based on this recognition, to determining experimentally the role of invasions in regulating and producing community structure. However, while many studies that attempt to assess the diversity of invasions acknowledge that the number of invaders is likely underestimated, there has been little attempt to formalize the sources of this underestimation, nor, more importantly, how such partial assessments of the scale of invasions (spatially and temporally) may influence our understanding of the evolution, ecology, history, and management of communities, or of our ability to predict invasions. Clearly, if we have only a partial view of the diversity of non-native species, this compromises our ability to predict what types of organisms can invade, to assess what environments and regions are more or less susceptible to invasions, and to understand invasion patterns over time and space.

The challenges – not mutually exclusive – in estimating the diversity of invaders in a community are shown in Box 2.1. Some of these, as noted below, are universal to any attempt to estimate alpha diversity, but are discussed here specifically as contributors to the underestimation of historical components of community assembly. Discussed here are 12 potential sources of error that have led to invader underestimation, together with some of the implications of such underestimations. The number of introduced species may, of course, also be overestimated, as discussed below as well.

2.2 Invader Underestimation – Systematics

2.2.1 Cryptogenic Species

Species that cannot be reliably demonstrated as being either introduced or native are cryptogenic (Carlton 1996). Most species that are now categorized as cryptogenic were previously treated as native; in turn, many species that should be recognized as cryptogenic are still regarded as native, the classic categorical default in biogeography and evolutionary biology. Few species previously treated as introduced have been converted to cryptogenic status, in part because the designation of a species as non-native has usually come about as the result of the application of conservative criteria.

Cryptogenic species have been recognized in marine habitats, to some extent in freshwater habitats, and rarely in terrestrial habitats. Cryptogenic species may include unidentified species if there is strong evidence that they may be introduced, but it is important to emphasize that not all unidentified species are cryptogenic (or native). Such evidence is reviewed by Chapman and Carlton (1991, 1994) and includes association with a nonindigenous facies, association with a "weedy" habitat, being a member of a genus or family that is otherwise not known from the region in question (but occurs elsewhere in the world), and other criteria.

Cryptogenic species are not introduced species of uncertain geographic origin, as used by Jensen and Knudsen (2005). No one term captures these "Flying Dutchmen," species that roam the seas on ships with no (as yet) certain home. Cryptogenic species are not introduced species whose mechanism of introduction is uncertain, as used by Englund (2002). The term to be used in this case is polyvectic (Carlton and Ruiz 2005). Cryptogenic species are not solely non-native species or "cryptic invaders," as used by Lockwood et al. (2007). By definition, it is not known if cryptogenic species are native or introduced.

Further, cryptogenic species do not automatically include species with a type locality outside of the region under study, such as the numerous taxa bearing European names on the Pacific coast of North America. In such cases, biogeographic and historical taxonomic considerations come into play. On the one hand, a substantial number of morphotaxa appear to drape naturally along the four northernmost coasts of the Pacific and Atlantic Oceans (northeastern Pacific, northwestern Pacific, northeastern Atlantic, and northwestern Atlantic) as a result of their Tertiary histories of evolution, expansion and contraction, but whether these remain the same genospecies (a taxon characterized by identical genomes throughout its range) is not known for most of these taxa. Transarctic species embrace a wide phyletic range: examples include the eelgrass *Zostera marina*, the sea anemone *Metridium senile*, the seaslugs *Alderia modesta* and *Dendronotus frondosus*, the barnacles *Balanus balanus* and *Semibalanus balanoides*, the priapulid *Priapulus caudatus*, and the mussels *Mytilus trossulus* and *Modiolus modiolus*. Coan et al. (2000) list more than 40 species of bivalve mollusks that are considered circumboreal.

On the other hand, nineteenth and early twentieth century monographs of European taxa ranging from foraminiferans to polychaetes to algae were used for many decades to identify species from around the world: the taxonomic residue of this globalization of euronomenclature remains with us in the form of perhaps thousands of undescribed (not introduced!) species bearing European names in

Africa, South America, North America, Asia, and Australasia. A rocky shore seaweed in California with a type locality of Italy would thus bear reexamination. Much of this early work served to lead to the misconception that many shallow-water marine organisms were "cosmopolitan". While many "cosmopolitan" species may in theory be considered cryptogenic, as their modern-day global distribution could be the result of centuries of international shipping, it is argued below that a more conservative (although vexing) approach is to consider many of these taxa to be undetermined – essentially now unidentified or unidentifiable species, pending systematic revisions.

2.2.2 Pseudoindigenous Species

An important but largely overlooked source of underestimating invasion diversity in a given region is the presence of *pseudoindigenes*, here defined as introduced species that are mistakenly considered as native (indigenous or endemic) to a location. Pseudoindigenes include four categories of taxa (Box 2.1). The level of sophistication of systematic and biogeographic knowledge predicts that there are more examples of category 1 than category 2, and more examples of category 2 than category 3, as explained below. Category 4 is conceptually distinct from the first three, and does not fall into this pattern.

Category 1: introduced species re-described as new after introduction, although already described in their native region Pseudoindigenes in this first category are common. Table 2.1 presents 94 introduced species redescribed as "new", following their introduction to a region. In many cases these species have

Box 2.1 Sources of error in estimating the number and thus importance of introduced species

SYSTEMATICS

Cryptogenic Species (*cryptogens*)

Species that cannot be reliably demonstrated as being either introduced or native.

Pseudoindigenous Species (*pseudoindigenes*)

Introduced species mistaken as native (indigenous or endemic) species:

1. Species redescribed as new after introduction, although already described in their native region (Table 2.1).
2. Species first described as new after introduction, and later found elsewhere (Table 2.2).

(continued)

Box 2.1 (continued)

3. Species described as new after introduction, and remaining unknown elsewhere (Table 2.3).
4. Species misidentified as previously known native species:
   A. Imperfect taxonomy
   B. Cryptic genospecies invasions

Unidentified Species, including many "**Cosmopolitan**" **Species**
Species that are not (unidentified) or cannot yet be (unidentifiable) identified to a level permitting biogeographic assessment.
Small Species
Species typically less than 1 mm in size that – *because of their size* – are considered naturally distributed ("smalls rule").
Uninvestigated Taxa
Species groups that are not or rarely studied.
Known but Unreported Taxa
Introduced species known to researchers but never published.
BIOGEOGRAPHY and COMMUNITY HISTORY
Widespread Intraoceanic Species
Species with *within*-ocean distributions that appear to mirror presumptive natural patterns.
Widespread Interoceanic Corridor Species
Species with *between*-ocean distributions that appear to mirror presumptive natural patterns.
Neritic Species with Presumptive Oceanic Dispersal
Species with planktonic life-history stages or living on floating habitats that are presumed to be amenable to dispersal on oceanic currents.
Resident Species
Introduced or cryptogenic species which, after a length of time, are grouped with the "native" component.
SAMPLING
Species in Underexplored Habitats and Associations

1. Undersampled parasitic, commensal, or symbiotic introduced species.
2. Undersampled introduced species in microhabitats and ecotonal habitats.

Incipient Invasions: Species with Small Population Sizes
Newly established introductions with small, restricted populations.

been described as new in several regions (up to 11 times for a single species), each time with a different name, to a total of 159 cases of mistaken re-descriptions. No systematic or focused search for pseudoindigenes in this category has been undertaken in any biota, and thus neither the regional nor global scale of this underestimation of introductions (and overdescription of biodiversity) is known. A few

Table 2.1 Examples of introduced species mistakenly redescribed as native species after their introduction or appearance in a new region

Species	Native to	Redescribed as (from)	Reference
Ciliophora (ciliates)			
Ancistrocoma pelseneeri Chatton & Lwoff, 1926	North Atlantic	*Parachaenia myae* Kofoid & Bush, 1936 (California)	Kozloff (1946)
Hydrozoa (hydroids)			
Obelia bidentata Clark, 1875	North Atlantic	*Obelia corona* Torrey, 1904 (California)	Cornelius (1975)
		Obelia multidentata Fraser, 1914 (Washington)	Cornelius (1975)
Cladonema radiatum Dujardin, 1843	North Atlantic	*Cladonema novae-zelandiae* Ralph, 1953 (New Zealand)	Schuchert (1996)
Pinauay crocea (Agassiz, 1862)	North Atlantic	*Parypha microcephala* Agassiz, 1865 (California)	Torrey (1902)
		Tubularia elegans Clark, 1876 (California)	Torrey (1902)
Cordylophora caspia (Pallas, 1771)	Ponto-Caspian	*Cordylophora lacustris* Allman, 1844 (Ireland)	Schuchert (2004)
		Cordylophora albicola Kirchenpauer, 1861 (Germany)	Schuchert (2004)
		Cordylophora americana Leidy, 1870 (Atlantic American coast)	Schuchert (2004)
		Cordylophora whiteleggi von Lendenfeld, 1887 (Australia)	Briggs (1931)
		Cordylophora fluviatilis Hamilton, 1928 (New Zealand)	Briggs (1931)
		Cordylophora lacustris otagoensis Fyfe, 1929 (New Zealand)	Schuchert (2004)
Pennaria disticha Goldfuss, 1820	North Atlantic	*Corydendrium splendidum* Boone, 1938 (Hawaii)	Cooke (1977)
Garveia franciscana (Torrey, 1902)	South Pacific-Indian Ocean	*Bimeria monodi* Billard,1927 (Cameroon)	Vervoort (1964) (all *Garveia franciscana*)
		Bimeria tunicata Fraser, 1944 (Louisiana)	
		Perigonimus megas Kinne, 1956 (Netherlands)	
Gonionemus vertens Agassiz, 1862	North Pacific	*Haleremita cumulans* Schaudinn, 1894 (Germany: Berlin Aquarium)	Edwards (1976)
		Gonionemus murbachi Mayer, 1901 (Massachusetts)	Kramp (1961)

		Gonionemus vindobonensis Joseph, 1918 (Adriatic: Trieste, in an aquarium)	Kramp (1961)
Nemopsis bachei Agassiz, 1849	Northwest Atlantic	*Nemopsis heteronema* Haeckel, 1879 (Norway)	Kramp (1961)
		Nemopsis crucifera (Forbes & Goodsir, 1853) (Scotland)	Kramp (1961)
Craspedacusta sowerbii Lankester, 1880	China	*Microhydra ryderi* Potts, 1885 (Pennsylvania)	Kramp (1961)
		Microhydra germanica Roch, 1924 (Germany)	Kramp (1961)
Anthozoa (sea anemones)			
Diadumene lineata (Verrill, 1873)	Japan	*Sagartia luciae* Verrill, 1898 (Connecticut)	Manuel (1988)
Turbellaria (flatworms)			
Rhynchodemus bilineatus (Mecznikov, 1866)	Europe	*Rhynchodemus sylvaticus* (Leidy, 1851) (Philadelphia, Pennsylvania)	Ogren (1989a)
		Rhynchodemus americanus Hyman, 1943 (Missouri)	Ogren (1989a)
Bipalium kewense Moseley, 1878	Indo-Pacific, Japan	*Bipalium manubriatum* Sharp, 1891 (Pennsylvania)	Ogren (1984)
		Bipalium costaricense Hyman, 1939 (Costa Rica)	Ogren (1984)
Polychaeta (worms)			
Myrianida pachycera (Augener, 1913)	Indo-West Pacific	*Myrianida crassicirrata* Hartmann-Schröder, 1965 (Oahu, Hawaiian Islands)	Nygren (2004)
Polydora cornuta Bosc, 1802	Northwest Atlantic	*Polydora amarincola* Hartman, 1936 (California)	Blake and Maciolek (1987)
Streblospio benedicti Webster, 1879	Northwest Atlantic	*Streblospio lutincola* Hartman, 1936 (California)	Carlton (1979)
Pseudopolydora kempi (Southern, 1921)	Japan	*Neopygopsio laminifera* Berkeley & Berkeley, 1954 (British Columbia)	Banse (1972)
Neanthes succinea (Frey & Leuckart, 1847)	Northwest Atlantic	*Nereis saltoni* Hartman, 1936 (California)	Carlton (1979)
Janua pagenstecheri (Quatrefages, 1865)	Europe	*Spirorbis unicornis* Bailey & Harris, 1968 (Galapagos Islands)	Knight-Jones et al. (1975)
		Spiroribis epichysis Bailey, 1970 (West Indies)	Knight-Jones et al. (1975)
Simplicaria pseudomilitaris (Thiriot-Quievreux, 1965)	Atlantic Ocean	*Spirorbis regalis* Bailey & Harris, 1968 (Galapagos Islands)	Vine et al. (1972)

(continued)

Table 2.1 (continued)

Species	Native to	Redescribed as (from)	Reference
Hydroides elegans (Haswell, 1883)	Australasia	*Hydroides pectinata* Lo Bianco, 1893 (Italy)	Zibrowius (1971)
		Hydroides trypanon Saint-Joseph, 1906 (Italy)	
		Hydroides pacificus Hartman, 1969 (California)	
Gastropoda (snails)			
Crepidula plana Say, 1822	Northwest Atlantic	*Crepidula sinuosa* Turton, 1825 (boat hull, Yorkshire, England)	Collin (2000)
Potamopyrgus antipodarum (Gray, 1843)	New Zealand	*Hydrobia jenkinsi* Smith, 1889 (England)	Ponder (1988)
Truncatella pulchella Pfeiffer, 1839	Northwest Atlantic	*Truncatella bairdiana* C. B. Adams, 1852 (Pacific Panama)	Clench and Turner (1948)
Corambe obscura (Verrill, 1870)	Northwest Atlantic	*Corambe batava* Kerbert, 1886 (Netherlands)	Swennen and Dekker (1995)
		Corambella carambola Marcus, 1955 (Brazil)	
Haminoea japonica (Pilsbry, 1895)	Japan	*Haminoea callidegenita* Gibson & Chia, 1989 (Washington)	Gosliner and Behrens (2006)
Siphonaria pectinata (Linnaeus, 1758)	Eastern Atlantic	*Siphonaria lineolata* d'Orbigny, 1841 (Cuba)	Carlton (1992)
		Siphonaria naufragum Stearns, 1872 (Florida)	Rosenberg (2005)
Ferrissia fragilis (Tryon, 1863)	North America	*Ferrissia wautieri* Mirolli, 1960 (Europe)	Walther et al. (2006)
Myosotella myosotis (Draparnaud, 1801)	Europe	*Melampus borealis* Conrad, 1832 (Rhode Island)	Martins (1996) (all *M. myosotis*)
		Auricula reflexilabris Orbigny, 1837 (Peru)	
		Alexia bermudensis H. & A. Adams, 1855 (Bermuda)	
		Auricula meridionalis Brazier, 1877 (Australia)	
		Auricularia meridionalis Brazier, 1877 (Australia)	
		Alexia acuminata Morelet, 1889 (South Africa)	
		Alexia pulchella Morelet, 1889 (South Africa)	

		Alexia myosotis marylandica Pilsbry, 1900 (Maryland)	
		Alexia subflava Fenaux, 1939 (Bermuda)	
Tarebia granifera (Lamarck, 1816)	Indo-Pacific	*Melania mauiensis* Lea, 1856 (Hawaii)	Cowie (1997)
Milax gagates (Draparnaud, 1801)	Europe	*Amalia babori* Collinge, 1897 (Hawaii)	Cowie (1997)
Meghimatium striatum Hasselt, 1824	Asia	*Philomycus australis* Bergh, 1870 (Hawaii)	Cowie (1997)
Allopeas gracile (Hutton, 1834)	Neotropics	*Bulimus junceus* Gould, 1846 (Hawaii)	Cowie (1997)
Paropeas achatinaceum (Pfeiffer, 1846)	Indo-Pacific	*Opeas henshawi* Sykes, 1904 (Hawaii)	Cowie (1997)
Laevicaulis alte (Ferussac, 1822)	Central Africa	*Vaginula leydigi* Simroth, 1889 (Australia)	Cowie (1997)
Hawaiia minuscula (Binney, 1841)	North America	*Helix kawaiensis* Reeve,1854 (Hawaii)	Cowie (1997)
Bivalvia (bivalves)			
Xenostrobus securis (Lamarck, 1819)	Australia/New Zealand	*Limnoperna fortunei kikuchii* Habe, 1981 (Japan)	Kimura et al. (1999)
Mytilus galloprovincialis Lamarck, 1819	Mediterranean	*Mytilus edulis diegensis* Coe, 1945 (California)	Coan et al. (2000)
Mytilopsis sallei (Recluz, 1849)	Eastern Pacific	*Mytilopsis allyneana* Hertlein & Hanna, 1949 (Fiji)	Morton (1981)
Mytilopsis leucophaeata Conrad, 1831	Northwest Atlantic	*Congeria cochleata* (Kickx, 1835) (Belgium)	Marelli and Gray (1983)
		Mytilopsis zeteki Hertlein & Hanna, 1949 (Panama Canal)	Rosenberg (2005)
Crassinella lunulata (Conrad, 1834)	Northwest Atlantic	*Crassinella oregonensis* Keen, 1938 (Oregon)	Coan (1979)
Mya arenaria Linnaeus, 1758	Northwest Atlantic	*Mya hemphilli* Newcomb, 1874 (California)	Carlton (1979)
Chama fibula Reeve, 1846	Philippines	*Chama hendersoni* Dall, Bartsch, & Rehder, 1938 (Hawaii)	Kay (1979)
Mercenaria campechiensis (Gmelin, 1791)	Northwest Atlantic	*Venus apodema* Dall, 1902 (Pacific Panama)	Keen (1971)
		Mercenaria kennicottii Dall, 1871 (Washington; probably adventitious shell [Carlton, 1979]).	Coan et al. (2000)
Venerupis philippinarum (A. Adams & Reeve, 1850)	Japan	*Paphia bifurcata* Quayle, 1938 (British Columbia)	Coan et al. (2000)
Hiatella arctica (Linnaeus, 1767)	Atlantic Ocean	*Saxicava hawaiensis* Dall, Bartsch, & Rehder, 1938 (Hawaii)	Kay (1979)
Martesia striata (Linnaeus, 1758)	South Pacific?	*Martesia hawaiensis* Dall, Bartsch, & Rehder, 1938 (Hawaii)	Kay (1979)

(continued)

Table 2.1 (continued)

Species	Native to	Redescribed as (from)	Reference
Lyrodus pedicellatus (Quatrefages, 1849)	South Pacific?	*Teredo chlorotica* Gould, 1870 (Masssachusetts, from ships from Pacific Ocean)	Turner (1966) (all *L. pedicellatus*)
		Teredo diegensis Bartsch, 1916 (California)	
		Teredo townsendi Bartsch, 1922 (California)	
		Teredo yatsui Moll, 1929 (Japan)	
		Teredo robsoni Roch, 1931 (South Africa)	
		Teredo tateyamensis Kuronuma, 1931 (Japan)	
		Teredo hawaiensis Dall, Bartsch & Rehder 1938 (Hawaii)	
		Teredo kauaiensis Dall, Bartsch & Rehder,1938 (Hawaii)	
		Teredo taiwanensis Taki & Habe, 1945 (Taiwan)	
		Teredo midwayensis Edmondson, 1946 (Midway Island)	
		Teredo honoluluensis Edmondson, 1946 (Hawaii)	
Psiloteredo megotara Hanley, 1848	Europe	*Teredo dilatata* Sowerby, 1851 (Massachusetts)	Turner (1966)
Bankia bipalmulata (Lamarck, 1801)	South Pacific ?	*Bankia hawaiiensis* Edmondson, 1942 *Bankia konaensis* Edmondson, 1942 (both, Hawaii)	Turner (1966)
Teredo navalis Linnaeus, 1758	South Pacific ?	*Teredo japonica* Clessin, 1893 (Japan)	Turner (1966) (all *Teredo navalis*)
		Teredo beachi Bartsch, 1921 (California)	
Teredo bartschi Clapp, 1923	South Pacific ?	*Teredo batilliformis* Clapp, 1924 (Bermuda)	Turner (1966) (all *Teredo bartschi*)
		Teredo aegyptia Roch, 1935 (Suez Canal)	
		Teredo hiloensis Edmondson, 1942 (Hawaii)	
Teredo clappi Bartsch, 1923	South Pacific ?	*Teredo trulliformis* Miller, 1924 (Hawaii)	Turner (1966)

Teredo furcifera von Marttens, 1894	South Pacific ?	*Teredo parksi* Bartsch, 1921 (Hawaii)	Turner (1966)
Copepoda (copepods)			
Mytilicola orientalis Mori, 1935	Japan	*Mytilicola ostreae* Wilson, 1938 (Washington)	Odlaug (1946)
Ostracoda (ostracodes)			
Eusarsiella zostericola Cushman, 1906	Northwest Atlantic	*Sarsiella tricostata* Jones, 1958 (California)	Kornicke (1967)
Spinileberis quadriaculeata (Brady, 1880)	Japan	*Spinileberis hyalinus* Watling, 1970 (California)	Watling (1975)
Cirripedia (barnacles)			
Balanus amphitrite Darwin, 1854	Indo-Pacific	*Balanus amphitrite hawaiiensis* Broch, 1927 (Hawaii)	Henry and McLaughlin (1975) (all *B. amphitrite*)
		Balanus amphitrite franciscanus Rogers, 1949	
		Balanus amphitrite herzi Rogers, 1949 (both, California)	
Elminius modestus Darwin, 1854	New Zealand	*Elminius cristallinus* Gruvel, 1907 (Azores)	Newman and Ross (1976)
Pollicipes polymerus Sowerby, 1833	Northeast Pacific	*Pollicipes polymerus madrasensis* Daniel, 1953 (India)	Newman and Killingley (1985)
Isopoda (isopods)			
Sphaeroma quoinaum Milne-Edwards,1840	New Zealand	*Sphaeroma pentodon* Richardson, 1904 (California)	Rotramel (1972)
Sphaeroma terebrans Bate, 1866	Indian Ocean	*Sphaeroma destructor* Richardson, 1897 (Florida)	Calman (1921)
Synidotea laevidorsalis (Miers, 1881)	Japan	*Synidotea laticauda* Benedict, 1897 (California)	Chapman and Carlton (1991)
		Synidotea marplatensis Giambiagi, 1922 (Argentina)	Chapman Carlton (1991)
Littorophiloscia culebrae (Moore, 1901)	North Atlantic	*Philoscia dartevellei* Brian, 1953 (Angola)	Taiti and Ferrara (1986)
Niambia capensis (Dollfus, 1895)	Southern Africa	*Porcellio littorina* Miller, 1936 (California)	Jass and Klausmeier (2000)
Porcellio dilatatus Brandt, 1833	Europe	*Porcellio spinicornis occidentalis* Miller, 1936 (California)	Van Name (1940)

(continued)

Table 2.1 (continued)

Species	Native to	Redescribed as (from)	Reference
Porcellio laevis Latreille, 1804	Europe	*Porcellio aztecus* Saussure, 1857 (Mexico)	Van Name (1936) (all *P. laevis*)
		Porcellio cubensis Saussure, 1857 (Cuba)	
		Porcellio mexicanus Saussure, 1857 (Mexico)	
		Porcellio interruptus Heller, 1861 (Chile)	
		Porcellio formosus Stuxberg, 1875 (California)	
		Porcellio parvicornis Richardson, 1902 (Bermuda)	
Porcellio scaber Latreille, 1804	Europe	*Porcellio nigra* Say, 1818 (Eastern United States)	Van Name (1936) (all *P. scaber*)
		Porcellio gemmulatus Dana, 1853 (United States)	
		Philoscia tuberculata Stimpson, 1856 (California)	
		Porcellio montezumae Saussure, 1857 (Mexico)	
Armadillidium vulgare (Latreille, 1804)	Europe	*Armadillo pilularis* Say, 1818 (United States)	Van Name (1936) (all *A. vulgare*)
		Armadillidium pilulare Stuxberg, 1875 (North America)	
		Uropodias bermudensis Richardson, 1902 (Bermuda)	
Amphipoda (amphipods)			
Caprella mutica Schurin, 1935	Japan	*Caprella macho* Platvoet, de Bruyne, & Meyling, 1995 (Netherlands)	Marellli (1981)
Caprella penantis Leach, 1814	North Atlantic?	*Caprella acutifrons neglecta* Mayer, 1890 (Hong Kong, off ship's bottom)	Stoddart and Lowry (2003)
Monocorophium insidiosum (Crawford 1937)	North Atlantic	*Corophium oaklandense* Shoemaker, 1949 (California)	Chapman (2007)

Stomatopoda (mantis shrimp)			
Gonodactylaceus falcatus (Forskal, 1775)	Indo-Pacific	*Gonodactylus aloha* Manning & Reaka, 1981 (Hawaii)	Ahyong (2001)
Decapoda (crabs)			
Rhithropanopeus harrisii (Gould, 1841)	Northwest Atlantic	*Pilumnus tridentatus* Maitland, 1874 (Netherlands)	Buitendijk and Holthuis (1949)
Carcinus maenas Linnaeus, 1758	Europe	*Carcinus granulatus* Say, 1817 (Long Island Sound)	Carlton and Cohen (2003)
Diptera (flies)			
Telmatogeton japonicus Tokunaga, 1933	Japan	*Telmatogeton remanei* Remmert, 1963 (Baltic Sea)	Kronberg (1986)
Heteroptera (water bugs)			
Trichocorixa reticulata (Guerin-Meneville, 1857)	North America	*Corixa blackburni* White, 1877 (Hawaii)	Zimmerman (1948)
Pycnogonida (sea spiders)			
Pigrogromitus timsanus Calman, 1927	southern hemisphere?	*Clotenopsa prima* Hilton, 1942 (Hawaii)	Stock (1968)
Kamptozoa (entoprocts)			
Urnatella gracilis Leidy, 1851	North America	*Urnatella indica* Seshaiya, 1946 (India)	Emschermann (1987)
Tunicata (sea squirts)			
Botryllus schlosseri (Pallas, 1766)	Europe?	*Botryllus gouldii* Verrill, 1871 (New York)	Van Name (1945)
Styela clava Herdman, 1881	Northwestern Pacific	*Styela mammiculata* Carlisle, 1954 (England)	Millar (1970)
Styela canopus (Savigny, 1816)	Northwestern Pacific	*Cynthia partita* (Stimpson, 1852) (Massachusetts)	Kott (1998) (all *S. canopus*)
		Styela variabilis Hancock, 1868 (England)	
		Cynthia stellifera Verrill, 1871 (Connecticut)	
		Styela canopoides Heller, 1877 (Adriatic Sea)	
		Styela partita bermudensis Van Name, 1902 (Bermuda)	

(continued)

Table 2.1 (continued)

Species	Native to	Redescribed as (from)	Reference
Styela plicata (Lesueur, 1823)	Northwestern Pacific	*Styela gyrosa* Heller, 1877 (Adriatic Sea)	Kott (1998)
		Styela pinguis Herdman, 1898 (Australia)	Kott (1998)
		Styela barnharti Ritter & Forsyth, 1917 (California)	Abbott and Johnson (1972)
Ascidiella aspersa (Muller, 1776)	Europe	*Ascidiella senegalensis* Michaelsen, 1915 (Senegal)	Monniot et al. (2001)
Ascidiella sydneiensis Stimpson, 1855	Australasia	*Phallusia longitubis* Traustedt, 1882 (West Indies)	Kott (1998)
Ascidia archaia Sluiter, 1890	Indo-West Pacific	*Ascidia corelloides* (Van Name, 1924) (Curacao)	Kott (1985); status as introduced to Caribbean herein
Ciona intestinalis (Linnaeus, 1767)	North Atlantic	*Ascidia diaphanaea* Quoy & Gaimard, 1834 (Australia)	Kott (1998)
		Ciona robusta Hoshino & Tokioka, 1967 (Japan)	Kott (1998)
Osteichthyes (fish)			
Omobranchus punctatus (Valenciennes, 1836)	Indo-Pacific	*Poroalticus sewalli* Foster, 1931 (Trinidad)	Springer and Gomon (1975)
Omobranchus ferox (Herre, 1927)	Philippines	*Omobranchus dealmeida* Smith, 1949 (Mozambique)	Springer and Gomon (1975)
Micropterus salmoides (Lacepède, 1802)	North America	*Pikea sericea* Fowler, 1938 (Hong Kong)	Robins and Böhlke (1960)
Hexanematichthys couma (Valenciennes 1840)	South America	*Sciadeichthys walrechti* Boeseman, 1954 (specimens washed ashore in Netherlands)	Marceniuk and Ferraris (2003)
Rhodophyta (red algae)			
Porphyra suborbiculata Kjellman, 1897	North Pacific	*Porphyra carolinensis* Coll & Cox, 1977 (Western Atlantic)	Broom et al. (2002)
Neosiphonia harveyi (Bailey, 1848)	Asia	*Polysiphonia havanensis insidiosa* J. Agardh, 1863 (Cuba)	Guiry (2003)
		Polysiphonia argentinica W. R. Taylor, 1939 (Argentina)	Guiry (2003)
Spermatophyta (seed plants)			
Zostera japonica (Ascherson & Graebner, 1907)	Western Pacific	*Zostera americana* den Hartog, 1970 (State of Wasington)	Bigley and Barreca (1982)

examples might suffice, but this longer list (although itself doubtless a small fraction of the total phenomenon) serves to illustrate the historical, taxonomic, and biogeographic breadth – the sheer ubiquity – of this category. Included are examples from aquatic (freshwater), marine, and terrestrial habitats.

With the onset of increased systematic work in the early nineteenth century, non-native species began to be mistakenly redescribed as natives (Table 2.1): the European green shore crab *Carcinus maenas* is described as a new species (*Carcinus granulatus*) in 1817 in North America, and, reciprocally, the American slipper limpet *Crepidula plana* is described as a new species (*Crepidula sinuosa*) in 1825 in Europe. The American dreissenid mussel *Mytilopsis leucophaeata* is also described as a new species (*Congeria cochleata*) in 1835 in Europe. As early as the 1830s, European species were being redescribed as native species in the Pacific Ocean, including the seasquirt *Ciona intestinalis* in Australia and the marsh snail *Myosotella myosotis* in Peru, ship-associated animals whose arrival in the Pacific Ocean could date back to the sixteenth century. Species commonly associated with global commerce are redescribed continuously: the Ponto-Caspian hydroid *Cordylophora lacustris* and the European synanthropic isopod *Porcellio laevis* are redescribed at least 6 times each, the European snail *Myosotella myosotis* is carried to South America, North America, Africa, and Bermuda, and redescribed as a new species 9 times, and the shipworm *Lyrodus pedicellatus*, not surprisingly, is redescribed at least 11 times from around the world.

Introductions continue to be redescribed as new taxa (Table 2.2): the Indo-Pacific mantis shrimp *Gonodactylaceus falcatus* was redescribed as a new species in 1981 from the Hawaiian Islands; the Australasian mussel *Xenostrobus securis* was redescribed as new, also in 1981, from Japan; the Indian Ocean seasquirt *Eusynstyela hartmeyeri* was redescribed as a new species from New Caledonia in 1991, and the Japanese skeleton shrimp (caprellid) *Caprella mutica* was redescribed as a new species from Europe in 1995. Kott (2004) described an abundant non-native ascidian (not seen or collected prior to the 1970s) in the New England fauna as an endemic species (*Didemnum vestum*), whose origin remains uncertain (indeed, it may fall into the second category of pseudoindigenes, below). All of these are examples only, and an unknown number of "new" species described from shallow coastal waters, especially harbors and estuaries subjected to the vectors that transport species around the world, are actually redescriptions of species already described from elsewhere in the world.

The lag time in recognizing that an introduced species has been mistakenly redescribed ranges from months to over 100 years. For example, the Japanese clam *Venerupis philippinarum* was mistakenly redescribed from British Columbia in 1938, and the Japanese ascidian *Styela clava* was mistakenly redescribed from England in 1954: in both cases, the error was recognized immediately. On the other hand, it took 125 years to show that a European seaslug, *Corambe batava* (long regarded as an endangered if not extinct species in Europe), was the common American seaslug *Corambe obscura*. Thus a great many pseudoindigenes remain

Table 2.2 Examples of introduced species first described from non-native regions (and later found elsewhere)

Species	Native to	First Described from	Reference
Turbellaria (flatworms)			
Taenioplana teredini Hyman, 1944	Southern hemisphere?	Hawaii	Prudhoe (1985)
Hydrozoa (hydroids)			
Blackfordia virginica Mayer, 1910	Ponto-Caspian	Virginia	Mills and Sommer (1995)
Garveia franciscana (Torrey, 1902)	South Pacific-Indian Ocean	San Francisco Bay	Vervoort (1964)
Pachycordyle navis (Millard, 1959)	Europe?	South Africa (from hull of resident ship)	Schuchert (2004)
Anthozoa (sea anemones)			
Nematostella vectensis Stephenson, 1935	North America	England	Pearson et al. (2002); Darling et al. (2004)
Carijoa riisei (Duchassaing & Michelotti, 1860)	Indo-West Pacific	Caribbean	Kahng (2005)
Clitella (oligochaete worms)			
Branchiura sowerbyi Beddard, 1892	Asia	Regent's Park, London	Cohen and Carlton (1995)
Limnodriloides monothecus Cook, 1974	Western Atlantic	Baja California	Cohen and Carlton (1995)
Gastropoda (snails)			
Babakina festiva (Roller, 1972)	Asia	California	Carlton (1979)
Bivalvia (bivalves)			
Mya arenaria Linnaeus, 1758	North America	North Sea	Coan et al. (2000)
Crassostrea angulata (Lamarck, 1819)	Asia	Europe	Ó Foighil et al. (1998)
Teredicola typica Wilson, 1942	Southern hemisphere?	Hawaii	McKoy (1975)

Isopoda (isopods)			
Iais californica (Richardson, 1904)	New Zealand	California	Rotramel (1972)
Limnoria tripunctata Menzies, 1951	Southern hemisphere	California	Cookson (1991)
Alloniscus oahuensis Budde-Lund, 1885	Indo-Pacific	Hawaii	Taiti and Howarth (1995)
Amphipoda (amphipods)			
Gammarus tigrinus Sexton, 1939	Atlantic North America	England	Bousfield (1973)
Diptera (flies)			
Canaceoides angulatus Wirth, 1969	Tropical Eastern	Hawaii	Englund et al. (2000)
Procanace williamsi Wirth, 1951	Asia	Hawaii	Englund et al. (2000)
Bryozoa (bryozoans)			
Victorella pavida Kent, 1870	Indian Ocean	England	Carlton (1979)
Rhodophyta (red algae)			
Neosiphonia harveyi (Bailey, 1848)	Asia	Connecticut	McIvor et al. (2001)
Spermatophyta (higher plants)			
Spartina alterniflora (Loisell, 1815)	North America	Bayonne, France	Mobberley (1956)

unrecognized, particularly in the face of declining taxonomic expertise required to ferret out such taxa.

The common early failure to recognize invasions as introductions can be understood in the context of the challenges of having access to sufficient literature, combined with an underappreciation for the potential of non-native species to appear at a systematist's doorstep. The failure to continue to do so in the twentieth and twenty-first centuries lies at several doorsteps, including the spread of species that are members of complexes difficult to tease apart morphologically, concomitant with increasing globalization and speed of commerce, leading to the requirement to be a master of the world's biota, a challenge for even specialized systematists. The lack of global knowledge of specific groups explains in large part the failure of many "environmental impact surveys" to report new invasions, as those workers involved in making routine identifications of marine invertebrates often do not possess global-scale expertise.

Whether newer or older invasions, the presence of anomalous clades in a biota does not always attract the attention of systematists. The "Californian" isopod *Sphaeroma pentodon* took many years to be accepted by North American workers as the Australian-New Zealand isopod *Sphaeroma quoianum*, despite the fact that the speciose genus *Sphaeroma* is centered in the western Pacific and Indo-Pacific, and no other species of *Sphaeroma* other than *S. pentodon* was known from the Northwest Pacific Ocean north of 25° north latitude. Monogeneric species occurring in oceans thousands of kilometers away from their closest evolutionary relatives or away from where the genus is otherwise represented by many species would bear reexamination relative to their biogeographic affinities and history, and thus perhaps their true identity.

Category 2: introduced species first described as new after introduction, and later found elsewhere As a result of species being transported from a region where the biota is poorly described to a region under greater biological and systematic scrutiny, a number of invasions have been first described in areas where they are not native. Twenty-one examples of such species that were then later discovered in their native, or other regions, are shown in Table 2.2. Fewer examples of these are available than those in the previous category, because the necessary "matches" require sophisticated global knowledge and, of course, sufficient exploration in native regions. Thus, the type locality of a species does not necessarily imply where a species is native. For the Hawaiian Islands, Cowie (1998) has noted that "It was only in the middle of the nineteenth century that naturalists really began to take note of the Hawaiian biota, describing many new, supposedly endemic species … Six of the ten introduced species first recorded between 1840 and 1889 were originally described from the Hawaiian Islands."

There may be equally long lag times in resolving where such species are native (introduced species that were described from a non-native region but that have not yet been found elsewhere are discussed in the next category). The common and widespread Atlantic North American filamentous red alga, *Neosiphonia harveyi* (= *Polysiphonia harveyi*), described in 1848 from Connecticut, was not recognized as native to Asia until the 1990s (McIvor et al. 2001). This species is further discussed below in two contexts, relative to biogeographic biases in detecting invasions, and relative to the concept of the "Missing 1000" invasions (Carlton 2000, 2002, and below).

One of the more famous invasions in this category (or perhaps the first, above) is the case of the appearance of an Asian oyster in the genus *Crassostrea* in the European theater about 500 years ago. Described by Lamarck in 1819 as a native southern European species, *Ostrea angulata* (now *Crassostrea angulata*), and long known as the "Portuguese oyster," investigators had concluded by the 1940s that it was identical with the Japanese *C. gigas*. However, recent work suggests that *C. angulata* may be a genetically distinct (although morphologically identical) sibling species of *C. gigas* (Ó Foighil et al. 1995, 1998; Boudry et al. 1998; Huvet et al. 2000, 2004; Lapegue et al. 2004; P. Gaffney, personal communication). If the two are not the same, *C. angulata* represents a case similar to that of *Mya arenaria* and *Spartina alterniflora*, noted below, although the recognition of *C. angulata* as a distinct genospecies in Asia has taken nearly 300 years.

Of interest is that two iconic marine organisms native to the American Atlantic coast were first described from Europe: the edible soft-shell clam *Mya arenaria* was described from the North Sea by Linnaeus in 1758 (what could not be known to Linnaeus was that it was introduced centuries earlier by the Vikings from North America), while the salt marsh cordgrass *Spartina alterniflora* was first described from France in 1815 (to where it had been introduced, perhaps by shipping, presumably in the eighteenth century or earlier). In both cases, the American biota was not yet well described, and, as noted above, these (and no doubt other) species first landed at the feet of European taxonomists. Similarly, the nineteenth century importation of exotic aquatic plants from Asia to the Kew Gardens of England (Desmond 1995) brought as yet undescribed associated species to the attention of British zoologists; examples include the freshwater jellyfish *Craspedacusta sowerbii* (Table 2.4 section B) and the freshwater worm *Branchiura sowerbyi* (Table 2.2).

For both this and the next category, an attendant risk is using a geographic name for a species, especially for taxa from harbors, ports, and other coastal zones influenced by human transport mechanisms (such as the Indian Ocean hydroid *Garveia franciscana*, or the New Zealand isopod *Iais californica*). Species with trivial names such as *californica*, *mexicana*, *mediterraneus*, and so forth, do not easily invite workers to look for the same species to be native on the other side of the world (see also an example from Brazil in Chap. 27, Ferreira et al.).

Admitted to Table 2.2, but an exception relative to the other species treated in this category, is the seaslug *Babakina festiva*. Although first described from California and not recognized as an introduction at the time, Roller (1972) noted that it had been known from Japan since at least 1956.

Unrelated to this phenomenon is the error of bestowing a geographic name on a species based on a mistaken source of the specimens being described. Thus the leucosiid (pebble) crab *Persephona mediterranea* (Herbst 1794) is a Western Atlantic species which does not occur in the Mediterranean (Williams 1984), while the purse oyster *Isognomon californicum* (Conrad, 1837) is a native Hawaiian species which does not occur in California (Kay 1979). A specimen of the Indo-West Pacific angelfish *Chaetodontophus mesoleucus*, mislabeled as coming from the Hawaiian Islands, was mistakenly redescribed as *Holacanthus bicolor oahuensis* Borodin, 1930 (Mundy 2005).

Category 3: introduced species described as new after introduction, and remaining unknown elsewhere The fewest examples of pseudoindigenes are in this category: if a species is not known from elsewhere, an assumption may be that it is native to the place where it is known. On the other hand, there may be clear but overlooked evidence that a species is not native. Such evidence may include localization to one or a few harbors along a well-explored coastline (perhaps combined with the demonstration that the species was not collected earlier and was not likely to be overlooked in previous centuries, when, in the same locations and habitats, extremely rare native species were collected), species whose only close morphological relatives are in another part of the world, species closely associated with a human transport vector, and so on. As noted elsewhere (Chapman and Carlton 1991, 1994), multiple sets of evidence, rather than relying on any one supporting criterion, strengthen such recognition.

Seven examples are shown in Table 2.3. Species that are restricted to one or a few harbors and are otherwise associated with introduced biotas include the sponge *Halichondria coerulea* and the crab *Pilumnus oahuensis*, both in Hawaii.

Table 2.3 Examples of introduced species described from non-Native regions and remaining unknown elsewhere

Species	Native to	Described from	Reference
Anthozoa (corals)			
Culicia rachelfitzhardingeae Cairns, 2006	Indo-Pacific?	Hawaii	Cairns (2006); Carlton and Eldredge (2009)
Porifera (sponges)			
Halichondria coerulea Bergquist, 1967	Indo-Pacific?	Hawaii	Carlton and Eldredge (2009)
Isopoda (isopods)			
Iais floridana Kensley & Schotte, 1999	Indian Ocean?	Florida	Herein
Caeijaera horvathi Menzies, 1951	Southern hemisphere?	California	Carlton and Eldredge (2009)
Ostracoda (ostracodes)			
Redekea californica De Vos & Stock,1956	Southern hemisphere?	California	Carlton (1979)
Decapoda (crabs)			
Pilumnus oahuensis Edmondson, 1931	Indo-Pacific/tropical East Pacific?	Hawaii	Carlton and Eldredge (2009)
Tunicata (sea squirts)			
Botrylloides diegensis Ritter & Forsyth, 1917	Indo-Pacific?	California	Carlton (2005)

The commensal isopod *Iais floridana*, described from Florida, is regarded as an introduction from the Indo-Pacific: its sole host is the introduced isopod *Sphaeroma terebrans*, and it is a close relative (if not a sibling species) of the Pacific Ocean *Iais singaporensis*.

The California botryllid seasquirt *Botrylloides diegensis*, described in 1917 from harbor pilings in San Diego, California, is regarded as introduced from the Western or South Pacific, where it is predicted it will be discovered. It is part of a clade of *Botrylloides* species all found in the Western or South Pacific; *B. diegensis* has no relatives in the Northeastern Pacific. It is suggested that it was carried in ship fouling to southern California in the nineteenth or earlier centuries.

Described as new, but recognized at the time as introduced Not strictly members of the above three categories are introduced species that while described as new were recognized at the time of description as non-native. Thus, these are not pseudoindigenes, as they were not mistaken as native species. Examples of these, which are presented here as a "tip of the hat" to the systematists and biologists who recognized their species as non-native, are shown in Table 2.4, divided into three categories, that parallel those above: species that were mistakenly redescribed (Table 2.4 section A), species first described from a non-native region and then found elsewhere (Table 2.4 section B), and species described from a non-native region and remaining unknown elsewhere (Table 2.4 section C). These investigators employed a variety of evidence to deduce that the species was not native; this evidence includes prior absence, association with a habitat created by human activity and dominated by exotic biota, and morphological similarity to autochthonous taxa. In the first case (Table 2.4 section A), there is the occasional temptation to describe introduced populations – although recognized as such! – as new subspecies (such as the barnacle *Balanus amphitrite saltonensis*, the amphipod *Caprella acanthogaster humboldtiensis*, the worm *Pseudopolydora kempi californica*, all from California, and all bestowed with regional names).

Cooper (1872), although recognizing the potential for ship-mediated transport of marine organisms, proceeded to describe the marsh snail *Alexia setifer* as a new species from San Francisco Bay, pointing out that the localities where it was found had been searched by collectors "for more than 20 years" prior to its discovery in 1871 – a rather strong assertion for how well those Bay shores were known in the 1850s and 1860s. Although unable to match it with a described Asian species, Cooper speculated that it may have been introduced with ships from China. Less than one year later Cooper (1873) recognized that it was the Atlantic snail *Phytia myosotis* (now *Myosotella myosotis*), although an erroneous reversal of nomenclatural fortune in the 1920s led to the use of the specific name *setifer* once again for the next 30 years.

The Japanese oyster-eating flatworm *Pseudostylochus ostreophagus* was first discovered in Puget Sound, Washington, in beds of imported Japanese oysters (*Crassostrea gigas*). Upon its discovery, fisheries biologists immediately went to Japan and discovered the flatworm there. The description of this worm as a new species was thus based on specimens from its native region (Hyman 1955). It is not listed in any of the tables here. This is a rare instance – and provides an important

Table 2.4 Examples of introduced species recognized at the time of their description as non-native or possibly non-native

A. Introduced species mistakenly redescribed from non-native regions

Species	Native to	Redescribed as (from)	Reference
Turbellaria (flatworms)			
Caenoplana coerulea Moseley, 1877	Australia	*Geoplana vaga* Hyman, 1943 (California)	Ogren (1989b)
Polychaeta (worms)			
Pseudopolydora kempi (Southern, 1921)	Indo-West Pacific	*Pseudopolydora kempi californica* (California)	Radashevsky and Hsieh (2000)
Cirripedia (barnacles)			
Balanus amphitrite Darwin, 1854	Indo-Pacific	*Balanus amphitrite saltonensis* Rogers, 1949 (Salton Sea, California)	Henry and McLaughlin, (1975); Flowerdew (1985); Raimondi (1992)
Amphipoda (amphipods)			
Caprella mutica Schurin, 1935	Japan	*Caprella acanthogaster humboldtiensis* Martin, 1977 (California)	Marelli (1981)
Mysidacea (mysids)			
Hyperacanthomysis longirostris Ii, 1936	Asia	*Acanthomysis bowmani* Modlin & Orsi, 1997 (California)	Fukuoka and Murano (2000)
Gastropoda (snails)			
Myosotella myosotis (Draparnaud, 1801)	Europe	*Alexia setifer* Cooper, 1872 (California)	Cooper (1872); Martins (1996)
Tunicata (sea squirts)			
Eusynstyela hartmeyeri Michaelsen, 1904	Red Sea, Indian Ocean	*Eusynstyela aliena* Monniot, 1991 (New Caledonia)	Monniot and Monniot (2001)

B. Species first described from non-native regions and subsequently found elsewhere

Species	Native to	Described from	Reference
Hydrozoa (hydroids)			
Craspedacusta sowerbii Lankester, 1880	China	Kew Gardens, London	Russell (1953)
Polychaeta (worms)			
Ficopomatus enigmaticus (Fauvel, 1923)	Australia	France	Cohen and Carlton (1995)
Copepoda (copepods)			
Oithona davisae Ferrari & Orsi, 1984	Asia	California	Ferrari and Orsi (1984)

(continued)

Table 2.4 Examples of introduced species recognized at the time of their description as non-native or possibly non-native

Isopoda (isopods)			
Gnorimosphaeroma rayi Hoestlandt, 1969	Japan	California	Hoestlandt (1973)
Bryozoa (bryozoans)			
Watersipora edmondsoni Soule & Soule, 1968	Indo-Pacific	Hawaii	Winston and Heimberg (1986)

C. Species first described from non-native regions and remaining unknown elsewhere

Species	Native to	Described from	Reference
Turbellaria (flatworms)			
Bdellocephala exotica (Hyman, 1953)	Asia?	Washington, D.C.	Hyman (1953)
Bipalium pennsylvanicum Ogren, 1987	Asia?	Pennsylvania	Ogren (1987)
Anthozoa (sea anemones)			
Diadumene franciscana Hand, 1956	Indo-Pacific/Asia?	California	Cohen and Carlton (1995)
Amphipoda (amphipods)			
Transorchestia enigmatica Bousfield & Carlton, 1969	New Zealand/Chile	California	Bousfield (2007)
Corophium alienense Chapman, 1988	Southeast Asia	California	Chapman (1988)
Mysidacea (mysids)			
Deltamysis holmquistae Bowman & Orsi, 1992	Asia	California	Bowman and Orsi (1992)
Decapoda (crabs)			
Acantholobulus pacificus (Edmondson, 1931)	Tropical Eastern Pacific	Hawaii	Felder and Martin (2003)

lesson in systematic biogeography – of a species being discovered in a non-native region, followed by biologists taking the unusual pro-active step (in this case economically motivated) to discover the origin of the species.

Category 4: introduced species misidentified as previously known native species Introduced species may be misidentified as native species because of insufficient taxonomic resolution, or because, despite seemingly adequate morphological taxonomy, cryptic invasions may occur that can only be revealed genetically.

Imperfect taxonomy Non-native species may be misidentified as described native species; we might say, "similar-looking native species," but such is not always the case. The introduced Atlantic gem clam *Gemma gemma* was misidentified as the quite distinct native Pacific clam *Transennella tantilla* (now *Nutricola tantilla*) throughout the American Pacific Northwest for many years (Carlton 1979). The introduced Atlantic amphipod *Ampelisca abdita* was referred to for many years in

San Francisco Bay as the Pacific amphipod *Ampelisca milleri*, despite the fact that, apart from clear morphological differences, the former occurred in estuarine muds in the Bay, and the latter was described from the open ocean, 75 m offshore of San Miguel Island, in southern California (Chapman 1988). The Japanese snail *Batillaria attramentaria* was misidentified as the native snail *Cerithidea californica* when first found in 1951 in Monterey Bay, California (Carlton 1979).

In more recent and better-known examples, the invasion of the Japanese seastar *Asterias amurensis* in Australia went unnoticed because it was misidentified for some years as the native seastar *Uniophora granulata* (Buttermore et al. 1994), and the first-observed specimens of the Caribbean barnacle *Chthamalus proteus* in Hawaii were identified as the native Hawaiian barnacle *Euraphia hembeli* (Zabin et al. 2007).

It may thus be predicted with some confidence that invasions have gone, or will go, unnoticed because of external morphological resemblances to similar-looking native species. The northeastern Pacific barnacle *Balanus glandula* is now well established in Argentina (Schwindt 2007) and in Japan (Kado 2003), but as a small white "acorn barnacle," may be overlooked elsewhere. Similarly, *Chthamalus proteus*, now abundant in the Hawaiian Islands (Southward et al. 1998) would be a challenge to detect if introduced to the warm coasts of California and Mexico (C. Zabin, personal communication), where other *Chthamalus* species occur. Godwin (2003) has noted that *C. proteus* survives on vessels on round-trip voyages between California and Hawaii.

Cryptic genospecies invasions Cryptic species (Bickford et al. 2007) provide one of the most difficult challenges in recognizing invasions, especially if a new invader appears to be morphologically identical to a native congener. A well-known example is the invasion of the Mediterranean mussel *Mytilus galloprovincialis* in southern California in the mid-twentieth century, which although first believed to be an invasion from Japan, was concluded to be a resurgence of the native mussel *Mytilus trossulus* (then known as *Mytilus edulis*, and specifically described as a new subspecies, *M. edulis diegensis*; Carlton 1979). As Geller (1999) has shown, this "resurgence" consisted of the invasion of the morphologically identical but genetically distinct non-native *M. galloprovincialis*.

In a parallel case, the marsh reed *Phragmites australis* "became" invasive in the nineteenth and twentieth centuries in Eastern North America; the invasion was caused by a non-native (but morphologically similar) genotype of *P. australis* (Saltonstall 2002).

Genetic analyses have also aided in revealing invasions of species of the jellyfish *Aurelia* (Dawson et al. 2005) and *Cassiopeia* (Holland et al. 2004), the polychaete worm *Myrianida* (Nygren 2004), the vermetid snail *Thylaeodus* (Strathmann and Strathmann 2006; Carlton and Eldredge 2009), the freshwater limpet *Ferrissia* (Walther et al. 2006), the bryozoans *Bugula* and *Watersipora* (Mackie et al. 2006), the brittlestar *Ophiactis* (Roy and Spooner 2002), and others. Many more such cases are to be expected.

Unidentified species, including many "cosmopolitan" species In most surveys of fauna and flora (sensu lato) some to many species cannot be identified for many reasons. These taxa should not default to being listed as "native," as is often the case. Thus these species are often not amenable to biogeographic assessment unless they are demonstrably undescribed taxa recognized as probable endemics to the region based upon habitat (such as deep cave endemics) or related taxa. Lee et al. (2003) refer to unidentified species as "indeterminate taxa." Clearly, introduced species may be amongst these, and perhaps commonly so. As noted above, when evidence can be mustered, some unidentified taxa can be considered cryptogenic, but taxa resolved only to phyletic, class, or ordinal levels should generally be considered as unidentified, and not cryptogenic (as used, for example, by Wasson et al. 2004).

Added to this category would be certain "species" whose only apparently available scientific name is one used for the same, or similar-looking, taxon around the world. These "cosmopolitan" taxa may include (1) species that have been globally spread by ships or other vectors (introductions), (2) a species-complex (including both native – and often undescribed – and introduced species), or (3) conceivably one naturally widespread species, although adequate mechanisms for global gene flow in ecological time that would prevent allopatric speciation are difficult to imagine, especially when no dispersal corridors appear to exist.

Thousands of species groups have not been adequately sorted into one of these three categories, and thus taxa with cosmonames should in many cases simply, but frustratingly, revert to being regarded as unidentified. This strategy would serve to strip away from a number of lists both introduced and cryptogenic species (for example, Cohen and Carlton 1995; Wasson et al. 2004). Marine examples include some "species" in the protist genera *Zoothamnium* and *Vorticella*, the sponge genus *Cliona*, hydroid genera such as *Plumularia, Sertularella, Campanularia, Obelia, Gonothyraea*, and *Dynamena*, bryozoan genera such as *Bugula* and *Bowerbankia*, and the caprellid amphipod genus *Caprella*.

An immediate derivative of this perspective is that assessments of regional biodiversity should divide all taxa into four categories: native, introduced, cryptogenic, and unidentified.

Small species Many microbial (Finlay 2002; Fenchel and Finlay 2004) and microscopic (Wyatt and Carlton 2002) species – essentially many taxa less than 1 mm in size – are considered naturally cosmopolitan, in part based upon the presumption that global dispersal for small organisms is naturally fluid and continuous, thus preventing allopatric speciation. Arguing against this is that wind and water do not act to homogenize the entire world; recent molecular evidence (Green and Bohannan 2006) thus suggests, not surprisingly, that greater provincial diversity exists among small organisms.

Because of severe taxonomic and biogeographic challenges, including the perception of natural cosmopolitanism noted above, small organisms are reported as introductions far less often than larger organisms. Newly-discovered small organisms– perhaps noticed because they have become common to abundant – are often assumed to be native. This rationale is based in part upon two arguments: (1) previously rare taxa may respond to environmental changes and become abundant (and thus detectable), and (2) new techniques permit the discovery of previously undetectable taxa. Relative to the first case, newly recognized taxa invoked as native were not simply previously rare – they were never previously recorded. In contrast, in most biotas, many extremely rare species may have been recorded over time. This argument thus requires that the species were so rare as to have survived no previous detection, and had no previous episodes of becoming common or abundant. Relative to the second case, improved microscopic and molecular techniques do permit fine-scale resolution of taxa previously undetectable, but previously undetectable taxa are not by default native – they were simply previously undetectable.

Carlton (in Wyatt and Carlton 2002) referred to this phenomenon as the "smalls rule of invasion ecology," defined as an inverse correlation of body size with the ability to be recognized as non-native (Carlton 2003). Thus small filamentous algae, other protists (such as foraminiferans, rhizopods, actinopods, ciliates, dinoflagellates (including *Pfiesteria* spp.), pelagophyceans (including *Aureococcus* spp.), diatoms, etc.), hydroids, flatworms, small nemerteans, rotifers, gastrotrichs, kinorhynchs, nematodes, oligochaetes, small polychaetes, copepods, mites, bryozoans, and a host of other taxa are rarely reported as invasions.

The same arguments that underlie the principles of natural dispersal of small organisms – ease of transport, the ability to form dormant life stages, large population sizes, and other characteristics (Fenchel and Finlay 2004; Green and Bohannan 2006) – equally support the possibility that many modern-day distributions of small organisms were created by the movement of terrestrial and aquatic media over the past centuries by human activity. Given this, it cannot logically be assumed that only natural dispersal has been in play.

Karling et al. (2000) examined the genetic variation (in the small subunit ribosomal RNA gene) of three species of planktonic foraminiferans that occur in both the Arctic and Antarctic, and identified at least one identical genotype in all three species, "indicating that trans-tropical gene flow must have occurred." Oceanographic phenomena that would lead to such gene flow in ecological time are speculative and have not been clearly demonstrated (Karling et al. 2000). In contrast, there has been clear potential for centuries of transtropical gene flow of planktonic foraminiferans – including between high-latitude waters – by means of steamship bath water in the nineteenth and early twentieth centuries, and by means of ballast water since the nineteenth century and continuing to date. Carlton (1985) has commented on the potential for the interchange of oceanic plankton by such human-mediated means as an alternative hypothesis to natural mixing.

The nearly complete absence of reports of introductions of such prominent planktonic and benthic taxa as diatoms and ciliate "protozoans" may be one of the larger gaps in introduced species diversity assessment. Indeed, in estuarine systems such as San Francisco Bay, it is possible that over 100 species of "protozoans" associated with soft and hard substrates, could be introduced, which, if so, alone would increase by more than one third the known invasions in the Bay. I comment upon the rarity of reports of introduced diatoms below.

Uninvestigated taxa A hallmark of modern censuses of marine life is that there are few or no systematists available to assist in the specialized collection and identification of an increasing number of taxa. The importance of the involvement of taxonomic experts in both field and laboratory work cannot be underestimated. Passive collections – samples taken by others and then provided to a systematist – often miss many small and cryptic taxa. Based upon my experience with "rapid assessment surveys" – in this case, biological surveys focused on boat harbor fouling communities along the coasts of North America and England – systematists working in the field often recover significantly more species in their specialty than do general field biologists. More importantly, there are often no available systematists in a region or country to assist either with such field work or with the identification of specimens.

Notable dearths in expertise now occur in "protozoans," sponges, hydrozoans, flatworms, nemerteans, nematodes, gastrotrichs, kinorhynchs, rotifers, kamptozoans, tardigrades, "oligochaetes," polychaetes, leeches, ostracodes, copepods, peracarid crustaceans in general (mysids, cumaceans, tanaids, isopods, amphipods), pycnogonids, mites, marine insects, bryozoans, and ascidians. In all of these groups, with regional exceptions, invasions are rarely reported. This category (uninvestigated taxa) overlaps with that of another, undersampled parasitic, commensal, or symbiotic associations, below.

Known but unreported taxa Interviews with systematists, naturalists, local biologists with many years of regional residence, and others, reveal that researchers are frequently aware of unpublished records of introduced species in a given region. Primary reasons cited for not publishing such records are lack of a perceived outlet to publish a paper on a new geographic record of one species, lack of time to write such papers, and, with some museum taxonomists, surprise that there would be interest in a new record of a small or "obscure" species.

2.3 Invader Underestimation – Biogeographic and Community History

2.3.1 Widespread Intraoceanic and Interoceanic Corridor Species

Widespread species within an ocean basin (intraoceanic) and between ocean basins (interoceanic) are almost always interpreted as natural distributions, in place for an undetermined length of time. When corridors are present – coastlines, continental shelves, islands, and so on – the interpretation of such distributions as natural and long-standing appears logical. On the other hand, absent paleontological and archeological evidence, we often have little to no understanding of the aboriginal distributions of many such widespread "corridor" species prior to the onset of global shipping, whose multi-millennial antiquity relative to invasions is clear (di Castri 1989; Leppäkoski et al. 2002; Wolff 2005).

What is the scale of this ancient potential human-mediated homogenization of biota, resulting in biogeographic patterns that mirror presumptive natural patterns? The Indo-Pacific marine biota serve as an example: here, presumably millions of species (most undescribed) occur in shallow waters from the Red Sea to Australasia, and often to outlying island groups, as far as the Hawaiian Archipelago, with less diverse groups reaching as far as the tropical and subtropical Eastern Pacific islands and coastlines. From the 1500s to the 1800s, ocean-going ships were floating zoos and botanical gardens of entrained species inside and outside their hulls (Carlton 1987, 1999a; Carlton and Ruiz 2003). Carlton (1999a) constructed an "imaginary assemblage of marine organisms on a wooden sailing vessel of 1750," excluding parasites, commensals, and other symbiotic species. Noting the scale of microhabitats available – hull fouling, hull boring, hull nestlers, the anchor, anchor chain, and anchor chain locker, sand and rock ballast – total species richness was calculated as easily over 150 species.

Carlton (1987) noted that this potential for ship-mediated homogenization "throughout the atolls and high islands of the central and South Pacific has rarely been considered." That a great many interoceanic and intraoceanic species had more restricted ranges prior to the onset of shipping – even if corridors were apparently available to such taxa – needs careful examination, a possibility that can now be tested with genetic analysis. In a similar fashion, Carlton and Hodder (1995) have argued that even along a moderately uniform coastline, ships transporting native species for centuries may have obscured original patterns of distribution.

In addition, species introduced to a coastline (or to an archipelago) can, over the centuries, become so widespread as to mimic natural patterns: we are often disinclined to suspect a species' natural status if it occurs from Alaska to Mexico, or from the Bay of Fundy to the Gulf of Mexico. Since many introduced species on these (and all) coasts have easily achieved such wide distributions, latitudinal breadth of occurrence can rarely alone be used to indicate either endemicity or aboriginal distributional patterns: the Asian alga *Neosiphonia harveyi* now ranges from Newfoundland to the Caribbean. Strasser (1999) has also noted that distributions created by human activity in modern time can recreate and parallel ancient distributions: the clam *Mya arenaria* was once widespread through the high latitudes of the North Atlantic and North Pacific Oceans until Pleistocene glaciation eliminated all but the Western North Atlantic populations: it was subsequently introduced (not reintroduced, having never been introduced in the first place) by humans to the Pacific Ocean and to Europe, recreating Tertiary patterns.

2.3.2 Neritic Species with Presumptive Oceanic Dispersal

As noted above, for many taxa there is a presumption that natural dispersal may play the greater role in the distribution of a species, even if such taxa are recorded from ship fouling and are unknown from ocean currents. Orensanz et al. (2002) and Castilla et al. (2005), presenting the first inventories of marine bioinvasions of Uruguay/Argentina, and Chile, respectively, thus excluded from consideration hydromedusae (and their hydroid polyps) and wood borers (such as gribbles, limnoriid isopods) and shipworms (teredinid bivalves), under the argument that their dispersal may also be natural on ocean currents. This will very likely lead to an underestimation of invader biodiversity.

Hydrozoan taxa involved represent harbor-dwelling fouling species in such genera as *Obelia, Campanularia, Sarsia, Ectopleura, Pinauay*, and *Plumularia*. Wood-boring taxa involved live in shallow bays, estuaries, and harbors; these include certain species of shipworms in the genera *Teredo* and *Lyrodus*, and certain species of gribbles in the genus *Limnoria*, known to infest wooden ships for centuries. None of the wood-boring species that are found in estuaries or harbors (or in wooden-hulled vessels), and none of the hydroids (polyps and medusae) found in fouling communities or open waters of marinas, ports, and harbors (or on ships' hulls or in ballast water) have been reported from ocean currents. Carlton (1999b) detailed the arguments, using shipworms as a model, that neritic taxa capable of floating or being floated require discovery in the open ocean before oceanic dispersal can be invoked. Relative to shipworms, the species found in floating wood at sea are, not surprisingly, a guild of neustonic, oceanic species, capable of living in that environment (and, in turn, are not the shipworm species found in harbors and ports).

2.3.3 Resident Species

As noted above, historical invasions are often difficult to detect, albeit less so now with the availability of genetic techniques. Not surprisingly, most invasions in the ocean have been recognized only since marine biologists appeared on the scene, even though vectors such as shipping had been in place for many centuries. Thus, European species were regularly transported to New England (and vice versa) on and in ships from the 1500s and on (and with more episodic earlier Viking voyages 500 years earlier). The documentation of the marine fauna and flora on both sides of the North Atlantic commenced 200–300 years later, in the 1700s and 1800s, with North American animals and plants being shipped back to European biologists for study and naming. Although larger fish, mollusks, and crustaceans reached European cabinets and universities by the late 1600s, most smaller taxa (such as bryozoans, sponges, hydroids, ascidians, small crustaceans, worms, and the remaining plethora of small invertebrates) were not specifically collected and shipped as such. The long history of invasions prior to collectors and biologists being present, combined with the lack of early investigations of many animal and plant groups, set the stage for the presumption of natural amphiatlantic distributions for many shallow-water taxa (Carlton 2003).

Stachowicz et al. (1999, 2002a, b) thus categorize certain fouling ascidians that occur both in Europe and New England as "native" species, including *Ciona intestinalis* and *Botryllus schlosseri*, as well as the fouling bryozoan *Cryptosula pallasiana*, and examine the interaction between these species and the introduced ascidians *Ascidiella aspersa, Diplosoma listerianum*, and *Botrylloides violaceus*. Stachowicz et al. (2002b) note that "Although the true status of the New England sea squirts as natives or invaders is difficult to resolve due to the poor fossil record of these soft-bodied organisms, all of these species have been present in New England for as long as humans have been studying these animals. Thus, these species form the resident community that current invaders encounter upon arrival."

Species status is not a dichotomy of native or introduced, and *Ciona, Botryllus* and *Cryptosula* can be considered cryptogenic, but not native. Genetic investigations will materially aid in sorting out these histories. Van Name (1945) considered *Botryllus schlosseri* to be introduced by ships from Europe to North America, and Carlton (2005) has proposed that it may in fact be native to the southwestern Pacific Ocean. Equally – or more – interesting, of course, is the interaction between a newer set of invaders and a previous set of invaders in terms of community history and development. Regardless, the concept of "resident" or "naturalized" is not a separate or distinct category of biogeographic, ecological, environmental, historical, or evolutionary status. Taxa capable of being identified to the species level are either native, introduced, or cryptogenic.

2.4 Invader Underestimation – Sampling

2.4.1 Species in Underexplored Habitats and Associations

All of the historical and taxonomic challenges noted above are further compounded when attempting to assess the systematics and biogeographic history of parasitic, commensal, and symbiotic organisms associated with either native species or demonstrably introduced taxa. In large part these challenges are rarely addressed since these associated taxa are rarely sampled in standard surveys. Monographic reports of invasions in marine and freshwater habitats from most areas of the world record the occasional parasitic or symbiotic species that were encountered by specialists, but the undersampling of this biotic component may rival the undersampling of microscopic free-living taxa.

Similarly, "microhabitats" that require specialized techniques, knowledgeable investigators, and experienced systematists, remain largely uninvestigated relative to invasions. Meiofaunal communities are a striking example. Ecotonal habitats, such as the supralittoral fringe (Carlton 2002) or oligohaline zones in estuaries, often have unique assemblages of species, but are the subject of few studies, and fewer still for invasions, as habitat-oriented ecologists often do not find themselves in transitional environments. Thus few terrestrial or marine ecologists have studied the intermediate maritime zone; similarly, freshwater and marine biologists have rarely studied the oligohaline zone between these two habitats.

2.4.2 Incipient Invasions: Species with Small Population Sizes

Newly invading species will, in the early stages of colonization, generally have small and restricted populations that are often difficult to detect through standard sampling programs, unless haphazardly stumbled upon. These are thus initially rare species, and fall into sampling challenges universal to assessing alpha diversity in a community (Rabinowitz et al. 1986; Chapman 1999). If vectors are present that lead to the continual inoculation (release) of propagules into a system, it is probable that at any given time some species are in the early stages of establishment, and may not be detected until several generations have reproduced. This is one of the most enduring challenges of quickly assessing the efficacy of vector management, as population lag times may lead to the detection of a new invader years after it is thought that a given vector is under some measure of control.

2.5 The Overestimation of Invader Diversity

Miscategorizing native species as introduced is rare, in large part because the classical default in systematics, evolutionary biology, and biogeography, is to assume that a species is native. Galil et al. (2002) note that the alpheid shrimp *Automate branchialis* Holthuis & Gottlie, 1958 was first considered to be a Lessepsian invader, but is in fact native to the Mediterranean Sea. Carlton and Eldredge (2009) note several cases of native Hawaiian mollusks, including *Bulla vernicosa, Vitularia miliaris*, and *Conus capitaneus*, mistaken as invasions. Understandably, based upon previous literature, Calcinai et al. (2004) treat the octocoral *Carijoa riisei* as an Atlantic species introduced to the Indo-Pacific, but it now appears to be native to the Pacific (Kahng 2005).

A classic and more complex case involves the history of the perception of the biogeography of the Atlantic South American xanthid crab *Pilumnoides perlatus* (Poeppig, 1836). Barnard (1950) noted its presence on ship-bottoms in South Africa; combined with its report of having been transported by ships to Britain, he suggested that it might be introduced to South Africa. However, Kensley (1981, p 10) speculated that its presence in both southern South America and in South Africa might be natural, explained by larval transport on the West Wind Drift, but he also believed that the question remained open as to whether it was introduced one way or the other (B. Kensley, personal communication).

Guinot and MacPherson (1987) then showed that the South African species, previously identified as *P. perlatus*, was in fact an undescribed taxon, naming it *Pilumnoides rubus*. In a further complication, they also then described the British specimens as a new native species, *Pilumnoides inglei*, rejecting the idea that it was introduced by ships. In addition to the fact that *P. inglei* had not been collected since 1913, they remarked that it was with "beaucoup d'hesitation que nous avons separe ce Crabe des cotes anglaises sous un nom distinct" [great hesitation that we have separated this crab from the British coasts under a distinct name]. They speculated that perhaps it was a cryptic species ("au biotope sans doute tres particulier" [without doubt a very special habitat]) in an attempt to explain why it had not been rediscovered. They noted it was, however, extremely close to the South American *P. perlatus*; despite describing it as a new species, they further noted that "nous n'avons pas releve de differences vraiment importantes entre ces deux especes" (!) [we have not noticed any truly important differences between the two species]). Ingle (1997) subsequently treated *P. inglei* as an introduced South American species, where, if it is distinct from *P. perlatus*, it remains unreported. As the systematics remains unsettled, it is not included here in Table 2.3, a category where *P. inglei* may eventually reside, returning to the original hypothesis that it was introduced.

2.6 Discussion

Taken as a whole, the combination of species that are cryptogenic, pseudoindigenous (including cryptic), unidentified, small, uninvestigated, unreported, pseudo-intraoceanic, pseudo-interoceanic, pseudo-oceanic, introduced "resident species," undersampled, and rare, potentially significantly alters our perception of the scale of invasions over time and space.

Perhaps no better examples of this are at hand than six recent independent studies in the Pacific and Atlantic Oceans (Table 2.5), summarizing known invasions in Denmark, the Azores, South Africa, Uruguay/Australia, Japan, and Chile. Each of these countries has experienced global shipping contact for many centuries. Despite the antiquity of external contact, no invasions are recognized in Denmark, Azores, South Africa, or Chile prior to the mid-nineteenth century (no earliest introduction dates are provided in the studies from Uruguay/Australia and Japan). The sole exception is the archeological recognition of the introduction by the Vikings of the North American soft-shell clam *Mya arenaria* to the European theatre in the thirteenth or fourteenth centuries.

Further, despite the potential scale of introductions, only a relatively few invasions are recognized from all of these locations, with a maximum of 33 species from the Azores, ranging down to only 18 in Denemark. In each of the countries shown in Table 2.5, we would expect significantly more invasions, at scales up to 5–10 times the numbers shown. The list of 22 species in South Africa (which has had European and transglobal shipping contact steadily since the 1500s) is rapidally expanding as more information is gathered; only 10 species were listed in the Robinson et al. (2005) publication vs the 22 in Chap. 23, Griffiths et al. Heavily obscured by the dark curtains of antiquity, the uncertainties of biogeographic interpretation, and sitting at the mercy of taxonomy, we simply do not yet know how many introduced species dominate the marine and estuarine environments of most coastlines of the world.

One of the central goals of ecology and evolution studies is to understand the patterns of the diversity, abundance, and distribution of species, and thus how communities came to be structured. Without an understanding of the history of communities, we cannot know the extent to which evolutionary processes have played fundamental roles in precipitating structures currently observed. The elegant summary of phytoplankton dynamics in San Francisco Bay, California, by Cloern and

Table 2.5 Estimates of number of introduced and cryptogenic marine species

Location	Number of introductions	Number of cryptogens	Earliest introduction	Reference
Atlantic Ocean				
Denmark	18	None listed	1895[a]	Jensen and Knudsen (2005)
Azores	33	18	1887	Cardigos et al. (2006)
South Africa	22	18	1955	Robinson et al. (2005); Chap. 23, Griffiths and Robinson
Uruguay/ Argentina	31	46	Not indicated	Orensanz et al. (2002)
Pacific Ocean				
Japan	25	None listed	Not indicated	Otani (2004)
Chile	51	None listed	1864	Castilla et al. (2005); Chap. 26, Castilla and Neill

[a] Except for the North American clam *Mya arenaria*, introduced circa 1250–1300

Dufford (2005) provides an example of the potential importance of understanding the history of invasions.

Cloern and Dufford (2005) report 500 distinct phytoplankton taxa in San Francisco Bay, with 396 of these identified to species level. The 81 most important species (by biomass) consist of a "community of cosmopolitan phytoplankton commonly observed in temperate estuaries and coastal waters globally," and "many key phytoplankton species in San Francisco Bay are the same taxa that develop blooms in the adjacent coastal upwelling systems, sugesting that phytoplankton diversity inside the estuary is influenced by exchanges with the coastal Pacific Ocean." Despite the predominance of phytoplankton in ballast water (McCarthy and Crowder 2000; Hulsman and Galil 2002), and despite the number of other additional vectors that have transported benthic diatoms to San Francisco Bay, no introduced diatoms, dinoflagellates, or other phytoprotists are recognized in San Francisco Bay, at either the morphospecies or genospecies level. Whether the presumably endemic coastal taxa informing San Francisco Bay populations are genetically the same as the estuarine taxa has by and large apparently not been established, although Cloern and Dufford (2005) note that "one mode of resilience to environmental variability is the occurrence within morphospecies of genetically distinct strains" (such as, we add here, might be introduced to estuarine environments). That diatoms in the open coastal zone can also be introduced is illustrated by the invasion of the (Australasian?) surf diatom *Attheya armatus* (= *Chaetoceros armatum*) around 1950 into the Pacific Northwest of North America (Schaefer and Lewin 1984).

Cloern and Dufford (2005) note that "phytoplankton photosynthesis is the primary energy supply to metazoan food webs of San Francisco Bay," and that growth and fecundity of invertebrates are strongly correlated with the HUFA (highly unsaturated fatty acids) of their food, particularly eicosapentaenoic acid (EPA) and docosahexaenoic acid (DHA). They note that 17 species comprise 89% of the phytoplankton biomass in the Bay, including two diatoms, two dinoflagellagtes, two cryptophytes, and *Mesodinium rubrum*; in combination these are rich in DHA and EPA and thus "the phytoplankton in San Francisco Bay are of high nutritional quality." Since the biogeographic history of the phytoflora of the Bay is unknown, whether this is a natural situation remains unknown.

In addition, phytoplankton size structure "influences energy allocation between the competing benthic and pelagic food webs because of differences in size-selective feeding between benthic suspension feeders (e.g. bivalve mollusks) and crustacean zooplankton (e.g. calanoid copepods)". Here again, which phytoplankters that contribute to the size spectrum in the Bay are or may be introduced is not known, so that any changing balance struck in influencing energy allocation over the decades is also not known. In turn it may be noted that a large proportion of the benthic suspension feeding community is composed of known introduced species in the Bay.

Cloern and Dufford (2005) further note that "92% of the [phytoplankton] biomass in San Francisco Bay comes from two algal divisions (diatoms and dinoflagellates) in which spore or cyst production is common" – benthic stages being a "mechanism to retain a species' genome within strongly advective systems such as estuaries." Cloern and Dufford (2005) suggest that this retentive mechanism may explain the predominance of cyst-forming species in the Bay – in turn, it may also explain the success of many of these species if they are introduced.

It seems clear that, in the absence of a phytoplankton history of the Bay, and given the fundamental role of phytoplankton in structuring estuarine trophodynamics, if a number of the species of abundant diatoms in San Francisco Bay were not there 100 years ago, a remarkble conversion in energy flow has occurred. While we use phytoplankton communities as a model here for the scale of what may have changed, the principles apply to all taxa, from bacteria to fish.

Carlton (2000, 2003), using the phrase the "Missing 1000," noted that the combination of shipping history and the lack of historical records in many parts of the world could have led to "nearly 1000 coastal species" being early introductions that have been overlooked and are now regarded as native: the late (1990s) recognition of the Asian alga *Neosiphonia harveyi* on the American Atlantic coast (there since the 1840s or earlier), and the even more recent discovery that the "Caribbean" octocoral *Carijoa riisei* (in the Atlantic since the 1850s or earlier) is native to the Indo-Pacific, only serve to underscore the scale at which such invasions may have occurred 100, 200, or more years ago. In retrospect, the number 1000 now seems too low, when the full suite of the sources of error in estimating non-native species is taken into account.

Ubi sumus? (Where are we?) Historical and continuing invasions preclude community equilibrium. In general, despite the number of macroinvasions that most marine ecologists are aware of around the world, the general sense of both the ubiquity and potential of invasions, and their consequences, appear to remain "off the radar" as a general principle in marine biogeography and community ecology. Thus reports seemingly as simple as a newly discovered amphipod, *Jassa marmorata*, on the Uruguay and Argentinean coast (Alonso de Pina 2005) are made without reference to the possibility of introduction, and yet *J. marmorata* is a potentially important predator structuring communities (Armsby and Tisch 2006). *J. marmorata* now occurs, distributed by ships over centuries, in many regions of the world (Conlan 1990), and yet there are no reports of its role outside the North Atlantic Ocean, because no ecological or experimental studies outside of its native region have been conducted – exactly the type of non-report that has led not a few recent investigators to conclude that most invasions have little to no impact in the communities to which they are introduced. *Ubi sumus*?, indeed.

2.7 The Way Forward: Solutions

Despite the erosion of systematic resources, in terms of the declining availability of taxonomic expertise, there are solutions. In the description of new species or the re-evaluation of old species, eyebrows need to be arched more highly to encompass all known global species within the genus or family of concern, even those from the most distant shores. Such action will reveal that newly-encountered species may well have names elsewhere; hints to potential source regions may arise from a having a finger on the pulse of the diversity and origin of the vectors in one's region that would import non-indigenous species. Re-examination of highly-localized, "endemic" species, especially in urbanized estuaries, will reveal that some of these are redescriptions of species from far-flung corners of the world.

Materially aiding in this endeavor is the application of molecular genetics. Morphological analyses remain the "bread and butter" of identifying species, as laboratories that would undertake genetic studies are not available to or affordable by all. However, genetic techniques in the twenty-first century will become less expensive and will more universally supplement (but not supplant) morphological studies, and we will thus expect significant breakthroughs in assessing both local and global biodiversity, as well as the biogeographic origins of species.

Finally, as a working rule, fewer assumptions should be made about the endemic or indigenous status of species, regardless of their apparently "natural" wide distribution, their size, or their presumptive methods of "natural" dispersal. Being more receptive to the potential scale of biogeographic complications that humans have wrought upon the Earth, long before biologists were present to observe the seeds of change, may reveal the depth and breadth of biotic transformations that commenced many centuries ago.

Acknowledgements I thank Alejandro Bortolus, Edwin Grosholz, and the late Brian Kensley for pointing out the papers of Mobberley (1956), Karling et al. (2000) and Barnard (1950), respectively.

Michelle Armsby shared with me her work on *Jassa marmorata* in the Gulf of Maine. I am as always grateful to Isabel Stirling (University of California, Berkeley) who provided copies of difficult-to-find papers. I thank Lu Eldredge for a copy of Moseley (1878), and the late Robert E. Ogren for conversations at the American Society of Zoologists meeting in December 1989 in Boston, and for subsequently providing copies of his papers on *Bipalium* and other land planarians. My thanks to Gil Rilov and Jeff Crooks for the invitation to participate in this collection.

References

Abbott DP, Johnson JV (1972) The ascidians *Styela barnharti, S. plicata, S. clava*, and *S. montereyensis* in Californian waters. Bull S Calif Acad Sci 71:95–105

Ahyong ST (2001) Revision of the Australian Stomatopod Crustacea. Rec Aust Mus Suppl 26:1–326

Alonso de Pina GM (2005) A new species of *Notopoma* Lowry & Bernard, 1996, and a new record of *Jassa marmorata* Holmes, 1903, from the southwestern Atlantic (Amphipoda: Corophiidea: Ischyroceridae). Proc Biol Soc Wash 118:528–538

Armsby M, Tisch N (2006) Intraguild predation and cannibalism in a size-structured community of marine amphipods. J Exp Mar Biol Ecol 333:286–295

Banse K (1972) On some species of Phyllodocidae, Syllidae, Nephtyidae, Goniadidae, Apistobranchidae, and Spionidae (Polychaeta) from the northeast Pacific Ocean. Pac Sci 26:191–222

Barnard KH (1950) Descriptive catalogue of South African decapod Crustacea (crabs and shrimps). Ann S Afr Mus 38:1–837

Bax N, Carlton JT, Mathews-Amos A, Haedrich RL, Howarth FG, Purcell JE, Rieser A, Gray A (2001) The control of biological invasions in the world's oceans. Conserv Biol 15:1234–1246

Bickford D, Lohman D, Sodhi NS, Ng PKL, Meier R, Winker K, Ingram KK, Das I (2007) Cryptic species as a window on diversity and conservation. Trends Ecol Evol 22:148–155

Bigley RE, Barreca JL (1982) Evidence for synonymizing *Zostera americana* den Hartog with *Zostera japonica*. Aschers Graebn Aquat Bot 14:349–356

Blake JA, Maciolek NA (1987) A redescription of *Polydora cornuta* Bosc (Polychaeta: Spionidae) and designation of a neotype. Bull Biol Soc Wash 7:11–15

Boudry PS, Heurtebise B, Collet F, Cornette, Gerard A (1998) Differentiation between populations of the Portuguese oyster, Crassostrea angulata (Lamarck) and the Pacific oyster, Crassostrea gigas (Thunberg), revealed by mtDNA RFLP analysis. J Exp Mar Biol Ecol 226:279–291

Bousfield EL (1973) Shallow-water gammaridean Amphipoda of New England. Cornell Univ Press, Ithaca New York

Bousfield EL (2007) Amphipoda: Gammaridea: Talitridae. In: Carlton J (ed) The Light & Smith manual: intertidal invertebrates from central California to Oregon, edn 4. University of California Press, Berkeley Los Angeles

Bowman TE, Orsi JJ (1992) *Deltamysis holmquistae*, a new genus and species of Mysidacea from the Sacramento – San Joaquin estuary of California (Mysidae: Mysinae: Heteromysini). Proc Biol Soc Wash 105:733–742

Briggs EA (1931) Notes on Australian athecate hydroids. Rec Aust Mus 18:279–282

Broom JE, Nelson WA, Yarish C, Jones WA, Aguilar Rosas R, Aguilar Rosas LE (2002) A reassessment of the taxonomic status of *Porphyra suborbiculata*, *Porphyra carolinensis* and *Porphyra lilliputiana* (Bangiales, Rhodophyta) based on molecular and morphological data. Eur J Phycol 37:227–235

Buitendijk AM, Holthuis LB (1949) Note on the Zuiderzee crab *Rhithropanopeus harrisii* (Gould) subspecies *tridentatus* (Maitland). Zool Meded 30:95–106

Butman CA, Carlton JT, Palumbi SR (1995) Whaling effects on deep-sea biodiversity. Conserv Biol 9:462–464

Buttermore RE, Turner E, Morrice E, Morrice MG (1994) The introduced Northern Pacific seastar *Asterias amurensis* in Tasmania. Mem Queensl Mus 36:21–25

Byers JE, Reichard S, Randall J, Parker I et al. (2002) Directing research to reduce the impacts of nonindigenous species. Conserv Biol 16:630–640

Cadotte MW, McMahon SM, Fukami T (eds) (2006) Conceptual ecology and invasion biology: reciprocal approaches to nature. Springer, Dordrecht

Cairns SD (2006) New records of azooxanthellate Scleractinia from the Hawaiian Islands. BP Bishop Mus Occas Pap 87:45–53

Calcinai B, Bavestrello G, Cerrano C (2004) Dispersal and association of two aliens species in the Indonesian coral reefs: the octocoral *Carijoa riisei* and the demosponge *Desmapsamma anchorata*. J Mar Biol Assoc UK 84:937–941

Calman WT (1921) Notes on marine wood-boring animals – II. Crustacea. Proc Zool Soc London 1920:215–220

Cardigos FT, Tempera S, Avila J, Goncalves A, Colaco, Santos RS (2006) Non-indigenous species of the Azores. Helgoland Mar Res 60:160–169

Carlton JT (1979) History, biogeography, and ecology of the introduced marine and estuarine invertebrates of the Pacific coast of North America. PhD dissertation, University of California Davis

Carlton JT (1985) Transoceanic and interoceanic dispersal of coastal marine organisms: the biology of ballast water. Oceanogr Mar Biol Annu Rev 23:313–371

Carlton JT (1987) Patterns of transoceanic marine biological invasions in the Pacific Ocean. Bull Mar Sci 41:452–465

Carlton JT (1992) Introduced marine and estuarine mollusks of North America: an end-of-the-20th-century perspective. J Shellfish Res 11:489–505

Carlton JT (1996) Biological invasions and cryptogenic species. Ecology 77:1653–1655

Carlton JT (1999a) The scale and ecological consequences of biological invasions in the world's oceans. In: Sandlund OT, Schei PJ, Viken Å (eds) Invasive species and biodiversity management. Kluwer Academic Publishers, Dordrecht, pp 195–212

Carlton JT (1999b) Molluscan invasions in marine and estuarine communities. Malacologia 41:439–454

Carlton JT (2000) Quo Vadimus Exotica Oceanica?: marine bioinvasion ecology in the twenty-first century. In: Pederson J (ed) Marine bioinvasions: Proceedings of the First National Conference. Massachusetts Institute of Technology, MIT Sea Grant College Program, MITSG 00-2, Cambridge, Massachusetts, pp 6–23

Carlton JT (2002) Bioinvasion ecology: assessing invasion impact and scale. In: Leppäkoski E, Gollasch S, Olenin S (eds) Invasive aquatic species of Europe. Distribution, impacts, and management. Kluwer Academic Publishers, Dordrecht, The Netherlands, pp 7–19

Carlton JT (2003) Community assembly and historical biogeography in the North Atlantic Ocean: the potential role of human-mediated dispersal vectors. Hydrobiologia 503:1–8

Carlton JT (2005) Setting ascidian invasions on the global stage. Keynote Address, 2005 International Invasive Sea Squirt Conference. Woods Hole Oceanographic Institution, Woods Hole MA. Abstract: http://www.whoi.edu/page.do?pid=11421&tid=282&cid=16303 (accessed April 2007)

Carlton JT, Cohen AN (2003) Episodic global dispersal in shallow water marine organisms: the case history of the European shore crabs *Carcinus maenas* and *Carcinus aestuarii*. J Biogeogr 30:1809–1820

Carlton JT, Hodder J (1995) Biogeography and dispersal of coastal marine organisms: experimental studies on a replica of a 16th-century sailing vessel. Mar Biol 121:721–730

Carlton JT, Eldredge LG (2009). Marine bioinvasions of Hawai'i: the introduced and cryptogenic marine and estuarine animals and plants of the Hawaiian archipelago. Bishop Museum Press, Honolulu, Hawaii

Carlton JT, Ruiz GM (2003) The magnitude and consequences of bioinvasions in marine ecosystems: implications for conservation biology. In: Norse EA, Crowder LB (eds) Marine conservation biology: the science of maintaining the sea's biodiversity. Island Press, Washington, DC, pp 123–148

Carlton JT, Ruiz GM (2005) Vector science and integrated vector management in bioinvasion ecology: conceptual frameworks. In: Mooney HA, Mack RN, McNeely JA, Neville LE, Schei PJ, Waage JK (eds) Invasive alien species: a new synthesis. Island Press, Covelo, California, pp 36–58

Castilla JC, Uribe M, Bahamonde N, Clarke M, Desqueyroux-Faundez R, Kong I, Moyano H, Rozbaczylo N, Santelices B, Valdovinos C, Zavala P (2005) Down under the southeastern Pacific: marine non-indigenous species in Chile. Biol Invas 7:213–232

Chapman JW (1988) Invasions of the northeast Pacific by Asian and Atlantic gammaridean amphipod crustaceans, including a new species of *Corophium*. J Crust Biol 8:364–382

Chapman JW (2007) Gammaridean amphipods. In: Carlton JT (ed) The Light & Smith manual: intertidal invertebrates from central California to Oregon, edn 4. University of California Press, Berkeley Los Angeles

Chapman JW, Carlton JT (1991) A test of criteria for introduced species: the global invasion by the isopod *Synidotea laevidorsalis* (Miers, 1881). J Crust Biol 11:386–400

Chapman JW, Carlton JT (1994) Predicted discoveries of the introduced isopod *Synidotea laevidorsalis* (Miers, 1881). J Crust Biol 14:700–714

Chapman MG (1999) Are there adequate data to assess how well theories of rarity apply to marine invertebrates? Biodivers Conserv 8:1295–1318

Clench WJ, Turner RD (1948) The genus *Truncatella* in the Western Atlantic. Johnsonia 2:149–164

Cloern JE, Dufford R (2005) Phytoplankton community ecology: principles applied in San Francisco Bay. Mar Ecol Progr Ser 285:11–28

Coan EV (1979) Recent Eastern Pacific species of the crassatellid bivalve genus *Crassinella*. Veliger 22:1–11

Coan EV, Scott PV, Bernard FR (2000) Bivalve seashells of western North America. Marine bivalve mollusks from Arctic Alaska to Baja California. St Barbara Mus Nat Hist Monogr 2

Cohen AN, Carlton JT (1995) Biological study. Nonindigenous aquatic species in a United States estuary: a case study of the biological invasions of the San Francisco Bay and Delta. A report for the United States Fish and Wildlife Service, Washington, D.C., and The National Sea Grant College Program, Connecticut Sea Grant, NTIS Rep PB96-166525

Collin R (2000) Phylogeny of the *Crepidula plana* (Gastropoda: Calyptraeidae) cryptic species complex in North America. Can J Zool 78:1500–1514

Conlan KE (1990) Revision of the crustacean amphipod genus *Jassa* Leach (Corophioidae: Ischyroceridea). Can J Zool 68:2031–2075

Cooke WJ (1977) Order Hydroida. In: Devaney DM, Eldredge LG (eds) Reef and shore fauna of Hawaii, Sect 1: Protozoa through Ctenophora. BP Bishop Museum Special Publication 64(1). Bishop Museum Press, Honolulu, Hawaii, pp 71–104

Cookson LJ (1991) Australasian species of Limnoriidae (Crustacea: Isopoda). Mem Mus Victoria 52:137–262

Cooper JG (1872) On new Californian Pulmonata. Proc Acad Nat Sci Phil 24:143–154

Cooper JG (1873) Note on *Alexia setifer* and its allies. Proc Calif Acad Sci Ser 1, 6:14–27

Cornelius PFS (1975) The hydroid species of *Obelia* (Coelenterata, Hydrozoa: Campanulariidae), with notes on the medusa stage. Bull Brit Mus Nat Hist 28:249–293

Cowie RH (1997) Catalog and bibliography of the nonindigenous nonmarine snails and slugs of the Hawaiian Islands. Bishop Mus Occas Pap 50

Cowie RH (1998) Patterns of introduction of non-indigenous non-marine snails and slugs in the Hawaiian Islands. Biodivers Conserv 7:349–368

Darling JA, Reitzel AM, Finnerty JR (2004) Regional population structure of a widely introduced estuarine invertebrate: *Nematostella vectensis* Stephenson in New England. Mol Ecol 13:2969–2981

Dawson MN, Gupta AS, England MH (2005) Coupled biophysical global ocean model and molecular genetic analyses identify multiple introductions of cryptogenic species. Proc Natl Acad Sci 102:11968–11973

Desmond R (1995) Kew: a history of the Royal Botanic Gardens. The Harvill Press, London, 466 pp
di Castri F (1989) History of biological invasions with special emphasis on the Old World. In: Drake JA, Mooney HA, di Castri F et al. (eds) Biological invasions. A global perspective. Wiley, New York, pp 1–30
Edwards C (1976) A study in erratic distribution: the occurrence of the medusa *Gonionemus* in relation to the distribution of oysters. Adv Mar Biol 14:251–284
Emschermann P (1987) Creeping propagation stolons – an effective propagation system of the freshwater entoproct *Urnatella gracilis* Leidy (Barentsiidae). Arch Hydrobiol 108:439–448
Englund RA (2002) The loss of native biodiversity and continuing nonindigenous species introductions in freshwater, estuarine, and wetland communities of Pearl Harbor, Oahu, Hawaiian Islands. Estuaries 25:418–430
Englund RA, Preston DJ, Wolff R, Coles SL, Eldredge LG, Arakai K (2000) Biodiversity of freshwater and estuarine communities in lower Pearl Harbor, Oahu, Hawaii with observations on introduced species. Bishop Mus Tech Rep 16
Felder DL, Martin JW (2003) Establishment of a new genus for *Panopeus bermudensis* Benedict & Rathbun, 1891, and several other xanthoid crabs from the Atlantic and Pacific oceans (Crustacea: Decapoda: Xanthoidea). Proc Biol Soc Wash 116:438–452
Fenchel T, Finlay BJ (2004) The ubiquity of small species: patterns of local and global diversity. Bioscience 54:777–784
Ferrari FD, Orsi JJ (1984) *Oithona davisae*, new species, and *Limnoithona sinensis* (Burckhardt, 1912) from the Sacramento-San Joaquin Estuary, California. J Crustac Biol 4:106–126
Finlay BJ (2002) Global dispersal of free-living microbial eukaryote species. Science 296:1061–1063
Flowerdew MW (1985) Indices of genetic identity and distance in three taxa within the *Balanus amphitrite* Darwin complex (Cirripedia, Thoracica). Crustaceana 49:7–15
Freeman AS, Byers JE (2006) Divergent induced responses to an invasive predator in marine mussel populations. Science 313:831–833
Fukuoka K, Murano M (2000) *Hyperacanthomysis*, a new genus for *Acanthomysis longirostris* Ii, 1936, and *A. brevirostris* Wang & Liu, 1997 (Crustacea: Mysidacea: Mysidae). Plankton Biol Ecol 47:122–128
Galil B, Froglia C, Noel P (2002) CIESM Atlas of exotic species in the Mediterranean. Crustaceans. Decapods and stomatopods. CIESM Publishers, Monaco
Geller JB (1999) Decline of a native mussel masked by sibling species invasion. Conserv Biol 13:661–664
Godwin LS (2003) Hull fouling of maritime vessels as a pathway for marine species invasions to the Hawaiian Islands. Biofouling 19(Suppl):123–131
Gosliner TM, Behrens DW (2006) Anatomy of an invasion: systematics and distribution of the introduced opisthobranch snail, *Haminoea japonica* Pilsbry, 1895 (Gastropoda: Opisthobranchia: Haminoeidae). Proc Calif Acad Sci (4) 57:1003–1010
Green J, Bohannan BJM (2006) Spatial scaling of microbial biodiversity. Trend Ecol Evol 21:501–507
Grosholz ED (2002) Ecological and evolutionary consequences of coastal invasions. Trend Ecol Evol 17:22–27
Grosholz ED (2005) Recent biological invasion may hasten invasional meltdown by accelerating historical introductions. Proc Natl Acad Sci 102:1088–1091
Guinot D, MacPherson E (1987) Revision du genre Pilumnoides Lucas, 1844, avec description de quatre especes nouvelles et creation de Pilumnoidinae subfam. nov. (Crustacea Decapoda Brachyura). Bull Mus Nat Hist Nat Paris (4) 9:211–247
Guiry M (2003) *Neosiphonia harveyi* (J Bailey). http://www.algaebase.org/ (accessed April 2007)
Henry DP, McLaughlin PA (1975) The barnacles of the *Balanus amphitrite* complex (Cirripedia, Thoracica). Zool Verh 141:254
Hoestlandt H (1973) Presence de *Gnorimosphaeroma rayi* Hoestlandt (Isopode, flabellifere) sur les cotes du Japon, de Siberie orientale et d'Hawaii, ainsi qu'indications sommaires de son polychromatisme genetique. C R Acad Sci Paris (D) 276:2817–2820

Holland BS, Dawson MN, Crow GL, Hofmann DK (2004) Global phylogeography of *Cassiopea* (Scyphozoa: Rhizostomeae): molecular evidence for cryptic species and multiple invasions of the Hawaiian Islands. Mar Biol 145:1119–1128

Hulsman N, Galil BS (2002) Protists – a dominant component of the ballast-transported biota. In: Leppäkoski ES, Gollasch, Olenin S (eds) Invasive aquatic species of Europe: distribution, impacts, and management. Kluwer Academic Publishers, Dordrecht, pp 20–27

Huvet AS, Lapègue A, Magoulas, Boudry P (2000) Mitochondrial and nuclear DNA phylogeography of *Crassostrea angulata*, the Portuguese oyster endangered in Europe. Conserv Genet 1:251–262

Huvet A, Fabioux C, McCombie H, Lapègue S, Boudry P (2004) Natural hybridization between genetically differentiated populations of *Crassostrea gigas* and *C. angulata* highlighted by sequence variation in flanking regions of a microsatellite locus. Mar Ecol Prog Ser 272:141–152

Hyman LH (1953) North American triclad Turbellaria. 14. A new, probably exotic, dendrocoelid. Am Mus Novit 1629:6

Hyman LH (1955) The polyclad flatworms of the Pacific coast of North America: additions and corrections. Am Mus Novit 1704:4–7

Ingle RW (1997) Crayfishes, lobsters, and crabs of Europe: an illustrated guide to common and traded species. Springer, Berlin Heidelberg New York

Jass J, Klausmeier B (2000) Endemics and immigrants: North American terrestrial isopods (Isopoda, Oniscidea) north of Mexico. Crustaceana 73:771–799

Jensen KR, Knudsen J (2005) A summary of alien marine benthic invertebrates in Danish waters. Oceanol Hydrobiol Stud 34(Suppl 1):137–162

Kado R (2003) Invasion of Japanese shores by the NE Pacific barnacle *Balanus glandula* and its ecological and biogeographical impact. Mar Ecol Prog Ser 249:199–206

Kahng S (2005) The invasion of a tropical coral reef ecosystem by an alien octocoral, *Carijoa riisei*. Fourth International Conference on Marine Bioinvasions, August 2005, Wellington, New Zealand, p 142 (Abstract)

Karling KF, Wade CM, Stewart IA, Kroon D, Dingle R, Brown AJL (2000) Molecular evidence for genetic mizing of Arctic and Antarctic subpolar populations of planktonic formanifers. Nature 400:43–47

Kay EA (1979) Hawaiian marine shells. Reef and shore fauna of Hawaii. Sect 4: Mollusca. B P Bishop Mus Spec Publ 64

Keen AM (1971) Sea shells of tropical west America; marine mollusks from Baja California to Peru, 2nd edn. Stanford University Press, Stanford, California

Kensley B (1981) On the zoogeography of southern African decapod Crustacea, with a distributional checklist of the species. Smith Contrib Zool 338:64

Kensley B, Schotte M (1999) New records of isopods from the Indian River Lagoon, Florida (Crustacea: Peracarida). Proc Biol Soc Wash 112:695–713

Kimura TM, Tabe M, Shikano Y (1999) *Limnoperna fortunei kikuchiii* Habe, 1981 (Bivalvia: Mytilidae) is a synonym of *Xenostrobus securis* (Lamarck, 1819): introduction into Japan from Australia and/or New Zealand. Venus 58:101–117

Knight-Jones P, Knight-Jones EW, Kawahara T (1975) A review of the genus *Janua*, including *Dexiospira* (Polychaeta: Spirorbidae). Zool J Linn Soc 56:91–129

Kolar CS, Lodge DM (2001) Progress in invasion biology: predicting invaders. Trends Ecol Evol 16:199–204

Kornicke LS (1967) A study of three species of *Sarsiella* (Ostracoda: Myodocopa). Proc U S Nat Mus 122, no 3594

Kott P (1985) The Australian Ascidiacea. Part 1, Phlebobranchia and Stolidobranchia. Mem Qd Mus 23:1–440

Kott P (1998) Tunicata. Zoological catalogue of Australia 34:51–252

Kott P (2004) A new species of *Didemnum* (Ascidiacea, Tunicata) from the Atlantic coast of North America. Zootaxa 732:1–10

Kozloff EN (1946) Studies on the ciliates of the family Ancistrocomidae Chatton and Lwoff (order Holotricha, suborder Thigmotricha). III. *Ancistrocoma pelseneeri* Chatton and Lwoff, *Ancistrocoma dissimilis* sp. nov., and *Hypocomagalma pholadidis* sp. nov. Biol Bull 91:189–199
Kramp PL (1961) Synopsis of the Medusae of the world. J Mar Biol Assoc U K 40
Kronberg I (1986) Riesenchromosomen und Artareal einer baltischen *Telmatogeton* – Art (Diptera: Chironomidae: Telmatogetoninae). Z Zool Syst Evolut 24:190–197
Lapegue S, Batista FM, Heurtebise S, Yu Z, Boudry P (2004) Evidence for the presence of the Portuguese oyster, *Crassostrea angulata*, in northern China. J Shellfisheries Res 23:759–764
Lee H, Thompson B, Lowe S (2003) Estuarine and scalar patterns of invasion in the soft-bottom benthic communities of the San Francisco estuary. Biol Invas 5:85–102
Leppäkoski E, Gollasch S, Olenion S (eds) (2002) Invasive aquatic species of Europe: distribution, impacts, and management. Kluwer Academic Publishers, Dordrecht
Lockwood JL, Hoopes MF, Marchetti MP (2007) Invasion ecology. Blackwell Publishing, Oxford
Lotze HK, Lenihan HS, Bourque BJ et al. (2006) Depletion, degradation, and recovery potential of estuaries and coastal seas. Science 312:1806–1809
Mackie JA, Keough MJ, Christidis L (2006) Invasion patterns inferred from cytochrome oxidase I sequences in three bryozoans, *Bugula neritina, Watersipora subtorquata*, and *Watersipora arcuata*. Mar Biol 149:285–295
Manuel RL (1988) British Anthozoa (Coelenterata: Octocorallia and Hexacorallia): keys and notes for the identification of the species, edn 2. Synopses of the British fauna (new series), no 18. Linnean Society of London
Marceniuk AP, Ferraris CJ Jr (2003) Ariidae (sea catfishes). In: Reis RE, Kullander SO, Ferraris CJ Jr (eds) Checklist of the freshwater fishes of South and Central America. Porto Alegre: EDIPUCRS, Brasil, pp 447–455
Marelli DC (1981) New records for Caprellidae in California. Proc Biol Soc Wash 94:654–662
Marelli DC, Gray S (1983) Conchological descriptions of *Mytilopsis sallei* and *Mytilopsis leucophaeata* of the brackish Western Atlantic. Veliger 25:185–193
Martins AMF (1996) Anatomy and systematics of the Western Atlantic Ellobiidae (Gastropoda: Pulmonata). Malacologia 37:163–332
McCarthy HP, Crowder LB (2000) An overlooked scale of global transport: phytoplankton species richness in ships' ballast water. Biol Invas 2:321–322
McIvor L, Maggs CA, Provan J, Stanhope MJ (2001) rbcL sequences reveal multiple cryptic introductions of the Japanese red alga *Polysiphonia harveyi*. Mol Ecol 10:911–919
McKoy JL (1975) Further records of *Teredicola typicus* C.B. Wilson, 1942 (Copepoda Cyclopoida) from shipworms in northern New Zealand. N Z J Mar Freshwater Res 9:417–421
Millar RH (1970) British ascidians. Synopses of the British fauna (new series), no 1. Linnean Society of London
Mills CE, Sommer F (1995) Invertebrate introductions in marine habitats: two species of hydromedusae (Cnidaria) native to the Black Sea, *Maeotias inexspectata* and *Blackfordia virginica*, invade San Francisco Bay. Mar Biol 122:279–288
Mobberley DG (1956) Taxonomy and distribution of the genus Spartina. Iowa State Coll J Sci 30:471–574
Modlin RF, Orsi JJ (1997) *Acanthomysis bowmani,* a new species, and *A. aspera* Ii, Mysidacea newly reported from the Sacramento-San Joaquin Estuary, California (Crustacea: Mysidae). Proc Biol Soc Washington 110:439–446
Monniot F, Monniot C (2001) Ascidians from the tropical western Pacific Ocean. Zoosystema 23:201–383
Monniot C, Monniot F, Griffiths CL, Schleyer M (2001) South African ascidians. Ann S A Mus 108
Mooney HA, Cleland EE (2001) The evolutionary impact of invasive species. Proc Natl Acad Sci 98:5446–5451

Morton B (1981) The biology and functional morphology of *Mytilopsis sallei* (Recluz) (Bivalvia: Dreissenacea) fouling Visakhapatnam Harbour, Andhra Pradesh, India. J Moll Stud 47:25–42

Mosely HN (1878) Description of a new species of land-planarian from the hothouses at Kew gardens. Ann Mag Nat Hist Ser 5 1:237–239

Mundy BC (2005) Checklist of the fishes of the Hawaiian Archipelago. Bishop Mus Bull Zool 6

Newman WA, Ross A (1976) Revision of the balanomorph. barnacles: including a catalog of the species. Mem San Diego Soc Nat Hist 9:1–108

Newman WA, Killingley JS (1985) The north-east Pacific intertidal barnacle *Pollicipes polymerus* in India? A biogeographical enigma elucidated by ^{18}O fractionation in barnacle calcite. J Nat Hist 19:1191–1196

Nygren A (2004) Revision of Autolytinae (Syllidae: Polychaeta). Zootaxa 680

Odlaug TO (1946) The effect of the copepod, *Mytilicola orientalis* upon the Olympia oyster, *Ostrea lurida*. Trans Am Microsc Soc 65:311–317

Ó Foighil DO, Gaffney PM, Hilbish TJ (1995) Differences in mitochondrial 16S ribosomal gene sequences allow discrimination among American [*Crassostrea virginica* (Gmelin)] and Asian [*C. gigas* (Thunberg) *C. ariakensis* Wakiya] oyster species. J Exp Mar Biol Ecol 192:211–220

Ó Foighil D, Gaffney PM, Wilbur AE, Hilbish TJ (1998) Mitochondrial cytochrome oxidase I gene sequences support an Asian origin for the Portuguese oyster, *Crassostrea angulata*. Mar Biol 131:497–503

Ogren RE (1984) Exotic land planarians of the genus *Bipalium* (Platyhelminthes: Turbellaria) from Pennsylvania and the Academy of Natural Sciences, Philadelphia. Proc Penn Acad Sci 58:193–201

Ogren RE (1987) Description of a new three-lined land planarian of the genus *Bipalium* (Turbellaria: Tricladida) from Pennsylvania, USA Trans Am Microsc Soc 106:21–30

Ogren RE (1989a) Identification features of the two-lined land planarian *Rhynchodemus sylvaticus*, with evidence that *Rhynchodemus americanus* is conspecific. Trans Am Microsc Soc 108:40–44

Ogren RE (1989b) Redescription and a new name for the blue land planarian *Geoplana vaga* Hyman now considered conspecific with *Caenoplana coerulea* Moseley from Australia (Turbellaria: Tricladida: Geoplanidae). J Penn Acad Sci 63:135–142

Orensanz JM, Schwindt E, Pastorino G et al. (2002) No longer the pristine confines of the world ocean: a survey of exotic marine species in the southwestern Atlantic Ocean. Biol Invas 4:115–143

Otani M (2004) Introduced marine organisms in Japanese coastal waters, and the processes involved in their entry. Jpn J Benthol 59:45–57 (English abstract, in Japanese; English translation available from author)

Pearson CV, Rogers AD, Sheader M (2002) The genetic structure of the rare lagoonal sea anemone, *Nematostella vectensis* Stephenson (Cnidaria, Anthozoa) in the United Kingdom based on RAPD analysis. Mol Ecol 11:2285–2293

Ponder WF (1988) *Potamopyrgus antipodarum* – a molluscan coloniser of Europe and Australia. J Moll Stud 54:271–285

Prudhoe S (1985) A monograph on Polyclad Turbellaria. British Museum (Natural History), Oxford University Press, Oxford, England

Rabinowitz D, Cairns S, Dillon T (1986) Seven forms of rarity and their frequency in the flora of the British Isles. In: Soule ME (ed) Conservation biology: the science of scarcity and diversity. Sinauer Publishers, pp 182–204

Radashevsky VI, Hsieh H-L (2000) *Pseudopolydora* (Polychaeta: Spionidae) species from Taiwan. Zoological Studies 39:218–235

Raimondi PT (1992) Adult plasticity and rapid larval evolution in a recently isolated barnacle population. Biol Bull 182:210–220

Ricciardi A, Rasmussen JB (1998) Predicting the identity and impact of future biological invaders: a priority for aquatic resource management. Can J Fish Aquat Sci 55:1759–1765

Robins CR, Böhlke JE (1960) *Pikea sericea*, a synonym of the American centrarchid fish, *Micropterus salmoides*. Copeia 1960(2):147

Robinson TA, Griffiths CL, McQuaid CD, Rius M (2005) Marine alien species of South Africa – status and impacts. Afr J Mar Sci 27:297–306

Roller RA (1972) Three new species of eolid nudibranchs from the west coast of North America. Veliger 14:416–423

Rosenberg G (2005) Malacolog 4.0: A database of Western Atlantic marine Mollusca. [WWW database (version 4.0.2)] URL http://data.acnatsci.org/wasp

Rotramel G (1972) *Iais californica* and *Sphaeroma quoyanum*, two symbiotic isopods introduced to California (Isopoda, Janiridae and Sphaeromatidae). Crustaceana(Suppl)3:193–197

Roy MS, Sponer R (2002) Evidence of a human mediated invasion of the tropical western Atlantic by the "world's most common brittlestar." Proc R Soc London B 269:1017–1023

Russell FS (1953) The Medusae of the British Isles. Cambridge University Press, Cambridge

Saltonstall K (2002) Cryptic invasion by a non-native genotype of the common reed, *Phragmites australis*, into North America. Proc Natl Acad Sci 99:2445–2449

Sax DF, Stachowicz JJ, Gaines SD (eds) (2005) Species invasions. Insights into ecology, evolution, and biogeography. Sinauer, Sunderland, Massachusetts

Schaefer CT, Lewin J (1984) Persistent blooms of surf diatoms along the Pacific coast, USA. Mar Biol 83:205–217

Schuchert P (1996) Athecate hydroids and their medusae (Cnidaria: Hydrozoa). N Z Ocean Inst Mem 106

Schuchert P (2004) Revision of the European athecate hydroids and their medusae (Hydrozoa, Cnidaria): Families Oceanidae and Pachycordylidae. Rev Suisse Zool 111:315–369

Schwindt E (2007) The invasion of the acorn barnacle *Balanus glandula* in the south-western Atlantic 40 years later. J Mar Biol Assoc UK 87:1219–1225

Southward AJ, Burton RS, Coles SL, Dando PR, DeFelice RC, Hoover J, Parnell PE, Yamaguchi T, Newman WA (1998) Invasion of Hawaiian shores by an Atlantic barnacle. Mar Ecol Prog Ser 165:119–126

Springer VG, Gomon MF (1975) Revision of the Blenniid fish genus *Omobranchus* with descriptions of three new species and notes on other species of the tribe Omobranchini. Smithsonian Contrib Zool 177:1–135

Stachowicz JJ, Whitlatch RB, Osman RW (1999) Species diversity and invasion resistance in a marine ecosystem. Science 286:1577–1579

Stachowicz JJ, Terwin JR, Whitlatch RB, Osman RW (2002a) Linking climate change and biological invasions: ocean warming facilitates nonindigenous species invasions. Proc Natl Acad Sci 99:15497–15499

Stachowicz JJ, Fried H, Osman RW, Whitlatch RB (2002b) Biodiversity, invasion resistance, and marine ecosystem function: reconciling pattern and process. Ecology 83:2575–2590

Stock JH (1968) Pycnogonida collected by the Galathea and Anton Bruun in the Indian and Pacific Oceans. Videnskabelige Meddelelser fra Dansk Naturhistorisk Forening 131:7–65

Stoddart HE, Lowry JK (2003) Zoological catalogue of Australia. Crustacea: Malacostraca: Peracarida: Amphipoda, Cumacea, Mysidacea. Volume 19.2B. CSIRO Publishing, Victoria, Australia

Strasser M (1999) *Mya arenaria* – an ancient invader of the North Sea coast. Helgolander Meeresuntersuchungen 52:309–324

Strathmann MF, Strathmann RR (2006) A vermetid gastropod with complex intracapsular cannibalism of nurse eggs and sibling larvae and a high potential for invasion. Pacific Sci 60:97–108

Straus SY, Lau JA, Carroll SP (2006) Evolutionary responses of natives to introduced species: what do introductions tell us about natural communities. Ecol Lett 9:357–374

Swennen C, Dekker R (1995) *Corambe batava* Kerbert, 1886 (Gastropoda: Opisthobranchia), an immigrant in the Netherlands, with a revision of the family Corambidae. J Moll Stud 61:97–107

Taiti S, Ferrara F (1986) Taxonomic revision of the genus *Littorophiloscia* Hatch, 1947 (Crustacea, Isopoda, Oniscidea) with descriptions of six new species. J Natl Hist 20:1347–1380

Taiti S, Howarth FG (1995) Terrestrial isopods from the Hawaiian Islands (Isopoda: Oniscidea). Occ Pap B P Bishop Mus 45:59–71

Torrey HB (1902) The Hydroida of the Pacific coast of North America. Univ Calif Publ Zool 1:1–104

Turner RD (1966) A survey and illustrated catalogue of the Teredinidae (Mollusca: Bivalvia). Museum of Comparative Zoology, Harvard Univ

Van Name WG (1936) The American land and fresh-water isopod Crustacea. Bull Am Mus Nat Hist 71:1–535

Van Name WG (1940) A supplement to the American land and fresh-water Isopod Crustacea. Bull Am Mus Natl Hist 77:109–142

Van Name WG (1945) The North and South American ascidians. Bull Am Mus Natl Hist 84:

Vervoort W (1964) Note on the distribution of *Garveia franciscana* (Torrey, 1902) and *Cordylophora caspia* (Pallas, 1771) in the Netherlands. Zool Mededel 39:125–146

Vine PJ, Bailey-Brock JH, Straughan D (1972) Spirorbinae (Polychaeta, Serpulidae) of the Hawaiian chain. Part 2, Hawaiian Spirorbinae. Pacific Sci 26:150–182

Walther AC, Lee T, Burch JB, Ó Foighil D (2006) Confirmation that the North American ancylid *Ferrissia fragilis* (Tryon, 1863) is a cryptic invader of European and East Asian freshwater ecosystems. J Moll Stud 72:318–321

Wasson K, Fenn K, Pearse JS (2004) Habitat differences in marine invasions of central California. Biol Invas 7:935–948

Watling L (1975) Remarks in: L. S. Kornicker, Spread of ostracodes to exotic environs on transplanted oysters. Bull Am Paleo 65:129–139

Williams AB (1984) Shrimps, lobsters, and crabs of the Atlantic coast of the eastern United States, Maine to Florida. Smithsonian Institution Press, Washington, D.C., 550 pp

Winston JE, Heimberg BF (1986) Bryozoans from Bali, Lombok, and Komodo. Am Mus Novit 2847:1–49

Wyatt T, Carlton JT (2002) Phytoplankton introductions in European coastal waters: why are so few invasions reported? In: CIESM (Commission Internationale pour l'Exploration Scientifique de la mer Mediterranée) Workshop Monographs no. 20, Monaco, pp 41–46

Wolff WJ (2005) Non-indigenous marine and estuarine species in The Netherlands. Zool Meded 79

Zabin CJ, Zardus J, Pitombo FB, Fread V, Hadfield MG (2007) A tale of three seas: consistency of natural history traits in a Caribbean-Atlantic barnacle introduced to Hawaii. Biol Invas 9:523–544

Zibrowius H (1971) Les especes méditerranéennes du genre *Hydroides* (Polychaeta, Serpulidae) remarques sur le pretendu polymorphisme de *Hydroides uncinata*. Tethys 2:691–746

Zimmerman EC (1948) Insects of Hawaii. Vol 3, Heteroptera. University of Honolulu Press, Honolulu

Chapter 3
Natural and Climate Change Mediated Invasions

Steve I. Lonhart

3.1 Introduction

Species distributions are constantly in flux. Biological and physical factors continually influence the rates of range expansions and contractions, altering the distribution of species in space and through time (MacArthur 1972; Brown 1995; Brown et al. 1996). Ranges expand as individuals colonize new areas and contract as populations become locally extinct. Understanding how organisms respond to environmental changes and describing the underlying mechanisms are key research components in the fields of ecology and biogeography. Knowing where populations occur—and where they are absent—provides insights into the ecological and physical factors that regulate patterns of density and distribution (see also Chap. 2, Carlton).

Historically, biological responses were due to natural processes and often occurred over long (geological) time scales. More recently, anthropogenic (i.e. human-mediated) processes have played an increasingly important role in driving patterns of density and distribution. In this chapter I will present biological invasions in the context of geographic range shifts, explore range shifts due to natural, anthropogenic, and artificial processes, and consider how climate change is already affecting species distributions.

3.2 The Geographic Range of a Species

The geographic range of a species is commonly defined as the known spatial extent of the species. Field guides often display range information as a map with polygons or shading to indicate species presence. Since species distributions are dynamic, a truly accurate assessment of the geographic range is nearly impossible. Instead, range maps represent estimates of distribution based on limited, often incomplete data and thus provide a general view of where a particular species occurs (Brown et al. 1996). Range limits, with the exception of

G. Rilov, J.A. Crooks (eds.) *Biological Invasions in Marine Ecosystems.*
Ecological Studies 204,

a few species, are likely either under- or over-estimates of the actual geographic range. This combination of historic, recent, and anomalous range records is not likely representative of a species' current range extent (Gaston 1994). For the majority of species, most of which receive little scientific attention, geographic ranges are likely underestimated due to inadequate sampling near range limits.

A suite of biological and physical factors regulates the distribution and density of each species. Physiological tolerance limits set thresholds beyond which an individual cannot reproduce, grow, or survive, and ecological interactions (e.g., competition, predation, mutualism) further modify these limits. Physical barriers can also prevent species from dispersing to all suitable habitats. For marine species, such barriers are often land masses, such as continents. For example, since completion of the Suez Canal in 1869, over 200 marine species have invaded the Mediterranean Sea from the Red Sea (Rilov et al. 2004; Chap. 27, Ferreira et al.). Similarly, the open ocean can act as a barrier to intertidal organisms with limited dispersal distances. In addition, there are numerous and often interacting environmental and physical factors driving individual and population-level responses that lead to local and regional fluctuations in density and spatial distribution. These natural responses make it nearly impossible to know the actual geographic range of a species at any given point in time.

3.3 Range Shifts

Understanding the causes and consequences of geographic range shifts assumes that new range records can be compared to existing range data that are both accurate and complete. Currently there are no widely accepted criteria used to determine what constitutes a range shift, let alone standard methods to measure and describe the geographic range of a species (Gaston 1996, 2003). Range shifts encompass contractions due to local extinction at the range edge and expansions as individuals invade beyond former range limits. But how long must a population be absent from a range edge to be considered locally extinct, warranting a range contraction? Conversely, when an individual is observed beyond its range limit, does that constitute a range expansion?

3.3.1 Factors that Influence our Understanding of the Geographic Range and Range Limits

"*I am never sure whether to be general or more detailed with distributions, as animal distribution records often tell us as much about the distribution of biologists as they do about the geographical range of a species.*" Dr. Bill Rudman, http://www.seaslugforum.net/

The geographic range of a species is dynamic, representing a moving target. In addition to the variation derived from environmentally driven and anthropogenic changes, there are also logistical constraints that limit our ability to describe effectively the geographic range of a species. Gaston (1994) noted that measures of geographic range are inaccurate or problematic for the following reasons: (1) the quality of data varies across the entire range since it is impossible to sample all areas equally; (2) the magnitude of error estimating the range varies as a function of true range size; (3) there is no standard way to deal with anomalous occurrences or transient or migratory species; and (4) the need to distinguish between historical and current range sizes.

Collecting new range data is not trivial and there are few biogeographic studies focused on detecting the range limits of marine species (Sagarin and Gaines 2002a). Instead, new range records are often collected serendipitously as a result of monitoring programs or surveys designed for other purposes. If we consider the most studied and accessible marine habitat—the rocky intertidal—we are still confronted with significant logistical barriers. For example, selection of rocky intertidal study sites is typically nonrandom and biased toward areas adjacent to marine laboratories or with relatively easy access. Sampling effort within and between sites may be unequal due to differences in exposure and available habitat. In addition, few sites are selected a priori as part of a study design focused on describing the range limits of a species (but see Gilman 2005).

With the exception of a few narrowly distributed species, comprehensive sampling of the entire geographic range is rarely feasible. This obstacle is reduced somewhat for intertidal species since the range is essentially linear (Sagarin and Gaines 2002b), but this is true only for intertidal obligates—many intertidal species also occur in the shallow subtidal. Surveys near range limits require intensive sampling effort to detect what should be a relatively rare occurrence. As such, these organisms are commonly undersampled at or near the range limit and therefore underestimate their spatial extent (Sagarin and Gaines 2002a). Spatial variation in abundance within the rocky intertidal makes it difficult to sample for certain species (Sagarin 2002). If the species is small or otherwise cryptic, the likelihood of detecting it declines. Furthermore, since distributions change over time, sampling the range requires repeated surveys on a regular basis. Such extensive on-site efforts are rarely undertaken for most species (but see Sagarin and Gaines 2002b; Gilman 2005).

Determining whether a new range record represents a "true" range expansion is difficult, requiring additional evidence. For example, it is not always possible to determine whether an extralimital range record represents a natural range expansion, a human-mediated invasion, or is a sampling artifact. The context of the discovery (i.e. species dispersal capabilities, site characteristics, and sampling history) provides additional information to evaluate the status of a new range record. Such records may not indicate that a species recently expanded into a new area, but instead may have been observed for the first time due to increased sampling effort or exploration of a new area. Without long-term, intensive sampling at a particular site, it is difficult to separate real range shifts from sampling artifacts.

There are also analytical issues that hinder accurate descriptions of geographic range. Existing range data can have multiple shortcomings. A range limit can be based either on a single, anomalous extralimital range record or the presence of a well-established population. As discussed earlier, anomalous, extralimital range records should not be used to determine geographic range. Conversely, an extralimital, well-established, self-sustaining population provides strong evidence of an expanded geographic range. There are also new range records that go unreported. During a search of several museum collections, Lonhart and Tupen (2001) discovered several "new" (i.e. unpublished) range records that exceeded recent field observations. However, unless sufficient metadata are included with museum voucher materials, it is difficult to differentiate extralimital range records from true range expansions. Furthermore, museum collections also represent a time-integrated view of the geographic range. Compiling range records that span decades, if not centuries, can be problematic when describing the current geographic range of a species.

Marine ecologists working in the field may lack the taxonomic expertise and natural history background to recognize extralimital species, and thus many potential new range records go unnoticed. In contrast, when taxonomic experts undertake expeditions to remote locations, it is not surprising that numerous new range records are reported (e.g., Vermeij et al. 1990; Bertsch et al. 2000). Ideally, coastlines would be systematically sampled, but this is not practical. Instead, targeted areas are sampled, and these are selected in a non-random manner. Thus the clustering of range limits at a particular site may be more indicative of where experts sampled than of the true limit of any particular species. Moerman and Estabrook (2006) describe a pattern where university botanists in North America have, in general, spent more time investigating areas near their university, resulting in higher local species richness than in counties more distant from their home institution, a phenomenon they call the 'botanist effect.'

Range maps may use some or all of these data to generate distribution polygons, and must extrapolate between the relatively few known data points (Gaston 1994; Brown et al. 1996). Thus, with the exception of a few species that have extremely limited distributions along the intertidal, the geographic range of a marine species cannot be known in great detail and is instead estimated using available information. While this level of detail is sufficient for biogeographic analyses that cluster endpoints at 1° latitude scales or larger (e.g., Roy et al. 1995), it may be insufficient to track invasions or ecological responses to climate change.

3.3.2 Natural Range Shifts

Natural range shifts require the establishment of extralimital populations without direct mediation by human activities. There are key spatial and temporal components to defining the validity of a proposed range shift, where a 'shift' may include range contraction and/or expansion. In the case of a range expansion, the new range record must occur beyond the known geographic range limits of the species. This is typically verified by consulting a field guide or other published reference that contains the spatial extent of the species in question. The second component, which receives little attention, is to determine whether the new range record represents an anomalous occurrence, or is part of a well-established population. If the new range record is based on one or a few individuals, temporarily found beyond the species' range limit, then the observation should be considered an extralimital range record. Relative to the geographic range of a species, extralimital range records should not constitute a basis for expanding the geographic range. To be considered a "true" range expansion, these extralimital individuals should be part of a self-sustaining, multi-generational and well-established population persisting beyond the previously known range limit.

This restrictive definition of range expansion excludes marginal-population that persist beyond the edge of the range for only a short period of time ('relict populations') or are sustained by external propagules ('sink-population'). Relict populations persist beyond their range limit for a single generation but fail to reproduce successfully, ultimately leading to local extinction. Such populations represent ephemeral range shifts. In contrast, sink populations persist beyond range limits, sustained by propagules derived from source populations within the established range limits. Although sink populations can persist for multiple generations, should dispersal from the source population cease, the sink population will become locally extinct.

Published geographic ranges may include data from anomalous extralimital range records and marginal populations. Without clear definitions for valid range records and criteria for including or excluding data from estimates of the geographic range, natural variation and sampling error may obscure our view of species' geographic ranges and their dynamics.

History and time Species range shifts occur for a number of reasons. Over the evolutionary history of a species, there are three general phases: initial expansion, equilibrium, and decline to extinction (Gaston 1996). During the initial expansion phase, a species invades new, suitable habitats. Expansion may be rapid or very slow, depending on the dispersal capabilities of the species and suitability of habitats. This is followed by a period of dynamic equilibrium, when the spatial extent of the species is stable and near its maximum. Finally, as the species declines and approaches global extinction, its range size diminishes by either contracting along the entire range towards the center or creating a patchwork of shrinking, isolated populations. Since these phases are not synchronous across taxa, each of the three phases is currently represented by a multitude of species. While these changes naturally take place at geologic time scales, anthropogenic processes have hastened the pace of species decline (e.g., habitat loss, pollution, over-fishing) and global spread (e.g., intentional and accidental introductions).

Response to natural changes Range shifts are also expected as species respond to natural changes in climatic conditions. Climatic external forcing occurs at various time scales, from seasonal to millennial (Root and Schneider 2002). Individuals and populations can respond to changes in weather and season, but these shifts are fine-scale and ephemeral, making them hard to detect and track. Furthermore, such shifts likely have little impact on general estimates of geographic range. Interannual shifts due to large-scale atmospheric and oceanographic changes, such as El Niño Southern Oscillation (ENSO) and North Atlantic Oscillation (NAO) events, can alter species distributions significantly (Harley et al. 2006). Although many of these range shifts are temporary, some can be lasting (Lonhart and Tupen 2001). At decadal scales, regime shifts such as the Pacific Decadal Oscillation (PDO) can dramatically alter patterns of species density and distribution (Chavez et al. 2003; McGowan et al. 2003). At millennial scales, paleontological records indicate shifts in species ranges in response to geologic processes and glacial-interglacial periods (Valentine and Jablonski 1993). These large-scale, low-frequency climatic changes lead to long-term, broad-scale range shifts.

3.3.3 Human-mediated Range Shifts

The pace and extent of species spread has rapidly accelerated as human modes of transportation (e.g., ships, planes) have increased in number, speed and distance traveled. Humans are moving species either intentionally or accidentally to all parts of the world. This accelerated global redistribution of species has lead to a phenomenon termed 'biotic homogenization', where community assemblages in different regions are becoming more similar to one another through the addition of cosmopolitan species (McKinney and Lockwood 1999).

Intentional species introductions occur for a variety of reasons (Chap. 5, Minchin et al.; Chap. 6, Hewitt et al.). In terrestrial systems, many species are brought to new environments for economic reasons (e.g., agriculture, silviculture), while other species are introduced as biocontrol agents to combat invasive pests. In aquatic systems, species are intentionally introduced for aquaculture, and in freshwater systems for commercial and recreational angling (Rahel 2000; Kolar and Lodge 2002). There is also increasing evidence that the release of pets from aquariums may be a significant source of species introductions (Semmens et al. 2004).

Not all introductions are intentional. Many species are accidentally introduced by human activities. In North America, Ruiz et al. (2000) conservatively estimate that just over half of the nearly 300 invasive species studied were introduced by shipping. Species can be transported on the hulls of vessels, in cargo, or in ballast tanks, either suspended in ballast water or in the tank's sediments. Hitchhikers are also found associated with commercially important species, such as oysters and abalone, whose shells serve as habitats for a myriad of invertebrates and algae (Culver and Kuris 2000; Wasson et al. 2001).

3.4 Climate Change and Range Shifts

The Intergovernmental Panel on Climate Change (IPCC) recently reported on observed climate changes to (1) atmospheric concentrations of greenhouse gases and aerosols, (2) the Earth's surface temperature and precipitation, (3) snow cover, sea and river ice, glaciers, and sea level, (4) climate variability, and (5) extreme climatic events (IPCC 2007). These changes are the result of natural, internal processes (e.g., El Niño Southern Oscillation, Pacific Decadal Oscillation), natural external forcing (e.g., Milankovitch cycles), and human-mediated external forcing (e.g., elevated CO_2 levels) (Beaugrand and Reid 2003). Describing the patterns and understanding the mechanisms that drive existing and predicted biological responses to climate change are active areas of research in the fields of ecology and biogeography.

Predicted and observed biological responses to global warming include changes in physiology, morphology, patterns of density and distribution, phenology, species interactions, and population genetics through local adaptation (Hughes 2000; McCarty 2001; Sagarin 2002; Beaugrand and Reid 2003; Helmuth et al. 2006; Parmesan 2006). A dynamic model of community response to climate change suggests species will respond individualistically rather than as a tightly linked species assemblage (Graham and Grimm 1990). Shifts will occur at the level of individuals, populations, and species—not at the level of communities—and will be limited by life history characteristics and phylogenetic constraints. Responses may be further limited by species interactions. For example, depending on coevolutionary relationships, the rate of range expansion for some species (e.g., parasites, mutualists, habitat specialists) will be limited by the rate of spread for an obligate host or habitat.

3.4.1 *Observed Biological Responses to Climate Change*

Organisms respond to climatic changes at various temporal scales. Many marine apex predators undergo seasonal migrations to forage in ephemeral but highly productive areas, to find mates, or give birth. Geographic range maps usually capture these temporary changes in distribution. At longer time scales (i.e. interannual, interdecadal, millennial), changes in distribution represent actual range shifts. The evidence for responses at these longer time scales is growing and several examples are presented below.

At interannual scales, atmospheric and oceanic processes act at large, basin-wide scales. In the Pacific Ocean, El Niño Southern Oscillation events, which persist for several months and occur every 2–7 years, increase sea surface temperature (SST) and alter equatorial and coastal current patterns (McGowan et al. 1998; Chavez et al. 1999). From coastal California there are multiple examples of new northern range records, spanning several taxa and nearly a century, that coincide with ENSO events (e.g., Hubbs and Schultz 1929; Glynn 1961; Richards and Engle 2001; Engle and Richards 2001). Poleward flow of coastal currents along the northeastern Pacific entrains the adults and larvae of subtropical and warm temperate species, depositing them well beyond their typical northern range limits. As ENSO conditions wane and SST drops, few of these extralimital individuals survive, although some may establish relict populations (Lonhart and Tupen 2001).

If SST serves as a key driver of change in coastal marine species, then during ENSO events populations of cold temperate species near their southern range limit should respond by either declining in abundance locally or contracting poleward. A recent study of the giant kelp, *Macrocystis pyrifera*, along the northeastern Pacific coast reported delayed recovery at its southern range limit in response to an ENSO event (Edwards and Hernández-Carmona 2005). During the 1982–1983 ENSO, high wave action, high SST and low nutrients decimated kelp populations at the southern limit. At these same areas the southern sea palm, *Eisenia arborea*, persisted during the 1982–1983 ENSO and quickly recruited at high densities into habitats devoid of *Macrocystis*. By coupling long-term monitoring data and field experiments, Edwards and Hernández-Carmona (2005) showed that the southern range shift of *Macrocystis* over a 20-year period was due to increased mortality and recruitment failure of *Macrocystis* after the ENSO event and the ability of the understory kelp *Eisenia* to competitively exclude *Macrocystis*.

Interdecadal regime shifts, a term used to describe significant and sustained changes in ecosystems responding to climate change (Hays et al. 2005), add yet another layer of complexity. The Pacific Decadal Oscillation occurs in the Pacific Ocean, lasts for 20–30 years, and is more pronounced at high latitudes (Mantua and Hare 2002). Long-term data sets are needed to track decadal changes, and in southern California the California Cooperative Oceanic Fisheries Investigations (CalCOFI) program has organized cruises to collect physical and biological data since 1949 (McGowan et al. 2003). During the winter of 1976–1977 there was an abrupt shift from the cooler 'anchovy regime' to the warmer 'sardine regime' (see Chavez et al. 2003). Following this regime shift, the offshore species composition and abundance of calanoid copepods changed, zooplankton phenology shifted, and the abundance and proportion of larval fishes changed (summarized in McGowan et al. 2003). Pelagic tunicates, significant members of zooplankton communities, also responded to the PDO regime shift (Lavaniegos and Ohman 2003). Of the 10 species studied, 4 were present in both cool and warm phases of the PDO, while 4 dropped below the limits of detection after the regime shift. During the warm phase, dramatic changes in biomass were not solely due to declines in abundance, but also to decreases in the size of individual zooids and colonies (Lavaniegos and Ohman 2003).

In the nearshore waters of southern California, Holbrook et al. (1997) documented substantial changes to the assemblages of reef fishes after the 1976–1977 regime shift and slight increase in SST (nearly 1 °C). At the two sites that were studied, species richness fell by up to 25% and dominance shifted from cold water to warm water species. In central California, Barry et al. (1995) re-sampled rocky intertidal invertebrates along a transect that had been initially sampled from 1931 to 1933. By locating the original bolts, they were able to replicate the original study and quantitatively compare abundances over the 60-year interval. Changes in abundance occurred for 32 of the 45 species analyzed, indicating a significant shift in community structure. When species were categorized by geographic range (i.e. southern/warm water, northern/cold water, or cosmopolitan), eight of nine southern species increased in abundance while five of eight northern species decreased significantly (Barry et al. 1995). Climate change was considered the primary driver of change, while alternative mechanisms such as habitat changes, anthropogenic effect, species interactions, ENSO events and upwelling variation were not considered as important (Sagarin et al. 1999). An unintended "experiment" on the California coast demonstrated how warming of seawater by the thermal outfall of a power-generating station caused dramatic changes in intertidal community structure through apparently cascading responses to changes in abundance of several key taxa, particularly habitat-forming foliose red algae (Schiel et al. 2004).

In the northeastern North Atlantic marine organisms have expanded northward concordant with warm water regime shifts and the North Atlantic Oscillation (NAO) (Hays et al. 2005). From 1960 to 1999 calanoid copepods (crustaceans) shifted up to 10° latitude northward as SST increased in part due to the NAO and climate change (Beaugrand et al. 2002). Physical and biological data suggest the northeastern North Atlantic is currently in a warm water dynamic regime and concomitant changes in the abundance, distribution, and diversity of phytoplankton and zooplankton communities may have contributed to the recent decline of Atlantic salmon stock (Beaugrand and Reid 2003). Southward et al. (1995) also report range shifts and population-level responses of plankton and intertidal barnacles and mollusks to increased SST in the western English Channel. During warm water phases, warm water species increased in abundance and expanded northward; the reverse was true during cool periods. Using 20 species with range limits in the North Sea, Perry et al. (2005) reported that during a period of increased SST half of the ranges shifted northward for warm water species and half of the ranges contracted for cold water species.

Prior to the acceleration of global change mediated by human activities (e.g., habitat loss, pollution, introduced species, and overfishing), changes in species-level distributions were relatively slow. Studies of the fossil record have added to our understanding of biological responses to climate change. During the Pleistocene, eastern Pacific marine mollusks indicate species range shifts and redistributions were common and driven by climatic changes associated with glacial-interglacial cycles (Valentine and Jablonski 1993; Roy et al. 1995, 1996). In fact, fossil evidence shows that species additions, deletions, and substitutions within marine communities are the rule rather than the exception (Valentine and Jablonski 1993).

3.5 Contrasting Natural Range Expansions and Biological Invasions

Range expansions as a result of natural processes (i.e. dispersal not aided by human activities) share many similarities with human-mediated biological invasions. Both result in the introduction of a species into a new habitat and expansion of the geographic range. However, there are also striking dissimilarities, and at least six have important ecological and biogeographic implications. First, the vectors and rates of transmission are very different. Biological invasions circumvent physical barriers to dispersal through human activities (e.g., shipping, aquaculture) and occur at an alarming and accelerating rate (Cohen and Carlton 1998). Natural range expansion of marine organisms often depends upon oceanic currents and occurs at longer, often geologic, time scales. Second, the rate of addition into communities is greater for introduced species than natives (Strauss et al. 2006). For example, marine biological invasion events often occur locally but at high frequencies, such as ships inoculating harbors on a daily basis (i.e. small spatial scale, high rate). In contrast, natural range expansions are driven by large-scale, low frequency climatic events (i.e. large spatial scale, slow rate). Third, many biological invasions fail because of very different environmental conditions between donor and recipient regions. Tropical species attached to the hull of a vessel are unlikely invaders of cold temperate regions. When a natural range expansion occurs, it is often because conditions in areas just beyond the current range limit change, becoming environmentally tolerable to the species in question. Fourth, invasive populations are disjunct from their native range in the donor region, often by very large distances. As a consequence, invasive populations can suffer Allee effects and become locally extinct unless there is a steady supply of new propagules. This is in contrast to natural range expansions, which are often contiguous with the main population. The 'rescue effect' (see Brown and Kodric-Brown 1977) may buffer extralimital populations from local extinction through immigration from the main, adjacent population. Fifth, responses to climate change will differ between native and invasive species (Carlton 2000), and the variation in response may be greatest for invasive species. As environmental conditions change in the recipient region, invasive species may increase or decline in abundance, depending on physiological tolerances. As SST increases due to climate change, selection by environmental conditions in the donor region will affect the ability of invasive species to persist in the recipient region: the abundance and distribution of warm water species should increase while cold water species could become locally extinct. Further, the likelihood of regional extinction is potentially higher for invasive species. Small and nascent invasive populations are more susceptible to the deleterious direct and indirect effects of climate change. For these vulnerable populations, a local extinction of an invasive population can also be a regional extinction. In contrast, native species responding to increasing SST may lead to poleward range expansions and range contractions away from the equator, but regional extinctions are unlikely. Finally, introduced species lack a co-evolutionary history with species in the recipient region (unless, by chance, there are other invasive species from the same donor region). Species undergoing a range expansion are likely moving into habitats and communities that share many of the same species. In spite of these differences, the response of established invasive species to recent global warming mirrors native species: on the Pacific coast of North America, all nine of the invasive species that have apparently responded to recent global warming moved poleward (Carlton 2000). Comparative studies on the responses of both natives and invasive species to climate change will improve our understanding of the biological and physical processes driving geographic range shifts and the success or failure of invasive species.

3.6 Conclusions

Without human activities to overcome physical barriers to dispersal, the ability of invasive species to spread long distances is minimal. In contrast, natives have a long history of opportunities to invade nearby areas, and the inability to expand further is due primarily to biological barriers affecting survival (e.g., physiological tolerances, species interactions) and not physical barriers to dispersal. The rate of change for natural range shifts is typically slow, occurring over decades and centuries and covering tens to hundreds of kilometers. In contrast, human-mediated invasions are occurring at an unprecedented rate, with species moved hundreds or thousands of kilometers in a matter of hours to days. The threat of biotic homogenization is significant: while the rate of climate change might be altered by human production of CO_2 and other greenhouse gases, and habitat loss can be reduced or even reversed, invasive species, once established, are rarely eradicated and therefore pose an ongoing threat with potentially severe ecological consequences.

References

Barry JP, Baxter CH, Sagarin RD, Gilman SE (1995) Climate-related, long-term faunal changes in a California rocky intertidal community. Science 267:672–675

Beaugrand G, Reid PC (2003) Long-term changes in phytoplankton, zooplankton and salmon related to climate. Global Change Biol 9:801–817

Beaugrand G, Reid PC, Ibañez F, Lindley JA, Edwards M (2002) Reorganization of North Atlantic marine copepod biodiversity and climate. Science 296:1692–1694

Bertsch H, Campillo OA, Arreola JL (2000) New distributional records of opisthobranchs from the Punta Eugenia region of the Baja California peninsula: a report based on 1997–1998 CONABIO-sponsored expeditions. Festivus 32:99–104

Brown JH (1995) Macroecology. University of Chicago Press, Chicago

Brown JH, Kodric-Brown A (1977) Turnover rates in insular biogeography: effect of immigration on extinction. Ecology 58:445–448

Brown JH, Stevens GC, Kaufman DM (1996) The geographic range: size, shape, boundaries and internal structure. Annu Rev Ecol Syst 27:597–623

Carlton JT (2000) Global change and biological invasions in the oceans. In:Mooney HA, Hobbs RJ (eds) Invasive species in a changing world. Island Press, Washington D.C., pp 31–53

Chavez FP, Strutton PG, Friederich GE, Feely RA, Feldman GC, Foley DG, McPhaden ML (1999) Biological and chemical response of the equatorial Pacific Ocean to the 1997–98 El Niño. Science 286:2126–2131

Chavez FP, Ryan J, Lluch-Cota SE, Niquen M (2003) From anchovies to sardines and back: multidecadal change in the Pacific Ocean. Science 299:217–221

Cohen AN, Carlton JT (1998) Accelerating invasion rate in a highly invaded estuary. Science 279:555–558

Culver CS, Kuris AM (2000) The apparent eradication of a locally established introduced marine pest. Biol Invas 2:245–253

Edwards MS, Hernández-Carmona G (2005) Delayed recovery of the giant kelp near its southern range limit in the North Pacific following El Niño. Mar Biol 147:273–279

Engle JM, Richards DV (2001) New and unusual marine invertebrates discovered at the California Channel Islands during the 1997–1998 El Niño. Bull South Calif Acad Sci 100:186–198

Gaston KJ (1994) Measuring geographic range sizes. Ecography 17:198–205

Gaston KJ (1996) Species-range-size distributions: patterns, mechanisms and implications. Trend Ecol Evol 11:197–201

Gaston KJ (2003) The structure and dynamics of geographic ranges. Oxford University Press, New York

Gilman S (2005) A test of Brown's principle in the intertidal limpet *Collisella scabra* (Gould, 1846). J Biogeogr 32:1583–1589

Glynn PW (1961) The first recorded mass stranding of pelagic red crabs, *Pleuroncodes planipes*, at Monterey Bay, California, since 1859, with notes on their biology. Calif Fish Game 47:97–101

Graham RW, Grimm EC (1990) Effects of global climate change on the patterns of terrestrial biological communities. Trend Ecol Evol 5:289–292

Harley CDG, Hughes AR, Hultgren KM, Miner BG, Sorte CJB, Thornber CS, Rodriguez LF, Tomanek L, Williams SL (2006) The impacts of climate change in coastal marine systems. Ecol Lett 9:228–241

Hays GC, Richardson AJ, Robinson C (2005) Climate change and marine plankton. Trend Ecol Evol 20:337–344

Helmuth B, Mieszkowska N, Moore P, Hawkins SJ (2006) Living on the edge of two changing worlds: forecasting the responses of rocky intertidal ecosystems to climate change. Annu Rev Ecol Evol Syst 37:373–404

Holbrook SJ, Schmitt RJ, Stephens JS (1997) Changes in an assemblage of temperate reef fishes associated with a climate shift. Ecol Appl 7:1299–1310

Hubbs CL, Schultz LP (1929) The northward occurrence of southern forms of marine life along the Pacific Coast in 1926. Calif Fish Game 15:234–241

Hughes L (2000) Biological consequences of global warming: is the signal already apparent? Trend Ecol Evol 15:56–61

IPCC (2007) Climate and biodiversity. Intergovernmental Panel on Climate Change technical paper

Kolar CS, Lodge DM (2002) Ecological predictions and risk assessment for alien fishes in North America. Science 298:1233–1236

Lavaniegos BE, Ohman MD (2003) Long-term changes in pelagic tunicate of the California Current. Deep-Sea Res II 50:2473–2498

Lonhart SI, Tupen JW (2001) New range records of 12 marine invertebrates: the role of El Niño and other mechanisms in southern and central California. Bull South Calif Acad Sci 100:238–248

MacArthur RH (1972) Geographical ecology: patterns in the distribution of species. Princeton University Press, New Jersey

Mantua NJ, Hare SR (2002) The Pacific decadal oscillation. J Oceanogr 58:35–44

McCarty JP (2001) Ecological consequences of recent climate change. Conserv Biol 15:320–331

McGowan JA, Cayan DR, Dorman LM (1998) Climate-ocean variability and ecosystem response in the northeast Pacific. Science 281:210–217

McGowan JA, Bograd SJ, Lynn RJ, Miller AJ (2003) The biological response to the 1977 regime shift in the California Current. Deep-Sea Res II 50:2567–2582

McKinney ML, Lockwood JL (1999) Biotic homogenization: a few winners replacing many losers in the next mass extinction. Trend Ecol Evol 14:450–453

Moerman DE, Estabrook GF (2006) The botanist effect: counties with maximal species richness tend to be home to universities and botanists. J Biogeogr 33:1969–1974

Parmesan C (2006) Ecological and evolutionary responses to recent climate change. Annu Rev Ecol Evol Syst 37:637–669

Perry AL, Low PJ, Ellis JR, Reynolds JD (2005) Climate change and distribution shifts in marine fishes. Science 308:1912–1915

Rahel FJ (2000) Homogenization of fish faunas across the United States. Science 288:854–856

Richards DV, Engle JM (2001) New and unusual reef fish discovered at the California Channel Islands during the 1997–1998 El Niño. Bull South Calif Acad Sci 100:175–185

Rilov G, Benayahu Y, Gasith A (2004) Prolonged lag in a population outbreak of an invasive mussel: a shifting-habitat model. Biol Invas 6:347–364

Root TL, Schneider SH (2002) Climate change: overview and implications for wildlife. In: Schneider SH, Root TL (eds) Wildlife responses to climate change: North American case studies. Island Press, Washington D.C., pp 1–56

Roy K, Jablonski D, Valentine JW (1995) Thermally anomalous assemblages revisited: patterns in the extraprovincial latitudinal range shifts of Pleistocene marine mollusks. Geology 23:1071–1074

Roy K, Valentine JW, Jablonski D, Kidwell SM (1996) Scales of climatic variability and time averaging in Pleistocene biotas: implications for ecology and evolution. Trend Ecol Evol 11:458–463

Ruiz GM, Fofonoff PW, Carlton JT, Wonham MJ, Hines AH (2000) Invasion of coastal marine communities in North America: apparent patterns, processes, and biases. Annu Rev Ecol Syst 31:481–531

Sagarin R (2002) Historical studies of species' responses to climate change: promises and pratfalls. In: Schneider SH, Root TL (eds) Wildlife responses to climate change: North American case studies. Island Press, Washington D.C., pp 127–163

Sagarin RD, Gaines SD (2002a) The 'abundant centre' distribution: to what extent is it a biogeographical rule? Ecol Lett 5:137–147

Sagarin RD, Gaines SD (2002b) Geographical abundance distributions of coastal invertebrates: using one-dimensional ranges to test biogeographic hypotheses. J Biogeogr 29:985–997

Sagarin RD, Barry JP, Gilman SE, Baxter CH (1999) Climate-related change in an intertidal community over short and long time scales. Ecol Monogr 69:465–490

Schiel DR, Steinbeck JR, Foster MS (2004) Ten years of induced ocean warming causes comprehensive changes in marine benthic communities. Ecology 85:1833–1839

Semmens BX, Buhle ER, Salomon AK, Pattengill-Semmens CV (2004) A hotspot of non-native marine fishes: evidence for the aquarium trade as an invasion pathway. Mar Ecol Progr Ser 266:239–244

Southward AJ, Hawkins SJ, Burrows MT (1995) Seventy years' observations of changes in distribution and abundance of zooplankton and intertidal organisms in the Western English Channel in relation to rising sea temperature. J Therm Biol 20:127–155

Strauss SY, Lau JA, Carroll SP (2006) Evolutionary responses of natives to introduced species: what do introductions tell us about natural communities? Ecol Lett 9:357–374

Valentine JW, Jablonski D (1993) Fossil communities: compositional variation at many time scales. In: Ricklefs RE, Schluter D (eds) Species diversity in ecological communities: historical and geographical perspectives. University of Chicago Press, Chicago, pp 341–348

Vermeij GJ, Palmer AR, Lindberg DR (1990) Range limits and dispersal of mollusks in the Aleutian Islands, Alaska. Veliger 33:346–354

Wasson K, Zabin CJ, Bedinger L, Diaz MC, Pearse JS (2001) Biological invasions of estuaries without international shipping: the importance of intraregional transport. Biol Conserv 102:143–153

Chapter 4
Modeling Marine Invasions: Current and Future Approaches

Marjorie J. Wonham and Mark A. Lewis

4.1 Introduction

This chapter focuses on how dynamical mathematical modeling has been and could be useful in understanding marine biological invasions. Mathematical models have long been central to the development of general ecological and invasion theory (e.g., Case 1990; Hastings et al. 2005; Lewis and Kareiva 1993; Neubert and Parker 2004; Shigesada and Kawasaki 1997). Although the dynamics of marine systems can be challenging to observe and model (e.g., deYoung et al. 2004; Kinlan et al. 2005), mathematical models are nonetheless beginning to provide insights into invasion dynamics in marine systems.

4.1.1 Why Marine Invasion Modeling?

Mathematical modeling is a tool, like natural history observations, field and laboratory experiments, and genetic analysis, that can provide insight into biological processes in general, and invasion dynamics in particular. The mathematical tools associated with ecological, epidemiological, evolutionary, and economic theory can all be brought to bear one way or another on problems of invasions. While models can provide new insights and perspectives into invasions, invasions can also motivate new ways to combine modeling approaches.

Before delving into this material, it is perhaps useful to consider two questions. First, is invasion modeling different from any other ecological modeling? Second, is marine invasion modeling different from any other invasion modeling?

Does invasion modeling differ from other ecological modeling? Current human-mediated invasions offer a dramatically sped up version of natural processes of colonization and extinction. At the community scale, the rapid accumulation of invaders requires us to consider in ecological time the global-scale dispersal and homogenization that traditionally have been the domain of paleontology and biogeography (e.g., Drake and Lodge 2004; Olden and Poff 2004). At the population scale, invasions prompt us to focus on the dynamics of small populations, and have

G. Rilov, J.A. Crooks (eds.) *Biological Invasions in Marine Ecosystems.* Ecological Studies 204,

spurred mathematical innovation in modeling spatial population dynamics in general, and Allee effects in particular (e.g. Hastings 1996). Invasions also encourage us to adopt and develop traditional conservation-oriented modeling tools, such as population viability analysis or fisheries harvest models, and turn them around to ask how best to eradicate, rather than to protect, a species. The striking impacts of selected invaders also provide a compelling incentive to combine ecological and economic modeling to help prevent and control invasions (e.g., Leung et al. 2002; Sharov 2004; Sharov and Liebhold 1998).

Does marine modeling differ from other invasion modeling? Marine systems differ in a multitude of ways from terrestrial ones, and can demand different modeling approaches. With respect to invasions, a conspicuous difference is the mobile nature of the habitat: water moves faster than continents. Ocean circulation operates in three dimensions, which can prove challenging to model. Depending on the question, marine invasion models may be nonspatial or may incorporate one or more spatial dimensions. One dimension may suffice for characterizing spread along a coastline, and two dimensions may adequately capture the movement of certain surface or benthic organisms. Three-dimensional models may be required to capture large-scale pelagic systems, or local settlement processes influenced by turbulent flow, eddies, and tidal exchange. As the number of spatial dimensions increases, so does the complexity of the model. Perhaps as a consequence of the habitat differences, models that effectively predict invasion-spread rates in terrestrial systems largely fail in marine systems (Grosholz 1996; Kinlan et al. 2005). This disparity highlights the need to focus on additional environmental processes when modeling marine invasions.

4.1.2 Scope of this Review

To contain this review, we have chosen to focus on dynamical mathematical models, which means we largely omit statistical models. We have also chosen to highlight how models have been applied to non-native species, which means we skip over many mathematically similar models that treat native species dynamics. We have organized the resulting collection of models in terms of biological invasion processes, mathematical model types, and the goals of the model.

Biologists and mathematicians may look at invasion modeling in somewhat different ways. A biologist may conceptualize invasions as a series of qualitatively distinct stages. A given invasion begins with a species being transported. The species will then establish and spread, interact with the resident community, have some degree of impact, and may or may not be subject to control or enhancement efforts. At a community scale, where multiple invasions occur, questions of invasibility, and the roles of disturbance, resource availability, and resident species diversity may arise.

A mathematician, in contrast, might categorize invasion processes according to the type of model that could be applied. For example, single-species population models can be used to ask questions about establishment and control, and multispecies models may provide insight into the role of interspecific interactions in establishment, impacts, and biological control. These models must be extended to a spatial context to address questions of invasion spread (Shigesada and Kawasaki 1997). A quite different category of bioeconomic cost-benefit analysis may be used to optimize invasion control efforts.

The goals of a model will determine its approach and structure. In any modeling enterprise, there is a tension between a model's tractability and its realism. Where a biologist may find a given model unrealistically simplified, a mathematician may find it excessively complex. The goal of a model can run the gamut from being highly strategic, intended to provide general insight into a certain kind of process, to highly tactical, intended to simulate and forecast the detailed operations of a particular system (Levins 1968). Where a model falls along this continuum likely dictates where it falls along a series of related spectra (Table 4.1). The success of a model can only be evaluated in the context of its goals: a highly strategic model is not intended to be very realistic, and a highly tactical model is not intended to be broadly applicable. A more detailed discussion of modeling philosophy and practice is beyond the scope of this chapter, but we refer in the Appendix to sources that we find helpful on these topics.

To review current marine invasion modeling, we mix and match freely between biological and mathematical perspectives of invasion biology, and also highlight differences in modeling goals. First, we look at models of human-mediated species transport, as this process sets the stage for the invasion dynamics that follow. Next, we turn to models that focus on the invader. In invasion biology terms, these models treat species establishment, impacts, and control. In modeling terms, they are population dynamics models that may be extended to include interspecific interactions, or integrated with bioeconomic cost-benefit analyses. Third, we delve into the very rich literature on invasion-spread modeling, which extends population dynamic models over one, two, or three spatial dimensions. Finally, we look at models that focus on the invaded community to explore questions of invasion resistance over space and time.

In each section, we briefly highlight relevant modeling approaches in general invasion biology and marine ecology. We then provide examples of marine invasion models that represent current and future directions in this area, and illustrate some of the contrasting goals of different models (Table 4.2). We end this review by identifying some promising areas for future mathematical modeling in the study of marine bioinvasions.

4.2 Invasion Pathway Models

Modeling an invasion pathway allows us to address questions of invader source and propagule pressure, the associated risk of species establishment, and invasion prevention (Jerde and Lewis 2007). In marine systems, the dominant invasion pathway is commercial shipping, with its associated ballast water, sediment, and hull fouling

Table 4.1 Modeling tradeoffs. Depending on its goals, a mathematical model may be more strategic, focusing on general insight into a certain kind of process, or more tactical, focusing on specific forecasting of a particular system. Broadly speaking, a model will tend to fall towards the left or the right side of these continua simultaneously, although there is room for mixing and matching approaches. The goal of a model is important to consider when evaluating its effectiveness

Tradeoffs	Continua	
Goal	Insight	Forecasting
Approach	Strategic	Tactical
Biology	Abstract	Realistic
Detail	Less	More
Uncertainty	Deterministic	Stochastic
Analysis	Analytical	Numerical
Applicability	General	Specific

communities. Two primary kinds of dynamic models have developed in this area: those that focus on a particular invasion vector, and those that model spatial patterns in regional or global species transport pathways.

Ballast-transport dynamics have been investigated using population models to compare the invasion potential of different species, and to evaluate prevention methods (e.g., MacIsaac et al. 2002; Wonham et al. 2005a, b). For marine and estuarine species, for example, Wonham et al. (2005b) used a simple population growth model to illustrate how the timing and level of open-ocean exchange can be optimized to reduce invasion risk for species with different salinity tolerances (Fig. 4.1). A more detailed population model applied to freshwater zooplankton resting stages in ballast sediments (Wonham et al. 2005a) could be extended to marine species as well.

A single invasion pathway rarely operates in isolation, and is typically connected to a larger spatial network of species transportation. Such networks can be modeled in a spatially implicit way using gravity models that represent the connections between pairs of points linked by a given transport pathway. This approach has been used regionally to model boat traffic and zebra mussel invasions, and globally to characterize shipping routes with high ballast water discharge (Bossenbroek et al. 2001; Drake and Lodge 2004; Leung et al. 2006).

In principle, invasion pathway models that predict propagule pressure over space and time could provide the initial conditions required for the population establishment and spread models treated in the following sections.

4.3 Population Models: Invasion Dynamics

In this section, we treat models that focus on a particular invader and its establishment, impacts, and control. These are all nonspatial models; spatial models, which use an additional set of mathematical tools, are considered in the next section. We begin with models of the invader dynamics alone, followed by models of invaders interacting with other species.

Table 4.2 Dynamical mathematical models of marine invasions reviewed here. Models classified by invasion stage and model focus. Some models are specifically developed for particular marine invaders; others are motivated by or applied to marine species, but are more general in their application

Invasion aspects	Physical and biological processes	Invader	Taxon	References
Pathways				
Delivery, prevention	Population growth	Multiple[a]	Invertebrates	Wonham et al. (2005b)
Delivery, establishment, prevention	-	Generic	Generic	Drake and Lodge (2004)
Population: invasion dynamics				
Single species				
Establishment	Chromosomal reversion, population growth	*Crassostrea ariakensis*	Bivalve (oyster)	Dew et al. (2003)
Establishment	Population and metapopulation growth	*Pseudodiaptomus marinus*	Crustacean (copepod)	Barry and Levings (2002)
Establishment, control	Population growth, removal	*Caulerpa taxifolia*	Macroalga	Ruesink and Collado-Vides (2006)
Control	Population growth, removal	*Ocinebrellus inornatus*	Gastropod (oyster drill)	Buhle et al. (2005)
Multiple species				
Establishment	Foodweb	*Mnemiopsis leidyi*	Ctenophore	Gucu (2002)
Establishment	Genetic fixation	Transgenes; *Oncorhynchus kisutch*	Transgene	Hedrick (2001)
Establishment, impact	Population growth, infection dynamics	Phocine distemper	Virus	Grenfell et al. (1992); Heide-Jørgensen and Härkönen (1992); De Koeijer et al. (1998); Harding et al. (2002, 2003); Lonergan and Harwood (2003)
Impact	Primary production, herbivory, predation	Generic; *Mnemiopsis leidyi*	Ctenophore	Morozov et al. (2005)

(continued)

Table 4.2 (continued)

Invasion aspects	Physical and biological processes	Invader	Taxon	References
Impact	Population growth (fish) + bioenergetic (jelly)	*Mnemiopsis leidyi*	Ctenophore	Knowler (2005)
Impact	Interspecific competition, parasitism	*Batillaria attramentaria*	Gastropod (mud snail)	Byers and Goldwasser (2001)
Impact	Foodweb	*Venerupis philippinarum*	Bivalve (clam)	Pranovi et al. (2003)
Impact, control	Bioeconomic	*Crepidula fornicata*	Gastropod (slipper shell)	Frésard and Boncoeur (2006)
Population: invasion spread				
Single species				
Establishment, prevention	Hydrodynamics, population growth, dispersal	Multiple[b]	Invertebrates	Drake et al. (2005)
Spread	Hydrodynamics, population growth, dispersal	Multiple[c]	Invertebrates	Grosholz (1996)
Spread	Hydrodynamics, dispersal	Multiple[d]	Invertebrates, macroalgae, fish	Kinlan et al. (2005)
Spread	Hydrodynamics, dispersal	*Phyllorhiza punctata*	Cnidarian (jellyfish)	Johnson et al. (2005)
Spread	Hydrodynamics, dispersal	*Asterias amurensis*	Echinoderm (sea star)	Parry et al. (2001)
Spread	Hydrodynamics, dispersal	*Crepidula fornicata*	Gastropod (slipper shell)	Viard et al. (2006)
Spread	Hydrodynamics, dispersal	*Theora lubrica, Musculista senhousia*	Bivalves (clam, mussel)	Inglis et al. (2006)
Spread	Population growth, dispersal	*Spartina alterniflora*	Plant (cordgrass)	Taylor et al. (2004)
Spread	Population growth, dispersal, control	*Spartina alterniflora*	Plant (cordgrass)	Taylor and Hastings (2004)

Spread, impact	Population growth, dispersal, habitat modification	*Spartina alterniflora*	Plant (cordgrass)	Cuddington and Hastings (2004)
Multiple species				
Spread, impact	Dispersal, infection dynamics	Phocine distemper virus	Virus	Swinton et al. (1998)
Spread	Population growth, dispersal	*Caulerpa taxifolia*	Macroalga	Hill et al. (1998); Aussem and Hill (1999, 2000)
Spread, impact	Hydrodynamics, dispersal, bioenergetics	*Mnemiopsis leidyi*	Ctenophore	Volovik et al. (1995)
Spread, impact	Hydrodynamics, dispersal, bioenergetics	*Mnemiopsis leidyi*	Ctenophore	Berdnikov et al. (1999)
Spread, impact	Hydrodynamics, dispersal, bioenergetics	*Mnemiopsis leidyi*	Ctenophore	Oguz et al. (2001)
Spread, impact	Hydrodynamics, dispersal, bioenergetics	*Mnemiopsis leidyi*	Ctenophore	Lebedeva and Shushkina (1994)
Spread, impact	Hybridization	Generic; *Spartina alterniflora*	Plant (cordgrass)	Hall et al. (2006)
Impact	Hydrodynamics, host-parasite dynamics	*Salmo salar*	Fish (salmon)	Krkošek et al. (2005, 2006)
Spread, impact, control	Population growth, dispersal, herbivory	*Elysia subornata*, *Caulerpa taxifolia*	Mollusc (sea slug), macroalga	Coquillard et al. (2000)
Impact	Hydrodynamics, foodweb	*Sabella spallenzani*	Annelid	Murray and Parslow (1999)
Community invasibility				
-	Growth, spread, competition	Multiple[e]	Invertebrates	Dunstan and Johnson (2005, 2006)
-	Foodweb	Multiple[f]	Invertebrates	Castillo et al. (2000)

(continued)

Table 4.2 (continued)

Invasion aspects	Physical and biological processes	Invader	Taxon	References
-	Species persistence	Multiple[g]	Invertebrates	Costello and Solow (2003); Solow and Costello (2004)
-	Species persistence	Multiple[h]	Invertebrates, macroalgae, fish, plants	Wonham and Pachepsky (2006)

[a]Wonham et al. (2005b). Model is applied to the marine invertebrates *Crassostrea gigas, Crassostrea virginica, Pinctada imbricata, Mytilus galloprovincialis, Venerupis philippinarum, Rapana venosa, Pacifastacus leniusculus, Carcinus maenas, Eriocheir sinensis, Dikerogammarus villosus, Eurytemora affinis,* and *Asterias amurensis*

[b]Drake et al. (2005). Model is applied generally to ctenophores, copepods, amphipods, decapods, and fish

[c]Grosholz (1996). Model is applied to the marine invertebrates *Botrylloides leachi, Carcinus maenas, Elminius modestus, Hemigrapsus sanguineus, Littorina littorea, Membranipora membranacea, Mytilus galloprovincialis, Perna perna, Philine auriformis,* and *Tritonia plebeia*

[d]Kinlan et al. (2005). Model is applied to a number of marine algae and invertebrates, and one fish species

[e]Dunstan and Johnson (2005, 2006). Model represents an epibenthic community in Tasmania, where at least two of the invertebrates (*Botrylloides leachi* and *Watersipora subtorquata*) are non-native (Hewitt et al. 2004) but are not singled out for analysis

[f]Castillo et al. (2000). Model represents a soft sediment community in Oregon, USA, which includes the introduced and cryptogenic invertebrates *Mya arenaria, Nippoleucon hinumensis, Heteromastus filiformis, Hobsonia florida, Pseudopolydora kempi,* and *Streblospio benedicti*

[g]Costello and Solow (2003); Solow & Costello (2004). Model is applied to marine invaders of San Francisco Bay

[h]Wonham & Pachepsky (2006). Model is applied to the marine invaders of the Baltic, Black, Caspian, Mediterranean, and North Seas; the Northeast Atlantic and Northeast Pacific Oceans; Hawai'i and Australia

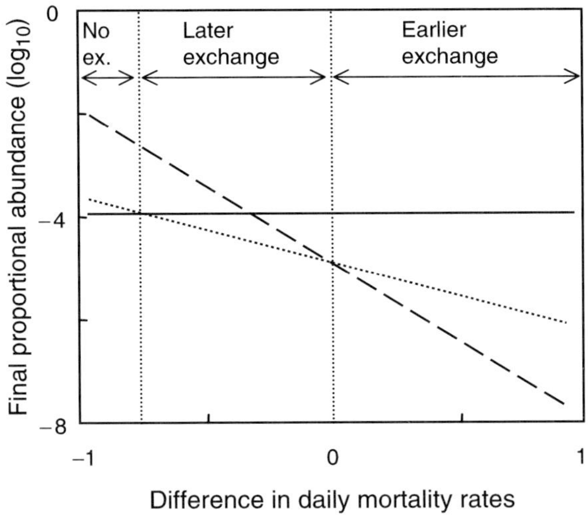

Fig. 4.1 Predicted effectiveness of three ballast-water exchange strategies. The *vertical axis* indicates the final abundance of organisms at the end of a ballast-water voyage, as a proportion of their initial abundance. The *horizontal axis* indicates the difference in a species' daily mortality rates in a ballast tank before and after exchange. Species with broader salinity tolerances would be near 0; those with narrower tolerances would be towards either end. Three scenarios are plotted, showing final abundance given later exchange (*dotted line*), earlier exchange (*dashed line*), and no exchange (*solid horizontal line*). *Vertical lines* separate the three regions in which each exchange strategy minimizes the final organism abundance. Redrawn from Wonham et al. (2005b)

4.3.1 Single-species Models

A dynamic model of a single invading population is generally the central element of an invasion modeling enterprise. Depending on the goal, such models can increase in complexity from simple exponential growth to include negative (intraspecific competition) and positive density dependence (Allee effects). Likewise, they can increase in detail from representing a homogeneous, to an age- or stage-structured, to an individual-based population.

In ecology and conservation, single-species models have been applied particularly effectively to population viability analysis (PVA) of threatened and endangered species (Holmes 2004; Morris and Doak 2003), and more recently to PVA of invasive species (Andersen 2005; Bartell and Nair 2003; McEvoy and Coombs 1999; Parker 2000; Shea and Kelly 1998; Shea and Possingham 2000). Most PVA work has been done with terrestrial species, but notable marine examples include cetaceans and turtles (e.g., Burkhart and Slooten 2003; Crowder et al. 1994). In these instances, single-species models can usefully identify key life stages for management actions. However, lessons learned from fisheries modeling illustrate the limitations of this isolated approach for modeling and managing complex systems of interacting species, and have led to more detailed multispecies and ecosystem scale modeling of marine systems (e.g., Butterworth and Plaganyi 2004; Fulton et al. 2003; Hollowed et al. 2000).

Only a small handful of marine invasion models use nonspatial single-species approaches alone. Most of these focus on the biological questions of population establishment and its inverse, invasion control. To investigate establishment at the genetic scale, Dew et al. (2003) developed an age-structured algorithm to model the risk of chromosomal reversion and establishment of outplanted triploid Suminoe oysters *Crassostrea ariakensis* in the northwest Atlantic. Barry and Levings (2002) implemented a stage-structured model of the copepod *Pseudodiaptomus marinus* in the northeast Pacific, to evaluate the establishment risk of a single population and of a spatially implicit metapopulation. The life history and component data presented by Rudnick et al. (2005) for the Chinese mitten crab (*Eriocheir sinensis*) could readily be formalized in a similar mathematical model. These models fall at the tactical end of the spectrum, in that they are focused on detailed predictions in a particular system.

Invasion control introduces additional elements into single species models. Ruesink and Collado-Vides (2006) used growth and recruitment data to parameterize a tactical model of the area occupied by the green alga *Caulerpa taxifolia* in the Mediterranean. They then numerically evaluated how the timing of control altered the total occupied area. In a more strategic approach Buhle et al. (2005) constructed and analyzed a matrix population model of the oyster drill *Ocinebrellus inornatus* in the northeast Pacific (Fig. 4.2). They then incorporated the predicted population growth rate into an economic cost-benefit analysis to determine which life stages offered more cost-effective control opportunities (Fig. 4.2). An intriguing spatial extension to these bioeconomic models involves long-term cost-benefit analysis of managing barrier zones adjacent to the population front (Leung et al. 2002; Sharov 2004; Sharov and Liebhold 1998) – an approach that could be applied to marine invasions as well.

There are many methods available for attempting invasion control, including physical, chemical, and biological means. In a single-species model, the control element can be formulated to represent the removal of a certain number or proportion of individuals. As such, it is analogous to the harvest element in simple fisheries models, and it best represents the effects of physical or chemical control. To model biological control, with feedback between the invader and the control species, takes us into the realm of multispecies models.

4.3.2 *Multi-species Models*

A number of classical modeling frameworks for species interactions, which have proved useful in developing ecological theory in general, have also been applied to understanding invasions. These include Lotka-Volterra competition models,

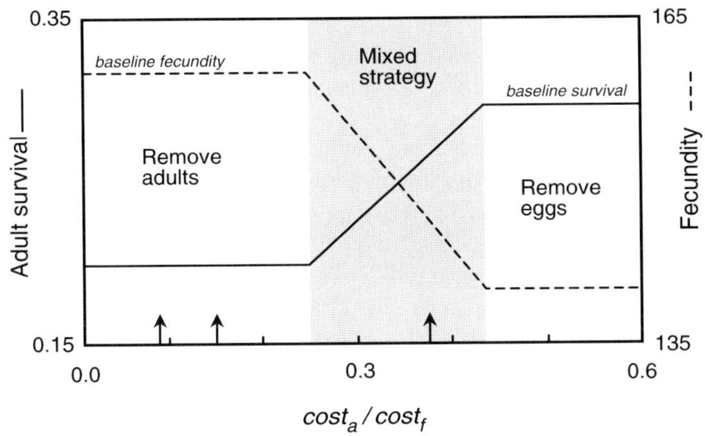

Fig. 4.2 Optimal control strategy determined from a bioeconomic model of invasive oyster drills, *Ocinebrellus inornatus*. The *x-axis* shows the ratio of costs of the same proportional reduction in adult survival and fecundity. The *y-axes* show the optimal target values for adult per capita survival (*left axis*) and fecundity (*right axis*) for controlling the invader. In this instance, the baseline adult survival probability is 0.3, and the baseline fecundity is 160 surviving offspring per adult. When removing adults is much cheaper than removing egg capsules ($cost_a \ll cost_f$), the optimal control strategy is to remove adults only, reducing survival to ~0.20. When the cost of removing adults begins to approach that of removing eggs, the optimal strategy becomes removing eggs only, reducing fecundity to ~140. At intermediate $cost_a/cost_f$ values, the optimal control strategy is a mixed one of removing both adults and eggs. *Arrows* indicate empirically estimated values of $cost_a/cost_f$ from different sites. Redrawn based on Buhle et al. (2005), and E. Buhle, personal communication to MJW

Volterra predation models, Nicholson-Bailey host-parasitoid models, Kermack-McKendrick epidemiological models, and their extensions and generalizations. (For an introduction to these models and their application to invasions, see Shigesada and Kawasaki 1997.)

In their simplest and most strategic formulations, these models remain theoretically tractable, and stability analysis can provide general insights into ecological invasion processes. Two particular biological questions can readily be addressed with these models: the ability of a species to invade an equilibrium population of one or more other species, and the impacts of its invasion on that equilibrium.

Generally, however, this strategic approach does not suffice to represent detailed processes of species interactions, or the combined interactions of a group of species at a community or ecosystem scale. As a result, many multispecies models of specific invasion systems move away from these classical strategic approaches to more detailed tactical formulations. Two marine invasion examples illustrate the interplay between these approaches.

Pathogen invasions are increasingly being reported from marine systems (e.g., Harvell et al. 1999), but classical epidemiological modeling (Anderson and May 1991; Kermack and McKendrick 1927) has rarely been applied to these diseases. However, certain long-established epidemiological models, known generally as *S-I* or Susceptible-Infectious models, have an impressive history of contributing to the understanding of infectious disease dynamics and control (Anderson and May 1991; Kermack and McKendrick 1927).

A series of three *S-I* models of phocine distemper virus (PDV) outbreaks in the northeast Atlantic illustrate a generally strategic approach (De Koeijer et al. 1998; Grenfell et al. 1992; Harding et al. 2002, 2003; Heide-Jørgensen and Härkönen 1992; Lonergan and Harwood 2003). Admittedly, PDV can perhaps only tangentially be considered a marine invasion, as its transmission is airborne. We mention it here since it can play a significant role in marine mammal population dynamics, and it illustrates an important class of models that can be applied to invasive infectious diseases.

These three PDV models illustrate how the same epidemic may be modeled using very different infection dynamics (mass action vs frequency dependent incidence functions), different treatments of time (continuous vs discrete), and different treatments of uncertainty (deterministic vs stochastic). One prediction from the first of these models was that another outbreak would not occur until the seal population had recovered for at least ten years after the 1988 outbreak (Grenfell et al. 1992). Indeed, a second outbreak occurred in 2002, prompting further modeling to explore the potential impact of recurring outbreaks on the population (Harding et al. 2002, 2003; Lonergan and Harwood 2003).

The second example is the invasion of the Black Sea by the comb jelly *Mnemiopsis leidyi*, which has been investigated with models that range from the strategic to the tactical. At the strategic end is a generic nutrient-phytoplankton-zooplankton (N-P-Z) model developed by Morozov et al. (2005) to investigate the impacts of adding a top predator to a marine plankton community. Although this exercise was motivated by the example of *Mnemiopsis*, its strategic generality makes it applicable, at least conceptually, to any such invasion.

Knowler (2005) took a somewhat more complex and detailed approach to developing a bioeconomic model of the impacts of *Mnemiopsis* on the Black Sea anchovy fishery. This study combined a Ricker stock-recruitment model of the anchovy population with a balance model of *Mnemiopsis* biomass and an economic cost-benefit model. It then used analytical methods to determine the optimal anchovy harvest policy in the post-invasion system.

At the tactical end of the spectrum is the mass balance model developed by Gucu (2002) to investigate factors contributing to the establishment of *Mnemiopsis*. This approach used the software package ECOPATH to develop a steady state model of the Black Sea marine ecosystem before and after the *Mnemiopsis* invasion. This detailed model helped visualize the differences in food web structure correlated with overfishing, eutrophication, and the subsequent comb jelly invasion (Gucu 2002).

An additional five case studies further highlight the difference in approach between simpler strategic and more complex tactical approaches to multispecies models of marine invasions. Hedrick (2001) used a very general algebraic framework to determine the invasion criteria and fitness impacts of transgenes. The model was developed nominally to consider competition between wild type and transgenes in coho salmon, for which reason we include it in our treatment of marine multispecies (or in this case, multigene) invasions, but it could apply broadly to transgene invasions from genetically modified organisms in any system (Hedrick 2001).

Also at the genetic level, Hall et al. (2006) developed a model of the population dynamics and genetics of hybridization between the Atlantic cordgrass *Spartina alterniflora* and its native congener, *S. foliosa*, in the northeast Pacific. Interestingly, they found that even without a selective advantage, the hybrid increased at the expense of the native and introduced genotypes (Hall et al. 2006).

To study a competitive interaction in detail, Byers and Goldwasser (2001) modeled the impact of the introduced mudsnail *Batillaria attramentaria* on a similar native species in the northeast Pacific (Fig. 4.3). They constructed an individual-based simulation model of the two snails and their respective resource conversion, parasitism, and mortality rates. The model was parameterized and validated with extensive field data, and then used to rank *Batillaria*'s competitive advantages and to predict the native snail's time to extinction (Fig. 4.3).

At an ecosystem scale, Pranovi et al. (2003) developed a complex, numerical mass balance model of the Manila clam *Venerupis* (=*Tapes*) *philippinarum* in the Venice lagoon. The clam has become an important commercial species, and is harvested with mechanical dredges that disturb the bottom sediments and associated community. In the so-called *Tapes* paradox, the clam is more abundant inside than outside fished areas. The model suggested that these positive feedback effects were somewhat limited, and predicted the degree to which eliminating the commercial clam harvest would increase the trophic level, total catch, and market value of the lagoon's other artisanal fisheries (Pranovi et al. 2003).

Finally, Frésard and Boncoeur (2006) conducted a cost-benefit analysis of controlling the slipper shell *Crepidula fornicata* on stocked commercial scallop beds in the northeast Atlantic. The costs of *Crepidula* to the fishery are direct, in that it must be removed from the shells of harvested scallops, and indirect, that it preemptively outcompetes settling scallops. The direct costs were estimated from the time devoted to scallop removal. In the absence of a *Crepidula* population dynamics model, and of competition coefficients between *Crepidula* and the scallops, the indirect costs were represented simply as a fixed reduction in harvestable area (Frésard and Boncoeur 2006). Thus, this model explores the impact of an invader on another species without explicitly having to model the population dynamics of either.

All the single and multispecies models considered thus far have been nonspatial, or have treated space implicitly. In the next section, we consider invasion models that incorporate space explicitly in one, two, or three dimensions.

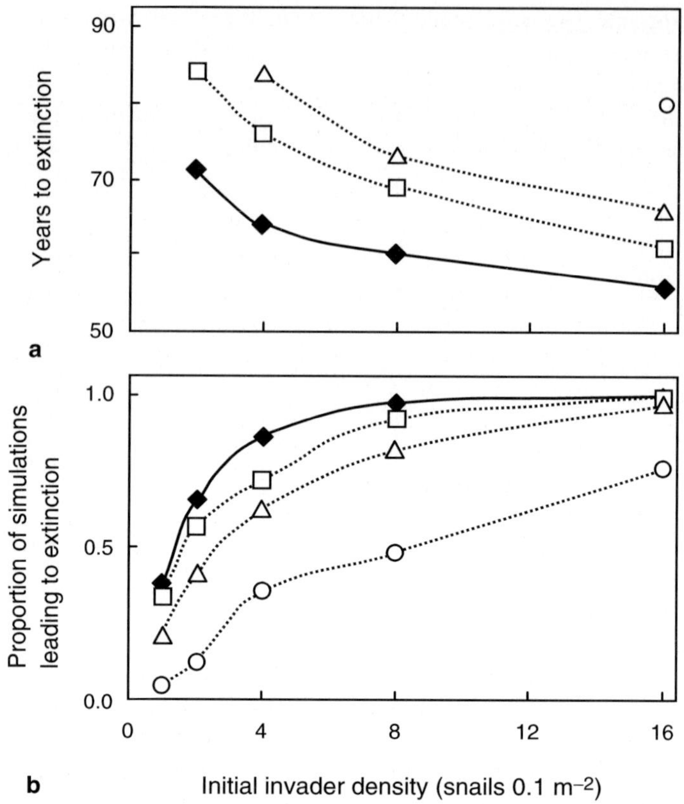

Fig. 4.3 Predicted impacts of the non-native mud snail *Batillaria attramentaria* on the native mud snail *Cerithidea californica*, based on a simulation model. **a** Median predicted years to extinction of the native snail after initial introduction of the invader at different densities. **b** Proportion of simulations that resulted in extinction of the native snail within 90 years. Outcomes shown for four scenarios: interspecific differences in parasitism and competition as observed in the field (*diamonds*), parasitism rates set equal for the two species (*squares*), competition parameters set equal (*triangles*), and both set equal (*circles*). In the last case, no median values are shown for lower invader densities in A because <50% of simulations led to extinction. Redrawn from Byers and Goldwasser (2001)

4.4 Population Models: Invasion Spread

A substantial proportion of the modeling work on invasions to date has focused on the rate at which the invader spreads. Marine environments, where the habitat itself is in motion, provide additional challenges for modeling spatial spread. We will first present models of a single invader spreading alone, followed by spatial models that incorporate multispecies interactions.

4.4.1 Single-species Models

The now-classic Fisher equation for modeling spread was originally developed to represent the spread of advantageous alleles through a population, and assumed logistic growth and random movement via diffusion (Fisher 1937). It was later adapted by Skellam (1951) to model the spread of invasive species, assuming exponential growth and diffusion. Both models lead to the same compact formula for the rate of spread, expressed in terms of the population's intrinsic growth rate and its so-called diffusion coefficient, which can be interpreted as measure of spatial movement (for details and extensions, see Shigesada and Kawasaki 1997).

The appeal of Fisher's formula lies not only in its simplicity, but also in the possibility of validation. Intrinsic growth rates and diffusion coefficients can be estimated from life history tables and mark-recapture studies, allowing spread rate predictions to be tested against historical data. This validation approach was pioneered by Andow et al. (1990) for terrestrial species. Here, Fisher's diffusion formula generally holds up well, although it tends to underestimate spread for invaders that exhibit long-distance non-diffusive jumps in space. For model extensions that incorporate such jumps using integrodifference equations see Kot et al. 1996; Neubert and Parker 2004; Lewis et al. 2005.

Marine and aquatic habitats differ notably from terrestrial ones in that they move on a relatively short timescale. Indeed, several recent reviews have highlighted the particular challenges of studying and modeling dispersal, particularly of larvae, in these systems (Kinlan et al. 2005; Kinlan and Hastings 2005; Levin 2006; Lutscher et al. 2005; Shanks et al. 2003; Siegel et al. 2003). These studies have emphasized the importance of local and regional hydrology and geomorphology, as well as larval physiology and behaviour, in influencing organism dispersal and spread.

Given the mobile nature of the habitat, we might expect simple diffusion models to perform poorly for most marine species. Specifically, we would expect them to underestimate downcurrent and overestimate upcurrent spread rates. Advection-diffusion models, which incorporate unidirectional current flow as well as diffusive spread, might be expected to work better. Advection rates would be expected, intuitively, to conspire with larval planktonic duration to influence a species spread rate: the longer an organism is in its planktonic stage, the farther it can spread, so the faster its invasion can progress.

Although the empirical relationship between larval duration and invasion rate has proved stubbornly elusive (Fig. 4.4), both reaction-diffusion and advection-diffusion models have been developed based on larval duration (Grosholz 1996, Kinlan et al. 2005). It is perhaps not surprising that these models have had negligible predictive power – curiously, though, they have an interesting tendency to overpredict spread rates for marine invertebrates and underpredict those for marine algae (Grosholz 1996, Kinlan et al. 2005).

This result leads us to ask what other factors could be at work to influence marine invasion spread rates. Likely candidates for slowing invasion rates are Allee effects and negative interspecific interactions, which are discussed in Sects. 4.4.2 and 4.4.3. Candidates for speeding up invasion rates are human-mediated transport, models which we discussed in Sect. 4.2, and positive interspecific interactions, which we mention briefly in Sect. 4.5. It is also important to consider that a species' spread rate may differ in different directions, and that models may have to be parameterized carefully to capture and predict this variation (e.g., Krkošek et al. 2007; Lubina and Levin 1988).

Although advection-diffusion models may not predict marine invasion rates very successfully, they may still provide insight into possible routes of organism transport. In this way, Johnson et al. (2005) used a modified surface advection component of the three-dimensional Princeton Ocean Model to hindcast possible sources of a jellyfish *Phyllorhiza punctata* bloom in the Mississippi Bight.

Similarly, Parry et al. (2001) modeled the dispersal of the sea star *Asterias amurensis* larvae, using an existing three-dimensional advection-diffusion model of Port Phillip Bay, Tasmania. The model predicted an overall *Asterias* distribution consistent with observed records, and local-scale incongruities were attributed to possible differences in predation pressure.

Viard et al. (2006) used a two-dimensional advection-diffusion model of the English Channel to predict the degree of larval slipper shell (*Crepidula fornicata*) exchange between populations. They found no correlation between predicted larval exchange and the observed genetic distance between populations, indicating high gene flow among populations.

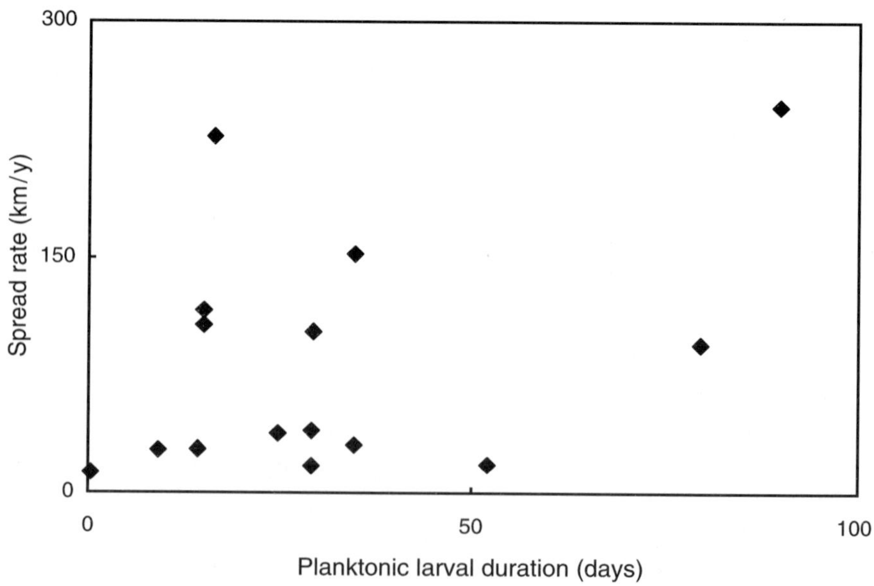

Fig. 4.4 Observed spread rates of marine invertebrate invaders as a function of observed planktonic larval duration. Redrawn from Kinlan et al. (2005)

Using a novel modeling approach, Inglis et al. (2006) combined a particle diffusion model with two different statistical habitat models to predict the distribution of bivalves *Theora lubrica* and *Musculista senhousia* in a New Zealand harbour. The diffusion component significantly improved the fit and accuracy of the habitat suitability index model, but only marginally improved the environmental regression model, and the improvement was more pronounced for *Theora* than for *Musculista* (Inglis et al. 2006).

4.4.2 Allee Effects

One mechanism that is well known to slow observed spread rates in terrestrial systems is an Allee effect (Hastings 1996). The hallmark of Allee dynamics is positive density dependence at low population levels. In other words, very small populations have lower per capita growth rates than slightly larger ones (Allee 1931; Gascoigne and Lipcius 2004). Demographic Allee effects – known in the fisheries literature as depensatory mortality – are likely to be particularly relevant at the establishment stage of a biological invasion.

In marine populations, Allee effects could arise from a wide range of mechanisms, and could be exacerbated or mitigated by the role of currents and eddies in dispersing or aggregating individuals. Although the empirical evidence for Allee effects in marine populations remains mostly indirect (Gascoigne and Lipcius 2004; Hutchings and Reynolds 2004), the potential interaction between harvesting and Allee effects (Gascoigne and Lipcius 2004; Hutchings and Reynolds 2004; Lundquist and Botsford 2004) suggests that invasion management actions could exploit Allee thresholds in controlling unwanted invaders.

An Allee effect is considered strong if the per capita growth rate becomes negative at a small population size, and weak if the growth rate decreases but remains positive. Mathematical extensions of Fisher's equation, in which logistic growth is replaced with Allee dynamics, make two predictions. First, both weak and strong Allee dynamics give rise to a slower invasion speed than the original model. Second, when Allee dynamics are strong, the initial colonization of invaders must exceed a threshold in both density and spatial extent if the invasion is to succeed (Kot et al. 1996; Lewis and Kareiva 1993; Wang and Kot 2001).

Allee effects have been shown empirically to play a role in the *Spartina alterniflora* invasion of the North American Pacific coast. This Atlantic cordgrass spreads rapidly across intertidal mudflats, filling in behind the invasion front to create dense meadows. The fecundity of individual *Spartina* is orders of magnitude higher for plants in established meadows than for isolated plants, which are limited by pollen availability and have reduced seed production. This leads to weak Allee dynamics. Taylor et al. (2004) developed and parameterized a spatially-explicit stochastic simulation model and a spatially-implicit deterministic model of this invasion in Willapa Bay, USA. By running the models with and without the Allee dynamics, they showed that even though the Allee effect is defined as weak, its effects are dramatic, almost halving the predicted spread rates (Fig. 4.5). From simulations, the Allee dynamics were deduced to confer heightened sensitivity of spread rates to the level of self-fertilization (Taylor et al. 2004).

Taylor and Hastings (2004) investigated the effectiveness of barrier zones for controlling the *Spartina alterniflora* invasion in the same location. They asked whether it was more efficient to prioritize removal of young, low-density outlier areas at the edge of an invasion, or older core population meadows. Their results indicated that *S. alterniflora* eradication was only possible if control of the faster growing low-density plants was prioritized. The most effective strategy, however, which would also require more resources, was to target the older core population areas as well. This was because, under the Allee dynamics seen in *Spartina*, eliminating high-density meadows lowered the risk of new propagule production far more than eliminating the lower-density outlying plants (Fig. 4.5).

Working in the same invasion system, Cuddington and Hastings (2004) developed a *Spartina alterniflora* spread model that included positive feedback dynam-

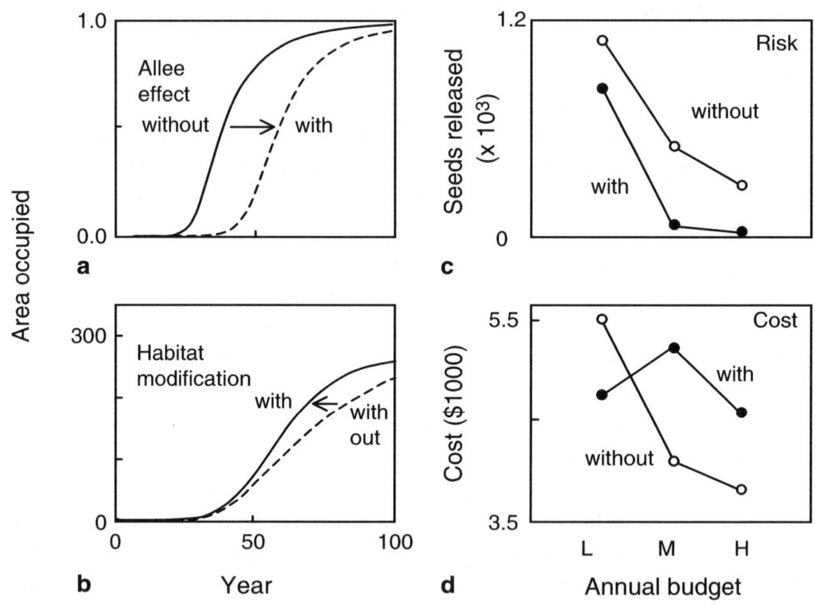

Fig. 4.5 Modeling analyses of the cordgrass *Spartina alterniflora* invasion with and without Allee effects. **a** Results of a simulation model showing that the empirically-observed Allee effect slows the rate of *Spartina* spread. Redrawn from Taylor et al. (2004). **b** Model results illustrating that habitat modification by spreading *Spartina* speeds up its invasion. Redrawn from Cuddington and Hastings (2004). **c** Predicted risk (thousands of potentially escaping seeds) associated with optimal *Spartina* control strategies, given low, medium, and high annual budgets, for a model with and without an Allee effect. Risk decreased with increasing budget, and when the Allee effect was incorporated. **d** Predicted cost (thousands of USD) for the same model. Allee effects led to cheaper eradication at a low budget, but more expensive eradication at higher budgets. Redrawn from Taylor and Hastings (2004)

ics driven by habitat engineering. As *Spartina* stands develop, they reduce water flow, trapping sediments and elevating the substratum into and eventually above the intertidal zone. In this way, the plant increases its available habitat. The model linked the area occupied by the invader and the distribution of habitat quality in a spatially implicit system of ordinary differential equations. The results illustrated that an engineer may enhance or reduce its own growth and spread rates, depending on the initial habitat quality distribution and the rate of habitat modification (Cuddington and Hastings 2004). Although this model is not explicitly spatial, we introduce it here because the positive feedback process shows intriguing similarities to Allee effects, in both its density dependence and its qualitative impacts on population spread (Fig. 4.5).

In a more general approach, Drake et al. (2005) developed a reaction-diffusion model with Allee dynamics to explore the chance of establishment for a range of planktonic marine organisms released from ballast water. Using this model, they found that for a given level of invasion risk, the acceptable release volume was far more sensitive to variation in the intrinsic rate of population growth than to variation in the strength of the Allee effect.

All the spread models treated so far are process oriented, in that specific terms represent movement, growth, reproduction, mortality, dispersal, and so forth. This makes it possible to evaluate the importance of each process in the invasion outcome. By way of contrast, if one is simply interested in forecasting a particular invasion, a less mechanistic approach can be used. For example, the goal of one series of papers was to predict the expansion of the invasive green alga *Caulerpa taxifolia* in the Mediterranean Sea (Aussem and Hill 1999, 2000; Hill et al. 1998).

Initially, a stochastic discrete event simulation model of the algal spread was developed and integrated with GIS habitat data (Hill et al. 1998). However, this model proved too computationally intensive to investigate the effects of various environmental variables, such as bathymetry, substrata and resident species, on the invader's spread. Accordingly, the authors developed a neural network metamodel designed to approximate the original simulation model but be more computationally efficient (Aussem and Hill 1999, 2000).

The neural network was trained by having it form rules connecting input (environmental variables) with output patterns (algal spread). The model was then tested by having it predict a sequence of historical spread independent of the data used to train the network, and was found to forecast accurately (Aussem and Hill 1999, 2000). Unlike the first and more mechanistic model, the *Caulerpa* metamodel is more phenomenological, in that the network rules do not necessarily have a biological interpretation. Both *Caulerpa* models are more tactical than strategic, in that accurate prediction rather than general insight is the goal.

Single-species spread models can incorporate aspects of the physical environment, such as advection, and intraspecific dynamics, such as Allee effects. However, no invader establishes in a vacuum: all interact to a greater or lesser extent with resident species, and some are targeted for biological control. This brings us to consider multispecies models of marine invasion spread.

4.4.3 Multi-species Models

Spatial multispecies models allow interspecific interaction dynamics to play out while species are dispersing across a domain via diffusion and advection. Many of the invasion systems studied with these models are extensions of nonspatial dynamics, derived by simply including diffusion and/or advection terms in the equations.

Lotka-Volterra competition dynamics have been extended to include random motion via diffusion, and used to model the spatial spread of one competitor into another's territory. The result is a system of nonlinear coupled reaction-diffusion equations. An early application of this model was to the spread of larger, introduced grey squirrels into areas occupied by native red squirrels in the United Kingdom (Okubo et al. 1989). Here a simple and compact formula, similar to that for Fisher's equation, was constructed. This formula described the predicted speed at which grey squirrels spread into, and displaced, the red squirrels, and could be calculated in terms of relative growth rates, competition coefficients and diffusion coefficients. Okubo et al. (1989) then measured the rates and coefficients and compared the predicted and observed spread rates, showing that they correlated closely. With both advection (unidirectional flow) and diffusion (random motion) included, competition models have been extended to aquatic systems, where they have been used to study mechanisms for species coexistence in river ecosystems (Lutscher et al. 2007).

Multispecies predator-prey models that include diffusion can be used to predict the rate of spread of an invasive predator into a native prey population, or, alternatively, the rate at which an invasive prey species can facilitate the spread of a predator that consumes it (Owen and Lewis 2001). This latter scenario has been analyzed in the context of biocontrol. If an invasive pest (prey) is spreading into a new environment, will a biocontrol agent (predator) be able to catch up to it and control its spread? Fagan et al. (2002) showed how to use coupled reaction-diffusion models to predict the biocontrol species attributes needed to catch up to the prey and control it effectively. Behind an invasion front, spatial predator-prey dynamics can become extremely complex and patchy (Sherratt et al. 1997): examples include the patchy distribution of marine phytoplankton and fish (Medvinsky et al. 2002) and of virally infected phytoplankton (Malchow et al. 2004).

As with the models considered above, most of the marine invasion systems studied with spatial models are extensions of nonspatial models we have already considered, including those for phocine distemper virus (PDV), *Mnemiopsis leidyi*, and *Caulerpa taxifolia*. Additional models treat the effects of increased filter feeding by the introduced annelid *Sabella spallenzani*, and the indirect effects of non-native farmed Atlantic salmon, *Salmo salar*.

The spatial spread of the 1998 PDV outbreak was modeled by Swinton et al. (1998), using an *S-I* type model distributed across a patchy network of seal sub-

populations. The authors found that the persistence of the epidemic depended on the number of patches when the patches were weakly coupled, and on the size of the patches when they were tightly coupled. They concluded that the 1988 outbreak died out because smaller, tightly coupled seal subpopulations could not sustain the infection (Swinton et al. 1998).

Three multispecies extensions of *Mnemiopsis* population modeling take a very different approach, coupling complex bioenergetic and hydrodynamic models to study the comb jelly invasion dynamics (Berdnikov et al. 1999; Oguz et al. 2001; Volovik et al. 1995). The goal of these models was to capture the observed pelagic foodweb dynamics before, during, and after the *Mnemiopsis* invasion. These tactical models, implemented as complex algorithms, are amenable to numerical but not analytical study. The models were assessed by qualitatively comparing their predictions to empirical data.

In a multispecies extension of *Caulerpa* spread models, Coquillard et al. (2000) focused on the potential reduction of the alga by the grazing gastropod *Elysia subornata*. In general, when biocontrol agents are introduced, a first objective is to optimize the release strategy to maximize the establishment of the agent (Shea and Possingham 2000). Another is to reduce the target species significantly by affecting vulnerable life-history stages (Shea and Kelly 1998). For a spreading invader, a third objective is to choose a control agent that will spread as quickly as the invader itself (Fagan et al. 2002). In the *Caulerpa-Elysia* model, all three objectives were analyzed by combining laboratory data and a spatially-explicit age-structured algorithmic model of the gastropod's growth to determine the optimal size, time, age, and number of releases for effective algal control (Coquillard et al. 2000).

Murray and Parslow (1999) developed a detailed bioenergetic and hydrodynamic model of Port Philip Bay, Australia. The model was used to consider the impacts of a number of environmental changes including nutrient loading and species invasions. As an example, the authors evaluated the potential impacts of the introduced annelid *Sabella spallenzani* by simulating an increase in filter-feeder biomass in the bay.

Finally, even without establishing in the wild, a non-native species can affect the community into which it is placed. For example, non-native farmed salmon, *Salmo salar*, serve as resident year-round hosts of parasitic sea lice in British Columbia estuaries (Krkošek et al. 2005). A spatially explicit model of sea lice population growth on juvenile salmon migrating past fish farms showed that farm hosts greatly amplified the natural infection levels and therefore juvenile salmon mortality (Fig. 4.6) (Krkošek et al. 2005, 2006).

In the previous two sections, we have looked at models that focus on an invading species and its establishment, spread, impacts, and control. The next section treats invasions from the community perspective, and examines models that focus on the dynamics of invasion success in different communities over space and time.

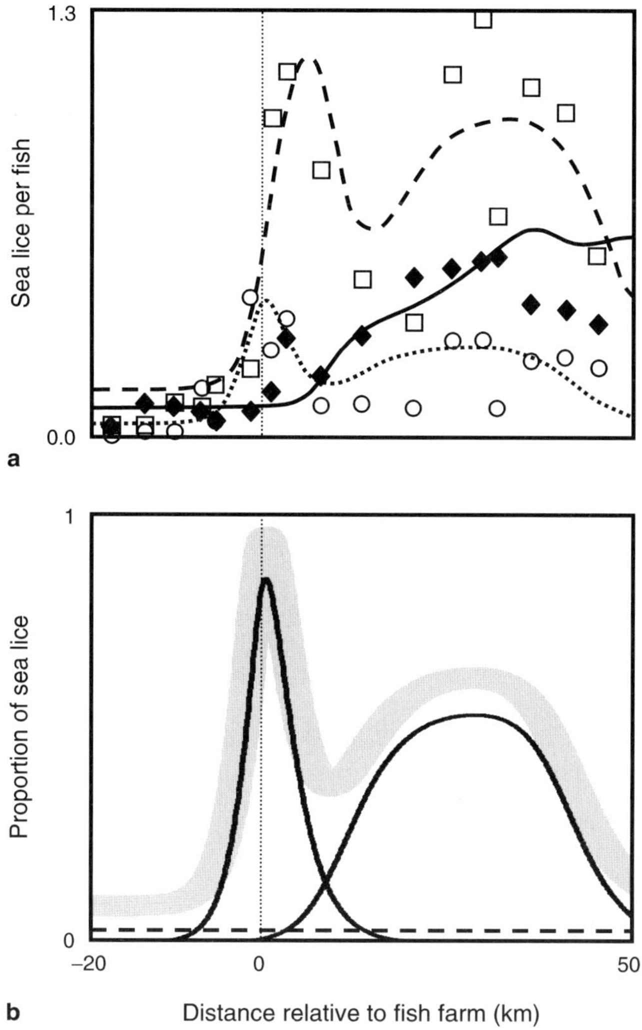

Fig. 4.6 Impacts of non-native farmed salmon, *Salmo salar*, on native host-parasite dynamics. **a** Observed (*points*) and predicted (*lines*) number of sea lice per native juvenile pink salmon along a coastal British Columbia migration route. Sea lice life stages are early juveniles (*circles, dotted line*), late juveniles (*squares, dashed line*), and adults (*diamonds, solid line*). **b** Planktonic distribution of sea lice early juveniles inferred from model. *Horizontal dashed line* is natural background lice level. *Solid black lines* are first generation (*left*) and second generation (*right*) lice originating from farm. *Thick gray line* is total early juvenile abundance in the plankton. *Vertical dotted lines* indicate farm location. Redrawn from Krkošek et al. (2005)

4.5 Community Invasibility Models

Spatial and temporal questions of community invasibility can be read, generally, under the longer-running ecological discussion of the connection between species diversity and community stability, both variously defined (e.g., Drake 1990; Post and Pimm 1983). In the invasion literature, this area has developed primarily as a discussion of community diversity and resistance to invasion – which can also be framed as a community's ability to repel or accommodate an invader, or as the likely success of an invader. The question of resilience in the face of invasion – which can also be framed as the scale of an invader's impact – has generally been treated separately in empirical work, but often simultaneously in modeling work.

Empirical studies, primarily in terrestrial plant communities, have led to a number of related conceptual models of the relationships among species richness, resource availability, disturbance, and invasibility (for recent reviews, see Davies et al. 2005; Richardson and Pysek 2006; Shea and Chesson 2002; Stachowicz and Byrnes 2006). Most of the abstract mathematical models exploring these relationships have focused on single-trophic level competitive interactions (Byers and Noonburg 2003; Case 1990, 1991; Hewitt and Huxel 2002; Melbourne et al. 2007; Mitchell et al. 2006; Rouget and Richardson 2003; Tilman 2004) and neutral dynamics (Fridley et al. 2004; Herben et al. 2004).

In marine systems, these questions have been investigated in a combined empirical and modeling study focused on competitive interactions in a marine fouling community (Dunstan and Johnson 2005, 2006). A spatially explicit individual-based model was parameterized with field data on growth, recruitment, and competitive interactions among 13 species in Tasmania, of which at least 3 were non-native (Hewitt et al. 2004). Community invasibility was defined in terms of species recruitment, whether native or non-native. The authors found that the relationship between species richness and invasibility was strongly mediated by the size of the patch being invaded. Larger patches had a more stable community composition over time, leading to less free space, and therefore lower recruitment and higher invasion resistance (Dunstan and Johnson 2006).

Compared to invasion resistance, comparatively few models have addressed the question of resilience, i.e., a community's response to invasion. Analytical exceptions are the patterns of resident species extinction reported for some community assembly models (Case 1990, 1991), and the biotic homogenization models developed in the context of freshwater fish invasions (Olden and Poff 2004).

In a marine case study, Castillo et al. (2000) constructed and analyzed a series of guild models of a soft-bottom invertebrate assemblage in the northeast Pacific. They found that for most of the models, the net feedback strength of the intertaxon interaction matrix was low, indicating that these systems would be expected neither to move away from nor to return to an equilibrium following a perturbation. The authors therefore suggested that this community was particularly able to accommodate the addition of invaders without experiencing species losses (Castillo et al. 2000).

One result that has emerged from both empirical and modeling approaches is the importance of spatial scale and invasion scenario in detecting and interpreting patterns of invasion resistance (Byers and Noonburg 2003; Hewitt and Huxel 2002). Since these empirical and theoretical approaches have so far primarily addressed

single trophic level and competitive interactions, it would be interesting to evaluate them, along with the related question of resilience, in the more complete context of a multi-trophic system.

Temporal patterns in the accumulation of invaders were initially highlighted by the striking marine example of San Francisco Bay (Cohen and Carlton 1998). Subsequent empirical and conceptual work connected this trend to the ideas of biotic resistance resulting from negative interactions, and invasional meltdown resulting from positive interactions (Ricciardi 2001; Simberloff and Von Holle 1999). Recent modeling has shown, however, that the observed pattern of acceleration in invasion numbers can be accounted for without any necessary increase in invasion rate or invasibility (Fig. 4.7) (Costello and Solow 2003; Solow and Costello 2004; Wonham and Pachepsky 2006). When null models reproduce observed patterns, it does not mean that other processes are not occurring. Rather, null models illustrate the patterns that would be expected in the absence of those processes, giving us a benchmark against which to compare empirical data.

4.6 Summary and Future Directions

The incorporation of modeling into the study of marine biological invasions is relatively recent, and holds exciting promise. In the sections above, we have touched on the range of existing marine invasion models. In this section, we briefly consider three areas that seem especially rich for future development.

4.6.1 Formalizing Conceptual Models Mathematically

The study of invasions has been characterized by an independent development of concepts and mechanisms that can, in many cases, be aligned with more general ecological concepts (e.g., Shea and Chesson 2002; Tilman 2004). On a broad scale, key invasion concepts of propagule pressure, species invasiveness, and community invasibility have largely been treated separately from each other in invasion biology. (For recent broad reviews, see Davies et al. 2005; Hails and Morley 2005; Melbourne et al. 2007; Mitchell et al. 2006; Richardson and Pysek 2006; Rouget and Richardson 2003; Shea and Chesson 2002; Stachowicz and Byrnes 2006; Tilman 2004.)

Recent invasion models, focused primarily on terrestrial plant systems, have offered general mathematical frameworks that begin to unite these concepts: Rouget and Richardson (2003) combined propagule pressure with environmental factors, Tilman (2004) connected success to resource competition, Mitchell et al. (2006) combined the effects of both abiotic and biotic factors, and Melbourne et al. (2007) examined the effects of a heterogeneous environment on the outcome of competition. More generally, ecological modeling that couples species dispersal

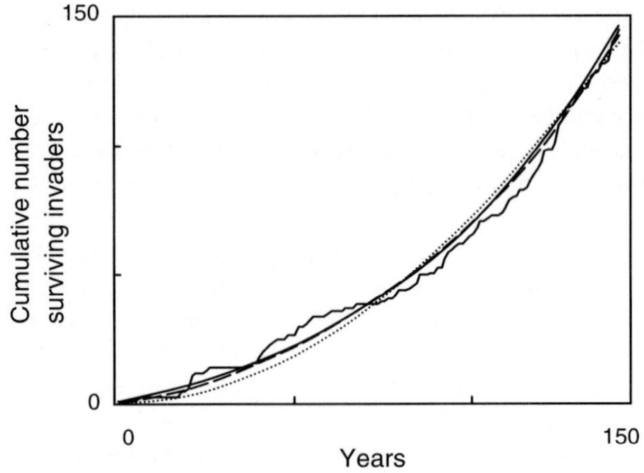

Fig. 4.7 Three different models capture the shape of the cumulative number of surviving invaders introduced into San Francisco Bay since the mid-1800s (*solid jagged line*). Predictions from a model incorporating introduction rate and discovery probability, for an increasing (*dashed curve*) and constant (*dotted curve*) introduction rate redrawn from Solow and Costello (2004). Predictions from a model incorporating introduction rate and survival probability, given constant introduction rate and constant survival probability (*solid curve*) redrawn from Wonham and Pachepsky (2006)

and species coexistence may be brought to bear on the problem of invasions (e.g., Fox and Srivastava 2006; He et al. 2005; MacArthur and Wilson 1967).

Mechanistic hypotheses toward explaining these general concepts include the influence on species establishment of propagule abundance, frequency, and quality, the influence on invasiveness of inherent traits, minimum residence time, enemy release, evolution of increased competitive ability, long distance dispersal, phenotypic plasticity, genetic drift, inbreeding, and hybridization for species invasiveness, and the influence on community invasibility of resource availability, disturbance, diversity, mutualisms, competition, predation, indirect interactions, spatial and temporal heterogeneity, and niche opportunities.

As invasion concepts are synthesized with each other and with classical ecology, they become increasingly amenable to formalization and testing through mathematical modeling. General ecological modeling that connects dispersal to species coexistence could also be brought to bear on the specific problem of invasions. For example, the nature and significance of propagule pressure (Colautti et al. 2006) has parallels in recruitment limitation and source-sink dynamics (e.g., Connolly et al. 2001; Levins 1969, 1970), which recent conceptual and mathematical syntheses can help explore (e.g., Amarasekare and Nisbet 2001; Hanski and Gaggiotti 2004; Holyoak et al. 2005). Patterns of invasion resistance and invasional meltdown (Elton 1958; Simberloff and Von Holle 1999) may be driven in part by the underlying processes of facilitation and inhibition first synthesized in classical

succession dynamics (Connell and Slatyer 1977). Recent modeling has scratched at the surface of these concepts and provides a framework that could be developed to explore these connections more explicitly (Costello and Solow 2003; Solow and Costello 2004; Wonham and Pachepsky 2006). It would be very interesting to explore the parallels between the notion of fluctuating resource availability (Davis et al. 2000), and the conceptual and mathematical formulations of the intermediate disturbance hypothesis (Connell 1978; Roxburgh et al. 2004).

A growing area of mathematical modeling in marine systems treats the design and impacts of marine protected areas (e.g., Gerber et al. 2003; Guichard et al. 2004). In so doing, it considers the problems of species dispersal and persistence, and connects to more general ecological and mathematical theories of minimum viable populations, habitat fragmentation, minimum available suitable habitat, and critical domain size (e.g., Pachepsky et al. 2005; With 2004), all of which apply also to invasion establishment and control. Other models that have provided insight into predicting marine population and community dynamics (e.g., Crowder et al. 1994; Wootton 2004) would be similarly informative in studying marine invasions. The considerable modeling tools developed in fisheries biology concerning harvest management (e.g., Kritzer and Sale 2004; Rose and Cowan 2003) could provide further insight into marine invasion control strategies.

4.6.2 Coupling Dynamical and Statistical Models

We have focused here on dynamical mathematical models, but there is a further wealth of statistical modeling of invasion patterns and processes. A number of examples concerning invasion transport and establishment illustrate ways in which these two modeling approaches could be coupled.

Statistical invasion risk analysis is extensively developed in the terrestrial realm, where it informs the International Plant Protection Convention, the international Agreement on the Application of Sanitary and Phytosanitary Measures, and numerous other international, national and regional policies regarding intentional and inadvertent imports and releases (Drake and Lodge 2006; Hayes 2003; Holt et al. 2006; Powell 2004; Wilson and Anton 2006). In marine systems (see also Chap. 20, Campbell), statistical risk analysis has been applied to organism transport in and on commercial and recreational vessels (Floerl et al. 2005, Hayes 2002a, b), and more generally to the accumulation of molluscan invaders, both terrestrial and marine, in the US (Levine and D'Antonio 2003). Any of these statistical frameworks could incorporate the kind of dynamic population modeling described in Sect. 2. Hayes (1998) outlines how this dual approach might be developed; the conceptual framework developed by Landis (2003) for green crab *Carcinus maenas* risk assessment would be amenable to this kind of mathematical formalization.

Statistical models of morphological and genetic population data (e.g., Bolton and Graham 2004; Daguin and Borsa 2000; Geller et al. 1997) and invader traits

(e.g., Kolar and Lodge 2002; Reichard and Hamilton 1997; Rejmánek and Richardson 1996) could also be coupled to dispersal models (e.g., Bossenbroek et al. 2001) to identify invasion sources. Recent developments in modeling invasion dynamics as stochastic processes (e.g. Drake and Lodge 2006; Jerde and Lewis 2007; Perrings 2005; Tilman 2004) also hold considerable promise for connecting to statistical analyses and their explicit treatment of uncertainty and variance.

A widely used statistical tool in species range predictions is environmental niche modeling. This approach has developed primarily in the context of predicting climate change impacts in terrestrial systems; there are a few examples of its use in forecasting invasions or describing their impacts (Herborg et al. 2007; Inglis et al. 2006; Peterson 2003; Peterson and Vieglais 2001; Vincent et al. 2006). The coupling of environmental niche modeling to population dynamics models (e.g., Akçakaya 2001; Akçakaya et al. 1995, 2004; Gutierrez et al. 2005), and hydrodynamic dispersal models (Inglis et al. 2006), holds promise for incorporating spatial variation in population dynamics and invasion speeds (e.g., Tobin et al. 2007).

4.6.3 Integrating Modeling and Empirical Work

Empirical and mathematical tools provide different kinds of insights into biological invasions. To a certain extent, their independent development can be profitable, but an interactive approach that incorporates both may generate the most understanding and predictive power.

Empirical and mathematical analysis complement and prompt each other in a number of ways. A model may generate a prediction that can be tested in the field, or a field study may generate a result that can be explored mathematically. Extensive empirical data – both observational and experimental – are required to parameterize and validate a model; models can help to synthesize empirical results across systems and scales. In some cases, modeling and empirical work can be combined in a single study, providing greater insight into local dynamics (e.g., Dunstan and Johnson 2005, 2006; Krkošek et al. 2005, 2006). In other cases, modeling may evolve in light of growing field data (e.g. Harding et al. 2002, 2003; Lonergan and Harwood 2003), or the synthesis of data may evolve with continued modeling (e.g., Byers and Noonburg 2003; Davies et al. 2005; Fridley et al. 2004). Neither the data nor the equations are an endpoint: it is the iteration between them that leads to evolution in understanding.

Many of the invasion hypotheses mentioned in Sect. 5.2 have arisen primarily from empirical observation, experimentation, and statistical analysis. Their current and future mathematical analysis holds exciting promise, and will in turn generate new ideas that can be examined again empirically. Marine systems provide additional challenges for both empirical and modeling work, and incorporating their unique elements is essential to a general understanding of marine invasions in particular, and invasion biology in general.

References

Akçakaya HR (2001) Linking population-level risk assessment with landscape and habitat models. Sci Total Environ 274:283–291

Akçakaya HR, McCarthy MA, Pearce JL (1995) Linking landscape data with population viability analysis – management options for the helmeted honeyeater *Lichenostomus melanops cassidix*. Biol Conserv 73:169–176

Akçakaya HR, Radeloff VC, Mlandenoff DJ, He HS (2004) Integrating landscape and metapopulation modeling approaches: viability of the sharp-tailed grouse in a dynamic landscape. Conserv Biol 18:526–537

Allee WC (1931) Animal aggregations: a study in general sociology. University of Chicago Press, Chicago

Amarasekare P, Nisbet RM (2001) Spatial heterogeneity, source-sink dynamics, and the local coexistence of competing species. Am Nat 158:572–584

Andersen MC (2005) Potential applications of population viability analysis to risk assessment for invasive species. Hum Ecol Risk Assess 11:1083–1095

Anderson RM, May (1991) Infectious diseases of humans. Oxford University Press, Oxford

Andow DA, Kareiva PM, Levin SA, Okubo A (1990) Spread of invading organisms. Landscape Ecol 4:177–188

Aussem A, Hill D (1999) Wedding connectionist and algorithmic modelling towards forecasting *Caulerpa taxifolia* development in the north-western Mediterranean sea. Ecol Model 120:225–236

Aussem A, Hill D (2000) Neural network metamodelling for the prediction of *Caulerpa taxifolia* development in the Mediterranean sea. Neurocomputing 30:71–78

Barry KL, Levings CD (2002) Feasibility of using the RAMAS metapopulation model to assess the risk of a non-indigenous copepod (*Pseudodiaptomus marinus*) establishing in Vancouver Harbour from ballast water. Can Tech Rep Fish Aquat Sci 2401

Bartell SM, Nair SK (2003) Establishment risks for invasive species. Risk Anal 24:833–845

Berdnikov SV, Selyutin VV, Vasilchenko VV, Caddy JF (1999) Trophodynamic model of the Black and Azov Sea pelagic ecosystem: consequences of the comb jelly, *Mnemiopsis leidyi*, invasion. Fish Res 42:261–289

Bolton TF, Graham WM (2004) Morphological variation among populations of an invasive jellyfish. Mar Ecol Progr Ser 278:125–139

Bossenbroek JM, Nekola JC, Kraft CE (2001) Prediction of long-distance dispersal using gravity models: zebra mussel invasion of inland lakes. Ecol Appl 11:1778–1788

Buhle ER, Margolis M, Ruesink JL (2005) Bang for buck: cost-effective control of invasive species with different life histories. Ecol Econ 52:355–366

Burkhart SM, Slooten E (2003) Population viability analysis for Hector's dolphin (*Cephalorhynchus hectori*): a stochastic population model for local populations. N Z J Mar Freshwater Res 37:553–566

Burnham KP, Anderson DR (1998) Model selection and inference: a practical information-theoretic approach. Springer, Berlin Heidelberg New York

Butterworth DS, Plaganyi EE (2004) A brief introduction to some approaches to multispecies/ecosystem modelling in the context of their possible application in the management of South African fisheries. Afr J Mar Sci 26:53–61

Byers JE, Goldwasser L (2001) Exposing the mechanism and timing of impact of nonindigenous species on native species. Ecology 82:1330–1343

Byers JE, Noonburg EG (2003) Scale dependent effects of biotic resistance to biological invasion. Ecology 84:1428–1433

Case TJ (1990) Invasion resistance arises in strongly interacting species-rich model competition communities. Proc Natl Acad Sci U S 87:9610–9614

Case TJ (1991) Invasion resistance, species build-up and community collapse in metapopulation models with interspecies competition. Biol J Linn Soc 42:239–266

Case TJ (1999) An illustrated guide to theoretical ecology. Oxford University Press, Oxford
Castillo GC, Li HW, Rossignol PA (2000) Absence of overall feedback in a benthic estuarine community: a system potentially buffered from impacts of biological invasions. Estuaries 23:275–291
Cohen AN, Carlton JT (1998) Accelerating invasion rate in a highly invaded estuary. Science 279:555–558
Colautti RI, Grigorovich IA, MacIsaac HJ (2006) Propagule pressure: a null model for biological invasions. Biol Invas 8:1023–1037
Connell JH (1978) Diversity in tropical rain forests and coral reefs. Science 199:1302–1310
Connell JH, Slatyer RO (1977) Mechanisms of succession in natural communities and their role in community stability and organization. Am Nat 111:1119–1144
Connolly SR, Menge BA, Roughgarden J (2001) A latitudinal gradient in recruitment of intertidal invertebrates in the northeast Pacific Ocean. Ecology 82:1799–1813
Coquillard P, Thibaut T, Hill DRC, Gueugnot J, Mazel C, Coquillard Y (2000) Simulation of the mollusc ascoglossa *Elysia subornata* population dynamics: application to the potential biocontrol of *Caulerpa taxifolia* growth in the Mediterranean Sea. Ecol Model 135:1–16
Costello CJ, Solow AR (2003) On the pattern of discovery of introduced species. Proc Natl Acad Sci U S 100:3321–3323
Crowder LB, Crouse DT, Heppell SS, Martin TH (1994) Predicting the impact of turtle excluder devices on loggerhead sea turtle populations. Ecol Appl 4:437–445
Cuddington K, Hastings A (2004) Invasive engineers. Ecol Model 178:335–347
Daguin C, Borsa P (2000) Genetic relationships of *Mytilus galloprovincialis* Lamarck populations worldwide: evidence from nuclear-DNA markers. In: Crame JA (ed) The evolutionary biology of the Bivalvia. Geological Society, London, pp 389–397
Davies KF, Chesson P, Harrison S, Inouye BD, Melbourne BA, Rice KJ (2005) Spatial heterogeneity explains the scale dependence of the native–exotic diversity relationship. Ecology 86:1602–1610
Davis MA, Grime JP, Thompson K (2000) Fluctuating resources in plant communities: a general theory of invasibility. J Ecol 88:528–536
De Koeijer A, Diekmann O, Reijnders P (1998) Modelling the spread of phocine distemper virus among harbour seals. Bull Math Biol 60:585–596
Dew JR, Berkson J, Hallerman EM, Allen KA Jr (2003) A model for assessing the likelihood of self-sustaining populations resulting from commercial production of triploid Suminoe oysters (*Crassostrea ariakensis*) in Chesapeake Bay. Fish Bull 101:758–768
deYoung B, Heath M, Werner F, Chai F, Megrey B, Monfray P (2004) Challenges of modeling ocean basin ecosystems. Science 304:1463–1466
Donovan T, Welden CW (2001a) Spreadsheet exercises in conservation biology and landscape ecology. Sinauer, New York
Donovan T, Welden CW (2001b) Spreadsheet exercises in ecology and evolution. Sinauer, New York
Drake JA (1990) The mechanics of community assembly and succession. J Theor Biol 147:213–233
Drake JM, Lodge DM (2004) Global hot spots of biological invasions: evaluating options for ballast-water management. Proc R Soc London Ser B 271:575–580
Drake JM, Lodge DM (2006) Allee effects, propagule pressure and the probability of establishment: risk analysis for biological invasions. Biol Invas 8:365–375
Drake JM, Lodge DM, Lewis M (2005) Theory and preliminary analysis of species invasions from ballast water: controlling discharge volume and location. Am Midl Nat 154:459–470
Dunstan PK, Johnson CR (2005) Predicting global dynamics from local interactions: individual-based models predict complex features of marine epibenthic communities. Ecol Model 186:221–233
Dunstan PK, Johnson CR (2006) Linking richness, community variability, and invasion resistance with patch size. Ecology 87:2842–2850
Elton CS (1958) The ecology of invasions by animals and plants. Methuen, London

Fagan WF, Lewis MA, Neubert M, van den Driessche P (2002) Invasion theory and biological control. Ecol Lett 5:148–157
Ferson S, Burgman M (eds) (2003) Quantitative methods for conservation biology. Springer, Berlin Heidelberg New York
Fisher RA (1937) The wave of advance of advantageous genes. Ann Eugen 7:355–369
Floerl O, Inglis GJ, Hayden BJ (2005) A risk-based predictive tool to prevent accidental introductions of nonindigenous marine species. Environ Manag 35:765–778
Fox JW, Srivastava D (2006) Predicting local/regional richness relationships using island biogeography models. Oikos 113:376–382
Frésard M, Boncoeur J (2006) Costs and benefits of stock enhancement and biological invasion control: the case of the Bay of Brest scallop fishery. Aquat Living Resource 19:299–305
Fridley JD, Brown RL, Bruno JF (2004) Null models of exotic invasion and scale-dependent patterns of native and exotic species richness. Ecology 85:3215–3222
Fulton EA, Smith ADM, Johnson CR (2003) Effect of complexity on marine ecosystem models. Mar Ecol Progr Ser 253:1–16
Gascoigne J, Lipcius RN (2004) Allee effects in marine systems. Mar Ecol Progr Ser 269:49–59
Geller JB, Walton W, Grosholz E, Ruiz G (1997) Cryptic invasions of the crab *Carcinus* detected by molecular phylogeography. Mol Ecol 6:256–262
Gerber LR, Botsford LW, Hastings A, Possingham HP, Gaines SD, Palumbi SR, Andelman S (2003) Population models for marine reserve design: a retrospective and prospective synthesis. Ecol Appl 13:S47–S64
Grenfell BT, Lonergan ME, Harwood J (1992) Quantitative investigations of the epidemiology of phocine distemper virus (PDV) in European common seal populations. Sci Total Environ 115:15–29
Grosholz ED (1996) Contrasting rates of spread for introduced species in terrestrial and marine systems. Ecology 77:1680–1686
Gucu AC (2002) Can overfishing be responsible for the successful establishment of *Mnemiopsis leidyi* in the Black Sea? Estuarine Coastal Shelf Sci 54:439–451
Guichard F, Levin SA, Hastings A, Siegel D (2004) Toward a dynamic metacommunity approach to marine reserve theory. Bioscience 54:1003–1011
Gutierrez AP, Pitcairn MJ, Ellis CK, Carruthers N, Ghezelbash R (2005) Evaluating biological control of yellow starthistle (*Centaurea solstitialis*) in California: a GIS based supply-demand demographic model. Biol Control 34:115–131
Haefner J (1996) Modeling biological systems: principles and applications. Chapman and Hall, New York
Hails RS, Morley K (2005) Genes invading new populations: a risk assessment perspective. Trends Ecol Evol 20:245–252
Hall RJ, Hastings A, Ayres DR (2006) Explaining the explosion: modelling hybrid invasions. Proc R Soc B 273:1385–1389
Hanski I, Gaggiotti OE (eds) (2004) Ecology, genetics, and evolution of metapopulations. Elsevier Academic Press
Harding KC, Härkönen T, Caswell H (2002) The 2002 European seal plague: epidemiology and population consequences. Ecol Lett 5:727–732
Harding KC, Härkönen T, Pineda J (2003) Estimating quasi-extinction risk of European harbour seals: reply to Lonergan and Harwood (2003). Ecol Lett 6:894–897
Harvell CD, Kim K, Burkholder JM, Colwell RR, Epstein PR, Grimes DJ, Hofmann EE, Lipp EK, Osterhaus ADME, Overstreet RM, Porter JW, Smith GW, Vasta GR (1999) Emerging marine diseases – climate links and anthropogenic factors. Science 285:1505–1510
Hastings A (1996) Models of spatial spread: a synthesis. Biol Conserv 78:143–148
Hastings A, Cuddington K, Davies KF, Dugaw CJ, Elmendorf S, Freestone A, Harrison S, Holland M, Lambrinos J, Malvadkar U, Melbourne BA, Moore K, Taylor C, Thomson D (2005) The spatial spread of invasions: new developments in theory and evidence. Ecol Lett 8:91–101

Hayes KR (1998) Ecological risk assessment for ballast water introductions: a suggested approach. ICES J Mar Sci 55:201–212
Hayes KR (2002a) Identifying hazards in complex ecological systems. Part 1: Fault-tree analysis for biological invasions. Biol Invas 4:235–249
Hayes KR (2002b) Identifying hazards in complex ecological systems. Part 2: Infection modes and effects analysis for biological invasions. Biol Invas 4:251–261
Hayes KR (2003) Biosecurity and the role of risk assessment. In: Ruiz G, Carlton JT (eds) Invasive species: vectors and management strategies. Island Press, Washington, DC, pp 382–414
He F, Gaston KJ, Connor EF, Srivastava DS (2005) The local-regional relationship: immigration, extinction, and scale. Ecology 86:360–365
Hedrick PW (2001) Invasion of transgenes from salmon or other genetically modified organisms into natural populations. Can J Fish Aquat Sci 58:841–844
Heide-Jørgensen M-P, Härkönen T (1992) Epizooitology of the seal disease in the eastern North Sea. J Appl Ecol 29:99–107
Herben T, Mandák B, Bímová K, Münzbergová Z (2004) Invasibility and species richness of a community: a neutral model and a survey of published data. Ecology 85:3223–3233
Herborg L-M, Jerde CL, Lodge DM, Ruiz GM, MacIsaac HJ (2007) Predicting invasion risk using measures of introduction effort and environmental niche models. Ecol Appl 17(3):663–674
Hewitt CL, Huxel GR (2002) Invasion success and community resistance in single and multiple species invasion models: do the models support the conclusions? Biol Invas 4:263–271
Hewitt CL, Campbell ML, Thresher RE, Martin RB, Boyd S, Cohen BF, Currie DR, Gomon MF, Keough MJ, Lewis JA, Lockett MM, Mays N, McArthur MA, O'Hara TD, Poore GCB, Ross DJ, Storey MJ, Watson JE, Wilson RS (2004) Introduced and cryptogenic species in Port Phillip Bay, Victoria, Australia. Mar Biol 144:183–202
Hilborn R, Mangel M (1997) The ecological detective: confronting models with data. Princeton University Press, Princeton, NJ
Hill D, Coquillard P, De Vaugelas J, Meinesz A (1998) An algorithmic model for invasive species: application to *Caulerpa taxifolia* (Vahl) C. Agardh development in the North-Western Mediterranean Sea. Ecol Model 109:251–265
Hollowed AB, Bax N, Beamish R, Collie JS, Fogarty M, Livingston PA, Pope J, Rice JC (2000) Are multispecies models an improvement on single-species models for measuring fishing impacts on marine ecosystems? ICES J Mar Sci 57:707–719
Holmes EE (2004) Beyond theory to application and evaluation: diffusion approximations for population viability analysis. Ecol Appl 14:1272–1293
Holt J, Black R, Abdallah R (2006) A rigorous yet simple quantitative risk assessment method for quarantine pests and non-native organisms. Ann Appl Biol 149:167–173
Holyoak M, Leibold MA, Holt R (2005) Metacommunities: spatial dynamics and ecological communities. Chicago University Press
Hutchings JA, Reynolds JD (2004) Marine fish population collapses: consequences for recovery and extinction risk. Bioscience 54:297–309
Inglis GJ, Hurren H, Oldman J, Haskew R (2006) Using habitat suitability index and particle dispersion models for early detection of marine invaders. Ecol Appl 16:1377–1390
Jerde CL, Lewis MA (2007) Waiting for invasions: a framework for the arrival of non-indigenous species. Am Nat 170:1–9
Johnson DR, Perry HM, Graham WM (2005) Using nowcast model currents to explore transport of non-indigenous jellyfish into the Gulf of Mexico. Mar Ecol Prog Ser 305:139–146
Kermack WO, McKendrick AG (1927) A contributions to the mathematical theory of epidemics. Proc R Soc A 115:700–721
Kinlan BP, Hastings A (2005) Rates of population spread and geographic range expansion: what exotic species tell us. In: Sax DF, Stachowicz J, Gaines SD (eds) Species invasions: insights into ecology, evolution, and biogeography. Sinauer, Sunderland, MA, p 495
Kinlan BP, Gaines SD, Lester SE (2005) Propagule dispersal and the scales of marine community process. Divers Distrib 11:139–148

Knowler D (2005) Reassessing the costs of biological invasion: *Mnemiopsis leidyi* in the Black Sea. Ecol Econ 52:187–199

Kolar CS, Lodge DM (2002) Ecological predictions and risk assessment for alien fishes in North America. Science 298:1233–1236

Kot M (2001) Elements of mathematical ecology. Cambridge University Press, Cambridge

Kot M, Lewis MA, van den Driessche P (1996) Dispersal data and the spread of invading organisms. Ecology 77:2027–2042

Kritzer JP, Sale PF (2004) Metapopulation ecology in the sea: from Levins' model to marine ecology and fisheries science. Fish Fish 5:131–140

Krkošek M, Lewis MA, Volpe JP (2005) Transmission dynamics of parasitic sea lice from farm to wild salmon. Proc R Soc London Ser B 272:689–696

Krkošek M, Lewis MA, Morton A, Frazer LN, Volpe JP (2006) Epizootics of wild fish induced by farm fish. Proc Natl Acad Sci U S 103:15506–15510

Krkošek M, Lauzon-Guay J-S, Lewis MA (2007) Relating dispersal and range expansion of California sea otters. Theoretical Population Biology: doi:10.1016/j.tpb.2007.1001.1008

Landis WG (2003) Ecological risk assessment conceptual model formulation for nonindigenous species. Risk Anal 24:847–858

Lebedeva LP, Shushkina EA (1994) Modeling the effect of *Mnemiopsis* on the Black Sea plankton community. Oceanology 34:72–80

Leung B, Lodge DM, Finoff D, Shogren JF, Lewis MA, Lamberti G (2002) An ounce of prevention or a pound of cure: bioeconomic risk analysis of invasive species. Proc R Soc London B 269:2407–2413

Leung B, Bossenbroek JM, Lodge DM (2006) Boats, pathways, and aquatic biological invasions: estimating dispersal potential with gravity models. Biol Invas 8:241–254

Levin LA (2006) Recent progress in understanding larval dispersal: new directions and digressions. Integr Comp Biol 46:282–297

Levine JB, D'Antonio CM (2003) Forecasting biological invasions with increasing international trade. Conserv Biol 17:322–326

Levins D (1968) Evolution in a changing environment. Princeton University Press, Princeton

Levins R (1969) Some demographic and genetic consequences of environmental heterogeneity for biological control. Bull Entomol Soc Am 15:237–240

Levins R (1970) Extinction. In: Gerstenhaber M (ed) Some mathematical problems in biology. American Mathematical Society, Providence, R.I., pp 75–107

Lewis MA, Kareiva P (1993) Allee dynamics and the spread of invading organisms. Theor Popul Biol 43:141–158

Lewis MA, Neubert MG, Caswell H, Clark J, Shea K (2005) A guide to calculating discrete-time invasion rates from data. In: Cadotte MW, McMahon SM, Fukami T (eds) Conceptual ecology and invasions biology: reciprocal approaches to nature. Springer, Berlin Heidelberg New York

Lonergan M, Harwood J (2003) The potential effects of repeated outbreaks of phocine distemper among harbour seals: a response to Harding et al. (2002). Ecol Lett 6:889–893

Lubina JA, Levin SA (1988) The spread of a reinvading species: range expansion in the California sea otter. Am Nat 131:526–543

Lundquist CJ, Botsford LW (2004) Model projections of the fishery implications of the Allee effect in broadcast spawners. Ecol Appl 14:929–941

Lutscher F, Pachepsky E, Lewis MA (2005) The effect of dispersal patterns on stream populations. SIAM J Appl Math 65:1305–1327

Lutscher F, McCauley E, Lewis MA (2007) Spatial patterns and coexistence mechanisms in systems with unidirectional flow Theor Popul Biol 71:267–277

MacArthur RM, Wilson EO (1967) The theory of island biogeography. Princeton University Press, Princeton, NJ

MacIsaac HJ, Robbins TC, Lewis MA (2002) Modeling ships' ballast water as invasion threats to the Great Lakes. Can J Fish Aquat Sci 59:1245–1256

Malchow H, Hilker FM, Petrovskii SV, Brauer K (2004) Oscillations and waves in a virally infected plankton system. Part I: The lysogenic stage. Ecol Complex 1:211–223
McEvoy PB, Coombs EM (1999) Biological control of plant invaders: regional patterns, field experiments, and structured population models. Ecol Appl 9:387–401
Medvinsky AB, Petrovskii SV, Tikhonova IA, Malchow H, Li B-L (2002) Spatiotemporal complexity of plankton and fish dynamics. SIAM Rev 44:311–370
Melbourne BA, Cornell HV, Davies KF, Dugaw CJ, Elmendorf S, Freestone AL, Hall RJ, Harrison S, Hastings A, Holland M, Holyoak M, Lambrinos J, Moore K, Yokomizo H (2007) Invasion in a heterogeneous world: resistance, coexistence or hostile takeover? Ecol Lett 10:77–94
Mitchell CE, Agrawal AA, Bever JD, Gilbert GS, Hufbauer RA, Klironomos JN, Maron JL, Morris WF, Parker IM, Power AG, Seabloom EW, Torchin ME, Vázquez DP (2006) Biotic interactions and plant invasions. Ecol Lett 9:726–740
Morozov AY, Nezlin NP, Petrovskii SV (2005) Invasion of a top predator into an epipelagic ecosystem can bring a paradoxical top-down trophic control. Biol Invas 7:845–861
Morris WF, Doak DF (2003) Quantitative conservation biology: theory and practice of population viability analysis. Sinauer Associates, Sunderland, MA
Murray AG, Parslow JS (1999) Modelling of nutrient impacts in Port Phillip Bay – a semi-enclosed marine Australian ecosystem. Mar Freshwater Res 50:597–611
Neubert MG, Parker IM (2004) Projecting rates of spread for invasive species. Risk Anal 24:817–831
Oguz T, Ducklow HW, Purcell JE, Malanotte-Rizzoli P (2001) Modeling the response of top-down control exerted by gelatinous carnivores on the Black Sea pelagic food web. J Geophys Res C 106:4543–4564
Okubo A, Maini P, Williamson MH, Murray JD (1989) On the spatial spread of the grey squirrel in Britain. Proc R Soc London B 238:113–125
Olden JD, Poff NL (2004) Ecological processes driving biotic homogenization: testing a mechanistic model using fish faunas. Ecology 85:1867–1875
Owen M, Lewis MA (2001) How predation can slow, stop or reverse a prey invasion. Bull Math Biol 63:655–684
Pachepsky E, Lutscher F, Nisbet RM, Lewis MA (2005) Persistence, spread and the drift paradox. Theor Popul Biol 67:61–73
Parker IM (2000) Invasion dynamics of *Cytisus scoparius*: a matrix model approach. Ecol Appl 10:726–743
Parry GD, Black KP, Hatton DN, Cohen BF (2001) Factors influencing the distribution of the exotic seastar *Asterias amurensis* during the early phase of its invasion of Port Phillip Bay, 1: Hydrodynamic factors. Mar Freshwater Resource Inst Rep 38, Queenscliff, Australia, 18 pp
Perrings C (2005) Mitigation and adaptation strategies for the control of biological invasions. Ecol Econ 52:315–325
Peterson AT (2003) Predicting the geography of species' invasions via ecological niche modeling. Q Rev Biol 78:419–433
Peterson AT, Vieglais DA (2001) Predicting species invasions using ecological niche modeling: new approaches from bioinformatics attack a pressing problem. Bioscience 51:363–371
Post WM, Pimm SL (1983) Community assembly and food web stability. Math Biosci 64:169–192
Powell MR (2004) Risk assessment for invasive plant species. Weed Technol 18S:1305–1308
Pranovi F, Libralato S, Raicevich S, Granzotto A, Pastres R, Giovanardi O (2003) Mechanical clam dredging in Venice lagoon: ecosystem effects evaluated with a trophic mass-balance model. Mar Biol 143:393–403
Reichard SH, Hamilton CW (1997) Predicting invasions of woody plants introduced into North America. Conserv Biol 11:193–203
Rejmánek M, Richardson DM (1996) What attributes make some plant species more invasive? Ecology 77:1655–1661
Ricciardi A (2001) Facilitative interactions among aquatic invaders: is an "invasional meltdown" occurring in the Great Lakes? Can J Fish Aquat Sci 58:2513–2525

Richardson DM, Pysek P (2006) Plant invasions: merging the concepts of species invasiveness and community invasibility. Prog Phys Geogr 30:409–431

Rose KA, Cowan JH (2003) Data, models, and decisions in US Marine Fisheries management: lessons for ecologists. Annu Rev Ecol Evol Syst 34:127–151

Rouget M, Richardson DM (2003) A semimechanistic model incorporating propagule pressure and environmental factors. Am Nat 162:713–724

Roughgarden J (1998) Primer of ecological theory. Prentice-Hall, Upper Saddle River, NJ

Roxburgh SH, Shea K, Wilson JB (2004) The intermediate disturbance hypothesis: patch dynamics and mechanisms of species coexistence. Ecology 85:359–371

Ruesink JL, Collado-Vides L (2006) Modeling the increase and control of *Caulerpa taxifolia*, an invasive marine macroalga. Biol Invas 8:309–325

Rudnick D, Veldhuizen T, Tullis R, Culver C, Hieb K, Tsukimura B (2005) A life history model for the San Francisco Estuary population of the Chinese mitten crab, *Eriocheir sinensis* (Decapoda: Grapsoidea). Biol Invas 7:333–350

Ruth M, Lindholm J (eds) (2002) Dynamic modeling for marine conservation. Springer, Berlin Heidelberg New York

Shanks AL, Grantham BA, Carr MH (2003) Propagule dispersal distance and the size and spacing of marine reserves. Ecol Appl 13:S159–S169

Sharov AA (2004) Bioeconomics of managing the spread of exotic pest species with barrier zones. Risk Anal 24:879–892

Sharov AA, Liebhold AM (1998) Bioeconomics of managing the spread of exotic pest species with barrier zones. Ecol Appl 8:833–845

Shea K, Chesson P (2002) Community ecology theory as a framework for biological invasions. Trend Ecol Evol 17:170–176

Shea K, Kelly D (1998) Estimating biocontrol agent impact with matrix models: *Carduus nutans* in New Zealand. Ecol Appl 8:824–832

Shea K, Possingham H (2000) Optimal release strategies for biological control agents: an application of stochastic dynamic programming to population management. J Appl Ecol 37:77–86

Sherratt JA, Eagan BT, Lewis MA (1997) Oscillations and chaos behind predator-prey invasion: mathematical artifact or ecological reality? Philos Trans R Soc London B 352:21–38

Shigesada N, Kawasaki K (1997) Biological invasions: theory and practice. Oxford University Press, Oxford, UK

Siegel D, Kinlan BP, Gaylord B, Gaines SD (2003) Lagrangian descriptions of marine larval dispersion. Mar Ecol Prog Ser 260:83–96

Simberloff D, Von Holle B (1999) Positive interactions in nonindigenous species: invasional meltdown? Biol Invas 1:21–32

Skellam JG (1951) Random dispersal in theoretical populations. Biometrika 38:196–218

Solow A, Costello C (2004) Estimating the rate of species introductions from the discovery record. Ecology 85:1822–1825

Stachowicz JJ, Byrnes JE (2006) Species diversity, invasion success, and ecosystem functioning: disentangling the influence of resource competition, facilitation, and extrinsic factors. Mar Ecol Prog Ser 311:251–262

Swinton J, Harwood J, Grenfell BT, Gilligan CA (1998) Persistence thresholds for phocine distemper virus infection in harbour seal *Phoca vitulina* metapopulations. J Anim Ecol 67:54–68

Taylor CM, Hastings A (2004) Finding optimal control strategies for invasive species: a density-structured model for *Spartina alterniflora*. J Appl Ecol 41:1049–1057

Taylor CM, Davis HG, Civille JC, Grevstad FS, Hastings A (2004) Consequences of an Allee effect in the invasion of a Pacific estuary by *Spartina alterniflora*. Ecology 85:3254–3266

Tilman D (2004) Niche tradeoffs, neutrality, and community structure: a stochastic theory of resource competition, invasion, and community assembly. Proc Natl Acad Sci U S 101:10854–10861

Tobin PC, Whitmire SL, Johnson DM, Bjørnstad ON, Liebhold AM (2007) Invasion speed is affected by geographical variation in the strength of Allee effects. Ecol Lett 10:36–43

Viard F, Ellien C, Dupont L (2006) Dispersal ability and invasion success of *Crepidula fornicata* in a single gulf: insights from genetic markers and larval-dispersal model. Helgol Mar Res 60:144–152

Vincent C, Mouillot D, Lauret M, Chi TD, Troussellier M, Aliaume C (2006) Contribution of exotic species, environmental factors and spatial components to the macrophyte assemblages in a Mediterranean lagoon (Thau lagoon, Southern France). Ecol Model 193:119–131

Volovik YS, Volovik AP, Myrzoyan ZA (1995) Modelling the *Mnemiopsis* sp. population in the Azov Sea. In: Harris R (ed) Zooplankton production. Proceedings of the ICES Symposium on Zooplankton Production, Plymouth (UK), 15–19 Aug 1994. Academic Press, London, pp 735–746

Wang MH, Kot M (2001) Speeds of invasion in a model with strong or weak Allee effects. Math Biosci 171:83–97

Wilson NLW, Anton J (2006) Combining risk assessment and economics in managing a sanitary-phytosanitary risk. A J Agric Econ 88:194–202

With KA (2004) Assessing the risk of invasive spread in fragmented landscapes. Risk Anal 24:803–815

Wonham MJ, Pachepsky E (2006) A null model of temporal trends in biological invasion records. Ecol Lett 9:663–672

Wonham MJ, Bailey SA, Lewis MA, MacIsaac HJ (2005a) Modelling the invasion risk of diapausing organisms transported in ballast sediments. Can J Fish Aquat Sci 62:2386–2398

Wonham MJ, Lewis MA, MacIsaac HJ (2005b) Minimizing invasion risk by reducing propagule pressure: a model for ballast-water exchange. Front Ecol Environ 3:473–478

Wootton JT (2004) Markov chain models predict the consequences of experimental extinctions. Ecol Lett 7:653–660

Appendix

For an introduction to the philosophy and practice of mathematical modeling in ecological systems, we find the texts by Case (1999), Haefner (1996), and Kot (2001) particularly helpful. Morris and Doak (2003) give a very accessible entree into population modeling, and the edited volume of Ferson and Burgman (2003) illustrates statistical and dynamical modeling case studies in conservation biology. For specific focus on likelihood methods applied to model selection, we recommend Burnham and Anderson (1998) and Hilborn and Mangel (1997). For a the mechanics of practical model building and analysis, including thoroughly worked computer exercises, Donovan and Weldon (2001a, b) provide ecology and conservation spreadsheet exercises in Microsoft Excel®, Roughgarden (1998) provides ecological examples and code in Matlab®, and Ruth and Lindholm (2002) investigate marine conservation problems using Stella®.

Section II
Invader Arrival

Chapter 5
Characterizing Vectors of Marine Invasion

Dan Minchin, Stephan Gollasch, Andrew N. Cohen, Chad L. Hewitt, and Sergej Olenin

5.1 Introduction

The arrival of an invasive species in a new region is the culmination of a set of relatively discrete steps, including the invader's initial association with a transport vector, its tolerance of environmental conditions encountered during transit, and its survival upon entering its new ecosystem (Ruiz and Carlton 2003). In the chapters that follow, a number of issues related to this process are presented. Chapter 6, Hewitt et al., discusses shipping, the most important of the marine invasion pathways. Chapter 7, Johnston et al., discusses the role of propagule pressure, how the quantity and quality of invader propagules determine invasion success. Chapter 8, Miller and Ruiz, follows with a framework for considering the distinct roles of source region, vector, and recipient region in assessing invasion success or failure within species pools. In addition, several vectors are discussed in relation to specific species and locales in the Geographic Perspectives section, which includes some assessments of temporal shifts in trading patterns (e.g. Chap. 24, Hayden et al.; Chap. 28, Fofonoff et al.). The importance of pathways, vectors, and modelling human activities is discussed in previous sections (Chap. 2, Carlton; Chap. 4, Wonham and Lewis).

5.2 Primary vs Secondary Introductions

A non-native organism arriving in a new location directly from its native region is called a primary introduction, while its subsequent spread from the founding site is considered to be a secondary introduction. This spread may occur through a combination of natural dispersal and human-associated transport mechanisms. As an introduced species expands its new range, further opportunities to spread by additional vectors may present themselves. Identifying how a species arrived, however, is not always possible (see Chap. 2, Carlton). On occasion, the arrival of an invader may result from a series of different vectors acting in a relay to convey that species (Minchin 2007).

G. Rilov, J.A. Crooks (eds.) *Biological Invasions in Marine Ecosystems.* Ecological Studies 204,

5.3 Principal Vectors of Marine Invasion

5.3.1 Shipping

Ships may transport non-native species both on the outer surface of the hull and inside the ship in solid or water ballast. Seaweeds, sponges, mussels, barnacles and other 'fouling' species can attach directly to ships' hulls, while mobile species may be present among them. These organisms may be carried over great distances (Minchin and Gollasch 2003). Surprisingly, small amounts of fouling growth on a ship's hull can increase drag and raise fuel consumption. Wooden-hulled ships, common during earlier periods and still in use in some regions today, can be colonized and extensively damaged by wood-boring marine organisms, primarily shipworms and certain isopod and amphipod species. These wood-borers have been carried to many parts of the world, resulting in broad geographic distributions.

Ships load ballast on-board to adjust buoyancy, provide stability and enhance maneuverability. Solid materials, including rocks and sand, were once used as ballast, laboriously packed into vessels by hand. Many ballast stones were stored for re-use in shore-side heaps where they could be accessed at different stages of the tide. Several species are believed to have traveled with solid ballast, either attached directly to ballast stones or otherwise loaded by the ballasting process, and survived transport in damp ballast holds. In addition to marine organisms, several species of plants (probably travelling as seeds) and insects are believed to have been introduced with solid ballast (Lindroth 1957). Starting around 1880, ships began to use water instead of solid ballast. This switch, which was largely complete by 1930, was aided by the construction of steel ships and the development of efficient engines for pumping water. The transport of ballast water provides opportunities for the spread of thousands of species (Carlton and Geller 1993), ranging from viruses to fishes (Gollasch et al. 2002). In addition, sediments that accumulate in the bottom of ballast tanks provide a refuge for infaunal species. As ships are an essential part of world trade, responsible for >90% of all cargo by weight, they will continue to provide many opportunities for the global redistribution of species (Minchin 2006).

5.3.2 Canals

A canal can enable the transfer of organisms between different biogeographical regions either via shipping or by the organisms themselves passing through the canal (Gollasch et al. 2006). Some canals have salinity or temperature barriers that may suppress the transfer of organisms. Many marine species have extended their ranges from the Red Sea to the Mediterranean Sea through the Suez Canals (see Chap. 31, Rilov and Galil). Fewer have migrated in either direction between the Caribbean and the eastern Pacific through the Panama Canals, where vessels and organisms must pass through fresh water (Cohen 2006). Some species with broad salinity tolerance limits have spread through in land waterways from the caspian and Black Seas and survived in brockish water environments of the northwestern european seas (Olenin 2002)

5.3.3 *Aquaculture*

Non-native species of fish, invertebrates and seaweeds are cultivated in many parts of the world. Species selected and imported for culture are normally hardy, because they need to survive confinement and exposure to a wide range of environmental conditions. Many aquaculture species are grown right out in the environment; others are intentionally released (ocean ranching); species held in cages, pens or ponds may regularly escape; their larvae, and their cultivated foods, may be released in discharges; and transportation accidents and other events will result in the release of others (Minchin 2007). Escaped or released aquaculture species can impact local species through competitive or other interactions, and by interbreeding with native stocks (Naylor et al. 2005). In addition, these aquaculture species may bring with them associated biota, including pests, parasites and diseases, which can also become established in the wild. Some of these will be previously unknown, and will only be recognized after they are introduced into new regions with aquaculture stock and their impacts on naive (previously unexposed) native host species become obvious (e.g. Kuris and Culver 1999; see also Chap. 2, Carlton). The economics of aquaculture operations often requires that cultured species be confined at high densities, and this may make them subject to frequent infestations. Escapees from pens and cages and stock movements may then spread these infestations widely. Even the release or outplanting of native aquaculture species to meet restoration goals may introduce non-native pathogens or parasites if the native species have been reared in the same facilities as infected host species, have been cross-contaminated through exposure to the same equipment, or have been returned from stocking experiments abroad (e.g. Engstrom 2001; Cohen 2002). The spread of associated non-native diseases, parasites and other organisms with the global transport of young oysters has been particularly notable (e.g., Ruesink et al. 2005; Chap. 23, Griffiths; Chap. 31, Rilov and Galil).

5.3.4 *Fisheries*

Fisheries development projects have introduced fishes, crustaceans and molluscs to boreal, temperate and tropical seas with the aim of establishing new populations. Some of these intentionally established populations have had unintended and harmful consequences for native species and habitats, including (again) the introduction of parasites and diseases (Minchin 2007; Chap. 32, Seo and Lee). In some cases, fisheries projects have introduced undesirable species that were misidentified as the target species. In addition, there is a substantial global trade in live baitfish, baitworms and other bait organisms that have the possibility (along with other organisms transported with them) of becoming established following release. Even non-living fisheries products (fresh or frozen) may carry pathogens, including viruses; these could be released into the wild if their dead hosts are used as bait or chum, released in drainage from fish processing plants.

5.3.5 Ornamental Species and Live Seafood

Aquarium species and live foods for human consumption are regularly imported through international airports and distributed to specialty shops, food markets or restaurants (e.g. Chapman et al. 2003). The subsequent release of these organisms, either intentionally or accidentally, can lead to their establishment. For example, the establishment of Pacific lionfish, *Pterois volitans*, on the east coast of North America (Whitfield et al. 2002), and of the green alga, *Caulerpa taxifolia,* in the Mediterranean Sea and southern California (Jousson et al. 2000; Chap. 31, Rilov and Galil), apparently resulted from releases from public or private aquaria.

5.3.6 Marine Leisure and Tourism

Small craft vary in design, speed and behaviour (Minchin et al. 2006), and this can influence the degree of fouling accumulating on hulls. Small craft can provide a significant mechanism for species dispersal within or between continents, including overland dispersal by trailered vessels. Other sporting activities, including fishing, diving, sailing, wind-surfing, and jet skiing, may involve the transfer of equipment, watercraft, or water contaminated with non-native species. Certain circumstances, including seasonal migrations of watercraft, may increase the potential for transporting organisms. Private craft offered for sale have a high probability of being transferred; very often these will lie at moorings or berths for some time before their sale and during this period can accumulate a large amount of fouling. Some specialized vessels act as 'floating dry-dock transporters,' where several watercraft, normally cruisers, 'swim' into the dock before it is dried out and are then carried across oceans and so may release biota to the area where these craft disembark.

5.3.7 Research and Education

Releases from research and academic institutions can occur from the escape, discarding or mercy release of experimental organisms, or from the discharge of organisms in flow-through seawater systems. They can also result from intentional plantings or releases of organisms for experimental purposes. For example, the invasion of the New Zealand mangrove, *Avicennia marina*, in a salt marsh in San Diego, California, occurred via the intentional planting of the species for physiological research (Callaway and Zedler 2004).

5.3.8 Habitat Restoration and Management

Plantings of non-native marine and estuarine angiosperms, such as cordgrasses in tidal marshes, have been undertaken to stabilize sediments, to prevent bank erosion, or to restore or 'improve' habitats (e.g., Daehler and Strong 1996). These have sometimes resulted in non-native plant populations that were subsequently deemed undesirable and became the target of costly and sometimes unsuccessful control efforts (see Chap. 17, Grosholz and Ruiz; Chap. 21, Hacker and Dethier). Recently, the transfer of oyster shell between bays for native oyster restoration projects in California has been assessed as a potential vector for introducing non-native species (Cohen and Zabin, in press).

5.4 Vector Management

It may prove difficult or impossible to eliminate or contain many introduced marine species, even if found soon after arrival. Consequently, preventing the arrival of harmful species, including controlling their transport with trade in advance of their inoculation, must be seen as the main objective for invasions management. With expanding international trade and greater opportunities for individuals to travel there are increased opportunities to spread non-native species. While management issues are treated more fully in Chap. 6, we highlight some management issues related to vectors here.

While some countries have legal requirements controlling the entry of live products such as fish and shellfish, there are usually few controls on imports of bait organisms, aquarium species or aquatic plants, and the implementation and enforcement of the regulations that do exist is often of limited scope and effectiveness. Also, unapproved movements of stock for angling or aquaculture purposes continue to take place. Greater public awareness and the full implementation of Codes of Practice, such as the ICES Code of Practice (ICES 2005), should reduce the risks from imported consignments. This code provides a practical risk-assessment approach for species involved in regular trade (see also Chap. 19, Hewitt et al.; Chap. 20, Campbell).

Although not developed specifically to control the spread of invasive species, there is a large body of work on the control of biofouling (e.g. Costlow and Tipper 1984). The most effective anti-fouling applications used organotins, which were in general use in paints from the 1970s to 2003. Since then, their use has been restricted because of serious impacts on a wide range of biota in and outside port areas. Though there are several less-effective alternatives in use or development, such as copper-based paints and non-toxic coatings, the loss of organotin paints makes controlling the transport of organisms in hull fouling even more challenging. It may help to focus on higher-risk circumstances, such as vessels that remain berthed, moored or anchored in one place for an extended period and accumulate a large amount of fouling, and are then moved over a long distance, as in the case of de-commissioned craft.

In concept, the management of organisms carried in ballast water and ballast sediments poses fewer challenges, as the organisms are isolated from the environment and can be manipulated and treated with greater freedom. At present, some jurisdictions require exchanges of port-loaded water for oceanic water at times when ships pass over deep water. This exchange is intended to purge the coastal species contained in the ballast water. If properly done, many of the organisms are extirpated in this way, though some, such as those associated with ballast sediments, may be little affected. Regulatory monitoring of high-seas ballast exchange is, however, difficult and very limited. Also, ballast exchanges are not always achievable on long distance routes, cannot reasonably be undertaken on short voyages, and may be unsafe during bad weather. Alternately, there are several treatment techniques at different stages of research and testing which may be effectively employed to remove organisms from or sterilize ballast water. The State of California has adopted regulations that would greatly limit the concentration of live organisms that could be discharged in ballast water after 2009–2016, which would necessitate the use of these treatment techniques. An International Maritime Organization Convention, if ratified and implemented, will impose discharge limits that are similar in form but not as strong.

The nature of primary and secondary introductions of many marine species is still poorly known, and an improved understanding would facilitate the development of more effective management approaches. Enough is known, however, to conclude that the main management emphasis should be placed on preventing primary introductions, since once a species has arrived and become established it is difficult and often impossible to control. Thoughtful regulations, if implemented and enforced effectively, will reduce the frequency of primary introductions, while monitoring for the arrival of species will in some cases enable early control of an invasion. Further developments in the growth of trade, including the expansion of the European Union and the continued development of bilateral and local trade agreements, are likely to result in further spread of marine species. However, rapid dissemination of information as an early warning instrument and greater public awareness may support management actions that could reduce this spread.

Acknowledgements Dan Minchin and Sergej Olenin were supported, in part, by the European Union 6th Framework grant ALARM (GOCE-CT-2003-506675) and Dan Minchin, Stephan Gollasch and Sergej Olenin by the grant DAISIE (SSPI-CT-2003-511202).

References

Callaway JC, Zedler JB (2004) Restoration of urban salt marshes: Lessons from southern California. Urban Ecosyst 7:133–150

Carlton JT, Geller JB (1993) Ecological roulette: the global transport of nonindigenous marine organisms. Science 261:78–82

Chapman JW, Miller TW, Coan EV (2003) Live seafood species as recipes for invasion. Conserv Biol 17(5):1386–1395

Cohen AN (2002) The release of pest species by marine aquaculture: lessons from a South African parasite introduced into California waters. In: Cohen AN, Webb SK (eds) Invasions in aquatic ecosystems: impacts on restoration and potential for control. Proceedings of a workshop, April 25, 1998, Sacramento, California. San Francisco Estuary Institute, Oakland, California, pp 9–13.

Cohen AN (2006) Species introductions and the Panama Canal. In: Gollasch S, Galil BS, Cohen AN (eds) Bridging divides: maritime canals as invasion corridors. Monogr Biol 83:127–199

Cohen AN, Zabin CJ (in press) Oyster shells as vectors for exotic organisms. J Shellfish Res

Costlow JC, Tipper RC (1984) Marine biodeterioration: an interdisciplinary study. Naval Institute Press, Annapolis, Maryland

Daehler CC, Strong DR (1996) Status, prediction and prevention of introduced cordgrass *Spartina* spp. invasions in Pacific estuaries, USA. Biol Conserv 78:51–58

Engstrom P (2001) Infected wild abalone stocks. State Department of Fish and Game. Santa Rosa Press Democrat, July 12, 2001

Gollasch S, Macdonald E, Belson S, Botnen H, Christensen J, Hamer J, Houvenaghel G, Jelmert A, Lucas I, Masson D, McCollin T, Olenin S, Persson A, Wallentinus I, Wetsteyn B, Wittling T (2002) Life in ballast tanks. In: Leppäkoski E, Gollasch S, Olenin S (eds) Invasive aquatic species of Europe: distribution, impacts and management. Kluwer Academic Publishers, Dordrecht, The Netherlands, pp 217–231

Gollasch S, Galil BS, Cohen AN (eds) (2006) Bridging divides: maritime canals as invasion corridors. Monogr Biol 83

ICES (2005) ICES Code of Practice on the introductions and transfers of marine organisms 2005. International Council for the Exploration of the Sea (ICES), Copenhagen, Denmark

Jousson O, Pawlowski J, Zaninetti L, Zechman FW, Dini F, Di Guiseppe G, Woodfield R, Millar A, Meinesz A (2000) Invasive alga reaches California. Nature 408:157–158

Kuris AM, Culver CS (1999) An introduced sabellid polychaete pest infesting cultured abalones and its potential spread to other California gastropods. Invert Biol 118:391–403

Lindroth, CH (1957) The faunal connections between Europe and North America. Wiley, New York, USA

Minchin D (2006) The transport and the spread of living aquatic species. In: Davenport J, Davenport JL (eds) The ecology of transportation, managing mobility for the environment. Springer, Berlin Heidelberg New York, pp 77–97

Minchin D (2007) Aquaculture and transport in a changing environment: overlap and links in ther spread of alien biota. Mar Pollut Bull 55:302–313

Minchin D, Gollasch S (2003) Fouling and ships' hulls: how changing circumstances and spawning events may result in the spread of exotic species. Biofouling 19:111–122

Minchin D, Floerl O, Savini D, Occhipinti-Ambrogi A (2006) Small craft and the spread of exotic species. In: Davenport J, Davenport JL (eds) The ecology of transportation, managing mobility for the environment. Springer, Berlin Heidelberg New York, pp 99–118

Naylor R, Hindar K, Fleming IA, Goldburg R, Williams S, Volpe J, Whoriskey F, Eagle J, Kelso D, Mangel M (2005) Fugitive salmon: assessing the risks of escaped fish from net-pen aquaculture. Bioscience 55(5):427–437

Olenin S (2002) Black Sea-Baltic Sea invasion corridors. In: Alien marine organisms introduced in the Mediterranean and Black Seas. CIESM workshops Monograph. F. Briand (ed) Commission Internationale pour l'exploration Scientifique de la Mediterranee Monaco: 29–33

Ruesink JL, Lenihan HS, Trimble AC, Heiman KW, Micheli F, Byers JE, Kay MC (2005) Introduction of non-native oysters: ecosystem effects and restoration implications.Ann Rev Ecol Evol Syst 36:643–689
Ruiz GR, Carlton JT (2003) Invasive species – vectors and management strategies. Island Press, Boca Raton, USA
Whitfield PE, Gardner T, Vives SP, Gilligan MR, Courtenay WR, Carleton-Ray G, Hare JA (2002) Biological invasion of the Indo-Pacific lionfish *Pterois volitans* along the Atlantic coast of North America. Mar Ecol Prog Ser 35:289–297

Chapter 6
The Vessel as a Vector – Biofouling, Ballast Water and Sediments

Chad L. Hewitt, Stephan Gollasch, and Dan Minchin

6.1 Introduction

Human-mediated marine bioinvasions have altered the way we view the marine environment – virtually all regions of the global oceans have experienced the introduction of marine species (e.g., Carlton 1979; Coles et al. 1999; Cranfield et al. 1998; Cohen and Carlton 1998; Hewitt et al. 1999, 2004; Orensanz et al. 2002; Leppäkoski et al. 2002; Lewis et al. 2003; Castilla et al. 2005; Wolff 2005; Gollasch and Nehring 2006; Minchin 2006), placing marine and coastal resources under increased threat. Humans have almost certainly transported marine species since early attempts to voyage by sea. These ancient transport vectors were slow, and for the most part restricted to small spatial scales. The beginning of significant exploration and subsequent expansion by Europeans (post 1500 AD) has resulted in the transport of many thousands of species across all world oceans (Crosby 1986; diCastri 1989; Carlton 2001).

The transport of species by human vectors was recognized by early workers (Ostenfeld 1908; Elton 1958), but it is only in the last few decades that significant progress on identifying patterns and processes has been made (e.g., Carlton 1985, 1996, 2001; Ruiz et al. 2000; Hewitt et al. 2004; Castilla et al. 2005; Minchin 2006). Numerous transport vectors have been identified and described (Carlton 2001; Chap. 5, Minchin et al.); however the majority of species appear to have been associated with vessel movements, either as exploratory, military, commercial or recreational vessels (e.g., Carlton 1985, 2001; Cohen and Carlton 1998; Hewitt et al. 1999; Gollasch et al. 2002, Minchin and Gollasch 2003).

The ship as a transport vector is comprised of several sub-vectors. These include (1) the hull and other 'niche' areas, such as the propeller, rudder, on exposed surfaces of water piping, seachests, and thruster tunnels, where accumulations of growths of organisms develop (typically known as hull fouling), (2) the boring of organisms into the structure of the vessel (primarily limited to wooden hulled vessels), and (3) the uptake of organisms in association with wet or dry ballast (Carlton 1985, 1996; Ruiz et al. 2000). Several of these ship sub-vectors are no longer active. Hull boring for example, virtually ceased to exist with the use of steel

G. Rilov, J.A. Crooks (eds.) *Biological Invasions in Marine Ecosystems.*
Ecological Studies 204,

as the primary ship-building material in merchant and naval vessels. However, many pleasure boats and fishing craft are still constructed of wood (Nagabhushanam and Sarojini 1997). Similarly, dry-ballast made up of sand, gravel and rock taken from littoral environments was replaced with water as ballast beginning in the late 1800s and had become phased out by 1950.

None of these sub-vectors is species-specific, and each is likely to transport entire assemblages of species. Each may also facilitate the transport of a differing suite of species with different physiological and ecological characteristics (see Table 6.1). Biofouling primarily transports species that have attached sedentary or sessile, benthic habits, or species associated with these communities (e.g., living in, between or on other organisms) (Minchin and Gollasch 2003). In contrast, ballast water transports species associated with the plankton either as holo-plankton (species that have their whole life-cycle in the water column), mero-plankton (species with a portion of their life-cycle in the water column), or tycho-plankton (species accidentally swept into the water column), and often include pelagic species. It is difficult to establish a firm link between an already established introduced species and the vector (or sub-vector) by which it arrived in the new location (Minchin 2007). Nevertheless, attempts at assigning linkages to sub-vectors based on life history modes, timing of invasions, and association between location of incursion and sub-vectors have been deduced by reasoned argument (e.g., Hewitt et al. 1999, 2004, in press; Fofonoff et al. 2003; Ruiz et al. 2000).

Table 6.1 Some comparative aspects of ballast water and associated sediments, and hull fouling

Item	Ballast water and sediments	Hull fouling
Amount ship ballast volume/hull surface area	ca 3–10 km^3 per year (Gollasch 2002b)	In 1982: 75,000 merchant vessels have 110 million m^2 (Olesen 1982)
Management physical/ chemical	Ballast purges/ exchange at sea	Antifouling agents/paints
		In water hull cleaning
	Flocculation of sediments	Dry docking/extended periods out of water
	Salinity changes	
	Filtration	Fresh/salt water immersion
	Irradiation	
	Centrifuge	
	Evolving chemical and physical treatments	
Relief	Baffles, platforms, supports, ledges and tanks walls	Smooth surfaces, some projections and cavities (sea chests, thruster tunnels)

(continued)

Table 6.1 (continued)

Item	Ballast water and sediments	Hull fouling
Temperature	Few sudden changes except during mid-ocean or port ballast exchanges	Fluctuations of temperature rapid and directly matching external environment during voyage
Salinity	Potential pulses of salinity associated with ballast exchanges	Salinity highly variable according to shipping routes and ports. Some species such as oysters can seal their shells for some days and avoid being exposed to changed salinity
Turbidity	Highly variable Periods of turbidity depending on ballast water uptake, port conditions (e.g., water depth, tug and dredge operations, tidal range) and sea state	Highly variable Periods of turbidity depending on port conditions (e.g., water depth, tug and dredge operations, tidal range, estuarine port) and sea state
Sedimentation	Accumulation on tank floors and on ledges	Very little except within highly fouled communities or hull pockets such as sea-chests
Turbulence	Variable, according tank position, often little flow	Extreme to moderate, according to weather conditions and ships speed
Light	Little or no light	Variable light, bright to shaded
Gas exchange	May be limited locally	Continuous
Uptake of organisms	Only at specific times of ballasting	At any time, most likely when in port – associated with port residency time
Biota	Mainly free living: mero-, holo- and tycho-planktonic organisms (Gollasch 2002a)	Mainly benthic Sessile, sedentary and some mobile species present (Gollasch 2002b)
Taxonomic range (refer to table of described taxa)	Viruses, micro-organisms, plants (largely planktonic and resting stages) and animals (Porifera to Teleosts) (Gollasch 2002a)	Viruses, micro-organisms, plants (all stages of single and multi-cellular plants) and animals (Porifera to Teleosts) (Gollasch 2002b)
Communities	Generally simple except in sediments	Development of complex communities possible
Life history characteristic	Predation, scavenging, deposit feeding, dormant/resting stages	Predation, herbivory, scavenging, filter/suspension feeding
Interactions with ambient communities	Restricted	Constant

(continued)

Table 6.1 (continued)

Item	Ballast water and sediments	Hull fouling
	Only following ballast water release	May be subject to predation and grazing, exposed to infestations, pollutants particularly when anchored or berthed
Excreta, exuvae, decaying remains	Retained	Generally lost, except barnacles and serpulids/spirorbids, some molluscs
Availability of 'food'	Restricted to organisms in ballast tank – live food declining according to voyage duration (detritus increasing during voyage), no or low levels of photosynthesis; potential to feed at all times	Food in ambient water variable (oceanic –low, coastal/estuarine – high), growth of 'fouling' community continuous, high to low levels of photosynthesis according to location on hull; feeding except at times during voyages or when physiologically challenged
Sexual reproduction	Micro-organisms and some Crustacea (copepods)	Most invertebrate phyla and algae
Asexual reproduction	Bacteria, protozoa, and diatoms (production of resting stages)	Some Anthozoa and planarians
Spawning/Sporulation	Not known	Many invertebrate and algal taxa – direct evidence for some molluscs, serpulids and bryozoa. May leave behind developing embryos in ports
Larval development	Larvae known in ballast tanks	Only brooded larvae

In this chapter we evaluate biofouling of the exposed surfaces and ballast water as sub-vectors of vessels by examining similarities and differences. We do not intend to undertake a comprehensive evaluation, but provide an indication of sub-vector activity and association with species together with the potential implications for management.

6.2 Biofouling

Much research has been focused on understanding the mechanisms of attachment and subsequent impacts on vessel performance of biological growth on the immersed hull surfaces of a vessel (e.g., Gollasch 2002a). It is now recognized that the drag generated by low levels of biofouling impair a vessel's efficiency, when measured as tonnage of fuel required to maintain speed. This has led to significant efforts to improve cost-effective anti-fouling methods that reduce biofouling growths during the operational periods between dry-dockings.

Early anti-fouling methods included tar mixed with horsehair, copper cladding affixed to the vessel hull, flat-headed nails to cover a wooden hull, the use of steel, paints with biologically active compounds using copper and later tri-butyl tin (TBT). Anti-fouling paints have proven extremely effective, particularly those based on organotins (see Minchin 2006). The advent and subsequent proliferation of TBT paints from the 1970s onwards, however, resulted in significant impacts to marine communities adjacent to ports, marinas and in busy shipping lanes. Impacts occur at very low concentrations, as these organotins also act as endocrine disrupters, causing sexual deformities in a variety of invertebrate species (see review by Fent, 1996). Concern over the impacts of TBT has resulted an internationally agreed-upon ban on the use of TBT under the International Convention on the Control of Harmful Anti-Fouling Systems on Ships of the International Maritime Organization (IMO), the United Nations body which deals with shipping (AFS 2001). This has involved the discontinuation of the application of organotin based anti-fouling paints since 2003. Port areas have generally shown declines of leached organotins since then, leading to improvements in water quality that may subsequently lead to port regions becoming more invasible (Nehring 2001). For example, it was assumed that the improved water quality has enabled a re-expansion of the zebra mussels in the port of Hamburg, Germany, in the River Elbe (Gollasch 2001).

Biofouling has typically been considered to be of historic significance (Carlton 2001) as many vessels will have been lost at sea during naval engagements and storms and will have spent long durations in port without the advantages of modern hull fouling controls. In contrast, today there are fast turn around times in ports and the use of efficient anti-fouling paints that reduce the opportunities for settlement. The increased speed of modern vessels en route will result in much of the attached biota becoming detached (see Minchin 2006). Yet even recently dry-docked and painted vessels can have significant biofouling. These can be observed in untreated areas where the ship is supported on blocks during dry dock and where paint cannot be applied, on slow service vessels such as barges where high quality paints are not used, and, where such vessels or structures are seldom dry-docked (e.g., Coutts 1999; Gollasch 2002a).

A calculation of the total wetted surface area, the permanently submerged hull surface even when the vessel is only partly loaded, for the active fleet in 1982 of 75,000 merchant ships was ~110 million m^2 (Olesen 1982). This will certainly be much greater today. In Germany alone, Bettelhäuser and Ulrich (1993) calculated that 290 vessels had a submerged hull surface area of 2.9 million m^2. These areas are largely anti-fouled, however Coutts (1999) estimated that up to 20% of the total wetted surface area of some vessels are untreated dry docking support strips. Similarly, Lenz et al. (2000) found that approximately 14% of the vessel's wetted surface area was fouled, despite the general usage of TBT as an antifoulant used at that time. The findings of Lenz et al. (2000) may have been affected by the age of antifouling paint on the vessels investigated. Nevertheless, it is highly probable that biofouling today, on account of implementation of the AFS Convention to phase out all organotins in ship antifoulants by 2008, is greater.

The situation for non-merchant vessels is likely to be much less positive. Whereas merchant vessels have a strong economic incentive to maintain their vessel hulls free of fouling, slow moving vessels such as oil platforms, oceanic barges, tugs, dredges, fishing vessels, and recreational craft, do not. In addition, many slow-moving vessels also have long port tenancies and are idle for significant periods of time, during which extensive fouling communities can accumulate. For example, recreational vessels offered for sale can remain idle and in-water for long durations. Once sold, these vessels are commonly transported to new locations by water or over-land without any attempt at cleaning.

Slow moving vessels have been implicated in a number of biofouling associated introductions. One fishing vessel, the *Yefim Gorbenko* was examined by researchers in New Zealand following its operation for several months in the New Zealand Exclusive Economic Zone (EEZ). The extent of biofouling on the vessel caused operational difficulties and was brought to dry-dock where it was found to have an accumulation of over 96 tonnes (wet weight) of biological material (Hay and Dodgshun 1997).

Leisure craft have not, until recently, been considered as significant contributors to the transport of species, however it would appear these are important in spreading some marine algae and invertebrates (Minchin et al. 2006; Hewitt et al., 2007). The incursion and successful eradication of the black striped mussel, *Mytilopsis sallei*, in Darwin, Northern Territory, Australia in 1999 was almost certainly associated with a recreational cruising yacht (Bax 1999; Willan et al. 2000). Similarly, the introduction of several ascidian species, such as a form of *Didemnum* (Fig. 6.1), and several tubeworms (Fig. 6.2), continue to spread worldwide (Zibrowius 1994; Çinar 2006; Valentine et al. 2007) and are thought to be spread by recreational craft (Minchin 2006).

Currently there are few easy detection or treatment methods available to biosecurity managers. High densities of biofouling can be readily observed from the surface, or by using in-water diver or remote camera inspections. Such monitoring requires the establishment of additional inspection regimes at ports of entry (Hewitt et al. 2004). In addition, once a vessel with significant fouling has been detected, options for management are limited. The vessel can be fined and/or sent on with directions to clean the hull prior to future re-entry; or directed to a dry-docking or hoisting facility should this be readily available. In case organisms are removed from the hull and other areas in dry-dock, measures should be taken to prevent their disposal into the water unless appropriately treated. Research efforts are currently underway to develop in-water cleaning methods that safely remove organisms from a vessel and vacuum or filter any dislodged material from the water column (B. Gould, Biosecurity New Zealand, personal communication).

Fig. 6.1 A didemnid tunicate fouling a hull of a recreational vessel in Ireland, 2005 (credit: D. Minchin)

Fig. 6.2 *Ficopomatus enigmatica*, an Indo-Pacific tubeworm, on a recreational vessel hull in Ireland, 2005 (credit: D. Minchin)

6.3 Ballast Water and Sediments

Ships' ballast water is taken on board and held within specialized tanks to maintain the trim and stability of ships, specifically when unladen. These tanks can be cargo holds temporarily used for holding ballast water; however the majority are purpose-built, complex structures designed to provide structural support for the ship. The tanks are typically distributed in different regions of a ship and are subjected to varying environmental conditions. For example, the forepeak tank, situated at the ship's bow, provides the most agitated conditions in poor weather, whereas those farther aft are not subjected to the same degree of disturbance. Almost all tanks used for holding ballast water accumulate sediments, the amount varying according to the ports vessels visit. Estuarine ports often have highly turbid water and any disturbance of the sea-bed can lead to plumes of sediment that may inadvertently be taken aboard (Fig. 6.3). These plumes may also contain biota, some of which may be in a resting state and could remain in this way, even in apparently unsuitable conditions for some months. Accumulations of sediments can become sufficient to support an infaunal community (e.g., Gollasch 2002a). These sediments are most usually silts and muds and carry a wide range of microbiota to fine web-like growths of slime-moulds (Hülsmann and Galil 2002) and more advanced metazoa (e.g., Gollasch 2002a; Gollasch et al. 2002).

Ostenfeld (1908) suggested that the use of water as ballast could form a transport mechanism for species transfers. Soon after, further evidence of ballast-water introductions took place with the first European appearance of the Chinese-mitten crab to the Aller River in Germany in ~1912 (Marquard 1926; Peters 1933). This concern has subsequently been confirmed with the identification of hundreds of species in the ballast tank environment (e.g., Carlton 1985; Carlton and Geller 1993; Gollasch et al. 2002). The uptake of water as ballast typically occurs in a port environment while the ship is at berth. As a consequence, any organism present in the water column either as a permanent member (holo-plankton and demersal species), or temporary member (mero- and tycho-plankton) will be entrained during ballast water uptake and subsequently transported with the vessel (e.g., Carlton and Geller 1993). Organisms from virtually all the major taxonomic and trophic groups have been detected in ballast water or in its accumulated sediments (Williams et al. 1988; Carlton and Geller 1993; Gollasch et al. 2002).

The dark conditions within tanks are unfavourable for photosynthesising plant stages and for those species that are dependant on sight for feeding, their numbers decline rapidly soon after ballasting. Other organisms also decline over time. Rarely have any accounts of species increasing their abundance been noted, as in the case of an harpacticoid copepod increasing its numbers in ballast tanks during a voyage, most likely due to reproduction during the voyage (Gollasch et al. 2000). The remains of those organisms that expire may become consumed by scavengers or are broken down by bacteria and fungi. Recent studies have shown that the micro-organisms that occur in ballast water are an important component of the ballast tank community, some of which can cause human diseases (Drake et al. 2001).

Fig. 6.3 Sediment disturbance in port due to vessel movement (credit: S. Gollasch)

Ballast water as a transport vector has attained the awareness of both the public and policy makers throughout the world. As discussed in Chap. 19, Hewitt et al., significant efforts at international, regional and national levels have been focused on developing appropriate management options for ballast water mediated introductions. Similar to biofouling, the IMO developed a Convention on ballast water management (BWM 2005; Gollasch et al. 2007; Chap. 19, Hewitt et al.). Existing methods include the exchange of coastal ballast water for open oceanic water while in transit. This ballast water exchange (BWE) may act to reduce the likelihood of species transport in some instances, but these activities may not dislodge any sediments that have accumulated. In addition, the practice of BWE may be both impractical and unsafe in specific locations and sea states and incorrect procedures have led to serious events at sea. At best BWE can only reduce the numbers of biota transferred by this process and is more effective if the properties of the exchanged water are different, such as in the case of fresh or brackish water being exchanged for oceanic water (Gollasch et al. 2007). For this reason, BWE can only be considered as a part-effective and temporary measure. In the meantime, technologies using varying treatment methods are being developed (e.g., Taylor et al. 2002). Unfortunately, many of these treatments take a long time and are only possible to use during long voyages and are not possible to use over short distance routes.

6.4 Discussion

Evaluations of various world regions indicates that shipping has been considered to be responsible for the majority of marine bioinvasions (e.g. Carlton 1985, 1996, 2001; Cohen and Carlton 1998; Cranfield et al. 1998; Hewitt et al. 1999, 2004; Ruiz et al. 2000; Gollasch 2002a; Leppäkoski et al. 2002; Fofonoff et al. 2003). The levels of certainty associated with assigning a specific vector to an invasion vary, in part, because few invasions are witnessed and are detected at some point following their arrival. This leads to attempts to assign responsibility to a vector as a consequence of evaluating life history characteristics, timing of arrival in relation to active vectors, and proximity to active vectors according to current use. Indeed, it appears that numerous species have the opportunity to be transported by either as biofouling or in ballast water (Hewitt et al. 1999, 2004; Minchin 2006) (Fig. 6.4).

Ruiz et al. (2000) considered that this ability to be transported at multiple stages of a life cycle might contribute to the invasion success by increasing inoculation pressure. The frequency of transmission along specific routes may also enhance the likelihood of successful establishment because of an increased opportunity of arriving at an optimal period for growth and reproduction. When considering the large amount of water carried worldwide, and the routes that many individual vessels undertake, it is perhaps of little surprise that organisms are carried to new regions. In addition, the extensively fouled surfaces of ships allow for the development of

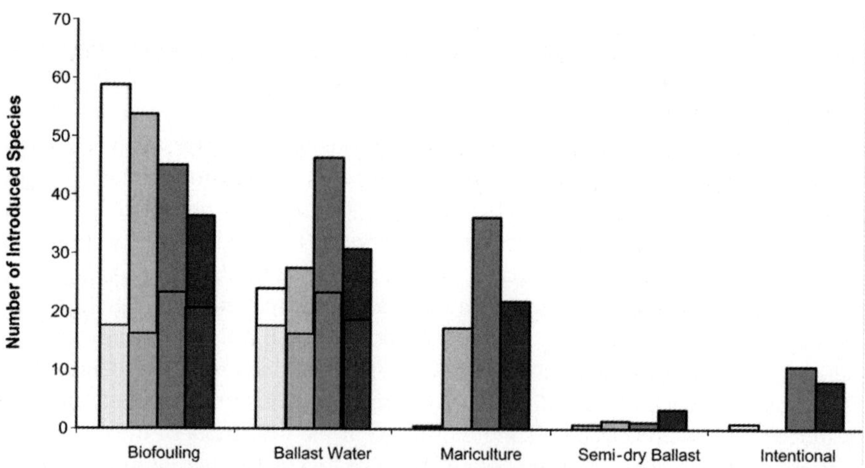

Fig. 6.4 An evaluation of historic marine bioinvasions according to five primary transport mechanisms: biofouling, ballast water, mariculture, semi-dry ballast and intentional: *white* – New Zealand (based on Cranfield et al. 1998); *light grey* – Port Phillip Bay (Hewitt et al. 1999, 2004); *dark grey* –North Sea (Jensen and Knudsen 2005; Wolff 2005; Gollasch and Nehring 2006; Minchin et al., in prep, Gollasch and Kerckhof, in prep); *black* – San Francisco Bay (Cohen and Carlton 1995); *stippled areas* represent species that can be both biofouling and ballast water

fouled species and all vessels also will have spread their compliment. While the great majority of routes for trade are now in general use, some new routes linking previously unconnected regions will undoubtedly be established as global trade increases. For example, a seasonal route is likely to evolve between the north Atlantic and north Pacific Oceans with the contraction of the Arctic ice-sheet as a consequence of global climate change.

It is of interest to managers whether biofouling or ballast water (and sediments) pose the greatest risk in distributing invasive species in the future (see Chap. 19, Hewitt et al. and Chap. 20, Campbell). With limited environmental management budgets and increasing pressures on the use of public funds, there is a need to target the most cost-effective and appropriate research and management activities that reduce the risks of costly invasions (Hewitt et al. 2004). In a recent analysis of incursions to Australia, New Zealand and the North Sea between 1995 and 2002, Australia had recorded 17 new incursions (Fig. 6.5A), New Zealand had recorded 18 incursions (Fig. 6.5B) and 20 new species were found in the North Sea (Fig. 6.5C). The Australian incursions could be divided into 13 biofouling alone (based on species' life history characteristics and locations of arrival), 2 to either biofouling or ballast water and 2 to others from non-ship vectors. In contrast, the New Zealand incursions included nine attributed to biofouling, three to ballast water, five to either biofouling or ballast water, and one to non-ship associated methods. Fifteen species reached the North Sea region with shipping (six with ballast water, six with biofouling and three in either biofouling or ballast water) and five with non-shipping vectors. In these studies it was assumed that all invasions were identified and that the appropriate vectors were correctly identified.

It is clear that both biofouling and ballast water are currently active and important vectors for the transport of marine species and will continue to spread species. Both vectors have special circumstances allowing for their transport according to the life-history stages and tolerances of biota. From a management perspective, these two sub-vectors associated with vessels require different regulatory frameworks and management responses, suggesting that further research into the relative risks of biofouling introductions vs ballast water introductions is needed to inform policy development better. Further, additional vector management options need to be developed to provide a suite of tools for appropriate management action. Examples include development of in-water hull cleaning devices for merchant and recreational vessels that will act in such a way that the removed biomass does not propagate and infect new regions; non-toxic antifouling paints to replace the current suite of organitin based paints; and ballast water treatment technologies to reduce invasion risk without increasing release of biologically active compounds in nearshore and coastal waters.

Additional work is needed on the transmission of parasites and diseases associated with biofouling species and in ballast water. These organisms are likely to create risks to aquaculture operations within and adjacent to ports and marinas. A number of commercially farmed molluscs or closely related species are found on the hulls of ships and capable of surviving long journeys (Minchin and Gollasch 2003). While these species are likely to have been transported and introduced to

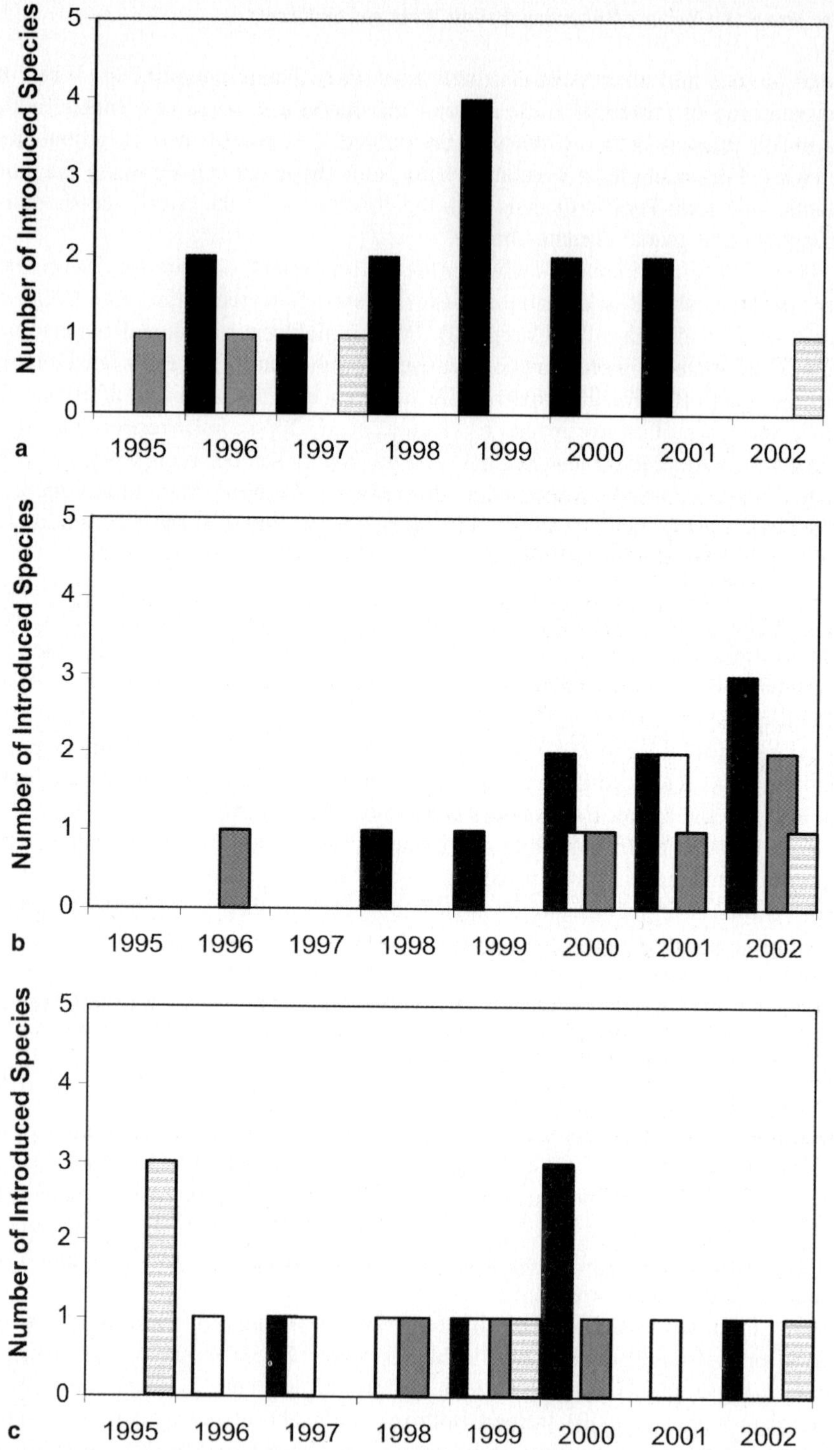

Fig. 6.5 An evaluation of recent Australian (**a**), New Zealand (**b**) and North Sea (**c**) incursions by sub-vector type based on life history characteristics, location of incursion, or observed vector associations: *black* – only hull fouling; *white* – only ballast water; *solid grey* – both hull fouling and ballast water; *stippled grey* – other non-shipping associated method

new regions, it is equally likely that parasites and pathogens were transported within these organisms. The further transport of these molluscs either through intentional stock transfers or with commercial or recreational vessel movements to new port regions may also lead to the further transfer of parasites and pathogens.

References

AFS (2001) International Convention on the Control of Harmful Anti-Fouling Systems on Ships. International Maritime Organization, London, United Kingdom; www.imo.org, accessed 30 January 2006

Bax NJ (1999) Eradicating a dreissenid from Australia. Dreissena! 10:1–5

Bettelhäuser F, Ullrich P (1993) Die Suche nach dem sozial- und umweltverträglichen Schiff - Die ersten Schritte. Arbeitskreis andere nützliche Objekte der Bremer Vulkan Werft, 1–92

BWM (2005) International Convention on the Control and Management of Ship's Ballast Water and Sediments. International Maritime Organization, London, United Kingdom; www.imo.org, accessed 30 January 2006

Carlton JT (1979) History, biogeography and ecology of the introduced marine and estuarine invertebrates of the Pacific Coast of North America. PhD Dissertation. University of California, Davis, California, USA

Carlton JT (1985) Transoceanic and interoceanic dispersal of coastal marine organisms: the biology of ballast water. Ocean Mar Biol Ann Rev 23:313–371

Carlton JT (1996) Pattern, process, and prediction in marine invasion ecology. Biol Cons 78:97–106

Carlton JT (2001) Introduced species in U.S. coastal waters: environmental impacts and management priorities. Pew Oceans Commission, Virginia, USA

Carlton JT, Geller JB (1993) Ecological roulette: the global transport of nonindigenous marine organisms. Science 261:78–82

Castilla JC, Uribe M, Bahamonde N, Clarke M, Desqueyroux-Faúndez R, Kong I, Moyano H, Rozbaczylo N, Santilices B, Valdovinos C, Zavala P (2005) Down under the southeastern Pacific: marine non-indigenous species in Chile. Biol Inv 7:213–232

Çinar ME (2006) Serpulid species (Polychaeta: Serpulidae) from the Levantine coast of Turkey (eastern Mediterranean), with special emphasis on alien species. Aquat Invas 4:223–240

Cohen AN, Carlton JT (1995) Biological study: non-indigenous aquatic species in a United States estuary: a case study of the biological invasions of the San Francisco bay and delta. US Fisheries and Wildlife and National Sea Grant College Program Report, NTIS Number PB96-166525, Springfield, Virginia, USA, 273 pp

Cohen AN, Carlton JT (1998) Accelerating invasion rate in a highly invaded estuary. Science 279:555–558

Coles SL, DeFelice RC, Eldredge LG, Carlton JT (1999) Historical and recent introductions of non-indigenous marine species into Pearl Harbor, Oahu, Hawaiian Islands. Mar Biol 135:147–158

Coutts ADM (1999) Hull fouling as a modern vector for marine biological invasions: investigation of merchant vessels visiting northern Tasmania. Masters Thesis, Australian Maritime College, Launceston, Tasmania, Australia

Cranfield HJ, Gordon DP, Willan RC, Marshall BA, Battershill CN, Francis MP, Nelson WA, Glasby CJ, Read GB (1998) Adventive marine species in New Zealand. NIWA Technical Report 34. National Institute of Water and Atmospheric Research, Wellington, New Zealand

Crosby AW (1986) Ecological imperialism: the biological expansion of Europe, 900–1900. Cambridge University Press, Cambridge

di Castri F (1989) History of biological invasions with specieal emphasis on the Old World. In: Drake JA, Mooney HA, di Castri F, Groves RH, Kruger FJ, Rejmánek M, Williamson M (eds) Biological invasions: a global perspective. SCOPE 37. Wiley, New York, pp 1–30

Drake LA, Choi K-H, Ruiz GM, Dobbs FC (2001) Global re-distribution of bacterioplankton and virioplankton communities. Biol Inv 3:193–199

Elton CS (1958) The ecology of invasions by animals and plants. Methuen, London, United Kingdom

Fent K (1996) Ecotoxicology of organotin compounds. Crit Rev Toxicol 26:1–117

Fofonoff PW, Ruiz GM, Steves B, Carlton JT (2003) In Ships or on ships? Mechanisms of transfer and invasion for non-native species to the coasts of North America. In: Ruiz GM, Carlton JT (eds) Invasive species: vectors and management strategies. Island Press, Washington, pp 152–182

Gollasch S (2001) German National Report. In: Report of the ICES Working Group on Introductions and Transfers of Marine Organism (WGITMO). Barcelona Meeting, pp 38–43

Gollasch S (2002a) The importance of ship hull fouling as a vector of species introductions into the North Sea. Biofouling 18(2):105–121

Gollasch S (2002b) Ballast water management in the north-east Atlantic. In: Tjallingii F (ed) Report to aid decision making on ballast water in the Oslo-Paris Commission (OSPAR). Biological Diversity Committee. North Sea Directorate, the Netherlands, p 48

Gollasch S, David M, Dragsund E, Hewitt CL, Fukuyo Y (2007) Critical review of the IMO International Convention on the Management of Ships' Ballast Water and Sediments. Harmful Algae 6: 585–600

Gollasch S, Lenz J, Dammer M, Andres HG (2000) Survival of tropical ballast water organisms during a cruise from the Indian Ocean to the North Sea. J Plankton Res 22:923–937

Gollasch S, MacDonald E, Belson S, Botnen H, Christensen JT, Hamer JP, Houvenaghel G, Jelmert A, Lucas I, Masson D, McCollin T, Olenin S, Persson A, Wallentinus I, Wetsteyn LPMJ, Wittling T (2002) Life in ballast tanks. In: Leppäkoski E, Gollasch S, Olenin S (eds) Invasive aquatic species of Europe distribution, impact and management. Kluwer Academic Publishers, The Netherlands, pp 217–231

Gollasch S, Nehring S (2006) National checklist for aquatic alien species in Germany. Aquat Invas 1(4):245–269

Gollasch S, Kerckhof F (in prep) National checklist of introduced aquatic alien species for Belgium

Hay C, Dodgshun T (1997) Ecosystem transplant? The case of the Yefim Gorbenko. Seafood N Z 13–15

Hewitt CL, Campbell ML, Thresher RE, Martin RB (eds) (1999) The introduced species of Port Phillip Bay, Victoria. Centre for Research on Introduced Marine Pests Technical Report No. 20. CSIRO Marine Research, Hobart

Hewitt CL, Campbell ML, Thresher RE, Martin RB, Boyd S, Cohen BF, Currie DR, Gomon MF, Keogh MJ, Lewis JA, Lockett MM, Mays N, McArthur MA, O'Hara TD, Poore GCB, Ross DJ, Storey MJ, Watson JE, Wilson RS (2004) Introduced and cryptogenic species in Port Phillip Bay, Victoria, Australia. Mar Biol 144:183–202

Hewitt CL, Campbell ML, Schaffelke B (2007) Introductions of marine macroalgae – accidental transfer pathways and mechanisms. Bot Mar 50:326–337

Hülsmann N, Galil BS (2002) Protists – a dominant component of the ballast – transported biota. In: Leppäkoski E, Gollasch S, Olenin S (eds) Invasive aquatic species of Europe: distribution, impact and management. Kluwer, Dordrecht, the Netherlands, pp 20–26

Jensen K, Knudsen J (2005) A summary of alien marine invertebrates in Danish waters. Oceanolog Hydrobiol Stud 34(Suppl 1):137–162

Lenz J, Andres H-G, Gollasch S, Dammer M (2000) Einschleppung fremder Organismen in Nord- und Ostsee: Untersuchungen zum ökologischen Gefahrenpotential durch den Schiffsverkehr. UBA Project Water: 102 04 250, Umweltbundesamt, Berlin, Texte 5

Leppäkoski E, Gollasch S, Olenin S (eds) (2002) Invasive aquatic species of Europe distribution, impact and management. Kluwer Academic Publishers, The Netherlands

Lewis PN, Hewitt CL, Riddle M, McMinn A (2003) Marine introductions in the Southern Ocean: an unrecognised hazard to biodiversity. Mar Pollut Bull 46:213–223

Marquard O (1926) Die Chinesische Wollhandkrabbe, *Eriocheir sinensis* Milne- Edwards, ein neuer Bewohner deutscher Flüsse. Fischerei 24:417–433

Minchin D (2006) The transport and the spread of living aquatic species. In: Davenport J, Davenport JL (eds) The ecology of transportation: managing mobility for the environment. Springer, Berlin Heidelberg New York, pp 77–97

Minchin D (2007) Aquaculture and transport in a changing environment: overlap and links in the spread of alien biota. Mar Pollut Bull 55(7/9):302–313

Minchin D, Bishop J, Cook E (in prep) Alien and cryptogenic species in brackish and marine regions of Britain

Minchin D, Gollasch S (2003) Fouling and ships' hulls: how changing circumstances and spawning events may result in the spread of exotic species. Biofouling 19:111–122

Minchin D, Gollasch S, Wallentinus I (2005) Vector pathways and the spread of exotic species in the sea. ICES Coop Res Rep 271

Minchin D, Floerl O, Savini D, Occhipinti-Ambrogi A (2006) Small craft and the spread of exotic species. In: Davenport J, Davenport JL (eds) The ecology of transportation: managing mobility for the environment. Springer, Berlin Heidelberg New York, pp 99–118

Nagabhushanam R, Sarojini R (1997) An overview of Indian research efforts on marine wood-boring and fouling organisms. In: Nagabhushanam R, Thompson M-F (eds) Fouling organisms of the Indian Ocean, biology and control technology. A.A. Balkema, Rotterdam, p 538

Nehring S (2001) After the TBT era: alternative antifouling paints and their ecological risks. Senckenbergiana Marit 31:341–351

Olesen G (1982) Antifouling paint – a source of pollution. Schiff Hafen 34(7):18–19

Oresanz JM, Schwindt E, Pastorino G, Bortolus A, Casas G, Darrigran G, Elías R, López Gappa JJ, Obenat S, Pascual M, Penchaszadeh P, Piriz ML, Scarabino F, Spivak ED, Vallarino EA (2002) No longer the pristine confines of the world ocean: a survey of exotic marine species in the southwestern Atlantic. Biol Inv 4:115–143

Ostenfeld CH (1908) On the immigration of Biddulphia sinensis Grev. and its occurrence in the North Sea during 1903–1907. Meddelelser Fra Kommissionen Fur Danmarks Fiskeri-og Havundersogelser. Serie Plankton 1, No 6

Peters N (1933) B. Lebenskundlicher Teil. In: Peters N, Panning, A (eds.) Die chinesische Wollhandkrabbe (Eriocheir sinensis H. MILNE-EDWARDS) in Deutschland. Akademische Verlagsgesellschaft mbH, Leipzig, pp 59–156

Ruiz GM, Fofonoff PW, Carlton JT, Wonham MJ, Hines AH (2000) Invasion of coastal marine communities in North America: apparent patterns, processes, and biases. Ann Rev Ecol Syst 31:481–531

Taylor A, Rigby G, Gollasch S, Voigt M, Hallegraeff G, McCollin T, Jelmert A (2002) Preventive treatment and control techniques for ballast water. In: Leppäkoski E, Gollasch S, Olenin S (eds) Invasive aquatic species of Europe: distribution, impacts and management. Kluwer Academic Publishers, Dordrecht, The Netherlands, p 583

Valentine PC, Carman MR, Blackwood DS, Heffron EJ (2007) Ecological observations on the colonial ascidian *Didemnum* sp. in a New England tide pool habitat. J Exp Mar Biol Ecol 342:109–121

Willan RC, Russell BC, Murfet NB, Moore KL, McEnnulty FR, Horner SK, Hewitt CL, Dally GM,Campbell ML, Bourke ST (2000) Outbreak of *Mytilopsis sallei* (Recluz, 1849) (Bivalvia: Dreissenidae) in Australia. Molluscan Res 20(2):25–30

Williams RJ, Griffiths FB, Van der Wal EJ, Kelly J (1988) Cargo vessel ballast water as a vector for the transport of non-indigenous marine species. Est Coast Shelf Sci 26:409–420

Wolff WJ (2005) Non-indigenous marine and estuarine species in The Netherlands. Zool Meded 79-1:1–116

Zibrowius H (1994) Introduced invertebrates: examples of success and nuisance in the European Atlantic and Mediterranean. In: Boudouresque CF, Briand F, Nolan C (eds) Introduced species in European coastal waters. European Commission, Luxembourg, pp 44–49

Chapter 7
The Role of Propagule Pressure in Invasion Success

Emma L. Johnston, Richard F. Piola, and Graeme F. Clark

7.1 Introduction

One of the core goals of invasion biology is the identification of factors that increase the risk of establishment success of non-native species. Historically, marine invasions have been investigated through observational studies and surveys (Cohen and Carlton 1998; Ruiz et al. 2000). These have guided ecologists towards the processes most relevant to invasion, but researchers are becoming increasingly aware of the limitations of observational studies alone. It is clear that different factors may influence invasion success at different stages of the invasion process (Kolar and Lodge 2001) and a major challenge is to quantify the relative importance of these factors. Understanding the intricacies of invasion dynamics requires a rigorous approach, in which potentially important factors can be controlled, manipulated and tested (Ruiz et al. 2000). Particularly strong calls have been made for the inclusion of propagule pressure or invader supply into our models, experiments and surveys, and this chapter reviews recent progress in elucidating the role of propagule pressure on invasion success in marine ecosystems.

7.2 Propagule Pressure

Propagule pressure describes a measure of the number of individuals released into an area to which they are not indigenous (Carlton 1996). In marine systems this can be the release of mature adult organisms but also, and perhaps more commonly, the release of early life-history stages such as larvae. Propagule pressure may be increased either through an increased number of arrival events (e.g. increased frequency of ship arrival) or an increased intensity of exposure during any one event (increased abundance of organisms released into the recipient water body from any one vector). Propagule pressure differs from settlement or recruitment because it represents the *potential* for introduction rather than a *realized* introduction. Propagules may be released but never join local populations. For example, millions

G. Rilov, J.A. Crooks (eds.) *Biological Invasions in Marine Ecosystems.*
Ecological Studies 204,

of non-native larvae may be released into an area (high propagule pressure), but most might die before settling out from the pelagic larval phase (settlement) or surviving to reproductive age (recruitment to the adult population).

Propagule pressure is well recognized as an important factor that may influence the success of species invasions (Lonsdale 1999; Ruiz et al. 2000; Puth and Post 2005; Drake and Lodge 2006). In part this may be because the basic relationship is rather intuitive: increasing propagule pressure should lead to an increased probability of invasion success (Leung et al. 2004). Recent models that incorporate propagule pressure are also proving to be more successful at explaining current distributions (Rouget and Richardson 2003). However, despite the profusion of models highlighting the theoretical importance of propagule pressure, there remain relatively few experimental studies that have measured or manipulated propagule pressure. Where studies have been conducted they report a clear positive effect of propagule pressure on invasion success in a range of taxa, including freshwater fish and invertebrates, beetles, birds, mice, voles and ungulates (Lockwood et al. 2005). What is perhaps surprising is the paucity of studies examining the role of propagule pressure on invasion in marine systems, relative to other habitats (Fig. 7.1).

The study of propagule pressure in invasion biology has many similarities to research in "supply-side ecology", which has long fascinated marine ecologists. Recent focus in this area has provided us with strong evidence for the stochastic, non-equilibrial and "open" nature of many marine systems (Underwood and Keough 2001). In such systems there are few circumstances in which propagule pressure would not influence population dynamics, even if the effects were difficult to detect (Caley et al. 1996). Any study of supply-side ecology may therefore shed some light on the invasibility of marine communities; however, very few explicitly claim to study invasion this way (Verling et al. 2005). The results of a literature search spanning the last decade of research in marine systems shows that the number of studies relating to supply-side ecology (including larval biology) outnumbers the number of supply-side invasion biology studies by forty to one (Fig. 7.1). However, attention to this issue appears to be increasing (Fig. 7.2).

The strong likelihood of propagule pressure effects in marine systems is complemented by an enormous applied interest in this area, particularly since it may be one of the few economically viable management options for the control and prevention of marine invasions. If the risk of invasion success can be modeled using basic information on vector types, arrival rates, pre- and post-border survival, fecundity and Allee effects, then we may have some success in developing much needed invasion risk assessment and management plans (Drake and Lodge 2006). Groundtruthing such models with successful experimental manipulations would then provide a strong theoretical framework with which to predict invasions, and help us to curb the increasingly rapid homogenization of the world's marine biota.

Here we review recent research on supply-side invasion in marine systems and identify productive avenues for future research. We start by describing studies that aim to explain invasion success through the characterization of certain biological characteristics of an invader, such as propagule resilience or propagule production. We then discuss how surveys can contribute to our understanding of supply-side

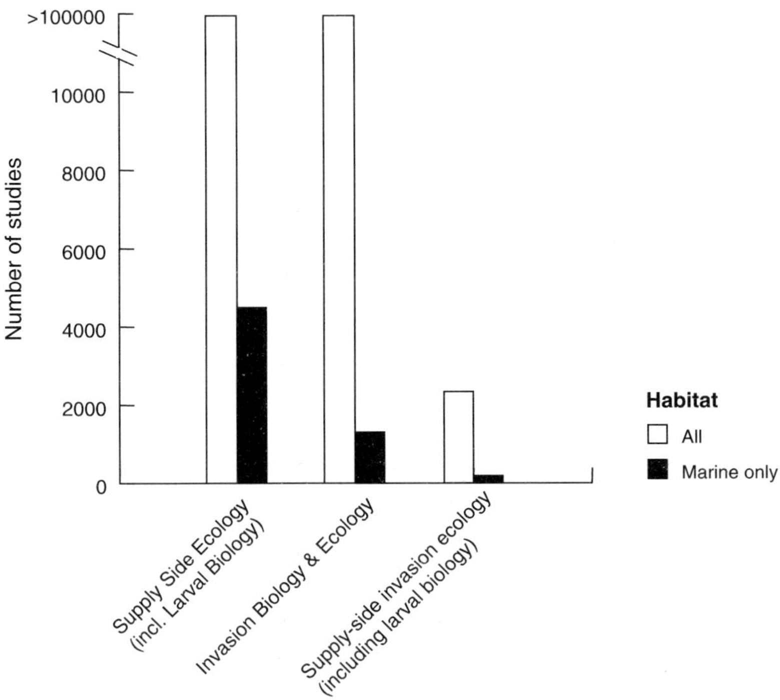

Fig. 7.1 Comparison of the number of studies of supply side ecology, invasion biology and supply side invasion ecology in all habitats vs marine systems. We conducted the literature search using the Web of Science for studies published between 1995 and 2007. Supply-side ecology papers were searched for using the terms larv*, propagule, supply-side ecology, supply side ecology, supply-side, supply side, bottom-up, and bottom*. Invasion biology studies were searched for using the terms invas*, invad*, exotic, alien, nonnative, non-native, nonindigenous, and non-indigenous. We also combined these search terms in order to compare the number of studies of supply-side ecology, invasion biology and supply-side invasion biology in terrestrial and marine systems. *Bars* represent the unrefined number of studies that were found in each search (see Figs. 7.2 and 7.3 for further analysis of the marine supply-side invasion biology studies)

invasion ecology. These studies include adult distribution and vector surveys, as well as investigations of population genetics. Finally we discuss studies that include manipulated invasive propagules in order to gauge community invasibility.

7.3 The Resilience of Propagules

Life history traits of individual taxa, attributes of source and recipient environments, and characteristics of transport vectors all contribute to determine which species become invaders (see Chap. 6, Hewitt et al.; Chap. 8, Miller and Ruiz;

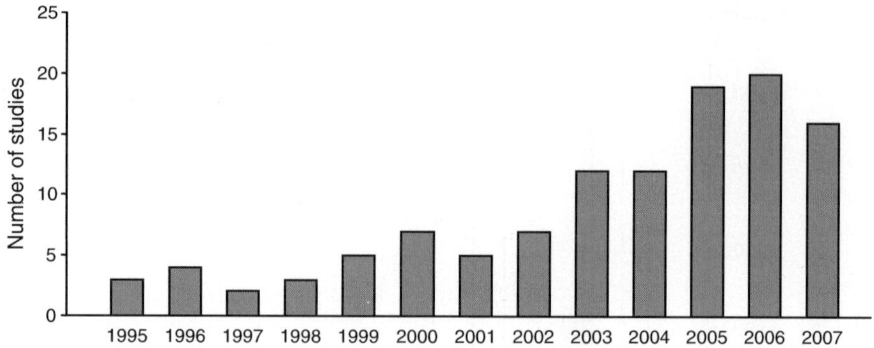

Fig. 7.2 Frequency of propagule studies by year for the last decade. Data was extracted from the literature search described in Fig. 7.1. A total of 187 abstracts were assessed and studies that were not about marine supply-side invasion biology were removed from the analysis. The remainder (109 studies) were categorized according to year of publication

Chap. 10, Smith; Chap. 12, Olyarnik et al.). They influence how invaders are transported, how many individuals survive transportation to be released (i.e. propagule pressure), and their eventual establishment success (Floerl and Inglis 2005; Wonham et al. 2005). Increased chance of invasion success should be associated with a high probability of being transported alive to a new region. Invasive propagules must therefore be entrained by vectors and able to survive distinct environments throughout the invasion process. These include the specific vector environment (i.e. the conditions experience during transportation), and the conditions of a specific recipient region(s) (for a detailed explanation see Chap. 8, Miller and Ruiz). Many common vectors that transport non-indigenous marine species present extremely stressful environments, and exert strong selective pressures on both target and non-target organisms. Organisms transported by ballast water for example, risk exposure to high concentrations of harmful metals, biocides and hypoxic conditions (Jelmert and Van Leeuwen 2000; Tamburri et al. 2002; Jones et al. 2006). Similarly, fouling organisms transported on the hulls of vessels are exposed to a range of chemical (Valkirs et al. 2003; Schiff et al. 2004; Srinivasan and Swain 2007) and physical stressors. The tolerance and life-history characteristics of larval and adult organisms entrained and transported under such hostile conditions may have a strong bearing on the propagule pressure they exert within recipient environments, and hence their invasive potential.

Adaptivity and tolerance to a range of biotic and abiotic stressors are important determinants of a species potential to colonize a new environment. Propagules at the larval stage are usually less resilient than adults (Connor 1972; Calabrese et al. 1973) and physical parameters such as temperature and salinity may strongly influence the survival of juvenile invaders. Determining the biological characteristics and environmental tolerances of larval stages of invasive species has been the subject of only a handful of recent studies (Fig. 7.3). These studies generally show that the larvae of successful invaders display wide tolerance to environmental conditions. Larvae of the extremely successful invasive marine crab *Carcinus maenas* are

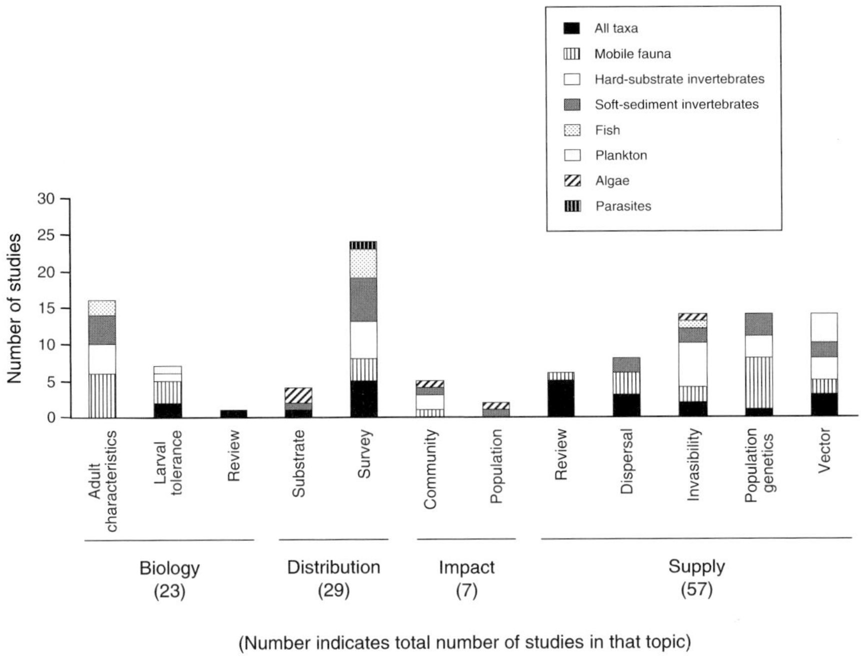

Fig. 7.3 Comparison of topics investigated in marine studies of supply-side invasion biology. Results of the search conducted in Fig. 7.1. were further categorised by organism and subject matter. A total of 187 abstracts were assessed and studies that were not marine or primarily about invasion biology were removed from the analysis. The remainder (109 studies) were categorized according to whether they were about (i) larval or adult invader characteristics (ii) surveys of existing invader distributions, (iii) studies of invader impacts and studies of invader propagule supply

able to survive and mature in salinities as low as 26 ppt (Bravo et al. 2007) and temperatures between 10 and 22.5 °C (deRivera et al. 2007), likely contributing to the continued spread of this invader along the North American coastline. Similarly, larvae of the invasive gastropod *Rapana venosa* from Chesapeake Bay have demonstrated prolonged tolerance to salinity levels as low as 7 ppt, facilitating its dispersal along large stretches of west coast United States (Mann and Harding 2003). Miller et al. (2007) determined that tolerance to low salinity conditions was one of the three most important attributes driving successful molluscan invasions in San Francisco Bay, along with developmental mode and population abundance. Phenotypic plasticity (Chap. 10, Smith) has even been demonstrated in the eggs and larvae of the grapsid crab *Chasmagnathus granulate*, allowing it to rapidly acclimate (and switch between) wide ranges of salinities (Charmantier et al. 2002). While *C. granulata* is not currently recognized as an invasive species, it is easy to envisage how such a trait may be an advantage to propagules of an introduced species across a wide range of environments.

Propagules of some invasive species even display significant tolerance to biocides and toxicants designed to prevent their dispersal. Toxic chemicals from homes, industry and agriculture are regularly released into bays and estuaries. Similar suites of toxicants are also used to try to prevent the recruitment and transport of marine organisms onto ship hulls, and a marine plant or animal that is capable of settling and surviving on these toxic paints may be transported around the globe (Hewitt et al. 1999). Copper and zinc are common constituents of antifouling paints used to prevent the spread of species on vessel and boat hulls. Field experiments examining recruitment to antifouling coated surfaces have demonstrated that larvae of the invasive bryozoan *Watersipora subtorquata* not only actively recruit and grow upon copper-treated surfaces, but facilitate the recruitment of other invasive species that use them as a refugia from the toxicant (Floerl et al. 2004). Laboratory studies of *W. subtorquata* and other invasive bryozoan larvae (including *Schizoporella errata* and *Bugula neritina*) show them to have copper tolerances up to five times greater than maximal levels observed under real-world conditions (Piola and Johnston 2006a).

The resilient nature of invasive propagules may be essential to their invasion success by not only increasing survival through the transportation process but also within the recipient environment. An example to highlight how the interplay between propagule traits, vector characteristics, and environmental conditions can influence invasion success is the relationship between pollution and invasion in marine systems. By their presence on antifouled surfaces some sessile invertebrates and algae clearly display a degree of resistance to certain contaminants, which could then become a desirable trait if the organism is transported to an environment containing the same (or similar) toxicants (e.g. a polluted harbor). This process of entrainment, resistance, transport and introduction may be even more successful if source and recipient environments experience similar levels of pollution. Invasive bryozoans display a high tolerance to the toxic heavy metal copper (Floerl et al. 2004; Piola and Johnston 2006b) and they are capable of developing and losing this tolerance depending on environmental conditions (Piola and Johnston 2006b). It is now clear that copper tolerance does confer a competitive advantage on some non-indigenous marine invertebrates (Dafforn et al. 2008; Piola and Johnston 2008; Crooks, unpublished data). We predict that this will also be the case for some organisms that are transported in ballast water, where the environment is likely to differ in many parameters, particularly salinity, temperature, dissolved oxygen and turbidity. Further research on the resilience of transported propagules to vector and recipient environments will help characterize the risk of transport of a range of marine invertebrates. Risk analysis frameworks that are based on biological characteristics of the invasive propagules must be interpreted with caution, however, since marine invasive species may also display rapid adaptation and/or evolution to new conditions (Sax et al. 2007).

7.4 Producing Propagules

Invasive species can also possess adult-stage characteristics that enhance their propagule supply and promote invasive success. Timing of reproductive maturity and numbers of propagules produced (fecundity) are two factors that often set successful invaders apart from competing native taxa. The rapa whelk *Rapana venosa* exhibits life history traits such as early reproductive maturity and annual production of large numbers of propagules (up to ~3600 embryos), which contribute to its success as an invader (Harding et al. 2007). Similarly, individuals of the round goby *Neogobius melanostomus* that have invaded brackish water habitats have been shown to invest more resources into reproduction and have higher fecundity than individuals of the same species living in marine habitats (Corkum et al. 2004). Through strong dispersal abilities and rapid adaptation (~3 years), they can readily become dominant in new environments. Analogous traits and high dispersal potential are also observed in the ubiquitous green mussel *Perna viridis* (Rajagopal et al. 2006) and the varnish clam *Nuttallia obscurata*, a recent invader to the NE Pacific ocean (Dudas and Dower 2006). Non-indigenous species with high propagule output often out-recruit native species and come to dominate local populations. For example, the introduced barnacles *Elminius modestus* (Watson et al. 2005) and *Chthamalus proteus* (Zabin et al. 2007) display high propagule production, high levels of recruitment back to parent populations, and, in the case of *C. proteus*, short larval development time – all of which enhance their dominance of invaded habitats.

Variable modes of propagule production and dispersal may also influence a species' invasion potential. By studying historical and modern day range limits of Californian marine bivalves, Roy et al. (2001, 2002) argue that bivalve invasions were generally driven by large bodied species/individuals, perhaps partly due to the fact that small bodied molluscs tend to brood their larvae resulting in reduced larval dispersal. In a previously undescribed gastropod in Hawaii, evidence of long-term sperm storage and extremely competent larvae at the time of hatching suggests a high potential for invasion in this species, as a single individual transported to a new region has the ability to fertilize and disperse propagules which have a high chance of survival (Strathmann and Strathmann 2006). The timing of propagule production may also influence invader success. By examining long term recruitment data, Stachowicz and Byrnes (2006) argue that the non-indigenous colonial ascidian *Botrylloides violaceus* increases recruitment success by spawning during periods of otherwise low recruitment. This effectively represents exploitation of a previously unoccupied ecological niche and points towards the importance of the timing of propagule supply in explaining invasion success (see Clark and Johnston 2005 for an experimental test of the timing of propagule arrival).

7.5 Inferring Propagule Supply from Current Distributions

Surveys of species distributions are a first and necessary step in identifying new species invasions. Although observational studies are inherently limited in their ability to determine causality, they can provide a basis for inference and give direction towards processes of interest. Surveys have proven instrumental in detecting the range expansion or initial occurrence of non-native species (e.g., Carriglio et al. 2004; Chap. 3, Lonhart). In isolation, these results can appear disparate, but growing syntheses of data from a wide range of locations has revealed some telling large-scale patterns (Ruiz et al. 2000; Chap.33, Preisler et al.). These generally show invasions to be most prevalent in anthropogenic environments such as ports and harbors, suggesting that invasion patterns are closely linked with the supply of invasive propagules through human-mediated inoculation vectors such as shipping (Ruiz et al. 2000). Long-term survey studies are also able to document the extreme degradation of environments by exotic species, such as has occurred in San Francisco Bay (Cohen and Carlton 1998) and Waitemata Harbour (Hayward et al. 1997).

A major shortcoming of observational studies, however, is their inability to separate effects of propagule supply from the invasibility of recipient communities (Lonsdale 1999). Unless there is a history of deliberate recorded introduction, then we can know little about the propagule pressure associated with introductions that failed to establish. Recent studies suggest that ports and harbours are threatened not only because of increased inoculation, but also because anthropogenic stresses compromise their ability to resist invasion. Such mechanisms are very difficult to ascertain with surveys alone, but could explain why some have detected significant invasion rates at sites distant from major transport hubs. Cohen et al. (2005), for example, found no significant difference in the numbers and proportion of introduced species at a series of disturbed estuarine sites with and without ports.

There are, however, sampling strategies that may increase our powers of inference. Studies that quantify recruitment as well as the incumbent communities can sometimes differentiate the importance of propagule supply vs competitive processes. Herbert et al. (2007) monitored the range expansion of intertidal barnacles on the English coast at the same time as recruitment patterns. They found that the distribution of populations was closely related to recruitment events, and inferred that the range limits would only be maintained given an ongoing supply of propagules. Surveys that aim to test hypotheses in addition to distributional patterns have also proved informative. This approach can reveal the small scale habitat preferences of invaders, such as *Codium fragile*, which tends to colonise and grow better on the sheltered sections of artificial structures compared to exposed sections (Bulleri et al. 2006). Importantly, the ability of surveys to identify relevant processes can often be scale dependent. Small-scale surveys infer the invasion history at particular sites (e.g. Marzano et al. 2003), but large scale, multi-factor surveys can show regional patterns of range, dispersal and habitat preference (Gust and Inglis 2006). Adding manipulative components to surveys can also help disentangle supply from survival. By including a manipulative component to an otherwise survey-based study, Piola and Johnston (2008) were able to show that some invasive species were competitively advantaged over natives in polluted conditions, suggesting that propagule pressure is not wholly responsible for the invasion of urban estuaries.

Molecular studies are a promising new tool to complement surveys and shed more light on supply-side invasion ecology. Several recent studies have used molecular information to attempt to reconstruct invasion pathways and infer at least something regarding the frequency or spatial extent of source populations. Current models would suggest that low levels of genetic variation will decrease the chance of population establishment, and Alee effects may be crucial to predicting invasion success (Drake and Lodge 2006). If true, then multiple inoculations separated in time and space, are likely to increase invasion success. Roman and Darling (2007) contend that low genetic diversity of invaders is usually overcome by multiple inoculations and thus founder effects are also generally overcome. We found six studies in the past decade (Fig. 7.3) that suggest that invader populations were genetically diverse and likely to result from multiple introductions (Dupont et al. 2003; Martel et al. 2004; Shefer et al. 2004; Zardus and Hadfield 2005; Roman 2006; Viard et al. 2006), and three that found evidence of low genetic diversity and strong founder effects, inferring restricted propagule supply (Patti and Gambi 2001; Mackie et al. 2006; Zardi et al. 2007). None were able to pinpoint explicitly source populations or the likely number of propagule arrivals.

Molecular studies of existing invasive populations tell us only about successful introductions that are currently at relatively large population sizes. Extrapolation of these population dynamics to the initial establishment is difficult (Leung et al. 2004) and there is a need to assess the genetic variability of transported individuals. Studies of the population genetics of organisms sampled from ballast water or hull fouling communities could be compared to that in the native and introduced ranges to illuminate the role of increased/decreased genetic variability in establishment success. Such studies will become increasingly feasible with the development of more sophisticated gene probes and similar detection technologies (e.g. Deagle et al. 2003; Gunasekera et al. 2005).

7.6 Vectors and Propagule Supply

Successful marine invasions are likely to be very difficult to reverse (Chap. 18, Hewitt et al.). While new management tools for the post-border control of invasives continue to be developed, preventing the transport of propagules through pre-border vector management remains the best option for controlling successful invasions. Of 109 recent papers we reviewed on invasion and propagule supply in marine systems (Fig. 7.3), approximately ten surveyed or discussed potential vectors. It is clear that propagule supply is therefore measured relatively rarely in marine environments, probably because of the difficulty in gaining access to commercial ships. Of the studies that did discuss vectors, half were relevant to marine invasions by ballast water organisms and half were relevant to fouling organisms. None related to invasions via the aquarium trade, fishing or aquaculture, which are other well recognized but less prevalent invasion vectors (Ruiz et al. 2000; Chap. 5, Minchin et al.). Ballast water studies either examined actual ballast water (Chu et al. 1997; Dickman and Zhang 1999; Verling et al. 2005), ballast sediment (Duggan et al. 2006; Radziejewska et al. 2006), or tested ballast management options such as irradiation and mid-ocean ballast exchange (Waite et al. 2003; Wonham et al. 2005). Fouling studies identified organisms likely to be transported on plastics (Barnes et al. 2004), antifouled surfaces (Floerl et al.

2004, 2005) and even the full-scale reproduction of early ship voyages (Carlton and Hodder 1995). In regions such as Australia, North America and Hawaii, it is estimated that between 55% and 85% of recorded marine NIS are fouling organisms introduced via fouling on vessel hulls or other floating structures (Wasson et al. 2001; Eldredge and Carlton 2002; Hewitt 2002), yet hull fouling remains largely unregulated in most countries (Gollasch 2002). Poor vessel maintenance, previously overlooked areas of ship hulls (e.g. sea-chest gratings), and even boat harbor design have all been shown to contribute to the increasing number of NIS propagules entering new regions and the frequency of their arrival (Floerl and Inglis 2003; Coutts and Taylor 2004; Floerl and Inglis 2005; Coutts and Dodgshun 2007; Chap. 6, Hewitt et al.).

In order to assess the risk of each transport vector, some measure of propagule pressure is necessary. Where direct measurements are difficult or costly (as described above) surrogates for propagule pressure are used. In terrestrial plant systems simple surrogates such as the number of visitors to nature reserves (Lonsdale 1999), or amount of trade (Thuiller et al. 2005) have been used. Drake and Lodge (2004) provide a useful model for assessing the risks of vector transfer of invasive propagules in ballast water. They constructed a ship transportation model and linked it to existing hotspots of marine invaders. They argue that reducing the average probability of an individual ship visit causing an invasion would be a more effective control measure than eliminating key ports that are significant epicenters of invasive propagule spread. Using proxies (such as shipping activity) for propagule pressure must be done with caution, however, since the number, diversity and quality of transported propagules will vary with factors such as vessel type and trip duration (Verling et al. 2005). Moreover, interregional transport may be possible without the obvious vectors such as commercial shipping (Wasson et al. 2001). There remains a clear paucity of studies of invasion vectors for marine systems relative to terrestrial ones (Verling et al. 2005), with particular need to survey actual large commercial vessels traveling rapidly and frequently across the globe (e.g. Coutts et al. 2006).

7.7 Manipulating Propagule Supply

Increasing the supply of invasive propagules is considered highly likely to increase invasion success, but the relative importance of this factor compared to others is not well understood. General discussions of invasion biology consistently emphasize the value of independently manipulating both the supply of invaders and the recipient community or environment in order to test the relative importance of each (e.g. Lonsdale 1999; Naeem et al. 2000). Manipulating two interacting components (i.e. invader and community) also permits interesting cross-factorial experiment designs. Propagules can be manipulated in regards to density, frequency, or the timing of inoculation relative to a disturbance, and crossed with any community treatment of interest. In particular, density and frequency are critical variables in supply-driven systems (Puth and Post 2005), and, if understood, should be considered in management plans for transport vectors such as shipping (Ruiz et al. 2000). Here, we describe an array of methods developed to experimentally manipulate invader propagules and their environment in marine systems. Terrestrial ecologists have enjoyed similar techniques for many decades, but marine examples have been less forthcoming due to the logistical difficulties of working with propagules in a fluid environment. Some of the techniques we describe have only been developed in recent years, and their potential applications are largely untapped. The suitability of each method will depend on the type of community, the invader, and the specific question being asked, but this summary aims to categorize and contextualize techniques so that marine ecologists may readily consider the range of tools at their disposal.

7.7.1 Manipulating Propagule at the Larval Stage

The most comprehensive tests of invasibility simulate invasion from the time propagules enter a recipient environment. Most marine organisms disperse in a planktonic life-history phase, and tests of propagule pressure for these species should ideally begin with the arrival of larvae or spores in the water column. This is akin to the seed-addition experiments conducted with terrestrial plants (e.g. Tilman 1997), which, although the most common form of invasion test in the terrestrial domain, have few marine analogies.

Over the past five years, however, several studies have developed novel methods of propagule manipulation. By injecting newly spawned larvae of the invasive bryozoan *Bugula neritina* into containers housing replicate sessile invertebrate assemblages (Fig. 7.4), Clark and Johnston (2005) were able to manipulate larval supply and community properties independently. They tested various hypotheses about effects of disturbance relative to a known pulse of propagules, and found that recruitment was largely explained by disturbance mediating resource availability. There are several advantages in using this technique compared to manipulating adult densities. It provides a more realistic test of invasion, since potential invaders are subject to biotic and abiotic filters at both the larval and adult life-history stages. Predation of larvae by incumbents, for example, may represent a significant component of invasion resistance in sessile invertebrates (Holloway and Keough 2002), but would be overlooked if we were only to consider invaders as adults. Additionally, many invertebrate larvae are known to exhibit active larval choice (Keough and Downes 1982), so allowing larvae to display small-scale habitat preference (as opposed to direct transplantation) increases their chance of survival and provides a more realistic invasion scenario. Controlled larval dosing has now been achieved with six species of bryozoa (Piola and Johnston, unpublished data), and is likely to be possible with many organisms that disperse in a larval phase.

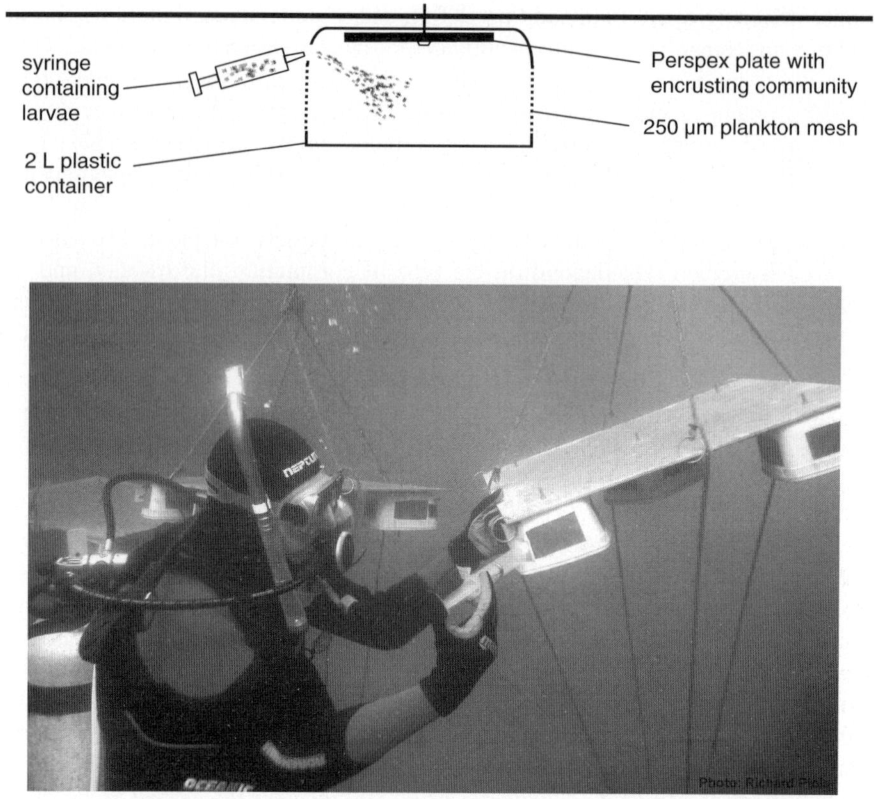

Fig. 7.4 Larval dosing technique developed by Clark and Johnston (2005). Recently spawned bryozoan larvae were drawn into a syringe and injected into a closed 2L plastic container. The container housed an experimental assemblage, and had mesh sides to allow water exchange. The bottom of the container was removed after 48 h, by which time most larvae had settled, so assemblages could experience natural environmental conditions

Valentine and Johnson (2003, 2005) took a different approach and enhanced the density of spores of the invasive algae *Undaria pinnatifida* by hanging mesh bags containing fertile sporophylls over experimental plots. The algae were replaced periodically to maintain a regular source of propagules. While this method didn't strictly control or quantify propagule supply, the authors were able to enhance supply over a relatively large area (16 m^2), which would have been impractical had they attempted to control supply precisely. Another study packaged seeds of marine grasses into biodegradable bags and buried them in the substrate (Dethier and Hacker 2005). When the bags degraded the seeds were sown, and they discovered that invasion patterns were more influenced by habitat quality than species interactions.

7.7.2 Manipulating Adult Invaders

Ecologists can ask different questions about an invader by manipulating established adults of the species. These studies typically look at growth of the invader under particular biotic and abiotic conditions, and/or impacts of the invader on the resident community (see Sect. 7.4). Such experiments may be the only manipulative option for organisms whose larvae are difficult or impossible to obtain. They are particularly relevant to the study of supply-side invasion biology in cases where invaders are likely to be transported as adults. This is possible for mobile hitch-hikers such as crustaceans, echinoderms and fish that may be transported in vessel cavities, or sessile species capable of regenerating from transported fragments (e.g. seaweeds and sponges). These studies generally aim to manipulate the densities of the invader population above natural levels, and generally use methods such as transplantation including caging.

Transplants are when species are relocated and assembled in a new environment, and are often used to manipulate densities of both the invader and native species. The methods used for relocation depend on the type of organism and the creativity of the researcher, but examples for sessile species include translocating panels with communities attached (Osman and Whitlatch 2004), gluing mussels onto panels (Stachowicz et al. 2002), and encouraging organisms to attach vegetatively to new substrates (Agius 2007). Vegetative attachment is a useful tool that gives the researcher much control over initial community composition. Rubber bands have been used to secure colonial ascidians (Aguis 2007) and sponges (Johnston and Clark 2007) to substrates while they attach, usually over a period of one to two weeks. Another technique for sessile species is to secure fragments onto a panel by stretching a hair-net over the surface (Fig. 7.5), which achieved a 50–90% survival rate in fragments of four encrusting bryozoan species (Piola and Johnston, unpublished data). Other methods include placing solitary ascidians inside cylinders in the field (Castilla et al. 2004), and manipulating densities of kelp and algae by attaching them to rope (Levin et al. 2002). Such studies test invasibility by how well a transplanted species survives and grows in a new location (e.g. Ruesink 2007) but may overestimate the success of any one propagule that would usually have arrived in the larval phase.

Many of the higher profile marine invaders are mobile organisms, particularly predators such as seastars and crabs. The mobility of these invaders adds several levels of difficulty to the already challenging task of manipulating the abundance of any marine invader. Several researchers have successfully manipulated mobile species usually by caging them within small patches of the marine environment or by tethering them to a patch. DeRivera et al. (2005) tethered the invasive green crab *Carcinus maenas* at 64 sites in 8 different bays on the eastern coast of North America. Each crab was fitted with a flexible steel halter and leash which was then attached to a lead weight. Biotic resistance in the form of predation by the native crab *Callincetes sapidus* was found to be a significant limiting factor on the abundance and geographic range of this species. Another recent study sheds much light on supply-side invasion

Fig. 7.5 Using hairnets to add non-indigenous species of sessile invertebrates. Fragments of four species of encrusting bryozoans were secured to a Perspex panel with a hairnet. After 45 days the colonies had vegetatively attached and grown to occupy most of the panel

ecology by manipulating recipient community diversity and the density of adult invaders of the porcelain crab (Hollebone and Hay 2007). The study added adult porcelain crabs to field-mesocosms and monitored larval recruitment of the invader through time. They found that the initial biotic resistance of diverse native assemblages was overwhelmed by increasing propagule pressure of invaders.

7.8 Conclusions

In this chapter we have discussed the role of propagule pressure in bioinvasions, and described recent attempts to manipulate propagule pressure in the marine environment. Research in terrestrial and freshwater systems indicates a clear positive relationship between propagule pressure and invasion success. Experimental work in marine systems suggests that the success of individual propagules will be enhanced by disturbance and increasing resource availability. Many of the recent marine studies represent significant advances in our knowledge of the invasion process, and have implications for the management and prevention of marine pest incursions. Nonetheless, our understanding of effects of variation in propagule pressure remains limited, and we now need to explore the processes in more detail. For example, are community properties that confer invasion resistance only important until propagule supply reaches a critical threshold? Over what range of community types and inoculation events is the relationship between supply and invasion linear, and when and how does it diverge? Further studies that manipulate invader densities from the very initial propagule stage are likely to provide us with even greater comprehension. We now echo the call of Ruiz and Carlton (2003) for more complex studies of propagule pressure that examine frequency, intensity, genetic and geographic variability of propagule arrival.

In caution, any test of invasibility that relocates an invasive species must be careful not to contribute to the range expansion of that organism. Most studies do this by deploying only one sex or only juveniles of sexually reproducing animals. For organisms that can reproduce asexually (e.g. the green alga *Caulerpa taxifolia*) it is probably only safe to manipulate invader densities in areas where the species is already known to occur. This may limit the ability to fully assess invader impacts but it does not negate the usefulness of experimental studies of invasion. With marine invasion frequency increasing rapidly, there has never been a more pressing time for us to think creatively about invasion biology, and to keep our feet a little wet.

References

Agius BP (2007) Spatial and temporal effects of pre-seeding plates with invasive ascidians: growth, recruitment and community composition. J Exp Mar Biol Ecol 342:30–39

Barnes DKA, Warren NL, Webb K, Phalan B, Reid K (2004) Polar pedunculate barnacles piggy-back on pycnogona, penguins, pinniped seals and plastics. Mar Ecol Prog Ser 284:305–310

Bravo MA, Cameron B, Metaxas A (2007) Salinity tolerance in the early larval stages of *Carcinus maenas* (Decapoda, Brachyura), a recent invader of the Bras d'Or Lakes, Nova Scotia, Canada. Crustaceana 80:475–490

Bulleri F, Abbiati M, Airoldi L (2006) The colonisation of human-made structures by the invasive alga *Codium fragile* ssp *tomentosoides* in the north Adriatic Sea (NE Mediterranean). Hydrobiologia 555:263–269

Calabrese A, Collier R, Nelson D, MacInnes J (1973) The toxicity of heavy metals to embryos of the American oyster *Crassostrea virginica*. Mar Biol 18:162–166

Caley M, Carr M, Hixon M, Hughes T, Jones G, Menge B (1996) Recruitment and the local dynamics of open marine populations. Annu Rev Ecol Syst 27:277–500

Carlton JT (1996) Biological invasions and cryptogenic species. Ecology 77:1653–1655

Carlton JT, Hodder J (1995) Biogeography and dispersal of coastal marine organisms – experimental studies on a replica of a 16th-century sailing vessel. Mar Biol 121:721–730

Carriglio D, Fanelli G, Rubino F (2004) First record of the alien gastropod *Melibe fimbriata* (Opistobranchia: Tethyidae) in the Taranto seas (Mediterranean Sea). J Mar Biol Assoc U K 84:1067–1068

Castilla JC, Guinez R, Caro AU, Ortiz V (2004) Invasion of a rocky intertidal shore by the tunicate *Pyura praeputialis* in the Bay of Antofagasta, Chile. Proc Natl Acad Sci USA 101:8517–8524

Charmantier G, Gimenez L, Charmantier-Daures M, Anger K (2002) Ontogeny of osmoremilation, physiological plasticity and larval export strategy in the grapsid crab *Chasmagnathus granulata* (Crustacea, Decapoda). Mar Ecol Prog Ser 229:185–194

Chu KH, Tam PF, Fung CH, Chen QC (1997) A biological survey of ballast water in container ships entering Hong Kong. Hydrobiologia 352:201–206

Clark GF, Johnston EL (2005) Manipulating larval supply in the field: a controlled study of marine invasibility. Mar Ecol Prog Ser 298: 9–19

Cohen AN, Carlton JT (1998) Accelerating invasion rate in a highly invaded estuary. Science 279:555–559

Cohen AN, Harris LH, Bingham BL, Carlton JT, Chapman JW, Lambert CC, Lambert G, Ljubenkov JC, Murray SN, Rao LC, Reardon K, Schwindt E (2005) Rapid assessment survey for exotic organisms in southern California bays and harbors, and abundance in port and non-port areas. Biol Invas 7:995–1002

Connor P (1972) Acute toxicity of heavy metals to some marine larvae. Mar Pollut Bull 3:190–192

Corkum LD, Sapota MR, Skora KE (2004) The round goby, *Neogobius melanostomus*, a fish invader on both sides of the Atlantic Ocean. Biol Invas 6:173–181

Coutts ADM, Dodgshun T (2007) A preliminary assessment of the nature and extent of fouling in ships' sea-chests in New Zealand. Mar Pollut Bull 54:875–886

Coutts ADM, Taylor MD (2004) A preliminary investigation of biosecurity risks associated with biofouling on merchant vessels in New Zealand. N Z J Mar Freshwater Res 38:215–229

Coutts ADM, Taylor MD, Hewitt CL (2006) Novel method for assessing the en route survivorship of biofouling organisms on various vessel types Mar Pollut Bull 54:97–100

Dafforn KA, Glasby TM, Johnston EL (2008) Differential effects of tributyltin and copper antifoulants on recruitment of non-indigenous species. Biofouling 24:23–33

Deagle BE, Bax N, Patil JG (2003) Development and evaluation of a PCR-based test for detection of *Asterias* (Echinodermata: Asteroidea) larvae in Australian plankton samples from ballast water. Mar Freshwater Res 54:709–719

deRivera CE, Ruiz GM, Hines AH, Jivoff P (2005) Biotic resistance to invasion: native predator limits abundance and distribution of an introduced crab. Ecology 86:3364–3376

deRivera CE, Hitchcock NG, Teck SJ, Steves BP, Hines AH, Ruiz GM (2007) Larval development rate predicts range expansion of an introduced crab. Mar Biol 150:1275–1288

Dethier MN, Hacker SD (2005) Physical factors vs. biotic resistance in controlling the invasion of an estuarine marsh grass. Ecol Appl 15:1273–1283

Dickman M, Zhang FZ (1999) Mid-ocean exchange of container vessel ballast water. 2: Effects of vessel type in the transport of diatoms and dinoflagellates from Manzanillo, Mexico, to Hong Kong, China. Mar Ecol Prog Ser 176:253–262

Drake JM, Lodge DM (2004) Global hotspots of biological invasions: evaluating options for ballast-water management. Proc R Soc Ser B 271:575–580

Drake JM, Lodge DM (2006) Alee effects, propagule pressure and the probability of establishment: risk analysis for biological invasions. Biol Invas 8:365–375

Dudas SE, Dower JF (2006) Reproductive ecology and dispersal potential of varnish clam *Nuttallia obscurata*, a recent invader in the Northeast Pacific Ocean. Mar Ecol Prog Ser 320:195–205

Duggan IC, Bailey SA, van Overdijk CDA, MacIsaac HJ (2006) Invasion risk of active and diapausing invertebrates from residual ballast in ships entering Chesapeake Bay. Mar Ecol Prog Ser 324:57–66

Dupont L, Jollivet D, Viard F (2003) High genetic diversity and ephemeral drift effects in a successful introduced mollusc (Crepidula fornicata: Gastropoda). Mar Ecol Prog Ser 253:183–195

Eldredge LG, Carlton J (2002) Hawaiian marine bioinvasions: a preliminary assessment. Pac Sci 56:211–212

Floerl O, Inglis GJ (2003) Boat harbour design can exacerbate hull fouling. Aust Ecol 28:116–127

Floerl O, Inglis GJ (2005) Starting the invasion pathway: the interaction between source populations and human transport vectors. Biol Invas 7:589–606

Floerl O, Pool TK, Inglis GJ (2004) Positive interactions between nonindigenous species facilitate transport by human vectors. Ecol Appl 14:1724–1736

Floerl O, Inglis GJ, Marsh HM (2005) Selectivity in vector management: an investigation of the effectiveness of measures used to prevent transport of non-indigenous species. Biol Invas 7:459–475

Gollasch S (2002) The importance of ship hull fouling as a vector of species introductions into the North Sea. Biofouling 18:105–121

Gunasekera RM, Patil JG, McEnnulty FR, Bax NJ (2005) Specific amplification of mt-COI gene of the invasive gastropod *Maoricolpus roseus* in planktonic samples reveals a free-living larval life-history stage. Mar Freshwater Res 56:901–912

Gust N, Inglis GJ (2006) Adaptive multi-scale sampling to determine an invasive crab's habitat usage and range in New Zealand. Biol Invas 8:339–353

Harding JM, Mann R, Ware Kilduff C (2007) The effects of female size on fecundity in a large marine gastropod *Rapana venosa* (Muricidae). J Shellfish Res 26:33–42

Hayward BW, Stephenson AB, Morley M, Riley JL, Grenfell HR (1997) Faunal changes in Waitemata Harbour sediments, 1930s-1990s. J R Soc N Z 27:1–20

Herbert RJH, Southward AJ, Sheader M, Hawkins SJ (2007) Influence of recruitment and temperature on distribution of intertidal barnacles in the English Channel. J Mar Biol Ass U K 87:487–499

Hewitt CL (2002) The distribution and biodiversity of tropical Australian marine bioinvasions. Pac Sci 56:213–222

Hewitt C, Campbell M, Thresher R, Martin R (1999) Marine biological invasions of Port Phillip Bay, Victoria. Technical Report No 20, CSIRO Marine Research, Centre for Research on Introduced Marine Pests, Hobart, Australia, p 344

Hollebone AL, Hay ME (2007) Propagule pressure of an invasive crab overwhelms native biotic resistance. Mar Ecol Prog Ser 342:191–196

Holloway MG, Keough MJ (2002) An introduced polychaete affects recruitment and larval abundance of sessile invertebrates. Ecol Appl 12:1803–1823

Jelmert A, Van Leeuwen J (2000) Harming local species or preventing the transfer of exotics? Possible negative and positive effects of using zinc anodes for corrosion protection of ballast water tanks. Water Res 34:1937–1940

Johnston EL, Clark GF (2007) Recipient environment more important than community composition in determining the success of an experimental sponge transplant. Restor Ecol 15

Jones A, Gensemer RW, Stubblefield WA, Van Genderen E, Dethloff GM, Cooper WJ (2006) Toxicity of ozonated seawater to marine organisms. Environ Toxicol Res 25:2683–2691

Keough MJ, Downes BJ (1982) Recruitment of marine invertebrates: the role of active larval choices and early mortality. Oecologia 54:348–352

Kolar CS, Lodge DM (2001) Progress in invasion biology: predicting invaders. Trends Ecol Evol 16:199–204

Leung B, Drake JM, Lodge DM (2004) Predicting invasions: propagule pressure and the gravity of Alee effects. Ecology 85:1651–1660

Levin PS, Coyer JA, Petrik R, Good TP (2002) Community-wide effects of nonindigenous species on temperate rocky reefs. Ecology 83:3182–3193

Lockwood JL, Cassey P, Blackburn T (2005) The role of propagule pressure in explaining species invasions. Trends Ecol Evol 20:223–228

Lonsdale WM (1999) Global patterns of plant invasions and the concept of invasibility. Ecology 80:1522–1536

Mackie JA, Keough MJ, Christidis L (2006) Invasion patterns inferred from cytochrome oxidase I sequences in three bryozoans, *Bugula neritina*, *Watersipora subtorquata*, and *Watersipora arcuata*. Mar Biol 149:285–295

Mann R, Harding JM (2003) Salinity tolerance of larval *Rapana venosa*: implications for dispersal and establishment of an invading predatory gastropod on the North American Atlantic Coast. Biol Bull 204:96–103

Martel C, Viard F, Bourguet D, Garcia-Meunier P (2004) Invasion by the marine gastropod *Ocinebrellus inornatus* in France. II. Expansion along the Atlantic coast. Mar Ecol Prog Ser 273:163–172

Marzano CN, Liaci SL, Fianchini A, Gravina F, Mercurio M, Corriero G (2003) Distribution, persistence and change in the macrobenthos of the lagoon of Lesina (Apulia, southern Adriatic Sea). Oceanol Acta 26:57–66

Miller AW, Ruiz GM, Minton MS, Ambrose RF (2007) Differentiating successful and failed molluscan invaders in estuarine ecosystems. Mar Ecol Prog Ser 332:41–51

Naeem S, Knops JM, Tilman D, Howe KM, Kennedy T, Gale S (2000) Plant diversity increases resistance to invasion in the absence of covarying extrinsic factors. Oikos 91:97–108

Osman RW, Whitlatch RB (2004) The control of the development of a marine benthic community by predation on recruits. J Exp Mar Biol Ecol 311:117–145

Patti FP, Gambi MC (2001) Phylogeography of the invasive polychaete Sabella spallanzanii (Sabellidae) based on the nucleotide sequence of internal transcribed spacer 2 (ITS2) of nuclear rDNA. Mar Ecol Prog Ser 215:169–177

Piola RF, Johnston EL (2006a) Differential resistance to extended copper exposure in four introduced bryozoans. Mar Ecol Prog Ser 311:103–114

Piola RF, Johnston EL (2006b) Differential tolerance to metals among populations of the introduced bryozoan *Bugula neritina*. Mar Biol 148:997–1010

Piola RF, Johnston EL (2008) Pollution reduces native diversity and increases invader dominance in marine hard-substrate communities. Diversity Distrib 14(2):329–342

Puth L, Post D (2005) Studying invasion: have we missed the boat? Ecol Lett 8:715–721

Radziejewska T, Gruszka P, Rokicka-Praxmajer J (2006) A home away from home: a meiobenthic assemblage in a ship's ballast water tank sediment. Oceanologia 48:259–265

Rajagopal S, Venugopala VP, van der Velde G, Jenner HA (2006) Greening of the coasts: a review of the *Perna viridis* success story. Aquat Ecol 40:273–297

Roman J (2006) Diluting the founder effect: cryptic invasions expand a marine invader's range. Proc R Soc B 273:2453–2459

Roman J, Darling JA (2007) Paradox lost: genetic diversity and the success of aquatic invasions. Trends Ecol Evol 22:454–464

Rouget M, Richardson DM (2003) Inferring process from pattern in plant invasions: a semimechanistic model incorporating propagule pressure and environmental factors. Am Nat 162:713–724

Roy K, Jablonski D, Valentine JW (2001) Climate change, species range limits and body size in marine bivalves. Ecol Lett 4:366–370

Roy K, Jablonski D, Valentine JW (2002) Body size and invasion success in marine bivalves. Ecol Lett 5:163–167

Ruesink JL (2007) Biotic resistance and facilitation of a non-native oyster on rocky shores. Mar Ecol Prog Ser 331:1–9

Ruiz GM, Carlton JT (2003) Invasion vectors: a conceptual framework for management. In: Ruiz G, Carlton JT (eds) Invasive species: vectors and management strategies. Kluwer Academic, Dordrecht, pp 529–547

Ruiz GM, Fofonoff PW, Carlton JT, Wonham MJ, Hines AH (2000) Invasion of coastal marine communities in North America: apparent patterns, processes, and biases. Annu Rev Ecol Syst 31:481–531

Sax DF, Stachowicz JJ, Brown JH, Bruno JF, Dawson MN, Gaines SD, Grosberg RK, Hastings A, Holt RD, Mayfield MM, O'Connor MI, Rice WR (2007) Ecological and evolutionary insights from species invasions. Trends Ecology Evolut 22:465–471

Schiff K, Diehl D, Valkirs AO (2004) Copper emissions from antifouling paint on recreational vessels. Mar Pollut Bull 48:371–377

Shefer S, Abelson A, Mokady O, Geffen E (2004) Red to Mediterranean Sea bioinvasion: natural drift through the Suez Canal, or anthropogenic transport? Mol Ecol 13:2333–2343

Srinivasan M, Swain GW (2007) Managing the use of copper-based antifouling paints. Environ Manag 39:423–441

Stachowicz JJ, Byrnes JE (2006) Species diversity, invasion success, and ecosystem functioning: disentangling the influence of resource competition, facilitation, and extrinsic factors. Mar Ecol Prog Ser 311:251–262

Stachowicz JJ, Fried H, Osman RW, Whitlatch RB (2002) Biodiversity, invasion resistance, and marine ecosystem function: reconciling pattern and process. Ecology 83:2575–2590

Strathmann MF, Strathmann RR (2006) A vermetid gastropod with complex intracapsular cannibalism of nurse eggs and sibling larvae and a high potential for invasion. Pac Sci 60:97–108

Tamburri MN, Wasson K, Matsuda M (2002) Ballast water deoxygenation can prevent aquatic introductions while reducing ship corrosion. Biol Conserv 103:331–341

Thuiller W, Richardson DM, Pysek P, Midgley G, Hughes G, Rouget M (2005) Niche-based modelling as a tool for predicting the risk of alien plant invasions at a global scale. Global Change Biol 11:2234–2250

Tilman D (1997) Community invasibility, recruitment limitation, and grassland biodiversity. Ecology 78:81–92

Underwood AJ, Keough MJ (2001) Supply-side ecology: the nature and consequences of variations in recruitment of intertidal organisms. In: Bertness M, Gaines S, Hay M (eds) Marine community ecology. Sinauer Associates, Sunderland, Maine, pp 183–200

Valentine JP, Johnson CR (2003) Establishment of the introduced kelp Undaria pinnatifida in Tasmania depends on disturbance to native algal assemblages. J Exp Mar Biol Ecol 295:63–90

Valentine JP, Johnson CR (2005) Persistence of the exotic kelp *Undaria pinnatifida* does not depend on sea urchin grazing. Mar Ecol Prog Ser 285:43–55

Valkirs AO, Seligman PF, Haslbeck E, Caso JS (2003) Measurement of copper release rates from antifouling paint under laboratory and in situ conditions: implications for loading estimation to marine water bodies. Mar Pollut Bull 46:763–779

Verling E, Ruiz GM, Smith LD, Galil B, Miller AW, Murphy KR (2005) Supply-side invasion ecology: characterizing propagule pressure in coastal ecosystems. Proc R Soc B 272:1249–1256

Viard F, Ellien C, Dupont L (2006) Dispersal ability and invasion success of *Crepidula fornicata* in a single gulf: insights from genetic markers and larval-dispersal model. Helgol Mar Res 60:144–152

Waite TD, Kazumi J, Lane PVZ, Farmer LL, Smith SG, Smith SL, Hitchcock G, Cap TR (2003) Removal of natural populations of marine plankton by a large-scale ballast water treatment system. Mar Ecol Prog Ser 258:51–63

Wasson K, Zabin CJ, Bedinger L, Diaz MC, Pearse JS (2001) Biological invasions of estuaries without international shipping: the importance of intraregional transport. Biol Conserv 102:143–153

Watson DI, O'Riordan RM, Barnes DKA, Cross T (2005) Temporal and spatial variability in the recruitment of barnacles and the local dominance of *Elminius modestus* Darwin in SW Ireland. Estuarine Coastal Shelf Sci 63:119–131

Wonham MJ, Lewis MA, MacIsaac HJ (2005) Minimizing invasion risk by reducing propagule pressure: a model for ballast-water exchange. Front Ecol Environ 3:473–478

Zabin CJ, Zardus J, Bettini Pitombo F, Fread V, Hadfield MG (2007) A tale of three seas: consistency of natural history traits in a Caribbean-Atlantic barnacle introduced to Hawaii. Biol Invas 9:523–544

Zardi GI, McQuaid CD, Teske PR, Barker NP (2007) Unexpected genetic structure of mussel populations in South Africa: indigenous *Perna perna* and invasive *Mytilus galloprovincialis*. Mar Ecol Prog Ser 337:135–144

Zardus JD, Hadfield MG (2005) Multiple origins and incursions of the Atlantic barnacle *Chthamalus proteus* in the Pacific. Mol Ecol 14:3719–3733

Chapter 8
Differentiating Successful and Failed Invaders: Species Pools and the Importance of Defining Vector, Source and Recipient Regions

A. Whitman Miller and Gregory M. Ruiz

8.1 Introduction

Attempts to understand the dynamics of biological invasions continue to abound in aquatic and terrestrial ecosystems. Identifying the biological attributes of successful invaders, or what makes a good invader, are among the most tantalizing questions still to be answered, especially in marine ecosystems. Numerous studies across a range of taxonomic groups have examined species' characteristics to determine whether certain species level factors strongly differentiate successful from failed invaders (see Rejmánek and Richardson 1996; Williamson and Fitter 1996; Reichard and Hamilton 1997; Miller 2000; Kolar and Lodge 2002; Prinzing et al. 2002; Cassey et al. 2004a,b; Miller et al. 2007). At the heart of these analyses is the comparison of successful and failed species pools, which are defined in various ways with specific consequences for the inferences that can result.

When trying to understand the effects of species characteristics on invasion outcome, most studies compare physiological tolerances, life history characteristics, and behavior of successful and failed invaders. Although it is certainly valid to compare any two groups to understand differences in their respective attributes, only a subset of such comparisons can answer questions about the invasion process. More specifically, invasions have a specific context and result from interactions among source regions, recipient regions, and transfer mechanisms (vectors). Thus, comparing invaders from one source region to non-invaders from a different source region may tell us little about attributes of successful invaders, because the latter group may not share the same opportunities for transfer and invasion, thereby introducing additional variables and confounding interpretation.

In this chapter we examine the role of source region, vector, and recipient region in evaluating successful vs unsuccessful invasions. First, we provide a general framework for identifying introduced species assemblages according to their vectors and sources. Second, we provide a conceptual model to illustrate explicitly the effects of controlling for vector, source region, and recipient region when comparing successful and failed invaders. Third, we review approaches used in some previous analyses to identify biological characteristics of successful invaders,

G. Rilov, J.A. Crooks (eds.) *Biological Invasions in Marine Ecosystems.*
Ecological Studies 204,

including multiple taxonomic groups and ecosystems. Throughout, our intention is to highlight the potential effects of specific types of comparisons on conclusions about invaders' attributes.

In this chapter, we consider a successful invasion to be the establishment of a self-sustaining population in a non-native region, regardless of abundance, geographic range, or impact. Thus, species are considered successful invaders, independent of any degree or classification of invasiveness (Prinzing et al. 2002, but see Richardson et al. 2000).

8.2 Identifying the Roles of Source Region, Recipient Region and Vector

Biological invasions result from multiple vectors and geographic sources. In addition to natural range extensions, recipient regions receive species from one or more regions via multiple human-mediated mechanisms, operating either simultaneously or at different times. For any given recipient region, the diversity of species transfers (i.e., the species richness of inoculants) and genetic diversity can be increased by drawing from (1) more than one source region, (2) more than one vector, and (3) more than one time period. For simplicity, we exclude the temporal dimension from further discussion in this paper and examine the implications of different sources and vectors on the analysis of invasion patterns.

For a single recipient region, inoculant diversity clearly results from the combined (summed) contributions across multiple source regions and vectors. In a simple framework, Fig. 8.1a illustrates how source regions and vectors yield unique, component species assemblages (pools) for delivery to a single recipient region. Thus, the species pool delivered from the first vector (V_1) and source region (S_1) will differ from that delivered by either V_1 operating from other source regions ($S_{2,3,n}$) or another vector ($V_{2,3,n}$) in the same source region (S_1). For example, species that were transferred to San Francisco Bay in ships' ballast water from Japan differ from those that arrived in ballast from China, and both differ from species that arrived on outer hulls of vessels or with live seafood (e.g., oysters) from Japan. Certainly there may be some overlap in species composition among cells in Fig. 8.1a, but each compartment represents a different pool of species.

In theory, understanding the full species pool delivered to a recipient region requires an accounting of all species inoculations according to vector and source (vector × source) combinations. In reality, such complete assessments are impossible. However, the additive but distinctive nature of inoculation across different compartments underscores several important aspects of propagule supply. First, the taxa transferred to one recipient region will differ by both vector and source region. Second, the importance (and even existence) of specific vectors and source regions will differ strongly among recipient regions ($R_{1,2,n}$). This latter aspect is portrayed by adding a third axis (Fig. 8.1b), whereby some component cells may be completely inactive in one source region but operational in others.

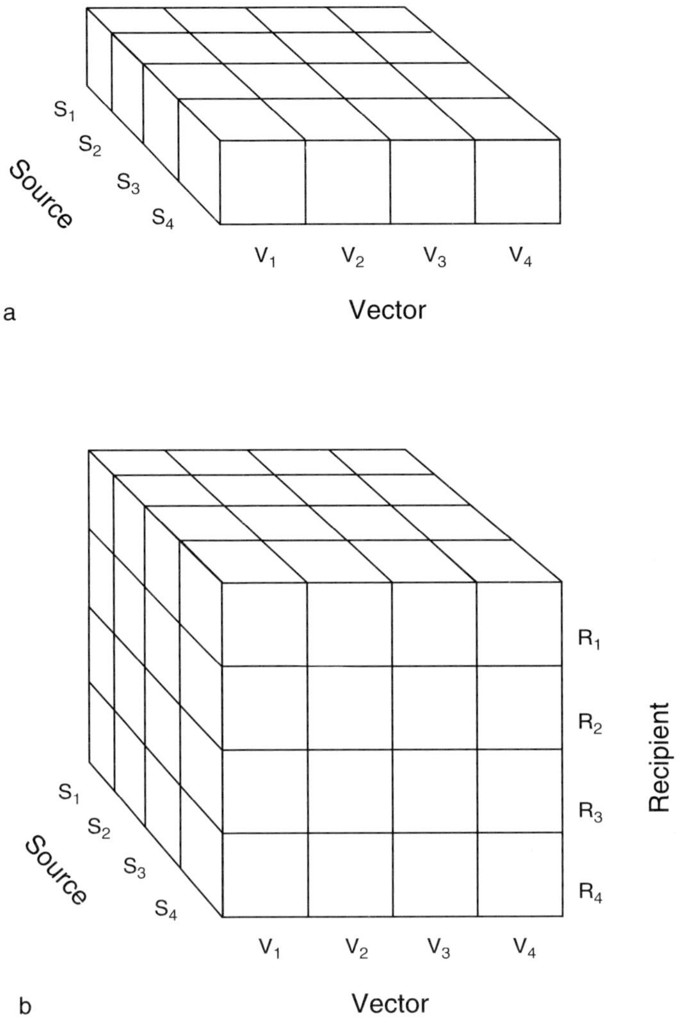

Fig. 8.1 Potential species assemblages entrained by vector (V) from source region (S) and introduced to recipient region (R). **a** Sixteen theoretical assemblages arising from four source regions and four vectors and introduced to a single recipient region. **b** How the same (source × vector) combinations can be introduced to more than one recipient region. For any given (source × vector) combination, time, distance, and degree of ecosystem matching will vary across recipient region, resulting in different patterns of invasion success and failure

Although this is perhaps an obvious accounting framework, it has some important implications for defining relevant species pools for analyses of invasion processes and patterns. For example, whereas one species entrained by a ballast water vector may survive transit only to die of salinity exposure after introduction, another species dies in transport. Each case ends in invasion failure, but the latter is an effect of the vector and the former a consequence of the recipient region. Still other species are never moved in ships' ballast water, so fail to colonize for yet another reason. Recognizing and disentangling the effects of sources, vectors, and recipient ecosystems is crucial for making informed comparisons among groups of species, especially when trying to identify biological attributes that lead to invasion success and failure. The remainder of this chapter is devoted to exploration and discussion of this issue.

8.3 Modeling Invasion State Space – Defining and Comparing Species Pools

When investigating biological attributes that correspond to invasion success and failure, it is vital to focus on the context and elements of the invasion process. Carlton (1979), Kolar and Lodge (2001) and others have described the invasion process as a series of sequential steps or filters whereby a species incrementally succeeds or fails. The generic invasion sequence includes: (1) vector entrainment, (2) transportation and survival, (3) introduction, (4) establishment and (5) subsequent spread. Several authors have specified the importance of treating the various stages of the invasion sequence separately (e.g., Kolar and Lodge 2001; Marchetti et al. 2004), since the attributes that enable a species to prevail at one stage may be quite different from those at another stage (e.g., establishment vs spread).

In their analyses of the biological attributes that lead to invasion success among parrots of the world, Cassey et al. (2004b) apply similar logic, arguing that invasion stages should be invoked to help define the discrete species pools used for comparison. For parrots, the putative biological and biogeographic factors correlated with invasion success differ, depending on whether the pool of successful parrot invaders (i.e., those that have successfully established self-sustaining populations beyond their native range) is compared with (1) those species released but which failed to establish, (2) species that were transported (with or without release) but which failed to establish, or (3) all parrot species of the world. By specifically modeling different species pool comparisons (i.e., comparison with (1) vs (2) or (3)) the authors demonstrate the genuine potential for drawing unsupported conclusions about invader attributes. Furthermore, Cassey et al. (2004a,b) explicitly constrained their analyses taxonomically by choosing all members of the parrot order. Additionally, they constrained the vector under consideration to intentional entrainment and transport beyond native ranges (i.e., the commercial pet trade), and the subsequent accidental release of parrots into the environment. Using a global-scale approach that includes all parrot species, regardless of source region (e.g., old world and new world parrots have equal weight in analyses), Cassey et al. (2004a) show that successful invaders tend to have broader diets and are more sedentary than failed invaders.

We agree with the type of explicit approach outlined above and suggest that invasion stages must always be considered in analyses that test for biological attributes of invaders. In general, studies investigating the contribution of biological attributes to invasion success should, when possible, constrain source and recipient regions and vector, ensuring appropriate comparisons that remove confounding (uncontrolled) variables. To understand more fully how biological attributes affect invasion success we must consider how a species, by dint of its biological attributes, interacts with: (1) its native environment – both the physical and biological aspects, (2) the vector by which the would-be invader is entrained, transported, and released into the receiving environment, and (3) the physical and biological aspects of the receiving environment.

To illustrate the consequence of selecting or constraining particular pathway parameters (source region, vector, and recipient region) for invasion analyses, we provide a conceptual model. Specifically, we identify theoretical pools of species for comparison and describe how different comparisons address different aspects of the invasion process. In practice, the exact membership of some species pools will often not be readily identifiable, and this is especially true for accidental introductions vs intentional introductions. Nevertheless, we believe the model has utility in that it highlights some of the limitations and pitfalls of incomplete knowledge, informing both the design and inference of analyses.

8.3.1 Species Pool Designations

The domain of potential introduced species to a recipient region was described above as the sum of species assemblages across all vectors and source regions (Fig. 8.1). Such a representation describes which species are theoretically associated with each compartment (vector × source), but it does not address the fate of these species at different stages of the invasion sequence (e.g., survival during transport or upon delivery). Disentangling the effects of source region, vector, and invasion stage is crucial to understanding both the invasion process and identifying key biological attributes that affect invasion success. Here we describe a conceptual model to identify specific successful and unsuccessful species pools at different stages of the invasion sequence, examining effects of source region, vector, recipient region.

Since individual invasions occur generally by inoculation from one discrete geographic region to another via some vector(s), we use Venn diagrams to define explicitly various sets of species that require consideration, or the potential invasion state space. These sets represent theoretical species pools whose intersections connote their relationship within the modeled invasion state space. Using this general approach, Fig. 8.2 graphically defines the complete domain of species (i.e., Total Source Diversity or SD) available in a single source region. SD_1 is the full set of species present in one source region, and its domain can be subdivided into the following: a set of species with biological attributes and capabilities of establishing self-sustaining populations (E_1) in a specific recipient region if introduced (R_1); the set of species that can be entrained by a particular vector (V_1); and the set of species with the potential to tolerate the rigors of transit in a particular vector (T_1).

SD_1 is not equal to S_1 of Fig. 8.1, as the latter represents those species that arrive by a particular vector to a particular recipient region, regardless of their survival in transit or following introduction. The sets or subdomains E_1, R_1, V_1, and T_1 are equivalent in Figs. 8.1–8.3.

Thus, potentially successful invaders that actually arrive to a recipient region are described by the intersection of sets V_1, T_1, and E_1 (i.e., $V_1 \cap T_1 \cap E_1$; Fig. 8.2, black). It is important to note that each of the subdomains V_1, T_1, and E_1 will overlap to a greater or lesser degree. For purposes of this and future discussion of this model, the following set notation will be used: a rectangle indicates the available universe of species in the source region (SD), the intersection of two sets is indicated with

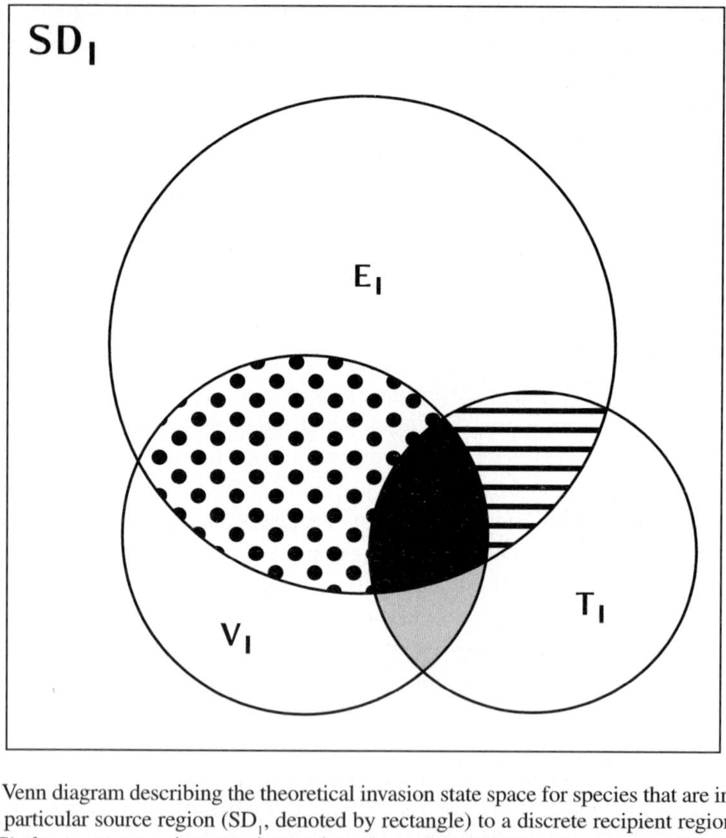

Fig. 8.2 Venn diagram describing the theoretical invasion state space for species that are introduced from one particular source region (SD_1, denoted by rectangle) to a discrete recipient region (R_1, not shown). *Circles* represent various species pools as they relate to the invasion process. E_1 refers to the theoretical pool of species residing in SD_1 that have the biological attributes and capacity to establish self sustaining populations in R_1, if introduced. V_1 represents the pool of species in SD_1 that are actually entrained by a particular vector. $E_1 \cap V_1$ is the subset of V_1 that could establish self-sustaining populations if introduced alive to a recipient region. The pool of species that could survive the rigors of a vector is denoted by T_1; however, only members of $V_1 \cap T_1$ are both entrained and can survive transit. The intersection all three species sets, $V_1 \cap T_1 \cap E_1$ (*black*), represent the theoretical pool of successful invaders that arrive to the recipient region

the symbol $\cap$, and a minus sign (–) denotes the complement of a set or intersection, i.e., the species excluded.

Table 8.1 provides a summary for how these variables combine and how these combinations are related to transfer opportunity and invasion outcome. For example, some species are entrained by a vector, tolerate transit, but fail to establish self-sustaining populations, $(V_1 \cap T1)–E_1$; Fig. 8.2, grey). Other species die in transit, $(V_1 \cap E_1)–T_1$ (Fig. 8.2, stippled). Another group of species has the capacity to survive transit and establish but is not ever entrained by V_1, $(T_1 \cap E_1)–V_1$ (Fig. 8.2, horizontal lines).

As noted by others (Prinzing et al. 2002; Cassey et al. 2004b; Pyšek et al. 2004), a crucial question is how "failed species" pools are designated for comparison with successful invaders (modeled here as $V \cap T \cap E$). Below, we explore various types of comparisons among species pools with our general conceptual model, adding increasing complexity from single to multiple source regions, vectors, and recipient regions. Throughout, our primary goal is to clarify the potential for confounding factors, and the inferences that can be drawn, when making specific comparisons among species assemblages.

8.3.2 *Species Pool Designations and Comparisons (Fixed Recipient, Source, and Vector)*

In this simple version of our model, which employs single subdomains, there are a variety of specific comparisons that seem most relevant. First, a comparison of successful invaders with the pool of species that are both entrained and introduced by the vector but which fail to establish self-sustaining populations, i.e., $V_1 \cap T_1 \cap E_1$ with $(V_1 \cap T_1)–E_1$ (Fig. 8.2, black vs grey, respectively). Because this analysis specifically constrains vector, source, and recipient region, it should yield biological differences associated with success and failure after introduction to the recipient region (R_1). A second comparison relates potentially successful invaders with species that were entrained, but which perished in transit before introduction (Fig. 8.2, black versus stippled). This comparison informs us of the effects of the vector (i.e., how species tolerate V_1). It should be noted that some members of $(V_1 \cap E_1)–T_1$ (Fig. 8.2, stippled) could be introduced via a different, non-lethal vector and go on

Table 8.1 Some expressions describing various states of entrainment (V_1), toleration of vector (T_1), and establishment (E_1), when species are taken from a source region (SD_1) and introduced to a recipient region (R_1)

Entrained, vector tolerated, established	Entrained, vector tolerated, not established	Entrained, vector not tolerated, not established	Not entrained, potential to tolerate vector, potential to establish	Invasion subdomains (V_1)	(T_1)	(E_1)
$V_1 \cap T_1 \cap E_1$				y	y	y
	$(V_1 \cap T_1)$-E_1			y	y	n
		$(V_1 \cap E_1)$-T_1		y	n	n
			$(T_1 \cap E_1)$-V_1	n	y	y

to establish self-sustaining populations in R_1. A third comparison, $V_1 \cap T_1 \cap E_1$ vs $(T_1 \cap E_1) - V_1$ (Fig. 2, black vs horizontal lines) contrasts successful invaders with a group of species that can establish self-sustaining populations in R_1 and survive transportation, but which are never entrained by the vector V_1. Importantly, members of $(T_1 \cap E_1) - V_1$ could also be successfully introduced to R_1 by a separate vector.

A somewhat coarser approach compares successful and failed invaders entrained by a particular vector, without regard to their ability to tolerate transport in the vector. In this comparison, the entire vector subdomain is compared with successful invaders $V_1 \cap T_1 \cap E_1$ (Fig. 8.2, black). This comparison does not distinguish whether failure to invade results from mortality in transit or upon arrival, but simply examines the joint outcome across both stages in the invasion sequence. In reality, detailed information may rarely exist for many vectors to measure such stage-specific effects, making this the best available approach.

For intentional introductions (e.g., fish stocking, aquaculture, horticulture, biocontrol) the selective actions of the vector on survivorship are largely neutralized, at least for the target species of interest. In these cases, the intersection of vector and survivorship is essentially complete (V=T), and a comparison of successful invaders ($V \cap T \cap E$ or $T \cap E$) vs introduced but failed species ($(V \cap T) - E$ or $T - E$) is a contrast for biological attributes that correlate with invasion success. This approach has been used with California fishes (Marchetti et al. 2004), although fish were derived from multiple source regions. While effective for examining traits associated with successful intentional introductions, there may still be differences by source region. A fully controlled comparison would explicitly contrast $E \cap T$ and $T - E$ separately for each source region (see below), to minimize any effect (weighting) by region, but this may present significant real-world challenges with respect to sample sizes needed for such analyses.

8.3.3 *Fixed Recipient and Source Regions, Multiple Vectors*

Depending on the number and types of vectors (V_1 to V_k) connecting SD_1 and R_1, the subset of potentially successful invaders will vary; however, the pool of species with the biological attributes and capabilities to succeed once introduced will remain constant for a given point in time E_1 (Fig. 8.3, black areas). Here we denote such species $E_{i,j}$, or the set of species from the *j*-th source region that could become established in the *i*-th recipient region, where $E_{1,1}$ represents the pool of species occurring in SD_1 with the biological capacity to establish in R_1. As the number of vectors connecting SD_1 with R_1 increases, a greater proportion of species with capacity to survive in R_1 will be sampled (Fig. 8.3). The species pool is described by the expression: $\sum_{k=1}^{k} E_{1,1} \cap V_k \cap T_k$, where K types of vectors are possible.

Since the vector is selective at two levels, both by the number and types of species it entrains and the rates of mortality it imposes on those species, vector type has direct influence on species introductions. Therefore, for each subset of successful

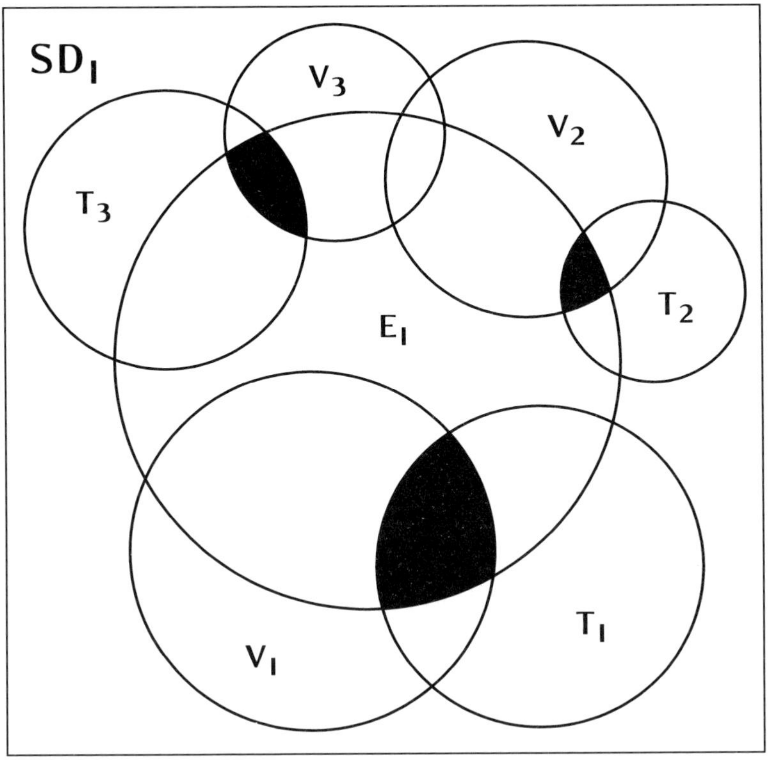

Fig. 8.3 Subsets of successful invaders (*black*) that emanate from a single source region (SD_1) and that are introduced via multiple vectors (V_1 to V_n) to a single recipient region (R_1). Associated abilities to tolerate vector-specific transits are depicted as (T_1 to T_n). The sum of successful invaders across vectors is described by the following expression: $\sum_{i=1}^{n} E_{1,1} \cap V_i \cap T_i$

invaders, individual comparisons should be constrained to subsets of "failed species" that correspond to like vectors, cumulatively denoted as $\sum_{k=1}^{k} (V_k \cap T_k) - E_{1,1}$, otherwise comparisons become confounded due to differential effects of multiple vectors on the delivery of viable organisms.

An uncontrolled approach is simply to identify those species that have invaded a particular region (R_1) from a specified source region, regardless of vector or time. In our model, successful invaders are defined as the intersection $V \cap T \cap E$ (Fig. 8.2, black); however, across multiple vectors invaders are illustrated by more than one such intersection (Fig. 8.3, black). When vector is not specified, successful invaders of R_1 are compared with the total species diversity of the source region, SD_1. On its own, this comparison may not reliably identify characteristics of successful invaders, because many species may simply never interface with a vector. Nevertheless, this approach may advance our understanding of invasion biogeography. For example, it may help identify (1) vector operation (historical and contemporary) and (2) those aspects of environmental matching that are crucial for invasion success.

8.3.4 *Fixed Recipient Region, Multiple Source Regions, Multiple Vectors*

For a single recipient region (R_1), alien species may arrive from numerous source regions (SD_1 to SD_k) and by numerous vectors (V_1 to V_k). Species with biological attributes and capabilities of establishing self-sustaining populations in R_1 are denoted as $E_{1,j}$. The sum of all such sets is limited by the number of vectors connecting R_1 and SD_j and is denoted $\sum_{j=1}^{j}\sum_{k=1}^{k} E_{1,j} \cap V_k \cap T_k$. Likewise, the complete collection of "failed invader" pools for comparison containing species that are both entrained and introduced by vectors but which fail to establish self-sustaining populations in R_1 are denoted $\sum_{j=1}^{j}\sum_{k=1}^{k} (V_k \cap T_k) - E_{1,j}$. Since comparisons seeking differences between successful and failed invaders are dependent on vector and source, corresponding pools must be parsed from the previous expressions and then compared individually to be meaningful.

In the Great Lakes, Kolar and Lodge (2002) compared successful fish invaders and failed introductions resulting from multiple source regions and multiple vectors. Biological attributes unique to the pool of successful invaders were used to parameterize a predictive model. The model was then applied to a group of 66 Ponto Caspian fish species to predict each species' likelihood to (1) establish, (2) spread, or (3) become a nuisance. This approach assumes the biological characteristics that conferred invasion success to the Great Lakes are similar across source regions (i.e., $E_{1,1}$ through $E_{1,j}$ are similar biologically). As in the analysis of California fish introductions by Marchetti et al. (2004), the effect of source region was not evaluated.

Perhaps more critically, the Great Lakes analysis appears confounded when considering vector. Here, attributes of failed and successful species pools were derived from multiple vectors and multiple source regions, being compared to characterize attributes of successful fish invaders. Nonetheless, these attributes were then applied in a predictive manner to a single vector, ballast water from a single region. This assumes there are no vector-effects, source region effects, or interaction between the two. The biological attributes of species that are capable of establishing when intentionally introduced may differ from those of species that can withstand the physiological rigors of a transatlantic voyage in a ballast tank. To our knowledge, no studies have yet quantitatively examined the simultaneous action of multiple vectors and/or source regions to a fixed recipient region.

8.3.5 Multiple Recipient Regions, Fixed Source Region, Multiple Vectors

A similar approach can be taken to describe potential species pools that hail from a single region (SD_1) but which are entrained and moved by one or more vectors to one or more recipient regions. Addressing the issue from this perspective allows one to consider comparisons of all the species being exported from a specific source to those species that actually have the biological capacity to succeed (i.e., E_1 to E_i) and establish self-sustaining populations in other locations. In this case, the theoretical sets of successful invaders of multiple recipient regions that arrive by multiple vectors are described by $\sum_{i=1}^{I}\sum_{k=1}^{K} E_{i,1} \cap V_k \cap T_k$. Again, to be instructive, each pool of successful species arriving by a particular vector must be treated individually for each vector and recipient region of interest. To illustrate, one could imagine focusing on the plankton assemblage available from SD_1 (e.g., port A) and then quantifying the entrainment and survivorship of such species in ballast tanks bound for ports B and C (i.e., R_1 and R_2) via onboard measurements. Theoretically, one could compare the collection of species successfully inoculated to ports B and C with those that (1) consistently perished in transit and (2) those species that have successfully established populations in ports B and C. Because of the polyvectic nature of invasion pathways (i.e., simultaneous operation of multiple vectors – see Carlton and Ruiz 2005), similar comparisons for other vectors linking ports A, B, and C would be necessary to understand the effects of the individual vectors that link a single source to one or more recipient regions.

To complicate the picture further, the following expression denotes the theoretical cumulative sum of species with the capacity to establish self-sustaining populations in multiple recipients regions, from multiple sources, and via multiple vectors: $\sum_{i=1}^{I}\sum_{j=1}^{J}\sum_{k=1}^{K} E_{i,j} \cap V_k \cap T_k$. Ironically, if not regrettably, this expression may be reflective of the world's increasingly connected network. One need only look to the complicated picture of commercial shipping to realize that ships link the ports of the world together to varying degrees and by a variety of simultaneously operating vectors ranging from ballast water, hull fouling, the biological content of sea chests, as well as organisms associated with cargo itself.

8.4 Some Recent Analyses of Invader Attributes Using Species Pool Comparisons

Species pool comparisons have been used extensively in studies of terrestrial faunal and floral invasions and more recently, in aquatic systems. Although source region, vector, and recipient region are broadly viewed as important to invasion success, they are rarely considered simultaneously or formally in analyses. In the literature, studies addressing invasion success versus failure typically focus on biological attributes associated with (1) recipient region, (2) vectors and pathways of transmission, or less commonly (3) source regions. Rarely have all three factors been considered in conjunction.

Table 8.2 summarizes some selected recent comparisons of successful and failed invaders' attributes, indicating which of the three pathway parameters, (source region, vector, and recipient region) were considered explicitly and controlled in analyses. This list is by no means exhaustive but is only meant to include some illustrative examples of terrestrial and aquatic studies across a broad taxonomic range. Each approach has merit, depending on the question being investigated, but it is also important to recognize the possible limitations and ramifications of excluding one or more of these invasion parameters when evaluating attributes of successful invaders. Among the studies reviewed here, six constrain a single invasion pathway parameter (three vector, three recipient region); seven constrain two parameters (five vector × recipient region, two vector × source); and two constrain all three parameters (source region × vector × recipient region – Table 8.2). Below, we discuss several of these examples in more detail, expanding on some references already presented.

A common approach constrains the species taxonomically and then compares successful with failed invaders generally, across many regions. In some instances vector is held constant and others not. For example, Rejmánek and Richardson (1996) employed this method in their study of pines, whereby they compared the biological attributes of cultivated (i.e., intentionally introduced) pine species.

Table 8.2 Invasion parameters constrained (indicated by "x") in selected recent analyses that attempt to differentiate successful and failed invaders according to their biological attributes

Source	Vector	Recipient	Taxa	Author
-	-	x	Fishes	Kolar and Lodge (2002)
-	x	x	Freshwater fishes	Marchetti et al. (2004)
-	x	-	Freshwater fishes	Ruesink (2005)
-	-	x	Marine bivalves	Roy et al. (2001)
x	x	x	Marine bivalves	Miller et al. (2002)
x	x	x	Marine gastropods/bivalves	Miller et al. (2007)
x	-	x	Plants	Goodwin et al. (1999)
-	-	-	Conifers	Richardson and Rejmánek (2004)
-	x	-	Pines	Rejmánek and Richardson (1996)
-	x	x	Woody plants	Reichard and Hamilton (1997)
-	-	x	Gymnosperms/angiosperms	Williamson and Fitter (1996)
x	-	x	Plants	Prinzing et al. (2002)
-	x	-	Parrots	Cassey et al. (2004a,b)
-	x	x	Birds	Duncan et al. (2001)
-	x	x	Birds	Duncan et al. (1999)
-	x	x	Birds	Veltman et al. (1996)

Using discriminant analyses they compared successful and failed pine species for a suite of biological characteristics and found that mean seed mass, minimum juvenile period and mean intervals between large seed crops differentiated the groups. No effort was made to control source or recipient region. Likewise, in their global survey of conifers, Richardson and Rejmánek (2004) compared the biological characteristics of invading conifers to those of other, non-invading, gymnosperms. Among freshwater fishes, Ruesink (2005) examined global introductions of freshwater fishes by humans, both with intention to establish non-native populations and for cultivation or use without intention to naturalize. Source and recipient regions were widespread and thus not controlled. Establishment was shown to increase in families with small body size, fish wish omnivorous diets, high endemism in recipient regions, and strength of introduction effort by humans. Such approaches can uncover much about the biology of invaders and non-invaders, but they may be limited in their ability to explain and predict invasion since they do not compare pools of successful and failed invaders that have followed similar invasion pathways.

A number of investigators have controlled vector by studying well documented, intentional introductions. Veltman et al. (1996) found that among 79 bird species intentionally introduced to New Zealand across 496 introduction events, 27 species succeeded and 52 failed. Invasion success was positively correlated with initial population size and negatively associated with native migratory behavior. This study held recipient region and vector constant, but drew from bird species hailing from more than one source. Duncan et al. (1999) correlated the life history characteristics of invading bird species of New Zealand with their invasion range size. Of 34 species, 17 were introduced intentionally from Britain and the remaining 17 were introduced from other source regions. In a separate study, Duncan et al. (2001) showed that of 52 bird species intentionally introduced to mainland Australia, introduction effort, habitat matching, invasion success elsewhere, and climate matching predicted invasion outcome. Again, however, vector and recipient region were constant, but source regions varied.

Among freshwater fishes, Marchetti et al. (2004) investigated the biological attributes of fish that were intentionally introduced to California watersheds, and thus constrained vector and recipient region, but not source region, since they compared fish species from numerous source areas. The biological attributes associated with success at each stage along the invasion pathway varied widely: establishment (parental care, size of native range, physiological tolerance, propagule pressure), spread (long-lived, regional origin, non-herbivores), and integration (a measure of abundance used as a proxy for degree of invasiveness – small, regional origin, non-invertebrate predators).

A pattern of divergent characters associated with invasion stages is also evident among introduced freshwater fishes in the Great Lakes, but the stage-specific characters are somewhat different from California invaders. According to Kolar and Lodge (2002) the hallmarks for establishment among the Great Lakes invaders were fast growth, broad temperature and salinity tolerance, and history of invasion elsewhere. In contrast, slow growth, low survivorship in high water temperatures, but an otherwise wide temperature tolerance, appear to be associated with the ability to spread within the recipient region.

The analyses by Kolar and Lodge (2002) constrained recipient region, but not source region or vector. To generate and parameterize a predictive model, Kolar and Lodge compared species characteristics from 24 species which failed to establish self-sustaining populations and 21 species that successfully established populations in the Great Lakes. As discussed earlier (see Sect. 8.3.4), the members of these species pools hailed from multiple source regions throughout the world, including regions in United States, East Asia, Northern Europe, the Ponto Caspian, Eurasia, South and Central America, and elsewhere. Furthermore, these species were introduced separately "via a variety of pathways including, but not limited to, intentional stocking, canals, natural waterways, release of baitfish and pets, escapes from aquaculture, and ballast water" (Kolar and Lodge 2002), which may have confounded their results.

Results from the California and Great Lakes studies, as well as the global freshwater fish analysis of Ruesink (2005), are difficult to compare, as the methods of analysis were not similar. Differential contributions of biological attributes could be reflections of (1) differences in recipient regions (e.g., California vs Great Lakes), (2) mixed vector effects (e.g., the ballast water vector may impose very different pressures on entrainment and en route survivorship than does intentional introduction), or (3) increased variability stemming from species pools drawn from diverse biogeographic origins, and distances.

In their study of European plant invaders of New Brunswick, Canada, Goodwin et al. (1999) controlled source and recipient regions, but not vector since invaders were introduced both intentionally and accidentally. Biological and distributional characteristics of successful invaders were compared with randomly chosen, non-invading congeners native to Europe. Successful invaders were shown to have significantly larger native geographic distributions than non-invaders, but the groups did not differ with respect to other life history characteristics. They concluded that larger native range size may reflect greater environmental tolerance and/or greater opportunity for accidental transport with humans. While both may be true, the conclusions are complicated because the invader and non-invaders species pools do not share the same introduction histories. Successful invaders comprise species that were introduced (1) intentionally or (2) accidentally, whereas non-invaders might be either failed invaders (i.e., entrained and/or introduced, but failed) or species that were never introduced by any means and thus never challenged by either vector or New Brunswick's physical and biological environment.

In an effort to control such uncertainties, Miller et al. (2002, 2007) held vector, source and recipient regions constant as a means for identifying pools of successful and failed species for comparison. The invasion pathway examined was the intentional introduction of live oysters *(Crassostrea virginica)* from the East coast of North America to San Francisco Bay. This pathway is viewed as responsible for the accidental introduction of scores of Western Atlantic invertebrates, including a selection of shelled mollusks (Cohen and Carlton 1995). By choosing a single vector, i.e., dredging and subsequent live shipment of eastern oysters by railroad, from a single source region (New York City and environs, the primary source of historical live oysters shipped to San Francisco Bay; see Carlton 1979 and references therein), and a single recipient region (San Francisco Bay), Miller et al. (2002, 2007) compiled a list of mollusk species that occurred with oysters and which would have been collected as by-catch in oyster dredges over the ~80-year period of vector operation. Statistical comparisons of successful San Francisco Bay invaders and failed invaders differentiated these groups according to tolerance of low salinity, reproductive mode, and abundance of species in the source region (Miller et al. 2002, 2007). It should be noted that, even in this case, mortality en route vs failure after inoculation into San Francisco Bay could not be evaluated. Although live oyster transport strived to maximize survivorship through the use of ice and refrigeration, measures that surely conferred success to some members of the entrained species pool, some species may have failed in transit, perhaps due to desiccation. In this case, without direct measures, it remains impossible to determine the exact cause of failure.

8.5 Discussion

Elucidating the interactions of species with their biotic and abiotic environments is of fundamental importance to ecologists. As non-native species are introduced beyond their historical ranges and cause environmental, economic, and human health impacts, these interactions become relevant to a much broader audience. From this perspective, there has been long-standing interest in understanding the biological profile of invasive species and predicting invasion success (Elton 1958; Baker 1965; Daehler and Strong 1993; Williamson and Fitter 1996). Species profiling, as an element of invasive species risk assessment, continues to be of great interest for the prediction and management of such species (Kolar and Lodge 2002; Ruesink 2005).

When trying to understand effects of species characteristics on invasion outcome, past research has often focused on biological attributes related to the invader's physiological tolerances, life history characteristics, and behavior. Beyond these, population level variability and biogeographic attributes are likely important to a species' ability to successfully establish self-sustaining populations beyond its native range. For example, the abundance of a species may have major implications for the number of propagules entrained, transported, and released in a recipient region. From a biogeographical standpoint, the size of a species' native range can influence the probability of a species' invasion success. This influence may be purely probabilistic, representing the level of human activity and thus the opportunity and likelihood of species entrainment, transport, and introduction (Simberloff 1989; Pyšek et al. 2004; Jeschke and Strayer 2005). Conversely, larger native ranges are sometimes interpreted as correlations of a species' ability to tolerate wider spectra of environmental/physiological conditions (Williamson and Fitter 1996; Goodwin et al. 1999). In both cases, knowledge of the physiological variability of the species across its native range (i.e., existence of specialized locally adapted biotypes) and the location of the donor population within a species' native range may also be crucial.

As already indicated, previous studies highlight the importance of species pool selections when attempting to accurately identify correlates of invasion success (Cassey 2004b; Pyšek et al. 2004). Prinzing et al. (2002) promoted the use of the so-called "source-area" approach as a means for identifying proper species pools to compare successful and failed invaders of particular source and recipient regions. They identified the need to constrain comparisons to species that originate in the same native region and which are introduced to a common recipient region. They used this approach to investigate plant species that occurred in Central Europe and subsequently succeeded or failed to establish populations in two provinces of Argentina. Although Pyšek et al. (2004) agree with the source-area approach in theory, they point to the importance of controlling for the size of invader's native range, and caution that such analyses should restrict source region species pools to native species rather than mixes of natives and non-natives, regardless of non-native residence time prior to re-introduction elsewhere. We agree with these assessments, but contend that analytical requirements should be even more stringent, and include explicit identification and treatment of vector when possible.

Finally, the distance of a source region from the recipient region may correlate with transit time, the degree of stress imposed by a vector, and ultimately the condition of the biota upon arrival to the recipient region (Pyšek et al. 2004; Carlton and Ruiz 2005). The interaction of vector and source region can affect survivorship during transportation and after introduction, depending on the duration and particular stresses of transit. Verling et al. (2005) have demonstrated differential zooplankton survivorship in ballast tanks based on voyage length. One limitation of our present treatment is that it does not explicitly capture the effects of time and distance during transport. For example, a species that can survive transit from source region A to recipient region B may experience much higher mortality during transit to a more distant recipient region C. In such cases, the physiological tolerance of organisms may limit their ability to survive transit.

Temporal aspects of invasion and species pool choice and comparison are not incorporated in our model; however, they are no doubt integral to invasions success. Time likely influences the invasibility of recipient regions as well as the supply of propagules from source regions. Seasonal variation and environmental shifts act to open and close windows of invasion opportunity (see Carlton 1996). Thus, analyses should strive to control or explicitly define the time period of study to counter temporal bias.

Despite numerous studies aimed at understanding the biological aspects of invasion success and failure, this remains a challenge, especially given the expanding diversity of source regions and vectors in play. Few studies have adequately controlled the effects of vector, or the interactions of vectors with source and recipient regions to enable robust comparisons of the biological attributes of successful and failed species. This is not an easy undertaking, and selection of appropriate species pools for comparison is essential if we are to understand the influence of biological attributes on invasion success.

Our goal in this chapter was to present a general framework for comparing traits among species pools to evaluate issues of invasion success. Clearly, this represents an ideal, where individual effects at each stage in the invasion sequence are measured. While the experimental effort required for unraveling all the interactions posed by multiple vectors that connect multiple regional species pools is surely beyond reach, our aim is to highlight, define, and parse such complexity in order to identify those parts that are most tractable to invasion scientists, either through comparative observations or by manipulative experiments.

Acknowledgements We would like to thank Richard Ambrose, Gail Ashton, Jeff Crooks, Richard Everett, Melissa Hagan, Steve Lonhart, Mark Minton, Gil Rilov, and Ray Wakeland, for their contributions to this chapter. We also wish to thank the Smithsonian Institution, National Sea Grant Program, and the United States Coast Guard for their assistance in funding this work.

References

Baker HG (1965) Characteristics and modes of origin of weeds. In: Baker HG, Stebbins GL (eds) The genetics of colonizing species. Academic Press, New York, New York, USA

Carlton JT (1979) History, biogeography, and ecology of the introduced marine and estuarine invertebrates of the Pacific coast of North America. PhD dissertation. University of California, Davis. 904 pp

Carlton JT (1996) Pattern, process, and prediction in marine invasion ecology. Biol Conserv 78:97–106

Carlton JT, Ruiz GM (2005) Vector science and vector management in bioinvasion ecology: conceptual frameworks. In: Mooney HA, Mack RN, McNeely JA et al. (eds) Invasive alien species, a new synthesis. Island Press, Covelo, California, USA, pp 36–58

Cassey P, Blackburn TM, Russell GJ, Jones KE, Lockwood JL (2004a) Influence on transport and the establishment of exotic bird species: an analysis of the parrots (Psittaciformes) of the world. Global Change Biol 10:417–426

Cassey P, Blackburn TM, Jones KE, Lockwood JL (2004b) Mistakes in the analysis of exotic species establishment: source pool designation and correlates of introduction success among parrots (Aves: Psittaciformes) of the world. J Biogeogr 31:277–284

Cohen AN, Carlton JT (1995) Nonindigenous aquatic species in a United States estuary: a case study of the biological invasions of the San Francisco Bay and Delta. A report for the U.S. Fish and Wildlife Service, Washington, D.C. and the National Sea Grant College Program Connecticut Sea Grant. 272 pp

Daehler CT, Strong DR (1993) Prediction and biological invasion. Trends Ecol Evol 10:380

Duncan RP, Blackburn TM, Veltman CJ (1999) Determinants of geographical range sizes: a test using introduced New Zealand Birds. J Anim Ecol 68:963–975

Duncan RP, Bomford M, Forsyth DM, Conibear L (2001) High predictability in introduction outcomes and the geographical range size of introduced Australian birds: a role for climate. J Anim Ecol 70:621–632

Elton CS (1958) The ecology of invasions of animals and plants. Methuen, London, UK

Goodwin BJ, McAllister AJ, Fahrig L (1999) Predicting invasiveness of plant species based on biological information. Conserv Biol 13:422–426

Jeschke JM, Strayer DL (2005) Invasion success of vertebrates in Europe and North America. Proc Natl Acad Sci USA 102:7198–7202.

Kolar CS, Lodge DM (2001) Progress in invasion biology: predicting invaders. Trends Ecol Evol 16:199–204

Kolar CS, Lodge DM (2002) Ecological predictions and risk assessment for alien fishes in North America. Science 298:1233–1236

Marchetti MP, Moyle PB, Levine R (2004) Invasive species profiling? Exploring the characteristics of non-native fishes across invasion stages in California. Freshwater Biol 49:646–661

Miller AW (2000) Assessing the importance of biological attributes for invasion success: Eastern oyster *(Crassostrea virginica)* introductions and associated molluscan invasions of Pacific and Atlantic coastal systems. D.Env. dissertation. University of California, Los Angeles 209 pp

Miller AW, Hewitt CL, Ruiz GM (2002) Invasion success: does size really matter? Ecol Lett 5:159–162

Miller AW, Ruiz GM, Minton MS, Ambrose RF (2007) Differentiating successful and failed molluscan invaders in estuarine ecosystems. Mar Ecol Prog Ser 332:41–51

Prinzing A, Durka W, Klotz S, Brandl R (2002) Which species become aliens? Evol Ecol Res 4:385–405

Pyšek P, Richardson DM, Williamson M (2004) Predicting and explaining plant invasions through analysis of source area floras: some critical considerations. Divers Distrib 10:179–187

Reichard SH, Hamilton CW (1997) Predicting invasions of woody plants introduced to North America. Conserv Biol 11:193–203

Rejmánek M, Richardson DM (1996) What attributes make some plant species more invasive? Ecology 77:1655–1661

Richardson DM, Rejmánek M (2004) Conifers as invasive aliens: a global survey and predictive framework. Divers Distrib 10:321–331

Richardson DM, Pyšek P, Rejmánek M, Barbour MG, Panetta FD, West CJ (2000) Naturalization and invasion of alien plants: concepts and definitions. Divers Distrib 6:93–107

Roy K, Jablonski D, Valentine JW (2001) Climate change, species range limits and body size in marine bivalves. Ecol Lett 4:366–370

Ruesink JL (2005) Global analysis of factors affecting the outcome of freshwater fish introductions. Conserv Biol 19:1883–1893

Simberloff D (1989) Which insect introductions succeed and which fail? In: Drake JA, Mooney HA, di Castri F, Groves RH, Kruger FJ, Rejmánek M, Williamson M (eds) Biological invasions: a global perspective. . Wiley, New York, pp 61–75

Veltman CJ, Nee S, Crawley MJ (1996) Correlates of introduction success in exotic New Zealand birds. Am Nat 147:542–557

Verling E, Ruiz GM, Smith LD, Galil B, Miller AW, Murphy KR (2005) Supply-side invasion ecology: characterizing propagule pressure in coastal ecosystems. Proc R S 272:1249–1257

Williamson MH, Fitter A (1996) The characters of successful invaders. Biol Conserv 78:163–170

Section III
Invader Establishment

Chapter 9
The Establishment of Invasive Species

Jeffrey A. Crooks and Gil Rilov

9.1 Introduction

In the continuum of invasion phases, establishment stands at the interface between the initial introduction of propagules and the integration of the invader into the ecological community. Although the edges of this transitional phase tend to blur, invader establishment is generally related to the survival of initially-transported individuals to form reproducing and expanding populations, influenced both by the characteristics of the invader and the receiving ecosystem. In practice, however, it is often difficult to distinguish between the factors operating in the arrival and establishment phases, because most of the information available is for invaders that have successfully negotiated both and have survived to form conspicuous incursions (Chap. 7, Johnston et al.; Chap. 8, Miller and Ruiz). We often know relatively little about how many invasions fail, when they fail, and why.

9.2 Factors Influencing Numeric and Geographic Growth of Invasive Populations

Despite the often limited ability to distinguish between factors operating across invasion phases, it is likely that some of the traits that increase probabilities of successful transit, such as tolerance to harsh and variable environmental conditions, might also better equip invaders to survive and reproduce after their association with the vector. It is also likely, however, that different processes will be operating in the newly encountered ecosystem. In the preceding section, Johnston et al. (Chap. 7) and Miller and Ruiz (Chap. 8) began to address some of the traits that characterize successful invaders. In this section, Smith (Chap. 10) continues to develop the theme of individual characteristics increasing likelihood of invasion, treating phenotypic plasticity associated with both transit and species establishment.

G. Rilov, J.A. Crooks (eds.) *Biological Invasions in Marine Ecosystems.*
Ecological Studies 204,

The environmental characteristics to which species must respond can be broadly broken down into abiotic and biotic factors. In general, any successful invader must pass through the "abiotic filter," representing the suite of physical and chemical properties in the receiving environment (Chap. 12, Olyarnik et al.). Understanding species tolerances and this abiotic filter informs efforts such as habitat- and climate-matching related to invasion success (e.g. Stachowicz et al. 2002; Stohlgren et al. 2005). In addition, there has been considerable discussion of how invaders respond to the quality of the environment, and there is often a noted propensity for invaders to do well in degraded or disturbed habitats. These relationships are discussed by Olyarnik et al. in this section (Chap. 12), as well as by Johnston et al. (Chap. 7) and Byers (Chap. 14) elsewhere in the book.

In order for incipient invasive populations to grow, they must also interact with biotic elements of their new ecosystem. For invaders that eventually become successful, this new ecosystem might be relatively benign if they leave their co-evolved predators and parasites behind during transit. This "enemy release hypothesis," as well as the broader role of parasites in biological invasions, is reviewed by Torchin and Lafferty (Chap. 11). Another key factor influencing invader success is the resident species pool, either native or previously introduced, that new invaders must interact with. Olyarnik et al. (Chap. 12) review the topic of invasibility and how organisms affect it. This biotic resistance to invasion, influenced by such factors as species diversity and redundancy, has received much attention in recent years, given its roots in basic ecological theory. There is also a substantial practical benefit to understanding controls on invader success, in that it should allow us to manage systems so that they are more effective at intrinsically resisting invasions.

Once invaders have passed through abiotic and biotic filters, the dynamics of the establishment phase itself are one of the best studied aspects of invasion (e.g. Hengeveld 1989; Shigesada and Kawasaki 1997). The quantitative dynamics of increases in density and range, for example, has been an area of active and fruitful modeling. These build upon established concepts of population growth (e.g. exponential) and spatial spread (e.g. reaction-diffusion), with increasing complexities related to the dynamics of biological invasion. One establishment-related concept which remains somewhat vexing, however, is the prolonged lag sometimes noted between initial establishment and subsequent numerical and geographic growth of invasive populations (Kowarik 1995; Rilov et al. 2004; Crooks 2005). These modeling-related concepts are reviewed in more depth by Wonham and Lewis (Chap. 4).

References

Crooks JA (2005) Lag times and exotic species: the ecology and management of biological invasions in slow motion. Ecoscience 12:316–329

Hengeveld, R (1989) Dynamics of biological invasions. Chapman and Hall, New York

Kowarik I (1995) Time lags in biological invasions with regard to the success and failure of alien species. In: Pysek P, Prach K, Rejmanek M, Wade M (eds) Plant invasions – general aspects and special problems. SPB Academic Publishing, Amsterdam, pp 15–38

Rilov G, Benayahu Y, Gasith A (2004). Prolonged lag in population outbreak of an invasive mussel: a shifting-habitat model. Biol Invas 6(3):347–364
Shigesada N, Kawasaki K (1997) Biological invasions: theory and practice. Oxford series in ecology and evolution. Oxford University Press, Oxford
Stachowicz JJ, Terwin JR, Whitlatch RB, Osman RW (2002) Linking climate change and biological invasions: ocean warming facilitates non-indigenous species invasion. Proc Natl Acad Sci USA 99:15497–15500
Stohlgren TJ, Crosier C, Chong GW, Guenther D, Evangelista P (2005) Life-history matching in invading non-native plant species. Plant Soil 277:7–18

Chapter 10
The Role of Phenotypic Plasticity in Marine Biological Invasions

L. David Smith

10.1 Introduction

The outcome of a species introduction depends, in large measure, on the abilities of the invader and species in the receiving community to respond to their new environments. A successful invader must survive changing environmental conditions at all stages leading to and following its introduction. Residents in the invaded community, in turn, must cope with environmental changes that result from the arrival of the new species. Adaptive responses (i.e., those that confer a fitness benefit) by either party have been viewed primarily to result from evolutionary changes in fixed traits in populations (Thompson 1998; Mooney and Cleland 2001; Cox 2004). Although intense selection can result in rapid phenotypic shifts across generations (Huey et al. 2000; Gilchrist et al. 2001; Reznick and Ghalambor 2001), this process does not encompass fully the dynamic nature of many invasions. A burgeoning literature indicates that individual organisms are capable of modifying ecologically important physiological, morphological, behavioral, and life-history features within a lifetime in response to environmental cues (Harvell 1986; Stearns 1989; Kingsolver and Huey 1998; Schlichting and Pigliucci 1998; West-Eberhard 2003; DeWitt and Scheiner 2004a). This phenomenon, known as phenotypic plasticity, provides a means by which an invader can respond relatively quickly to its new biotic or abiotic environment. Similarly, phenotypic plasticity may allow resident species to mitigate changes wrought by the invader. The role of adaptive phenotypic plasticity in biological invasions, however, has been largely ignored in marine settings.

Our understanding and interpretation of marine biological invasions will be incomplete on several counts if we fail to acknowledge or test for the potential influence of phenotypic plasticity. First, phenotypic plasticity can provide a mechanistic explanation to understand and predict (1) why and how some individuals or species invade and others do not, (2) what the ecological effects and eventual ranges of the invader might be, and (3) how native species might respond to the introduction. In particular, knowledge of the type, direction, and magnitude of induced responses is critical if we are to decipher direct and indirect ecological effects stemming from species introductions. Second, recognition of phenotypic plasticity's

G. Rilov, J.A. Crooks (eds.) *Biological Invasions in Marine Ecosystems.*
Ecological Studies 204,

influence should aid in our interpretation of spatial patterns or temporal changes that develop following marine introductions. In particular, changes in characteristics of populations following an invasion cannot necessarily be assumed to reflect selection (rapid or otherwise). Rather, studies need to test for environmental and genetic influences and their interactive contribution to phenotype across ecologically relevant environments. Such studies will give a better indication of factors driving short- and long-term evolutionary responses to marine invasions.

To gain a better understanding of plasticity's role in marine invasions, we need to (1) be cognizant of the different forms of plasticity and of the selective factors that favor adaptive plasticity, (2) identify environmental cues and phenotypic responses that are likely to influence the outcome of marine invasions, (3) determine circumstances specific to marine introductions in which plasticity might be a contributing factor, and (4) design appropriate studies to test for plasticity. To encourage research efforts to address these issues, I have divided this chapter into five sections. First, I provide a brief overview of phenotypic plasticity and its relationship to biological invasions. Second, I address why we should expect phenotypic plasticity to be important in marine invasions. Third, based on general patterns emerging from studies of marine introductions, I suggest when and where phenotypic plasticity might be expected to play a significant role in marine invasions. Fourth, I provide examples where phenotypic plasticity has had multiple ecological, and perhaps evolutionary, effects in marine invasions. Finally, I identify larger questions and suggest methods to test for phenotypic plasticity in marine systems.

10.2 Phenotypic Plasticity

10.2.1 A Brief Overview

Excellent overviews of phenotypic plasticity exist (Stearns 1989; Gotthard and Nylin 1995; Kingsolver and Huey 1998; Pigliucci 2001; DeWitt and Scheiner 2004b), so I will only briefly describe key points here. In the last two decades, research has shown unequivocally that phenotypic plasticity is an important adaptive mechanism for coping with environmental variation. Phenotypic plasticity is distinct from other evolved responses to environmental heterogeneity, such as intermediate phenotypes (generalists) or bet hedging (several phenotypes produced in a single environment), in that different environments generate different phenotypes (DeWitt and Langerhans 2004). For a single genotype, phenotypic plasticity is also distinct from developmental noise in that variation generated by the former is directional in nature, while variation arising from the latter is directionally random. Adaptive phenotypic plasticity is indicated when a genotype responds to a specific environmental signal with a repeatable, directional change in phenotype that increases fitness (Travis 1994; Doughty and Reznick 2004). Environmental cues that trigger phenotypic responses can be biological or physical in nature and may vary and interact over different temporal or spatial scales. For example, induced defensive responses by prey to highly localized signals (e.g., chemical cues from a predator) may be modified by broader scale environmental heterogeneity (e.g., water flow, temperature gradients).

Phenotypic plasticity can be manifested in physiological, behavioral, morphological, or life-history traits. Depending on the trait and taxon in question, responses may be discrete (i.e., polyphenisms) or continuous, and reversible or irreversible (David et al. 2004). While each two-way combination is possible, the distinction between a discrete and continuous response is not always clear-cut, and reversibility is seldom tested. Examples of discrete, irreversible phenotypic responses include temperature-related sex determination in reptiles (Janzen 1995) and use-induced handedness in lobsters (Govind 1987). Discrete, but reversible, changes include predator-induced diel vertical migration in zooplankton (Bollens and Frost 1991) and seasonal changes in bird plumage (Piersma and Drent 2003). Continuous, reversible phenotypic responses are evident in many physiological and morphological traits. Examples include physiological acclimation to seasonal temperature change in ectotherms and size adjustments in trophic structures and digestive systems of bivalves, echinoderms, and birds (Piersma and Drent 2003). Finally, continuous, but irreversible, changes occur in many life history characteristics. For example, many molluscs, echinoderms, and amphibians alter the timing of metamorphosis to various environmental cues (e.g., predators, diet, conspecific density) (Hadfield and Strathmann 1996; Doughty and Reznick 2004).

A major conclusion based on diverse studies is that phenotypic plasticity has a genetic basis and, thus, the potential to evolve (Berrigan and Scheiner 2004; Windig et al. 2004). In the case of a continuous trait, a plastic genotype expresses a range of phenotypes as a function of the environment. This relationship is the genotype's norm of reaction (Fig. 10.1) (Stearns 1989). Both the magnitude and direction of reaction norms can vary among genotypes. Where genetic variation for plasticity exists [i.e., the genotype-environment (G × E) interaction is significant], plasticity is potentially subject to selection (Weis and Gorman 1990; Thompson 1991; DeWitt and Scheiner 2004b; Fordyce 2006). In such cases, the relative fitness of each genotype will depend on the particular environment in which it is measured (Fig. 10.1). Whether genes for plasticity exist and are targets of selection or whether plasticity is a by-product of selection on trait means in different environments remains a source of debate (Via et al. 1995). Regardless, selection for adaptive phenotypic plasticity is expected in instances where (1) the environment is sufficiently fine-grained that heterogeneity is experienced by the organism or, in some cases, its maternal parent (Agrawal et al. 1999), (2) environmental cues are reliable, (3) correlated responses improve fitness and are, to some extent, heritable, and (4) some costs (e.g., reduced fecundity) or limits (e.g., ability to predict future conditions) to plasticity exist (Travis 1994; DeWitt et al. 1998; Weinig 2000; Berrigan and Scheiner 2004). With regard to the last point, constraints on plasticity are assumed (although rarely tested); otherwise, evolution should select for genotypes capable of always matching the best phenotype of possible alternatives to the appropriate environment (DeWitt et al. 1998).

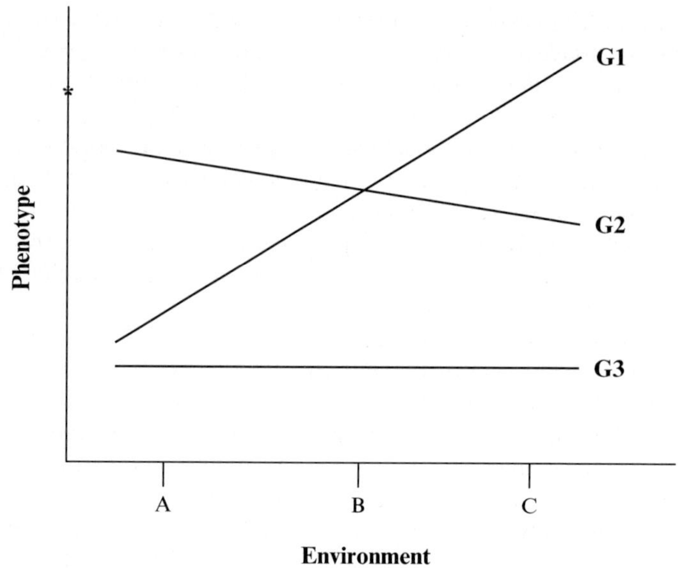

Fig. 10.1 Reaction norms for three hypothetical genotypes: G1, G2 and G3. G1 has the most plastic phenotypic response across environments A, B, and C, as indicated by the steepness of its slope. G2 is less responsive phenotypically to its environment and opposite in direction to that of G1. G1 and G2 show a significant G × E interaction. G3 shows no phenotypic plasticity across environments. If * represents an adaptive optimum, then relative fitness will vary among genotypes as a function of the environment, with G1<G2 in environment A; G1=G2 in B; and G1>G2 in C. G3 is at a relative disadvantage in all three environments

10.2.2 General Importance to Invasions

A number of authors have identified phenotypic plasticity as a characteristic of a good invader (Bradshaw 1965; Rejmánek and Richardson 1996; Daehler 2003; Clements et al. 2004; Richards et al. 2006). At the most basic level, as an organism moves outside its native range, plasticity should give it the flexibility, within limits, to match its phenotype to its new environment. If the organism's fitness is enhanced, then its chances for survival and successful introduction increase. For any organism, the benefit of plasticity will depend on "how often the correct phenotype-environment matches are made" (DeWitt and Scheiner 2004b). In the context of invasions, organisms are removed from their normal selective regime; consequently, opportunities for correct phenotype-environment matches may become less predictable. For example, a genotype that has evolved to respond to cues that are highly specific to its home range may fail to recognize or respond optimally to unfamiliar cues in the invaded habitat. In such cases, plasticity may confer little or no benefit. Other more generic cues may be recognizable regardless of location, and the adaptive responses may be universally beneficial. For example, kairomones (i.e., interspecific chemical messengers that benefit the recipient) may be species-specific in some cases and not in others

(Gilbert 1999). Dzialowski et al. (2003) documented defensive morphological responses by an introduced cladoceran to kairomones released by native insect and fish predators, which suggests cue generality. Work with other taxa, however, indicates kairomone specificity (Tollrian and Harvell 1999). At present, our understanding of cue recognition is too rudimentary to predict how well introduced or native species will respond to signals in their new environments (Smith and Jennings 2000; Miner et al. 2005; Freeman and Byers 2006).

Phenotypic plasticity's influence on post-invasion ecological processes, while limited in effect initially, may become increasingly important with time. For example, phenotypic plasticity may 'buy' time for the invader to integrate and adapt evolutionarily to its new surroundings in the period immediately after an introduction (Sexton et al. 2002; Schlichting 2004; Fordyce 2006). As the introduced species becomes established, phenotypic plasticity has the potential to alter the outcome of ecological interactions between populations of introduced and resident species by giving one or the other an advantage in resource acquisition or reproductive success (Daehler 2003). The extent to which native and non-native species respond reciprocally to one another is largely unknown (Agrawal 2001). If bidirectional responses prove common, then each invasion has the potential to set off a dynamic process in which phenotypic change in one participant alters the 'environment' of the other, and vice versa. The degree to which these reciprocal interactions escalate will depend on the relative symmetry in responses by each species; the duration of contact; and the reversibility of the response (Agrawal 2001).

The ecological consequences of plasticity in an invasion may extend beyond the direct effects of the introduced and resident species on each other's abundance. Recent studies have shown that modification of a trait in one species in response to another can indirectly affect the abundances of other species in the community (Callaway et al. 2003; Dill et al. 2003; Trussell et al. 2003; Werner and Peacor 2003). These trait-mediated indirect interactions (TMIIs) are now thought to be ubiquitous in ecological communities. While studies have tended to focus on indirect effects stemming from behavioral plasticity, morphological plasticity can also lead to TMIIs and have long-term effects on community structure. For example, Raimondi et al. (2000) demonstrated that predator-induced changes in the form of intertidal barnacles indirectly increased algal cover. In their system, juvenile barnacles responded to cues from a predatory whelk by developing a predator-resistant 'bent' morphology. In contrast to 'normal' conical barnacles, bent morphs inhibited recruitment of mussels, and in so doing, permitted encrusting alga to occupy space. Because the induced changes in barnacle form were irreversible, the alteration to community structure was potentially long lasting. This and other examples indicate that plastic responses following an introduction could have multiple and cascading ecological consequences for the community.

As the invader spreads in the new habitat, phenotypic plasticity could generate variation over broad, as well as narrow, spatial or temporal scales and potentially lead to evolutionary change. Numerous studies have argued that changes in phenotype following an introduction demonstrate rapid evolutionary change (Thompson 1998; Ellstrand and Schierenbeck 2000; Cox 2004; Freeman and Byers 2006). In a number of cases, associated genetic changes have been documented (Carroll and Dingle 1996; Gilchrist et al. 2001). Phenotypic plasticity might also explain observed temporal shifts or spatial differences. For example, morphological shifts following two different introductions were attributed initially to rapid selection: one case involved divergence in hind limb lengths of *Anolis* lizards experimentally introduced to islands (Losos et al. 1997); the other involved increases in shell thickness of snails following the introduction of a predatory crab (Vermeij 1982a,b; Seeley 1986). Subsequent studies revealed that the phenotypic changes could be induced and were adaptive (Losos et al. 2000; Trussell and Smith 2000). While evidence of phenotypic plasticity in these studies does not preclude rapid evolutionary change after an invasion, it suggests that selection may be acting on reaction norms rather than on fixed traits. In considering whether phenotypic shifts in post-invasion populations represent adaptive change, we need to tease apart the relative contributions of genetic, environmental, and G × E effects experimentally and test for fitness benefits and costs over a range of relevant environments (Lee and Petersen 2002; Lee CE et al. 2003; Miner et al. 2005). Ultimately, short-term phenotypic responses may produce long-term evolutionary changes in post-invasion communities that result in diversification (Day et al. 1994; West-Eberhard 2003; Schlichting 2004).

10.2.3 Empirical Work in Non-marine Systems

Studies that have explicitly examined phenotypic plasticity in the context of biological invasions have been conducted primarily in terrestrial and freshwater systems. These studies generally took one of four directions. One approach, which typically focused on consumer-prey interactions, examined responses of the native species to cues from non-native species or vice versa (Dzialowski et al. 2003; Moore et al. 2004). A second tack, often involving plants, compared phenotypic responses in non-native vs native species in response to various environmental cues (e.g., nutrient gradients) (Milberg et al. 1999; Schweitzer and Larson 1999; Niinemets et al. 2003). A third category of studies compared phenotypic differences or performances of the invasive species in their native vs introduced habitats typically to find a causal explanation for post-introduction changes in or success of the invader (Leclaire and Brandl 1994; Willis et al. 2000). A fourth group of studies tested for phenotypic responses of invasive species over various environmental gradients in the invaded habitat to reveal characteristics that might give the invader a fitness advantage or to distinguish genetic vs environmental contributions to observed spatial variation (Byers and Quinn 1998; Parker et al. 2003; Kollmann and Banuelos 2004). Overall, plasticity was observed in most of these studies, and responses were in a direction assumed to be adaptive. Although hypothesized fitness benefits were seldom tested explicitly, this body of work illustrates the diverse effects of phenotypic plasticity in invasions of terrestrial and freshwater systems and points to useful avenues of study in marine systems.

10.3 Phenotypic Plasticity in Marine Systems

Over the last two decades, research has demonstrated that phenotypic plasticity is common among marine organisms. Its prevalence may relate to at least three factors. First, although the marine realm is more constant physically than either freshwater or terrestrial habitats, environmental heterogeneity nonetheless exists at many levels and is pronounced at the margins of oceans (e.g., estuaries, intertidal zones, surface waters). Physical factors, such as salinity, temperature, pressure, oxygen level, light level, water flow, and wave stress can vary over a range of temporal and spatial scales. In addition, biotic factors (e.g., densities of predators, competitors, parasites, prey) can also fluctuate, even in places where physical conditions are relatively stable. Second, the sessile nature of many adult benthic marine organisms may have selected for phenotypic plasticity. Bradshaw (1965) noted that phenotypic plasticity was common in terrestrial plants and suggested that such a strategy might be necessary given that quick escape from environmental stresses is impossible. The same selective pressure should hold for post-settlement stages of benthic macrophytes and invertebrates that fix themselves onto or burrow into the substratum. Finally, planktonic dispersal of algal spores, plant seeds, and larvae of marine organisms by coastal or ocean currents is common (Kinlan and Gaines 2003). These dispersing forms will encounter a range of environmental conditions as they develop in the water column or recruit to the benthos, so phenotypic plasticity should be favored.

A summary of representative examples of phenotypic plasticity in marine systems illustrates the taxonomic breadth and the range of environmental cues and phenotypic responses associated with the phenomenon (Table 10.1). In almost all examples listed, specific cues were identified and manipulated to test for plasticity. I did not include field demonstrations of phenotypic plasticity (e.g., reciprocal transplants) if they did not attempt to isolate environmental cues. To date, phenotypic plasticity has been documented in marine protists, algae, vascular plants, and invertebrate and vertebrate animals in the benthos, plankton, and nekton. Phenotypic plasticity occurs in larval as well as post-larval life stages and in diverse ecological settings including rocky intertidal habitats, subtidal soft bottom communities, temperate estuaries, tropical coral reefs, coastal waters, and oceanic pelagic zones. Environmental cues that trigger plastic responses include food resources (prey quantity and type), physical factors (salinity, temperature, light, oxygen, water flow or wave energy), consumers (chemical, tactile, visual cues or simulated damage), and competitors.

I classified the types of induced responses broadly as involving physiological, behavioral, morphological, or life history traits (Table 10.1). The specific characters that were modified, however, were quite variable. For example, difference in food *quantity* altered the length of ciliated feeding arms in echinoid larvae (Hart and Strathmann 1994), reproductive effort in clams (Ernande et al. 2004), and body size in marine iguanas (Wikelski and Thom 2000). Differences in the *quality* of food in the diet resulted in modification of claw size in crabs (Smith and Palmer 1994),

Table 10.1 Examples of phenotypic plasticity in marine systems

General cue	Inducer	Response type[a]	Characters	Taxon	Life Habit[b]	Reference[c]
Food resource	Quantity	LH	Egg size, development time	Gastropoda, Asteroidea	B, P	Chester (1996); George (1999)
		LH	Reproductive effort, growth rate	Bivalvia	B	Ernande et al. (2004)
		M	Trophic morphology	Echinoidea, Ophiuroidea	B, P	Ebert (1996); Hart and Strathmann (1994); Podolsky and McAlister (2005)
		M	Body size	Reptilia[d]	B	Wikelski and Thom (2000)
	Quality	LH	Development time	Echinoidea	P	Heyland and Hodin (2004)
		M	Trophic morphology	Bivalvia, Decapoda, Gastropoda, Osteichthyes	B, N	Drent et al. (2004); Eastman and DeVries (1997); Padilla (1998); Smith and Palmer (1994); Baldridge (2006)[d]
		P	Digestive chemistry	Polychaeta	B	Bock and Mayer (1999)
Physical	Salinity	M	Body form, growth rate	Osteichthyes	N	Corti et al. (1996)
		P	Osmoregulatory capacity	Decapoda	P	Charmantier et al. (2002)
		P	Salinity tolerance	Anthophyta	B	Castillo et al. (2005)
	Temperature	M	Body size	Euphausiacea	P	Marinovic and Mangel (1999)
		P	Metabolic rate	Gastropoda	B	Sokolova and Portner (2003)
		P	Heat-shock proteins	Bivalvia	B	Hamdoun et al. (2003)
	Light	P	Photosynthetic rate	Bacillariophyta, Chlorophyta	B	Gallagher et al. (1984); Raniello et al. (2004)
		M	Branch morphology	Rhodophyta	B	Monro and Poore (2005)
	Water flow or wave energy	M	Trophic morphology	Cirripedia	B	Arsenault et al. (2001); Marchinko (2003)
		M	Growth rate, shell shape	Bivalvia	B	Steffani and Branch (2003)
		M	Pedal foot size, growth rates	Gastropoda[e]	B	Etter (1988, 1996); Trussell (1997)
		M	Body stiffness	Porifera	B	Palumbi (1984)
		M	Shell thickness	Gastropoda	B	Trussell (1996)

	Substratum	M	Growth form	Chlorophyta	B	Garbary et al. (2004)
Consumer[f]	Chemical	B	Diel vertical migration	Copepoda	P	Bollens and Frost (1991)
		M	Spines	Bryozoa	B	Harvell (1986)
		M	Shell thickness, growth rate, byssal thread #	Bivalvia	B	Caro and Castilla (2004); Leonard et al. (1999)
		M, B	Shell thickness, growth & feeding rate	Gastropoda	B	Appleton and Palmer (1988); Palmer (1990); Dalziel and Boulding (2005)
		M, B	Shell growth, avoidance	Gastropoda	B	Behrens Yamada and Boulding (1998)
		P	Phlorotannin, polyphenolic production	Phaeophyta	B	Toth and Pavia (2000); Van Alstyne (1988)
	Chemical, tactile or visual[g]	B	Burial response	Bivalvia	B	Whitlow et al. (2003)
		B	Foraging rate	Gastropoda	B	Trussell et al. (2003)
		M	Spine length	Decapoda	B	Davis et al. (2005)
		M	Body form	Cirripedia	B	Lively et al. (2000)
		M	Shell thickness	Bivalvia, Gastropoda	B	Smith and Jennings (2000); Trussell (1996); Trussell and Smith (2000)
Damage[h]		LH	Reproductive timing	Anthophyta, Bryozoa	B	Bortolus et al. (2004); Harvell and Helling (1993)
		M	Spicule concentration	Porifera	B	Hill and Hill (2002)
		M	Sclerite size and density	Anthozoa	B	West (1997)
		M	Toughness/tensile strength	Phaeophyta	B	Lowell et al. (1991)
		P	Polyphenolic production	Phaeophyta	B	Yates and Peckol (1993); Peckol et al. (1996)
Competitor	Density	LH	Reproductive timing	Bryozoa	B	Harvell and Helling (1993)
		M	Stolons	Bryozoa	B	Harvell and Padilla (1990)

(continued)

Table10. 1 (continued)

General cue	Inducer	Response type[a]	Characters	Taxon	Life Habit[b]	Reference[c]
		M	Giant cannibal morph, lateral wings	Ciliata	P	Kuhlmann et al. (1999)
		M	Shell form	Gastropoda	B	Boulding and Hay (1993)
		M	Sweeper tentacles	Anthozoa	B	Miles (1991); Harvell (1999)
		M, B	Gender	Osteichthyes	N	Munoz and Warner (2003)

[a] Response types: B, Behavioral, LH, Life history; M, Morphological; P, Physiological
[b] Life habit: B, benthos; N, nekton; P, plankton
[c] Reference in bold & italicized if any study organism not native to system
[d] Temperature is or is likely to be an additional cue
[e] Foraging time modified, so food quantity an additional cue
[f] Consumer: predator or herbivore
[g] Unable to distinguish between inducers
[h] Damage usually simulated by experimenter

digestive chemistry in polychaetes (Bock and Mayer 1999), and development time in echinoid larvae (Heyland and Hodin 2004). The presence of consumers triggered diel vertical migration in copepods (Bollens and Frost 1991), production of polyphenolic defenses in brown algae (van Alstyne 1988), spine formation in bryozoans (Harvell 1986), and increases in shell thickness in gastropod snails (Appleton and Palmer 1988). Responses to variation in the physical factors included physiological acclimation and morphological changes. Plasticity may often reflect responses to multiple factors (Etter 1996; Peckol et al. 1996).

10.4 Where Phenotypic Plasticity Might Play a Role in Marine Invasions

The ubiquity of environmentally induced changes in marine systems strongly suggests that the phenomenon could play an important role in marine invasions by either facilitating the introduction of a non-native marine species or damping its impact should native species respond. Few studies have tested for plasticity's role in marine invasions, but some general patterns from studies on marine invasions are beginning to emerge. These suggest that future studies might focus on particular habitats, taxonomic groups, stages in the invasion sequence, or vectors that transport marine species.

10.4.1 Habitats

Organisms must adjust to prevailing biotic and abiotic conditions in the new habitat if they are to survive. For the invader, the 'new' habitat includes all environments encountered during the relatively short periods of uptake and transfer (discussed below) as well as conditions experienced over longer time frames after release. Plasticity would be expected to contribute to post-release success in variable habitats (or microhabitats) provided (1) the fluctuations are sufficiently fine-grained for the organism to experience them and (2) the organism has the capacity to change its phenotype at the appropriate time scale (Miner and Vonesh 2004; Miner et al. 2005). At a gross level, plasticity may be more critical for organisms released into estuaries, which often vary extensively in salinity, temperature, and turbidity, than in the more uniform conditions of the open ocean. Because potential invaders are most likely to be released in ports or other developed coastal areas, plasticity for physiological tolerance to pollutants or hypoxic conditions might be particularly advantageous (see Chap. 7, Johnston et al.). Plasticity might also play a greater role in latitudes (or depths) that are more diurnally or seasonally variable than in those that are thermally constant. These spatial scales, however, may be too broad; all environments vary at some scale, and plasticity, in its many manifestations, has the potential to influence invasion success where variation exists.

10.4.2 Taxonomic Groups

The list of marine organisms known either to exhibit phenotypic plasticity or to have been introduced somewhere in the world's oceans is taxonomically diverse. Plasticity studies on those groups that are transported or introduced most frequently, however, might yield useful taxonomic generalities. For example, in a survey of marine coastal introductions in North America, crustaceans, molluscs, annelids and algae (in descending order) were responsible for approximately 70% of the nearly 300 invasions (Ruiz et al. 2000). Although few studies have tested for plasticity in marine crustaceans, recent findings of diet- and flow-related changes in trophic structures (Smith and Palmer 1994; Arsenault et al. 2001; Marchinko and Palmer 2003) and predator-induced spine elongation and diel vertical migration (Bollens and Frost 1991; Frost and Bollens 1992; Davis et al. 2005) suggest that plasticity may contribute to their invasion success. Plastic responses by planktonic copepods deserve particular scrutiny, because they are the most abundant metazoan in ballast water (Lavoie et al. 1999; Smith et al. 1999). In contrast to crustaceans, plasticity in molluscs is better known and modifications in their behavior, shell form or life history have been documented in a number of invasions (Trussell and Smith 2000; Trussell et al. 2003; Whitlow et al. 2003; Ernande et al. 2004). Marine algae are extremely plastic, both physiologically and morphologically, and this flexibility may have contributed to the spread of the green alga *Caulerpa racemosa* in the Mediterranean (Raniello et al. 2004). Finally, diatoms and dinoflagellates are common constituents of the ballast water biota (and undoubtedly overlooked as invaders; see Chap. 2, Carlton). Plasticity in response to nutrient and light levels occurs in both groups (Gallagher et al. 1984; Lee ES et al. 1999) and could profoundly influence their survival during ballast water transport.

10.4.3 Steps in the Invasion Sequence

The degree to which plasticity influences an introduction may differ depending on the particular stage of an invasion sequence. A successful introduction consists of an organism's uptake by a vector, transport from the donor to the receiving region, release into the new habitat, establishment of a reproducing population, and geographic spread and integration. Studies investigating plasticity's role in terrestrial and freshwater invasions have focused on post-introduction (stages 4 and 5) phenotypic change. This sequential skew probably reflects multiple factors including the difficulty and uncertainty associated with predicting which species will invade; long-standing interest in elucidating critical post-introduction ecological consequences; the practical necessity of control or eradication; and the relative ease with which later steps can be examined. These considerations aside, phenotypic plasticity has the potential to increase survival at any step in the invasion sequence. With mechanized travel, technological advances, and improved shipping practices, the uptake, transport, and release of species occur more quickly now than in the past. As a consequence, adaptive responses to environmental variation in the initial stages of an invasion may tend to be behavioral or physiological in nature (but see shipping, below). The later stages (establishment, spread, and integration) play out over longer time frames (weeks to centuries); consequently, plasticity later in the invasion sequence may also encompass morphological changes or life history shifts and selection on reaction norms.

10.4.4 Vectors

The relative importance of phenotypic plasticity in the invasion sequence will almost certainly depend on the type of vector and purpose for moving the species. For example, the importance of plasticity early in the invasion sequence may depend on which organisms, if any, are targeted for transport; whether the objective is to promote their survival; and the extent to which environmental conditions fluctuate during transport. Primary mechanisms for moving marine species include: shipping (ballast water, hull fouling); mariculture, seafood, bait, and marine ornamentals industries (see Chap. 5, Minchin et al.; Chap. 6, Hewitt et al.); research and educational operations; and coastal restoration projects (Weigle et al. 2004). Of these, shipping is the most important vector for transporting marine species both in terms of numbers of organisms transported and geographic scope (Carlton and Geller 1993; Ruiz et al. 2000). Shipping is also the least selective vector in terms of uptake of organisms, provides a largely uncontrolled environment during transport, and releases organisms abruptly into the receiving environment. Thus, organisms transported by ships may rely to a greater extent on plastic responses than for any other vector.

10.4.4.1 Shipping Pathways

Ships transfer organisms in ballast water and on the surface of the hull, and plasticity could influence which individuals or taxa are taken up. Behavioral plasticity, in particular, could make some organisms more prone than others to being sampled or captured. Sih et al. (2004) discussed the potential importance of behavioral syndromes to plasticity and invasions. Behavioral syndromes are suites of correlated behaviors expressed consistently among individuals in multiple situations. Studies have shown that behavioral characteristics that favor high dispersal rates, such as boldness, aggressiveness, and high activity, are often correlated. Plasticity in this 'aggressive' syndrome could determine which individuals make the trip or not.

The rapid and often extreme changes in environmental conditions encountered during a ship's transit undoubtedly stress organisms physiologically and, in doing so, may favor those individuals capable of short-term adaptive responses.

Selective pressures experienced in transit are probably intense given the high mortality rates that have been recorded (e.g., up to 92% for voyages ~16–22 days; Verling et al. 2005). Once inside a ballast tank, organisms must cope with an absence of light, probable reductions in nutrients or available prey over time, changes in temperature and oxygen levels, and competition with other organisms entrained in the water parcel (Carlton 1985; Chap. 6, Hewitt et al.). Adaptive phenotypic responses to these environmental changes would enhance survival over short voyages, particularly for vulnerable early life history stages, and may be critical for all life history stages as voyage duration increases. For survivors of the transit, short-term responses to environmental variation could also have longer-term (i.e., post-invasion) repercussions. For example, food limitation in a variety of marine invertebrates can affect fecundity of adults, the degree to which adults provision eggs, and the size, morphology, and development time of larvae (Hadfield and Strathmann 1996).

Fluctuations in water temperature, salinity, and flow speed encountered by organisms that foul ship's hulls or sea chests (Coutts et al. 2003) should be more extreme than those experienced by organisms in ballast tanks. Hull-fouling organisms must cope with ambient water conditions as ships move between tropical and temperate regions or between high and low salinity ports. Many species are capable of rapid physiological acclimation to changing temperature or salinity (Gandolfi et al. 2001; Charmantier et al. 2002), and these should be favored during transit. Organisms attached to the hull must also survive a range of flow speeds. In response to variable flow rates, suspension-feeding barnacles fine-tune the length of their cirri (Marchinko and Palmer 2003), sponges modify stiffness of the body wall (Palumbi 1984), and snails alter foot size (Etter 1996). Although such morphological responses take time, hitchhikers on ships' hulls and in sea chests may travel oceans for years.

Once organisms are released into the new environment, phenotypic plasticity may permit them to adjust to abrupt transitions and stressful conditions of the receiving waters or extend time in the plankton until suitable settlement sites are found. Extreme differences in environmental conditions will exceed all physiological responses, but plasticity should become increasingly important to survival as physical-chemical differences narrow. For example, Smith et al. (1999) examined survivorship of various invertebrate taxa that were collected from ballast tanks and maintained experimentally at different combinations of temperature and salinity for two weeks. Not surprisingly, highest mortality was documented in instances in which temperatures and salinities of the ballast water and experimental conditions differed the most. Survivorship improved as temperature and salinity differences narrowed, and survivorship patterns, although complex, revealed substantial tolerance in the polychaetes, bivalves, and crustaceans tested.

10.4.4.2 Non-shipping Pathways

Phenotypic plasticity may play a less critical role during the uptake and transit stages of non-shipping pathways than it does for the shipping pathway. First, compared to the indiscriminate sampling of the water column during ballasting uptake, non-shipping pathways generally target specific taxa for transport. Plastic behavioral responses to environmental cues are unlikely to affect which baitworms or soft-shell clams are taken, although behavioral plasticity could conceivably increase the likelihood of capture of certain target or non-target species (e.g., more aggressive ornamental fish; crabs hiding among harvested oysters). Second, organizations involved in non-shipping pathways want their products to survive transit, so they strive to keep environmental conditions as stable as possible. Better handling and packaging methods and faster shipments should reduce environmental fluctuations for organisms moved by researchers or the seafood, bait, aquaculture, biological supply, or marine ornamentals trades (see Chap. 6, Hewitt et al.). Many of these species, however, may have been selected for trade in the first place because they were physiologically plastic and, thus, better able to tolerate environmental changes during transit.

The importance of phenotypic plasticity to organisms during or after release by non-shipping pathways may hinge on whether their survival in the wild (vs in contained systems) is desired or not. If survival in the wild is desired, then the entity (e.g., a mariculturist) handling the organism has probably sought to minimize differences between source and receiving environments. As a consequence, plasticity may become less critical. If, however, release is either accidental or a by-product of an activity (e.g., dumping unused bait worms, unwanted pets, or holding water) (Weigle et al. 2004), plasticity may be more important for survival. As in the case of ballast water release, an individual's ability to adjust physiologically, behaviorally, or ecologically should improve as environmental differences between source and receiving waters narrow. Thus, marine ornamental species, which are typically tropical in origin, will almost certainly die if released accidentally or deliberately into cold New England waters. In contrast, chances of survival would improve greatly if tropical marine ornamentals were released in the warm waters of the southeastern U.S.

10.5 Examples of Phenotypic Plasticity's Effect in Marine Invasions

10.5.1 Post-invasion Effects

The role of phenotypic plasticity in marine invasions is assessed most easily in the period after an introduction, when the invader is known and the time frame for study is longer. Results of several studies indicate that plasticity could contribute to the survival and spread of a marine species after its introduction (Table 10.1). For example, Ernande et al. (2004) found shifts in resource allocation in life history traits (survival, growth, reproductive effort) of the Pacific oyster *Crassostrea gigas* in response to variation in food abundance. In the same species, Hamdoun et al. (2003) documented increased expression of heat-shock proteins and greater thermal tolerance as water temperatures rose in summer. In each instance, phenotypic responses helped to explain the oyster's ability to cope with local environmental heterogeneity and might reasonably have been expected to contribute to the establishment and spread of this species. Neither study, however, focused on the fact that the species was introduced. Other workers have documented adaptive plasticity in species in their native range (e.g., bryozoan *Membranipora membranacea*) (Harvell 1986) that coincidentally have invaded elsewhere. Without direct tests, we can only speculate as to whether fitness advantages in the native habitat would be similar in the new environments.

Only a few studies have studied plasticity explicitly in the context of marine invasions. For example, Castillo et al. (2005) examined physiological tolerance of the invasive cordgrass *Spartina densiflora* to a range of salinities in greenhouse experiments. Their results showed high acclimation capacity and good performance between 0.5 and 20 ppt, which helped to explain the rapid colonization of brackish areas and slower expansion rates in high salinity areas in Europe. Raniello et al. (2004) demonstrated photosynthetic plasticity in the invasive tropical alga *Caulerpa racemosa* in response to light and temperature. Their results suggested that *C. racemosa*'s colonization success resulted from its ability to exploit low light conditions found underneath native seagrass canopies and to acclimate to winter temperatures in the Mediterranean. Garbary et al. (2004) linked invasion success of the alga *Codium fragile* in eelgrass-dominated beds in eastern Canada to advantageous growth forms induced in the alga by the substratum.

10.5.2 The Invasion of *Carcinus maenas* in the Gulf of Maine

To date, the most comprehensive examination of phenotypic plasticity's post-invasion ecological repercussions involve studies of the European green crab *Carcinus maenas* in northeastern North America. First introduced to the mid-Atlantic coast of North America from Europe in the early 1800s, *C. maenas* expanded its range north of Cape Cod, Massachusetts into the Gulf of Maine around 1900, and reached its present northern extent in the Canadian Maritime provinces by the early 1950s (Scattergood 1952; Welch 1968). The presence of this shell-crushing predator has been shown experimentally to induce behavioral and morphological shifts in native molluscan prey in both soft- and hard-bottom habitats. For example, soft-shell clams *Mya arenaria* burrowed deeper in sediments (Whitlow et al. 2003), and two intertidal snails limited their foraging and sought refuge in the presence of *C. maenas* (Trussell et al. 2003). The green crab's presence also induced snails and mussels to increase shell thickness (and, in the latter, produce more byssal threads) (Trussell 1996; Leonard et al. 1999; Smith and Jennings 2000; Trussell and Smith 2000). In each case, the phenotypic response was either hypothesized or demonstrated to reduce prey vulnerability. Costs of the defensive responses generally involved reduced feeding efficiency or slower growth rates.

Trait modifications induced by *Carcinus maenas* can also affect abundances of other species in the immediate community. For example, Trussell et al. (2003) demonstrated in laboratory experiments that increased risk of predation by *C. maenas* suppressed drilling by dogwhelks on barnacles and grazing by periwinkles on fucoid algae. Similarly, in field experiments, the presence of *C. maenas* appeared to limit grazing by juvenile periwinkles and resulted in higher densities of fucoid algae on experimental tiles than in treatments without crabs (Trussell et al. 2002). The authors concluded that the presence of *C. maenas* could shape community composition indirectly by inducing behavioral changes in their prey.

Carcinus maenas' ability to modify prey morphology has led to reinterpretation of the mechanism driving an historical shift and a biogeographic pattern in shell form in the Gulf of Maine. In the century after the green crab's arrival, shell thickness in the periwinkle *Littorina obtusata* increased between 50–82% and in the dogwhelk *Nucella lapillus* by 12%. These shifts were initially attributed to rapid selection for thick-shelled morphs (Vermeij 1982b; Seeley 1986); however, results of a more recent reciprocal transplant experiment suggest that predator-induced plasticity can account for such change (Trussell and Smith 2000). Phenotypic plasticity may also account for a present-day latitudinal cline in shell thickness in *L. obtusata*. Experimental work suggests that thicker shelled snail populations in the southern Gulf of Maine may be responding to the combined effects of warmer waters and higher *C. maenas* abundances (Trussell 2000; Trussell and Smith 2000). In contrast, thinner shelled *L. obtusata* populations in the northern Gulf of Maine reflect the induced effects of colder water temperatures (which limit shell calcification) and fewer predators. Reciprocal transplant experiments have revealed significant source × rearing location effects in morphological responses of northern and southern snail populations (Trussell 2000; Trussell and Etter 2001). These G × E interactions indicate genetic variation in plasticity between populations although maternal effects or effects induced early in development cannot be discounted.

Recent findings suggest that *Carcinus maenas*' range expansion in the Gulf of Maine may have initiated an ecological arms race. Claw sizes of *Carcinus maenas* vary in a manner consistent with the pattern of breakage resistance in *Littorina obtusata* (Smith 2004). Crabs from the southern Gulf of Maine, where thicker shelled *L. obtusata* were found, had relatively larger crusher (but not cutter) claws than their northern counterparts. Laboratory experiments suggest this spatial pattern may reflect a temperature-dependent induced response to prey defenses (Baldridge 2006). Crabs that were raised at warm temperature on a diet of thicker shelled *L. obtusata* exhibited greater increases in crusher (but not cutter) claw size than did crabs reared on thinner shelled *L. obtusata*. In contrast, crabs reared in colder water (typical of summertime temperatures in Maine) showed reduced foraging activity, no differences in diet treatment effects on crusher or cutter claw size, and longer intermolt periods. Colder water temperatures at the northern end of the *C. maenas*' range, then, may limit claw change and give snails there an advantage that is absent further south. Perhaps the most important conclusion to arise from these and earlier studies is that induced responses following a marine invasion can generate broad scale spatial and historical patterns of phenotypic variation.

Taken together, the ecophenotypic responses associated with the *Carcinus maenas* invasion illustrate the far-reaching and often complex consequences of phenotypic plasticity in a marine invasion. Given that the green crab has invaded Australia, Japan, South Africa, and the west coast of North America at different time points in the last two centuries (Grosholz and Ruiz 1996), we can test for the generality of plasticity's effects at these other sites. Comparable investigations of adaptive plasticity in other marine invasions are essential if we are to understand the extent of its influence on invasion process and pattern.

10.6 Future Research Directions

Biological invasions provide a useful tool to examine questions pertaining to phenotypic plasticity's ecological and evolutionary importance (e.g., does plasticity influence niche construction, population stability or species co-existence?), in part, because we know when the species-environment interactions begin. Our understanding of plasticity's significance for biological invasions, however, is at an embryonic stage (Richards et al. 2006). A better understanding of phenotypic plasticity's role may help in our interpretation of many of the key questions in invasion biology. Does phenotypic plasticity contribute to biotic resistance to invasions? Does it enhance survival of the invader in disturbed habitats? Does it help explain post-invasion lag times or increased rates of dispersal? We may benefit by framing our investigations to address questions about phenotypic plasticity's role in: (1) the *processes* that result in an invasion, (2) the changes in *patterns* that are generated by an invasion, and (3) the *predictions* that can be made concerning impacts and range limits of invasive species.

In terms of process, we should examine *how* various expressions of phenotypic plasticity facilitate the introduction or spread of invasive species and the extent to which phenotypic responses by native species impede the invader. For example, our understanding of the invasion process will be improved if we test for phenotypic plasticity's effects during the transfer period. In particular, phenotypic responses by holoplanktonic organisms and early life history stages of meroplankton to the ballast water environment may greatly influence an organism's energetic reserves, time to metamorphosis, post-metamorphic size, and survival, in transit and after release.

Where phenotypic plasticity is tested, investigators would do well to determine the nature and magnitude of the response(s), whether responses are in fact adaptive, and costs associated with plasticity. The most direct methods to test these questions are to subject experimental organisms to different levels of an environmental cue, look for a phenotypic response in a fitness-related trait, and test for associated effects on performance or survival. Although rarely performed, a complete assessment of the adaptive nature of the plastic response would test for fitness in all phenotype-environment combinations (i.e., induced forms in original and new environments, non-induced forms in original and new environments) (Doughty and

Reznick 2004). Other experiments are needed to assess (1) the capacity of both the invader and the native species to recognize cues in their new environments, (2) the potential for reciprocal phenotypic responses, and (3) the extent to which trait changes mediate abundances of other members of the community.

Common to all of these questions is the difficult problem of assessing the effects of multiple cues on phenotypic variation. For example, what is the cumulative effect on an organism's phenotype when it is exposed to variation in food resources, competitors, and predators in different temperature and flow regimes? Such a multi-factor experiment would be logistically impossible, but experiments do need to move toward testing subsets of relevant cues over biologically realistic ranges. At a minimum, researchers need to examine how an organism's response to an environmental cue of interest interacts with variation in at least one other important biotic or abiotic cue in their system. For example, latitudinal gradients in water temperature or additional competitor species could influence the magnitude of phenotypic responses in an invader and either limit or enhance its ecological impact or geographic extent. In designing experiments or interpreting results, researchers need to be cognizant of the complex interplay between (1) the nature and reliability of the cue, (2) the type, time course, and reversibility of the phenotypic response, (3) taxon-specific constraints (e.g., in patterns of growth), and (4) how each of the aforementioned might be modified by additional environmental cues.

As illustrated by the *Carcinus maenas* invasion, an assessment of phenotypic plasticity is essential if we are to interpret patterns of phenotypic variation generated by an invasion. Common garden experiments can be used to test whether observed phenotypic variation is largely genetic or environmental (Parker et al. 2003). Reciprocal transplant experiments can be used to distinguish the relative contributions of the source population and the rearing environment to phenotypic variation and determine whether a population is most fit in a given environment. Ethical considerations should preclude reciprocal transplant experiments in the field if invasive species are to be moved outside their existing range. In such situations, simulated conditions in the laboratory will have to suffice. The genetic contribution to phenotypic variation can be determined by designing the experiment with some level of family structure (clones, half-sibs, full-sibs). While this type of experimental design is well suited to clonally reproducing marine organisms (e.g., bryozoans, cnidarians), it may be logistically infeasible for non-clonal organisms or ones that have long-lived planktonic phases or complex development. If genetic and environmental contributions can be determined, then experiments can test for selection on reaction norms (Kingsolver and Huey 1998; Lee CE et al. 2003). Similarly, with this knowledge, we may interpret the extent to which environmental and genetic influences are working in concert (co-gradient variation) or in opposition (counter-gradient variation) to generate large-scale spatial patterns (Conover and Schultz 1995; Trussell and Etter 2001).

Finally, information on phenotypic plasticity's contribution to invasion process and pattern should allow for better predictions of the ecological impacts of an invasion. Controlled experiments can be used to test relative fitness or performance in environmental conditions inside and outside an invader's current range. Depending on the direction and magnitude of the phenotypic response, current and future impacts of the invasive species on native species might be predicted with some confidence. Such information, in combination with knowledge of the potential for evolutionary change, should greatly assist biologists in determining both the short- and long-term risks associated with a given invasion.

Acknowledgments I thank E. Callaghan, D. Lello, and an anonymous reviewer for their editorial comments.

References

Agrawal AA (2001) Phenotypic plasticity in the interactions and evolution of species. Science 294:321–326

Agrawal AA, Laforsch C, Tollrian R (1999) Transgenerational induction of defenses in animals and plants. Nature 401:60–63

Appleton RD, Palmer AR (1988) Water-borne stimuli released by predatory crabs and damaged prey induce more predator-resistant shells in a marine gastropod. Proc Natl Acad Sci U S A 85:4387–4391

Arsenault DJ, Marchinko KB, Palmer AR (2001) Precise tuning of barnacle leg length to coastal wave action. Proc R Soc London B 268:2149–2154

Baldridge A (2006) The role of phenotypic plasticity in the interaction between an invasive crab predator, *Carcinus maenas,* and its native snail prey, *Littorina obtusata*. Masters thesis. Smith College, Northampton, Massachusetts

Behrens Yamada S, Boulding EG (1998) Claw morphology, prey size selection and foraging efficiency in generalist and specialist shell-breaking crabs. J Exp Mar Biol Ecol 220:191–211

Berrigan D, Scheiner SM (2004) Modeling the evolution of phenotypic plasticity. In: DeWitt TJ, Scheiner SM (eds) Phenotypic plasticity: functional and conceptual approaches. Oxford University Press, New York, p 82–97

Bock MJ, Mayer LM (1999) Digestive plasticity of the marine benthic omnivore *Nereis virens*. J Exp Mar Biol Ecol 240:77–92

Bollens SM, Frost BW (1991) Diel vertical migration in zooplankton: rapid individual-response to predators. J Plankton Res 13:1359–1365

Bortolus A, Laterra P, Iribarne O (2004) Crab-mediated phenotypic changes in *Spartina densiflora* Brong. Estuarine Coastal Shelf Sci 59:97–107

Boulding EG, Hay TK (1993) Quantitative genetics of shell form of an intertidal snail: constraints on short-term response to selection. Evolution 47:576–592

Bradshaw AD (1965) Evolutionary significance of phenotypic plasticity in plants. Adv Genet 13:115–155

Byers DL, Quinn JA (1998) Demographic variation in *Alliaria petiolata* (Brassicaceae) in four contrasting habitats. J Torrey Bot Soc 125:138–149

Callaway RM, Pennings SC, Richards CL (2003) Phenotypic plasticity and interactions among plants. Ecology 84:1115–1128

Carlton JT (1985) Transoceanic and interoceanic dispersal of coastal marine organisms: the biology of ballast water. Oceanogr Mar Biol Annu Rev 23:313–371

Carlton JT, Geller JB (1993) Ecological roulette: the global transport of nonindigenous marine organisms. Science 261:78–82

Caro AU, Castilla JC (2004) Predator-inducible defenses and local intra-population variability of the intertidal mussel *Semimytilus algosus* in central Chile. Mar Ecol Prog Ser 276:115–123

Carroll SP, Dingle H (1996) The biology of post-invasion events. Biol Conserv 78:207–214

Castillo JM, Rubio-Casal AE, Redondo S, Alvarez-López AA, Luque T, Luque C, Nieva FJ, Castellanos EM, Figueroa ME (2005) Short-term responses to salinity of an invasive cordgrass. Biol Invas 7:29–35

Charmantier G, Giménez L, Charmantier-Daures M, Anger K (2002) Ontogeny of osmoregulation, physiological plasticity and larval export strategy in the grapsid crab *Chasmagnathus granulata* (Crustacea, Decapoda). Mar Ecol Prog Ser 229:185–194

Chester CM (1996) The effect of adult nutrition on the reproduction and development of the estuarine nudibranch, *Tenellia adspersa* (Nordmann, 1845). J Exp Mar Biol Ecol 198:113–130

Clements DR, DiTommaso A, Jordan N, Booth BD, Cardina J, Doohan D, Mohler CL, Murphy SD, Swanton CJ (2004) Adaptability of plants invading North American cropland. Agric Ecosyst Environ 104:379–398

Conover DO, Schultz ET (1995) Phenotypic similarity and the evolutionary significance of countergradient variation. Trends Ecol Evol 10:248–252

Corti M, Loy A, Cataudella S (1996) Form changes in the sea bass, *Dicentrarchus labrax* (Moronidae: Teleostei), after acclimation to freshwater: an analysis using shape coordinates. Environ Biol Fish 47:165–175

Coutts ADM, Moore KM, Hewitt CL (2003) Ships' sea-chests: an overlooked transfer mechanism for non-indigenous marine species? Mar Pollut Bull 46:1504–1515

Cox GW (2004) Alien species and evolution. Island Press, Washington, DC

Daehler CC (2003) Performance comparisons of co-occurring native and alien invasive plants: implications for conservation and restoration. Annu Rev Ecol Evol Syst 34:183–211

Dalziel B, Boulding EG (2005) Water-borne cues from a shell-crushing predator induce a more massive shell in experimental populations of an intertidal snail. J Exp Mar Biol Ecol 317:25–35

David JR, Gibert P, Moreteau B (2004) Evolution of reaction norms. In: DeWitt TJ, Scheiner SM (eds) Phenotypic plasticity: functional and conceptual approaches. Oxford University Press, New York, pp 50–63

Davis JLD, Eckert-Mills MG, Young-Williams AC, Hines AH, Zohar Y (2005) Morphological conditioning of a hatchery-raised invertebrate, *Callinectes sapidus*, to improve field survivorship after release. Aquaculture 243:147–158

Day T, Pritchard J, Schluter D (1994) A comparison of two sticklebacks. Evolution 48:1723–1734

DeWitt TJ, Langerhans RB (2004) Integrated solutions to environmental heterogeneity: theory of multimoment reaction norms. In: DeWitt TJ, Scheiner SM (eds) Phenotypic plasticity: functional and conceptual approaches. Oxford University Press, New York, pp 98–111

DeWitt TJ, Scheiner SM (eds) (2004a) Phenotypic plasticity: functional and conceptual approaches. Oxford University Press, New York

DeWitt TJ, Scheiner SM (2004b) Phenotypic variation from single genotypes: a primer. In: DeWitt TJ, Scheiner SM (eds) Phenotypic plasticity: functional and conceptual approaches. Oxford University Press, New York, pp 1–9

DeWitt TJ, Sih A, Wilson DS (1998) Costs and limits of phenotypic plasticity. Trends Ecol Evol 13:77–81

Dill LM, Heithaus MR, Walters CJ (2003) Behaviorally mediated indirect interactions in marine communities and their conservation implications. Ecology 84:1151–1157

Doughty P, Reznick DN (2004) Patterns and analysis of adaptive phenotypic plasticity in animals. In: DeWitt TJ, Scheiner SM (eds) Phenotypic plasticity: functional and conceptual approaches. Oxford University Press, New York, pp 126–150

Drent J, Luttikhuizen PC, Piersma T (2004) Morphological dynamics in the foraging apparatus of a deposit feeding marine bivalve: phenotypic plasticity and heritable effects. Funct Ecol 18:349–356

Dzialowski AR, Lennon JT, O'Brien WJ, Smith VH (2003) Predator-induced phenotypic plasticity in the exotic cladoceran *Daphnia lumholtzi*. Freshwater Biol 48:1593–1602

Eastman JT, Devries AL (1997) Biology and phenotypic plasticity of the Antarctic nototheniid fish *Trematomus newnesi* in McMurdo Sound. Antarct Sci 9:27–35

Ebert TA (1996) Adaptive aspects of phenotypic plasticity in echinoderms. Oceanol Acta 19:347–355

Ellstrand NC, Schierenbeck KA (2000) Hybridization as a stimulus for the evolution of invasiveness in plants? Proc Natl Acad Sci U S A 97:7043–7050

Ernande B, Boudry P, Clobert J, Haure J (2004) Plasticity in resource allocation based life history traits in the Pacific oyster, *Crassostrea gigas*. I. Spatial variation in food abundance. J Evol Biol 17:342–356

Etter RJ (1988) Asymmetrical developmental plasticity in an intertidal snail. Evolution 42:322–334

Etter RJ (1996) The effect of wave action, prey type, and foraging time on growth of the predatory snail *Nucella lapillus* (L). J Exp Mar Biol Ecol 196:341–356

Fordyce JA (2006) The evolutionary consequences of ecological interactions mediated through phenotypic plasticity. J Exp Biol 209:2377–2383

Freeman AS, Byers JE (2006) Divergent induced responses to an invasive predator in marine mussel populations. Science 313:831–833

Frost BW, Bollens SM (1992) Variability of diel vertical migration in the marine planktonic copepod *Pseudocalanus newmani* in relation to its predators. Can J Fish Aquat Sci 49:1137–1141

Gallagher JC, Wood AM, Alberte RS (1984) Ecotypic differentiation in the marine diatom *Skeletonema costatum*: influence of light intensity on the photosynthetic apparatus. Mar Biol 82:121–134

Gandolfi A, Todeschi EBA, Van Doninck K, Rossi V, Menozzi P (2001) Salinity tolerance of *Darwinula stevensoni* (Crustacea, Ostracoda). Ital J Zool 68:61–67

Garbary DJ, Fraser SJ, Hubbard C, Kim KY (2004) *Codium fragile*: rhizomatous growth in the *Zostera* thief of eastern Canada. Helgol Mar Res 58:141–146

George SB (1999) Egg quality, larval growth and phenotypic plasticity in a forcipulate seastar. J Exp Mar Biol Ecol 237:203–224

Gilbert JJ (1999) Kairomone-induced morphological defenses in rotifers. In: Tollrian R, Harvell CD (eds) The ecology and evolution of inducible defenses. Princeton University Press, Princeton, NJ, pp 127–141

Gilchrist GW, Huey RB, Serra L (2001) Rapid evolution of wing size clines in *Drosophila subobscura*. Genetica 112/113:273–286

Gotthard K, Nylin S (1995) Adaptive plasticity and plasticity as an adaptation: a selective review of plasticity in animal morphology and life history. Oikos 74:3–17

Govind CK (1987) Muscle and muscle fiber type transformation in clawed crustaceans. Am Zool 27:1079–1098

Grosholz ED, Ruiz GM (1996) Predicting the impact of introduced marine species: lessons from the multiple invasions of the European green crab *Carcinus maenas*. Biol Conserv 78:59–66

Hadfield MG, Strathmann MF (1996) Variability, flexibility and plasticity in life histories of marine invertebrates. Oceanol Acta 19:323–334

Hamdoun AM, Cheney DP, Cherr GN (2003) Phenotypic plasticity of HSP70 and HSP70 gene expression in the Pacific oyster (*Crassostrea gigas*): implications for thermal limits and induction of thermal tolerance. Biol Bull 205:160–169

Hart MW, Strathmann RR (1994) Functional consequences of phenotypic plasticity in echinoid larvae. Biol Bull 186:291–299

Harvell CD (1986) The ecology and evolution of inducible defenses in a marine bryozoan: cues, costs, and consequences. Am Nat 128:810–823

Harvell CD (1999) Complex biotic environments, coloniality, and heritable variation for inducible defenses. In: Tollrian R, Harvell CD (eds) The ecology and evolution of inducible defenses. Princeton University Press, Princeton, NJ, pp 231–244

Harvell CD, Helling R (1993) Experimental induction of localized reproduction in a marine bryozoan. Biol Bull 184:286–295

Harvell CD, Padilla DK (1990) Inducible morphology, heterochrony, and size hierarchies in a colonial invertebrate monoculture. Proc Natl Acad Sci U S A 87:508–512

Heyland A, Hodin J (2004) Heterochronic developmental shift caused by thyroid hormone in larval sand dollars and its implications for phenotypic plasticity and the evolution of nonfeeding development. Evolution 58:524–538

Hill MS, Hill AL (2002) Morphological plasticity in the tropical sponge *Anthosigmella varians*: responses to predators and wave energy. Biol Bull 202:86–95

Huey RB, Gilchrist GW, Carlson ML, Berrigan D, Serra L (2000) Rapid evolution of geographic cline in size in an introduced fly. Science 287:308–309

Janzen FJ (1995) Experimental evidence for the evolutionary significance of temperature-dependent sex determination. Evolution 49:864–873

Kingsolver JG, Huey RB (1998) Evolutionary analyses of morphological and physiological plasticity in thermally variable environments. Am Zool 38:545–560

Kinlan BP, Gaines SD (2003) Propagule dispersal in marine and terrestrial environments: a community perspective. Ecology 84:2007–2020

Kollmann J, Banuelos MJ (2004) Latitudinal trends in growth and phenology of the invasive alien plant *Impatiens glandulifera* (Balsaminaceae). Divers Distrib 10:377–385

Kuhlmann H-W, Kusch J, Heckmann K (1999) Predator-induced defenses in ciliated protozoa. In: Tollrian R, Harvell CD (eds) The ecology and evolution of inducible defenses. Princeton University Press, Princeton, NJ, pp 142–159

Lavoie DM, Smith LD, Ruiz GM (1999) The potential for intracoastal transfer of non-indigenous species in the ballast water of ships. Estuarine Coastal Shelf Sci 48:551–564

Leclaire M, Brandl R (1994) Phenotypic plasticity and nutrition in a phytophagous insect: consequences of colonizing a new host. Oecologia 100:379–385

Lee CE, Petersen CH (2002) Genotype-by-environment interaction for salinity tolerance in the freshwater-invading copepod *Eurytemora affinis*. Physiol Biochem Zool 75:335–344

Lee CE, Remfert JL, Gelembiuk GW (2003) Evolution of physiological tolerance and performance during freshwater invasions. Integr Comp Biol 43:439–449

Lee ES, Lewitus AJ, Zimmer RK (1999) Chemoreception in a marine cryptophyte: behavioral plasticity in response to amino acids and nitrate. Limnol Oceanogr 44:1571–1574

Leonard GH, Bertness MD, Yund PO (1999) Crab predation, waterborne cues, and inducible defenses in the blue mussel, *Mytilus edulis*. Ecology 80:1–14

Lively CM, Hazel WN, Schellenberger MJ, Michelson KS (2000) Predator-induced defense: variation for inducibility in an intertidal barnacle. Ecology 81:1240–1247

Losos JB, Warheit KI, Schoener TW (1997) Adaptive differentiation following experimental island colonization in *Anolis* lizards. Nature 387:70–73

Losos JB, Creer DA, Glossip D, Goellner R, Hampton A, Roberts G, Haskell N, Taylor P, Ettling J (2000) Evolutionary implications of phenotypic plasticity in the hindlimb of the lizard *Anolis sagrei*. Evolution 301–305

Lowell RB, Markham JH, Mann KH (1991) Herbivore-like damage induces increased strength and toughness in a seaweed. Proc R Soc London B 243:31–38

Marchinko KB (2003) Dramatic phenotypic plasticity in barnacle legs (*Balanus glandula* Darwin): magnitude, age dependence, and speed of response. Evolution 57:1281–1290

Marchinko KB, Palmer AR (2003) Feeding in flow extremes: dependence of cirrus form on wave-exposure in four barnacle species. Zoology 106:127–141

Marinovic B, Mangel M (1999) Krill can shrink as an ecological adaptation to temporarily unfavourable environments. Ecol Lett 2:338–343

Milberg P, Lamont BB, Perez-Fernandez MA (1999) Survival and growth of native and exotic composites in response to a nutrient gradient. Plant Ecol 145:125–132

Miles JS (1991) Inducible agonistic structures in the tropical corallimorpharian, *Discosoma sanctithomae*. Biol Bull 180:406–415

Miner BG, Vonesh JR (2004) Effects of fine grain environmental variability on morphological plasticity. Ecol Lett 7:794–801

Miner BG, Sultan SE, Morgan SG, Padilla DK, Relyea RA (2005) Ecological consequences of phenotypic plasticity. Trends Ecol Evol 20:685–692

Monro K, Poore AGB (2005) Light quantity and quality induce shade-avoiding plasticity in a marine macroalga. J Evol Biol 18:426–435

Mooney HA, Cleland EE (2001) The evolutionary impact of invasive species. Proc Natl Acad Sci U S A 98:5446–5451

Moore RD, Griffiths RA, O'Brien CM, Murphy A, Jay D (2004) Induced defenses in an endangered amphibian in response to an introduced snake predator. Oecologia 141:139–147

Muñoz RC, Warner RR (2003) Alternative contexts of sex change with social control in the bucktooth parrotfish, *Sparisoma radians*. Environ Biol Fish 68:307–319

Niinemets U, Valladares F, Ceulemans R (2003) Leaf-level phenotypic variability and plasticity of invasive *Rhododendron ponticum* and non-invasive *Ilex aquifolium* co-occurring at two contrasting European sites. Plant Cell Environ 26:941–956

Padilla DK (1998) Inducible phenotypic plasticity of the radula in *Lacuna* (Gastropoda: Littorinidae). Veliger 41:201–204

Palmer AR (1990) Effect of crab effluent and scent of damaged conspecifics on feeding, growth, and shell morphology of the Atlantic dogwhelk *Nucella lapillus* (L.). Hydrobiologia 193:155–182

Palumbi SR (1984) Tactics of acclimation: morphological changes of sponges in an unpredictable environment. Science 225:1478–1480

Parker IM, Rodriguez J, Loik ME (2003) An evolutionary approach to understanding the biology of invasions: local adaptation and general-purpose genotypes in the weed *Verbascum thapsus*. Conserv Biol 17:59–72

Peckol P, Krane JM, Yates JL (1996) Interactive effects of inducible defense and resource availability on phlorotannins in the North Atlantic brown alga *Fucus vesiculosus*. Mar Ecol Progr Ser 138:209–217

Piersma T, Drent J (2003) Phenotypic flexibility and the evolution of organismal design. Trends Ecology Evol 18:228–233

Pigliucci M (2001) Phenotypic plasticity: beyond nature and nurture. Johns Hopkins University Press, Baltimore, Maryland

Podolsky RD, McAlister JS (2005) Developmental plasticity in *Macrophiothrix* brittlestars: are morphologically convergent larvae also convergently plastic? Biol Bull 209:127–138

Raimondi PT, Forde SE, Delph LF, Lively CM (2000) Processes structuring communities: evidence for trait-mediated indirect effects through induced polymorphisms. Oikos 91:353–361

Raniello R, Lorenti M, Brunet C, Buia MC (2004) Photosynthetic plasticity of an invasive variety of *Caulerpa racemosa* in a coastal Mediterranean area: light harvesting capacity and seasonal acclimation. Mar Ecol Progr Ser 271:113–120

Rejmánek M, Richardson DM (1996) What attributes make some plant species more invasive? Ecology 77:1655–1661

Reznick DN, Ghalambor CK (2001) The population ecology of contemporary adaptations: what empirical studies reveal about the conditions that promote adaptive evolution? Genetica 112/113:183–198

Richards CL, Bossdorf O, Muth NZ, Gurevitch J, Pigliucci M (2006) Jack of all trades, master of some? On the role of phenotypic plasticity in plant invasions. Ecol Lett 9:981–993

Ruiz GM, Fofonoff PW, Carlton JT, Wonham MJ, Hines AH (2000) Invasion of coastal marine communities in North America: apparent patterns, processes and biases. Annu Rev Ecol Syst 31:481–531

Scattergood LW (1952) The distribution of the green crab, *Carcinides maenas* (L.) in the northwestern Atlantic. Report No. Fisheries Circular 8, Maine Department of Sea and Shore Fisheries, Augusta, ME

Schlichting CD (2004) The role of phenotypic plasticity in diversification. In: DeWitt TJ, Scheiner SM (eds) Phenotypic plasticity: functional and conceptual approaches. Oxford University Press, New York, pp 191–200

Schlichting CD, Pigliucci M (1998) Phenotypic evolution: a reaction norm perspective. Sinauer Associates, Sunderland, MA

Schweitzer JA, Larson KC (1999) Greater morphological plasticity of exotic honeysuckle species may make them better invaders than native species. J Torrey Bot Soc 126:15–23
Seeley RH (1986) Intense natural selection caused a rapid morphological transition in a living marine snail. Proc Natl Acad Sci U S A 83:6897–6901
Sexton JP, McKay JK, Sala A (2002) Plasticity and genetic diversity may allow saltcedar to invade cold climates in North America. Ecol Appl 12:1652–1660
Sih A, Bell A, Johnson JC (2004) Behavioral syndromes: an ecological and evolutionary overview. Trends Ecol Evol 19:372–378
Smith LD (2004) Biogeographic differences in claw size and performance in an introduced crab predator *Carcinus maenas*. Mar Ecol Prog Ser 276:209–222
Smith LD, Jennings JA (2000) Induced defensive responses by the bivalve *Mytilus edulis* to predators with different attack modes. Mar Biol 136:461–469
Smith LD, Palmer A (1994) Effects of manipulated diet on size and performance of brachyuran crab claws. Science 264:710–712
Smith LD, Wonham MJ, McCann LD, Ruiz GM, Hines AH, Carlton JT (1999) Invasion pressure to a ballast-flooded estuary and an assessment of inoculant survival. Biol Invas 1:67–87
Sokolova IM, Pörtner H-O (2003) Metabolic plasticity and critical temperatures for aerobic scope in a eurythermal marine invertebrate (*Littorina saxatilis*, Gastropoda: Littorinidae) from different latitudes. J Exp Biol 206:195–207
Stearns SC (1989) The evolutionary significance of phenotypic plasticity. BioScience 39:436–445
Steffani CN, Branch GM (2003) Growth rate, condition, and shell shape of *Mytilus galloprovincialis*: responses to wave exposure. Mar Ecol Prog Ser 246:197–209
Thompson JD (1991) Phenotypic plasticity as a component of evolutionary change. Trends Ecol Evol 6:246–249
Thompson JN (1998) Rapid evolution as an ecological process. Trends Ecol Evol 13:329–332
Tollrian R, Harvell CD (1999) The evolution of inducible defenses: current ideas. In: Tollrian R, Harvell CD (eds) The ecology and evolution of inducible defenses. Princeton University Press, Princeton, NJ, pp 306–322
Toth GB, Pavia H (2000) Water-borne cues induce chemical defense in a marine alga (*Ascophyllum nodosum*). Proc Natl Acad Sci U S A 97:14418–14420
Travis J (1994) Evaluating the adaptive role of morphological plasticity. In: Wainwright PC, Reilly SM (eds) Ecological morphology: integrative organismal biology. University of Chicago Press, Chicago, pp 99–122
Trussell GC (1996) Phenotypic plasticity in an intertidal snail: the role of a common crab predator. Evolution 50:448–454
Trussell GC (1997) Phenotypic plasticity in the foot size of an intertidal snail. Ecology 78:1033–1048
Trussell GC (2000) Phenotypic clines, plasticity, and morphological trade-offs in an intertidal snail. Evolution 54:151–166
Trussell GC, Etter RJ (2001) Integrating genetic and environmental forces that shape the evolution of geographic variation in a marine snail. Genetica 112:321–337
Trussell GC, Smith LD (2000) Induced defenses in response to an invading crab predator: an explanation of historical and geographic phenotypic change. Proc Natl Acad Sci U S A 97: 2123–2127
Trussell GC, Ewanchuk PJ, Bertness MD (2002) Field evidence of trait-mediated indirect interactions in a rocky intertidal food web. Ecol Lett 5:241–245
Trussell GC, Ewanchuk PJ, Bertness MD (2003) Trait-mediated effects in rocky intertidal food chains: predator risk cues alter prey feeding rates. Ecology 84:629–640
van Alstyne KL (1988) Herbivore grazing increases polyphenolic defenses in the intertidal brown alga *Fucus distichus*. Ecology 69:655–663
Verling E, Ruiz GM, Smith LD, Galil B, Miller AW, Murphy KR (2005) Supply-side invasion ecology: characterizing propagule pressure in coastal ecosystems. Proc R Soc B 272: 1249–1257

Vermeij GJ (1982a) Environmental change and the evolutionary history of the periwinkle (*Littorina littorea*) in North America. Evolution 36:561–580

Vermeij GJ (1982b) Phenotypic evolution in a poorly dispersing snail after arrival of a predator. Nature 299:349–350

Via S, Gomulkiewicz R, Jong GD, Scheiner SM, Schlichting CD, Tienderen PV (1995) Adaptive phenotypic plasticity: consensus and controversy. Trends Ecol Evol 10:212–217

Weigle SM, Smith LD, Carlton JT, Pederson J (2004) Assessing the risk of introducing exotic species via the live marine species trade. Conserv Biol 19:213–223

Weinig C (2000) Plasticity versus canalization: population differences in the timing of shade-avoidance responses. Evolution 54:441–451

Weis AE, Gorman WL (1990) Measuring selection on reaction norms: an exploration of the *Eurosta-Solidago* system. Evolution 44:820–831

Welch WR (1968) Changes in the abundance of the green crab, *Carcinus maenas* (L.), in relation to recent temperature changes. Fish Bull 67:337–345

Werner EE, Peacor SD (2003) A review of trait-mediated indirect interactions in ecological communities. Ecology 84:1083–1100

West JM (1997) Plasticity in the sclerites of a gorgonian coral: tests of water motion, light level, and damage cues. Biol Bull 192:279–289

West-Eberhard MJ (2003) Developmental plasticity and evolution. Oxford University Press, New York

Whitlow WL, Rice NA, Sweeney C (2003) Native species vulnerability to introduced predators: testing an inducible defense and a refuge from predation. Biol Invas 5:23–31

Wikelski M, Thom C (2000) Marine iguanas shrink to survive El Nino. Nature 403:37–38

Willis AJ, Memmott J, Forrester RI (2000) Is there evidence for the post-invasion evolution of increased size among invasive plant species? Ecol Lett 3:275–283

Windig JJ, De Kovel CGF, De Jong G (2004) Genetics and mechanics of plasticity. In: Dewitt TJ, Scheiner SM (eds) Phenotypic plasticity: functional and conceptual approaches. Oxford University Press, New York, pp 31–49

Yates JL, Peckol P (1993) Effects of nutrient availability and herbivory on polyphenolics in the seaweed *Fucus vesiculosus*. Ecology 74:1757–1766

Chapter 11
Escape from Parasites

Mark E. Torchin and Kevin D. Lafferty

11.1 Introduction

In betting circles, the odds-on favorite of a sporting match can depend on the health of the star players; a pulled hamstring or bad flu can determine the winner. Introduced species are often in contest with native species and their impacts are directly proportional to their demographic performance in the novel environment. Demographic performance can encompass population level parameters, such as densities, abundances, and biomass as well as individual level parameters, such as growth rate, survivorship and fecundity. One indication of an invader's demographic performance is size because individuals that grow fast or live long can become large. On average, marine invaders attain larger sizes compared to populations in their native range (Grosholz and Ruiz 2003). This increased performance, if it translates into increased standing biomass, should positively correlate with an invader's impact (Crivelli 1983). Exploring reasons for invasion success will not only accelerate our understanding of species interactions, but will strengthen our ability to manage invasions.

Debate as to what factors facilitate invasion success has led to quantitative evaluation of a long-standing explanation for successful introduced species: the enemy release hypothesis (ERH). One prediction of the ERH is that introduced populations lack natural enemies compared to populations within their original range (Williams 1954; Elton 1958). Another prediction of the ERH is that introduced species should benefit from enemy-mediated competition because they are less likely to be affected by natural enemies than their native competitors (Elton 1958; Keane and Crawley 2002). Here, the first prediction requires a comparison of the same species across native and introduced populations. The second prediction requires a comparison of populations of an introduced species with populations of one or more native species which coexist in the same community as the invader (generally competitors). These fundamentally different predictions are sometimes confused and lead to misleading evaluations of the ERH. The two predictions are not comparable or mutually exclusive. For example, a population of an introduced species might be similarly affected by natural enemies compared to sympatric pop-

G. Rilov, J.A. Crooks (eds.) *Biological Invasions in Marine Ecosystems.*
Ecological Studies 204,

ulations of native species. However, the invading population may still experience enemy release compared to conspecific populations in the native range. The effects of natural enemies on a species are a function of the total number of natural enemies, their mean abundance and their individual impacts at the population level (Torchin and Mitchell 2004). Therefore, release can occur through losing natural enemies (enemy escape), or having increased resistance or tolerance to existing natural enemies. Below, we consider parasites as the main category of natural enemies, recognizing that other natural enemies are important. We generally limit our comparisons to the number of parasite species per host species. Although the number of parasite species is the easiest metric to use for evaluating broad patterns, Torchin and Mitchell (2004) note that other aspects of parasitism such as pathology and the prevalence or intensity of infection can lead to important variation in the effects of parasitism among parasite species. Developing research strategies to broadly evaluate these factors will ultimately enable a more robust examination of the ERH.

The number of parasite species that attack an introduced species is a function of the number of parasites it has in its native range, minus the number it escapes during the introduction, plus the number it acquires in the introduced range. The difference between the number of parasite species in populations of hosts in the native and introduced ranges is perhaps the most simplistic measure of parasite release. A more comprehensive measure of parasite release is a function of number of parasite species, their mean abundance and their virulence, accounting for possible interspecific interactions amongst them (Torchin and Mitchell 2004). Evidence from plants (Fenner and Lee 2001; Wolfe 2002; Mitchell and Power 2003; DeWalt et al. 2004) and animals (Cornell and Hawkins 1993; Torchin et al. 2003) indicates that introduced populations experience a net reduction in parasitism (both in terms of species richness and prevalence within a host population) compared to conspecific populations from the native range. Studies of terrestrial species provide the bulk of the insight on how introduced species escape their parasites; based on a literature search using the ISI Web of Science,18 of the 19 recent papers specifically mentioning the enemy release hypothesis concerned terrestrial plants. Still, information is emerging for some marine invasions (Calvo-Ugarteburu and McQuaid 1998a, b; Torchin et al. 2001, 2002, 2005; Bachelet et al. 2004). Of the 26 animal species Torchin et al. (2003) examined, aquatic species (n = 15, both marine and freshwater) experienced a greater release from parasites than terrestrial species (Fig. 11.1). Release may not be permanent. Parasites can accumulate on introduced populations over time as more opportunities for parasites to invade or adapt to the invader increase (Cornell and Hawkins 1993).

Enemy release should usually benefit an invader in its new range, but for it to lead to a net demographic advantage relative to other species in the community depends on the direct and indirect effects of parasites on native competitors and the relative strength of competition. Both native and introduced parasite species can attack a native competitor (introduced parasites are an important consideration we do not explore in this chapter). The parasite advantage for an invader should be proportional to the impact of parasitism on a competing native host species minus

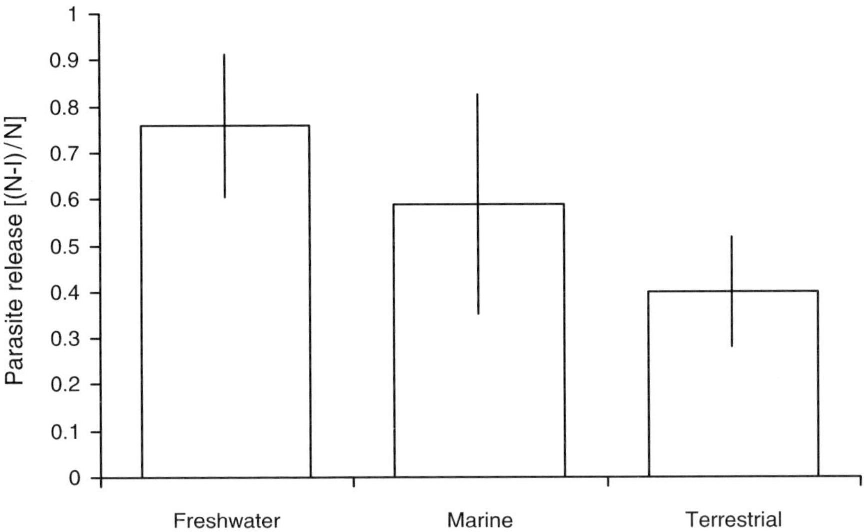

Fig. 11.1 Parasite release (species richness) experienced by introduced species in different habitats. Data are from Torchin et al. (2003). Release is the proportion (N–I)/N, where N is the number of parasite species in the native range and I is the number of parasite species in the introduced range. *Error bars* are 95% confidence intervals calculated using the standard deviation of the mean

the impact of parasites on the introduced host species (Torchin and Mitchell 2004). Sometimes, as evidenced by literature on plants, natural enemies can have similar impacts on sympatric native and introduced species (Blaney and Kotanen 2001; Agrawal and Kotanen 2003). This may be particularly true if invaders are sympatric with closely related native species, as host shifting may be more common, increasing the accumulation of parasites on invaders (Torchin and Mitchell 2004).

11.2 Reasons for Parasite Escape

The invasion process can filter out parasites in several ways (Fig. 11.2). First, species invasions, and the establishment of introduced populations, are often the result of a few colonists arriving and reproducing in the new environment. This, and the fact that parasites are generally not homogenously distributed among host populations, reduces the likelihood of introducing parasitized hosts (Fig. 11.2a). This is similar to the "bottle-necking" mechanism that can reduce the genetic diversity of invading population (Tsutsui et al. 2000). Second, even if parasitized hosts invade a new location, the low density of founder populations may impair parasite transmission. Thus, invaders may escape directly transmitted parasites, which require a minimum host density for transmission (Fig. 11.2b). Third, many

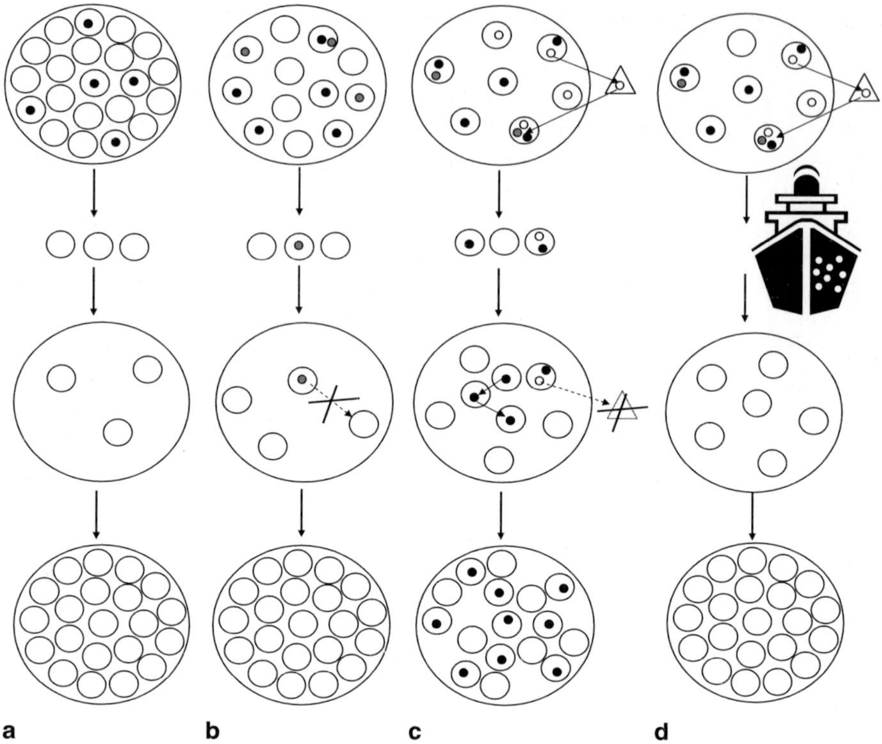

Fig. 11.2 Diagrammatic representation of the reasons (mechanisms) for parasite escape: *Large circles* are host populations, *medium circles* are host individuals, *small circles* represent parasite species: **a** few colonists and heterogeneously distributed parasites reduce the chances that parasitized hosts will be introduced; **b** small founder populations with host densities too low for parasite transmission prevent successful establishment of co-introduced parasites; **c** parasites with complex life cycles requiring more than one host (triangle is additional required host species) will not establish if suitable hosts are not present in new range – however, for directly transmitted parasites, when introduced population densities exceed the threshold for parasite transmission introduced parasites may establish; **d** introduction of uninfected larval stages, such as those introduced by ballast water will exclude parasites from transferring to novel location. Ballast water introduction may be particularly potent means for marine species to escape parasites

parasites have complex life cycles requiring the presence of more than one host. If suitable hosts are not present in the novel environment, complex life cycle parasites will not establish (Fig. 11.2c). Finally, introduction of uninfected life-history stages, such as larvae or seeds, will preclude most parasites which are infectious to adults. While these stages can harbor parasites, parasites of larvae are often lost post-recruitment and generally do not infect adult stages (Rigby and Dufour 1996; Cribb et al. 2000). This mechanism may be particularly important for marine and aquatic species that arrive as larvae in ballast water (Fig. 11.2d) (Lafferty and Kuris 1996).

11.3 Consequences of Parasite Release

For parasite release to confer an advantage requires that parasites negatively affect host populations. Mathematical models indicate that the spread and impact of a directly transmitted infectious disease agent through a population increases with the density of susceptible and infectious hosts. This means that parasites can be density-dependent sources of mortality or reproductive impairment – a prerequisite for being able to regulate populations (Anderson and May 1979). Laboratory experiments where parasites are added to an uninfected host population or where parasites are removed from an infected host population have demonstrated that parasites can limit host populations (Greenwood et al. 1936; Park 1948; Stiven 1964; Keymer 1981; Lanciani 1982; Anderson and Crombie 1984; Scott and Anderson 1984 Scott 1987). Experiments with a parasitic nematode and red grouse (Hudson and Dobson 1989; Dobson and Hudson 1992; Hudson et al. 1998) demonstrate that parasites can regulate host populations in nature. The importance of parasites in driving host population demographics and ecosystem function is becoming evident (reviewed in Hudson et al. 2006). Understanding the role of parasites in natural systems is a key step in determining the importance of their absence for some invasive species.

11.4 Fishes

Although limited, studies examining parasitism in introduced and native populations of marine and estuarine fishes are beginning to emerge. The rabbitfish, *Siganus rivulatus*, provides one example of the potential for fish to escape natural enemies and experience enemy release. Twenty-two species parasitize the rabbitfish in its native Red Sea (Diamant et al. 1999). Rabbitfish have migrated to the Mediterranean through the Suez Canal, and the lack of genetic differentiation between introduced and native rabbitfish populations suggests that many individuals made the trip (Bonhomme et al. 2003; Hassan et al. 2003). It is no surprise, therefore, that eight species of rabbitfish parasites (mostly protozoans) have also invaded the Mediterranean. None of these have complex life cycles (Diamant et al. 1999), suggesting that a lack of appropriate intermediate hosts may limit some Red Sea parasites from establishing in the Mediterranean (Fig. 11.2c). Only one new parasite (a ciliate) has successfully colonized rabbitfish in the Mediterranean (Diamant et al. 1999). The rabbitfish is now abundant in the Mediterranean (Bonhomme et al. 2003), perhaps thanks to an escape from parasitism.

A euryhaline invader, the round goby, *Neogobius melanostomus*, is native to the Ponto-Caspian region where it is parasitized by 50 species with typical prevalences of 100% and intensities of 1000 parasites per fish (Pronin et al. 1997; Corkum et al. 2004). It has now invaded the Baltic Sea and the Laurentian Great Lakes, presumably via ballast water or movement through shipping lanes (Corkum et al. 2004). Goby populations in the Great Lakes have fewer parasites (10 species, 76% prevalence) compared to native populations (Pronin et al. 1997). Baltic populations have even fewer parasites (four species, 4–20% infection prevalence). Pronin et al. (1997) suggests the lack of parasites in the Great lakes may be partly due to limited parasitic infections in the zebra mussel (*Dreissena polymorpha*), another introduced species originating from the same location as the round goby. The zebra mussel serves as an intermediate host for several of the parasites infecting the round goby in its native range (Pronin et al. 1997), but introduced populations are generally less parasitized compared to native populations (Molloy 1998). This situation may not be stable. Co-invasion by two potential hosts opens the door to a variety of Ponto-Caspian parasites with complex life cycles.

11.5 Molluscs

Marine species introduced via aquaculture provide a unique vector for accidental introductions, including parasites. Historic aquaculture practices often introduced large quantities of adult individuals of many species (both wanted as well as unwanted hitchhikers) (Naylor et al. 2001; Chap. 5, Minchin et al.). Large quantities and repeated introductions likely facilitate transfer and establishment of introduced parasites better than other introduction pathways (Torchin et al. 2002, 2003, 2005).

The Asian mud snail, *Batillaria attrementaria* is one such aquaculture introduction. Introduced to the west coast of North America as a byproduct of Pacific oyster (*Crassostrea gigas*) aquaculture from Japan (Bonnot 1935; Barrett 1963), *B. attrementaria* populations now occur from Southern Canada to Central California (Byers 1999). While in its native range, *B. attrementaria* is infected with at least eight different morphologically distinct trematode species (Shimura and Ito 1980; Rybakov and Lukomskaya 1988; Harada and Suguri 1989; Torchin et al. 2005), only a single trematode species (apparently from Japan) infects introduced populations in North America (Torchin et al. 2005). This introduced trematode was the most geographically widespread and abundant species within Japan (Torchin et al. 2005; Miura et al. 2005), evidence that common parasite species are the ones most likely to invade (Torchin et al. 2003). In some invaded locations, *B. attrementaria* encounters a very similar native competitor, *Cerithidea californica* (Byers 2000). This native snail has a rich trematode fauna, including a species that uses it as a first and second intermediate host (*B. attrementaria* only serves as a first intermediate host to its introduced parasite) (Torchin et al. 2005). None of these species are able to infect *B. attrementaria*.

Batillaria attrementaria is outcompeting the native *C. californica* where they co-occur (Byers 2000). From a traditional food-web perspective, replacement of one species by an ecological analog should have no detectable effect on food-web topology. However, Torchin et al. (2005) posit that the replacement of the native snail would also result in the replacement of ten or more native trematodes with one

Japanese trematode. This has implications for parasite diversity and could also reduce parasitism in the invertebrate community. For example, the trematodes that parasitize the native snail have life cycles that use crabs, polychaetes, mollusks and fishes as second intermediate hosts, while the Japanese trematode only uses fishes. In a food web that considers parasites and their effect on trophic interactions, Lafferty and Kuris (in press) found that the high host specificity of trematodes for the snail hosts, the low diversity of trematodes in the introduced snail and the local extirpation of the native snail results in a large decrease in connectance in the estuarine food web.

As an aside, it is worth noting that the ability to document parasite release can be a function of the capability to find and identify parasites. Recent molecular investigations found multiple cryptic trematode species in *B. attrementaria* in Japan (Miura et al. 2005) and a complex of three cryptic species in North America (Miura et al. 2006). Regardless, compared to native populations in Japan, introduced populations still experience a significant reduction in parasite species richness.

Another snail to invade California along with oysters is the Atlantic mud snail, *Ilyanassa obsoleta* (Demond 1952; Carlton 1999). Similarly, this snail has escaped many of its parasites. There are five trematode species reported to infect *I. obsoleta* in its introduced range (Grodhuas and Keh 1958) – about half the number recorded from populations where this snail is native (Stunkard 1983; Curtis 1997). One of these is probably an introduced trematode, *Austrobilharzia variglandis*, which causes swimmer's itch in humans (Miller and Northup 1926; Stunkard and Hinchliffe 1952; Grodhuas and Keh 1958). The remaining four unidentified species (Grodhuas and Keh 1958) are possibly a subset of *I. obsoleta's* native suite of parasites as well (Torchin et al. 2002; Torchin and Kuris 2005), but this awaits study.

Another recently established gastropod species, *Cyclope neritea* in the Bay of Biscay, was likely accidentally introduced through shellfish culture practices (Bachelet et al. 2004). In its native range in the Mediterranean, *C. nerita* can be heavily parasitized by trematodes (Bachelet et al. 2004), but appears to have escaped parasitism in the introduced range (1% prevalence with one trematode species). This likely gives it a parasite advantage over a native competitor, *Nassarius reticulatus*, (6–18% prevalence with six trematode species) because infected *N. reticulatus* exhibit reduced survivorship (Bachelet et al. 2004).

Introduced to South Africa, the Mediterranean mussel, *Mytilus galloprovincialis* competes with the native mussel, *Perna perna* and has a parasite advantage. While two pathogenic trematode species infect the native mussel, the introduced mussel remains uninfected (Calvo-Ugarteburu and McQuaid 1998a, b). These parasites reduce the native mussel's competitive ability against the invader and may partly explain the invasion success of the Mediterranean mussel in South Africa (Calvo-Ugarteburu and McQuaid 1998a, b). Despite the parasite advantage, it is not clear whether this an example of escape from natural enemies. Although native populations of *M. galloprovincialis* harbor trematodes (Cheng 1967), they are not common in northern Spain (Calvo-Ugarteburu and McQuaid 1998b) where the invasion may have originated.

11.6 Crabs

In its native Japan, the intertidal crab, *Hemigrapsus sanguineus*, is infected by a parasitic castrator (rhizocephalan barnacle, up to 64% prevalence) (Yamaguchi et al. 1994) and trematode metacercariae (A. Kuris and K. Lafferty, unpublished data). The crab recently invaded the East Coast of the USA where it is large and abundant (Lohrer et al. 2000). Parasitological examination of over 1000 introduced crabs from New Jersey failed to recover any parasites (McDermott 1998) from introduced *H. sanguineus*, but, Torchin et al. (2001) found one out of 27 introduced crabs from New England infected with a nematode. A similar parasitological comparison of the New Zealand pie crust crab, *Cancer novaezelandiae*, found abundant trematode metacercariae in a native New Zealand population and no parasites in Tasmania where the crab has been introduced (Kuris and Gurney 1997).

The European green crab, *Carcinus maenas* has invaded several locations around the world. Torchin et al. (2001) examined several populations of the green crab both in its native range in Europe and introduced populations around the world. In Europe, the crab's demographic performance (individual size and biomass) declines with the prevalence of parasitic castrators (which block reproduction and stop growth). Parasitic castrators explain 64% of the variation in crab size and 36% of the variation in crab biomass. Parasitic castrators do not infect introduced populations. As one would expect from the association between parasitism and performance in Europe, introduced populations of the green crab are significantly larger in body size and population biomass than native populations with parasites. The introduced populations are most similar to the few European populations where parasitism is rare. A few generalist and relatively non-pathogenic parasite species infect some of the introduced populations. This tends to occur in the populations that invaded a relatively long time ago. Parasite species richness increases in older introductions (Table 1 in Torchin et al. 2001) and time since introduction accounts for over 80% of the variance in the number of parasite species found in the introduced populations (Fig. 11.3).

11.7 Conclusion

Evidence that introduced marine species escape parasites is emerging. The extent to which they accumulate parasites in the novel region generally does not make up for the loss on ecological time scales. This leads to enemy release and makes it more likely that they will have a parasite advantage over competitors. The extent to

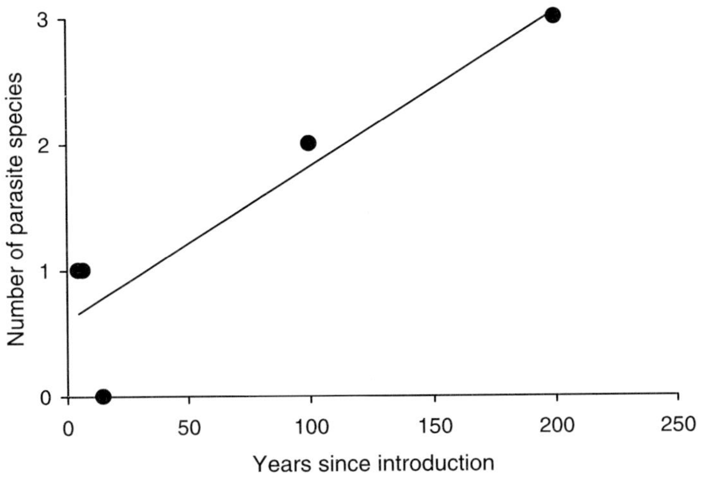

Fig. 11.3 Association between number of parasites found (Torchin et al. 2001) and time since introduction for introduced populations of the European green crab, *Carcinus maenas*. y= 0.01x + 0.6, $R^2 = 0.83$, $P = 0.03$

which this facilitates their demographic expansion and success in their introduced range remains an important question.

References

Agrawal AA, Kotanen PM (2003) Herbivores and the success of exotic plants: a phylogenetically controlled experiment. Ecol Lett 6:712–715

Anderson RM, Crombie JA (1984) Experimental studies of age-prevalence curves for *Schistosoma mansoni* infections in populations of *Biomphalaria glabrata*. Parasitology 89:79–105

Anderson RM, May RM (1979) Population biology of infectious diseases. Part 1. Nature 280:361–367

Bachelet G, Simon-Bouhet B, Desclaux C, Garcia-Meunier P, Mairesse G, de Montaudouin X, Raigne H, Randriambao K, Sauriau PG, Viard F (2004) Invasion of the eastern Bay of Biscay by the nassariid gastropod Cyclope neritea: origin and effects on resident fauna. Mar Ecol Prog Ser 276:147–159

Barrett EM (1963) The California oyster industry. Fish Bull 123

Blaney CS, Kotanen PM (2001) Effects of fungal pathogens on seeds of native and exotic plants: a test using congeneric pairs. J Appl Ecol 38:1104–1113

Bonhomme F, Baranes A, Golani D, Harmelin-Vivien M (2003) Lack of mitochondrial differentiation between Red Sea and Mediterranean populations of the Lessepsian rabbitfish, Siganus rivulatus (Perciformes : Siganidae). Sci Mar 67:215–217

Bonnot P (1935) The California oyster industry. Calif Fish Game 21:65–80

Byers JE (1999) The distribution of an introduced mollusc and its role in the long-term demise of a native confamilial species. Biol Invas 1:339–353

Byers JE (2000) Competition between two estuarine snails: Implications for invasions of exotic species. Ecology 81:1225–1239

Calvo-Ugarteburu G, McQuaid CD (1998a) Parasitism and invasive species: effects of digenetic trematodes on mussels. Mar Ecol Prog Ser 169:149–163

Calvo-Ugarteburu G, McQuaid CD (1998b) Parasitism and introduced species: epidemiology of trematodes in the intertidal mussels *Perna perna* and *Mytilus galloprovincialis*. J Exp Mar Biol Ecol 220:47–65

Carlton JT (1999) Molluscan invasions in marine and estuarine communities. Malacologia 41:439–454

Cheng TC (1967) Marine molluscs as hosts for symbioses – with a review of known parasites of commercially important species. Adv Mar Biol 5:1–300

Corkum LD, Sapota MR, Skora KE (2004) The round goby, *Neogobius melanostomus*, a fish invader on both sides of the Atlantic Ocean. Biol Invas 6:173–181

Cornell HV, Hawkins BA (1993) Accumulation of native parasitoid species on introduced herbivors: a comparison of hosts as natives and hosts as invaders. Am Nat 141:847–865

Cribb TH, Pichelin S, Dufour V, Bray RA, Chauvet C, Faliex E, Galzin R, Lo CM, Lo-Yat A, Morand S, Rigby MC, Sasal P (2000) Parasites of recruiting coral reef fish larvae in New Caledonia. Intl J Parasitol 30:1445–1451

Crivelli AJ (1983) The destruction of aquatic vegitation by carp. Hydrobiologia 106:37–41

Curtis LA (1997) Ilyanassa obsoleta (Gastropoda) as a host for trematodes in Delaware estuaries. J Parasitol 83:793–803

Demond J (1952) The Narissariidae of the west coast of North America between Cape San Lucas, Lower California, and Cape Flattery, Washington. Pac Sci 6:300–317

DeWalt SJ, Denslow JS, Ickes K (2004) Natural-enemy release facilitates habitat expansion of the invasive tropical shrub *Clidemia hirta*. Ecology 85:471–483

Diamant A, Banet A, Paperna I, von Westernhagen H, Broeg K, Kruener G, Koerting W, Zander S (1999) The use of fish metabolic, pathological and parasitological indices in pollution monitoring – II. The Red Sea and Mediterranean. Helgol Mar Res 53:195–208

Dobson AP, Hudson PJ (1992) Regulation and stability of a free-living host-parasite system, *Trichostrongylus tenuis* in red grouse. II. Population models. J Anim Ecol 61:487–500

Elton CS (1958) The ecology of invasions by animals and plants. Methuen, London

Fenner M, Lee WG (2001) Lack of pre-dispersal seed predators in introduced Asteraceae in New Zealand. N Z J Ecol 25:95–99

Greenwood M, Bradford-Hill A, Topely WWC, Wilson J (1936) Experimental epidemiology. Med Res Counc Spec Rep 209:204

Grodhaus G, Keh B (1958) The marine, dermatitis-producing cercaria of *Austrobilharzia variglandis* in California (Trematoda: Schistosomatidae). J Parasitol 44:633–638

Grosholz ED, Ruiz GM (2003) Biological invasions drive size increases in marine and estuarine invertebrates. Ecol Lett 6:700–705

Harada M, Suguri S (1989) Surveys on cercariae in brackish water snails in Kagawa Prefecture, Shikoku, Japan. Jpn J Parasitol 38:388–391

Hassan M, Harmelin-Vivien M, Bonhomme F (2003) Lessepsian invasion without bottleneck: example of two rabbitfish species (*Siganus rivulatus* and *Siganus luridus*). J Exp Mar Biol Ecol 291:219–232

Hudson PJ, Dobson AP (1989) Population biology of *Trichostrongylus tenuis*, a parasite of economic importance for red grouse management. Parasitol Today 5:283–291

Hudson PJ, Dobson AP, Newborn D (1998) Prevention of population cycles by parasite removal. Science 282:2256–2258

Hudson PJ, Dobson AP, Lafferty KD (2006) Is a healthy ecosystem one that is rich in parasites? Trends Ecol Evol 21:381–385

Keane RM, Crawley MJ (2002) Exotic plant invasions and the enemy release hypothesis. Trends Ecol Evol 17:164–170

Keymer AE (1981) Population dynamics of *Hymenolepis diminuta*: the influence of infective-stage density and spatial distribution. Parasitology 79:195–207

Kuris AM, Gurney R (1997) Survey of Tasmanian crabs for parasites: a progress report. Proceedings of the first international workshop on the demography, impacts and management of the introduced populations of the European crab, *Carcinus maenas*. Centre for Research on Introduced Marine Pests. Technical Report 11:92–94

Lafferty KD, Kuris AM (1996) Biological control of marine pests. Ecology 77:1989–2000

Lafferty KD, Kuris AM In press. Parasites reduce food web robustness because they are sensitive to secondary extinction as illustrated by an invasive estuarine snail. Philosophical Transactions of the Royal Society B.

Lanciani CA (1982) Parasite-mediated reductions in the survival and reproduction of the backswimmer *Buenoa scimitra* (Hemiptera: Notonectidae). Parasitology 85:593–603

Lohrer AM, Whitlatch RB, Wada K, Fukui Y (2000) Home and away: comparisons of resource utilization by a marine species in native and invaded habitats. Biol Invas 2:41–57

McDermott JJ (1998) The western Pacific brachyuran (*Hemigrapsus sanguineus: Grapsidae*), in its new habitat along the Atlantic coast of the United States: geographic distribution and ecology. ICES J Mar Sci 55:289–298

Miller HM, Northup FE (1926) The seasonal infestation of *Nassa obsoleta* (Say) with larval trematodes. Biol Bull 50:490–508

Mitchell CE, Power AG (2003) Release of invasive plants from fungal and viral pathogens. Nature 421:625–627

Miura O, Kuris AM, Torchin ME, Hechinger RF, Dunham EJ, Chiba S (2005) Molecular genetic analyses reveal cryptic species of trematodes in the intertidal gastropod, Batillaria cumingi (Crosse). Intl J Parasitol 35:793–801

Miura O, Torchin ME, Kuris AM, Hechinger RF, Chiba S (2006) Introduced cryptic species of parasites exhibit different invasion pathways. Proc Natl Acad Sci 103:19818–19823

Molloy DP (1998) The potential for using biological control technologies in the management of Dreissena spp. J Shellfish Res 17:177–183

Naylor RL, Williams SL, Strong DR (2001) Ecology – aquaculture – a gateway for exotic species. Science 294:1655–1656

Park T (1948) Experimental studies of interspecies competition. 1. Competition between populations of the flour beetles, *Tribolium-Confusum Duval* and *Tribolium-Castaneum Herbst*. Ecol Monogr 18:265–307

Pronin NM, Fleischer GW, Baldanova DR, Pronina SV (1997) Parasites of the recently established round goby (*Neogobius melanostomus*) and tubenose goby (*Proterorhinus marmoratus*) (Cottidae) from the St. Clair River and Lake St. Clair, Michigan, USA. Folia Parasitologica 44:1–6

Rigby MC, Dufour V (1996) Parasites of coral reef fish recruits, *Epinephelus merra* (Serranidae), in French Polynesia. J Parasitol 82:405–408

Rybakov AV, Lukomskaya OG (1988) On the life cycle of *Acanthoparyphium macracanthum* sp.n. (Trematoda, Echinostomatidae). Parazitologiya 22:224–229

Scott ME (1987) Regulation of mouse colony abundance by *Heligmosomoides polygyrus* (Nematoda). Parasitology 95:111–129

Scott ME, Anderson AM (1984) The population dynamics of *Gyrodactylus bullatarudis* (Mongenea) on guppies (*Poecilia reticulata*). Parasitology 89:159–194

Shimura S, Ito J (1980) Two new species of marine cercariae from the Japanese intertidal gastropod, *Batillaria cumingii* (Crosse). Jpn J Parasitol 29:369–375

Stiven AE (1964) Experimental studies on the host parasite system hydra and *Hydramoeba hydroxena* (Entz.). II. The components of a single epidemic. Ecol Monogr 34:119–142

Stunkard HW (1983) The marine cercariae of the Woods Hole Massachusetts region, a review and a revision. Biol Bull 164:143–162

Stunkard HW, Hinchliffe MC (1952) The morphology and life-history of *Microbilharzia variglandis* (Miller and Northrup 1926) Stunkard and Hinchliffe, 1951, avian blood flukes whose larve cause "swimmer's itch" of ocean beaches. J Parasitol 38:248–265

Torchin ME, Kuris AM (2005) Introduced marine parasites. In: Rhode K (ed) Marine parasitology. CSIRO Publishing

Torchin ME, Mitchell CE (2004) Parasites, pathogens and invasions by plants and animals. Front Ecol Environ 2:183–190

Torchin ME, Lafferty KD, Kuris AM (2001) Release from parasites as natural enemies: increased performance of a globally introduced marine crab. Biol Invas 3:333–345

Torchin ME, Lafferty KD, Kuris AM (2002) Parasites and marine invasions. Parasitology 124: S137–S151

Torchin ME, Lafferty KD, Dobson AP, McKenzie VJ, Kuris AM (2003) Introduced species and their missing parasites. Nature 421:628–630

Torchin ME, Byers JE, Huspeni TC (2005) Differential parasitism of native and introduced snails: replacement of a parasite fauna. Biol Invas 7:885–894

Tsutsui ND, Suarez AV, Holway DA, Case TJ (2000) Reduced genetic variation and the success of an invasive species. Proc Natl Acad Sci USA 97:5948–5953

Williams JR (1954) The biological control of weeds. Report of the Sixth Commonwealth Entomological Congress, London, UK, pp 95–98

Wolfe LM (2002) Why alien invaders succeed: support for the escape-from-enemy hypothesis. Am Nat 160:705–711

Yamaguchi T, Tokunaga S, Aratake H (1994) Contagious infection by the rhizocephalan parasite *Sacculina* sp. In the grapsid crab *Hemigrapsus sanguineus* (De Haan). Crustacean Res 23:89–101

Chapter 12
Ecological Factors Affecting Community Invasibility

Suzanne V. Olyarnik, Matthew E.S. Bracken, Jarrett E. Byrnes, A. Randall Hughes, Kristin M. Hultgren, and John J. Stachowicz

12.1 Introduction

What makes a community invasible? For over a century ecologists have sought to understand the relative importance of biotic and abiotic factors that determine community composition. The fact that we are still exploring this topic today hints at both its importance and complexity. As the impacts from harmful non-native species accumulate, it has become increasingly urgent to find answers to the more applied aspects of this question: what makes a habitat vulnerable to invasion by additional species, and which species are likely to invade? Answers to these questions will not only aid in targeting conservation efforts but will also advance our understanding of marine community ecology.

Although the relative importance of abiotic vs. biotic factors in making a habitat invasible varies, abiotic factors undoubtedly serve as the first "filter" to invasions, limiting establishment of non-native (=exotic) species to conditions approximating their native ranges. As obvious examples, tropical corals will not establish in boreal waters, and temperate rocky intertidal species will not colonize tropical shores. Similarly, species cannot invade a community if propagules do not arrive at the site. Other chapters in this volume cover the influence of abiotic factors and propagule supply (Chap. 7, Johnston et al.; Chap. 8, Miller and Ruiz; Chap. 19, Hewitt et al.), so we only briefly review these factors. In this chapter we focus on the question of predicting invasion success of non-native species that are (1) transported to the habitat in question (i.e., propagule supply is not extremely limiting) and (2) physiologically capable of surviving in the climatic regime. We begin with the observation that even in areas of suitable habitat within the current range of an introduced species, there is often dramatic variation in the density, presence, and overall success of the invader. We seek to explain this variation in terms of processes that control the availability of resources. These include not only abiotic and physical factors that determine base resource levels, but also interactions between species or between organisms and their environment that increase resource availability (through disturbance) or decrease resource availability (through competitive processes), or create new resources (through facilitation) (Fig. 12.1).

G. Rilov, J.A. Crooks (eds.) *Biological Invasions in Marine Ecosystems.*
Ecological Studies 204,

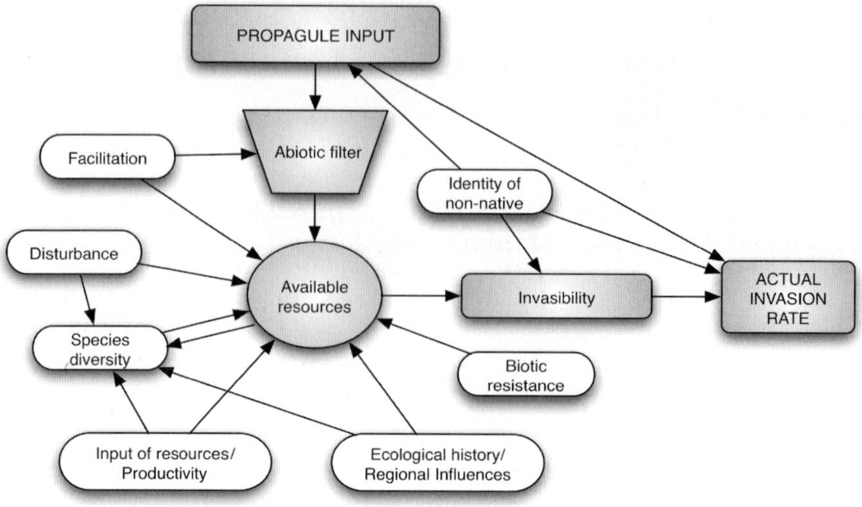

Fig. 12.1 Factors that have been shown to affect invasions of non-native species into marine communities. Assuming exotic propagules are present and able to survive the initial abiotic "filter" of the new habitat (see Sect. 12.2), their ability to invade the community can be determined by several, potentially interacting, factors that affect resource availability (Sects. 12.3–12.7), as well as the rate of propagule input and the characteristics of the exotic relative to the resident species already present (Rejmanek et al. 2005). Figures 12.3 and 12.4 illustrate how some of these factors interact to affect invasibility using specific case studies

Over the past few decades, we have seen substantial progress in understanding biological invasions, including the identification of several factors that affect community invasibility. While there is a growing body of research on invasibility in marine systems, the majority of studies are from terrestrial systems (see Fig. 12.2). Where available, we review results from marine studies; where these are scarce, we draw attention to this limitation and supplement our review with what, if anything, is known from other systems.

12.2 The Abiotic Filter

The combination of abiotic conditions under which an organism can sustain populations without immigration ("the fundamental niche") has long been recognized as a constraint on the distribution of a species, and most non-native species are at least somewhat limited to habitats ecologically similar to their ancestral ranges (Holt et al. 2005; Peterson 2003). While this concept has been incorporated into climate- and/or habitat-matching methods aimed at predicting what habitats would be ecologically appropriate for potential invaders (Peterson 2003; Peterson and Vieglais

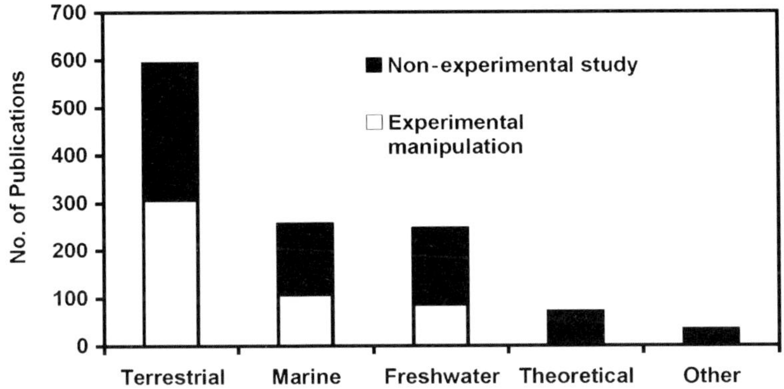

Fig. 12.2 Bioinvasion papers from the ecology literature from 1995 to 2005 by system and study type. Published papers from terrestrial systems outnumbered those from marine two to one. Of these, approximately half of the terrestrial studies involved an experimental manipulation, while only 42% of the marine and 35% of the freshwater studies did. The *white portion of each bar* represents the number of studies that included an experimental manipulation; the *black portion* represents all other types of studies. The system category "other" includes studies that either cut across all systems or did not specify a system. We searched the Web of Science using the following search terms: invas*, invad*, exotic, alien, nonnative, non-native, nonindigenous, non-indigenous. The analysis was restricted to 14 journals: Ecology, Science, Nature, Oecologia, OIKOS, Ecological Applications, Biological Invasions, Ecology Letters, Journal of Ecology, Journal of Animal Ecology, Proceedings of the National Academy of Sciences of the United States of America, Marine Ecology Progress Series, Journal of Experimental Marine Ecology and Biology, and Hydrobiologia

2001), the relative efficiency of modeling abiotic niche requirements (or similar methods of predicting invasions) depends on whether the abiotic requirements are conserved over time (Holt et al. 2005) or whether non-natives undergo rapid changes in their abiotic requirements in their invaded ranges (Quinn et al. 2000; Stockwell and Ashley 2004; Wares et al. 2005). Some studies do indicate widespread niche conservatism over time (Holt et al. 2005). For example, the crab *Hemigrapsus sanguineus* has nearly identical abiotic habitat requirements in its invaded and ancestral ranges (Lohrer et al. 2000). In contrast, other studies have found that abiotic habitat characteristics vary among spatially independent invasions, as is the case for the European green crab *Carcinus maenas* (Grosholz and Ruiz 1996). The degree to which a species' abiotic requirements are conserved or can change during the course of invasions is still not well understood, but it depends on factors such as propagule delivery, initial genetic diversity, and the historical pattern of the species' introductions (Wares et al. 2005).

Given that a non-native species has passed through the initial abiotic "filter" (Fig. 12.1), community ecology theory can provide insights on when abiotic or biotic factors may be most important in determining the course (and success) of the subsequent invasion. Abiotic factors might be of primary importance in determining the abundance of invaders, for example in physically stressed habitats where the effects of competition and predation are typically weaker (see also Chap. 7, Johnston et al.). In Puget Sound salt marsh communities, Dethier and Hacker (2005) found that abiotic factors such as salinity and sediment type primarily controlled invasion success of the English cordgrass *Spartina anglica,* whereas biotic factors had no effect.

Variation in abiotic effects may also mediate biotic interactions, such as competition and predation, between natives and non-natives. Byers (2002b) showed that an abiotic factor (sediment type) mediated biotic impacts (crab predation) on the non-native clam *Nuttallia*. In salt marshes, relaxation of salinity stress via increased rainfall and other freshwater inputs increases growth and competitive advantage of non-native plants relative to natives (Dethier and Hacker 2005; Kuhn and Zedler 1997; Minchinton 2002). Conversely, biotic interactions can modify abiotic conditions by buffering abiotic (physical) stress (Bertness and Hacker 1994), and this modification might either facilitate or inhibit invasions (for further discussion see Sect. 12.5.2). Ultimately, it is the interplay of abiotic and biotic factors in a community (via resource availability, disturbance, facilitation, and other factors) that affects the amount of resources available for non-natives and, consequently, the invasibility of the community (Fig. 12.1).

12.3 Resource Levels

Assuming once again that a non-native species has passed through the initial abiotic “filter”, increased availability of limiting resources can influence the invasibility of a habitat. In this section, we focus on how additions of resources to resource-limited systems can facilitate invasions by alleviating resource limitation (Davis MA et al. 2000). In subsequent sections, we review how competitive interactions (Sect. 12.4), facilitation (Sect. 12.5) and disturbance (Sect. 12.6) can each alter the actual level of available resources for exotic species attempting to establish.

Because a non-native species will require certain limiting resources in order to establish and spread, it will be more successful in habitats where competition for these resources is reduced (Davis MA et al. 2000; Davis MA and Pelsor 2001; Stachowicz et al. 2002a; Stohlgren et al. 2002). Enhanced resource availability can occur either when resource use by the existing species in a community declines (e.g., due to disturbance or consumption of competitors) or when the supply of limiting resources increases faster than the native members of the community can sequester those resources (e.g., eutrophication). When resource utilization decreases or total resource supply increases, a community can become more susceptible to invasion.

Direct evidence for the effects of enhanced resource availability on marine and aquatic bioinvasions is limited to only a few case studies, though high productivity has been implicated in promoting the spread of non-native species in marine systems (Branch and Steffani 2004). The results of a nutrient addition experiment in super-littoral rock pool microcosms suggest that adding nutrients enhances invasibility by increasing limiting resources (Romanuk and Kolasa 2005). Similarly, adding nutrients to an aquatic plankton community stimulated bacterial production, which in turn promoted invasion by ciliates due to decreased competition for resources (Jiang and Morin 2004). Conversely, limiting resource availability can decrease invasion success: because phytoplankton availability in the water column is reduced inside eelgrass beds due to decreased water flow, eelgrass limits the growth and survival of the suspension-feeding non-native mussel *Musculista senhousia*, reducing invasion compared to areas where eelgrass meadows are fragmented or absent and rates of food delivery are higher (Allen and Williams 2003). If we consider resource availability more broadly to include space, there is clear evidence from marine systems that if native species are unable to consume or utilize the primary limiting resource, a community may be more susceptible to invasion (Stachowicz et al. 1999, 2002a). Finally, at larger scales, nutrient runoff from the land is increasing the distribution and abundance of an exotic vascular plant (*Phragmites australis*) in New England salt marshes (Bertness et al. 2002).

Most theory and evidence supporting the role of resource availability in mediating invasibility comes from studies of terrestrial primary producers. For example, nutrient additions to nutrient-poor grasslands can promote the invasion and eventual dominance of non-native annual species (Burke and Grime 1996; Huenneke et al. 1990). In contrast, long-term observational data suggest that more productive systems may actually be characterized by reduced invasion success, likely because (within the "normal" limits of a system) native species respond to increases in resource availability by increasing growth rates, allowing them to sequester the additional resources and preventing the establishment and growth of non-natives (Cleland et al. 2004). Thus, predicting invasibility based simply on the productivity of a system without measuring unused limiting resources is problematic.

12.4 Biotic Resistance and Species Diversity

Biotic resistance refers to the ability of a community of resident species to repel invaders as a result of species interactions, and it includes a variety of mechanisms. It can encompass the idea that native consumers or pathogens reduce invader fitness, but is most often conceived of as a product of competition with natives within the same trophic level. In terms of competition, the biotic resistance paradigm usually assumes that communities are both resource-limited and strongly structured by interspecific competition and resource partitioning (May and MacArthur 1972). A key aspect of this paradigm requires a comparison between the resource requirements of a potential invader and those required by the existing community. If the new species has optimum performance at resource levels that overlap minimally with the residents, then it will have a higher probability of successful invasion compared to other species with resource requirements more similar to the residents' (Tilman 2004). For example, the dramatic success of *Hemigrapsus sanguineus* in New England has been attributed to the absence of ecological equivalents as no other grapsid crabs are present in the intertidal there (Lohrer et al. 2000). However, this sort of attribution of causation is most usually done post-hoc. Determining the degree to which resource requirements of exotics that have not yet arrived will overlap with native species is not always straightforward: resource requirements of the invader may change in the invaded range, and biotic interactions with native species in the new habitat may alter access to limiting resources. Comparing a potential invader's performance in its native range (including resource uptake rates, resilience to varying conditions, and competitive abilities) to the performance of natives in a community at risk of invasion may increase our ability to predict whether that particular exotic will be successful in invading the community.

A single resident species may play a strong role in excluding particular invaders (Lennon et al. 2003; Lohrer et al. 2000; McGrady-Steed et al. 1997) when the characteristics of the invader and resident are sufficiently similar. In other cases a single predator or herbivore may exert a dominant influence on the abundance of non-natives (deRivera et al. 2005; Lohrer et al. 2000). A recent meta-analysis of the role of biotic resistance in plant invasions (Levine et al. 2004) suggests that these forces can be strong, but they rarely enable communities to resist invasion completely, instead acting to constrain the abundance of exotic species once they have successfully established. The results of an experiment that examined effects of the native community on an introduced oyster were consistent with this review at larger scales (across wave exposure), but not at smaller scales (across tidal elevation) (Ruesink 2007). Other experiments in marine systems that formally test the strength of biotic resistance forces are lacking. In a different meta-analysis, Parker et al. (2006) found native herbivores suppressed exotic plants, but exotic herbivores facilitated both the abundance and species richness of exotic plants. The increasing, widespread replacement of native with exotic herbivores is likely to facilitate invasions and may trigger an invasional meltdown (see Sect. 12.5.3 for further discussion).

Because there are a variety of mechanisms (competition, predation, disease) that can promote biotic resistance, it is likely that different resident species will have different effects on invaders, and those effects may interact. Ruesink (2007) simultaneously manipulated native predators and neighbors (competitors) and found these two different guilds affected introduced oysters differently. Predators reduced oyster survival, but neighbors both reduced growth and improved survival at some wave-exposed sites. Although invasion biology is beginning to address interactive effects of resident species on invaders, it is clear that we need to perform additional studies to examine the mechanisms involved and to consider the consequences of changing diversity (extinctions and invasions) on these interactions.

A corollary to the idea of biotic resistance is that, all else being equal, having more species should result in lower total resource availability, decreasing the success of new species (Case 1990; Elton 1958; Fargione et al. 2003; Fargione and Tilman 2005; MacArthur 1970, 1972; Naeem et al. 2000; Tilman 2004). Because of its intuitive appeal and the lure of a simple measure of invasion resistance, a large number of studies have addressed the question of whether more diverse communities are more resistant to invasion. The meta-analysis mentioned above estimated the effect of species diversity itself as equal in magnitude to strong individual species effects (Levine et al. 2004). However, the analysis was limited to experimental studies, and experimental and observational approaches and small- and large-scale studies have often reached opposing conclusions, sometimes leading to contentious debate. In the remainder of this section, we focus on the relationship between resident diversity and invasibility.

12.4.1 Experimental vs Observational Approaches

Contributing to the apparent paradox between results from experimental and observational diversity-invasibility studies are the fundamentally different questions the two approaches pose, although this difference is not always recognized. Manipulative experiments ask what are the consequences of species loss for the likelihood of invasion. Surveys, on the other hand, tell us locations with species-rich native communities are characterized by conditions that promote exotic species richness. These issues are partially reviewed in Stachowicz and Byrnes (2006), and more fully in Fridley et al. (2007). Both of these questions are important from basic and applied standpoints, and a synthesis of these approaches can provide considerable insights into the conditions under which diversity is and is not an important determinant of invasion success.

Mathematical models (Case 1990, 1991; Tilman 2004; Chap. 4, Wonham and Lewis) and manipulative experiments in both marine and terrestrial systems (Britton-Simmons 2006; Kennedy et al. 2002; Levine 2000; Naeem et al. 2000; Stachowicz et al. 1999, 2002a; White and Shurin 2007) generally (but not unequivocally—see Arenas et al. 2006) support the idea that increasing diversity decreases invasibility. Stachowicz et al. (1999, 2002a) manipulated the species richness of these communities and challenged them with different potential invaders. They found that in all cases invasion success decreased with increasing resident species richness because individual species were complementary in their temporal patterns of space occupation. Individual species all fluctuated in abundance, but these fluctuations were out of phase. Thus, at least one species was always abundant and occupying space in the high-diversity treatments, whereas there were periods of high space availability in the low-diversity treatments. This mechanism appears to operate in the field, as surveys at small and medium scales showed a negative relationship between native and non-native richness (Stachowicz et al. 2002a) and showed that complementary temporal niches arise from seasonal differences in recruitment patterns at larger scales (Stachowicz and Byrnes 2006). Even a brief window of open space (2 weeks) (Stachowicz et al. 2002a) can increase invasion success by an order of magnitude (see Davis MA and Pelsor 2001 for a terrestrial example). Such seasonal or temporal niches may drive diversity effects on invasion resistance in other communities. For example, the biomass of mobile and sessile invertebrate invaders in experimental seagrass mesocosms decreased with increasing species richness of resident mobile invertebrates (France and Duffy 2006). Grazers in this system do show seasonal abundance patterns (Duffy et al. 2001;

Parker et al. 2001), which should produce more complete resource use throughout the season and may contribute to this effect.

In contrast to these findings, an experimental study of marine algae found that native algal functional group richness did not affect invasion success of other, non-resident natives, and that functional group identity most strongly affected invasion instead (Arenas et al. 2006). They found that resource availability did control invasion success, but algal identity (and not richness) controlled resource availability. However, the clever experimental design, which assembled polycultures from small monocultures of each functional group chipped from the rock, did not include all possible functional groups (ephemeral green algae, one of the native invaders, was not included). The design also decreased the likelihood of canopy layering, in which understory species (which had low space availability but high light levels) and canopy species (which had low light levels but high bare substrate) might have acted to complementarily reduce overall resource levels and invasion success. In fact such multivariate complementarity was found in a study of algal invasion into communities of varying functional group richness on the west coast of North America (Britton-Simmons 2006). A third macroalgal study found an overall negative effect of species richness on invasion, even though algal richness enhanced invader settlement by facilitation; apparently this was due to the extremely low survival of settlers in diverse patches (White and Shurin 2007). Thus on balance, experimental marine studies from a diverse suite of taxa (sessile inverts, mobile inverts and algae) generally concur with terrestrial experiments (Fridley et al. 2007) in finding a negative effect of increasing diversity on invasion success, and that this is mediated in large part by complementary resource use among taxa.

Unlike controlled experiments that manipulate only one factor (richness), observational studies, while they cannot unambiguously assign causation, enable an assessment of whether the mechanistic effects of richness identified in experiments are sufficiently strong to generate patterns in the context of natural variation in other important factors. Compared with terrestrial systems, there have been surprisingly few observational studies of resident diversity and invasion in the sea. A survey of sessile marine invertebrates in Tasmania found a positive correlation between the number of native species and the species richness (and to a lesser extent the abundance) of settlers (including both native and non-native species; Dunstan and Johnson 2004). They attributed the positive correlation to a combination of interspecific facilitation and low richness communities being dominated by a few large colonies, which were difficult to displace (Dunstan and Johnson 2004). Likewise, a similar study performed across several spatial scales found that the strength and direction of the relationships between native and exotic plant richness and cover in estuarine plant communities varied among sites and sampling scale (Bruno et al. 2004). Both these studies suggest that any negative effects of diversity are weak relative to other factors, and thus do not generate a negative relationship between native and invader richness in the field. In contrast, several studies have found negative correlations between native richness and invader abundance (White and Shurin 2007) or invader richness (Stachowicz et al. 2002a). Both of these studies also found that reduced diversity increased invasion success in experiments, suggesting that native richness can be strong enough to generate patterns in some cases but can also be overwhelmed by other factors.

A paired experimental and observational approach to this question can clearly produce important insights, but experiments may be limited in their ability to rigorously manipulate multiple factors. Using a multiple regression approach, Stachowicz and Byrnes (2006) examined the context dependency of richness effects on invasion. They found that substrate heterogeneity and the availability of primary space markedly influenced the slope relationship. Specifically, the substrate heterogeneity and additional settlement space generated by a structurally complex bryozoan (*Watersipora subtorquata*) caused the native-invader richness relationship to shift from negative to positive (Fig. 12.3). Thus a negative relationship was only found when facilitators were rare and space was limiting, suggesting that the conditions

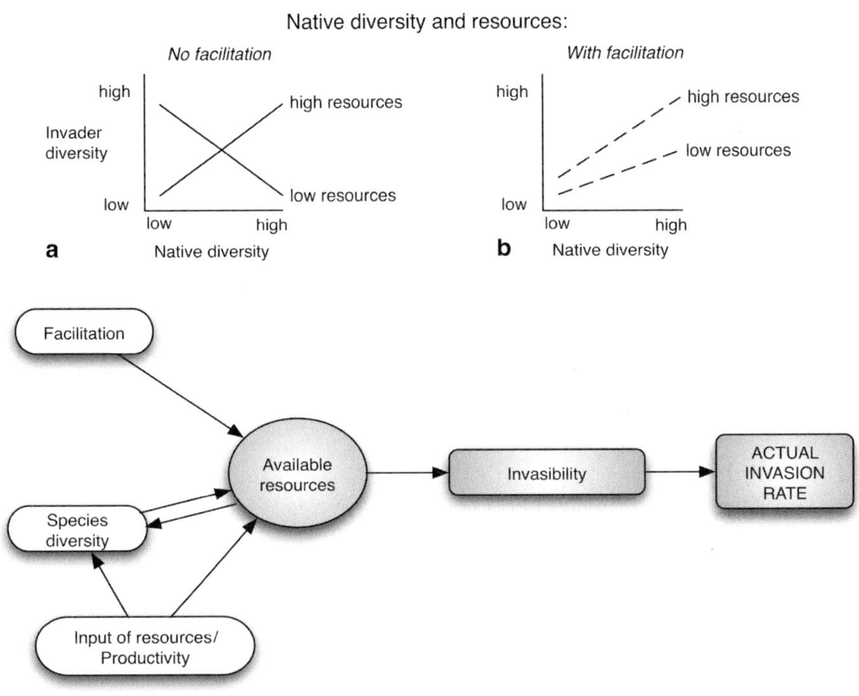

Fig. 12.3 Native diversity can interact with the resource pool to determine invasibility of a system. This concept derives from experiments in sessile invertebrate communities, where space is an important resource. **a** In a community where the amount of available space is limited, a more diverse assemblage of native species more fully co-opts the limiting resource, reducing invasion success through competition and biotic resistance. Where there is abundant space or very high levels of propagule supply, competition and biotic resistance are relatively unimportant and both native and invader diversity are determined by extrinsic factors (e.g. recruitment, disturbance, flow), creating a positive correlation (Levine 2000; Stachowicz et al. 2002a; Stachowicz and Byrnes 2006). **b** Where facilitation increases the availability of limiting resources (e.g., by providing space for secondary settlement), resource limitation is alleviated, overwhelming the negative effects of diversity on invasibility (Stachowicz and Byrnes 2006; see Sect. 12.5.1)

under which the effect of species richness on invasion is dominant are somewhat restricted. Terrestrial studies often agree, finding that the positive effects of heterogeneity or resource levels on both native and exotic richness drive the native-exotic richness correlation, particularly at larger spatial scales, whereas smaller scale negative relationships are often found, reflective of the limited heterogeneity at that scale (e.g., Shea and Chesson 2002; Davies et al. 2005; Fridley et al. 2007).

12.4.2 Large- vs Small-scale Studies

The change in the slope of the native-exotic richness relationship from negative to positive as one moves from small to large-scale observational studies is a bit more vexing than the difference between experimental and observational studies. There appear to be multiple, non-mutually exclusive explanations for the shift from a nearly universally positive slope at large spatial scales to variable slopes at smaller scales. Competition-based models suggest that at small scales, species interactions are strong, resulting in biotic resistance; across large scales, increased resources reduce interaction strengths, resulting in a higher probability of invasion success (Byers and Noonburg 2003). Indeed most studies of interspecific competition in sedentary or sessile organisms would support the idea that the effects of competition occur primarily at a neighborhood level. In contrast, Fridley et al. (2004) caution that random assembly of communities of neutral species also produce a negative relationship between native and exotic diversity at the smallest scales and a positive relationship at larger scales, so biological mechanisms are not required to produce these results.

Davies et al. (2005), building on the arguments of Shea and Chesson (2002), explained the invasion paradox as a consequence of spatial environmental heterogeneity. They showed that native and exotic richness at large scales were both positively influenced by variance in soil resources (such as soil depth), suggesting that shifts in competitive dominance of different species over a range of resource levels favor the coexistence of natives and exotics on scales where such heterogeneity emerges (Chesson 2000; Huston and Deangelis 1994; Tilman 2004). This theory and the temporally-based analogue of Davis J et al. (2002) are specific cases of more general ideas relating to the persistence of rare species (i.e., an invader) in a variable environment (Chesson 1994). The result is a consistently positive native-exotic richness relationship at large scales, where heterogeneity is virtually guaranteed. At smaller scales plots may be homogenous or heterogeneous, leading to higher variability in diversity-invasibility patterns. In their study of landscape scale patterns in shoreline plant communities, Bruno et al. (2004) found native and exotic richness were negatively correlated within some, but not most, sites (small scale), and were positively correlated across sites (larger scale). They suggested three mechanisms: (1) exotics could be facilitated by residents (see Sect. 12.5), (2) the factors that promote high native richness (e.g., high propagule supply, resource availability and favorable environmental conditions) could also enhance success of exotics (see Sect. 12.4.1 and Fig. 12.3a), and (3) sampling artifacts could be driving positive native-exotic relationships (this section). The authors point out that none of these mechanisms preclude biotic resistance by native species, therefore a positive relationship should not necessarily lead to the rejection of the diversity-invasibility paradigm. Further, the shoreline habitats of New England are subject to frequent disturbance, which often acts to increase available resources and reduce competition, increasing invasibility (see Sect. 12.6).

A major cause of variation in small-scale heterogeneity may be the presence or absence of biogenic habitat structure (see also Sect. 12.5). In small-scale field surveys of native and invader richness, the slope of the relationship between native and invader diversity can change with the availability of resources (space) and the presence of habitat-forming foundation species (Stachowicz and Byrnes 2006). In these systems, negative relationships between native and exotic richness are only found when primary space is limiting and structure-forming species like mussels and head-forming bryozoans that provide habitat heterogeneity and additional settlement space are rare. We further explore the role of facilitation and resource provision in the next section.

12.5 Facilitation

So far we have focused on species interactions that decrease resource availability, but by their presence most species also create some resources. There are obvious examples that cross trophic levels, such as the presence of a plant providing food for herbivores, but others are more subtle and can act within guilds or trophic levels. For example, while a mussel holds space, its hard shell provides additional attachment surface that can alleviate space limitation for other sessile suspension feeders. Invasion biology, like ecology in general, has often focused on negative, resource-consuming interactions like competition and predation to the exclusion of facilitative interactions among species (Bertness and Callaway 1994; Bruno 2005; Bruno et al. 2003; Rodriguez 2006). However, native and exotic species can provide habitat or offer a refuge from physical stress or predation and thus could either directly facilitate non-native species or inhibit them by facilitating native competitors or predators (see also Chap. 16, Crooks; Chap. 17, Grosholz and Ruiz). In this section, we examine the direct and indirect effects of the presence of different types of positive interactions on available resources for exotic species in marine communities.

12.5.1 Increasing Resource Availability

Marine systems provide numerous examples of foundation species enhancing biological diversity by providing structure and substrate, and by reducing the effects of disturbance for organisms (Bruno and Bertness 2001; Bruno et al. 2003;

Rodriguez 2006; Stachowicz 2001). Both in subtidal and intertidal habitats, native mussels provide additional surface area for sessile invertebrates and a dense structural matrix for many smaller mobile invertebrates (Paine and Suchanek 1983). Mussel and oyster farms have severe problems with non-native fouling organisms such as the solitary sea squirts *Ciona intestinalis* (Carver et al. 2003) and *Styela clava*, as the hard shells provide ideal substrate for colonization by sessile invertebrates. Native kelp in the Gulf of Maine provides a substrate for the non-native bryozoan *Membranipora membranacea* (Berman et al. 1992) that normally resides on kelp in the Pacific (Dixon et al. 1981; Hepburn and Hurd 2005). The alga *Chondrus crispus* is host to a variety of non-native bryozoans and tunicates (Stachowicz and Whitlatch 2005). These relationships are commonly assumed to be parasitic or competitive, but they can be mutualistic, as the epibionts can provide associational defenses (Wahl and Hay 1995) or provide nutrients via their excretions (Hepburn and Hurd 2005; Hurd et al. 1994). Native species that support invaders as epibionts facilitate invasions in the same way that the construction of artificial jetties, breakwaters, oil platforms, and marinas and docks do – by adding available resources (space).

Provision of habitat for exotic species by already established non-native species follows similar patterns. The non-native mudsnail *Battilaria attramentaria* provides settlement substrate for the non-native slippershell *Crepidula convexa* and the invasive anemone *Diadumene lineata* in the mudflats of tidal estuaries in Washington State (Wonham et al. 2005). Often, these exotics facilitate the establishment of other non-native species in habitats where they could not have otherwise existed (see Fig. 12.3b). On boats covered with anti-fouling paint, the non-native bryozoan *Watersipora subtorquata* can settle in areas where the paint has chipped, and then create large foliose heads that provide settlement substrate and a means of dispersal for other non-native species (Floerl and Inglis 2005). The non-native worm *Ficopomatus enigmaticus* creates rocky reefs in the middle of soft-sediment estuaries, enabling a variety of other exotic species to enter an otherwise inaccessible habitat (Schwindt and Iribarne 2000).

In general, when exotic, habitat-forming species invade habitats lacking other foundation species, the consequences for community invasibility will be dramatic. However, when other native facilitators are present, the outcome is less clear. For example, although the invasive mussel *Mytilus galloprovincialis* serves the role of a foundation species in Sydney Harbor, it does not contain as high of an abundance or richness as adjacent native algal turfs (Chapman et al. 2005), nor does it contain any species that the turfs lack. In contrast, native-exotic hybrids of *Spartina* cordgrass in San Francisco Bay produce taller and denser stems, greatly decreasing habitat suitability for native infaunal species (Brusati and Grosholz 2006). Because interactions between habitat-forming species and their associates are relatively generalized, we see no reason why exotics or natives should, as a rule, benefit more from such interactions. However, even subtle differences in the types of habitat structure provided can dramatically alter composition of associated faunal communities, so comparative studies of the relative effects of native and non-native foundation species on the success of exotic (and native) species is clearly an area in need of greater attention.

12.5.2 Ameliorating Physical Stress

Species that have large individuals or that form dense aggregations often buffer the physical environment, making it more hospitable for smaller or less stress-tolerant species. For example, in physically stressful intertidal habitats canopy-forming algae and plants are known to facilitate the persistence of many species by reducing heat, desiccation, and/or salt stress (Bertness and Callaway 1994; Bertness and Hacker 1994; Bertness and Shumway 1993). Similarly, the non-native alga *Caulerpa taxifolia* experiences higher growth in the presence of native seagrasses due to lower water velocity that reduces blade fragmentation (Ceccherelli and Cinelli 1998, 1999). But as *Caulerpa* grows it can become larger than the seagrass bed that facilitated it, eventually overgrowing it, suggesting that these sorts of facilitative interactions may be most important during the early establishment phase of an invasion, as is true more broadly of the "nurse plant" effect (Callaway 1995). In many cases, exotic foundation species can stabilize the substrate (e.g., seagrass, marsh cordgrass or mat-forming bivalves), fundamentally altering water flow and disturbance regimes among many other effects. We provide examples of how these negatively affect natives and may positively affect exotics in Sect. 12.6.

12.5.3 Facilitation Cascades and "Invasional Meltdown"

There is some evidence that facilitation can lead to positive feedbacks among invaders that progressively increase the invasion susceptibility of a community. Resident organisms can make a habitat more easily invaded if they suppress the densities of potential predators or competitors of a newly arrived, non-native species. These types of facilitators can reduce the effect of both diversity and biotic resistance on invasion resistance. These facilitation interactions tend to be more likely in ecosystems that have been heavily affected by anthropogenic activities or are already heavily invaded, leading to "invasional meltdown" (Simberloff and Von Holle 1999).

For example, one of the major factors enabling the invasion of the Japanese alga *Codium fragile* ssp. *tomentesoides* in the Gulf of Maine was a dramatic increase in the native urchin *Strongylocentrotus droebachiensis* (Harris and Tyrrell 2001; Sumi and Scheibling 2005). Rising urchin densities were caused at least in part by declines in urchin predators due to overfishing (Steneck et al. 2002). *Codium* is a low-preference food for these urchins and is avoided until other native algae, particular the dominant bed-forming *Laminaria saccharina*, have been dramatically reduced or eliminated (Sumi and Scheibling 2005). This state change was facilitated by a second non-native species, the epiphytic bryozoan *Membranipora membranacea*, which encrusts native kelp, increasing its palatability to herbivores and its susceptibility to mechanical breakage (Dixon et al. 1981; Levin et al. 2002), facilitating *Codium* establishment. Interestingly, once established, *Codium* provides substrate for a variety of non-native epibionts, propagating a cycle of invasional meltdown in the Gulf of Maine (Harris and Jones 2005; Mathieson et al. 2003).

A single species can also have impacts by both altering the competitive landscape and removing direct predators under different scenarios. The exotic European green crab *Carcinus maenas* has been implicated in at least two scenarios of facilitation of other exotic species. Much like urchins in the previous example, on the west coast of North America *Carcinus* has been shown to reduce the density of native clams, *Nutricola* sp., thereby releasing the formerly suppressed, non-native clam *Gemma gemma* (Grosholz 2005). On the east coast of North America, instead of reducing competitors, *Carcinus* removes a variety of small gastropods that would, in its absence, reduce the abundance of non-native fouling organisms through predation (Osman and Whitlatch 1998, 2004; Stachowicz and Whitlatch 2005).

12.6 Disturbance

Ecologists have long recognized that disturbances can facilitate non-native species (Byers 2002b; Clark and Johnston 2005; Elton 1958; Grosholz 2002; Occhipinti-Ambrogi and Savini 2003; Chap. 7, Johnston et al.). Although disturbance can be defined in many ways, we focus here on those forces (natural or human-caused) that remove native biomass and/or change environmental conditions, thereby increasing or creating available resources.

Despite the well-documented effects of natural disturbance on marine communities (Connell 1978; Dayton 1971; Sousa 1979), there are only a few documented cases of natural disturbances facilitating non-native invasions in marine systems (Nichols et al. 1990; Valentine and Johnson 2003). In contrast, there are numerous examples of anthropogenic disturbance such as physical disturbance, over-harvesting, and the presence of non-native species contributing to invasion success (Byers 2002a). It has been suggested that coastal areas such as estuaries, bays and lagoons are hotspots for invasions due to not only to the high rate of human-mediated transfer between these areas but also high rates of disturbance that either remove native biomass or alter physical conditions (e.g., pollutants, dredging, development, etc.) in such a way to facilitate exotic species (see Fig. 12.4; also see Chap. 33, Preisler et al.). As a corollary, one explanation for the relatively small number of invaders along the coast of Chile is a scarcity of sheltered areas that are more likely to attract human development and associated disturbance (Castilla et al. 2005).

To understand the role of disturbance in invasion success, it is important to identify whether the mechanism by which disturbance acts is direct (changing environmental suitability) or indirect (increased resource availability by reduction of native abundance). We first review examples of disturbance increasing available resources. Returning to an earlier example, seagrass beds are often fragmented as a result of physical disturbance from dredging and development; the non-native mussel, *Musculista senhousia*, grows best in disturbed eelgrass beds as compared to continuous

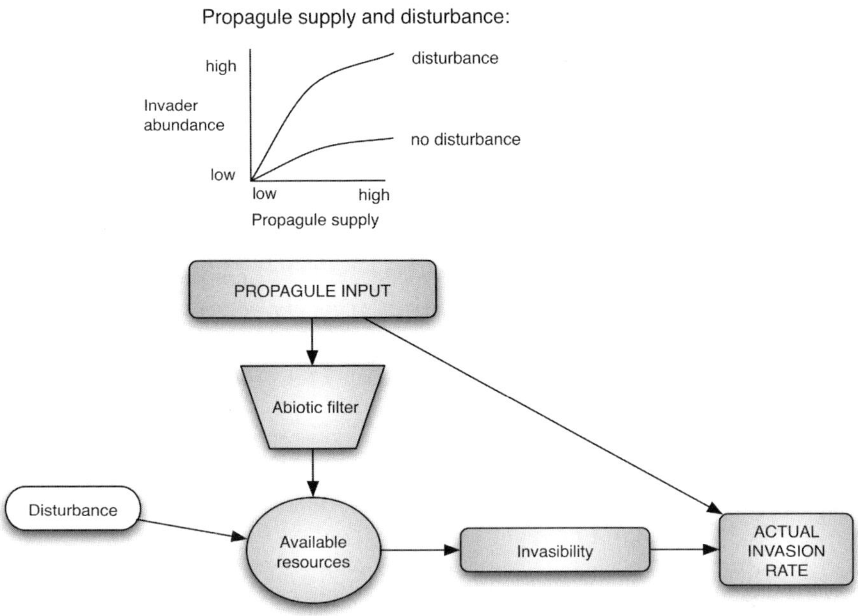

Fig. 12.4 Disturbance and propagule supply can interact to modify invader success. Invasion increases with propagule supply, but in the absence of disturbance or other factors that release resources, high levels of propagule input are needed for successful invasion because individual survival rates are low. Where disturbance is stronger, fewer propagules may be required for successful establishment. For example, in an experiment that controlled propagule supply, physical disturbance was shown to increase the invasion success of the introduced species *Bugula neritina* in sessile, marine invertebrate communities (see Sect. 12.6) (Clark and Johnston 2005; Chap. 7, Johnston et al.)

beds with slower flow rates and reduced food availability (Allen and Williams 2003; Reusch and Williams 1999). In another experiment that simulated anthropogenic impacts, physical disturbance increased space availability and the recruitment success of the non-native bryozoan *Bugula neritina*, although chemical pollution did not (Clark and Johnston 2005). In addition, non-native species can be a cause of disturbance to natives, facilitating the establishment of other non-natives, as discussed previously for the bryozoan, *Membranipora membranacea* (see Sect. 12.5.3).

In addition to removing species, natural or human-mediated disturbances can change habitat conditions and result in mismatch between native species and their environment (Byers 2002a). Such a situation can lead to a competitive advantage for non-native species if they are successful under the new conditions (Byers 2002a). In an experiment that controlled nonnative propagule supply, Crooks et al. (unpublished data) found that native diversity declined significantly with increasing concentrations of copper, a common pollutant from boat-bottom paint, while exotics did not. Although there are few experimental data, evidence from marine systems suggests that human-mediated eutrophication enhances the survival and growth of non-native species by creating conditions that are unfavorable for native species (Allen and Williams 2003; Byers 2002a; Occhipinti-Ambrogi and Savini 2003). In addition to disturbing native species directly (see above), the presence of other non-native species can also result in abiotic and/or biotic environmental change that is detrimental to native species (Byers 2002a; Grosholz 2002; Levin et al. 2002; Neira et al. 2005). In particular, non-native foundation species such as the *Spartina alterniflora-S. foliosa* hybrid in San Francisco Bay, (Brusati and Grosholz 2006; Neira et al. 2005) or the mussel *Musculista senhousia* (Crooks and Khim 1999) can greatly alter habitat characteristics (e.g., sediment organic content, light availability, physical and structural complexity) reducing the fitness or abundance of natives, and in some cases, facilitating exotics. A major source of disturbance (anthropogenic and natural) that can directly facilitate invasions is rising ocean temperatures as a result of global climate change (Harley et al. 2006). Temperature increases from climate change have facilitated range expansions of many tropical and subtropical organisms into temperate areas (Barry et al. 1995; Beare et al. 2004; Perry et al. 2005; Southward et al. 1995). In many instances, the success of non-native species may depend on the magnitude of temperature *extremes* (minimum and maximum) in certain seasons, instead of increases in *mean* temperature. For example, warmer *winter* temperatures accelerated the seasonal timing of recruitment of invasive tunicates in New England, such that in warm years exotics recruited before natives, whereas in colder years the reverse was true (Stachowicz et al. 2002b). Temperature extremes had similarly opposing effects on the magnitude of recruitment of native and exotic species. In no cases was there any effect of annual mean temperature on recruitment of natives or exotics. In concert, the results suggest that decreasing severity of New England winters over the past two decades has likely facilitated the shift in dominance of fouling communities to recent invaders. Finally, interactions between human-driven alterations in biotic pressure (e.g. overfishing of top predators) and climate (ocean warming) is also thought to have accelerated shifts in entire communities, for example the dramatic shift to an exotic-dominated community assemblage in the subtidal areas of the Gulf of Maine discussed above (Sect. 12.5.3) (Harris and Tyrrell 2001).

Another important consideration for predicting habitat invasibility is the timing of disturbance in relation to the arrival of invader propagules (Clark and Johnston 2005; Chap. 7, Johnston et al.). For example, experimental removal of the native algal canopy just prior to recruitment of the non-native kelp *Undaria* facilitated the introduction to a greater degree than when the disturbance occurred after the recruitment pulse (Valentine and Johnson 2003). Presumably early life history stages of many invaders cannot persist long in established communities of natives, but may thrive when native biomass is decreased. Interestingly, disturbance is not critical for the persistence of *Undaria* in this system; once established, *Undaria* can persist even in the absence of grazing urchins (Valentine and Johnson 2005). Prolonged disturbance can reduce natives to sufficiently low abundance that even if the disturbance ceases, recolonization by natives is slow and perhaps recruitment limited, so systems remain dominated by exotic species unless natives are artificially restored (Seabloom et al. 2003).

We close this section with several generalizations that have recently emerged from work on the disturbance – invasibility relationship in freshwater and terrestrial environments that have yet to be fully examined in marine systems (Lozon and MacIsaac 1997). First, non-native species may be more likely to be associated with human disturbances than natural ones (Lozon and MacIsaac 1997), which is supported by the qualitative trend in the marine literature. Second, disturbance is more important to the success of plant than animal invasions in terrestrial systems (Lozon and MacIsaac 1997); similar comparisons might hold for mobile vs sessile organisms in marine communities, but few comparative data are available. Third, assessing causation to disturbance, per se, is difficult because disturbance often co-varies with many other factors that also influence resource supply and niche space (e.g., biotic resistance, species diversity, propagule supply (Clark and Johnston 2005; Connell 1978; Davis MA et al. 2002; Klein et al. 2005; Naeem et al. 2000) and because invaders themselves may cause significant changes in disturbance regimes. Careful, multifactorial experiments will be needed to help address this issue.

12.7 Historical Context

In this chapter we have reviewed how abiotic conditions (resource availability and disturbance) and biotic interactions (competition and facilitation) and the interaction between the two affect the relative abundance of invaders in marine communities. We would be remiss, however, if we did not emphasize that a community's history – on both geologic (Daehler and Strong 1996; Vermeij 1991a, 2005) and ecological timescales (Davis MA and Pelsor 2001; Grosholz 2005; Vitousek et al. 1996) can affect the probability and course of an invasion.

Large-scale geological invasions of species (i.e., biotic interchanges) between two or more regions have occurred several times in the past, for instance the Trans-Arctic interchange between the North Pacific and North Atlantic via the Bering Strait 3.5 Ma (Vermeij 1991a, b). Vermeij observed that most of these interchanges involved asymmetrical movements of species from one geographic donor region to another receiver region, and he found that the number of invaders to a region was proportional to the number of species that had gone extinct there (the "ecological opportunity" hypothesis). Species poor regions were disproportionately invaded by species from species rich invasions. For example, the rocky-shore fauna of the Northwest Atlantic suffered a catastrophic extinction event in the early Pliocene, dramatically reducing species diversity. During the subsequent Trans-Arctic interchange the majority of species involved (for example, 261/295 of mollusc fauna) migrated from the North Pacific to the Northwest Atlantic (Vermeij 1991a, b). On a shorter geologic timescale, many have noted that the exceptionally high invasion rate of U.S. Pacific coast estuaries may be related to the fact that the estuaries are geologically "young" (i.e., they formed less than 10,000 years ago) and are therefore species poor (Cohen and Carlton 1998; Daehler and Strong 1996). And on an even shorter timescale, the degree of "saturation" of a community (or lack thereof) appears to be relevant for ecological timescales as well. Moyle and Light (1996) noted that freshwater fish invaders were more successful in streams that had recently suffered a catastrophe (e.g., a major flooding event), and human-caused extinctions and local extirpations have also been shown to make a community susceptible to invasion (as discussed in Sect. 12.6).

12.8 Conclusion

Most of the studies we have reviewed focus on one or perhaps two of the various factors affecting habitat invasibility; studies that consider the interaction among factors are rare. Yet Fig. 12.1 clearly illustrates that many factors simultaneously affect resource availability both directly and indirectly. A major priority for invasion biology (and community ecology) is to assess the relative roles of these factors in determining community composition, and how these relative roles change across environmental gradients. A promising avenue for untangling these relationships is to conduct factorial experiments with factors that are often related, for example, disturbance and propagule supply (Clark and Johnston 2005). Where multifactorial experiments are not feasible, or where larger spatial scales are desirable, careful correlative studies that measure a range of covariates can be very useful (e.g., Stachowicz and Byrnes 2006).

A number of the relationships in Fig. 12.1 have been depicted as unidirectional when in fact they may be bi-directional, leading to the potential for feedback. For example, disturbance affects diversity but diversity may also affect the susceptibility of communities to disturbance (Sect. 12.6). Given that each of the factors alone can affect invasibility in different ways, it seems important to understand how their interactive effects might differ from their effects in isolation (see Figs. 12.3 and 12.4 for conceptual models).

Finally we suggest that a diversity of approaches and study systems will aid in developing more robust generalities in invasion biology. Our focus on resource availability is driven in part by the bias in invasion ecology toward the study of sessile or sedentary taxa; if mobile taxa are less resource-limited, then perhaps other factors drive differential patterns of invasion in these groups (but see France and Duffy 2006). Similarly, more explicit recognition of the benefits as well as the limitations of both experimental and observational approaches will clarify our thinking and reduce misunderstandings. Indeed, the literature in both invasion biology and general ecology shows that studies that combine these approaches often reach robust conclusions (Levine 2000; Stachowicz et al. 2002a; White and Shurin 2007). We urge experimentalists to complement their studies with larger scale experiments, and vice-versa. Simultaneous application of both approaches in the same system should enhance our ability to connect pattern and mechanism in invasion biology.

Acknowledgements We thank L. Rodriguez and two anonymous reviewers for comments on a previous draft of the manuscript and the editors for the invitation to participate in this project. We also acknowledge support from the National Science Foundation Biological Oceanography Program (to JJS) and Graduate Research Fellowship program (to SVO, ARH, and KMH).

References

Allen BJ, Williams SL (2003) Native eelgrass *Zostera marina* controls growth and reproduction of an invasive mussel through food limitation. Mar Ecol Prog Ser 254:57–67

Arenas F, Sanchez I, Hawkins SJ, Jenkins SR (2006) The invasibility of marine algal assemblages: Role of functional diversity and identity. Ecology 87:2851–2861

Barry JP, Baxter CH, Sagarin RD, Gilman SE (1995) Climate related, long-term faunal changes in a California rocky intertidal community. Science 267:672–675

Beare DJ, Burns F, Greig A, Jones EG, Peach K, Kienzle M, McKenzie E, Reid DG (2004) Long-term increases in prevalence of North Sea fishes having southern biogeographic affinities. Mar Ecol Prog Ser 269–278

Berman J, Harris L, Lambert W, Buttrick M, Dufresne M (1992) Recent invasions of the Gulf of Maine: Three contrasting ecological histories. Conserv Biol 6:435–441

Bertness MD, Callaway R (1994) Positive interactions in communities. Trends Ecol Evol 9:191–193

Bertness MD, Hacker SD (1994) Physical stress and positive associations among marsh plants. Am Nat 144:363–372

Bertness MD, Shumway SW (1993) Competition and facilitation in marsh plants. Am Nat 142:718–724

Bertness MD, Ewanchuk PJ, Silliman BR (2002) Anthropogenic modification of New England salt marsh landscape. Proc Natl Acad Sci 99:1395–1398

Branch GM, Steffani CN (2004) Can we predict the effects of alien species? A case-history of the invasion of South Africa by *Mytilus galloprovincialis* (Lamarck). J Exp Mar Biol Ecol 300:189–215

Britton-Simmons KH (2006) Functional group diversity, resource preemption and the genesis of invasion resistance in a community of marine algae. OIKOS 113:395–401

Bruno JF (2005) Insights into biotic interactions from studies of species invasions. In: Sax DF, Gaines SD, Stachowicz JJ (eds) Species invasions: insights into ecology, evolution, and biogeography. Sinauer Associates, Sunderland, Massachusetts, pp 13–40

Bruno JF, Bertness MD (2001) Habitat modification and facilitation in marine benthic communities. In: Bertness MD, Gaines SD, Hay ME (eds) Marine community ecology. Sinauer Associates, Sunderland, MA, pp 201–220

Bruno JF, Stachowicz JJ, Bertness MD (2003) Inclusion of facilitation into ecological theory. Trends Ecol Evol 18:119–125

Bruno JF, Kennedy CW, Rand TA, Grant M-B (2004) Landscape-scale patterns of biological invasions in shoreline plant communities. Oikos 107:531–540

Brusati ED, Grosholz ED (2006) Native and introduced ecosystem engineers produce contrasting effects on estuarine infaunal communities. Biol Invas 8:683–695

Burke MJW, Grime JP (1996) An experimental study of plant community invasibility. Ecology 77:776–790

Byers JE (2002a) Impact of non-indigenous species on natives enhanced by anthropogenic alteration of selection regimes. Oikos 97:449–458

Byers JE (2002b) Physical habitat attribute mediates biotic resistance to non-indigenous species invasion. Oecologia 130:146–156

Byers JE, Noonburg EG (2003) Scale dependent effects of biotic resistance to biological invasion. Ecology 84:1428–1433

Callaway RM (1995) Positive interactions among plants. Bot Rev 61:306–349
Carver CE, Chisholm A, Mallet AL (2003) Strategies to mitigate the impact of Ciona intestinalis (L.) biofouling on shellfish production. J Shellfish Res 22:621–631
Case TJ (1990) Invasion resistance arises in strongly interacting species-rich model competition communities. Proc Natl Acad Sci 87:9610–9614
Case TJ (1991) Invasion resistance species build-up and community collapse in metapopulation models with interspecies competition. Biol J Linn Soc 42:239–266
Castilla JC, Uribe M, Bahamonde N, Clarke M, Desqueyroux-Faundez R, Kong I, Moyano H, Rozbaczylo N, Santelices B, Valdovinos C, Zavala P (2005) Down under the southeaster Pacific: marine non-indigenous species in Chile. Biol Invas 7:213–232
Ceccherelli G, Cinelli F (1998) Habitat effect on spatio-temporal variability in size and density of the introduced alga Caulerpa taxifolia. Mar Ecol Prog Ser 163:289–294
Ceccherelli G, Cinelli F (1999) Effects of *Posidonia oceanica* canopy on *Caulerpa taxifolia* size in a north-western Mediterranean bay. J Exp Mar Biol Ecol 240:19–36
Chapman MG, People J, Blockley D (2005) Intertidal assemblages associated with natural corallina turf and invasive mussel beds. Biodivers Conserv 14:1761–1776
Chesson P (1994) Multispecies competition in variable environments. Theor Popul Biol 45:227–276
Chesson P (2000) General theory of competitive coexistence in spatially-varying environments. Theor Popul Biol 58:211–237
Clark GF, Johnston EL (2005) Manipulating larval supply in the field: a controlled study of marine invasibility. Mar Ecol Prog Ser 298:9–19
Cleland EE, Smith MD, Andelman SJ, Bowles C, Carney KM, Horner-Devine MC, Drake JM, Emery SM, Gramling JM, Vandermast DB (2004) Invasion in space and time: non-native species richness and relative abundance respond to interannual variation in productivity and diversity. Ecol Lett 7:947–957
Cohen AN, Carlton JT (1998) Accelerating invasion rate in a highly invaded estuary. Science 279:555–558
Connell JH (1978) Diversity in tropical rain forests and coral reefs. Science 199:1302–1310
Crooks JA, Khim HS (1999) Architectural vs. biological effects of a habitat-altering, exotic mussel, *Musculista senhousia*. J Exp Mar Biol Ecol 240:53–75
Daehler CC, Strong DR (1996) Status, prediction and prevention of introduced cordgrass *Spartina* spp. invasions in Pacific estuaries, USA. Biol Conserv 78:51–58
Davies KF, Chesson P, Harrison S, Inouye BD, Melbourne BA, Rice KJ (2005) Spatial heterogeneity explains the scale dependence of the native-exotic diversity relationship. Ecology 86:1602–1610
Davis J, Levin L, Walther S (2002) Artificial armored shorelines: sites for open-coast species in a southern California bay. Mar Biol 140:1249–1262
Davis MA, Pelsor M (2001) Experimental support for a resource-based mechanistic model of invasibility. Ecol Lett 4:421–428
Davis MA, Grime JP, Thompson K (2000) Fluctuating resources in plant communities: a general theory of invasibility. J Ecol 88:528–534
Dayton PK (1971) Competition, disturbance, and community organization: the provision and subsequent utilization of space in a rocky intertidal community. Ecol Monogr 41:351–389
deRivera CE, Ruiz GM, Hines AH, Jivoff P (2005) Biotic resistance to invasion: native predator limits abundance and distribution of an introduced crab. Ecology 86:3364–3376
Dethier MN, Hacker SD (2005) Physical factors vs. biotic resistance in controlling the invasion of an estuarine marsh grass. Ecol Appl 15:1273–1283
Dixon J, Schroeter SC, Kastendiek J (1981) Effects of the encrusting bryozoan, *Membranipora membranacea*, on the loss of blades and fronds by the giant kelp, *Macrocystis pyrifera* (Laminariales). J Phycol 17:341–345
Duffy J, Macdonald K, Rhode J, Parker J (2001) Grazer diversity, functional redundancy, and productivity in seagrass beds: an experimental test. Ecology 82:2417–2434
Dunstan PK, Johnson CR (2004) Invasion rates increase with species richness in a marine epibenthic community by two mechanisms. Oecologia 138:285–292

Elton CS (1958) The ecology of invasions by animals and plants. Methuen, London
Fargione JE, Tilman D (2005) Diversity decreases invasion via both sampling and complementarity effects. Ecol Lett 8:604–611
Fargione JE, Brown CS, Tilman D (2003) Community assembly and invasion: an experimental test of neutral versus niche processes. Proc Natl Acad Sci USA 100:8916–8920
Floerl O, Inglis GJ (2005) Starting the invasion pathway: the interaction between source populations and human transport vectors. Biol Invas 7:589
France K, Duffy J (2006) Consumer diversity mediates invasion dynamics at multiple trophic levels. OIKOS 113:515–529
Fridley JD, Brown RL, Bruno JF (2004) Null models of exotic invasion and scale-dependent patterns of native and exotic species richness. Ecology 85:3215–3222
Fridley JD, Stachowicz JJ, Naeem S, Sax DF, Seabloom EW, Smith MD, Stohlgren TJ, Tilman D, Von Holle B (2007) The invasion paradox: reconciling pattern and process in species invasions. Ecology 88
Grosholz ED (2002) Ecological and evolutionary consequences of coastal invasions. Trends Ecol Evol 17:22–27
Grosholz ED (2005) Recent biological invasion may hasten invasional meltdown by accelerating historical introductions. Proc Natl Acad Sci 102:1088–1091
Grosholz ED, Ruiz GM (1996) Predicting the impact of introduced marine species: lessons from the multiple invasions of the European green crab *Carcinus maenas*. Biol Conserv 78:59–66
Harley CDG, Hughes AR, Hultgren KM, Miner BG, Sorte CJB, Thornber CS, Rodriquez LF, Tomanek L, Williams SL (2006) The impacts of climate change in coastal marine systems. Ecol Lett 9:228–241
Harris LG, Jones AC (2005) Temperature, herbivory and epibiont acquisition as factors controlling the distribution and ecological role of an invasive seaweed. Biol Invas 7:913–924
Harris LG, Tyrrell MC (2001) Changing community states in the Gulf of Maine: synergism between invaders, overfishing and climate change. Biol Invas 3:9
Hepburn CD, Hurd CL (2005) Conditional mutualism between the giant kelp *Macrocystis pyrifera* and colonial epifauna. Mar Ecol Prog Ser 302:37–48
Holt RD, Barfield M, Golmulkiewicz R (2005) Theories of niche conservatism and evolution. In: Sax DF, Stachowicz JJ, Gaines SD (eds) Species invasions: insights into ecology, evolution, and biogeography. Sinauer Associates, Sunderland, Mass
Huenneke LF, Hamburg SP, Koide R, Mooney HA, Vitousek PM (1990) Effects of soil resources on plant invasion and community structure in Californian serpentine grassland. Ecology 71:478–491
Hurd CL, Durante KM, Chia FS, Harrison PJ (1994) Effect of bryozoan colonization on inorganic nitrogen acquisition by the kelps *Agarum fimbriatum* and *Macrocystis integrifolia*. Mar Biol 121:167–173
Huston MA, Deangelis DL (1994) Competition and coexistence: the effects of resource transport and supply rates. Am Nat 144:954–977
Jiang L, Morin PJ (2004) Productivity gradients cause positive diversity-invasibility relationships in microbial communities. Ecol Lett 7:1047–1057
Kennedy TA, Naeem S, Howe KM, Knops JMH, Tilman D, Reich P (2002) Biodiversity as a barrier to ecological invasion. Nature 417:636–638
Klein J, Ruitton S, Verlaque M, Boudouresque C-F (2005) Species introductions, diversity and disturbances in marine macrophyte assemblages of the northwestern Mediterranean Sea. Mar Ecol Prog Ser 290:79–88
Kuhn NL, Zedler JB (1997) Differential effects of salinity and soil saturation on native and exotic plants of a coastal salt marsh. Estuaries 20:391–403
Lennon JT, Smith VH, Dzialowski AR (2003) Invasibility of plankton food webs along a trophic state gradient. OIKOS 103:191–203
Levin PS, Coyer JA, Petrik R, Good TP (2002) Community-wide effects of nonindigenous species on temperate rocky reefs. Ecology 83:3182–3193
Levine JM (2000) Species diversity and biological invasions: Relating local process to community pattern. Science 288:852–854

Levine JM, Adler PB, Yelenik SG (2004) A meta-analysis of biotic resistance to exotic plant invasions. Ecol Lett 7:975–989

Lohrer AM, Whitlatch RB, Wada K, Fukui Y (2000) Home and away: comparisons of resource utilization by a marine species in native and invaded habitats. Biol Invas 2:41–57

Lozon JD, MacIsaac HJ (1997) Biological invasions: are they dependent on disturbance? Environ Rev 5:131–144

MacArthur RH (1970) Species-packing and competitive equilibrium for many species. Theor Popul Biol 1:1–11

MacArthur RH (1972) Geographical ecology: patterns in the distribution of species. Harper and Row, New York

Mathieson AC, Dawes CJ, Harris LG, Hehre EJ (2003) Expansion of the Asiatic green alga *Codium fragile* subsp *tomentosoides* in the Gulf of Maine. Rhodora 105:1–53

May RM, MacArthur RH (1972) Niche overlap as a function of environnmental variability. Proc Natl Acad Sci 69:1109–1113

McGrady-Steed J, Harris PM, Morin PJ (1997) Biodiversity regulates ecosystem predictability. Nature (London) 390:162–165

Minchinton TE (2002) Precipitation during El Nino correlates with increasing spread of *Phragmites australis* in New England, USA, coastal marshes. Mar Ecol Prog Ser 305–309

Moyle PB, Light T (1996) Fish invasions in California: do abiotic factors determine success? Ecology 77:1666–1670

Naeem S, Knops JMH, Tilman D, Howe KM, Kennedy T, Gale S (2000) Plant diversity increases resistance to invasion in the absence of covarying extrinsic factors. OIKOS 91:97–108

Neira C, Levin LA, Grosholz ED (2005) Benthic macrofaunal communities of three sites in San Francisco Bay invaded by hybrid *Spartina*, with comparison to uninvaded habitats. Mar Ecol Prog Ser 292:111–126

Nichols FH, Thompson JK, Schemel LE (1990) Remarkable invasion of San Francisco Bay (California, USA) by the Asian clam *Potamocorbula amurensis*. II. Displacement of a former community. Mar Ecol Prog Ser 66:95–101

Occhipinti-Ambrogi A, Savini D (2003) Biological invasions as a component of global change in stressed marine ecosystems. Mar Pollut Bull 46:542–551

Osman RW, Whitlatch RB (1998) Local control of recruitment in an epifaunal community and the consequences to colonization processes. Hydrobiologia 376:113–123

Osman RW, Whitlatch RB (2004) The control of the development of a marine benthic community by predation on recruits. J Exp Mar Biol Ecol 311:117–145

Paine RT, Suchanek TH (1983) Convergence of ecological processes between independently evolved competitive dominants – a tunicate-mussel comparison. Evolution 37:821–831

Parker J, Duffy J, Orth R (2001) Plant species diversity and composition: experimental effects on marine epifaunal assemblages. Mar Ecol Prog Ser 224:55–67

Parker JD, Burkepile DE, Hay ME (2006) Opposing effects of native and exotic herbivores on plant invasions. Science 311:1459–1461

Perry AL, Low PJ, Ellis JR, Reynolds JD (2005) climate change and distribution shifts in marine fishes. Science 308:1912–1915

Peterson AT (2003) Predicting the geography of species' invasions via ecological niche modeling. Q Rev Biol 78:419–433

Peterson AT, Vieglais DA (2001) Predicting species invasions using ecological niche modeling: new approaches from bioinformatics attack a pressing problem. BioScience 51:363–371

Quinn TP, Unwin MJ, Kinnison MT (2000) Evolution of temporal isolation in the wild: Genetic divergence in timing of migration and breeding by introduced chinook salmon populations. Evolution 54:1372–1385

Rejmanek M, Richardson DM, Pysek P (2005) Plant invasions and invasibility of plant communities. In: Van Der Maarel E (ed) Vegetation ecology. Blackwell, Oxford, UK, pp 332–355

Reusch TBH, Williams SL (1999) Macrophyte canopy structure and the success of an invasive marine bivalve. OIKOS 84:398–416

Rodriguez LF (2006) Can invasive species facilitate native species? Evidence of how, when, and why these impacts occur. Biol Invas 8:927–939
Romanuk TN, Kolasa J (2005) Resource limitation, biodiversity, and competitive effects interact to determine the invasibility of rock pool microcosms. Biol Invas 7:711–722
Ruesink JL (2007) Biotic resistance and facilitation of a non-native oyster on rocky shores. Mar Ecol Prog Ser 331:1–9
Schwindt E, Iribarne OO (2000) Settlement sites, survival and effects on benthos of an introduced reef-building polychaete in a SW Atlantic coastal lagoon. Bull Mar Sci 67:73–82
Seabloom EW, Harpole WS, Reichman OJ, Tilman D (2003) Invasion, competitive dominance, and resource use by exotic and native California grassland species. Proc Natl Acad Sci U S A 100:13384–13389
Shea K, Chesson P (2002) Community ecology theory as a framework for biological invasions. Trends Ecol Evol 17:170–176
Simberloff D, Von Holle B (1999) Positive interactions of nonindigenous species: invasional meltdown? Biol Invas 1:21
Sousa WP (1979) Distubance in marine intertidal boulder fields: the nonequilibrium maintenance of species diversity. Ecology 60:1225–1239
Southward AJ, Hawkins SJ, Burrows MT (1995) Seventy years' observations of changes in distribution and abundance of zooplankton and intertidal organisms in the western English Channel in relation to rising sea temperature. J Therm Biol 20:127–155
Stachowicz JJ (2001) Mutualism, facilitation, and the structure of ecological communities. BioScience 51:235–246
Stachowicz JJ, Byrnes JE (2006) Species diversity, invasion success and ecosystem functioning: disentangling the influence of resource competition, facilitation, and extrinsic factors. Mar Ecol Prog Ser 311:251–262
Stachowicz JJ, Whitlatch RB (2005) Multiple mutualists provide complementary benefits to their seaweed host. Ecology 86:2418–2427
Stachowicz JJ, Whitlatch RB, Osman RW (1999) Species diversity and invasion resistance in a marine ecosystem. Science 286:1577–1579
Stachowicz JJ, Fried H, Osman RW, Whitlatch RB (2002a) Biodiversity, invasion resistance, and marine ecosystem function: reconciling pattern and process. Ecology 83:2575–2590
Stachowicz JJ, Terwin JR, Whitlatch RB, Osman RW (2002b) Linking climate change and biological invasions: ocean warming facilitates nonindigenous species invasions. Proc Natl Acad Sci 99:15497–15500
Steneck RS, Graham MH, Bourque BJ, Corbett D, Erlandson JM, Estes JA, Tegner MJ (2002) Kelp forest ecosystems: biodiversity, stability, resilience and future. Environ Conserv 29:436–459
Stockwell CA, Ashley MV (2004) Rapid adaptation and conservation. Conserv Biol 18:272–273
Stohlgren TJ, Chong GW, Schell LD, Rimar KA, Otsuki Y, Lee M, Kalkhan MA, Villa CA (2002) Assessing vulnerability to invasion by nonnative plant species at multiple spatial scales. Environ Manag 29:566–577
Sumi CBT, Scheibling RE (2005) Role of grazing by sea urchins *Strongylocentrotus droebachiensis* in regulating the invasive alga *Codium fragile* ssp *tomentosoides* in Nova Scotia. Mar Ecol Prog Ser 292:203–212
Tilman D (2004) Niche tradeoffs, neutrality, and community structure: a stochastic theory of resource competition, invasion, and community assembly. Proc Natl Acad Sci 101:10854–10861
Valentine JP, Johnson CR (2003) Establishment of the introduced kelp *Undaria pinnatifida* in Tasmania depends on disturbance to native algal assemblages. J Exp Mar Biol Ecol 295:63–90
Valentine JP, Johnson CR (2005) Persistence of the exotic kelp *Undaria pinnatifida* does not depend on sea urchin grazing. Mar Ecol Prog Ser 285:43–55
Vermeij GJ (1991a) Anatomy of an invasion: the trans-Arctic interchange. Paleobiology 17:281–307

Vermeij GJ (1991b) When biotas meet: understanding biotic interchange. Science 253:1099–1104

Vermeij GJ (2005) Invasion as expectation: a historical fact of life. In: Sax DF, Stachowicz JJ, Gaines SD (eds) Species invasions: insights into ecology, evolution, and biogeography. Sinauer Associates, Sunderland, Mass

Vitousek PM, D'Antonio CM, Loope LL, Westbrooks R (1996) Biological invasions as global environmental change. Am Sci 84:468–478

Wahl M, Hay ME (1995) Associational resistance and shared doom: effects of epibiosis on herbivory. Oecol Hist Arch 102:329

Wares JP, Hughes AR, Grosberg RK (2005) Mechanisms that drive evolutionary change. In: Sax DF, Stachowicz JJ, Gaines SD (eds) Species invasions: insights into ecology, evolution, and biogeography. Sinauer Associates, Sunderland, Massachusetts

White L, Shurin J (in press2007) Diversity effects on invasion vary with life history stage in marine macroalgae. OIKOS 116(7):1193–1203

Wonham MJ, O'Connor M, Harley CDG (2005) Positive effects of a dominant invader on introduced and native mudflat species. Mar Ecol Prog Ser 289:109–116

Section IV
Invader Integration into Ecosystems

Chapter 13
The Integration of Invasive Species into Marine Ecosystems

Gil Rilov

13.1 Introduction

This section of the book deals largely with what happens once invasive species have become established, whereas the previous two sections dealt primarily with how the arrival and establishment of these species occurs. Invader interactions can take many forms and it is expected that most species would interact with their biotic and abiotic environment in more than one way. Indeed, a truly comprehensive list of the impacts of invaders would be identical to a list of the role of any species in an ecosystem. Although the kinds of interactions may not be unique between exotics and natives, what is likely to differ is the degree to which effects occur, due to such factors as differing evolutionary histories and disruptions to habitats or species that may be concomitant with invasions. It is these ecological interactions that lie at the heart of the concern about invasive species. In a recent cover-page article in NEWSWEEK (January 15, 2007) it was estimated that globally bioinvasion toll (terrestrial and aquatic combined, including pests and pathogens) on the economy and the environment is close to US$1.4 trillion a year, and will only increase!

13.2 Are Most Marine Invasions Ecologically Harmless?

Despite all this, out of the thousands of introduced marine species, we only hear of relatively few that have caused notable impacts on their new environment. Why is that? Is it because most of them are harmless? Is it because we have not measured or monitored their effects properly or persistently? Or is it because most of them are recent invaders, and given enough time many will become harmful? Furthermore, as Carlton suggests in Chap. 2, it is possible that many early invasions of species that are now thought to be indigenous (in specific regions) or cosmopolitan have actually caused considerable alteration to native communities but we just don't know about it because nobody recorded that change.

G. Rilov, J.A. Crooks (eds.) *Biological Invasions in Marine Ecosystems.*
Ecological Studies 204,

The question of whether most invaders are ecologically harmless touches on a very basic ecological topic: does the existence or well-being of one species always come at the expense, however minute, of some other species in the same habitat? The honest answer to this question is that we don't really know because nature is very complex and we cannot measure everything. For example, in the case of possible competition, we cannot measure all the resources that a particular species is using and may be competed for by an invader. In reality, we might expect the full range of effects from huge to minute, and sometimes the effect of an invader can in fact be "positive", in terms of increasing the fitness of a native species. Ecosystem engineers (Chap. 16, Crooks), for instance, can facilitate the proliferation of both native and invasive species by supplying a suitable habitat. Is facilitation "good" or "bad" in this context? Or is any change that is not natural "bad"? After all, once a new species occurs unnaturally in a specific region, that region is no longer "ecologically pristine" because the biodiversity in it has been altered.

In many cases, biodiversity (at least in terms of species richness) regionally increases by invasions because, as of yet, we are not aware of any marine example of species that went globally extinct due to an invasion (Briggs 2007). This means that, with invasions, the total number of species in a region has overall increased ("extinctions" on a very localized scale still occur). For example, the eastern Mediterranean is naturally relatively impoverished in species, and thus species invasions, mostly from the Red Sea through the Suez Canal (see Chap. 31, Rilov and Galil) has dramatically increased the total number of species in the region. Today, one can have a glimpse of some of the glorious richness of the Red Sea (e.g., several colorful coral reef fishes) by diving on Levantine reefs, as this region is becoming more "tropical" in its species composition. Some people might view this increase in species richness as positive, but from an ecological stand point, the system is off-balance. What the invaded systems are really loosing is not biodiversity; it is what we could call their "biological uniqueness" or integrity, as the world biomes become more homogenized.

Although to the best of our knowledge native species in the marine environment have not gone extinct (yet) due to invasions, we know that they can "suffer" heavy losses when they interact, directly or indirectly, with invasive species. We know that the establishment and proliferation of some species have been on the expense of native species that have dwindled considerably as a result. So far, rigorous experimentation that tests the mechanism of such interactions is limited, as most authors in this section conclude. We can see a system changing profoundly, but in most cases we are not exactly sure what the mechanism is. However, the situation is not all bleak. In the last decade or so, increasing numbers of good lab and field experiments have been conducted, which certainly expanded our understanding of how invasive species interact with their receiving environment. Many of these studies are mentioned in this section at least with some degree of detail. Several high profile marine invaders, such as the European Green Crab, were studied from a multitude of perspectives related to their post-establishment interactions, i.e., as a strong competitor, predator and facilitator (mainly for the establishment of other invasive species).

13.3 Ecological Interactions of Marine Invaders

In the first chapter in this section (Chap. 14), Byers examines the literature for evidence of competitive interactions between invasive and native species. He found 49 invasive species that have been studied in this context, over 50% of them being molluscs or algae, more than 80% of the studies being conducted in Europe or North America and all being coastal (mostly intertidal). The majority of these studies showed that competition was important and in some cases can bring native species to localized extinction. However, Byers concludes that the taxonomic and geographic bias in the literature makes generalization and real evaluation of the magnitude of importance of competition in invasions, especially on the community and long term scale, difficult if not impossible. More studies are needed for that.

In Chap. 15, Rilov examines studies on predator-prey interactions in the context of invasions. Rilov found less than 30 publications that directly look at predator-prey interactions of marine invasions. Forty four invader-native species interactions and three invader-invader interactions were identified. Some invaders were studied as both predator and prey in the invaded systems. Interestingly, the most studied taxonomic group of invasive predators is pelagic and coastal zooplankton, while the most-studied invasive prey are molluscs. Rilov focuses on three major topics: the invasion of predatory zooplankton, predation as a biological control of invasions, and predation by exotic species as a facilitator for further invasions. He concludes that (1) invasive zooplankton can have major, ecosystem-level, effects on their environment, (2) highly selective predators can potentially be very effective in eliminating invaders, and (3) predators can have the capacity to change the environment to facilitate further invasions.

Chapter 16 deals with invaders as ecosystem engineers. Crooks defines how engineering is different from other types of influences inflicted on the environment by invaders, describes the different types of engineering effects and gives examples of prominent engineers. Some engineering effects are quite easily defined, for example the addition of structure (e.g., a mussel shell or a tubeworm reef) that supply habitat for other species, whether native or invasive. Other engineering effects may be more difficult to distinguish from non-engineering effects, for example indirect effects on the invaded food web. Crooks provides a research framework with which ecologists should investigate an invaded systems to separate the different processes.

In the final chapter of the section (Chap. 17), Grosholz and Ruiz take a broad perspective that encompasses some the interactions described in the previous chapters, and focus on invader impacts that spread across trophic levels, whether by trophic, competitive, or non-trophic mechanisms. They found that well-documented trophic cascades in marine systems were rare, but evidence of other multitrophic effects of invaders is relatively abundant. The magnitude of the effects clearly varies by species and location, however. They also develop a new perspective for viewing invasions as a human-mediated disturbance, which can be used as a tool for understanding and predicting the potential for multitrophic level effects. Using a comparison between the effects of an invasion of an exotic crab and an immense recruitment event of a native crab in the same region, they suggest that the speed at which the disturbance occurs may be equally important to the size of the disturbance. The authors propose that an invasion that occurs slowly may allow the system sufficient time to rebound via indirect effects reducing the potential for impacts to cascade across multiple trophic levels.

13.4 Conclusion

What becomes evident from the synthesis and analysis included in all chapters is that ecologists have made some major strides in the understating of the integration processes involved in marine bioinvasion, but that the gaps in knowledge are still quite large. Hopefully the perspectives offered in these chapters can be used to help fill these gaps in the coming years.

Reference

Briggs JC (2007) Marine biogeography and ecology: invasions and introductions. J Biogeogr 34(2):193–198

Chapter 14
Competition in Marine Invasions

James E. Byers

14.1 Introduction

Competition is a negative interaction between two or more species that utilize the same shared, limiting resource (Connell 1983). Although competition can have large local, immediate effects, (e.g. on demography, resource use, etc.), competition in many marine systems is often assumed to have minimal effect on population persistence, primarily due to characteristics of the dominant life histories of marine organisms. Notably, a large proportion of marine species have pelagic larvae and thus often reside in open populations where the supply of progeny is decoupled from progeny production. Thus, although competition can still affect adults, future generations are supplied from distant populations that can "rescue" populations of inferior competitors from being excluded. Even in relatively closed marine habitats, e.g., bays or estuaries, a constant influx of larvae in ballast water (Verling et al. 2005) may make many populations effectively open, subsidizing populations of species that would otherwise be excluded. The open nature of larval production and delivery applies to food resources as well. The preponderance of filter feeders, which feed on a food resource that is typically replenished frequently (e.g., with tidal cycle) and whose supply is often decoupled from consumptive pressure by resident organisms, may reduce the occurrence of resource competition.

Bringing evidence to bear on the frequency and strength of competition in marine species is not easy. Experimental manipulations are usually logistically difficult. For example, planktonic species are extremely hard to track because of their small size and fluidity. Also, dramatic ontogenetic changes and concomitant dietary and habitat shifts are common as well, meaning that even if competition between some life stages can be elucidated, its relative importance on populations overall may be difficult to assess. Thus, particularly in marine systems where logistical and common life history characteristics can make competition hard to study, it is important to assess what has been done, how well it has been done, and what future research needs are.

To illuminate the larger issues regarding the commonality of competition in marine invasion, how central a role it plays, and biases in its study, I reviewed the

G. Rilov, J.A. Crooks (eds.) *Biological Invasions in Marine Ecosystems.*
Ecological Studies 204,

marine literature. I used the resulting database to address not only how often competition with exotic species significantly affects native species, but also the responses typically measured to index competitive effects, the taxa commonly studied for competitive interactions, and the marine habitats and regions represented. Furthermore, I examined the database to see whether some characteristics of marine lifestyles (e.g., open systems, filter feeding) mitigate competition's role in marine invasions.

14.2 A Review of the Competition Literature on Marine Invasive Species

I searched the peer-reviewed literature in ISI Web of Science (covering 1977 through June 2005) and Aquatic Sciences and Fisheries Abstracts (covering 1971–June 2005). I used the search terms: (marsh or estuar* or bay or sea or coast* or marine or ocean*) and compet* and (inva* or introduc* or alien or nonnative or non-indigenous or non-native or non-indigenous). Such a broad, inclusive search allowed even studies that only mentioned "competition" (or any root of the word) to be examined. Initial searches produced nearly 300 papers on competition in marine invasions. Several studies on anadromous fish, e.g. salmon, were excluded because all the competition work was done in freshwater habitats and life stages. Articles were indexed according to the nonindigenous species examined, habitat, feeding mode, geographic location of the study, response variables measured, whether appreciable competition was concluded by the study, and what type of evidence was brought to bear on this conclusion. The evidence comprising this latter category was classified as one of five types: experimental; natural experiment or sampling of natural pattern; component parts demonstrated (e.g., dietary overlap or resource conversion efficiency relative to a similar native species); correlational or observational measures; or assumed or stated as background information (often by relying on previous studies). Studies in the last category were excluded from the database; however, the original sources upon which these studies often relied were usually already contained in the database.[1]

[1] Although these evidence categories somewhat reflect the strength of the evidence for competitive interactions, there can be exceptions. For example, although papers under the category "competition assumed" were excluded from analyses, the category of course does not necessarily indicate that the inference of competition is incorrect. An exotic species that forms a dense monospecific stand soon after invading is often presumed to be a superior competitor. Common examples in the literature include *Mytilus galloprovincialis*, *Spartina* sp., and *Caulerpa taxifolia* where researchers cite the fast dominance and loss of similar native species as evidence of competition. For some species, such a conclusion may be well accepted without much direct experimental evidence. Nonetheless, experimental documentation is still the gold standard for competition since negative covariation in native and exotic species' abundance can also be driven by underlying environmental factors, like climate shifts, disturbance, etc. Experiments explicitly document the mechanisms of interaction and success.

To enhance the independence of the database, multiple studies by the same lead author on the same species were only counted once. Specifically, I included the study that used the most rigorous approach to examine the presence of competition. Careful review of all papers yielded 80 that met the described criteria and these were analyzed for three primary aspects. First, to examine the frequency with which various nonindigenous taxa were studied for competition, each nonindigenous species in each competition study was counted. That is, if multiple exotic species were examined within a single study, each species was tallied. Then, to determine how broadly a given taxonomic group had been studied, I next tallied the occurrence of nonindigenous species in the database only once, regardless of how many times it was studied. Second, to understand where marine invader competition studies were conducted, I recorded the geographical regions of each study; such a *per study* basis avoided over-weighting studies that had examined multiple species.[2] I also tallied the marine habitats examined in the studies, counting each habitat only once per species, thus avoiding multiple counts of habitats that re-occurred for commonly studied species. Third, I calculated the overall frequency of competition on a *per study* basis.

14.2.1 What Exotic Taxa are Studied for Competitive Interactions?

The top two taxonomic groups studied for competitive effects were molluscs and algae, which accounted for greater than 60% of all studies in the database (Fig. 14.1). After adding the tunicates, arthropods (represented almost exclusively by crabs), and marsh grasses to this assemblage, these top five taxonomic groups comprised nearly 90% of all competition studies on nonindigenous species. Again, these numbers are conservative because they exclude multiple studies done by a lead author on the same species.

Because competitive interactions of several nonindigenous species were examined in multiple studies, an alternative way of examining the database is to look at how many distinct nonindigenous species are represented in each taxonomic group. This approach counts each species only once, irrespective of how often it may have been examined by multiple studies. In other words, it addresses whether a well-studied taxonomic group in the previous analysis (Fig. 14.1a) is composed of a few studies on many species, or many studies on a few species. For the most part, the greater the number of studies, the greater the number of nonindigenous species represented in a taxonomic group (Fig. 14.1). Thus, the relative rankings of the taxonomic groups was essentially unchanged from the previous analysis with the

[2] Berman et al. (1992) was treated as three separate studies because it compared and contrasted patterns of three very distinct invaders that drew upon independent datasets and approaches to evaluate each species.

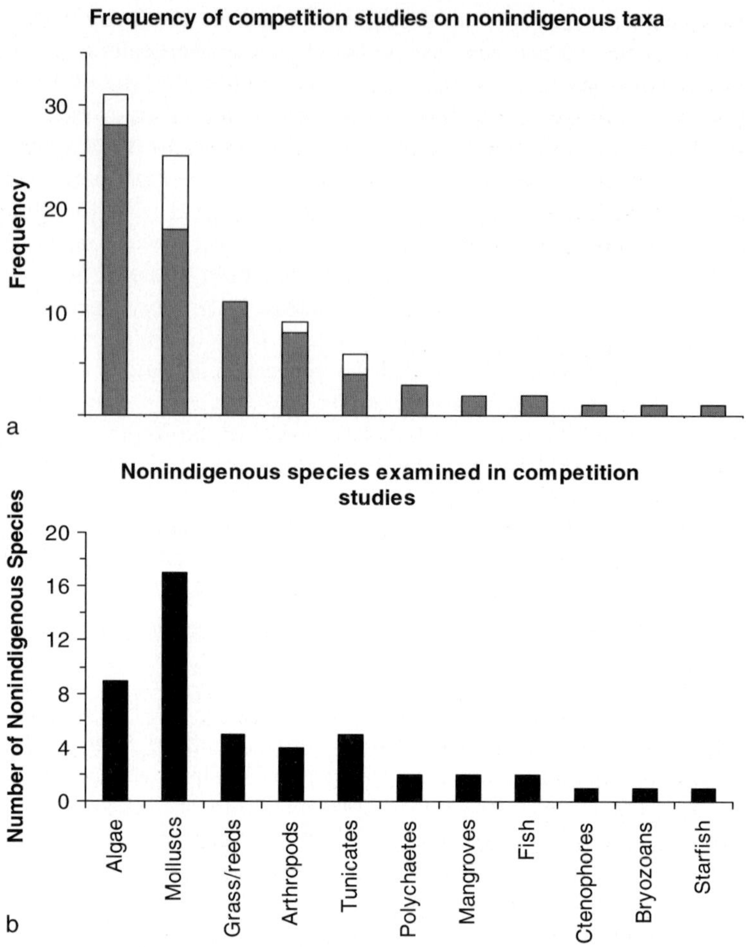

Fig. 14.1 Total number of competition studies (by distinct first authors) on nonindigenous marine species categorized by taxonomic group. Four studies examined competition between exotic species; however, the grand majority examined effects of one or more exotic species on one or more native species. If multiple nonindigenous species were examined in a single study, each species was tallied. **a** *Open white bars* represent competition studies where no competition was detected with a native species; *solid bars* represent competition studies where competition was concluded to be occurring. **b** Number of different nonindigenous species examined for competition within each taxonomic group. This figure reflects how many different species are represented in the competition studies depicted in **a**

notable exception of molluscs surpassing algae as the most broadly studied group (Fig. 14.1b). That is, molluscs are represented by many species with a few studies each, compared to algal species, which are represented by many studies on a few species. As a case in point, the algae *Caulerpa taxifolia* and *Sargassum muticum* had 9 and 10 competition studies, respectively, accounting for >60% of all exotic algal competition studies. In total, 49 exotic marine species have been studied for competitive effects (and only 26 have been studied experimentally), with molluscs and algae representing >50% of these.

14.2.2 What Regions and Habitats are Studied?

Geographically, 83% (66/80) of the marine invader competition studies were conducted in Europe and North America. Even here the effort was not even, because almost all of studies in North America come from the northeastern United States and the US Pacific coast. Of the sole nine studies from the southern hemisphere, four were from Australia and three from South Africa (and all three focused on the same species—*Mytilus galloprovincialis*, see Chap. 24, Hayden et al.). Although some of the nonindigenous species studied for competition were tropical in origin (e.g., *Caulerpa taxifolia*), only three competition studies on exotic species have been performed in the tropics—two in Hawaii (Zabin and Hadfield 2002; Krauss and Allen 2003) and one in Guam (Braley 1984). This lack of competitive studies may be partially attributable to a paucity of invaders in the tropics or because we have only sparse data on the extent of marine invasions in the tropics (e.g., Coles et al. 1999; Englund 2002).

All studies were conducted near shore. However, author-defined habitats indicated a reasonably even spread of studies throughout nearshore habitats (Fig. 14.2). Studies were not exclusively confined to sheltered habitats that might at least superficially seem to offer more tractable study conditions. About half of the competition studies were performed on coasts, seas, or fjords/sounds. Very few studies were explicitly subtidal (e.g., Britton-Simmons 2004; Ross et al. 2004). The few studies that concluded no significant competition was occurring were spread fairly proportionately throughout the habitat types.

14.2.3 How Often Does Competition Occur and How Are Its Effects Measured/Indexed?

Initial searches produced nearly 300 papers on competition in marine invasions. However, more than two-thirds of these were excluded because they only assumed competition, or stated it as background information by relying on previous studies. Most of the studies where competition was assumed were for species that were not necessarily more difficult to measure or manipulate. Several were

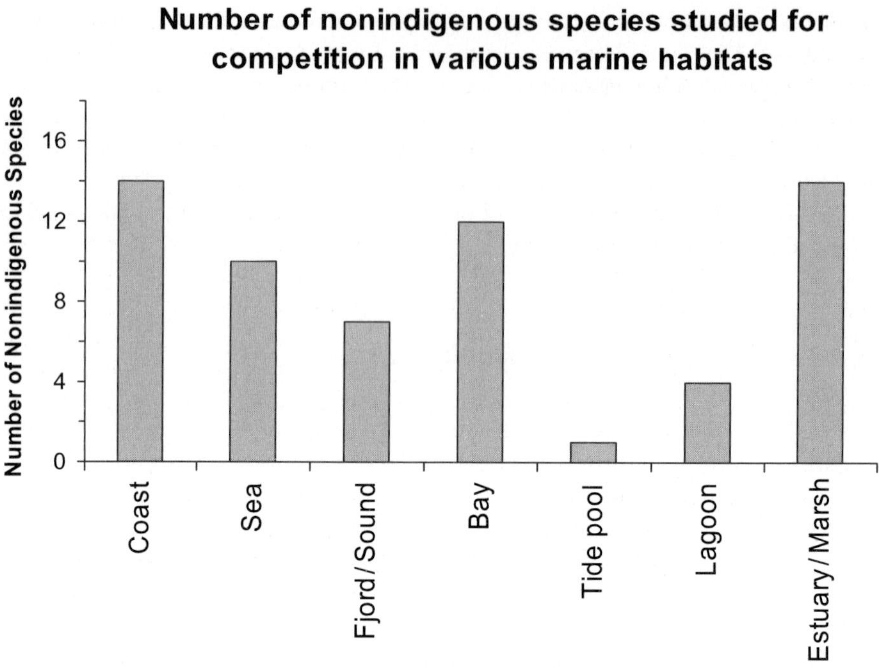

Fig. 14.2 The number of nonindigenous species studied for competition in various marine habitats. These habitats have overlapping definitions, but I used the authors' designations as best I could. The graph is based on the number of distinct species instead of studies to avoid taxonomic inflation that could result from having multiple studies of the same species in the same habitat type. However, this analysis is not completely independent of taxonomy since some species were studied in more than one habitat, e.g. oysters, and were counted for each of those habitat types. Habitats are subjectively ranked in order of openness from most to least. (Although tide pools are extremely closed habitats at low tide, this is offset by their connectedness to the larger adjacent habitat at high tide

species that authors cited had been already documented in other studies as highly competitive (e.g. *Caulerpa taxifolia*, *Mytilus galloprovincialis*). Many others (e.g., Pipitone et al. 2001) were largely descriptive studies merely suggesting reasons for an invader's success.

Of the studies in the database where competition was directly addressed, 80% (64 of 80) concluded significant competition between exotic and native species was occurring, in some cases with dramatic effects like local exclusion. Five of these studies concluded that competition was occurring only with some studied native species, while other native species in the study were not affected. For example, Crooks (2001) showed that byssal mats formed by the Asian mussel *Musculista senhousia*, significantly affected growth and survivorship of the surface dwelling, suspension feeding clam, *Clione* sp., while the deep dwelling deposit feeding clam, *Macoma nasuta*, was unaffected (see Chap. 18, Hewitt et al.). Four additional studies in the database (5%) did not involve native species and examined competition between two exotic species (Piazzi and Cinelli 2001; Piazzi and Ceccherelli 2002; Krauss and Allen 2003; Ross et al. 2004); significant competition was concluded in all four.

Of studies in the database, 15% concluded that no appreciable competition between nonindigenous species and native species was occurring. This percentage was essentially identical (16%) if just looking at studies where competition was experimentally examined (n=37). These studies that found no significant competition were spread fairly proportionately throughout the habitat types. I explored the 12 studies on 13 species that found no competition to determine if there were any unifying characteristics. Of the 13 studied species, 10 were animals, 8 of which were filter feeders (6 molluscs and 2 tunicates). Filter feeders have been proposed to experience reduced competition because the planktonic food resource on which they feed is often not a limiting factor (Levinton 1972; Peterson 1979; Byers 2005). Especially in areas with appropriate tidal flushing, food can replenish quickly. However, despite the expectation that filter feeders may have one less resource for which to compete, reducing the overall likelihood or magnitude of competition, filter feeders were not overly represented in the "no competition" subset of studies relative to their overall frequency in the database.

Typical of competition studies in general, growth, fecundity, survivorship, and avoidance behavior were commonly recorded response metrics. Many studies used several of these response variables in combination. For example, Britton-Simmons (2004) in Washington USA showed that native brown and red algal species were more abundant and native kelp grew twice as fast where the Asian alga *Sargassum muticum* was experimentally removed. Complementary measurements in experimental plots identified shading as the mechanism of competitive impact on natives rather than changes to water flow, sedimentation, or nutrient availability. Also, in Washington USA Byers (2005) examined whether increasing densities of the Asian clam, *Venerupis philippinarum*, influenced growth, survivorship, and fecundity of a confamilial native clam, *Protothaca staminea*. Even at experimental densities 50% higher than any measured in field, *Venerupis* has no direct effect on itself or *Protothaca*. Differences in the clams' burial depths apparently minimized competition for space and exploitative competition between these filter feeders was minimal.

Some studies, particularly those that examined algae, showed local competitive exclusion, at least over small spatial scales. For example, in New England USA, Bertness et al. (2002) quantified the percentage of shoreline developed at the terrestrial edge of salt marshes, which strongly influenced nitrogen run-off. In turn, nitrogen availability explained 45% of the variation in the dominance of the invasive marsh reed *Phragmites australis*. Thus, nutrient enrichment associated with shoreline development gave *Phragmites* competitive advantage over traditional high-marsh vegetation. In some areas *Phargmites* expansion (coupled with native *Spartina alterniflora* expansion from the low marsh) reduced plant species richness fivefold, creating species-poor monocultures.

14.3 Should Competition be Stronger in Human Mediated Invasions?

The collective conclusion of these studies was that competition was frequent and strong. In this section I outline three key ways that humans have altered colonization events that may enhance the role of competition in invasions, especially compared to historical invasion processes, i.e., natural colonization.

First, rapid rates of human-mediated invasions are several orders of magnitude greater than natural background rates (Cohen 2005). With the accelerated pace of species invasion mediated by humans, co-evolution and competitive exclusion have little time to re-equilibrate a community after an invasion before the next invader arrives. The rapidly assembled (or amended) communities have insufficient time for co-evolution to mitigate species interactions like competition, e.g. through character displacement. At least one theoretical work that has treated natural and human mediated invasions as distinct demonstrated a difference in the way that colonization mechanisms structure communities. Rummel and Roughgarden (1985), using a Lotka-Volterra framework, compared communities with the same resource distribution that were coevolution-structured (i.e., analogous to natural, well-spaced colonization events) and invasion-structured (i.e., analogous to human-mediated, rapid introductions). Both model communities were the same except that in coevolutionary trials after every invasion the community was allowed to coevolve to a new equilibrium. The alternation of invasion and coevolution continued until further invasions were not possible or until a cycle developed. In contrast, invasion-structured communities were continually subjected to invasions with no coevolution occurring between invasions. Invasio-structured communities ended with more species and with smaller niche separation than the coevolution-structured communities. The close packing of species in invasion communities resulted in higher competition coefficients and caused invasion structured communities to be less stable. This theory could be extended to predict, for example, that the 260+ nonindigenous marine and estuarine species in San Francisco Bay may still be far from asymptoting, and the Bay may be able to attain higher species diversity in modern times full of invasive species than it did of natives historically.

Second, human mediated transport of propagules causes a breakdown of one of the key assumptions of succession theory and community development—the trade-off exhibited by each species in its colonization and competition abilities. Theory predicts, and some empirical work shows, that the best colonizers and dispersers are the weedy species, which in turn compete poorly during successional processes when the more slowly arriving, but superior competitor species displace them (e.g., Clements 1916; Huston and Smith 1987; Tilman 1990). Poor competitors survive in the system through their dispersal advantage. Humans now artificially disperse large number of species, increasing the opportunity for establishment and causing a breakdown in the classic tradeoff. A species can now be a great competitor and have its dispersal placed on equal footing with the naturally good dispersers (but poor competitors). For example, tens of millions of metric tons (MT) of ballast water are discharged yearly into US ports, with each MT on average containing ~1,000–10,000 zooplankton organisms (Verling et al. 2005). With invasive propagule pressure (Chap. 10, Smith) now boosted enormously by humans in many places, colonization of novel environments is seldom determined by a species' own dispersal abilities. Given the magnitude of these new, human-mediated vectors, good competitors should be showing up far more frequently than they would naturally.[3]

Third, sharp increases in anthropogenic disturbances may alter the receptiveness of a community to colonizers (see also Chap. 7, Johnston et al.; Chap. 12, Olyarnik et al.). Specifically, such disturbances may override a previous ability of a native species to exclude invaders preemptively, thus increasing the ability of a colonizing nonindigenous species to compete its way into a community. Theoretically, one might assume that resident species should have the upper hand when competing with invasive species because the residents, with thousands of years of incumbency, should be best adapted to the local environment (Vermeij and Dudley 2000). However, environments that have been seriously altered by anthropogenic disturbance outside the magnitude, duration, or frequency of natural disturbances that affect communities, can create a mismatch between traits of the native species and the environmental conditions to which they have long adapted—a phenomenon termed selection regime modification (Byers 2002). Nonindigenous species may have equal opportunity when these environments suddenly become just as novel for the native species as they are for the exotics (Byers 2002). Invasive species should on average have enhanced establishment success and enhanced competitive impacts in these modified areas (Wasson et al. 2005; Tyrrell and Byers 2007). As anthropogenic impacts continue at unprecedented rates in nearly all ecosystems on earth (Janzen 1998; Vitousek et al. 1997), this mechanism may be increasingly influential and may also help to explain the observed strong correlation of invasions in disturbed environments (see reviews by D'Antonio et al. 1999; Dukes and Mooney 1999). Selection regime modification has certainly occurred in many nearshore marine habitats which are heavily disturbed. Salt marshes, for example, are heavily filled and channelized, and receive high quantities of freshwater runoff and contaminants (e.g., Cairnes 1993; Kennish 2001; Valiela et al. 2004) and also contain relatively large numbers of invasive species.

[3] Ballast subsidies not only dismantle the competition-colonization trade-off allowing strong competitors to overcome dispersal limitation and invade, but additionally, poor competitors, once arrived, can often persist artificially through the help of such human-mediated propagule subsidies.

14.4 Caveats to the Conclusion of Strong and Frequent Competition in the Marine Invasion Literature

In general reviews of competition in aquatic and terrestrial habitats (not just reviews of invasions), significant competition was also found to occur in a majority of studies that explicitly examined its presence (Connell 1983; Schoener 1983; Gurevitch et al. 1992; Bruno et al. 2005). However, due to publishing biases, where only positive results are published, or studies are conducted on species where competition is thought to be occurring, competition's common occurrence is not supported by thorough, unbiased literature. Competition studies of marine invaders share this general bias. Additionally, inventories of nonindigenous species (which thus dictate the candidate species for nonindigenous competition studies) are themselves biased because of historical records and modern sampling efforts that favor detection of exotic species that are larger and are economically or culturally important (Ruiz et al. 2000; Byers, in press). Furthermore, competition studies on marine invaders do not draw proportionately from the biased nonindigenous species inventories and are themselves skewed further toward larger, conspicuous, easily-accessed taxa. Given the young age of invasion biology as a discipline and the field's relevance to applied issues, it is perhaps unsurprising that published studies focus on species that are abundant, problematic, conspicuous, and easy to access.

On the positive side, molluscs and algae are the second and forth most documented exotic taxa, respectfully, in coastal North America (Ruiz et al. 2000). Twelve species (18%) of molluscs and four species of algae (16.7%) identified as established nonindigenous species by Ruiz et al. (2000) have been studied for competition.[4] Tunicates are also well represented in competition studies with four species examined (21% of the list of Ruiz et al. 2000). However, for all other exotic taxonomic groups in North America, fewer than 10% of the species comprising them have been studied for competitive effects. The most troublesome discrepancy is with the crustaceans, which despite being the largest taxonomic group of nonindigenous species in North America (85 established exotics), has had only 3 species (~3.5%) studied for competitive effects. Annelids are the third largest nonindigenous group in North America (28 species), with one of these species having been studied for competition elsewhere in the world where it is also nonindigenous. The skew in studied taxa may be especially important to overall generalizations about competition gleaned from the literature because certain life history characteristics that may influence competition (e.g., brooder vs broadcast spawner; filter vs deposit feeder) are often phylogentically correlated.

---

[4] One of the four algal species, *Caulerpa taxifolia*, had not invaded North America at the time of the compendium of Ruiz et al. (2000). It was included in this tally because of its subsequent high profile invasion of southern California. For molluscs, *Ovatella myosotis* was included; it is a known invader of US West coast (Berman and Carlton 1991), but in bays other than those compiled by Ruiz et al. (2000). Also, three molluscs included in this count are nonindigenous to North America, but were studied for competition elsewhere in the world where they are also nonindigenous.

The geographic skew is also particularly acute, with almost all studies carried out in temperate regions of the Northern Hemisphere; however, this is certainly where most invaders have been inventoried and studied in general. It is encouraging that species were not all studied in the most self-enclosed/confined habitats (e.g., marshes, lagoons) (Fig. 14.2); however they do almost exclusively favor the most tractable life stages (e.g., benthic adults). Finally, certain species, e.g., *Mytilus galloprovincialis*, but especially *Sargassum muticum* and *Caulerpa taxifola*, become focal points of competition research because of real or perceived impacts on native species.

Only a handful of studies in the database reveal competitive exclusion (e.g., Bertness et al. 2002; Steffani and Branch 2005). These studies were all done at smaller spatial scales. A paucity of experimental demonstrations of invaders driving competitive exclusion of native species was also found in larger, general literature reviews that incorporated freshwater and terrestrial habitat (Simberloff 1981; Gurevitch and Padilla 2004; Bruno et al. 2005). Most studies in the database did not last more than 1–3 years and demonstrate short-term effects on demography, behavior, or reduction in biomass. However, negative demographic effects and projections of current interaction coefficients suggest that local competitive exclusion is an expected endpoint after several decades for many of the impacted native species in these studies (e.g., Byers 2000; Sebastian et al. 2002), and evidence of actual exclusion would likely increase with an increase in studies' durations. However, future competition studies should examine exclusion explicitly because presently when species compete, but no exclusion occurs, it is difficult to discern whether the lack of competitive exclusion reflects a time lag due to the length of the process (i.e., an extinction debt – cf. Tilman et al. 1994) or species coexistence due to tighter species packing in non-coevolved communities (Rummel and Roughgarden 1985). Ironically, for several of the studies where competition was only assumed (and which were therefore not included in the formally analyzed database), exclusion seems a bigger problem. This may be because competition and exclusion by certain invaders (e.g., *Caulerpa taxifola*, *Mytilus galloprovincialis*) appear so obvious that one does not need a formal study.

Theoretically one might expect that heavy study of native-exotic competitive interactions in marshes, estuaries, and bays (Fig. 14.2) would boost the conclusion of a large role for competition overall. Populations of organisms in these semi-enclosed habitats tend to be closed and there is a high proportion of tractable, benthic species. Even planktonic species and larvae are often retained in these environments. Within closed populations competitive effects are coupled between adults and the recruiting population, making competitive exclusion likely compared to open populations where even inferior competitor species can be continuously subsidized by larval inputs from distant sources. Also because marshes in particular are often heavily altered by humans, they are a prime place for selection regime modification and thus high competitive impact by exotic species. The database however did contain a fairly wide representation of species life history attributes (e.g., broadcast spawners and direct developers) and, although many studies were in closed habitats, roughly half (36/80) were done in relatively open marine habitats (Fig. 14.2). Also, competition was still commonly found in these open habitats perhaps because populations there can still be closed, and most importantly because many competition studies examine immediate effects of demography, fecundity, behavior, and population trends of current residents independent of recruitment (i.e., many studies did not look at long-term persistence of populations). I suspect that with more competition studies, less publication bias for positive results, and more emphasis on population level responses, enclosed habitats would have disproportionately higher incidence of competitively influential exotic species.

Although open populations of broadcast spawning organisms were represented in the database, there were still no studies in the open ocean, and few from subtidal habitats. This representation of habitats in the competition literature, although skewed, essentially reflects that few exotic species have been documented offshore (although admittedly this area is seldom examined in most nonindigenous species surveys) (e.g., Ruiz et al. 1997; Cohen and Carlton 1998; Hewitt et al. 1999). A strong habitat focus on marshes, estuaries, and bays may be appropriate since it seems that exotic marine species are more abundant in these habitats (Ruiz et al. 1997; Chap. 33, Preisler et al.). Wasson et al. (2005) documented 527 invertebrates in Elkhorn Slough, CA – 444 natives, 58 exotics, and 25 cryptogens (11% exotics). The surrounding rocky intertidal open coast contained 588 species – 567 natives, 8 exotics, and 13 cryptogens (1% exotics). Exotic species in the estuary were not only more diverse, but also more abundant than on the open coast. Byers and Pringle (2006) demonstrate that the advection typical of open coastlines make retention and thus establishment difficult and may be largely responsible for the dearth of invasive species there. If so, the very factor—a retentive environment—that likely makes population-level competitive effects more common in semi-enclosed habitats, may also be responsible for increased exotic establishment there.

14.5 Future Directions

Rather than studies which single out individual biotic mechanisms like competition, perhaps the greatest need is studies that examine the relative importance of various biotic interactions enabling invasion and governing post-invasion interactions (Bruno et al. 2005). Apparent competition in particular is one such biotic interaction whose importance is increasingly recognized in exotic-native interactions that should be considered among the typical biotic interactions like competition, predation, and facilitation. Apparent competition can result when a shared parasite or predator species mediates interactions between exotic and native competitors. Two theoretical papers have demonstrated how this interaction may substantially influence exotic species success and impact (Courchcamp et al. 2002; Noonburg and Byers 2005). Certainly modern biological control practitioners have recognized the potency of apparent competition, in large part explaining why host specificity screening is a vital step before introducing potential control agents. Despite the need to examine this important interaction, a search of ISI Web of Science using the terms: "apparent competition and (non-native or introduced or nonindigenous or invasion or invader or invasive or exotic)" yielded only 64 papers—and only one of these was a marine study (Byers 2005). Although potentially more complex because more species are involved, a theoretical expectation of the importance of apparent competition and the paucity of studies suggest a ripe avenue for future research.

The literature search presented here looked primarily at the frequency with which competition was concluded. Ideally a meta-analysis of competitive interaction strengths would be very informative; however, only 26 species in the database were examined experimentally and these experiments frequently used different response variables that make quantitative comparisons difficult. Standardization of data collected, e.g., per capita interaction strength (Wootton and Emmerson 2005), would enable future generalizations and comparisons. For example, once the quantitative invasion literature is better and more evenly developed, a formal comparison of competition between native species and human-mediated invaders vs competition between species in studies not focused on invaders (e.g., Connell 1983; Schoener 1983) would be illustrative. Comparison of interaction strengths between these two groups of studies would help to address whether competition is stronger and more frequent in human-mediated invasions.

14.6 Conclusion

Competition, or its absence, is often assumed to fundamentally influence invasion success. Several key aspects of the current, human-mediated spread of exotics suggest that competitive interaction strengths between invader and resident biota should be stronger compared to historical invasions driven by colonization events of natural means and rates. The marine literature reveals that significant competition between natives and exotics is very common. However, this synopsis is couched with caution because it is not based on a large number of data-driven studies. Fewer than half the studies in the database were experimental, and the database itself excluded many dozens of studies that only assumed competition. Furthermore, empirical studies mostly address algae and molluscs in nearshore habitats in temperate North America and Europe. While competition seems important and common in marine invasions that have examined it empirically, we must be cognizant of these biases and avoid overextending conclusions of its importance. Human-mediated invasions may provide an unprecedented opportunity to examine strong competition in action; however, to truly evaluate it well, we should gauge its strength relative to other biotic interactions and abiotic factors that determine invasion success and impact.

Acknowledgements I thank Irit Altman, April Blakeslee, Tyra Byers, John Meyer, Gil Rilov, and two anonymous reviewers for their helpful advice and editing to shape this chapter.

References

Berman J, Carlton JT (1991) Marine invasion processes: interactions between native and introduced marsh snails. J Exp Mar Biol Ecol 150:267–281

Berman J, Harris L, Lambert W, Buttrick M, Dufresne M (1992) Recent invasions of the Gulf of Maine: three contrasting ecological histories. Conserv Biol 6:435–441

Bertness MD, Ewanchuk PJ, Silliman BR (2002) Anthropogenic modification of New England salt marsh landscapes. Proc Natl Acad Sci U S A 99:1395–1398

Braley RD (1984) Mariculture potential of introduced oysters *Saccostrea cucullata* tuberculata and *Crassostrea echinata*, and a histological study of reproduction of *Crassostrea echinata*. Aust J Mar Freshwater Res 35:129–141

Britton-Simmons KH (2004) Direct and indirect effects of the introduced alga *Sargassum muticum* on benthic, subtidal communities of Washington State, USA. Mar Ecol Prog Ser 277:61–78

Bruno JE, Fridley JD, Bromberg KD, Bertness MD (2005) Insights into biotic interactions from studies of species invasions. In: Sax DF, Stachowicz JJ, Gaines SD (eds) Species invasions: insights into ecology, evolution, and biogeography. Sinauer, Sunderland, MA, p 495

Byers JE (2000) Competition between two estuarine snails: Implications for invasions of exotic species. Ecology 81:1225–1239

Byers JE (2002) Impact of non-indigenous species on natives enhanced by anthropogenic alteration of selection regimes. Oikos 97:449–458

Byers JE (2005) Marine reserves enhance abundance but not competitive impacts of a harvested nonindigenous species. Ecology 86:487–500

Byers JE (in press) Invasive animals in marshes: biological agents of change. In: Silliman BR, Bertness MD, Grosholz ED (eds) Salt marshes under global siege. University of California Press

Byers JE, Pringle JM (2006) Going against the flow: retention, range limits and invasions in advective environments. Mar Ecol Prog Ser 313:27–41

Cairnes J (1993) Is restoration ecology practical? Restor Ecol 1:3–7

Clements FE (1916) Plant succession. Carnegie Institute, Washington

Cohen AN (2005) Basis for a standard based on the natural rate of invasion. Memo to the Ballast Water Treatment Standards Committee (August 7, 2005). Appendix 5. In: Ballast water discharge standards: report and recommendation of the California Advisory Panel on ballast water performance standards

Cohen AN, Carlton JT (1998) Accelerating invasion rate in a highly invaded estuary. Science 279:555–558

Coles SL, DeFelice RC, Eldredge LG, Carlton JT (1999) Historical and recent introductions of non-indigenous marine species into Pearl Harbor, Oahu, Hawaiin islands. Mar Biol 135:147–158

Connell JH (1983) On the prevalence and relative importance of interspecific competition: evidence from field experiments. Am Nat 122:661–696

Courchamp F, Langlais M, Sugihara G (2000) Rabbits killing birds: modelling the hyperpredation process. J Anim Ecol 69:154–164

Crooks JA (2001) Assessing invader roles within changing ecosystems: historical and experimental perspectives on an exotic mussel in an urbanized lagoon. Biol Invas 3:23–36

D'Antonio CM, Dudley TL, Mack M (1999) Disturbance and biological invasions: direct effects and feedbacks. In: Walker L (ed) Ecosystems of disturbed ground. Elsevier, Amsterdam, The Netherlands, pp 413–452

Dukes JS, Mooney HA (1999) Does global change increase the success of biological invaders? Trends Ecol Evol 14:135–139

Englund RA (2002) The loss of native biodiversity and continuing nonindigenous species introductions in freshwater, estuarine, and wetland communities of Pearl Harbor, Oahu, Hawaiian Islands. Estuaries 25:418–430

Gurevitch J, Padilla DK (2004) Are invasive species a major cause of extinctions? Trends Ecol Evol 19:470–474

Gurevitch J, Morrow LL, Wallace A, Walsh JS (1992) A meta-analysis of competition in field experiments. Am Nat 140:539–572
Hewitt CL, Campbell ML, Thresher RE, Martin RB (1999) Marine biological invasion of Port Phillip Bay, Victoria. Centre for Research on Introduced Marine Pests, Hobart, Australia
Huston M, Smith T (1987) Plant succession - life history and competition. Am Nat 130:168–198
Janzen D (1998) Gardenification of wildland nature and the human footprint. Science 279:1312–1313
Kennish MJ (2001) Coastal salt marsh systems in the US: a review of anthropogenic impacts. J Coast Res 17:731–748
Krauss KW, Allen JA (2003) Influences of salinity and shade on seedling photosynthesis and growth of two mangrove species, *Rhizophora mangle* and *Bruguiera sexangula*, introduced to Hawaii. Aquat Bot 77:311–324
Levinton J (1972) Stability and trophic structure in deposit-feeding and suspension-feeding communities. Am Nat 106:472–486
Noonburg EG, Byers JE (2005) More harm than good: when invader vulnerability to predators enhances impact on native species. Ecology 86:2555–2560
Peterson CH (1979) Predation, competitive exclusion, and diversity in the soft-sediment benthic communities of estuaries and lagoons. In: Livingston RJ (ed) Ecological processes in coastal and marine systems. Plenum, New York, pp 233–263
Piazzi L, Ceccherelli G (2002) Effects of competition between two introduced *Caulerpa*. Mar Ecol Prog Ser 225:189–195
Piazzi L, Cinelli F (2001) Distribution and dominance of two introduced turf-forming macroalgae on the coast of Tuscany, Italy, northwestern Mediterranean Sea in relation to different habitats and sedimentation. Bot Mar 44:509–520
Pipitone C, Badalamenti F, Sparrow A (2001) Contribution to the knowledge of *Percnon gibbesi* (Decapoda, Grapsidae), an exotic species spreading rapidly in Sicilian waters. Crustaceana 74:1009–1017
Ross DJ, Johnson CR, Hewitt CL, Ruiz GM (2004) Interaction and impacts of two introduced species on a soft-sediment marine assemblage in SE Tasmania. Mar Biol 144:747–756
Ruiz GM, Carlton JT, Grosholz ED, Hines AH (1997) Global invasions of marine and estuarine habitats by non- indigenous species: mechanisms, extent, and consequences. Am Zool 37:621–632
Ruiz GM, Fofonoff PW, Carlton JT, Wonham MJ, Hines AH (2000) Invasion of coastal marine communities in North America: Apparent patterns, processes, and biases. Annu Rev Ecol Syst 31:481–531
Rummel JD, Roughgarden J (1985) A theory of faunal buildup for competition communities. Evolution 39:1009–1033
Schoener TW (1983) Field experiments on interspecific competition. Am Nat 122:240–285
Sebastian CR, Steffani CN, Branch GM (2002) Homing and movement patterns of a South African limpet *Scutellastra argenvillei* in an area invaded by an alien mussel *Mytilus galloprovincialis*. Mar Ecol Prog Ser 243:111–122
Simberloff D (1981) Community effects of introduced species. In: Nitecki MH (ed) Biotic crises in ecological and evolutionary time. Academic Press, New York, pp 53–81
Steffani CN, Branch GM (2005) Mechanisms and consequences of competition between an alien mussel, *Mytilus galloprovincialis*, and an indigenous limpet, *Scutellastra argenvillei*. J Exp Mar Biol Ecol 317:127–142
Tilman D (1990) Constraints and tradeoffs: toward a predictive theory of competition and succession. Oikos 58:3–15
Tilman D, May RM, Lehman CL, Nowak MA (1994) Habitat destruction and the extinction debt. Nature 371:65–66
Tyrell MC, Byers JE (2007) Do artificial substrates favor nonindigenous fouling species over natives? J Exp Mar Biol Ecol 342(1):54–60
Valiela I, Rutecki D, Fox S (2004) Salt marshes: biological controls of food webs in a diminishing environment. J Exp Mar Biol Ecol 300:131–159

Verling E, Ruiz GM, Smith LD, Galil B, Miller AW, Murphy KR (2005) Supply-side invasion ecology: characterizing propagule pressure in coastal ecosystems. Proc R Soc B 272:1249–1256

Vermeij GJ, Dudley R (2000) Why are there so few evolutionary transitions between aquatic and terrestrial ecosystems? Biol J Linn Soc 70:541–554

Vitousek PM, Aber JD, Howarth RW, Likens GE, Matson PA, Schindler DW, Schlesinger WH, Tilman DG (1997) Human alteration of the global nitrogen cycle: sources and consequences. Ecol Appl 7:737–750

Wasson K, Fenn K, Pearse JS (2005) Habitat differences in marine invasions of Central California. Biol Invas 7:935–948

Wootton JT, Emmerson M (2005) Measurement of interaction strength in nature. Annu Rev Ecol Evol Syst 36:419–444

Zabin C, Hadfield MG (2002) Do locals rule? Interactions between native intertidal animals and a Caribbean barnacle in Hawai'i. Pac Sci 56:235–236

Chapter 15
Predator-Prey Interactions of Marine Invaders

Gil Rilov

15.1 Introduction

Predator-prey interactions are among the most fundamental processes shaping the structure and function of ecological communities, particularly in marine systems. In the past several decades, it has become clear that humans are interfering considerably with these interactions in many marine systems, mainly by removing top predators via harvesting (Myers and Worm 2003), but also through biological introductions. Most introduced species that have become established in their new environment should be expected to integrate in some way into the food web acting as consumers (predators, herbivores, detritivores) or as prey. Surprisingly, there is a relative paucity of studies that have examined the ecological effects of exotic species on predator-prey interactions, and the potential consequences of these effects for local communities. In this chapter, I will briefly review progress in understanding predator-prey interactions in marine systems, and examine the existing evidence for bioinvasion impacts on these interactions, focusing on relatively well-studied examples.

15.2 The Role of Predation in Marine Communities

It was Joseph Connell, in the mid-1950s, who first demonstrated experimentally that predators can dramatically affect the distribution and abundance of their prey, and therefore community structure as a whole (Connell 1961). While working on the intertidal rocks of the Isle of Cumbrae in Scotland, he noticed that barnacles are mostly restricted to the mid- and high shore levels of the intertidal zone, while seaweeds dominate the low shore level. This suggested to him that barnacles either don't settle or don't survive well on the low shore. Connell's simple but elegant experiments showed that by excluding predatory whelks from the low intertidal zone using cages, barnacles recruit and survive well and, over time, outcompete seaweeds.

G. Rilov, J.A. Crooks (eds.) *Biological Invasions in Marine Ecosystems.*
Ecological Studies 204,

In studies involving the removal of sea stars off specific areas on the wave-exposed Washington rocky shore, Bob Paine in the mid-1960s demonstrated the profound influence of a "keystone" predator, the ochre sea star *Pisaster ochraceus,* on marine benthic communities (Paine and Schindler 2002). Paine showed that after the removal of sea stars, the mussel *Mytilus californianus* rapidly extended its vertical distribution from its previous mid shore level to the low shore zone. When present, the ochre sea star ferociously feeds on the dominant space competitor, mussels, leaving cleared areas for the establishment of a more diverse assemblage of lesser competitors including seaweeds. When sea stars are absent, however, these other species are outcompeted by the mussels. This experiment thus demonstrated the indirect effect of predation on community structure – how a predator can facilitate the coexistence of many species by removing a dominant competitor. Such indirect effects of predation have been a major topic in ecological studies in the past two decades (e.g., Bruno and O'Connor 2005). Paine later demonstrated a similar phenomenon on the other side of the Pacific Ocean by removing sea stars on the shores of New Zealand (Paine and Schindler 2002). With their innovative science, the two pioneering naturalists, Connell and Paine, launched the era of experimentation in marine ecology.

Soon after their initial discoveries it has become clear that the magnitude of predation effects as well as other species interactions can vary greatly from one place to another. It has been suggested (Menge and Sutherland 1987) and later tested and modeled (Connolly and Roughgarden 1999; Menge and Farrell 1989) that species interaction effects in the rocky intertidal are highly dependent on both the magnitude of environmental stress and the rates of recruitment.

Although initial ecological experimental studies focused mainly on the rocky intertidal, more evidence quickly emerged that predation can be a strong ecological force in other coastal systems. For example, Peterson in the early 1980s showed that predation by crabs can dramatically affect population size structure and species interactions of two bivalves in lagoonal sand and mudflats in California (Peterson 1982). Recent experimental manipulations of the dominant salt-marsh grazer (the periwinkle, *Littoraria irrorata*) and its predators (e.g., blue crabs and terrapins) have demonstrated a strong top-down control of marsh plant production (Silliman and Bertness 2002). Without the presence of predators, periwinkle grazing can convert a salt marsh into a barren mudflat in less than a year (a clear example of a trophic cascade; see also Chap. 17, Grosholz and Ruiz). This means that over-harvesting of snail predators such as blue crabs could be a contributing factor to the massive die-off of salt marshes across the southeastern United States (Silliman et al. 2005).

On coastal subtidal reefs, predation has also been shown to be an important process in shaping ecological communities. For example, in South Africa, hordes of whelks, *Burnupena* spp., in some places can control lobster populations to such an extent that mussels are freed from their main predator, the lobster. In a classic study, Barkai and McQuaid (1988) noticed that two islands in South Africa, only 4 km apart, have very different communities. On one island the benthic community was dominated by high densities of rock lobsters and seaweeds, while other organisms were relatively scarce. There, the abundant lobsters feed on both mussels and whelks. On another island, the community is dominated by beds of mussels and large populations of sea cucumbers, sea urchins, and whelks. On this island the numerous whelks mass-attack any juvenile lobster that tries to recruit to the population, thereby excluding lobsters as effective predators on the island and subsequently allowing the existence of mussel beds.

Another example comes from the western Atlantic shores of North America. In an unusual event during the mid-1990s, the mussel *Mytilus edulis* massively recruited to subtidal rocks across the southwest Gulf of Maine (Witman et al. 2003). It took about a year for this episodic increase in food resources to be expressed in dramatically elevated levels of recruitment, biomass, and density of their main predators, sea stars, *Asterias* spp., and rock crabs, *Cancer* spp. Shortly after, these predators practically eliminated most of those newly created beds of juvenile mussels. These results clearly demonstrate the existence of coupling between the supply of prey (bottom-up control) to the system and the degree of predation on this prey (top-down control) in this environment (Witman et al. 2003). A few recent studies have also demonstrated the potential effects of subtidal predators, such as lobsters (Robles et al. 1990; Robles 1997) and fish (Rilov and Schiel 2006a), on intertidal prey, and how this subtidal-intertidal trophic linkage can be highly context (e.g. seascape, bottom topography) dependent (Rilov and Schiel 2006b).

In kelp forests, predation and herbivory can be very important as well. Sea otters, sea urchins and kelp live in a delicate balance of coexistence. In areas where sea otters have been previously exterminated, sea urchins, the otters' favorite prey, have become so abundant that they reduced the kelp forest to bare rock by chewing on their holdfast until the entire plant detached from the substrate (Estes et al. 1998, 2004). After their recovery in southwest Alaska (years after their hunting stopped), otter populations declined again, probably due to increased predation by killer whales as the result of declines in other whales, which are their favorite prey. In response to the otters' decline, urchin populations exploded again, decimating the kelp forest. Estes et al. (2004) suggested that the dynamics of kelp forests in southwest Alaska have thus been influenced by an ecological chain reaction (or a trophic cascade) that encompassed numerous species and large scales of space and time. Numerous other studies have found similar results in other coastal systems, notably so in coral reefs, were predation was shown to be a strong population and community-shaping force (e.g., Hixon and Carr 1997; Hixon et al. 2002; Holbrook and Schmitt 2002).

To date, most studies on consumer interactions were conducted in coastal systems, which are relatively accessible to humans. We know little of how useful the concepts that emerged from these studies are to communities inhabiting the vast continental shelves and the open ocean. Studying predator-prey interactions offshore and in deep waters is a tremendous challenge, and investigators must turn to more correlative studies to examine these relationships or use unintended human "experiments" (fishing as a predator-removal experiment) to test them. The importance of predators in these systems usually become apparent after they are lost to the system. The collapse of many stocks of the Atlantic Cod (*Gadus morhua*) due to overfishing (Myers et al. 1997) serves as a good example. Meta-analysis on population abundances of the cod and its prey, the northern shrimp (*Pandalus borealis*), in different regions of the Atlantic continental shelf showed that eight out of nine regions demonstrated negative correlations of cod and shrimp biomass, suggesting that when cod populations collapsed the populations of its prey the shrimp, increases, and supporting a "top-down" view of the system (Worm and Myers 2003). Food-webs in pelagic systems may be even more difficult to study, therefore making predator-prey interactions more elusive. However, there is growing evidence that fishing the big predators is shifting the food web structure to greater numbers of smaller prey species (Pauly et al. 1998; Pauly and Palomares 2005; Scheffer et al. 2005).

The above examples clearly demonstrate that predator-prey interactions play a pivotal role in marine ecological communities. As the number of marine invasions continues to increase, it is important to ask: how does the arrival and establishment of new predators and prey change consumer-prey interactions and the structure of these communities?

15.3 Predator-prey Interactions in Invaded Systems: A Literature Review

I searched the literature with the ISI Web of Science search engine (between 1996 and 2006) using the following string of keywords: "(marine or sea or ocean) and (invasiv* or exotic* or nonindigenous) and (predat*)" to look for studies that were published on predator-prey interactions in invaded marine systems. This search resulted in 99 entries. Of them, only 29 were relatively related to the topic. Of these 29 papers, 7 were mostly speculative, general reviews or were describing initial stages of food-web model development that were not useful for the purpose of this analysis and thus were omitted from it. Several more papers that were not picked-up by the search engine or were published earlier were added to the remaining 22 papers for the analysis.

The following categories were used to analyze the literature:

1. Origin of invader.
2. Taxonomic group to which the invader and the native organism belong (e.g., fish/crab/clam etc.). In some studies the natives were treated as a group (e.g. zooplankton, shorebirds) and therefore lumped together in both the summary table and the analysis.
3. Classification as predator or prey.
4. Habitat where the interaction occurred using two major categories, biogeographical region (temperate/tropical waters) and vertical zone (intertidal/subtidal/pelagic).
5. Substratum where interaction occurred (soft/hard/open water).
6. Invaded region.
7. Study type – observational/experimental or a combination of those two.
8. Experiment type – field, lab or both.
9. Major findings; e.g., native/invader found in diet, effect on prey abundance, preference for a specific prey.

15.3.1 Descriptive Statistics of Predator-prey Interactions in Invaded Systems

I found 21 exotic species for which some aspect of predator-prey interactions have been studied in the invaded system (Table 15.1). Forty four invader-native species interactions and three invader-invader interactions were identified. Some exotic species such as the whelk *Rapana venosa* in Chesapeake Bay, the copepod *Pseudodiaptomus inopinus*, the cladoceran *Cercopagis pengoi* in the Baltic Sea, and the comb jellyfish *Mnemiopsis leidyi* in the Black Sea were studied as both predator and prey in the invaded system.

Interestingly, the most studied taxonomic group of invasive predators is pelagic and coastal zooplankton. The five species studied are two copepods (*Pseudodiaptomus inopinus* and *Tortanus dextrilobatus*), a cladoceran (*Ceropagis pengi*), and two comb jellyfishes (*Mnemiopsis leidyi* and *Beroe ovata*). Four of the invasive species studied are intertidal or shallow subtidal macro-benthic predators such as decapods (European green crab *Carcinus maenas*, Asian shore crab *Hemigrapsus sanguineus*), a sea star (*Asterias amurensis*), and a gastropod (rapa whelk, *Rapana venosa*). Another invertebrate predator is of terrestrial origin, the fire ant, *Solenopsis invicta*, that infests sea turtle nests on the beaches of Florida. The rest of the predators are vertebrates: the American mink, *Mustela vison*, that feeds on fish on the Spanish shores, and the ship rat, *Rattus rattus*, that preys on shore bird nests and other marine organisms on islands. Invasive prey includes mostly mollusks: the clams *Nuttallia obscurata* and *Anadara demiri*, *Gemma gemma*, and the mussels, *Brachidontes pharaonis* and *Musculista senhousia*, but also a hydrozoan, *Moerisia lyonsi*, and a tunicate, *Pyura praeputialis*, and the predatory zooplankton mentioned above that serve as prey for larger predators such as fish. These species have been introduced across a broad range of geographic regions including the east and west coast of the North America, Chile, islands off the Scotland shores, the Baltic Sea, the Black Sea, western and eastern Mediterranean and southern Australia.

The majority of the interactions studied had some component of experimental work in them (Fig. 15.1), although some were very basic, such as prey preference with no quantitative information on population level effects. This is understandable in most cases because of the difficulty of testing such effects. For example, the effects of predation by zooplankton could realistically only be estimated using correlative and modeling approaches or using mesocosms. Quite a few predator-prey interactions were tested in the field (Fig. 15.1); most of them involve macro-predators. For example, effects of predation by the European green crab *Carcinus maenas* (Grosholz 2005; Grosholz and Ruiz 1995; Grosholz et al. 2000) were tested on native and invasive prey, identifying potential native predators on the invasive soft-sediment dweller *Musculista senhousia* (Reusch 1998). All interaction studies were conducted in temperate waters, largely in the shallow littoral, either in the intertidal zone, the subtidal zone or in both zones (and on both soft and hard bottoms), but a few were also conducted in open water habitats (Fig. 15.1). Why interactions have not been studied in tropical or polar ecosystems is unclear. Because this finding is

Table 15.1 Predator prey interactions between invasive-native and invasive-invasive species found in the literature search

	Invader		Native								
Origin	Taxa and Species	pd/pry	Taxa and Species	pd/pry	Habitat	Substrate	Region	obs/exp	F/L	Effect	References
Brazil	Fire ant *Solenopsis invicta*	pd	Sea turtles	pry	Temperate intertidal	Soft	Florida	obs/exp	F/L	Infestation of sea turtle nests; ants effect freshwater sea turtle hatchlings in lab	Sakai et al. (2001)
Ponto-Caspian	Cladoceran *Cercopagis pengoi*	pd	Zooplankton *Bosmina*	pry	Temperate pelagic	Open water	Gulf of Riga (Baltic Sea)	obs	F/L	Relationship between *Ceropagis* increase and zooplankton decrease	Ojaveer et al. (2004)
Ponto-Caspian	Cladoceran *Cercopagis pengoi*	pd	Meso-zooplankton	pry	Temperate pelagic	Open water	Northern Baltic Sea	obs	L	Stable isotopes and models show shift in food web structure	Gorokhova et al. (2005)
Ponto-Caspian	Cladoceran *Cercopagis pengoi*	pry	Fish *Clupea harengus mebras*	pd	Temperate pelagic	Open water	SE Finland (Baltic Sea)	obs	F/L	Correlation between abundance in water and in herring's diet	Antsulevich and Valipakka (2000); Gorokhova et al. (2004)
Ponto-Caspian	Cladoceran *Cercopagis pengoi*	pry	Fish *Sprattus sprattus*	pd	Temperate pelagic	Open water	Northern Baltic Sea	obs	F/L	Correlation between abundance in water and in sprate's diet	Gorokhova et al. (2004)
Asia	Clam *Nuttallia obscurata*	pry	Crab *Cancer productus*	pd	Temperate intertidal	Soft	San Juan Islands	exp	F	Substrate dependent predation effects; where calm cannot burry efficiently it is absent	Byers (2002)
Indian Ocean	Clam *Anadara demiri*	pry	Sea star *Astropecten irregularis*	pd	Temperate subtidal	Soft	Italy/Adriatic Sea	obs	F	Predation on juvenile clams in the winter (the calm is third most abundant prey in its stomach)	Morello et al. (2004)

South East Asia	Clam *Nuttallia obscurata*	pry	Crab *Cancer magister*	pd	Temperate intertidal	Soft	Barkley Sound, British Columbia	exp	L	Crab preferred the invader clam over local clams *Protothaca staminea* and *Tapes philippinarum,* if sediment was shallow enough (where the invader could not borrow deep)	Dudas et al. (2005)
	Clam *Nuttallia obscurata*	pry	Crab *Cancer productus*	pd	Temperate intertidal	Soft	Barkley Sound, British Columbia	exp	L	Crab preferred the invader clam over local clams *Protothaca staminea* and *Tapes philippinarum,* if sediment was shallow enough (where the invader could not borrow deep)	Byers (2002); Dudas et al. (2005)
West Atlantic	Comb jelly *Mnemiopsis leidyi*	pd	Ichthyo-plankton and meso-zooplankton	pry	Temperate pelagic	Open water	Black Sea	obs	F	Strong correlation between *mnemiopsis* increase and zooplankton decrease	Shiganova (1998); Shiganova et al. (2001b)
West Atlantic	Comb jelly *Mnemiopsis leidyi*	pd	Tunicate *Oikopleura dioica*	pry	Temperate pelagic	Open water	Black Sea	obs	F	Abundance of the tunicate reduced with the appearance of *Mnemiopsis* but are rebounding with the introduction of *Beroe*	Shiganova (2005)
Asia	Copepod *Tortanus dextrilobatus*	pd	Copepod *Acartiura sp.*	pry	Temperate pelagic	Open water	San Francisco Estuary	exp	L	Significant effect of predation on native zooplankton	Hooff and Bollens (2004)

(continued)

Table 15.1 (continued)

	Invader		Native								
Origin	Taxa and Species	pd/pry	Taxa and Species	pd/pry	Habitat	Substrate	Region	obs/exp	F/L	Effect	References
Asia	Copepod *Tortanus dextrilobatus*	pry	Fish	pd	Temperate pelagic	Open water	San Francisco Estuary	obs	F/L	In diet, likely to be important in the flow of material and energy in the invaded systems	Bollens et al. (2002)
Asia	Copepod *Pseudodiaptomus inopinus*	pry	Fish (flounder, perch, gobies)	pd	Temperate pelagic	Open water	Washington and Oregon estuaries	obs	F/L	Found in diet and preyed upon in similar rates as the native copepod *Acartiura*	Bollens et al. (2002)
Asia	Copepod *Pseudodiaptomus inopinus*	pry	Shrimp *Crangon franciscorum*	pd	Temperate pelagic	Open water	Washington and Oregon estuaries	obs	F/L	Found in diet and preyed upon in similar rates as the native copepod *Acartiura*	Bollens et al. (2002)
Asia	Copepod *Pseudodiaptomus inopinus*	pry	Mysid *Neomysis mercedis*	pd	Temperate pelagic	Open water	Washington and Oregon estuaries	obs	F/L	Found in diet and preyed upon in similar rates as the native copepod *Acartiura*	Bollens et al. (2002)
Europe	Crab *Carcinus maenas*	pd	Clam *Nutricola tantilla*	pry	Temperate intertidal	Soft	Central California	obs/exp	F/L	Five- to tenfold density declines within three years of arrival; strong preference for the native *Nutricola* spp. vs. the invader clam *Gemma gemma*. The native clam abundance reduced as the invasive increased; indirect invasive-invasive positive effect	Grosholz et al. (2000); Grosholz (2005)

Europe	Crab *Carcinus maenas*	pd	Clam *Nutricola confusa*	pry	Temperate intertidal	Soft	Central California	obs/exp	F/L	Five- to tenfold density declines within three years of arrival; strong preference for the native *Nutricola* spp. vs. the invader clam *Gemma gemma*. The native clam abundance reduced as the invasive increased; indirect invasive-invasive positive effect	Grosholz et al. (2000); Grosholz (2005)
Europe	Crab *Carcinus maenas*	pd	Crab *Hemigrapsus oregonensis*	pry	Temperate intertidal	Soft	Central California	exp	F/L	Five- to tenfold density declines within three years of arrival	Grosholz et al. (2000)
Europe	Crab *Carcinus maenas*	pd	Mussel *Mytilus edulis*	pry	Temperate intertidal	Hard	Long Island Sound	exp	F	*Carcinus* had higher per-capita predation effect than *Hemigrapsus*	Lohrer and Whitlatch (2002b)
Europe	Crab *Carcinus maenas*	pd	Clam *Katelysia scalarina*	pry	Temperate subtidal	Soft	Tasmania	exp	F	Strong predation effects mostly on juveniles	Walton et al. (2002)
Europe	Crab *Carcinus maenas*	pd	Clam *Fulvia tenuicostata*	pry	Temperate subtidal	Soft	Tasmania	exp	F	Strong predation effects mostly on juveniles	Ross et al. (2004)
Europe	Crab *Carcinus maenas*	pd	Clam *Mya arenaria*	pry	Temperate intertidal	Soft	Maine	exp	F	Strong predation effects, and effects on burrowing depth	Whitlow et al. (2003)
Europe	Crab *Carcinus maenas*	pd	Snail *Littorina obtusata*	pry	Temperate intertidal	Hard	Northern Maine and Massachusetts	exp	F/L	Induced defenses in prey – thicker shell; Where snails had thicker shells crabs had larger, stronger crasher claws	Trussell and Nicklin (2002); Smith (2004)

(continued)

Table 15.1 (continued)

Origin	Invader		Native		Habitat	Substrate	Region	obs/exp	F/L	Effect	References
	Taxa and Species	pd/pry	Taxa and Species	pd/pry							
Europe	Crab *Carcinus maenas*	pd	Crab *Cancer magister*	pry	Temperate intertidal	Hard/soft	Washington State	exp	F/L	Adult green crab prey on juvenile *Dungeness* crab	McDonald et al. (2001)
Asia	Crab *Hemigrapsus sanguineus*	pd	Mussel *Mytilus edulis*	pry	Temperate intertidal	Hard	Long Island Sound	exp	F/L	Higher per population predation effect than *Carcinus;* male and female crabs consumed mussels	Lohrer and Whitlatch (2002b); Bourdeau and O'Connor (2003)
Asia	Crab *Hemigrapsus sanguineus*	pd	Clam *Mercenaria mercenaria*	pry	Temperate intertidal	Hard	Massachusetts	exp	L	Females did not open clams, but ate flesh from previously opened clams	Bourdeau and O'Connor (2003)
Asia	Crab *Hemigrapsus sanguineus*	pd	Snail *Littorina littorea*	pry	Temperate intertidal	Hard	Massachusetts	exp	L	Very few *L. littorea* were consumed	Bourdeau and O'Connor (2003)
Asia	Whelk *Rapana venosa*	pd	Clam *Mercenaria mercenaria*	pry	Temperate intertidal	Soft	Chesapeake Bay	exp	L	Prey preference for *M. mercenaria* over natives: *Crassostrea virginica, Mya arenaria*, or *Mytilus edulis*	Harding and Mann (1999)
Asia	Whelk *Rapana venosa*	pry	Crab *Callinectes sapidus*	pd	Temperate intertidal	Soft	Chesapeake Bay	exp	L	The crabs readily feeds on juvenile rapa whelks <55 mm	Harding (2003)
Black Sea and Middle East	Hydrozoan *Moerisia lyonsi*	pry	Jellyfish *Chrysaora quin-quecirrha*	pd	Temperate subtidal	Hard	Chesapeake Bay	exp	L	The invasive medusa is consumed by a local jellyfish which may control its distribution limits	Ma and Purcell (2005)

Red Sea	Mussel *Brachidontes pharaonis*	pry	Whelk *Stramonita haemastoma*	pd	Temperate intertidal	Hard	Israel	exp	F/L	The invasive mussel became the preferred food item in the native gastropod's diet	Rilov et al. (2002)
Asia	Mussel *Musculista senhousia*	pry	Fish *Umbrina roncador*	pd	Temperate littoral	Soft	California	obs/exp	F	Found in guts	Crooks (2002)
Asia	Mussel *Musculista senhousia*	pry	Fish *Roncador stearnsii*	pd	Temperate littoral	Soft	California	obs/exp	F	Found in guts	Crooks (2002)
Asia	Mussel *Musculista senhousia*	pry	Fish *Anisotremus davidsonii*	pd	Temperate littoral	Soft	California	obs/exp	F	Found in guts	Crooks (2002)
Asia	Mussel *Musculista senhousia*	pry	Shorebirds (willet, godwit)	pd	Temperate littoral	Soft	California	obs/exp	F	Observed feeding, strong effects on experimental mussels	Crooks (2002)
Asia	Mussel *Musculista senhousia*	pry	Whelk *Pteropurpura festiva*	pd	Temperate littoral	Soft	California	obs/exp	F	Decimated transplanted mussels quickly, the invasive mussel was favorite prey	Reusch (1998)
North Pacific	Sea star *Asterias amurensis*	pd	Clam *Fulvia tenuicostata*	pry	Temperate littoral	Soft	SE Tasmania	exp	F	Strong predation effects mostly on adults	Ross et al. (2004)
North Pacific	Sea star *Asterias amurensis*	pd	Clam *Katelysia scalarina*	pry	Temperate subtidal	Soft	SE Tasmania	exp	F	Strong predation effects mostly on juveniles	Ross et al. (2004)
North Pacific	Sea star *Asterias amurensis*	pd	Soft sediment assemblage	pry	Temperate subtidal	Soft	SE Tasmania	exp	F	Alteration of assemblage by predation mainly on shallow dwelling bivalves	Ross et al. (2003)

(continued)

Table 15.1 (continued)

	Invader		Native								
Origin	Taxa and Species	pd/pry	Taxa and Species	pd/pry	Habitat	Substrate	Region	obs/exp	F/L	Effect	References
Australia	Tunicate *Pyura praeputialis*	pry	Sea stars	pd	Temperate intertidal	Hard	Chile	exp	F	Predation on *Pyura* juveniles by starfish and snails constitutes a regulatory mechanism for the setting of its low intertidal limit	Castilla et al. (2004)
Australia	Tunicate *Pyura praeputialis*	pry	Snails *Concholepas concholepas*	pd	Temperate intertidal	Hard	Chile	exp	F	Predation on *Pyura* juveniles by starfish and snails constitutes a regulatory mechanism for the setting of its low intertidal limit	Castilla et al. (2004)
America	American mink *Mustela vison*	pd	Crabs, fish *Anadara demiri*	pry	Temperate intertidal	Hard	Spain	obs	F	Many in diet, strong predation effects assumed, mostly on fish	Delibes et al. (2004)
mainland	Ship rat *Rattus rattus*	pd	Seabirds	pry	Temperate intertidal	Hard	Outer Hebrides, Scotland	obs	F/L	Strong predation effects on shore bird nests, feed also on other marine food sources	Stapp (2002)
	Invader 1		**Invader 2**								
Asia vs. Europe	Crab *Hemigrapsus sanguineus*	pd	Crab *Carcinus maenas*	pry	Temperate intertidal	Hard	Southern New England	obs/exp	F/L	Strong predation effect on juvenile *Carcinus*	Lohrer and Whitlatch (2002a)

Asia vs. Asia	Copepod *Tortanus dextrilobatus*	pd	Cyclopoid *Oithona davisae*	pry	Temperate pelagic	Open water	San Francisco estuary	exp	L	*Tortanus* prefers to feed on larger native copepods over *Oithona* Moderate effect of predation on non-native zooplankton	Hooff and Bollens (2004)
Mediterranean Sea vs. West Atlantic	Comb jelly *Beroe ovata*	pd	Comb jelly *Mnemiopsis leidyi*	pry	Temperate pelagic	Open water	Black Sea	obs.	F	In Black Sea increase in fish larval food and in fish larvae since the introduction of *Beroe*;	Gordina et al. (2005)
							Caspian Sea	exp	L	*Beroe* feeds on *Mnemiopsis* but larvae do not survive in Caspian Sea conditions	Kideys et al. (2004)

pd = predator, pry = prey, obs= observational, exp = experimental, F = field study, L = lab study

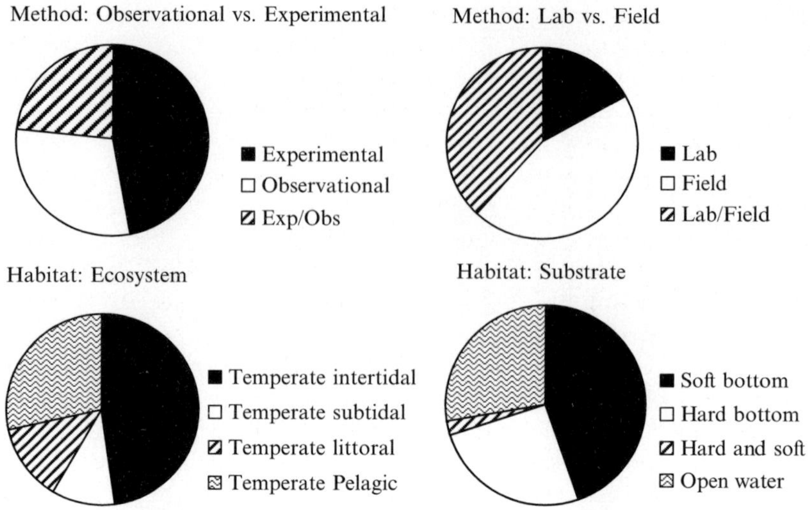

Fig. 15.1 Proportion of the different methods and habitats in studies on predator-pry interaction of marine invasions.

generally true for marine bioinvasion studies on other topics as well, the absence of such studies is discussed elsewhere in this book (Chap. 14, Byers).

Some invasive predators were shown to consume only specific size classes of the native prey population. For example, in Tasmania, the green crab, though a generalist, prefers to prey on juvenile clams, *Fulvia tenuicostata*, whereas the invasive sea star *Asterias amurensis* in the same region prefers mostly adult *F. tenuicostata*, suggesting that the combined effect of the two invaders on the clam may be greater than that due to each predator alone (Ross et al. 2004).

Predator-prey interactions are mentioned or discussed in other chapters of this book, primarily in next chapter on multitrophic-level effects of invasions as well as in a biogeographical context in Sect. VI. Here I will discuss three topics related to predator-prey interactions in the invasion context that have not been addressed directly or have only been discussed briefly in the other chapters. These include: the invasion of predatory zooplankton, predation as a biological control of invasions, and predation by exotic species as a facilitator for further invasions (invasional meltdown).

15.3.2 Invasions of Predatory Zooplankton

In their review, Bollens et al. (2002) report that invasions of marine and estuarine holoplankton (species that spend all their life as plankton) are more numerous (32 taxa combined) than freshwater invasions (27 taxa), but that more studies have been conducted on the freshwater invasions (134 vs 102). Of these studies, few have focused on the biology and ecology of these invasions, including predator-prey interactions. Nonetheless, the fact that 5 of the 12 invasive predators identified in my literature search are zooplankton is intriguing. Is it because they are very conspicuous (small but occur in billions and thus relatively easy to detect), have strong impacts on entire ecosystems by radically affecting both lower and higher trophic levels or just because researchers interested in this taxonomic group are paying more attention to invasion effects than other researchers? At this point it is difficult to answer this question, but it could be a combination of all these reasons.

Some of these invasions have already been shown to affect dramatically the structure of zooplankton communities in the invaded systems; these can result in cascading effects on other trophic levels. The comb jelly fish *Mnemiopsis leidyi*, by devouring almost everything smaller than itself, has severely impacted zooplankton communities in the Black Sea for almost a decade, but now seems to be controlled by another comb jellyfish invader, *Beroe ovata* that feeds almost exclusively on *M. leidyi* (see below, and Chap. 30, Leppäkoski et al.). *M. leidyi* invasion effects were so dramatic that it quickly led to fisheries collapses, mainly anchovy. This is because *M. leidyi* eats fish eggs, fish larvae and also the food that larvae and adults fish prey upon, leading to starvation. By feeding heavily on meroplankton (larvae of benthic species that spend their early life in the open water) as well, *M. leidyi* has also affected benthic communities.

In the northeast Pacific Ocean, two Asian copepods—*Pseudodiaptomus inopinus* in Washington and Oregon coastal estuaries, and *Tortanus dextrilobatus* in San Francisco Bay—have increased in abundance by orders of magnitude in the past decade (Hooff and Bollens 2004). Following these invasions, the entire zooplankton community structure has shifted dramatically, with overwhelming domination by the newcomers in some locations. The invasive copepods have been incorporated into the food web by becoming a dominant food source, and, being a large predatory copepod itself, *T. dextrilobatus* has also became a potentially important predator in the system. Hooff and Bollens (2004) suspect that *P. inopinus*—because it follows the same vertical migration as its dominant bentho-pelagic invertebrate predators (e.g. mysids and shrimp), daytime on the bottom and nighttime in the water above it—is creating changes in the relative magnitudes of energy flow in the benthic vs pelagic food webs in invaded estuaries of the Pacific Northwest. This hypothesis, though interesting and plausible, needs to be tested in future experimental studies. *Tortanus dextrilobatus* have shown a preference for large native copepods of San Francisco Bay over smaller exotic ones that also exist in this system. This raises the question of whether it is possible that by feeding preferentially on native species, *T. dextrilobatus* facilitates the establishment of other exotic copepods via the reduction of competition for resources? An example of such a process with invasive macro-predators and prey is described in the next section.

Cercopagis pengoi is a predatory cladoceran native to the Ponto-Caspian region that invaded the Baltic Sea in the 1990s and, later, the North American Great Lakes. Its effects on the pelagic food web in the Baltic were measured by examining stomach contents of zooplanktivorous fish (herring and sprat) and, by modeling the food web structure using stable isotope analysis (Gorokhova et al. 2004, 2005). These studies demonstrated that *C. pengoi* is a major component (60–70%) of young-of-year fish diets, but that *C. pengoi* also competes with young fish for the same shared food (mostly mesozooplankton). The models also show that the trophic level of herring shifted from 2.6 to 3.4 following the invasion of the predatory zooplankton. The fact that *C. pengoi* serves as both prey and competitor for pelagic fish makes it difficult to evaluate the net trophic outcome of this invasion. However, the invasion of this predator demonstrates again the extraordinary capability of zooplankton invasions to alter entire ecosystems.

One thing to note when looking at the case studies of zooplankton invasions in the marine environment, and their enormous influence, is that they all come from systems that are mostly enclosed: the Baltic Sea, the Black and Caspian Seas and estuaries on the US West Coast. This observation again leads to the question of why we find more marine invasions in enclosed seas and in estuarine areas, a subject that is discussed in detail in Chap. 33, Preisler et al.

15.3.3 Predation as an Invasion-control Mechanism

Using biological control to combat aggressive invaders has been discussed as a potential management tool, although it is subject to much controversy and has been yet to be substantially implemented in marine systems (Lafferty and Kuris 1996; Kuris and Lafferty 2000; Kuris 2003; Secord 2003; Messing and Wright 2006: Chap. 18, Hewitt et al.). Currently, biocontrol is often considered using parasites or parasitoids as control agents, but not predators (as in classic biocontrol). Although intentionally introducing exotic predators is highly debatable because of potentially large and unpredictable impacts on non-target species and the ecosystem, some insight into the role of predators in keeping invasive populations in check can be gained by examining the role of accidentally introduced or native predators.

One unplanned but intriguing example has been the very recent "Black Sea experiment" of biological control of one invader by another. As mentioned above, the densities of the ctenophore *Mnemiopsis leidyi* have been dwindling considerably since the first blooms of the new ctenophore *Beroe ovata* occurred in 1999 (Finenko and Romanova 2000; Finenko et al. 2003; Shiganova and Bulgakova 2000). Unlike *M. leidyi*, the newcomer is a specialized predator that feeds mainly on *Mnemiopsis* and is very effective in controlling its populations levels (Finenko et al. 2001; Kideys 2002; Shiganova et al. 2001a, 2003). This specialized predation is thought to be, at least partly, related to the signs of recovery of the Black Sea ecosystem (Kideys 2002; Shiganova et al. 2001b). When *Mnemiopsis* appeared in the Caspian Sea (Ivanov et al. 2000), it immediately caused the same disastrous effects on zooplankton communities and fisheries as it did in the Black Sea the previous decade. Therefore, investigators suggested that perhaps through an intentional introduction of *B. ovada* to the Caspian Sea, populations of *Mnemiopsis* could be controlled. To evaluate its feasibility for mitigating the *Mnemiopsis* invasion in the Caspian Sea, *B. ovata* was transported from Turkish waters to laboratories in Iran. Experiments were performed to determine rates of survival, feeding and respiration (in the low salinity waters of the Caspian Sea) of *B. ovata* (Kideys et al. 2004). Kideys et al. (2004) showed that *B. ovata* survives well in the Caspian waters and also feeds voraciously on *M. leidyi*. It can also reproduce in these waters; however its larvae, at least in the experimental laboratory conditions in this study, did not survive more than a few hours. These results indicate that unless new data and field experiments suggest otherwise, poor survival of larvae may hinder a successful establishment of *B. ovata* in the Caspian Sea and therefore compromise the potential use of this species as a controlling agent of *M. leidyi* in this sea.

An example for a possible natural control of an invasion via predation by a native species is that of blue crabs (*Callinectes sapidus*) in the Chesapeake Bay preying on the Asian veined rapa whelk (*Rapana venosa*). The large (up to 15 cm shell length) rapa whelk was first noticed in the Chesapeake Bay in 1998 and since then many individuals of varying sizes, including viable egg cases, were found in the Bay (Harding 2003). It is clear that such a large whelk has the potential to exert considerable effects on local invertebrate prey populations in general and that of bivalves in particular. Early studies have shown a dietary preference by the whelk for the clam *Mercenaria mercenaria* over the oyster *Crassostrea virginica*, the clam *Mya arenaria*, or the mussel *Mytilus edulis*, although other prey would be consumed in the absence of *Mercenaria* (Harding and Mann 1999). In the Chesapeake Bay, there is a suite of native predators that are likely to feed on whelks. These include the channeled whelks (*Busycotypus canaliculatus*), knobbed whelks (*Busycon carica*), blue crabs (*C. sapidus*), mud crabs (*Eurypanopeus sp.*), and spider crabs (*Libinia emarginata*) (Harding and Mann 1999). All crabs were indeed shown to feed on *R. venosa* (Harding and Mann 2003), and blue crabs were specifically tested in the laboratory for whelk size preference, and predation techniques by different-sized crabs (Harding 2003). Results showed that all tested sizes of blue crabs prey upon rapa whelks <35 mm shell length, but even the big crabs cannot consume whelks >55 mm, indicating a size refuge for the invasive whelk in the Chesapeake Bay. Harding (2003) suggested that blue crab predation on juvenile rapa whelks may serve as a natural control on population growth and range expansion in estuarine areas both in the Chesapeake Bay and along the North American Atlantic coast.

The potential harmful effect to the ecology of a system by rapa whelks are exemplified in the Black Sea. Rapa whelk (*Rapana thomasiana*) invasions in the Black Sea are thought to be responsible for drastic declines of bivalve populations there, including the decimation of oyster populations (Harding 2003; Zolotarev 1996). In this ecosystem there are no apparent potential predators to control the invasive whelk populations, which may explain its decimating effects.

In a recent laboratory study in western Canada, Dudas et al. (2005) showed that local crabs, dungeness crab, *Cancer magister* and red rock crab, *Cancer productus* preferred a recent clam invader, the varnish clam, *Nuttallia obscurata*, over the co-occurring native littleneck clams, *Protothaca staminea* and the Japanese littleneck *Venerupis philippinarumm,* introduced in the early 1900s. This was attributed to lower handling time, higher pick-up success and increased profitability of consuming varnish clams compared to the local clams. When feeding trials were conducted in substratum (allowing the clams to bury) this preference switched to the littleneck species, likely due to their shallower burial depth (i.e. up to 10 cm depth compared to 30 cm for varnish clams). Dudas et al. (2005) suggested that morphology (i.e. thin shell, compressed shape) of the invader increases its vulnerability to predation, but burial depth provides a predation refuge. Because of the importance of burial depth, substratum type also plays an important role in the extent of this invasion.

Byers (2002) also investigated the influence of substratum type and burial depth on predation vulnerability of *N. obscurata* to the red rock crab. His results indicated that substratum type, because of its influence on burial depth, may limit the varnish clams' distribution within a beach and may exclude it from beaches with improper sediment characteristics. This is another demonstration of how native predators could keep an invasion in check, at least at specific habitats.

What do these three examples tell us about the potential of predators, be they invaders or natives, as controlling agents on aggressive and potentially harmful invaders? They show that highly selective predators can potentially be very effective in decimating invaders, even if the invader's populations can be counted in the billions. Especially promising is the potential for facilitating already-occurring species (e.g. Reusch 1998), given the ramifications of attempting to introduce intentionally a non-native predator for the sake of control. It should be noted, however, that even working with native predators will have potential implications for other resident species, and this must bear consideration before any control program is implemented.

15.3.4 Predation by Invaders as a Facilitative Force for Further Invasions

The process of "invasional meltdown" can occur through various ecological mechanisms, some of which have been discussed elsewhere in this book. Predation of an invader on native species is one mechanism that could potentially facilitate the establishment of new exotic species. It can also lead to an outburst of old but latent invasions if the new predator selectively preys on native species and therefore reduce competition for the old, rare invader prey, allowing it to increase its numbers. This latter process is well exemplified with the invasion of the European green crab in US West Coast estuaries. For over a decade, in Bodega Bay, California, there have been studies of the abundance and distribution patterns of the green crab *Carcinus maenas*, the native soft-sediment intertidal clams *Nutricola tantilla* and *N. confusa*, and the biologically similar eastern gem clam *Gemma gemma* which existed in the area since at least the 1960s but in low numbers (Grosholz 2005; Chap. 17, Grosholz and Ruiz). The potential feeding preference of the green crab on the native vs the invasive clams, and the potential competitive interactions among the clams, were also experimentally examined. Grosholz (2005) found that the appearance of the green crab in the bay was associated with a remarkably rapid increase in the abundance of *G. gemma* that coincided with the dramatic decline in the relative dominance of *Nutricola* spp (Fig. 15.2). Since the declines of *Nutricola* spp. in 1996, *G. gemma* has also rapidly expanded its distribution in Bodega Harbor. In the laboratory, green crabs strongly preferred consuming the native over the invasive species. This preference

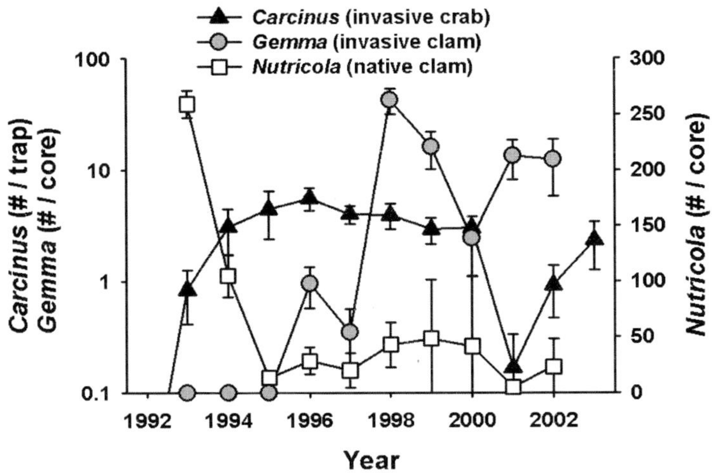

Fig. 15.2 Changes in the abundance of native clams (*Nutricola spp.*), invasive clams (*G. gemma*), and invasive European green crabs (*Carcinus maenas*) in Bodega Bay during 1993–1994 after the invasion of green crabs. *Error bars* represent 1 SEM. Copied with permission from Grosholz ED (2005) Recent biological invasion may hasten invasional meltdown by accelerating historical introductions. Proceedings of the National Academy of Sciences of the United States of America 102:1088–1091. Copyright (2005) National Academy of Sciences, U.S.A

for *Nutricola* spp. was largely driven by size differences, as crabs preferred the larger native over the smaller invasive clam. Competition experiments in the field showed that at densities similar to those measured prior to green crab invasion, the native *N. confusa* exert strong interspecific competitive effects on growth of the introduced *G. gemma*. At current densities of the native clam, however, no such effects were evident. This study is the first to demonstrate the population level consequences of a new marine invasive predator that rapidly transformed a historically benign introduction into an aggressively expanding invasion due to positive indirect interactions among invaders.

15.4 Conclusion

Predation has been shown repeatedly to be a critical process in shaping ecological communities, with effects ranging from weak to very strong (e.g. keystone), and including both direct and indirect. Predation effects were also shown to have the potential to cascade through many levels of the food web and can vary considerably, depending on local conditions such as environmental stress and rates of larval supply of prey species (bottom-up/top-down coupling). The importance of predation by invasive species as a force altering ecological communities has long been recognized in terrestrial environments. On oceanic islands, predators such as the brown tree snake in Guam, or the Australian brush-tailed possum in New Zealand have decimated, and even brought to extinction, numerous species. A recent meta-analysis demonstrated that in interactions among terrestrial invertebrates alien predators are more "dangerous" than native ones (Salo et al. 2006). The studies discussed above, and those mentioned in Chap. 17, clearly demonstrate that predator-prey interactions can also be extremely important when considering marine bioinvasions, and a similar comparison between the impacts of invasive and native predators would be a next necessary step once enough experimental studies become available. Of course, the life history of the organisms involved and their degree of dispersal can modulate the potential impacts of these interactions. As populations of most marine organisms are considered more "open" than those of most terrestrial organisms (although there is a growing body of evidence that suggest that they may be much more "closed" than previously thought; Mora and Sale 2002), one would expect that complete annihilation of native populations by invasive predators is less likely (Briggs 2006). However, the case of the invasion of the comb jellyfish *Mnemiopsis leidyi* in the Black Sea is an alarming example of the potentially devastating effects of massive invasions of predators. Likewise, the appearance of *Beroe ovata,* that later contributed to the removal of most of *Mnemiopsis* from that environment, demonstrates the potentially rapid progression of such effects. The rates at which these interactions progress and change highlight the importance of studies that follow the invasion of predator and prey species closely, and carefully test the ecological processes involved in their interactions with native and with other invasive species. We should also increase research efforts on taxonomic groups of invaders other than the ones discussed above, such as fish (75% of the species in the Mediterranean are carnivores or omnivores; see Chap. 31, Rilov and Galil) and polychaetes, and that can also be important predators in marine systems.

Although the line of evidence is still limited, studies so far suggest that the strongest impacts by marine invading predators would occur in enclosed or semi-enclosed waters such as estuaries or small seas. In such habitats, prey cannot escape far or migrate out of the region when exposed to intense predation and thus its populations can dwindle rapidly. Along the same line, invasive predators are usually more mobile than their native prey (for example, crabs or sea stars preying on bivalves), meaning that a single predator can attack many prey individuals and thus have the potential to exert strong predation effects on the prey population. Furthermore, we can expect that invasive predators that are generalists (such as the European green crab, with eight prey species—though in different biogeographical regions—appearing on the list in Table 15.1) would have a wide impact on many species but would not drive any particular species to local extinction. Successful invasive predators that are more specialists in their diet are expected to have stronger direct effects on a single or a few species and possible wider indirect effects if the native prey is a key species in the invaded habitat. The flip side of this is that effective control of invasive prey by native predators would depend on strong preference towards the invader over native prey. We have seen such an example with native predatory crabs (*Cancer magister* and *C. productus*) and their control of invasive clams (*Nuttallia obscurata*) in U.S. West

Coast estuaries (Byers 2002; Dudas et al. 2005). However, this example also demonstrates that such control can be highly context-depended; the degree of impact of crab predation on the clams was greatly dependent on sediment depth because depth affected the clam's ability to avoid predation by burrowing: the deeper the sediment the lower the predation pressure.

Finally, whether an invasive predator can bring native prey to (at least local) extinction is an intriguing open question. As most massive invasions of marine predators are fairly recent (the last 2–3 decades) it would be premature to conclude that because no species has gone extinct yet, extinction is not a plausible scenario. It is expected that if extinctions do occur, they would take place in enclosed or semi-enclosed habitats where native populations are more isolated and cannot escape, rather than on the open coast. Extinction would most probably be caused by a specialist rather than a generalist predator, because predation would be more targeted and thus effective. Potential exterminators would also be either large, highly-mobile and long-lived species, or small but with very high fecundity and population growth rates. Meta-analysis of existing case studies that compare effect size of different invasive predators, as well as modeling and comparative-experimental approaches across species, habitats and regions, would help to increase our understanding of the importance of predation by and on invaders and predict their future impacts on marine ecosystems.

References

Antsulevich A, Valipakka P (2000) Cercopagis pengoi – new important food object of the Baltic herring in the Gulf of Finland. Intl Rev Hydrobiol 85:609–619

Barkai A, McQuaid C (1988) Predator-prey role reversal in a marine benthic ecosystem. Science 242:62–64

Bollens SM, Cordell JR, Avent S, Hooff R (2002) Zooplankton invasions: a brief review, plus two case studies from the northeast Pacific Ocean. Hydrobiologia 480:87–110

Bourdeau PE, O'Connor NJ (2003) Predation by the nonindigenous Asian shore crab Hemigrapsus sanguineus on macroalgae and molluscs. Northeast Nat 10:319–334

Briggs JC (2006) Marine biogeography and ecology: invasions and introductions. J Biogeogr: doi:10.1111/j.1365-2699.2006.01632.x

Bruno JF, O'Connor MI (2005) Cascading effects of predator diversity and omnivory in a marine food web. Ecol Lett 8:1048–1056

Byers JE (2002) Physical habitat attribute mediates biotic resistance to non-indigenous species invasion. Oecologia 130:146–156

Castilla JC, Guinez R, Caro AU, Ortiz V (2004) Invasion of a rocky intertidal shore by the tunicate Pyura praeputialis in the Bay of Antofagasta, Chile. Proc Natl Acad Sci U S A 101:8517–8524

Connell JH (1961) Effect of competition, predation by *Thais lapillus* and other factors on natural populations of the barnacle *Balanus balanoides*. Ecol Monogr 31:61–104

Connolly SR, Roughgarden J (1999) Theory of marine communities: competition, predation, and recruitment-dependent interaction strength. Ecol Monogr 69:277–296

Crooks JA (2002) Predators of the invasive mussel *Musculista senhousia* (Mollusca: Mytilidae). Pac Sci 56:49–56

Delibes M, Clavero M, Prenda J, Blazquez MD, Ferreras P (2004) Potential impact of an exotic mammal on rocky intertidal communities of northwestern Spain. Biol Invas 6:213–219

Dudas SE, McGaw LJ, Dower JF (2005) Selective crab predation on native and introduced bivalves in British Columbia. J Exp Mar Biol Ecol 325:8–17

Estes JA, Tinker MT, Williams TM, Doak DF (1998) Killer whale predation on sea otters linking oceanic and nearshore ecosystems. Science 282:473–476

Estes JA, Danner EM, Doak DF, Konar B, Springer AM, Steinberg PD, Tinker MT, Williams TM (2004) Complex trophic interactions in kelp forest ecosystems. Bull Mar Sci 74:621–638

Finenko GA, Romanova ZA (2000) Population dynamics and energetics of the ctenophore Mnemiopsis leidyi in Sevastopol Bay. Oceanology 40:677–685

Finenko GA, Anninsky BE, Romanova ZA, Abolmasova GI, Kideys AE (2001) Chemical composition, respiration and feeding rates of the new alien ctenophore, Beroe ovata, in the Black Sea. Hydrobiologia 451:177–186

Finenko GA, Romanova ZA, Abolmasova GI, Anninsky BE, Svetlichny LS, Hubareva ES, Bat L, Kideys AE (2003) Population dynamics, ingestion, growth and reproduction rates of the invader Beroe ovata and its impact on plankton community in Sevastopol Bay, the Black Sea. J Plankton Res 25:539–549

Gordina AD, Zagorodnyaya JA, Kideys AE, Bat L, Satilmis HH (2005) Summer ichthyoplankton, food supply of fish larvae and impact of invasive ctenophores on the nutrition of fish larvae in the Black Sea during 2000 and 2001. J Mar Biol Assoc U K 85:537–548

Gorokhova E, Fagerberg T, Hansson S (2004) Predation by herring (*Clupea harengus*) and sprat (*Sprattus sprattus*) on *Cercopagis pengoi* in a western Baltic Sea bay. Ices J Mar Sci 61:959–965

Gorokhova E, Hansson S, Hoglander H, Andersen CM (2005) Stable isotopes show food web changes after invasion by the predatory cladoceran Cercopagis pengoi in a Baltic Sea bay. Oecologia 143:251–259

Grosholz ED (2005) Recent biological invasion may hasten invasional meltdown by accelerating historical introductions. Proc Natl Acad Sci U S A 102:1088–1091

Grosholz ED, Ruiz GM (1995) Spread and potential impact of the recently introduced European Green Crab, *Carcinus Maenas*, in Central California. Mar Biol 122:239–247

Grosholz ED, Ruiz GM, Dean CA, Shirley KA, Maron JL, Connors PG (2000) The impacts of a nonindigenous marine predator in a California bay. Ecology 81:1206–1224

Harding JM (2003) Predation by blue crabs, *Callinectes sapidus*, on rapa whelks, *Rapana venosa*: possible natural controls for an invasive species? J Exp Mar Biol Ecol 297:161–177

Harding JM, Mann R (1999) Observations on the biology of the veined Rapa whelk, *Rapana venosa* (Valenciennes, 1846) in the Chesapeake Bay. J Shellfish Res 18:9–17

Harding JM, Mann R (2003) Rapa whelks (*Rapana venosa*) as prey items for Chesapeake bay fauna: natural controls for an invasive species? In: Proceedings of the Third International Conference on Marine Bioinvasions (La Jolla, California), p 53

Hixon MA, Carr MH (1997) Synergistic predation, density dependence, and population regulation in marine fish. Science 277:946–949

Hixon MA, Pacala SW, Sandin SA (2002) Population regulation: historical context and contemporary challenges of open vs. closed systems. Ecology 83:1490–1508

Holbrook SJ, Schmitt RJ (2002) Competition for shelter space causes density-dependent predation mortality in damselfishes. Ecology 83:2855–2868

Hooff RC, Bollens SM (2004) Functional response and potential predatory impact of Tortanus dextrilobatus, a carnivorous copepod recently introduced to the San Francisco Estuary. Mar Ecol Prog Ser 277:167–179

Ivanov VP, Kamakin AM, Ushivtzev VB, Shiganova T, Zhukova O, Aladin N, Wilson SI, Harbison GR, Dumont HJ (2000) Invasion of the Caspian Sea by the comb jellyfish *Mnemiopsis Leidyi* (Ctenophora). Biol Invas 2:255–258

Kideys AE (2002) Fall and rise of the Black Sea ecosystem. Science 297:1482–1484

Kideys AE, Finenko GA, Anninsky BE, Shiganova TA, Roohi A, Tabari MR, Youseffyan M, Rostamian MT, Rostami H, Negarestan H (2004) Physiological characteristics of the ctenophore *Beroe ovata* in Caspian Sea water. Mar Ecol Prog Ser 266:111–121

Kuris AM (2003) Eradication of introduced marine pests. In: Rapport DJ, Lasley WL, Rolston DE et al. (eds) Managing for healthy ecosystems. Lewis Publishers, New York, pp 549–556

Kuris AM, Lafferty KD (2000) Can biological control be developed as a safe and effective mitigation against established introduced marine pests? In: Peterson J (ed) Marine bioinvasions: proceedings of a conference, January 24–27, 1999. MIT Sea Grant College Program, Cambridge, pp 102–106

Lafferty KD, Kuris AM (1996) Biological control of marine pests. Ecology 77:1989–2000

Lohrer AM, Whitlatch RB (2002a) Interactions among aliens: apparent replacement of one exotic species by another. Ecology 83:719–732

Lohrer AM, Whitlatch RB (2002b) Relative impacts of two exotic brachyuran species on blue mussel populations in Long Island Sound. Mar Ecol Prog Ser 227:135–144

Ma XP, Purcell JE (2005) Effects of temperature, salinity, and predators on mortality of and colonization by the invasive hydrozoan *Moerisia lyonsi*. Mar Biol 147:215–224

McDonald PS, Jensen GC, Armstrong DA (2001) The competitive and predatory impacts of the nonindigenous crab Carcinus maenas (L.) on early benthic phase Dungeness crab Cancer magister Dana. J Exp Mar Biol Ecol 258:39–54

Menge BA, Farrell TM (1989) Community structure and interaction webs in shallow marine hard-bottom communities: tests of an environmental stress model. Adv Ecol Res 19:189–262

Menge BA, Sutherland JP (1987) Community regulation: variation in disturbance, competition, and predation in relation to environmental stress and recruitment. Am Nat 130:730–757

Messing RH, Wright MG (2006) Biological control of invasive species: solution or pollution. Front Ecol Environ 4:132–140

Mora C, Sale PF (2002) Are populations of coral reef fish open or closed? Trend Ecol Evol 17:422–428

Morello E, Solustri BC, Froglia C (2004) The alien bivalve *Anadara demiri* (Arcidae): a new invader of the Adriatic Sea, Italy. J Mar Biol Assoc U K 84:1057–1064

Myers RA, Worm B (2003) Rapid worldwide depletion of predatory fish communities. Nature 423:280–283

Myers RA, Hutchings JA, Barrowman NJ (1997) Why do fish stocks collapse? The example of cod in Atlantic Canada. Ecol Appl 7:91–106

Ojaveer H, Simm M, Lankov A (2004) Population dynamics and ecological impact of the non-indigenous Cercopagis pengoi in the Gulf of Riga (Baltic Sea). Hydrobiologia 522: 261–269

Paine RT, Schindler DE (2002) Ecological pork: novel resources and the trophic reorganization of an ecosystem. Proc Natl Acad Sci U S A 99:554–555

Pauly D, Palomares ML (2005) Fishing down marine food web: it is far more pervasive than we thought. Bull Mar Sci 76:197–211

Pauly D, Christensen V, Dalsgaard J, Froese R, Torres F (1998) Fishing down marine food webs. Science 279:860–863

Peterson CH (1982) The importance of predation and intra- and interspecific competition in the population biology of two infaunal suspension-feeding bivalves, *Protothaca staminea* and *Chione undatella*. Ecol Monogr 52:437–475

Reusch TBH (1998) Native predators contribute to invasion resistance to the non-indigenous bivalve Musculista senhousia in southern California, USA. Mar Ecol Prog Ser 170:159–168

Rilov G, Schiel RD (2006) Trophic linkages across seascapes: subtidal predators limit effective mussel recruitment in rocky intertidal communities. Mar Ecol Prog Ser 327:83–93

Rilov G, Schiel RD (2006) Seascape-dependent subtidal-intertidal trophic linkages. Ecology 87(3):731–744

Rilov G, Gasith A, Benayahu Y (2002) Effect of an exotic prey on the feeding pattern of a predatory snail. Mar Environ Res 54:85–98

Robles CD (1997) Changing recruitment in constant species assemblages: Implications for predation theory in intertidal communities. Ecology 78:1400–1414

Robles C, Sweetnam D, Eminike J (1990) Lobster predation on mussels – shore-level differences in prey vulnerability and predator preference. Ecology 71:1564–1577

Ross DJ, Johnson CR, Hewitt CL (2003) Variability in the impact of an introduced predator (*Asterias amurensis*: Asteroidea) on soft-sediment assemblages. J Exp Mar Biol Ecol 288:257–278

Ross DJ, Johnson CR, Hewitt CL, Ruiz GM (2004) Interaction and impacts of two introduced species on a soft-sediment marine assemblage in SE Tasmania. Mar Biol 144: 747–756

Sakai AK, Allendorf FW, Holt JS, Lodge DM, Molofsky J, With KA, Baughman S, Cabin RJ, Cohen JE, Ellstrand NC, McCauley DE, O'Neil P, Parker IM, Thompson JN, Weller SG (2001) The population biology of invasive species. Annu Rev Ecol Syst 32:305–332

Salo P, Korpima E, Banks PB, Nordstro M, Dickman CR (2006) Alien predators are more dangerous than native predators to prey populations. Proc R Soc B 274:1237–1243

Scheffer M, Carpenter S, de Young B (2005) Cascading effects of overfishing marine systems. Trends Ecol Evol 20:579–581

Secord D (2003) Biological control of marine invasive species cautionary tales and land-based lessons. Biol Invas 5:117–131

Shiganova TA (1998) Invasion of the Black Sea by the ctenophore Mnemiopsis leidyi and recent changes in pelagic community structure. Fish Oceanogr 7:305–310

Shiganova TA (2005) Changes in appendicularian *Oikopleura dioica* abundance caused by invasion of alien ctenophores in the Black Sea. J Mar Biol Assoc U K 85:477–494

Shiganova TA, Bulgakova YV (2000) Effects of gelatinous plankton on Black Sea and Sea of Azov fish and their food resources. Ices J Mar Sci 57:641–648

Shiganova TA, Mirzoyan ZA, Studenikina EA, Volovik SP, Siokou-Frangou I, Zervoudaki S, Christou ED, Skirta AY, Dumont HJ (2001a) Population development of the invader ctenophore Mnemiopsis leidyi, in the Black Sea and in other seas of the Mediterranean basin. Mar Biol 139:431–445

Shiganova TA, Bulgakova YV, Volovik SP, Mirzoyan ZA, Dudkin SI (2001b) The new invader Beroe ovata Mayer 1912 and its effect on the ecosystem in the northeastern Black Sea. Hydrobiologia 451:187–197

Shiganova TA, Musaeva EI, Bulgakova YV, Mirzoyan ZA, Martynyuk ML (2003) Invaders ctenophores Mnemiopsis leidyi (A. Agassiz) and Beroe ovata Mayer 1912, and their influence on the pelagic ecosystem of northeastern Black Sea. Biol Bull 30:180–190

Silliman BR, Bertness MD (2002) Atrophic cascade regulates salt marsh primary production. Proc Natl Acad Sci U S A 99:10500–10505

Silliman BR, van de Koppel J, Bertness MD, Stanton LE, Mendelssohn IA (2005) Drought, snails, and large-scale die-off of southern US salt marshes. Science 310:1803–1806

Smith LD (2004) Biogeographic differences in claw size and performance in an introduced crab predator *Carcinus maenas*. Mar Ecol Prog Ser 276:209–222

Stapp P (2002) Stable isotopes reveal evidence of predation by ship rats on seabirds on the Shiant Islands, Scotland. J Appl Ecol 39:831–840

Trussell GC, Nicklin MO (2002) Cue sensitivity, inducible defense, and trade-offs in a marine snail. Ecology 83:1635–1647

Walton WC, MacKinnon C, Rodriguez LF, Proctor C, Ruiz GA (2002) Effect of an invasive crab upon a marine fishery: green crab, Carcinus maenas, predation upon a venerid clam, Katelysia scalarina, in Tasmania (Australia). J Exp Mar Biol Ecol 272:171–189

Whitlow WL, Rice NA, Sweeney C (2003) Native species vulnerability to introduced predators: testing an inducible defense and a refuge from predation. Biol Invas 5:23–31

Witman JD, Genovese SJ, Bruno JF, McLaughlin JW, Pavlin BI (2003) Massive prey recruitment and the control of rocky subtidal communities on large spatial scales. Ecol Monogr 73:441–462

Worm B, Myers RA (2003) Meta-analysis of cod-shrimp interactions reveals top-down control in oceanic food webs. Ecology 84:162–173

Zolotarev V (1996) The Black Sea ecosystem changes related to the introduction of new mollusc species. Mar Ecol (Pubblicazioni Della Stazione Zoologica Di Napoli I) 17:227–236

Chapter 16
The Role of Exotic Marine Ecosystem Engineers

Jeffrey A. Crooks

> *How, for instance, would you classify the effect of the common European periwinkle,* Littorina littorea, *an intertidal snail, on the shores of New England? By grazing on the algae on the rocks and on the rhizomes of marsh grass, this gastropod has shifted the coastal landscape from mud flats and salt marshes to a rocky shore.*
>
> (Williamson 1996)

16.1 Ecosystem-Level Impacts

Of all invader impacts, those likeliest to have the most wide-reaching consequences are alterations to ecosystems, as they can essentially "change the rules of existence" for broad suites of resident biota (Vitousek 1990). One often-considered class of ecosystem-level effects of exotics is the disruption of energy or material fluxes (Chap. 17, Grosholz and Ruiz). For example, the initiation of trophic cascades, which can be triggered by events such as the invasion of new predators, can dramatically alter energy flow within ecosystems (e.g., Spencer et al. 1991). Similarly, the cycling of nutrients through biogeochemical pathways can be affected by exotics (e.g., Larned 2003). This can occur through the invasion of species that differ from natives in their utilization of nutrients, such as when nitrogen-fixing plants invade nitrogen-poor soils (Vitousek et al. 1987). In addition to directly affecting the cycling of energy or nutrients, exotics also can alter the actual physical or chemical nature of the ecosystem itself. Such organisms have been called ecosystem engineers (Jones et al. 1994, 1997).

Reduced to its essence, ecosystem engineers affect other biota via alterations to the abiotic environment (Fig. 16.1). These species create, destroy, or otherwise modify habitats, and thereby affect resources or stressors (e.g., living space, sediment, and ambient temperature) that affect other organisms (Jones et al. 1994; Crooks 2002). The beaver is the classic example of an ecosystem engineer. By creating dams out of trees, beavers dramatically change the nature of the ecosystem by converting forests to ponds, thereby benefiting aquatic species at the expense of terrestrial ones. Beavers have effects beyond pond creation, however. The trees they fell are themselves ecosystem engineers that create shade, provide structure for nests, and dampen winds, and the loss of these engineering functions alters forest areas.

G. Rilov, J.A. Crooks (eds.) *Biological Invasions in Marine Ecosystems.*
Ecological Studies 204,

Although there has been some debate about ecosystem engineering (e.g., Brown 1995; Power 1997; Jones and Gutiérrez 2007), the concept has provided a unifying theme for both basic and applied ecology as it captures the myriad ways in which species can alter the physical or chemical environment (Coleman and Williams 2002; Crooks 2002; Cuddington et al. 2007; Hastings et al. 2007), As a case in point, it clearly answers Williamson's question (above) about how to possibly consider the habitat-altering role of an herbivorous snail. Beyond categorization, a key element in the development of this concept has been the need to address an important body of ecological interactions that has not easily fallen into current theoretical or empirical constructs such as models of food web dynamics or mass-balanced biogeochemical cycles. To this end, new approaches incorporating engineering dynamics are emerging (e.g., Cuddington and Hastings 2004; Byers et al. 2006; Cuddington et al. 2007). Finally, the study of non-native engineers has offered valuable (if unfortunate) opportunities to investigate fundamental links between species and ecosystems in a way that might not be afforded by examining long-integrated native species (Vitousek 1990; Crooks 2002).

In this chapter, I will explore engineering by invaders more fully, focusing on marine exotic species. I will start this review with a relatively broad treatment of engineering in relation to the other activities of organisms, using an illustrative example. This will extend earlier work (Crooks 2002) comparing engineering to the two other major classes of ecosystem-level effects: changing the flow of energy through

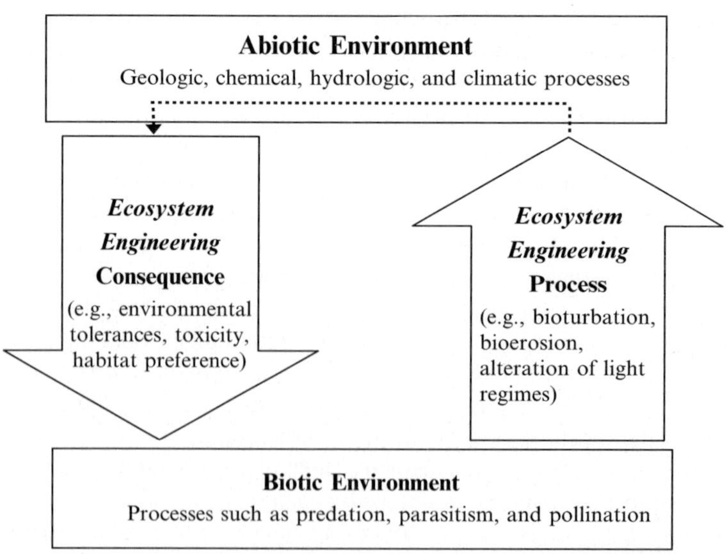

Fig. 16.1 Ecosystem engineering, representing the biogenic modification of the abiotic environment, and subsequent abiotic effects on biota (see Jones and Gutiérrez 2007 for more on engineering process and consequence)

food webs, and altering nutrient/biogeochemical cycling. I will then illustrate the diversity of effects of invasive marine engineers, and conclude with a treatment of the broader implications of the engineering concept for marine conservation.

16.2 Engineering

The first step in applying the engineering concept is to define clearly what it is, and what it is not, especially in relation to other ecosystem-level effects of species. In some cases, the distinctions between engineering and non-engineering processes are clear, whereas in other cases the differences are more subtle. To highlight this, I will examine the effects of an invasive bivalve in an ecosystem (Fig. 16.2). Although this example is ultimately a hypothetical composite of several different species, it will draw on the relatively large body of research on the role of bivalves (and particularly invaders) in systems (e.g., Darrigran 2002; Ojaveer et al. 2002; Gutierrez et al. 2003; Newell 2004; Crooks 2006a; Chap. 23, Griffiths et al.).

The key element of engineering is that it requires two major steps (Fig. 16.1): the modification of the abiotic environment by the engineer, and the response of other biota to this abiotic alteration (engineering process and consequence, respectively; see Jones and Gutiérrez 2007). In the bivalve example (Fig. 16.2), all of the effects on the environment stem from only four basic activities: shell creation, biofiltration of the water column, biodeposition of material, and respiration (production of carbon dioxide). These activities can be placed in three classes (Fig. 16.2), representing: (1) engineering (black lines), either directly affecting physical resources (solid lines) or indirectly affecting trophic or nutrient resources (dashed lines), (2) non-engineering activities or consequences, through either trophic or nutrient pathways (white lines), or (3) a combination of engineering and non-engineering (gray lines).

16.2.1 Autogenic vs Allogenic Engineering

These effects on the physical environment can take two primary forms; what has been termed *autogenic* and *allogenic engineering* (Jones et al. 1994). First, in *autogenic engineering*, the physical environment is changed by the bodies of the engineers themselves (Jones et al. 1994). In the case of the bivalve, the creation of shells represents autogenic engineering (Crooks 1998; Gutiérrez et al. 2003). This alteration of physical structure provides an important abiotic resource, living space, to which other species respond (Fig. 16.2). For example, small benthic macrofauna can achieve high abundances within the shell matrices created by exotic bivalves (Stewart and Haynes 1994; Crooks 1998; Crooks and Khim 1999). Similarly, the trees mentioned in the example above, before being felled by beavers, performed autogenic engineering in that their roots, trunks, branches, and leaves all modify the abiotic environment and affect other forest biota.

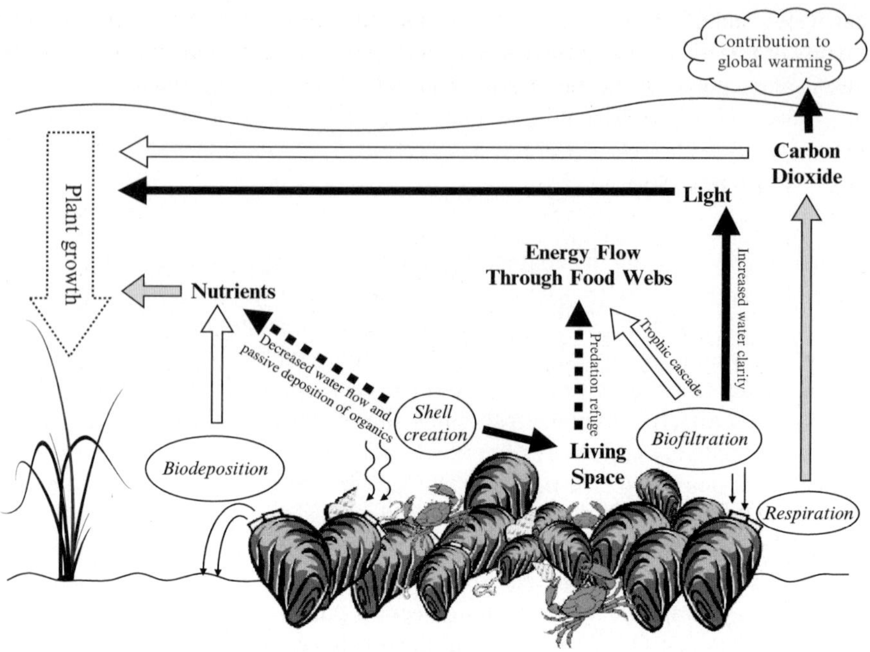

Fig. 16.2 Examples of ecosystem-level effects of an invasive bivalve. *Black lines* represent engineering activities, including direct effect on physical resources (*solid lines*) and indirect effects on biogeochemical and tropic resources (*dashed lines*). *White lines* represent non-engineering activities (i.e., biogeochemical and trophic flows). *Gray lines* represent a combination of engineering and non-engineering. (adapted from Crooks 2006b)

The other primary class of biogenic habitat modification is *allogenic engineering*. Here, the engineer changes the physical state of the ecosystem via mechanical, chemical, or other means (Jones et al. 1994). The bivalves filter water, and this clears suspended materials from the water column and increases water clarity (Fig. 16.2). This increased light penetration can then facilitate plant growth at depth (Phelps 1994; Strayer et al. 1999). This type of engineering does not involve the organism itself as a structural element, as does autogenic engineering. The activities of the beavers also represent allogenic engineering. It is important to note, as demonstrated with the bivalves, that organisms can simultaneously perform autogenic and allogenic engineering.

The biological activities that give rise to allogenic engineering are varied, including burrowing (Talley et al. 2001; Talley and Crooks 2007) and bulldozing sediments (Bertness 1984). In some cases, allogenic engineering can result from feeding activities, such as grazing down vegetative structure by herbivores (de Vos et al. 1956; Ford and Grace 1998; Talley and Crooks 2007), but engineering encompasses a different ecological pathway and does not include the flow of energy via trophic interactions. For example, one important consequence of bivalve biofiltration could be a decrease in planktonic standing stocks (Fig. 16.2), which can induce trophic cascades (Kimmerer et al. 1994; Vander Zanden et al. 1999; Chap. 17, Grosholz and Ruiz; Chap. 15, Rilov). However, this is strictly a biotic interaction that does not need to involve any abiotic element, and there is a well-developed body of theory devoted to this sort of direct flow of energy through food webs. Conversely, the degree to which biofiltration increases light penetration and allows plant growth (Phelps 1994; Strayer et al. 1999) is not captured with traditional theory – neither the bivalve nor the plankton is in the plant's food web (Fig. 16.2). It would take different models, incorporating both abiotic and biotic interactions, to capture this effect. This highlights an important principal: *the same biological activity can give rise to both engineering and non-engineering effects.*

16.2.2 Indirect Effects of Engineering on Food Webs and Nutrient Cycling

Although engineering itself does not directly affect energy flow through food webs, it can give rise to indirect effects which subsequently affect trophic flows. For example, it is possible that the small macrofauna within the shell matrix might be responding to a refuge from predation from carnivorous fish unable to eat the bivalve or its associated biota (e.g., Mayer et al. 2001). If the removal of this macrofaunal food source affects energy flow through food webs it would be an indirect energy flow consequence of the modification of habitats achieved via engineering (Fig. 16.2). Again, this distinction can be emphasized by recognizing that the bivalve might be a key driver of energy flow but would not even be in the food web containing the small macrofauna and the fish.

Another principal ecosystem-level role of species is to directly affect biogeochemical cycling, which is not engineering (Crooks 2002). For example, active biodeposition of nitrogenous wastes of bivalves can affect nutrient availability and the subsequent responses of other species, like plants (Reusch and Williams 1998). Like feeding, these fluxes of materials, such as nutrients, oxygen, or carbon dioxide, through biogeochemical cycles can be distinguished from engineering. Unlike feeding, however, biogeochemical cycling involves transitions into the abiotic realm and then back to the biotic, making this distinction between engineering and non-engineering more subtle. The difference lies in how the transition back to the biotic realm occurs. For example, bivalves produce carbon dioxide through respiration (Fig. 16.2). If this is directly assimilated by the plant through photosynthesis, this is a direct, stoichiometric transfer of materials that is analogous to the transfer of energy through foodwebs. There is little value in thinking of this as engineering (Jones et al. 1997). However, the action of CO_2 within the abiotic realm to change habitats, such as acting as a greenhouse gas, could lead to effects on the distribution and abundance of other organisms (Chauvaud et al. 2003). This would be engineering in its broadest sense.

The modification of habitats via engineering can indirectly affect nutrient cycling, as was seen for trophic flows (Fig. 16.2). The shells produced by the bivalve can slow water flow and passively increase the flow of material to the benthos (Crooks and Khim 1999). This might affect the direct use of nutrients by organisms such as plants. This is engineering, however, in that this passive contribution to the nutrient pool does not involve active participation by the bivalve, but rather occurs via the modification of habitats. As discussed earlier, the animal also directly affects nutrients via active biodeposition (Fig. 16.2). This raises another important principal which will have ramifications for assessing engineering effects: *the same resource pool can be affected by an organism's engineering and non-engineering activities.*

16.3 Detecting Engineering Effects

The examples above distinguish between three different classes of ecosystem-level processes: effects on trophic flows, effects on biogeochemical cycling, and ecosystem engineering (Crooks 2002). However, there are many other sorts of interactions between an invasive engineer and other biota. They might eat or be eaten (without substantially affecting energy flow within the ecosystem), be parasitized, or directly compete for food. All such factors will work in concert to constitute the integrated effect of an invader on other species. Given this complexity, how does one detect and quantify the net effects due to engineering alone?

Assessing the role of engineering can be complex, but it essentially involves making observations, and often intervening, at different stages in the cycle of engineering process and consequence (Jones and Gutiérrez 2007). In general terms, one can observe or modify some subset or all of the following: the engineer itself, its effects on the abiotic environment, the interactions within the abiotic realm, and the abiotic effects on biota (Fig. 16.1). In practice, there are many possible approaches and combinations of approaches that can be brought to bear on the problem, but some common ones are highlighted below.

16.3.1 Effects on Abiotic Environment Alone

One common means of assessing the potential impact of engineers in systems is to assess only their effects on the abiotic environment, and then extrapolate to effects on other species via this abiotic alteration. For example, the isopod *Sphaeroma quoyanum* has been demonstrated to burrow into mud banks of salt marshes and increase erosion rates, effectively converting marsh to mudflat (Talley et al. 2001; Talley and Crooks 2007). Although the cascading biotic effects of this increased marsh erosion rate have not been quantified, in qualitative terms it seems clear that mudflat organisms would benefit and marsh organisms would not. Although an indirect inference, this highlights one of the powers of engineering concept. Basic ecological research over the years has offered insight into how the abiotic environment influences biota, such as the role of light, temperature, or sediment properties in shaping the distribution and abundance of species. If one can demonstrate the effect of the invader on the abiotic realm, general inferences and perhaps even quantitative predictions can be made related to biotic effects.

16.3.2 Comparisons of Invaded and Uninvaded Areas

In terms of direct, quantitative assessments, probably the most common way in which the biotic effects to engineers are (putatively) assessed is to compare biotic assemblages in areas with engineers to areas without, separated in either space or time (e.g., Crooks 1998; Wikström and Kautsky 2004; Sax et al. 2005). However, the simple comparison of such assemblages does not give a direct assessment of the relative strength of engineering. For example, in the case of increased plant growth observed in areas with invasive filter feeding bivalves (Strayer et al. 1999), it is likely that engineering via increasing water clarity is important. However, it is also possible that the species are responding to bivalve's direct increase in nutrient regeneration (e.g., Fig. 16.2), which is not engineering. Further study would be necessary to determine the degree to which engineering is important. Despite these limitations, documentation of patterns observed under natural conditions has utility, and is a logical (and often necessary) starting point for more detailed investigations. Moreover, increased predictive power might be gained if comparisons across other species with similar engineering effects are made (e.g., Crooks 2002).

16.3.3 Experimental Approaches

In order to assess engineering in more depth, experiments are often employed. One fruitful approach is to replicate the structural modification created by the engineer. For organisms that have effects via autogenic engineering, impacts can be assessed by using structural equivalents of the organism, and then comparing the effects generated by these to those generated by the living organisms (e.g., Holloway and Keough 2002). For organisms with hard exoskeletons (e.g., bivalves, corals, tube worms, and vermetid gastropods), the actual structure produced by the organism can be used as the structural equivalent, whereas in other cases one might need to use artificial structure that mimics the living organism (e.g., plastic strips for eelgrass or soda straws for polychaete tubes). For an organism with strong engineering effects (and barring any unforeseen interactions), we would expect the structural equivalent alone to account for much of the effect the living organism, the latter of which represents the structure plus the biological activities. For example, in the study of an invasive mussel on a tidal flat, the effects of the mussel shell and a byssal mat mimic accounted for much of the effect produced by the living organism, and structural effects were determined to greatly outweigh biological effects alone (Crooks and Khim 1999). In the case of an invasive spionid polychaete, however, a tube mimic had no affect on amphipod behavior, suggesting the presence of some non-structurally mediated interaction (Neideman et al. 2003).

While the use of structural equivalents is very effective at assessing engineering effects mediated by structural changes, engineering also can occur through other means. For example, if an experiment assessing the effect of bivalve engineering on seagrass growth found that shells alone had little effect, it could not be stated that engineering in general was not occurring. The active biofiltration by the bivalve has engineering effects (increased light penetration) that could not be assessed by using shells alone. As discussed above, however, biofiltration of water leading to increased light penetration (engineering) and biodeposition of nutrients (non-engineering), both involve the activities of the living bivalve (Fig. 16.2). Experimentally discriminating between the relative importance of such effects can be more complicated, but could involve attempting to manipulate the resources which the eelgrass responds, such as increasing shade or artificially supplying nutrients (Allen and Williams 2003).

Many other approaches to assessing engineering are also possible, but the key in all of them is to assess what is being examined and realize its limitations. For example, it is possible to recreate an engineering process and compare responses in those areas to controls. If one were assessing the engineering impacts of snail grazing, one could compare biotic communities in artificially cleared patches to uncleared, control patches. In such cases, however, it would be important to distinguish between those strictly biotic effects due to consumption of plant material, and those due to the removal of the engineering plants and their effects.

16.4 Effects of Exotic Engineers on Abiotic Factors and Processes

The role of marine species in affecting the abiotic environment, and the consequent effect for other biota, has long been part of marine ecological thinking, especially for benthic systems. There is, therefore, a well-developed context for examining the more formalized concept of engineering in the marine realm. Furthermore, the high invasion rates in coastal systems offer opportunities for examination of many different types of exotic engineers (see also Wallentinus and Nyberg 2005). Below I will identify some general effects of engineers that modify properties of the water column and benthos, as well as the more synthetic effects of the broad suite of organisms that affect structure and habitat complexity. Although not meant to be exhaustive, when taken with the bivalve examples presented above, this will highlight some of the major roles of exotic engineers within invaded ecosystems.

16.4.1 Water Column and Hydrodynamic Conditions

Although the water column is intrinsically less vulnerable to the effects of engineers (due to large volumes and the paucity of physical structure), invaders can affect water column properties and processes. For example, there has been increasing interest and concern related to the ballast-water transport of toxic species responsible for harmful algal blooms (Hallegraeff 1998; Van Dolah 2000; Hopkins 2002). The toxic effects of these species, often dinoflagellates, can result from several different mechanisms. One common pathway is for other species to ingest the toxic plankton, causing stress or death for these organisms or those further up the food chain. This type of interaction is not best considered as engineering, as it involves direct trophic interactions without an abiotic intermediary. However, some blooms do result in alterations to the environment, often through mechanical disruption of cells (e.g., through wave action), effectively creating an abiotic environment detrimental to other species (Van Dolah 2000). Similarly, dead biomass associated with blooms can lead to microbially-induced anoxic conditions, with further effects for resident biota (Van Dolah 2000). Another form of environmental alteration by exotic bloom species includes the production of copious mucus, which can coat the seafloor (Hopkins 2002).

Benthic species also can affect water column properties, such as through the filter-feeding activities of the bivalves described above (Fig. 16.2). Another exotic filter-feeder, the tube-worm *Ficopomatus enigmaticus*, can affect water quality and light penetration (often in a positive way) by filtering large volumes of water within enclosed harbors (Davies et al. 1989). On smaller scales, canopy producing exotics can locally affect available light, such as has been observed with the invasive macroalga *Sargassum muticum* on hard substrates (Staehr et al. 2000; Wallentinus 2002; Britton-Simmons 2004).

Exotic engineers also can alter the hydrodynamic regime experienced by other species. The slipper limpet, *Crepidula fornicata*, can increase drag on the species to which it is attached, increasing dislodgment (Thieltges 2005; Chap. 29, Gollasch et al.). Similarly, the alga *Codium fragile* has earned the name "oyster thief," as it too increases drag on the bivalves to which it attaches and can lead to the loss of these commercially important species from beds (Steneck and Carlton 2001).

16.4.2 Sedimentation and Benthic Conditions

When exotic engineers remove material from the water column, either actively and/or passively, the flux of material to the seabed is increased, thereby altering benthic conditions (e.g., Fig. 16.2). For example, in addition to increasing local shading, *Sargassum* can alter the physical environment by slowing water flow and increasing sedimentation rates, thus favoring sediment-dwelling species such as harpacticoid copepods (Viejo 1999; Wallentinus 2002). Two exotic red algal turf species also increase sediment accumulation, making conditions unfavorable for other algae (Piazzi and Cinelli 2000). Introduced vascular plants, such as *Spartina* spp. on the west coast of the United States, are also noted for their ability to increase sedimentation rates, especially compared to vegetation-free sediments (Daehler and Strong 1996; Chap. 17, Grosholz and Ruiz). Replacing a native plant with an exotic also can affect benthic conditions. This been observed with the *Phragmites australis* invasion of *Spartina alterniflora* marshes on the east coast of the U.S., where the invaded habitats have lower soil salinities, less microtopographic relief, and higher redox potentials (Gratton and Denno 2005).

As exemplified by the exotic bivalves (Fig. 16.2), exotic animals can affect benthic habitats both passively and actively. Other examples include the invasive riverine amphipod, *Cheliorophium* (=*Corophium*) *curivispinum*, which creates networks of tubes that increase mud on hard substrates (van den Brink et al. 1993). The tubeworm *Ficopomatus enigmaticus* can increase carbonate sediments within invaded lagoons through the creation and subsequent breakdown of calcareous tubes (Schwindt et al. 2002). Slipper limpets (*Crepidula*), in addition to increasing drag, biodeposit feces and pseudofeces, which can smother bivalves and make the sediment too muddy for successful recruitment by oysters (Thieltges 2005; Chap. 29, Gollasch et al.). Burrowing exotics also can affect sedimentary conditions. In the Baltic, the bioturbating worm *Marenzelleria viridis* burrows more deeply than native species, thus increasing the depth of the oxic zone and the area habitable by other macrofauana (Zmudzunski 1996; Chap. 30, Leppäkoski et al.).

16.4.3 Plants and Animals as Agents of Structural Change

As has been highlighted above, many of the engineering effects within invaded ecosystems broadly result from some modification to physical structure. However, it is often difficult to tease apart the proximate mechanisms giving rise to effects observed when structure is modified in a system. Adding, removing, or modifying structure in marine systems can simultaneously affect key resources or stressors such as water flow, food availability, light, temperature, ambient environmental conditions, enemy-free space, and surfaces for attachment. Also, if the structural elements are organisms themselves (as opposed to structures created out of abiotic materials), resident biota might eat the engineer and be responding to the direct provision of food resources, which is not engineering. For example, *Sargassum* is successfully colonized by many of the ephiphytes found on native macroalgae, but some of this pattern is certainly due to the food value of the alga (Viejo 1999; Wernberg et al. 2004). Despite these potential difficulties, the net consequences of such alterations can be profound and it is worth characterizing the roles of structure producing plants, algae, and animals in aquatic systems, as they can differ from the roles of these species in terrestrial habitats.

One fundamental difference between engineering in aquatic and terrestrial systems is that animals in water play a much more varied role in relation to structure than they do on land. Unlike on land, animals in the water can perform substantial autogenic engineering. There are numerous examples of marine animals with large effects related to autogenic engineering. In addition to the bivalves discussed above (Fig. 16.2), exotic structure-producers, such as barnacles, worms, tunicates, and snails, can create biogenic matrices and structural resources utilized by other species (e.g., Crooks 1998; Olenin and Leppäkoski 1999; Castilla et al. 2004; Wonham et al. 2005). A striking example is the tube-building polychaete worm, *Ficopomatus enigmaticus*, which is able to form massive reefs within invaded lagoons, with dramatic consequences for physical properties and processes as well as resident biota (Carlton 1979; Davies et al. 1989; Schwindt et al. 2002, 2004).

Animals in marine systems also play their more typical role as agents of physical disturbance. For example, exotic grazers in Gulf of Mexico marshes, such as nutria and wild boar, can inhibit soil-building processes, potentially contributing to and exacerbating marsh submergence (Ford and Grace 1998). Other invasive engineers that destroy structure include bioeroders such as isopods (Talley et al. 2001; Talley and Crooks 2007) and mitten crabs (Rudnick et al. 2003; Chap. 29, Gollasch et al.), and disease agents that eventually destroy structures produced by autogenic engineers such as oysters (Ruiz et al. 1999).

As in terrestrial systems, large autotrophs are often the dominant structural elements in the nearshore environment. The invasion of seagrasses such as *Zostera japonica* (Posey 1988), mangroves such as *Rhizophora mangle* (Demopoulos 2004), marsh plants such as *Spartina* spp. (Daehler and Strong 1996; Neira et al. 2005; Chap. 17, Grosholz and Ruiz), and macroalgae such as *Undaria pinnatifida* (Curiel et al. 2001; Casas et al. 2004; Hewitt et al. 2005) and *Caulerpa taxifolia* (Meinesz 1999; Levi and Francour 2004) can alter habitats (the community-level effects of such species will be discussed in more detail below). Invasions of the upland transition zone by terrestrial plants also can affect marine systems (e.g. Whitcraft et al. 2007). For example, Australian pines can grow on sandy, treeless coastlines (e.g., in Florida), and fallen trees on the beach can inhibit sea turtle nesting, but standing trees can actually facilitate nesting by blocking city lights (Salmon et al. 1995).

16.4.4 Net Effects of Exotic-induced Changes to Habitat Complexity

Some general patterns emerge in the examination of exotic marine engineers that affect structure and complexity. In general, the type of biotic effect depends on how complexity is affected. Similar to patterns observed for native species, those organisms that increase complexity often facilitate suites of resident biota, while those that decrease complexity have the opposite effect (Crooks 2002). When a structure-producing organism invades an area with little structure ("discrete trait invaders": see Chapin et al. 1996), dramatic changes can result. For example, Japanese eelgrass (*Zostera japonica*) has invaded the Pacific Northwest of the US, and as it can live higher in the intertidal than the native eelgrass, it converts bare mudflats to eelgrass meadows (Posey 1988). This increased structure actually benefits small macrofauna, which tend to live in higher densities within the meadows of the invasive plants. A similar macrofaunal facilitation has been observed with the invasions of mangroves onto tidal flats in Hawaii (Demopoulos 2004), tubeworms and oysters in Argentina (Schwindt et al. 2002; Escapa et al. 2004), and tunicates in Chile (Castilla et al. 2004). Conversely, when complexity is decreased due to invasion, such as through the activities of grazing animals in marshes (de Vos et al 1956; Ford and Grace 1998), negative effects are often found (Crooks 2002).

When an invasive engineer is compared to a native engineer with similar structural features, their effects are often less dramatic (e.g., Wikström and Kautsky 2004). Examination of invasive *Phragmites australis* and native *Spartina alterniflora* stands in Chesapeake Bay revealed some small (but perhaps important) differences in many aquatic species (Posey et al. 2003; Weis and Weis 2003; but see Gratton and Denno 2005). For the invasive cordgrass *Spartina anglica* in Australia, infaunal communities in the exotic vegetation were similar to that found in native cordgrass, but different than that found on mudflats (Hedge et al. 2000).

Although the patterns above, where the degree of complexity is a primary driver of faunal response, appear relatively robust, there are of course exceptions, and these are important in highlighting general principles (Crooks 2002). For example, comparisons of faunal communities in meadows of a native-exotic hybrid cordgrass to both vegetated and unvegetated sediments in San Francisco Bay yielded variable results, including some instances of decreased densities and diversities within vegetated habitats (Chap. 17, Grosholz and Ruiz). These differences were suggested to arise from factors related to the age of the invasion, ambient environmental conditions, and life histories of resident organisms (Neira et al. 2005). The importance of environmental context and the types of organisms examined also is emphasized by the effects of the aforementioned grazing snail *Littorina littorea* (Bertness 1984). This snail converts soft-sediment habitats to rocky shores through its grazing and bulldozing activities, and was found to benefit rocky shore biota at the expense of soft-sediment biota. While this is not surprising, it does emphasize that *engineers typically do not destroy "habitat" per se – they transform it from one habitat type to another.* Those organisms able to live within the new habitat type will benefit, while those associated with the old habitat type will not (Jones et al. 1997; Crooks 2002; Talley and Crooks 2007).

16.4.5 Scale

Another key factor involved in faunal responses to engineers is scale. Temporal scales will be important, as patterns will likely change related to the age of the invasion (Holloway and Keough 2002; Crooks 2005; Neira et al. 2005). Spatially, at broad scales, landscapes with patches of affected by invasive engineers and invader-free patches will likely have higher diversities than more uniform landscapes (Wright et al. 2002). Scale will also play a role in biotic responses within patches. A good example of this is the soft-sediment dwelling mussel, *Musculista senhousia. Musculista* can form dense mats, constructed with its sticky byssus, on intertidal and subtidal soft sediments (Crooks 1998). Typical densities are 5,000–10,000 m^{-2}, although densities of over 100,000 m^{-2} have been reported (Crooks 1996; Crooks and Soulé 1999). Descriptive studies demonstrate that at small spatial scales, the mussel appears to benefit many small organisms that occur in higher abundances within the complex mat matrix when compared to nearby, mat-free areas (Creese et al. 1997; Crooks 1998). This is predominantly a structural effect, as experiments testing shells and mats alone account for much of the effects seen in natural mats (Crooks and Khim 1999). However, for larger organisms not able to live within the dense mat matrix, such as bivalves and eelgrass, the mussel can have detrimental effects likely related to both habitat modification and competition for food (Reusch and Williams 1998; Crooks 2001; Mistri 2004).

16.5 Implications

Although continued study of invasive marine engineers will lead to a better understanding of the consequences of anthropogenic biological invasions, there are other important implications of exotic engineers. For example, although engineering focuses on alterations of the natural environment, the modification of man-made structures by exotic engineers can have important economic and management implications. Examples include crustaceans burrowing into levees (Talley et al. 2001; Rudnick et al. 2003), shipworms and gribbles destroying wooden ships, piers, and docks (Carlton 1979; Chap. 29, Gollasch et al.), and fouling organisms clogging water intakes into power plants (Nalepa and Schloesser 1993) and increasing drag on ships (Chap. 6, Hewitt et al.). Exotic engineers can alter the local environment in ways that might have public health effects, and it is known that some dinoflagellates produce aerosolized toxics that can affect humans (Van Dolah 2000). The study of exotic engineers can also provide insight into the effects of natural invasions of engineers, such as kelp-grazing urchins that create barrens and crown-of-thorn starfish that destroy reefs. Understanding the consequences of adding habitat-altering species can also lend insight into what happens when desirable engineers are removed from systems (e.g., Newell 1988).

When viewed in total, engineering, a conceptual construct that integrates from the biotic to the abiotic and back again, provides a valuable context for much theoretical and applied thinking. It is essentially an indirect interaction that is decoupled from the biota doing the actual work, and our relatively good understanding of abiotic influences on biotic properties and processes can be used to help predict the effects of engineers. Although we have been slower to appreciate the extent to which biota can complete the loop by affecting the abiotic realm, this is increasingly coming into focus.

Acknowledgements I thank the NCEAS working group on Modelling Ecosystem Engineers: Jeb Byers, Kim Cuddington, Clive Jones, Theresa Talley, John Lambrinos, Alan Hastings, and Will Wilson, for the opportunity to explore and share ideas associated with this topic. Two anonymous reviewers provided valuable comments on the manuscript, and I also thank Gil Rilov for his work on this volume. This publication was supported in part by the National Sea Grant College Program of the U.S. Department of Commerce's NOAA Grant R/CZ 190C through the CA Sea Grant College Program, the San Diego Foundation, and the Western Regional Panel of the Aquatic Nuisance Species Task Force.

References

Allen BJ, Williams SL (2003) Native eelgrass *Zostera marina* controls growth and reproduction of an invasive mussel through food limitation. Mar Ecol Prog Ser 254:57–67

Bertness MD (1984) Habitat and community modification by an introduced herbivorous snail. Ecology 65:370–381

Britton-Simmons KH (2004) Direct and indirect effects of the introduced alga *Sargassum muticum* on benthic subtidal communities of Washington State, USA. Mar Ecol Prog Ser 277:61–78

Brown JH (1995) Organisms as engineers: a useful framework for studying effects on ecosystems? Trends Ecol Evol 10:51–52

Byers JE, Jones CG, Cuddington K, Talley TS, Hastings A, Lambrinos JG, Crooks JA, Wilson WG (2006) Using ecosystem engineers to restore ecological systems. Trends Ecol Evol 23:493–500

Carlton JT (1979) History, biogeography, and ecology of the introduced marine and estuarine invertebrates of the Pacific coast of North America. PhD Thesis, University of California, Davis

Casas G, Scrosati R, Luz Piriz M (2004) The invasive kelp *Undaria pinnatifida* (Phaeophyceae, Laminariales) reduces native seaweed diversity in Nuevo Gulf (Patagonia, Argentina). Biol Invas 6:411–416

Castilla JC, Lagos NA, Cerda M (2004) Marine ecosystem engineering by the alien ascidian *Pyura praeputialis* on a mid-intertidal rocky shore. Mar Ecol Prog Ser 268:119–130

Chapin FS, Reynolds H, D'Antonio CM, Eckhart V (1996) The functional role of species in terrestrial ecosystems. In: Walker B, Steffen W (eds). Global change in terrestrial ecosystems. Cambridge University Press, Cambridge, pp 403–428

Chauvaud L, Thompson JK, Cloern JE, Thouzeau G (2003) Clams as CO_2 generators: the *Potamocorbula amurensis* example in San Francisco Bay. Limnol Oceanogr 48: 2086–2092

Coleman FC, Williams SL (2002) Overexploiting marine ecosystem engineers: potential consequences for biodiversity. Trends Ecol Evol 17:40–44

Creese RSH, DeLuca S, Wharton W (1997) Ecology and environmental impact of *Musculista senhousia* (Mollusca: Bivalvia: Mytilidae) in Tamaki Estuary, Auckland, New Zealand. N Z J Mar Freshwater Res 31:225–236

Crooks JA (1996) The population ecology of an exotic mussel, *Musculista senhousia*, in a southern California bay. Estuaries 19:42–50

Crooks JA (1998) Habitat alteration and community-level effects of an exotic mussel, *Musculista senhousia*. Mar Ecol Prog Ser 162:137–152

Crooks JA (2001) Assessing invader roles within changing ecosystems: historical and experimental perspectives on an exotic mussel in an urbanized lagoon. Biol Invas 3:23–36

Crooks JA (2002) Characterizing ecosystem-level consequences of biological invasions: the role of ecosystem engineers. Oikos 97:153–166

Crooks JA (2005) Lag times and exotic species: the ecology and management of biological invasions in slow-motion. Ecoscience 12:316–329

Crooks JA (2006a) *Musculista senhousia*. In: Boersma PD, Reichard S, Van Buren A (eds) Invasive species of the Pacific northwest. University of Washington Press, Seattle, pp 112–113

Crooks JA (2006b) The arrival, establishment, and integration of an invasive marine mussel into foreign ecosystems. In: Koike F, Clout MN, Kawamichi M, De Poorter M, Iwatsuki K (eds) Proceedings of the International Conference on Assessment and Control of Invasion Risks. Shoukadoh Book Sellers, Kyoto, Japan, and IUCN Press, Gland, Switzerland, pp 113–115

Crooks JA, Khim HS (1999) Architectural vs. biological effects of habitat-altering, exotic mussel, *Musculista senhousia*. J Exp Mar Biol Ecol 240:53–75

Crooks JA, Soulé ME (1999) Lag times in population explosions of invasive species. In: Sandlund OT, Schei PJ, Viken A (eds) Invasive species and biodiversity management. Kluwer Academic Publishers, Dordrecht, The Netherlands, pp 103–125

Cuddington K, Hastings A (2004) Invasive engineers. Ecol Model 178:335–347

Cuddington K, Byers JE, Wilson WG, Hastings A (2007) Ecosystem engineers. Plants to protists. Elsevier, San Diego, USA

Curiel D, Guidetti P, Bellemo G, Scattolin M, Marzocchi M (2001) The introduced alga *Undaria pinnatifida* (Laminariales, Alariceae) in the lagoon of Venice. Hydrobiologia 477:209–219

Daehler CC, Strong DR (1996) Status, prediction and prevention of introduced cordgrass *Spartina* spp. invasions in Pacific estuaries, USA. Biol Conserv 78:51–58

Darrigran G (2002) Potential impact of filter-feeding invaders on temperate inland freshwater environments. Biol Invas 4:145–156

Davies BR, Stuart V, de Villiers M (1989) The filtration activity of a serpulid polychaete population (*Ficopomatus enigmaticus* (Fauvel)) and its effects on water quality in a coastal marina. Estuarine Coastal Shelf Sci 29:613–620

de Vos A, Manville RH, Van Gelder RG (1956) Introduced mammals and their influence on native biota. Zoologica 41:163–194

Demopoulos AWJ (2004) Aliens in paradise: a comparative assessment of introduced and native mangrove benthic community composition, food-web structure, and litter-fall production. PhD Thesis. University of Hawaii, Honolulu, HI

Escapa M, Isaacch JP, Daleo JAP, Iribarne O, Borges M, Dos Santos EP, Gagliardini DA, Lasta M (2004) The distribution and ecological effects of the introduced Pacific oyster *Crassostrea gigas*. J Shellfish Res 23:765–772

Ford MA, Grace JB (1998) Effects of vertebrate herbivores on soil processes, plant biomass, litter accumulation and soil elevational changes in a coastal marsh. J Ecol 86:974–982

Gratton C, Denno RF (2005) Restoration of arthropod assemblages in a *Spartina* salt marsh following removal of the invasive plant *Phragmites australis*. Restor Ecol 13:358–372

Gutierrez JL, Jones CG, Strayer DL, Iribarne O (2003) Mollusks as ecosystem engineers: the role of shell production in aquatic habitats. Oikos 101:79–90

Hallegraeff GM (1998) Transport of toxic dinoflagellates via ships ballast water: bioeconomic risk assessment and efficacy of possible ballast water management strategies. Mar Ecol Prog Ser 168:297–309

Hastings A, Byers JE, Crooks JA, Cuddington K, Jones CG, Lambrinos JG, Talley TS, Wilson WG (2007) Ecosystem engineering in space and time. Ecol Lett 10:153–164

Hedge P, Kriwoken LK (2000) Evidence for effects of *Spartina anglica* invasion on benthic macrofauna in Little Swanport estuary, Tasmania. Aust Ecol 25:150–159

Hewitt CL, Campbell ML, McEnnulty F, Moore KM, Murfet NB, Robertson B, Schaffelke B (2005) Efficacy of physical removal of a marine pest: the introduced kelp *Undaria pinnatifida* in a Tasmanian Marine Reserve. Biol Invas 7:251–263

Holloway MG, Keough MJ (2002) Effects of an introduced polychaete, *Sabella spallanzanii*, on the development of epifaunal assemblages. Mar Ecol Prog Ser 236:137–154

Hopkins CCE (2002) Introduced marine organisms in Norwegian waters, including Svalbard. In: Leppäkoski E, Gollasch S, Olenin S (eds) Invasive aquatic species of Europe. Kluwer Academic Publishers, Dordrecht, The Netherlands, pp 240–252

Jones CG, Gutiérrez JL (2007) On the purpose, meaning, and usage of the physical ecosystem engineering concept. In: Cuddington K, Byers JE, Wilson WG, Hastings A (eds) Ecosystem engineers. Plants to protists. Elsevier, San Diego, USA, pp 3–24

Jones CG, Lawton JH, Shachak M (1994) Organisms as ecosystem engineers. Oikos 689: 373–386

Jones CG, Lawton JH, Shachak M (1997) Positive and negative effects of organisms as physical ecosystem engineers. Ecology 78:1946–1957

Kimmerer WJ, Gartside E, Orsi JJ (1994) Predation by an introduced clam as the likely cause of substantial declines in zooplankton of San Francisco Bay. Mar Ecol Prog Ser 113:81–93

Larned ST (2003) Effects of the invasive, nonindigenous seagrass *Zostera japonica* on nutrient fluxes between the water column and benthos in a NE Pacific estuary. Mar Ecol Prog Ser 254:69–80

Levi F, Francour P (2004) Behavioural response of *Mullus surmuletus* to habitat modification by the invasive macroalga *Caulerpa taxifolia*. J Fish Biol 64:55–64

Mayer CM, Rudstam LG, Mills EL, Cardiff SG, Bloom CA (2001) Zebra mussels (*Dreissena polymorpha*), habitat alteration, and yellow perch (*Perca flavescens*) foraging: system-wide effects on behavioural mechanisms. Can J Fish Aquat Sci 58:2459–2467

Meinesz A (1999) Killer algae: the true tale of biological invasion. University of Chicago Press, Chicago

Mistri M (2004) Effect of *Musculista senhousia* mats on clam mortality and growth: much ado about nothing? Aquaculture 241:207–218

Nalepa TF, Schloesser SW (1993) Zebra mussels. Biology, impacts, and control. Lewis Publishers, Boca Raton, FL

Neideman R, Wenngren J, Olafsson E (2003) Competition between the introduced polychaete *Marenzelleria* sp. and the native amphipod *Monoporeia affinis* in Baltic soft bottoms. Mar Ecol Prog Ser 264:49–55

Neira C, Levin LA, Grosholz ED (2005) Benthic macrofaunal communities of three sites in San Francisco Bay invaded by hybrid Spartina, with comparison to uninvaded habitats. Mar Ecol Prog Ser 292:111–126

Newell RIE (1988) Ecological changes in Chesapeake Bay: are they the result of overharvesting the American oyster, *Crassostrea virginica*. In: Lynch MP, Krome EC (eds) Understanding the estuary: advances in Chesapeake Bay Research. Chesapeake Research Consortium Publication 129, Baltimore, Maryland, pp 536–546

Newell RIE (2004) Ecosystem influences of natural and cultivated populations of suspension-feeding bivavle molluscs: a review. J Shellfish Res 23:51–61

Ojaveer H, Leppäkoski E, Olenin S, Ricciardi A (2002) Ecological impact of Ponto-Caspian invaders in the Baltic Sea, European inland waters and the Great Lakes: an inter-ecosystem comparison. In: Leppäkoski E, Gollasch S, Olenin S (eds) Invasive aquatic species of Europe. Kluwer Academic Publishers, Dordrecht, The Netherlands, pp 412–425

Olenin S, Leppäkoski E (1999) Non-native animals in the Baltic Sea: alteration of benthic habitats in coastal inlets and lagoons. Hydrobiologia 393:233–243

Phelps HL (1994) The Asiatic clam (*Corbicula fluminea*) invasion and system-level ecological change in the Potomac River Estuary near Washington, D.C. Estuaries 17:614–621

Piazzi L, Cinelli F (2000) Effets de l'expansion des Rhodophyceae introduites *Acrothamnion preissii* et *Womersleyella setacea* sur les communautes algales des rhizomes de *Posidonia oceanica* de Metiterranee occidentale. Cryptogamie Algologie 21:291–300

Posey M (1988) Community changes associated with the spread of an introduced seagrass, *Zostera japonica*. Ecology 69:974–983

Posey M, Alphin TD, Meyer DL, Johnson JM (2003) Benthic communities of common reed *Phragmites australis* and marsh cordgrass *Spartina alterniflora* marshes in Chesapeake Bay. Mar Ecol Prog Ser 261:51–61

Power ME (1997) Estimating impacts of a dominant detritivore in a neotropical stream. Trends Ecol Evol 12:47–49

Reusch TBH, Williams SL (1998) Variable responses of native eelgrass *Zostera marina* to a non-indigenous bivalve *Musculista senhousia*. Oecologia 113:428–441

Rudnick DA, Hieb K, Grimmer KF, Resh VH (2003) Patterns and processes of biological invasion: The Chinese mitten crab in San Francisco Bay. Basic Appl Ecol 4:249–262

Ruiz GM, Fofonoff P, Hines AH (1999) Non-indigenous species as stressors in estuarine and marine communities: assessing invasion impacts and interactions. Limnol Oceanogr 44:950–972

Salmon M, Reiners R, Lavin C, Wyneken J (1995) Behavior of loggerhead sea turtles on an urban beach. I. Correlates of nest placement. J Herpetol 29:560–567

Sax DF, Kinlan BP, Smith KF (2005) A conceptual framework for comparing species assemblages in native and exotic habitats. Oikos 108:457–464

Schwindt E, Bortolus A, Iribarne O (2002) Invasion of a reef-builder polychaete: direct and indirect impacts on the native benthic community structure. Biol Invas 3:137–149

Schwindt E, Iribarne OO, Isla FI (2004) Physical effects of an invading reef-building polychaete on an Argentinean estuarine environment. Estuarine Coastal Shelf Sci 59(1):109–120

Spencer CN, McClelland BR, Stanford JA (1991) Shrimp stocking, salmon collapse, and eagle displacement. Cascading interactions in the food web of a large aquatic ecosystem. BioScience 41:14–21

Staehr PA, Pedersen MF, Thomsen MS, Wernberg T, Krause-Jensen D (2000) Invasion of *Sargassum muticum* in Limfjorden (Denmark) and its possible impact on the indigenous macroalgal community. Mar Ecol Prog Ser 207:79–88

Steneck RS, Carlton JT (2001) Human alterations of marine communities: students beware! In: Bertness M, Gaines S, Hay M (eds) Marine community ecology. Sinauer, Sunderland, MA, pp 455–468

Stewart TW, Haynes JM (1994) Benthic macroinvertebrate communities of southwestern Lake Ontario following invasion of *Dreissena*. J Great Lakes Res 20:479–493

Strayer DL, Caraco NF, Cole JJ, Findlay S, Pace ML (1999) Transformation of freshwater ecosystems by bivalves. Bioscience 49:19–27

Talley TS, Crooks JA (2007) Habitat conversion associated with bioeroding marine isopods. In: Cuddington K, Byers JE, Wilson WG, Hastings A (eds) Ecosystem engineers. Plants to protists. Elsevier, San Diego, USA, pp 185–202

Talley TS, Crooks JA, Levin LA (2001) Habitat utilization and alteration by the burrowing isopod *Sphaeroma quoyanum* in California salt marshes. Mar Biol 138:561–573

Thieltges DW (2005) Impact of an invader: epizootic American slipper limpet *Crepidula fornicata* reduces survival and growth in European mussels. Mar Ecol Prog Ser 286:13–19

van den Brink FWB, van der Velde G, bij de Vaate A (1993) Ecological aspects, explosive range extension and impact of a mass invader, *Corophium curvispinum* Sars, 1895 (Crustacea: Amphipoda), in the Lower Rhine (The Netherlands). Oecologia 93:224–232

Van Dolah FM (2000) Marine algal toxins: origins, health effects, and their increased occurrence. Environ Health Perspect 108(Suppl 1):133–141

Vander Zanden MJ, Casselman JM, Rasmussen JB (1999) Stable isotope evidence for the food web consequences of species invasions in lakes. Nature 401:464–467

Viejo RM (1999) Mobile epifauna inhabiting the invasive *Sargassum muticum* and two local seaweeds in northern Spain. Aquat Bot 64:131–149

Vitousek PM (1990) Biological invasions and ecosystem processes: towards an integration of population biology and ecosystem studies. Oikos 57:7–13

Vitousek PM, Walker LR, Whiteaker LD, Mueller-Dombois D, Matson PA (1987) Biological invasion by *Myrica faya* alters ecosystem development in Hawaii. Science 238:802–804

Wallentinus I (2002) Introduced marine algae and vascular plants in European aquatic environments. In: Leppäkoski E, Gollasch S, Olenin S (eds) Invasive aquatic species of Europe. Distribution, impacts, and management. Kluwer Academic Publishers, Dordrecht, The Netherlands, pp 27–52

Wallentinus I, Nyberg CD (2005) Introduced marine organisms as habitat modifiers. Mar Pollut Bull 55:323–332

Weis JS, Weis P (2003) Is the invasion of the common reed, *Phragmites australis*, into tidal marshes of the eastern US an ecological disaster? Mar Pollut Bull 46:816–820

Wernberg T, Thomsen MS, Staehr PA, Pedersen MF (2004) Epibiota communities of the introduced and indigenous macroalgal relatives *Sargassum muticum* and *Halidrys siliquosa* in Limfjorden (Denmark). Helgol Mar Res 58:154–161

Whitcraft CR, Talley DM, Crooks JA, Boland J, Gaskin J (2007) Invasion of tamarisk (*Tamarix* spp.) in a southern California salt marsh. Biol Invas 9:875–879

Wikstrom SA, Kautsky L (2004) Invasion of a habitat-forming seaweed: effects on associated biota. Biol Invas 6:141–150

Williamson M (1996) Biological invasions. Chapman and Hall, London

Wonham MJ, O'Connor M, Harley CDG (2005) Positive effects of a dominant invader on introduced and native mudflat species. Mar Ecol Prog Ser 289:109–116

Wright JP, Jones CG, Flecker AS (2002) An ecosystem engineer, the beaver, increases species richness at the landscape scale. Oecologia 132:96–101

Zmudzinski L (1996) The effect of the introduction of the American species *Marenzelleria viridis* (Polcyhaeta: Spionidae) on the benthic ecosystem of Vistula lagoon. Mar Ecol 17:221–226

Chapter 17
Multitrophic Effects of Invasions in Marine and Estuarine Systems

Edwin D. Grosholz and Gregory M. Ruiz

17.1 Introduction

Invasions in marine and estuarine systems are no longer news to the coastal ecologists who have been documenting the numerous introduced species appearing in coastal systems over the last several decades. We are now faced with a dramatically altered benthic landscape that includes a diversity of both native and non-native species. As ecologists, we seek to understand not only how these altered systems function, but predict future changes despite the fact that the species assemblage is continually shifting.

In a background of seemingly constant change, one of the top priorities for ecologists is to understand why some introductions result in large changes to native communities, changes that alter ecosystem function or services and that have large impacts on the natural and/or human sphere (Parker et al. 1999; Ruiz et al. 1999; Grosholz et al. 2000). We wish to understand what the consequences of invasions are and how we can learn from these to predict future impacts. Why is it that some invasions, perhaps not many, cause dramatic changes, whereas others appear to result in comparatively little change?

This chapter focuses on invasions that have resulted in measurable changes at multiple trophic levels in marine and estuarine systems. We begin by examining multiple trophic level changes in response to invasion in several ways. First, we discuss the evidence for multitrophic level changes resulting from species introductions. Second, we provide some illustrative examples of multispecies impacts on non-native species from our own work in Pacific estuaries. This focuses particular attention on an ecosystem engineer (see also Chap. 16, Crooks), including effects on ecosystem processes as well as food webs, and also major predator. Third, we compare changes from an introduction to those resulting from natural disturbances. Using concepts from disturbance theory, we suggest a framework to better describe the circumstances under which we might expect invasions to result in changes across multiple trophic levels.

Ultimately, we conclude this investigation by identifying common features of invasions that have produced changes across multiple trophic levels. We do this in

G. Rilov, J.A. Crooks (eds.) *Biological Invasions in Marine Ecosystems*.
Ecological Studies 204,

the hope that this may help prioritize strategies and efforts, by identifying invasion pathways that deliver high-impact species and the targets for eradication, control or other management intervention in the event that these species do become introduced.

17.2 Multitrophic Impacts

Whether introduced species create changes that rebound across multiple levels is part of a larger debate about the generality of multitrophic level processes, such as trophic cascades (Carpenter et al. 1985; Carpenter and Kitchell 1993). A trophic cascade is traditionally defined as a system of three or more trophic levels in which consumers at higher trophic levels, by reducing consumers at intermediate trophic levels, produce increases in biomass of organisms at lower trophic levels, typically primary producers. We follow earlier authors by defining trophic cascades as "community cascades," rather than the more ubiquitous "species cascades," which are simply changes among species within a trophic level without broad scale changes across the entire trophic level (Polis 1999).

Trophic cascades appear to be common in aquatic systems although by no means ubiquitous in nature (Strong 1992; Pace et al. 1999; Polis et al. 2000; Shurin et al. 2002). Several trophic cascades have been documented in marine systems showing stronger top down control than in most other systems (Shurin et al. 2002). Most trophic cascades in marine systems have involved urchins grazing macroalgae (including kelps) on hard substrate (Shurin et al. 2002), although experiments involving other grazers and other systems including mesocosms suggest that trophic cascades may be more common in benthic marine systems (Bruno and O'Connor 2005; Duffy et al. 2005; Byrnes et al. 2006). Given the relative rarity of well-documented trophic cascades in marine systems (Shurin et al. 2002; Silliman and Bertness 2002), we consider a much broader range of multitrophic level processes that may result from species introductions.

There are many examples of introduced species where impacts on multiple trophic levels have been reasonably demonstrated (Table 17.1). These include species whose effects are mediated through trophic impacts (see also Chap. 15, Rilov), such as filter feeding bivalve molluscs *Corbula amurensis, Corbicula fluminea,* and *Dreissena* spp. that consume water column primary producers and consumers. Also included are species that act through non-trophic mechanisms and affect habitat structure, causing changes at many trophic levels (see also Chap. 16, Crooks). Often referred to as ecosystem engineers (Jones et al. 1994, 1997; see Chap. 18, Hewitt et al.), these species have been shown to influence ecosystem processes in a variety of ways (Vitousek 1990; Williamson 1996; Parker et al. 1999; Crooks 2002). Ecosystem engineers in aquatic systems can influence light availability, alter wave motion and water flow, modify sediments and influence the storage, availability and recycling of nutrients. They can also result in changes across many trophic levels by increasing or decreasing habitat complexity (Crooks 2002). Examples of

Table 17.1 Examples of multitrophic impacts of invading species in marine and estuarine environments

Invasive species	Location	Trophic level	Trophic groups affected	Reported effects	References
Phragmites australis –Common reed	New Jersey estuaries	Primary producer	Benthic macroalgae, benthic invertebrates, omnivorous fish	Shading, decreased benthic microalgal biomass, changes in benthic invertebrate abundance, herbivorous insect habitat, altered marsh topography and hydroperiod, decreased abundance of *Fundulus heteroclitus*	Angradi et al. (2001), Able and Hagen (2003), Currin et al. (2003), Raichel (2003), Gratton and Denno (2005)
Mnemiopsis leidyi – Ctenophore	Black Sea, Caspian Sea	Planktonic predator	Phytoplankton, herbivorous zooplankton, planktivorus fishes	Reduction of mesozooplankton abundance, increased phytoplankton biomass, reduction of planktivorous fish abundance	Shiganova (1998, 2001), Kideys (2002), Bilio and Nierman (2004)
Beroe ovata – Ctenophore	Black Sea	Planktonic predator (feeds only on *Mnemiopsis*)	Predatory zooplankton (*Mnemiopsis*) herbivorous zooplankton, planktivorous fishes	Reduction of *Mnemiopis* abundance, increased mesozooplankton biomass, reduction of planktonic fish abundance, increased planktivorous abundance	Shiganova (2001), Kideys (2002)
Ficopomatus engimaticus – polychaete tubeworm	Coastal lagoons, Argentina	Suspension feeder (primarily phytoplankton)	Predatory crabs, deposit-feeding polychaetes	Calcareous *Ficopomatus* reefs add structure to mudflats, attracting predatory crabs (*Cyrtograpsus angulatus*), resulting in a reduction of soft-bottom polychaetes	Schwindt (2001)
Corbula amurensis – Asian brackishwater clam	San Francisco Bay	Suspension feeder (primarily phytoplankton)	Phytoplankton, zooplankton, native bivalves, deposit-feeding macrobenthos, fishes, molluscivorous waterfowl	Decreased phytoplankton biomass, direct predation on zooplankton, decreased abundance of most zooplankton, increased water clarity, general decrease in fish abundance, growth, and gut fulness, valuable food item for molluscivorous waterfowl, but a source of contaminants	Alpine and Cloern (1992), Kimmerer et al. (1994), Feyrer et al. (2003), Richman and Lovvorn (2004)

(continued)

Table 17.1 (continued)

Invasive species	Location	Trophic level	Trophic groups affected	Reported effects	References
Corbicula fluminea – Asian freshwater clam	Potomac River estuary	Suspension feeder (primarily phytoplankton)	Phytoplankton, submerged vegetation, small planktivorous-benthivorous fishes, predatory fishes, waterfowl	Decreased phytoplankton biomass, increased water clarity, recovery of submerged vegetation, increased abundance of small fishes and large predatory fishes, increased abundance of herbivorous, molluscivorous, and piscivorous waterfowl	Cohen et al. (1984), Killgore et al. (1989), Phelps (1994)
Invasive species	Location	Trophic level	Trophic groups affected	Reported effects	References
Dreissena polymorpha – Zebra mussel	Hudson River estuary	Suspension feeder (primarily phytoplankton)	Phytoplankton, zooplankton, native bivalves, deposit-feeding macrobenthos	Decreased phytoplankton biomass, decreased abundance of some zooplankton (*Bosmina* spp.), increased water clarity, recovery of submerged vegetation, increased abundance in nearshore fishes, decreased abundance of open-water fishes	Strayer et al. (1999, 2004)
Musculista senhousia – Senhouse's date mussel	Mission Bay, California	Suspension feeder (primarily phytoplankton)	Seagrasses, native bivalves, deposit-feeding macrobenthos	Replacing seagrass (*Zostera marina*) beds, creating mats on the surface of mudflats, providing habitat for macrobenthos, especially surface-feeders, competing with native bivalves	Crooks (1998), Dexter and Crooks (2000), Williams et al. (2005)
Cercopagis pengoi – Fishook cladoceran	Eastern Baltic Sea	Planktonic predator	Herbivorous zooplankton, planktivorous fishes	Reduction of *Bosmina maritima* populations, changes in seasonal dynamics of copepods. Alteration of planktivorous fish diets (*Clupea haerengus, Sprattus sprattus, Osmerus eperlanus*), food overlap with planktivorous fishes and their larvae	Pollumae and Valjataga (2004), Ojaveer et al. (2004), Gorokhova et al. (2005)
Carcinus maenas – Green crab	Nahant, Massachusetts Bay	Predator	Herbivorous snails, macroalgae	Free-roaming *Carcinus maenas* or caged crabs, producing chemical cues, resulted in emigration of *Littorina littorea* from tidepools, resulting in reduced grazing and increased growth of green algae (*Ulva, Enteromorpha*)	Trussell et al. (2004)

such ecosystem engineers in Table 17.1 include the emergent marsh plants *Phragmites australis*, reef-forming polychaetes *Ficopomatus enigmaticus*, and mat-forming mussels *Musculista senhousia*.

Table 17.1 focuses primarily on predation and habitat alteration, where effects are perhaps best documented, but this is not intended as a comprehensive list. Certainly there are other cases within these categories as well as other modes of interaction where non-native species may affect multiple trophic levels. For example, changes brought about by a bryozoan (*Membranipora membranacea*) in the southern Gulf of Maine (Levin PS et al. 2002), in concert with a newly introduced alga *Codium fragile* ssp. *tomentosoides*, has reduced the abundance of kelps (*Laminaria* spp.) with negative implications for fishes dependent on kelp habitats. The introduced gastropod *Littorina littorea* is presumed to have changed marginal soft-sediment habitats to rocky, cobble coastlines (Bertness 1984). Additionally, parasites like digenetic trematodes utilize multiple host species and trophic levels, where they may have effects. It is thought that the competitive displacement of the native California mud snail *Cerithidea californica* by the introduced mud snail *Batillaria attramentaria* could affect multiple trophic levels (Byers 2000; Torchin et al. 2005). The changes in snail species may have affects on both lower and higher trophic levels. For the latter, the native snail is host to 17 species of larval digenetic trematodes (Sousa 1984), but the introduced species is host to only one morphospecies (Torchin et al. 2005). The trematodes also use a variety of vertebrates (including fish, birds, and mammals) as definitive hosts, but effects at these higher trophic levels have not been demonstrated.

As we go beyond the functional identity of the species in Table 17.1, it is evident that species from a wide range of taxonomic groups and habitat types can produce change across trophic levels in estuarine and marine food webs. Our list highlights plants, ctenophores, cladocerans, polychaetes, clams, mussels, crabs among others. It also includes examples from salt marshes, rocky shores, oligohaline estuaries, coastal lagoons, and inland seas.

Thus, we surmise that the potential scope for multi-trophic impacts associated with non-native species is indeed great, but our understanding of extent (frequency) and circumstances for such effects is still limited to a few well-studied examples. For the vast majority of invasions, few quantitative data exist to evaluate direct and indirect effects (Ruiz et al. 1999). Although many invasions are associated with conspicuous changes, it is frequently difficult to discern cause-effect relationships in the absence of explicit studies. This is especially challenging due to many confounding factors in coastal systems, which are rapidly changing and subject to multiple stresses from eutrophication, overexploitation, sedimentation, contaminant inputs, and habitat alteration. For example, with *Corbicula* in the Potomac River, reduction of nutrient inputs roughly coincided with the *Corbicula* introduction and therefore both factors likely played a role in the increasing water clarity and cover of submerged aquatic vegetation (SAV) (Phelps 1994). For these and many observed changes coincident with other invasions, the relative importance of the invasions compared with other stressors has yet to be evaluated.

In the following sections, we highlight two examples in more detail, drawing upon our own research. These are intended as illustrative case histories, one for an ecosystem engineer and the other for a predator, in which there exists considerable quantitative data to evaluate invasion effects at multiple trophic levels.

17.3 Case History #1: The Ecosystem Engineer *Spartina* in West Coast Estuaries

We examine the impacts of invasive *Spartina* on various trophic levels in two estuaries in western North America: San Francisco Bay, CA and Willapa Bay, WA. *Spartina* acts as an ecosystem engineer through extensive modification of the above and below ground environment, which produces predictable physical and chemical changes. The total effects on community and ecosystem processes are a function of the magnitude of change in a range of properties including vascular plant biomass both above-ground and below-ground, the rates of production and decomposition of detritus, photosynthesis and productivity of microalgae, and the biogeochemistry of the surrounding sediments. All of these processes can strongly affect the abundances of organisms at several different trophic levels (Fig. 17.1).

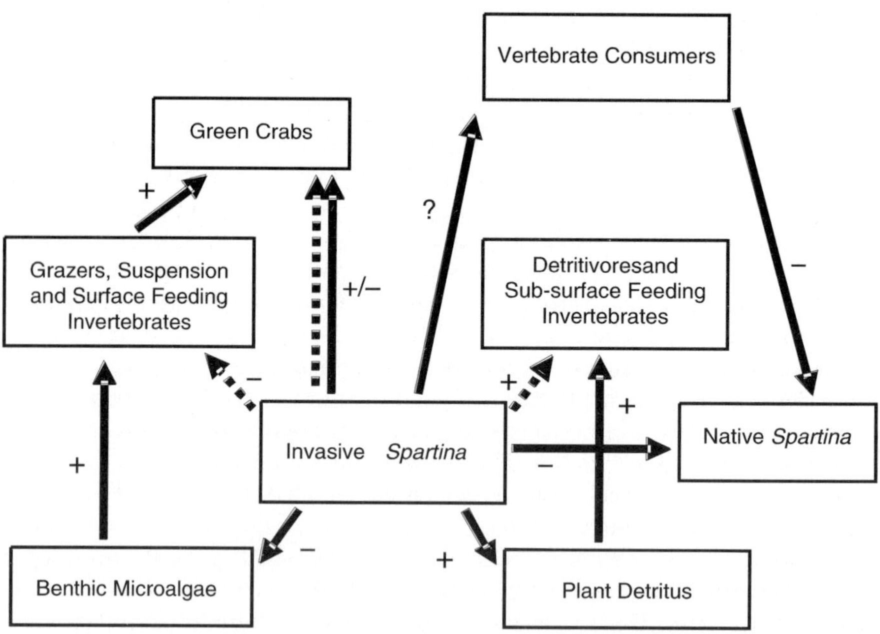

Fig. 17.1 San Francisco Bay food web showing the impacts of the *Spartina* hybrid (*S. alterniflora* × *S. foliosa*) on different trophic levels. Direct effects are indicated by *solid black lines* and indirect effects are indicated by *dashed black lines*

The invasions of Willapa Bay and San Francisco Bay by *Spartina alterniflora* have very different histories but very similar impacts. The invasion of *Spartina alterniflora* in San Francisco Bay began in 1975 as an intentional introduction from the eastern U.S. by the Army Corp of Engineers for salt marsh restoration (Ayres et al. 2003, 2004). Then *S. alterniflora* hybridized with the native *S. foliosa* sometime after the initial introduction (Daehler and Strong 1997). The hybrid form has displaced both parent species and has a broader tidal range than other native plants (Ayres et al. 2003, 2004). In San Francisco Bay, hybrid *Spartina* has successfully colonized many hundreds of hectares in the central and southern portions of the bay, where it has invaded previous unvegetated habitat at the lower end of its tidal distribution and has excluded native plants towards the upper end of its distribution. In Willapa Bay, *Spartina alterniflora* was accidentally introduced around 1890 (Feist and Simenstad 2000; Davis et al. 2004; Civille et al. 2005) and now occupies more than 1500 ha throughout the bay. There is no native *Spartina* in Willapa Bay; thus, no hybridization occurred. However, just as in San Francisco Bay, *Spartina* is colonizing open mud flat transforming a historically unvegetated area to a densely vegetated habitat.

Despite the differences in the circumstances of the two invasions, the consequences for the food webs in the two sites have been similar, so we discuss the impacts generally for invasive *Spartina* in the two bays. The most important impacts across trophic levels occur as a consequence of the large amount of above ground biomass produced by invasive *Spartina*. The *Spartina* canopy significantly reduces light penetration, restricts water flow, and produces greater sediment accretion compared with the surrounding unvegetated habitat (Neira et al. 2005, 2006). The higher rate of sediment accretion together with the buildup of below ground biomass and peat also increases the tidal elevation of invaded areas. Unvegetated sites in areas of even moderate flow experience changes in elevation of many centimeters over periods of in some cases just a few weeks in comparison with very gradual elevation changes in adjacent vegetated areas (E. D. Grosholz, unpublished data).

The reduction in light reaching the sediment surface can significantly affect photosynthesis of benthic microalgae. In west coast estuaries, benthic microalgal production provides a substantial proportion of overall primary production in the system (Deegan and Garritt 1997; Kwak and Zedler 1997; Page HM 1997). *Spartina* consistently reduces microalgal primary productivity relative to open mudflats in San Francisco Bay (Grosholz et al., in press, A. C. Tyler et al., in preparation). Overall, the *Spartina* invasion reduced microalgal productivity with important implications for higher trophic levels.

Spartina also reduced the biomass and diversity of infaunal invertebrates (Fig. 17.1). In some sites in San Francisco Bay, invertebrate densities declined by as much 75% relative to unvegetated mudflat (Neira et al. 2005). Species richness also showed a significant 25% decline in *Spartina* invaded areas compared with unvegetated areas (Neira et al. 2005, 2006; Levin LA et al. 2006; Grosholz et al., in press) and with native *Spartina foliosa* areas (Brusati and Grosholz 2006).

Similarly, in Willapa Bay, invertebrate species richness was lower in *Spartina alterniflora* areas (Grosholz et al., in press).

In addition, these experimental studies demonstrated that the changed infaunal density and diversity were the result of the combined effects of reduced survival associated with the preemption of substantial amounts of below ground habitat by *Spartina*, changes in the food supply and predation pressure as well as the physical and chemical changes to sediments and porewater (Neira et al 2005, 2006; Levin LA et al. 2006; Grosholz et al., in press). The hybrid is affecting predatory epifaunal decapods as well. Data from San Francisco Bay show that the European green crab is three to five times more abundant in the hybrid *Spartina*, particularly smaller size classes, compared with the adjacent mudflat (Neira et al. 2006). Predatory green crabs, in turn, have a significant impact on the invertebrate taxa that are negatively affected by *Spartina* invasion. Therefore, their changing abundance may reinforce the impacts of the invasion (Neira et al. 2006). Data for Willapa Bay suggest a similar pattern with *Spartina* facilitating younger size classes of green crabs (Yamada 2001) (Fig. 17.1). Furthermore, Neira et al. (2006) experimentally demonstrated that green crab foraging also had impacts on lower trophic levels by reducing levels of chl *a* and increasing sediment organic matter (see section on Green Crabs above).

The structure of benthic food webs has also shifted following *Spartina* invasion. Based on experiments in both San Francisco Bay and Willapa Bay that traced the fate of isotopically labeled *Spartina* detritus and microalgae, the *Spartina* invasion has resulted in a shift from an infaunal invertebrate community dominated by omnivores and surface feeders (bivalves, amphipods) that consume benthic microalgae, to a community dominated by below ground feeders that primarily consume plant detritus (capitellid polychaetes, oligochaetes) (Levin LA et al. 2006; Grosholz et al., in press) (see Fig. 17.1).

The invasion of west coast estuaries by *Spartina* may also have important implications for wintering populations of migratory shorebirds. Many shorebirds require open mudflats for foraging and will not use vegetated areas (Page GW et al. 1999; Stenzel et al. 2002). So when *Spartina* invades normally unvegetated mudflats, this is a de facto loss of foraging habitat for some bird species. Also, existing data show that the biomass of invertebrates is greater on tidal flat areas at higher tidal elevations (Christiansen et al., in review). Therefore, as the *Spartina* invasion expands to occupy a greater portion of these higher elevation areas, shorebirds will be increasingly forced to forage at lower tidal elevations, which are not only exposed for shorter periods of time, but also contain lower densities of invertebrate prey.

In summary, the impacts of invasive *Spartina* highlight a range of impacts that ecosystem engineers can have in estuarine and marine systems. This invasive plant changes many aspects of the physical environment including light, water flow, sediment accretion, detrital accumulation, and biogeochemical processes. The outcome of these extensive changes is significant changes in primary producers, detrivores, herbivores, omnivores, as well as secondary and top consumers. Invasive plants like this represent some of the most important threats to maintaining native assemblages and the ecosystem services provided by native communities.

17.4 Case History #2: Contrasting Trophic Effects of a Non-Native and Native Predator in a Central California Estuary

The European crab *Carcinus maenas* is one of the most widely introduced estuarine species, having established populations in North America, Africa, Asia and Australia (Yamada 2001) and most recently South America (Hildago et al. 2005). In western North America, it was first introduced in San Francisco Bay around 1989 and has colonized virtually all major bays and estuaries from Monterey Bay, CA to Gray's Harbor, WA (Grosholz et al. 2000; Yamada 2001).

The reserve of the University of California Bodega Marine Laboratory has been invaded by green crabs since 1994 (Grosholz and Ruiz 1995). The reserve is part of a 2-km^2 harbor that is a largely marine embayment with mostly sandy mud substrate and limited freshwater input except during winter storm events. Benthic invertebrates have been studied intermittently in Bodega Harbor over the past nearly 50 years (see Grosholz 2005). For the past 20 years, the distribution and abundance of at least two dozen species of invertebrates have been tracked annually (Grosholz et al. 2000). Annual census data for a dozen species of shorebirds has also been collected in the harbor during this period. We used these long term data together with field and lab experiments to assess the impact of green crabs on three functional groups: (1) infaunal suspension- and deposit-feeding molluscs and polychaetes, (2) epifaunal decapod crustaceans, and (3) wintering shorebirds.

Populations of native shore crabs *Hemigraspus oregonensis* and small native clams *Nutricola confusa* and *N. tantilla* declined by 80–90% over a three year period, following the invasion in 1994 (Grosholz et al. 2000; Grosholz 2005; see Chap. 15, Rilov). Other invertebrate taxa also declined including gammaridean amphipods, while at the same time taxa such as tanaids and syllid and spionid polychaetes increased in abundance. The taxa that declined most strongly were the ones selectively preyed upon by green crabs in lab and field trials (Grosholz et al. 2000). By contrast, populations of wintering shorebirds remained statistically unchanged during this period (Grosholz et al. 2000) (Fig. 17.2). We also found no evidence for the birds choosing habitats within the harbor that were less affected by green crabs.

Green crabs also appear to have influenced primary production. Manipulative experimental studies of green crabs conducted in nearby San Francisco Bay have demonstrated reductions in chlorophyll *a* and increases in total sediment organics in the presence of green crabs (Neira et al. 2006). In Bodega Harbor, using a time series of aerial photos over a 10-year period since the green crab invasion, the cover of macroalgae (*Ulva* spp. and *Graciliariopsis sjoestedtii*) increased significantly in areas at lower tidal heights (E. D. Grosholz and G. M. Ruiz, unpublished data).

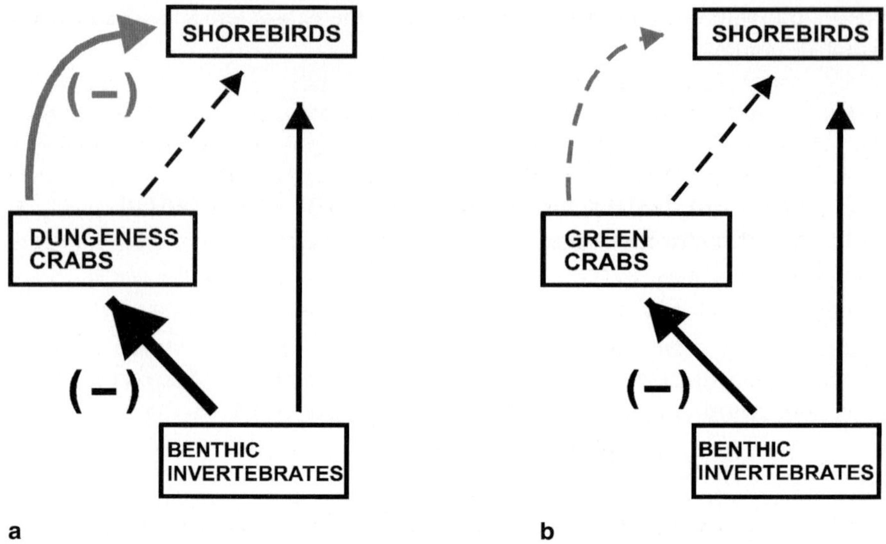

Fig. 17.2a,b Bodega Harbor food web showing the impacts of: **a** Dungeness crabs; **b** European green crabs on different trophic levels. Direct effects are indicated by *black straight lines* and indirect effects are indicated by *gray curved lines*. Significant interactions are indicated by *solid lines* and insignificant interactions are indicated by *dashed lines*

Ten years before the arrival of *C. maenas*, Bodega Harbor was colonized in force by an extremely large number of the juvenile Dungeness crab *Cancer magister*. Although native to central California, this species shows enormous variation in recruitment, with the crab being rare to undetectable in most years within Bodega Harbor. In 1985, postlarval Dungeness crabs occurred in excess of 4000 crabs per pitfall trap per day, and the harborwide population was estimated in excess of 100 million crabs (Ruiz 1987; Everett 1989). This cohort of crabs was gone within one year, moving to deeper water along the outer coast, and comparable trapping efforts in most years yield no crabs. An event of this magnitude appears to occur less than once every 20 or 30 years (McConnaughey et al. 1992). Such unusual recruitment events are likely driven by winds and currents outside of the Bodega Harbor ecosystem (McConnaughey et al. 1992; Wing et al. 1995).

Immediately following this massive and sudden recruitment of Dungeness crabs, the abundances of many invertebrates plummeted several orders of magnitude. The largest effects were documented for the same species of clams and native crabs as affected by the subsequent European crab invasion. Following the invertebrate declines was a significant drop in the abundance of shorebird in Bodega Harbor (Fig. 17.2). Banding data for birds such as dunlin (*Calidris alpina*) support the conclusion that shorebirds dispersed to nearby bays that had not been affected by mass recruitment of Dungeness crabs (Ruiz 1987). Shorebird dispersal was attributed to food limitation, resulting from direct effects of Dungeness crab predation on primary shorebird prey species such as clams (*Nutricola* spp.) (Ruiz 1987; Ruiz et al. 1989).

The impacts of the Dungeness crab recruitment event shows some striking similarities and stark contrasts to the green crab invasion. Both species caused a dramatic reduction in the abundances of benthic invertebrates in Bodega Harbor. Declines in the most preferred prey such as bivalves, small crabs and amphipods were significant in both events. However, the rapid decline in invertebrate prey following the native Dungeness crab recruitment was followed by a corresponding decline in shorebird populations. In contrast, shorebirds did not decline following the green crab invasion (Fig. 17.2).

17.5 Invasions as Disturbance: A Framework for Understanding Multitrophic-level Impacts

The overall goal of this chapter is to describe and understand when, where and why species introductions in coastal systems do or do not result in changes at multiple trophic levels. For species that produce trophic impacts, it is important to understand the impact that invasion has on these systems and how they respond. Here we consider invasions as a type of disturbance, which can be characterized by a suite of attributes with varying degrees of impact. Our objective is to apply the conceptual framework established for natural disturbance to invasions, in order to develop more accurate predictions regarding the impacts of invasions across trophic levels.

Ecologists have been studying naturally occurring disturbances in coastal systems for much longer than they have been studying invasions (Dayton 1971; Sousa 1979, 1984; Connell et al. 1997). A substantial body of theory has developed around disturbance in natural systems and much of this has emphasized how spatial and temporal characteristics of disturbances can influence their impacts (Sousa 1984, 2001; Pickett and White 1985). Characteristics of the recipient system are also important, but for the purposes of comparison, we will focus on features of disturbance such as the magnitude, duration, frequency and spatial extent that have been shown to be important in determining the impacts of a given disturbance (Bender et al. 1984; Sousa 1984, 2001; Pickett and White 1985).

To argue that invasion can fairly be considered as a disturbance, it important to clarify the definition of disturbance including physical and biological disturbance. Physical disturbance includes forces such as hurricane winds, storm waves, sediment accumulation, drifting logs, ice scour, freshwater flooding, and anoxia (Pickett and White 1985; McGuiness 1987; Keough and Quinn 1998; Sousa 2001). Biological disturbance can result from accumulations of algal or salt marsh plant wrack, whiplash by algae, bioturbation, allelopathy, and red tides. Forces such as insect outbreaks and disease outbreaks are recognized as biological disturbances in various systems (Pickett and White 1985). In addition, agents of disturbance such as sedimentation, flooding, euthrophication, logs, anoxia, and temperature events can be significantly influenced or mediated by human activities.

As with other forms of disturbance, biological invasions result in discrete events, each of which has a particular mode of action, magnitude, and spatial and temporal scale. This recognizes that biological invasions represent a continuum of disturbance depending on the invasion. For example, there could be an initial invasion event with continued low effects and periodic fluctuations. The magnitude of these fluctuations may be extreme, as with the "outbreaks" that occur at low frequency for the Chinese mitten crab (Rudnick et al. 2003). There could also be an initial event with a modest duration after which the invader fails to become established such as *Tritonia* in Massachusetts (Allmon and Sebens 1988), possibly resulting in ephemeral effects. Alternatively, there could be an initial event with a sustained population and continued high persistent effects, as with European green crabs in California (Grosholz et al. 2000; Grosholz 2005). As with the temporal dynamics, invasions also vary widely in spatial extent.

In general, we wish to explore such a disturbance framework for understanding the circumstances under which invasions will or will not result in changes at multiple trophic levels. However, we also suggest that this approach may allow comparisons with other forms of disturbance, since in most coastal systems changes caused by species invasions occur against a background of other human mediated stressors and natural disturbances. In some situations, introduced species invasions may differ qualitatively from natural disturbances in ways that may allow us to understand and predict their impacts. For instance, we know that many invading species show delays in their rate of population growth and spread after they first get established in the new range (Lubina and Levin 1988; Andow et al. 1990; Hastings 1996). Therefore, the impacts of many of these invaders may not be immediate, but may require several years before the impacts on native species are acute. As we argue below, delays in the build up of impacts on native species may determine whether the impacts will extend across multiple trophic levels.

We illustrate this approach by comparing the invasion of the European green crab (above) with the large recruitment of the native Dungeness crab. As discussed in the previous section, both events occurred in Bodega Harbor, affecting the same soft-sediment community but in different ways. Although *C. magister* is native, it exhibits extreme variation in abundance, representing an event perhaps analogous to an invasion event or outbreak of a species like the mitten crab discussed above. Other unusual recruitment events, for example the Crown-of-Thorns sea star and the damage it caused on Pacific coral reefs, have also represented a tremendous disturbance (Moran 1986). Thus, for this same community, we are able to examine two separate events, using a disturbance framework.

The Dungeness crab disturbance was of short "duration" (Fig. 17.3a), with juvenile crabs emigrating to deeper water and out of the harbor within one year, and these crabs have been rare to absent since 1985. In contrast, green crabs have been conspicuously and continuously present each year since 1994 (>10 years). In this sense, the green crab invasion represents a long-term "press" disturbance, while the Dungeness recruitment represents a short-term "pulse" disturbance (Bender et al. 1984; Attayde and Hansson 2001). The differences in duration of the two disturbances (how long the disturbance persisted for) appears not to be a factor in

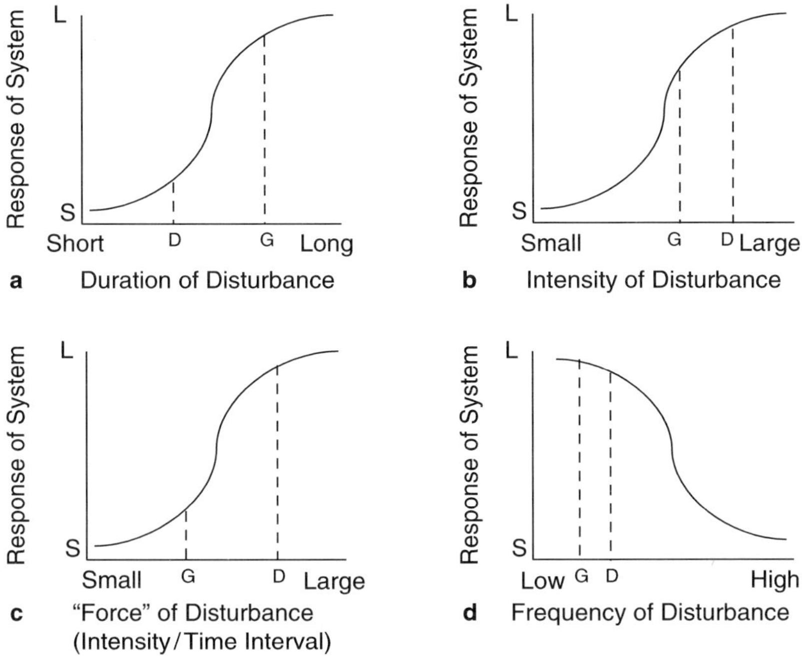

Fig. 17.3. a Plot of duration of the disturbance vs response of the system. For all plots, the shapes of the lines are arbitrary and are meant to describe a generalized positive relationship. Also throughout, D refers to Dungeness crabs and G refers to green crabs. **b** Plot of magnitude or "intensity" of the disturbance vs the response of the system. **c** Plot of the "force" of the disturbance vs response of the system. Force is defined as the disturbance intensity divided by the time interval required to reach the maximum disturbance intensity. **d** Plot of frequency of disturbance vs response of the system

determining the impacts on shorebirds, since the short-lived disturbance (Dungeness crabs) resulted in significant impacts at higher trophic levels, while the long-term disturbance (green crabs) did not (Fig. 17.3a).

We suggest that two factors contributed to the divergent impacts on higher trophic levels (shorebirds). The first factor was the difference in the "intensity" of the two disturbances (Fig. 17.3b). Intensity is typically defined as the strength of the disturbing force (e.g. wind speed in the case of a hurricane) as distinct from the "severity" of a disturbance, which defines the degree of damage or impact (direct effects) on the organism, community, or ecosystem (Sousa 1984, 2001; Pickett and White 1985). For invasions, this may relate to the abundance and mode of action of the organism. In this case, both crabs had a similar mode of action, preying on the same community. However, the density of Dungeness crab juveniles was three orders of magnitude greater than the density of green crabs at the peak of each disturbance. This greater intensity clearly contributed to the different impacts; however, the more moderate predation intensity of green crabs was still sufficient to reduce the abundance of shorebird prey by more than 90%. Therefore, there must be other factors contributing to the higher trophic level impacts.

The time scale of the two events suggests a second factor also contributed to the different impacts on shorebirds, which we define as the "force" of the disturbance (Fig. 17.3c). We define "force" as the ratio of the intensity of the disturbance (from Fig. 17.3b) over the *time interval* between the first occurrence of the disturbance and the point at which the disturbance reached its maximum intensity. This "time interval" is *not* the duration of the disturbance, which is how long the disturbance persists for, but instead is the time required for the event to "build up" to the point of maximum disturbance. The force of disturbance is analogous to the physical definition of force and describes the rate of change. For example, the force from a hammer blow will be greater if the mass of the hammer is applied over a fraction of a second (a sharp strike) than if it is slowly applied over 10 s (slowly pushing down). With the Dungeness crab disturbance, the maximum intensity was reached within a few weeks. By comparison, the green crab disturbance built up slowly to maximum intensity over two years after the initial introduction.

The differences in the force of the two disturbances resulted in very different indirect effects that are likely responsible for the dispersal of shorebirds. The greater force of the Dungeness disturbance (greater intensity over a shorter time interval) resulted in severe declines of all prey species for shorebirds over a short time interval. There was no opportunity for recruitment of invertebrate populations to replace what had been consumed by the millions of juvenile Dungeness crabs; thus, shorebirds were left with little to eat and correspondingly dispersed to other harbors (Ruiz 1987). In contrast, the lesser force of the green crab disturbance (lesser intensity developing over a much greater time interval) permitted the indirect effects of predation to become manifest. The slower rate of bivalve decline in response to the more gradual green crab invasion may have permitted alternative, though less preferred, food sources (e.g. polychaetes) to increase (Grosholz et al. 2000). This indirect effect of crab predation may also have allowed opportunistic shorebirds to slowly switch to the alternative prey instead of dispersing away from Bodega Harbor (Estelle 2005).

The idea that multitrophic level processes like trophic cascades become attenuated over time has been developed previously (Persson 1999; Polis 1999), and we now extend this concept to invasions, including the component of "rapid effects" with one of "strong effects." Not only was the Dungeness recruitment event a much stronger disturbance (larger hammer) than the green crab invasion, but more importantly, the "strike" occurred much faster than with the green crab invasion.

Other temporal characteristics such as the frequency and the timing of the two disturbances differed somewhat between the two events. With respect to frequency, disturbance theory predicts that the impact of the disturbance on the recipient system will decrease as the frequency of the disturbance increases (Fig. 17.3d). In this system, there are data to support the idea that the circumstances creating large recruitment events for native crabs may recur every 20 or 30 years (McConnaughey et al. 1992).

This contrasts with the single event frequency that accompanies the invasion of a non-native species, an event that is not typically repeated. Therefore, disturbance theory would predict that the rare, but more frequently occurring, Dungeness crab recruitment event would have a smaller impact on the system, which is the opposite from what is observed. However, it is also noteworthy that many of the impacted invertebrate taxa are effectively annuals, such that Dungeness crab recruitments may span >20–30 generations. The point at which such rare events differ from unique events is not clear and deserves further consideration.

Our results suggest that a key to predicting whether an invasion will result in multitrophic level impacts is not only the size of the invasion and mode of action, but how rapidly the invasion builds up. In general, the intensity of the impacts of introduced species may be comparable to rare natural disturbances. However, we emphasize that when invasions are viewed as a disturbance to the system, short-term changes may not be as important or ultimately as dramatic as the longer-term changes to natural systems. We conclude that application of disturbance theory to analyze the consequences of introduced species invasions may provide a valuable approach to better resolve and predict the impacts of invasions on native species, and it may also facilitate direct comparisons with both naturally and human-mediated disturbances.

17.6 Conclusions

The evidence for multitrophic level impacts in estuarine and marine systems is still limited compared to terrestrial systems. This is likely due in large part to the paucity of studies that investigate impacts beyond the immediate circumstances of the invasion, including the availability of pre-invasion data collected over appropriate temporal and spatial scales that is needed to infer cause-effect relationships. It is nonetheless clear that several species likely produced broad-scale changes over several trophic levels.

As indicated in Table 17.1, invasions that are known to produce multitrophic level effects include: (1) filter feeders controlling phytoplankton and zooplankton abundance, (2) predators with a capacity for rapid growth that reduce important key "foundation" species, and (3) ecosystem engineers, which alter primary production, nutrient cycling or other key ecosystem processes. There are likely other mechanisms that result in multitrophic effects.

We hypothesize that effects of marine invasions may commonly extend to multiple trophic levels, although the magnitude of the effects will clearly vary by species and location. Viewing invasions as a human-mediated disturbance may provide a useful tool for understanding and predicting the potential for multitrophic level effects, especially in considering impacts to have multiple attributes that reside along a continuum instead of existing as a binary state. Our results suggest that not only is the size of the disturbance important, but also the speed at which the disturbance occurs may be equally important. If the invasion occurs slowly, there may be sufficient time for the system to rebound via indirect effects reducing the potential for impacts to cascade across multiple trophic levels.

Although the current examples of multitrophic and large scale impacts fall into a few general categories (Table 17.1; Ruiz et al. 1999), it is premature to draw conclusions about where and when such impacts will occur. This is clearly a high priority both for science and management. As a minimum, our results indicate a premium for preventing and removing species that are filter-feeders, generalist predators, and ecosystem engineers. However, this should not be viewed as an exclusive list, as such impacts may exist in other functional groups (e.g., disease organisms and parasites), and a rigorous, quantitative assessment is still lacking in this area. It is our hope that a disturbance framework may prove useful in advancing this evaluation.

Acknowledgements The authors wish to express their sincere thanks to J. Crooks and G. Rilov for their invitation to contribute to this volume. We wish to thank W. P. Sousa for discussions of natural disturbance and the application to invasive species. We thank P. Fofonoff for creation of the table and assistance with references and table summarizing species associated with cascades and engineers. We would like to thank M. Torchin and external reviewers for their comments that helped strengthen this chapter. This work was supported by funding from the National Science Foundation (DEB-9322797 CRB, DEB-0083583 Biocomplexity), Sea Grant-NOAA Partnership Program, Pacific States Marine Fisheries Commission and the Smithsonian Institution.

References

Able KW, Hagan SM (2003) Impact of common reed, *Phragmites australis*, on essential fish habitat: influence on reproduction, embryological development, and larval abundance of mummichog (*Fundulus heteroclitus*). Estuaries 26:40–50

Allmon RA, Sebens K (1988) Feeding biology and ecological impact of an introduced nudibranch, *Tritonia plebeia*, New England, USA. Mar Biol 99:375–385

Alpine AE, Cloern JE (1992) Trophic interactions and direct physical effects control phytoplankton biomass and production in an estuary. Limnol Oceanogr 37:946–955

Andow DA, Kareiva PM, Levin SA, Okubo A (1990) Spread of invading organisms. Landscape Ecol 4:177–188

Angradi TR, Hagan SM, Able KW (2001) Vegetation type and the intertidal macroinvertebrate fauna of the brackish marsh: *Phragmites* vs. *Spartina*. Wetlands 21:75–92

Attayde JL, Hansson LA (2001) Press perturbation experiments and the indeterminacy of ecological interactions: effects of taxonomic resolution and experimental duration. Oikos 92:235–244

Ayres DR, Strong DR, Baye P (2003) *Spartina foliosa* (Poaceae): a common species on the road to rarity. Madroño 50:209–213

Ayres DR, Smith DL, Zaremba K, Klohr S, Strong DR (2004) Spread of exotic cordgrasses and hybrids (*Spartina* sp.) in the tidal marshes of San Francisco Bay. Biol Invas 6:221–231

Bender EA, Case TJ, Gilpin ME (1984) Perturbation experiments in community ecology: theory and practice. Ecology 65:1–13

Bertness MD (1984) Habitat and community modification by and introduced herbivorous snail. Ecology 65:370–381

Bilio M, Niermann U (2004) Is the comb jelly really to blame for it all? *Mnemiopsis leidyi* and the ecological concerns about the Caspian Sea. Mar Ecol Prog Ser 269:173–183

Bruno JF, O'Connor MI (2005) Cascading effects of predator diversity and omnivory in a marine food web. Ecol Lett 8:1048–1056
Brusati ED, Grosholz ED (2006) Native and introduced ecosystem engineers produce contrasting effects on estuarine infaunal communities. Biol Invas 8:683–695
Byers JE (2000) Competition between two estuarine snails: implications for invasions of exotic species. Ecology 81:1225–1239
Byrnes J, Stachowicz JJ, Hultgren KM, Hughes AR, Olyarnik SV, Thornber CS (2006) Predator diversity strengthens trophic cascades in kelp forests by modifying herbivore behaviour. Ecol Lett 9:61–71
Carpenter SR, Kitchell JF (1993) The trophic cascade in lake ecosystems. Cambridge University Press
Carpenter SR, Kitchell JF, Hodgson JR (1985) Cascading trophic interactions and lake productivity. Bioscience 35:634–639
Christiansen NC, Grosholz ED, Rosso P (in review) Quantifying the potential impact of the *Spartina* invasion on invertebrate food resources for foraging shorebirds in San Francisco Bay. Proceedings of the 3rd Annual Invasive Spartina Conference. Cambridge Publications, Cambridge, UK
Civille JC, Sayce K, Smith SD, Strong DR (2005) Reconstructing a century of *Spartina alterniflora* invasion with historical records and contemporary remote sensing. Ecoscience 12:330–338
Cohen RRH, Dresler PV, Phillips EJP, Cory RL (1984) The effect of the Asiatic clam, *Corbicula fluminea*, on phytoplankton of the Potomac River, Maryland. Limnol Oceanogr 29:170–180
Connell JH, Hughes TP, Wallace CC (1997) A 30-year study of coral abundance, recruitment, and disturbance at several scales in space and time. Ecol Monogr 67:461–488
Crooks JA (1998) Habitat alteration and community-level effects of an exotic mussel, *Musculista senhousia*. Mar Ecol Prog Ser 162:137–152
Crooks JA (2002) Characterizing ecosystem-level consequences of biological invasions: the role of ecosystem engineers. Oikos 97:153–166
Currin CA, Wainwright SC, Able KW, Weinstein MP, Fuller CM (2003) Determination of food web support and trophic position of the Mummichog, *Fundulus heteroclitus* in New Jersey Smooth Cordgrass (*Spartina alterniflora*), Common Reed (*Phragmites australis*) and restored salt marshes. Estuaries 26:495–510
Daehler CC, Strong DR (1997) Hybridization between introduced smooth cordgrass (Spartina alterniflora; Poaceae) and native California cordgrass (*S-foliosa*) in San Francisco Bay, California, USA. Am J Bot 84:607–611
Davis HG, Taylor CM, Civille JC, Strong DR (2004) An Allee effect at the front of a plant invasion: *Spartina* in a Pacific estuary. J Ecol 92:321–327
Dayton PK (1971) Competition, disturbance, and community organization: the provision and subsequent utilization of space in a rocky intertidal environment. Ecol Monogr 41: 351–389
Deegan LA, Garritt RH (1997) Evidence for spatial variability in estuarine food webs. Mar Ecol Prog Ser 147:31–47
Dexter DM, Crooks JA (2000) Benthic communities and the invasion of an exotic mussel in Mission Bay, San Diego: a long-term history. Bull Calif Acad Sci 99:128–146
Duffy JE, Richardson JP, France KE (2005) Ecosystem consequences of diversity depend on food chain length in estuarine vegetation. Ecol Lett 8:301–309
Estelle VB (2005) Impacts of an introduced marine invertebrate on wintering shorebirds. PhD Dissertation, Colorado State University, Ft Collins, CO
Everett RA (1989) The effects of seasonal blooms of macroalgae and an unusual settlement of predaceous crabs on the structure of an intertidal infaunal assemblage. PhD Dissertation. University of California, Berkeley, p 242
Feist BE, Simenstad CA (2000) Expansion rates and recruitment frequency of exotic smooth cordgrass, *Spartina alterniflora* (Loisel), colonizing unvegetated littoral flats in Willapa Bay, Washington. Estuaries 23:267–274

Feyrer F, Herbold B, Matern SA, Moyle PB (2003) Dietary shifts in a stressed fish assemblage: consequences of a bivalve invasion in the San Francisco estuary. Environ Biol Fish 67:277–288

Gorokhova E, Hansson S, Hoeglander H, Andersen CM (2005) Stable isotopes show food web changes after invasion by the predatory cladoceran *Cercopagis pengoi* in a Baltic Sea bay. Oecologia 143:251–259

Gratton C, Denno RF (2005) Restoration of arthropod assemblages in a *Spartina* salt marsh following removal of the invasive plant *Phragmites australis* 13:358–372

Grosholz ED (2005) Recent biological invasion may hasten invasional meltdown by accelerating historical introductions. Proc Natl Acad Sci USA. 102:1088–1091

Grosholz ED, Ruiz GM (1995) Spread and potential impact of the recently introduced European green crab, *Carcinus maenas*, in central California. Mar Biol 122:239–247

Grosholz ED, Ruiz GM, Dean CA, Shirley KA, Maron JL, Connors PG (2000) The impacts of a non-indigenous marine predator in a California Bay. Ecology 81:1206–1224

Grosholz ED, Levin LA, Tyler AC, Neira C (2009) (in press) Changes in community structure and ecosystem function following *Spartina alterniflora* invasion of Pacific estuaries. In: Silliman B, Bertness MD, Grosholz ED (eds) Anthropogenic changes in salt marsh ecosystems. University of California Press, Berkeley, CA

Hasting AM (1996) Models of spatial spread: is the theory complete? Ecology 77:1675–1679

Hidalgo FJ, Baron PJ, Orensanz JM (2005) A prediction come true: the green crab invades the Patagonian coast Biol Invas 7:547–552

Jones CG, Lawton JH, Shachak M (1994) Organisms as ecosystem engineers. Oikos 69:373–386

Jones CG, Lawton JH, Shachak M (1997) Positive and negative effects of organisms as physical ecosystem engineers. Ecology 78:1946–1957

Keough MJ, Quinn GP (1998) Effects of periodic disturbances from trampling on rocky intertidal algal beds. Ecol Appl 8:141–161

Kideys AE (2002) Fall and rise of the Black Sea ecosystem. Science 297:1482–1483

Killgore KJ, Jack RP, Morgan II, Rybicki NB (1989) Distribution and abundance of fishes associated with submersed aquatic plants in the Potomac River. North Am J Fish Manag 9:101–111

Kimmerer W, Gartside E, Orsi JJ (1994) Predation by an introduced clam as the likely cause of substantial declines in zooplankton of San Francisco Bay. Mar Ecol Prog Ser 113: 81–93

Kwak TJ, Zedler JB (1997) Food web analysis of southern California coastal wetlands using multiple stable isotopes. Oecologia 110:262–277

Levin LA, Neira C, Grosholz ED (2006) Invasive cordgrass modifies wetland trophic function. Ecology 87:419–432

Levin PS, Coyer JA, Petrik R, Good TP (2002) Community-wide effects of nonindigenous species on temperate rocky reefs. Ecology 83:3182–3193

Lubina JA, Levin SA (1988) The spread of a reinvading species: range expansion in the California sea otter. American Naturalist 131:526–543

McConnaughey RA, Armstrong DA, Hickey BM, Gunderson DR (1992) Juvenile Dungeness crab (*Cancer magister*) recruitment variability and oceanic transport during the pelagic larval phase. Can J Fish Aquat Sci 49:2028–2044

McGuinness KA (1987) Disturbance and organisms on boulders. I. Patterns in the environment and the community. Oecologia 71:409–419

Moran PJ (1986) The *Acanthaster* phenomenon. Oceanogr Mar Biol Annu Rev 24:379–480

Neira C, Levin L, Grosholz ED (2005) Benthic macrofaunal communities of three sites in San Francisco Bay invaded by hybrid *Spartina*, with comparison to uninvaded habitats. Mar Ecol Prog Ser 292:111–126

Neira C, Grosholz ED, Levin LA, Blake R (2006) Mechanisms generating modification of benthos following tidal flat invasion by a *Spartina* (*alterniflora X foliosa*) hybrid. Ecol Appl 16:1391–1404

Ojaveer H, Simm M, Lankov A (2004) Population dynamics and ecological impact of the non-indigenous *Cercopagis pengoi* in the Gulf of Riga (Baltic Sea). Hydrobiologia 522:261–269

Pace ML, Cole JJ, Carpenter SR, Kitchell JF (1999) Trophic cascades revealed in diverse ecosystems. Trends Ecol Evol 14:483–488

Page GW, Stenzel LE, Kjelmyr JE (1999) Overview of shorebird abundance and distribution in wetlands of the Pacific Coast of the contiguous United States. Condor 101:461–471

Page HM (1997) Importance of vascular plant and algal production to macro-invertebrate consumers in a southern California salt marsh. Estuarine Coastal Shelf Sci 45:823–834

Parker IM, Simberloff D, Lonsdale WM (1999) Impact: toward a framework for understanding the ecological effects of invaders. Biol Invas 1:3–19

Persson L (1999) Trophic cascades: abiding heterogeneity and the trophic level concept at the end of the road. Oikos 85:385–397

Phelps HL (1994) The Asiatic clam (*Corbicula fluminea*) invasion and system-level ecological change in the Potomac River estuary near Washington, DC Estuaries 17:614–621

Pickett SA, White PS (1985) The ecology of natural disturbance and patch dynamics. Academic Press, Orlando, FL

Polis GA (1999) Why are parts of the world green? Multiple factors control productivity and the distribution of biomass. Oikos 86:3–15

Polis GA, Sears ALW, Huxel GR, Strong DR, Maron J (2000) When is a trophic cascade a trophic cascade? Trends Ecol Evol 15:473–475

Pollumae A, Valjataga K (2004) *Cercopagis pengoi* (Cladocera) in the Gulf of Finland: environmental variables affecting its distribution and interaction with *Bosmina coregoni maritima*. Proceedings of the Estonian Academy of Sciences: Biol Ecol 53:276–282

Raichel DL (2003) The influence of *Phragmites* (Common Reed) on the distribution, abundance, and potential prey of a resident marsh fish in the Hackensack Meadowlands, New Jersey. Estuaries 26:511–512

Richman SE, Lovvorn JR (2004) Relative foraging value to lesser scaup ducks of native and exotic clams from San Francisco Bay. Ecol Appl 14:1217–1231

Rudnick DA, Hieb K, Grimmer KF, Resh VH (2003) Patterns and processes of biological invasion: the Chinese mitten crab in San Francisco Bay. Basic Appl Ecol 4:249–262

Ruiz G (1987) Interactions among shorebirds, crab, and their invertebrate prey populations. Dissertation. University of California, Berkeley, CA, USA

Ruiz GM, Connors PG, Griffen SE, Pitelka FA (1989) Structure of a wintering dunlin population. Condor 91:562–570

Ruiz GM, Fofonoff P, Hines AH, Grosholz ED (1999) Non-indigenous species as stressors in estuarine and marine communities: assessing invasion impacts and interactions. Limnol Oceanogr 44:950–972

Schwindt E (2001) Invasion of a reef-builder polychaete: direct and indirect impacts on the native benthic community structure. Biol Invas 3:137–149

Shiganova TA (1998) Invasion of the Black Sea by the ctenophore *Mnemiopsis leidyi*, and recent changes in pelagic community structure. Fish Oceanogr 7:305–310

Shiganova TA (2001) The new invader *Beroe ovata* and its effect on the ecosystem in the Northeastern Black Sea. Hydrobiologia 471:187–197

Shurin JB, Borer ET, Seabloom EW, Anderson K, Blanchette CA, Broitman B, Cooper SD, Halpern BS (2002) A cross-ecosystem comparison of the strength of trophic cascades. Ecol Lett 5:785–791

Silliman BR, Bertness MD (2002) Atrophic cascade regulates salt marsh primary production. Proc Natl Acad Sci USA 99:10500–10505

Sousa WP (1979) Disturbance in marine inter-tidal boulder fields – the non-equilibrium maintenance of species-diversity. Ecology 60:1225–1239

Sousa WP (1984) The role of disturbance in natural communities. Annu Rev Ecol Syst 15:353–391

Sousa WP (2001) Disturbance in marine systems. Marine conservation biology. Sinauer Assoc, Sunderland, MA

Stenzel LE, Hickey CM, Kjelmyr JE, Page GW (2002) Abundance and distribution of shorebirds in the San Francisco Bay area. West Birds 33:69–98
Strayer DL, Caraco NF, Cole JJ, Pace ML (1999) Transformation of freshwater ecosystems by bivalves: a case study of zebra mussels in the Hudson River. BioScience 49:19–27
Strayer DL, Hattala KA, Kahnle AW (2004) Effects of an invasive bivalve *Dreissena polymorpha* on fish in the Hudson River estuary. Can J Fish Aquat Sci 61:924–941
Strong DR (1992) Are trophic cascades all wet – differentiation and donor-control in speciose ecosystems. Ecology 73:747–754
Torchin ME, Hechinger HF, Huspeni TC, Whitney KL, Lafferty KD (2005) Ecology of the introduced ribbed mussel (*Geukensia demissa*) in Estero de Punta Banda, Mexico: interactions with the native cord grass, *Spartina foliosa*. Biol Invas 7:607–614
Trussell GC, Ewanchuk PJ, Bertness MD, Silliman BR (2004) Trophic cascades in rocky shore tide pools: distinguishing lethal and nonlethal effects. Oecologia 139:427–432
Vitousek PM (1990) Biological invasions and ecosystem processes – towards an integration of population biology and ecosystem studies. Oikos 57:7–13
Williams SL, Ebert TA, Allen BJ (2005) Does the recruitment of a non-native mussel in native eelgrass habit explain their disjunct adult distributions? Divers Distrib 11:409–416
Williamson MH (1996) Biological invasions. Chapman and Hall, London
Wing SR, Botsford LW, Largier JL, Morgan LE (1995) Spatial structure of relaxation events and crab settlement in the northern California upwelling system. Mar Ecol Prog Ser 128:199–211
Yamada SY (2001) Global invader: the European green crab. Oregon Sea Grant, Oregon State University

Section V
Management Perspectives

Chapter 18
Marine Bioinvasion Management: Structural Framework

Chad L. Hewitt, Richard A. Everett, Naomi Parker, and Marnie L. Campbell

18.1 Introduction

Significant global change has occurred through the accidental and intentional human mediated introductions of species in regions outside of their evolutionary origins can no longer be disputed (e.g., Lubchenco et al. 1991; Carlton 2001; Pimentel 2002). This change is well documented in a variety of terrestrial and freshwater ecosystems (e.g., Drake and Mooney 1989) and is becoming increasingly apparent in marine and estuarine habitats in all of the world's oceans (e.g., Carlton 2001; Chap. 2, Carlton). Documenting the scale and rates of marine introductions and the subsequent changes to invaded systems has captured much of the marine invasion ecology effort during the last 25 years (e.g., Grosholz et al. 2000; Carlton and Ruiz 2004).

While the lessons that can be learned about evolution, ecosystem function, community dynamics, and species biology and ecology from the study of biological introductions are fascinating (e.g., Harper 1965; Carlton and Ruiz 2004), the challenge "what should we and/or what can we do?" remains. The options appear to be simple, however the details of implementation are difficult: *we can choose to do nothing or we can choose to act.*

18.2 Inaction vs Action

Those who suggest that we do nothing generally fall into two categories. The first suggests that human mediated invasions are merely an extension of natural processes. This is generally phrased as "after all, humans are natural and therefore their actions are natural and indeed, introduced species are merely a special case of 'invaders'." Many of these researchers have discussed the apparent xenophobia regarding treatment of non-native species (e.g., Egler 1942; Davis 2003) and the overt use of negative syntax in discussions of biological introductions (Eser 1998; Sagoff 1999) with no recognition of the beneficial uses of non-native species

G. Rilov, J.A. Crooks (eds.) *Biological Invasions in Marine Ecosystems.*
Ecological Studies 204,

(Kangas 2004). Indeed, many in the group debate the veracity of the larger scale claims of impact, specifically the contribution of introduced species to the Sixth Extinction (Leakey and Lewin 1996) through human mediated global homogenization (e.g., Lubchenco et al. 1991; Chapin et al. 2000). Indeed, the suggestion that the global exchange of species will lead to increased speciation events in the future (Rozensweig 2001; Sax and Gaines 2003) has been used to refute some of the potential impacts of invasions.

A second group suggests that fiscal pragmatism dictates a "do nothing" strategy – we cannot afford to prevent future invasions by restricting current trade activities *without scientific evidence*, nor is it an appropriate use of public monies to address existing introductions *until valid evidence is presented demonstrating sufficient harm*. These arguments align with the World Trade Organisation framework, however they are opposed to the Precautionary Approach agreed to by Parties to the Convention on Biological Diversity (1993). As a consequence, the burden of proof is vested with the management agency to demonstrate risk (e.g., Hewitt et al. 2004; Anderson 2007; see also Campbell, Chap. 20).

While there are several who continue to debate a do nothing strategy, the global community has chosen action, driven in large part by the increasing demonstrable effects of non-native species on environmental, economic, social (including human mortality and morbidity) and cultural values. Action, however, is a continuum. We can act to prevent new introductions by erecting stringent border controls through effective management of the transport pathways and vectors and choose to accept what has been currently introduced as a *fait accompli,* or we can act to prevent new introductions, while simultaneously managing, controlling and ultimately eradicating those species that have already been introduced to a region.

18.3 Marine Biosecurity – Prevention, Regulation, and Management of Invasions

Effective marine biosecurity is the science-based protection of native marine biodiversity and marine ecosystems that provide environmental, economic, social and cultural values to society through effective management and control of non-native species (e.g., Biosecurity Council 2003; Hewitt et al. 2004). This requires a goal, preferably stated in a policy context, which is clear, concise and widely accepted. This goal can and should be stated at multiple levels, local (e.g., state, province), national, regional and global in order to create both understanding and concerted effort. Marine invaders, more so than their freshwater and terrestrial counterparts, frequently transcend geopolitical boundaries once introduced to a region, consequently marine invasions cannot be managed in isolation of neighbouring states.

This section discusses several options available for the prevention, regulation, management and control of human-mediated biological introductions. We currently have a limited suite of tools in the toolbox for marine biosecurity delivery. Here we present chapters that provide a broad overview of regulatory frameworks at national, regional and international levels addressing marine bioinvasions (Chap. 19, Hewitt et al.), and address the use of risk assessment for marine biosecurity (Chap. 20, Campbell). The implications of non-native species eradication on the long-term recovery of native ecosystems, with some discussion of *Spartina* management (Chap. 21, Hacker and Dethier), are also considered.

Hewitt et al. (Chap. 19) provide a broad overview of the development of regulatory frameworks at national, regional and international scales, with illustrative examples drawn from the three authors experience in the policy and regulatory environments of their central government agencies in Australia, New Zealand, and the USA. They identify a number of international and regional agreements that are capable of addressing marine introductions and discuss where such action is occurring. Lastly, they identify issues at their national levels which provide an insight to the gaps and inconsistencies that may exist. Embedded within their discussion is a theme of the role that science plays in addressing this pressing issue.

Campbell's chapter (Chap. 20) provides a discussion of how the operational tool of risk assessment can be employed to address marine biosecurity issues. Her discussion outlines the generic risk assessment framework and identifies methods that are currently being employed at the Species level for pre-border assessments of intentional importation to post-border analyses of species level risk once an incursion is detected; the Vector level to determine which shipments or vessels pose the greatest risk; and the Pathway level to aid in targeting either high risk trading routes (source regions) or high risk receiving locations (ports, marinas). Campbell provides multiple examples of assessments at these three levels and specifically includes both pre-border and post-border demonstrations. Campbell emphasises the role of risk assessment as a tool to aid the transparency and consistency of decision making in biosecurity management.

Hacker and Dethier (Chap. 21) address the important topic of what to do with an ecosystem once the invader is controlled. Often, invasive species management efforts are just that – focused on the invader. However, the larger goal of an invader management effort should typically be to recover natural species and processes. Hacker and Dethier discuss how the removal of an introduced species that acts as an ecosystem engineer (e.g., alters or modifies the habitat or energy flow of the ecosystem; see Chap. 16, Crooks) can require habitat restoration to facilitate recovery of the native ecosystem. They present a modified version of the Alternate Stable State model and suggest that this model may provide predictive ability to anticipate the consequences of eradication attempts. They provide several examples of how eradication of *Spartina* in differing habitat contexts is likely to alter the biosecurity outcomes that the management action achieves. Their final note is a caution for managers to consider the post-eradication implications of species removal.

18.4 Other Management Options

This section provides an initial glimpse at management options including regulatory frameworks. Management issues are also addressed in other chapters throughout the book. Wonham and Lewis (Chap. 4) discuss some management-related modelling, including such topics as risk assessment, Marine Protected Areas, and invasion resistance. The relationship between environmental quality and invader success is also discussed in several chapters (e.g. Chap. 12, Olyarnik et al.; Chap. 7, Johnston et al.), with the implication being that improving degraded environmental conditions may increase the ability of systems to resist invaders – this is a topic that merits more attention (see also Ruiz and Crooks 2001; Crooks and Suarez 2006). Regional approaches to management and regulations are also covered in the section that follows.

These are by no means the sum total of management options available for implementation. Although some management of vessel-related vectors is highlighted in Chap. 6 (Hewitt et al.), we do not fully address the significant suite of research that has been undertaken on the management and control of species in ballast water and biofouling of ships hulls. A significant technical effort has occurred on a global scale to research and develop acceptable and functional methods of removing species from ballast water. For reference to the recent work in these areas we point the reader to the International Maritime Organization (www.imo.org) for a current update on the variety of research and development activities to address these problems. We provide a limited suite of starter references below (see Additional Literature below).

Similarly, the physical, chemical and biological control options available (realistically or theoretically) for implementation in the marine environment have not been addressed in a comprehensive fashion (but see McEnnulty et al. 2001; Bax et al. 2001). Physical and chemical tools are being developed in a variety of contexts around the globe and range from physical removal by divers (e.g., Culver and Kuris 2000; Hewitt et al. 2005), heat treatment (e.g., Wotton and Hewitt 2004), and chemical control (Bax 1999; Willan et al. 2000).

Of greater controversy however is the use of classical and neo-classical biological control in the marine environment (e.g., Simberloff and Stiling 1996; Lafferty and Kuris 1996; Kuris 2003; Thresher and Kuris 2004; Chap. 15, Rilov). These techniques, coupled with genetic technologies that provide the ability to increase specificity, have been widely hailed as the solution to well-established species (Myers et al. 2000; Kuris 2003; Thresher and Kuris 2004); however, the negative public perceptions of biological control in many communities has precluded much needed work in this arena (e.g., Follett and Duan 2000; Thresher and Kuris 2004). While no single source provides a clear indication of the current state of biological control work for use in the marine environment, we refer the reader to a number of references (see Additional Literature below) that provide some access to this rich literature.

References

Anderson LWJ (2007) Control of invasive seaweeds. Bot Mar 50:418–437

Bax NJ (1999) Eradicating a dreissenid from Australia. Dreissena! 10(3):1–5

Bax NJ, Carlton JT, Mathews-Amos A, Haedrich RL, Howarth FG, Purcell JE, Rieser A, Gray A (2001) The control of biological invasions in the World's oceans. Conserv Biol 15(5):1234–1246

Biosecurity Council (2003) Tiakina Aotearoa – Protect New Zealand: The biosecurity strategy for New Zealand. Wellington, New Zealand, Ministry of Agriculture and Forestry. 63 pp <<www.maf.govt.nz/biosecurity-strategy>> accessed 30 January 2006

Carlton JT (2001) Introduced species in U.S. coastal waters: environmental impacts and management priorities. Pew Oceans Commission, Arlington, Virginia, 28 pp

Carlton JT, Ruiz GM (2004) The magnitude and consequences of bioinvasions in marine ecosystems: implication for conservation biology. In: Norse EA, Crowder LB (eds) Marine conservation biology: the science of maintaining the sea's biodiversity. Island Press, Washington, USA, pp 123–148

Chapin FS, Zavaleta ES, Eviner VT, Naylor RL, Vitousek PM, Reynolds HL, Hooper DU, Lavorel S, Sala OE, Hobbie SE, Mack MC, Diaz S (2000) Consequences of changing biodiversity Nature 405(6783):234–242

Crooks JA, Suarez A (2006) Hyperconnectivity, invasive species, and the breakdown of natural barriers to dispersal. In: Crooks KR, Sanjayan MA (eds) Connectivity conservation: maintaining connections for nature. Cambridge University Press, New York, pp 451–478

Culver CS, Kuris AM (2000) The apparent eradication of a locally established introduced marine pest. Biol Invas 2:245–253

Davis MA (2003) Biotic globalization: does competition from introduced species threaten biodiversity? Bioscience 53:481–489

Drake J, Mooney HA, Di Castri F, Groves R, Kruger FJ, Rejmánek M, Williamson M (eds) (1989) Biological invasions: a global perspective. Wiley, Chichester

Egler FE (1942) Indigene versus alien in the development of arid Hawaiian vegetation. Ecology 23:14–23

Eser U (1998) Assessment of plant invasions: theoretical and philosophical fundamentals. In: Starfinger U, Edwards K, Kowarik I, Williamson M (eds) Plant invasions: ecological mechanisms and human responses. Backhuys Publishers, Leiden, The Netherlands, pp 95–107

Follett PA, Duan JJ (eds) 2000. Nontarget effects of biological control. Kluwer Academic Publishers, Boston

Grosholz ED, Ruiz GM, Dean CA, Shirley KA, Maron JL,Connors PG (2000) The impacts of a nonindigenous marine predator in a California bay. Ecology 81:1206–1224

Harper JL (1965) Establishment, aggression, and cohabitation in weedy species. In: Baker HG, Stebbins GL (eds) The evolution of colonizing species. Academic Press, New York, NY, pp 245–265

Hewitt CL, Willing J, Bauckham A, Cassidy AM, Cox CMS, Jones L, Wotton DM (2004) New Zealand Marine Biosecurity: delivering outcomes in a fluid environment. N Z J Mar Freshwater Res 38:429–438

Hewitt CL, Campbell ML, McEnnulty F, Moore KM, Murfet NB, Robertson B, Schaffelke B (2005) Efficacy of physical removal of a marine pest: the introduced kelp *Undaria pinnatifida* in a Tasmanian Marine Reserve. Biol Invas 7(2):251–263

Kangas PC (2004) Ecological engineering principles and practice. Lewis Publishers, New York

Kuris AM (2003) Eradication of introduced marine pests. In: Rapport DJ, Lasley BL, Rolston, DE et al. (eds) Managing for healthy ecosystems. Lewis Publishers, New York, pp 549–556

Lafferty KD, Kuris AM (1996) Biological control of marine pests. Ecology 77:1989–2000

Leakey R, Lewin R (1996) The sixth extinction: patterns of life and the future of humankind. Doubleday and Company

Lubchenco JA, Olson M, Brubaker LB, Carpenter SR, Holland MM, Hubbell SP, Levin SA, MacMahon JA, Matson PA, Melillo JM, Mooney HA, Peterson CH, Pulliam HR, Real LA, Regal PJ, Risser PG (1991) The sustainable biosphere initiative: an ecological research agenda. Ecology 72:371–412

McEnnulty FR, Bax NJ, Schaffelke B, Campbell ML (2001) A review of rapid response options for the control of ABWMAC listed introduced marine pest species and related taxa in Australian waters. CRIMP Tech Rep Ser 23. Commonwealth Scientific and Industrial Research Organisation, Hobart, Australia

Myers JH, Simberloff D, Kuris AM, Carey JR (2000) Eradication revisited: dealing with exotic species. Trends Ecol Evol 15:316–320

Pimentel D (2002) Biological invasions: economic and environmental costs of alien plant, animal, and microbe species. CRC Press, Boca Raton, USA. 369 pp

Rosenzweig ML (2001) The four questions: what does the introduction of exotic species do to diversity. Evol Ecol Res 3:361–367

Ruiz GM, Crooks JA (2001) Marine invaders: patterns, effects, and management of non-indigenous species. In: Gallagher P, Bendell-Young L (eds) Waters in peril, Kluwer Academic Publishers, pp 3–17

Sagoff M (1999) What's wrong with alien species? Report of the Institute for Philosophy and Public Policy, University of Maryland, College Park, Maryland

Sax DF, Gaines SD (2003) Species diversity: from global decreases to local increases. Trends Ecol Evol 18:561–566

Simberloff D, Stiling P (1996) How risky is biological control? Ecology 77:1965–1974

Thresher RE, Kuris AM (2004) Options for managing invasive marine species. Biol Invas 6:295–300

Willan RC, Russell BC, Murfet NB, Moore KL, McEnnulty FR, Horner SK, Hewitt CL, Dally GM, Campbell ML, Bourke ST (2000) Outbreak of *Mytilopsis sallei* (Recluz, 1849) (Bivalvia: Dreissenidae) in Australia. Mollusc Res 20(2): 25–30

Wotton DM, Hewitt CL (2004) Marine biosecurity post-border management: developing incursion response systems for New Zealand. N Z J Mar Freshwater Res 38:553–559

Additional Literature

Ballast Water

BWM (2005) International Convention on the Control and Management of Ship's Ballast Water and Sediments, International Maritime Organization, London, United Kingdom <<www.imo.org>> accessed 30 January 2006

Carlton JT (1985) Transoceanic and interoceanic dispersal of coastal marine organisms: the biology of ballast water. Oceanogr Mar Biol Annu Rev 23:313–371

Carlton JT, Navarret AM, Mann R (1982) Biology of ships' ballast water: the role of ballast water in the transoceanic dispersal of marine organisms. Final Project Report, National Science Foundation, Division of Applied Research (Biology), DAR-8008450, 161 pp

Gollasch S (2002) Ballast water management in the North-East Atlantic. In: Tjallingii F (ed) Report to aid decision making on ballast water in the Oslo-Paris Commission (OSPAR), Biological Diversity Committee. North Sea Directorate, the Netherlands, 48 pp

Gollasch S, MacDonald E, Belson S, Botnen H, Christensen JT, Hamer JP, Houvenaghel G, Jelmert A, Lucas I, Masson D, McCollin T, Olenin S, Persson A, Wallentinus I, Wetsteyn LPMJ, Wittling T (2002) Life in ballast tanks. In: Leppäkoski E, Gollasch S, Olenin S (eds) Invasive aquatic species of Europe: distribution, impact and management. Kluwer Academic Publishers, The Netherlands. pp 217–231

Gollasch S, David M, Dragsund E, Hewitt CL, Fukuyo Y (2007) Critical review of the IMO International Convention on the Management of Ships' Ballast Water and Sediments. Harmful Algae 6:585–600

Hülsmann N, Galil BS (2002) Protists – a dominant component of the ballast – transported biota. In: Leppäkoski E, Gollasch S, Olenin S (eds) Invasive aquatic species of Europe: distribution, impact and management. Kluwer, Dordrecht, the Netherlands, pp 20–26

Matheickal JT, Raaymakers S (eds) (2004) 2nd International Ballast Water Treatment R&D Symposium, IMO London, 21–23 July 2003: Proceedings. GloBallast Monogr Ser No 15. IMO London

Raaymakers S (ed) (2002) Baltic Regional Workshop on Ballast Water Management, Tallinn, Estonia, 22–24 Oct 2001: Workshop Report. GloBallast Monogr Ser No 2. IMO London

Raaymakers S (2003) 1st International Workshop on Guidelines and Standards for Ballast Water Sampling, Rio de Janeiro, Brazil, 7–11 April 2003: Workshop Report. GloBallast Monogr Ser No 9. IMO London

Raaymakers S, Gregory C (2003) 1st East Asia Regional Workshop on Ballast Water Control & Management, Beijing, China, 31 Oct–2 Nov 2002: Workshop Report. GloBallast Monogr Ser No 6. IMO London

Taylor A, Rigby G, Gollasch S, Voigt M, Hallegraeff G, McCollin T, Jelmert A (2002) Preventive treatment and control techniques for ballast water. In: Leppäkoski E, Gollasch S, Olenin S (eds) Invasive aquatic species of Europe: distribution, impacts and management. Kluwer Academic Publishers, Dordrecht, The Netherlands, 583 pp

Williams RJ, Griffiths FB, Van der Wal EJ, Kelly J (1988) Cargo vessel ballast water as a vector for the transport of non-indigenous marine species. Est Coast Shelf Sci 26:409–420

Biofouling

AFS (2001) International Convention on the Control of Harmful Anti-Fouling Systems on Ships, International Maritime Organization, London, United Kingdom <<www.imo.org>> accessed 30 January 2006

Gollasch S (2002) The importance of ship hull fouling as a vector of species introductions into the North Sea. Biofouling 18(2):105–121

Hay C, Dodgshun T (1997) Ecosystem transplant? The case of the *Yefim Gorbenko*. Seafood N Z May:13–15

Hewitt CL, Campbell ML, Schaffelke B (2007) Introductions of marine macroalgae – accidental transfer pathways and mechanisms. Bot Mar 50:397–417

Lewis PN, Hewitt CL, Riddle M, McMinn A (2003) Marine introductions in the Southern Ocean: an unrecognised hazard to biodiversity. Mar Pollut Bull 46:213–223

Minchin D (2006) The transport and the spread of living aquatic species. In: Davenport J, Davenport JL (eds) The ecology of transportation: managing mobility for the environment. Springer, Berlin Heidelberg New York, pp 77–97

Minchin D, Floerl O, Savini D, Occhipinti-Ambrogi A (2006) Small craft and the spread of exotic species. In: Davenport J, Davenport JL (eds) The ecology of transportation: managing mobility for the environment. Springer, Berlin Heidelberg New York, pp 99–118

Biological Control

Anderson LWJ (2003) A review of aquatic weed biology and management research conducted by the United States Department of Agriculture—agricultural research service. Pest Manag Sci 59:801–813

Byrne M, Cerra A, Nishigaki T, Hoshi M (1997) Infestation of the testes of the Japanese sea star *Asterias amurensis* by the ciliate *Orchitophyra stellarum*: a caution against the use of the ciliate for biological control. Dis Aquat Organisms 28:235–239

Coquillard PT, Thibaut T, Hill DRC, Gueugnot J, Mazel C, Coquillard Y (2000) Simulation of the mollusc Ascoglossa *Elysia subornata* population dynamics: application to the potential biocontrol of *Caulerpa taxifolia* growth in the Mediterranean Sea. Ecol Model 135:1–16

Kuris AM, Lafferty KD (1992) Modeling crustacean fisheries – effects of parasites on management strategies. Can J Fish Aquat Sci 49:327–336

Kuris AM, Lafferty KD (2000) Can biological control be developed as a safe and effective mitigation against established introduced marine pests? In: Peterson J (ed) Marine bioinvasions: proceedings of a conference, January 24–27, 1999. MIT Sea Grant College Program, Cambridge, pp 102–106

Kuris AM, Lafferty KD, Grygier MJ (1996) Detection and preliminary evaluation of natural enemies for possible biological control of the northern Pacific seastar, *Asterias amurensis*. CSIRO Centre Res Introduced Mar Pests Tech Rep 3:1–17

Kuris AM, Lafferty KD, Torchin ME (2005) Biological control of the European green crab, *Carcinus maenas*: natural enemy evaluation and host specificity. In: Hoddle MS (ed) International symposium on biololgical control of arthropods abstract, vol 1. USDA Forest Service, Davos, Switzerland

Messing RH, Wright MG (2006) Biological control of invasive species: solution or pollution. Front Ecol Environ 4:132–140

Secord D (2003) Biological control of marine invasive species cautionary tales and land-based lessons. Biol Invas 5:117–131

Thibaut T, Meinesz A (2000) Are the Mediterranean ascoglossan molluscs *Oxynoe olivacea* and *Lobiger serradifalci* suitable agents for a biological control against the invading tropical alga *Caulerpa taxifolia*? C R Acad Sci Ser III Sci Vie 323:477–488

Thibaut TA, Meinesz A, Burtaire L, Charrier S, Mangialajo L, Ierardi S, Vidal V (1998) Biological control of *Caulerpa taxifolia* in the Mediterranean Sea: use of tropical and Mediterranean ascoglossans. In: Boudouresque CF, Meinesz A, Gravez V (eds) Proceedings of the 3rd International Workshop on Caulerpa taxifolia, Marseille, France, 19–20 September 1997, pp 105–111

Thibaut TA, Meinesz A, Amade P, Charrier S, De Angelis K, Ierardi S, Mangialajo L, Melnick J, Vidal V (2001) *Elysia subornata* (Mollusca) a potential control agent of the alga *Caulerpa taxifolia* (Chlorophyta) in the Mediterranean Sea. J Mar Biol Assoc U K 81:497–504

Chapter 19
Examples of Current International, Regional and National Regulatory Frameworks for Preventing and Managing Marine Bioinvasions

Chad L. Hewitt, Richard A. Everett, and Naomi Parker

19.1 Introduction

There is a growing understanding at public and policy levels that bioinvasions represent a significant threat to the environment, economic, social and cultural values (e.g., Lubchenco et al. 1991; Pimentel et al. 2000a, b; Carlton 2001; Pimentel 2002). In terrestrial environments this has resulted in the adoption of regulatory frameworks at national, regional and international scales for the maintenance of quarantine and biosecurity protection of human health and economy. The recent outbreaks of Bovine Spongiform Encephalitus (BSE) and Avian Influenza have galvanised efforts throughout the globe to put in place appropriate measures for the protection and maintenance of our societal values, specifically human health and economy.

In contrast to the terrestrial (and to a lesser degree freshwater) situation, the awareness of marine bioinvasions as a societal threat has been slow to develop. In part this may be due to the relatively recent advent of marine bioinvasion research as a discipline. The seminal treatment of bioinvasions by Elton (1958) had less than 10% (14 pages) allocated to marine invasions. Similarly, The Scientific Committee on Problems of the Environment (SCOPE) evaluations of bioinvasions during the 1980s and 1990s had no explicit focus on marine systems (e.g., Drake et al. 1989; Mooney et al. 2005). Over the last 25 years, however, a significant volume of work has been established that clearly identifies the global scale of marine bioinvasions (e.g., Carlton 1996, 2001; Hewitt et al. 1999; Hewitt 2003; Ruiz et al. 2000), with the development of theoretical frameworks that underpin explicit calls for research (e.g., Vermeij 1996; Carlton 2001; Byers et al. 2002; Ruiz and Hewitt 2002).

A second, and perhaps more significant reason for reduced awareness of marine bioinvasions by policy makers is the focus of the threat (see Reiser et al. 2004). Unlike terrestrial systems where bioinvasions have caused significant damage to economic interests (e.g., agriculture, forestry, animal husbandry), the majority of demonstrable marine bioinvasion impacts appear to be primarily on native biodiversity and ecosystem health (e.g. Grosholz et al. 2000; Carlton and Ruiz 2004; Schaffelke and Hewitt, 2007) with few demonstrable direct impacts on economic values (but see Pimentel 2002). Similarly, the kinds of devastating impacts that are

G. Rilov, J.A. Crooks (eds.) *Biological Invasions in Marine Ecosystems.*
Ecological Studies 204,

so readily observed on land, are seen by only a few members of the public when they occur in the marine environment – many marine invasions remain unnoticed for years if not decades (see discussion in Cohen and Carlton 1998).

The previous chapters in this volume demonstrate that marine bioinvasions research has progressed from initial descriptive analyses, to more comprehensive and synthetic development of predictive models. As a consequence of this growing understanding, public and government awareness has significantly increased, resulting in the development of regulatory frameworks at national, regional and international scales. Several frameworks provide turn-key responses to the bioinvasions threat and target a single vector or impacted industry. Others have attempted a more comprehensive response to the issue and have placed marine bioinvasions in the context of quarantine response and control.

It is beyond the scope of this chapter to undertake a comprehensive review of regulatory frameworks at global (international), regional (multi-lateral) and national scales. Instead, we attempt to provide a discussion at the three scales, and provide representative examples of the current situation with an eye to the gaps and inconsistencies that arise. All three authors have operated in the policy development and regulatory implementation environments within the central government of three different countries and provide their operational insights to this evaluation rather than strict legal interpretations.

19.2 International Frameworks

A number of international instruments (conventions and treaties) and non-binding agreements exist that address bioinvasions. These include conventions and treaties with general obligations for signatory Parties; conventions and treaties vested within the broader context of international quarantine agreements; conventions and treaties associated with protection of biological diversity, specifically the Convention on Biological Diversity (CBD 1992), and, conventions that support the internationally consistent management of specific transport vectors (e.g., International Maritime Organisation).

19.2.1 General Obligations

The rights and responsibilities agreed in the United Nations Convention on the Law of the Sea (UNCLOS 1982) have created the legal basis for subsequent marine legal regimes. UNCLOS explicitly places a general requirement for Parties to take measures "to prevent, reduce and control pollution of the marine environment resulting from…the intentional or accidental introduction of species alien or new, to a particular part of the marine environment, which may cause significant and harmful changes thereto" (Article 196). This obligation applies explicitly to waters under the jurisdiction of a Party, and therefore is restricted to the proclaimed Exclusive Economic Zone (EEZ).

Several instruments focus on the regimes for management and protection of specific locations or environments. The Protocol on Environmental Protection to the Antarctic Treaty identifies general obligations on Parties to prevent the introduction of alien species (intentionally or accidentally) into the Antarctic Treaty area. This obligation has been historically applied to terrestrial environments and resulted in the removal of sled dogs from the Antarctic. Only recently has there been a recognition that marine bioinvasions are of significance to higher latitudes (e.g. Lewis et al. 2003; Hines et al. 2000; Niimi 2004). As a matter of importance, the 28th Antarctic Treaty Consultative Meeting (June 2005; CEP VIII 2005) considered the issues of ballast water and biofouling transport into the Antarctic Treaty area and the possible regulatory mechanisms to be implemented by Parties. It was proposed that consideration be given to voluntary application of the BWM Convention within Antarctic waters. In addition, the Scientific Committee for Antarctic Research (SCAR) presented information concerning biofouling at the 29th Antarctic Treaty Consultative Meeting in 2006. The issues of marine invasions in the Antarctic region have now been identified as a matter of urgency at the CEP and SCAR.

Similarly, Parties of the World Heritage Convention (UNESCO 1972), who act to establish World Heritage Sites in the marine environment, assume obligations to protect the values for which the site has been identified (e.g., important and significant habitats for conservation of biological diversity). The World Heritage listing of Heard and Macdonald Islands was largely based on their unique status as an example of sub-Antarctic Islands without bioinvasions. Unlike Heard and Macdonald Islands, other World Heritage sites are often subject to multiple-use, including commercial activities such as shipping, ports, recreational SCUBA diving, boating and fishing (e.g., Shark Bay, Western Australia; see Wyatt et al. 2005). Inaction by a Party to respond to invasive alien species, including failure to make provisions through national management plans, legislation and regulations, could impair the values of a property, possibly resulting in the removal of the property from World Heritage listing.

Non-binding agreements, including voluntary codes, that incorporate requests to limit or minimize the spread of alien species include the FAO Code of Conduct on Responsible Fisheries (Article 9 on Aquaculture Development; FAO 1995), and the International Council for the Exploration of the Seas (ICES) Code of Practice on the Introduction and Transfers of Marine Organisms (ICES 2005). Both codes provide a risk management framework for operational implementation to provide surety to neighbouring coastal states that intentional introductions follow acceptable guidelines. In addition, Parties of the Ramsar Convention (Convention on Wetlands 1971) are urged to ensure that measures are in place to prevent or control invasive alien species (Resolution VIII.18).

19.2.2 International Quarantine Agreements

A number of international agreements oriented towards maintaining a standardised set of quarantine arrangements for the protection of human health, and economically important plant and animal health have been established over the last 60 years.

These include the International Plant Protection Convention (IPPC 1951; revised 1997), the World Animal Health Organisation (Office International des Épizooties; OIE), and the Codex Alimentarius (food standards). These three organizations are recognized under the World Trade Organisation (WTO) as the only acceptable standard setting bodies within the Sanitary and Phytosanitary Standards (SPS) Agreement. Of these, the IPPC and OIE are of greatest importance.

The IPPC was established in 1952 to prevent damage to plants of economic importance through introductions of disease, parasites or alien species. As a consequence, national phytosanitary quarantine systems have been established by member states through a series of agreed standards which include: establishment of import regulations and compliance systems including the establishment of Pest Risk Assessments to international standards; establishment and maintenance of surveillance systems with timely reporting to associate states; the use of eradication and control systems to manage detected pests; and an export certification system to meet the needs of trading partners. The WTO have recently agreed that the IPPC is the appropriate responsible treaty for establishing Sanitary and Phytosanitary Standards for plant protection.

The IPPC however, has a much broader scope, including protection of the natural environment from many harmful non-indigenous species (Hedley 2004). Hedley (2004) insists that consideration of both direct and indirect impacts to natural environmental flora must be considered at both international and national levels. Similarly, while the IPPC has rarely been applied to the protection of the natural environment, consideration must be given to the IPPC application to protection of marine flora. Recent discussions and evaluations indicate that there is no limitation within the Convention to application in a marine context (UNEP/CBD/COP/8/3 2005).

Similarly, the OIE was created to regulate the spread of pathogens and parasites of domestic stock animals and has since expanded its scope to encompass the transfer of pathogens, parasites and pests of animals. Unlike the IPPC, it is clear that the OIE does not have a mandate to regulate or set standards for bioinvasions that affect the natural environment. The OIE operates with a black-list of species to determine mandatory reporting requirements and obligations for surveillance regimes, response preparedness and certification schemes. Much like IPPC however, the OIE provides a system under which a national quarantine system can operate with trading partners.

The OIE currently has listed a number of pathogens of commercial marine acquaculture species (salmonids and molluscs). The application of the OIE to marine systems is unquestionable, it will however require significant efforts to increase the listed species or to shift the OIE philosophy in order to encompass pests of natural systems.

19.2.3 Convention on Biological Diversity

The Convention on Biological Diversity (CBD 1992) is a comprehensive and legally binding Convention for the protection of biodiversity. It is one of the few instruments that explicitly addresses the obligations on Parties to manage alien species through the prevention or minimization of introductions, spread and impacts. A large number of countries are party to the Convention (179), significantly however, the United States has not become a Party to the convention, and participates as an observer in most fora.

The CBD places a number of obligations on Parties (as paraphrased below):

- To ensure that activities within their jurisdiction or control do not cause damage to the environment of other States or of areas beyond the limits of national jurisdiction. (Article 3).
- To prevent the introduction of, control or eradicate those alien species which threaten ecosystems, habitats or species (Article 8(h)).
- To ensure that the environmental consequences of its programmes and policies that are likely to have significant adverse impacts on biological diversity are duly taken into account (Article 14.1).

In addition, the Conference of Parties of the CBD has identified non-binding actions that it recommends to Parties. These include Decision VII/5 on marine biological diversity, that recommends Parties and other Governments use native species and subspecies in marine aquaculture (paragraph 45(g)), and expressed support for regional and international collaboration to address transboundary impacts of marine aquaculture on biodiversity, such as spread of disease and invasive alien species (paragraph 51).

The CBD, while not a standard setting instrument, is a facilitating body through which balance between economic growth (and international trade) can be sought with protection for biological values. The CBD has established an extensive programme of work and supports the Global Invasive Species Program (GISP).

19.2.4 Vector-based International Instruments

Ships have been identified as an important vector of both marine and freshwater invasive alien species (Carlton 1996; Hewitt et al. 1999; Ruiz et al. 2000; Chap. 4, Wonham and Lewis; Chap. 5, Minchin et al.). Ships, both commercial and recreational, carry organisms by a number of modes, the most important of which are ballast water and fouling (Gollasch 2002; Chap. 6, Hewitt et al.). No single International Instrument addresses both of these modes, but each is the subject of a specific Convention managed by the International Maritime Organization (IMO).

19.2.4.1 Ballast Water

The IMO within its Marine Environmental Protection Committee (MEPC) undertook assessment of the aquatic bioinvasions issue in the late 1980s.

It was internationally recognized in the mid-1980s that shipping was contributing to the transport and establishment of harmful aquatic species, specifically through ballast water. Initially an Assembly Resolution was adopted to provide guidance to vessels on appropriate management efforts to minimize the likelihood of aquatic bioinvasions. A Ballast Water Working Group (BWWG) was established to help identify the appropriate instrument for an international regulatory framework.

After more than 13 years of deliberation, the International Convention on the Control and Management of Ships' Ballast Water and Sediments was adopted on 13 February 2004 (BWM 2005). Its purpose is "to prevent, minimise and ultimately eliminate the risks to the environment, human health, property and resources arising from the transfer of harmful aquatic organisms and pathogens through the control and management of ships' ballast water and sediments". The BWM Convention fundamentally relies on a number of Guidelines that are currently under development, as well as technological advances to allow appropriate and safe treatment of ballast water.

While the BWM Convention does not provide complete coverage of vessels (military and small vessels are excluded) and has a long phase-in period, in the long term, all discharges will undergo some form of management to an internationally agreed standard, including on the high seas. Of greatest concern in the short-term is the inability for the BWM Convention or technology to resolve the issue of Short Sea Shipping routes. These routes typically remain inside the 200-nm (or even 50-nm) restrictions listed in the BWM Convention and are often of insufficient duration for Ballast Water Exchange to occur without deviation and/or delay of the vessel. As a consequence, several regional coalitions (e.g. Mediterranean, Baltic) are considering agreements to minimize the risks posed by these vessels and routes.

19.2.4.2 Biofouling

Currently no international instrument explicitly addresses the bioinvasion aspect of managing biofouling (including fouling on the vessel hull, sea chests, internal piping and other 'niche' areas such as around the propeller). However, the International Convention on the Control of Harmful Anti-Fouling Systems on Ships (AFS 2001) may inadvertently increase the bioinvasions risks associated with biofouling by removing the use of tri-butyl organotin (TBT) paints due to the deleterious environmental impacts.

The risks associated with biofouling are influenced by the type of vessel, the surfaces susceptible to biofouling and the nature of vessel movements; e.g., vessels which move slowly (e.g., mobile drilling rigs or barges) are likely to harbour more extensive biofouling communities and therefore may pose greater risks than fast moving vessels (Chap.5 , Minchin et al.; Chap. 6, Hewitt et al.).

Several national delegations to IMO MEPC have indicated that they will call on the IMO to consider mechanisms to minimise biofouling as a matter of urgency. In the meantime, the CBD Conference of Parties has called on IMO to action development of a biofouling regulatory framework. Simultaneously, the 28th Antarctic Treaty Consultative Meeting has considered biofouling risks as a matter of urgency and the Scientific Committee for Antarctic Research made a submission to the 29th Antarctic Treaty Consultative Meeting in 2006 on this issue (CEP VIII 2005).

19.2.4.3 Aquaculture and Intentional Stocking for Fisheries

Currently no international instrument explicitly addresses the use of non-native species for establishing new aquaculture industries or fisheries. Several Codes have been developed as voluntary guidelines on this issues, such as the Food and Agriculture Organization (FAO) voluntary Code of Conduct on Responsible Fisheries (Article 9 on Aquaculture Development; FAO 1995) and the ICES Code of Practice Concerning Introductions and Transfers of Marine Species (ICES 1994). The FAO has promoted the CCRF as a best-practice guide to the management and maintenance of fisheries and aquaculture facilities and is supported by several technical guidelines. Two technical guidelines address the use of Alien Species in aquaculture: FAO Technical Guideline Number 2 (FAO 1996) presents guidance on the application of the precautionary approach in capture fisheries and species introductions; and, FAO Technical Guideline Number 5 (FAO 1997) discusses each CCRF Article. Of these articles, Article 9.1.2 identifies the potential genetic impacts of introduced (alien) species through introgression and competition with native stocks. Article 9.2.3 explicitly discusses the need for consultation with neighbouring states when considering the introduction of Alien Species into a transboundary system. Article 9.3 (and all sub-articles) identifies the need to minimise the adverse effects of alien species to genetic resources and ecosystem integrity and encourage the use of native species whenever possible, the application of standard quarantine procedures and the establishment (or adoption) of codes of practice for approvals and management of introduced species.

19.3 Regional Multi-lateral Frameworks

Numerous regional multi-lateral treaties, conventions, and agreements are in place that currently address or are considering the issues of marine bioinvasions. These include regional conventions such as the Barcelona Convention (Mediterranean Sea), the Helsinki Commission (Baltic Sea), the OSPAR Commission (North-East Atlantic including the North Sea), and regional cooperative agreements such as the UNEP regional Seas programs, the South Pacific Regional Environmental Program (SPREP), and the Asia-Pacific Economic Cooperation (APEC). Here we present two examples, the activities within the Helsinki Commission and APEC.

19.3.1 Helsinki Commission

The Helsinki Commission (HELCOM – www.helcom.fi), launched in 1974, is the governing body of the "Convention on the Protection of the Marine Environment of the Baltic Sea Area" also known as Helsinki Convention. The overall objective of the convention is to protect the marine environment of the Baltic Sea from all sources of pollution through intergovernmental co-operation between Denmark, Estonia, Finland, Germany, Latvia, Lithuania, Poland, Russia, Sweden and the European Community. A Memorandum of Understanding exists between the International Council for the Exploration of the Sea (ICES) and HELCOM with the aim to cooperate more closely on matters of mutual interest.

HELCOM has been working to improve the Baltic marine environment through approximately 200 Recommendations. HELCOM deals with (selection relevant to biological invasions) the maritime transport (including ballast water matters), environmental impacts of fishery management and practices and protection and conservation of marine and coastal biodiversity. The precautionary principle is the overall approach (see Chap. 20, Campbell).

To help prevent the spread of alien species, HELCOM is also supporting the International Convention for the Control and Management of Ships' Ballast Water and Sediments and supports also country inventories of introduced species, the Baltic Sea Regional Programme, the "Database on Alien Species in the Baltic Sea" (http://www.ku.lt/nemo/mainnemo.html), regional and national workshops relevant to introduced species. One of the recent HELCOM activities regarding biological invasions is the development of a risk assessment based Regional Strategic Action Plan on Ballast Water Management for the Baltic Sea Area. As a first tool, a risk assessment approach of ballast water mediated species introductions is currently being worked out (Leppäkoski and Gollasch, in preparation).

19.3.2 Asia-Pacific Economic Cooperation

The Asia-Pacific Economic Cooperation (APEC) is a key regional forum, which promotes economic growth and prosperity in the region and strengthens the Asia-Pacific community. In preparation for the first Meeting of Oceans Ministers (MoM), the Fisheries Working Group and the Marine Resource Conservation Working Group (MRC WG) embarked on a joint project in 2001 to identify gaps in marine biosecurity delivery in APEC member economies. This project, sponsored by Australia and Chile, was designed to address the threat of introduced species that can become pests within APEC economies. The project entailed characterization of each of the economies' strategies to control marine pests and to draft elements of a regional risk management framework for possible use by APEC member economies (Williamson et al. 2002). As a direct result, the MoM (APEC 2002) identified invasive alien species as a priority for developing coordinated management efforts and informed the development of the Seoul Oceans Declaration.

APEC has also held a meeting in Beijing, China hosted jointly by the US State Department and China to evaluate the coordination of alien invasive species activities by APEC across all Working Groups and ecosystems (APEC 2005).

19.4 National Examples

Several nations have established regulatory frameworks for the prevention and management of intentional and accidental marine bioinvasions (see Williamson et al. 2002; Miller and Fabian 2004). Most nations manage intentional introductions in keeping with standard quarantine practices for purposes of managing trade activities. Here we present examples from Australia, New Zealand and the USA to illustrate three similar, but different approaches to the problem of unintentional invasions.

19.4.1 Australia

Significant and high profile marine invasions have occurred in Australia resulting in serious environmental and economic impacts (see also Chap. 25, Sliwa et al.). Of particular note are the North Pacific Seastar, *Asterias amurensis* – a voracious predator now well established in Tasmania and Port Phillip Bay (Victoria) (see Hewitt et al. 1999) the Black-striped Mussel, *Mytilopsis sallei* – a fast-growing fouling species successfully eradicated from Darwin Harbour in 1999 (Bax 1999), and a number of dinoflagellates (*Gymnodinium catenatum*, and several *Alexandrium* species) toxic to humans (Hallegraeff 1993; Bolch and de Salas 2007).

These species, their recognised impacts, and the need for identification of prevention and eradication methods, has led to the development of a taskforce for the prevention and management of marine pest incursions. The taskforce recommended the development of a National System for the Prevention and Management of Marine Pest Incursions to address all potential marine pest vectors underpinned by a risk assessment framework and to specifically establish arrangements for prevention, emergency preparedness and response, and ongoing management and control.

The development of the National System, due to be implemented from October 2006 has been coordinated by the Department of Agriculture, Fisheries and Forestry and collaboratively developed through a National Introduced Marine Pests Coordination Group consisting of representation from all Australian States and the Northern Territory, marine industries (shipping, ports, fishing, aquaculture), conservation groups and researchers. The National System is supported by an Intergovernmental Agreement between the States/NT and the federal government that was signed by all but one state in 2005 that outlines governance arrangements, roles and responsibilities. Transport, primary industry and environmental interests have been involved in the development of the National System.

Prevention measures under the National System encompass both prevention of introductions into Australia and prevention of translocations around Australia. At the international border, Australia has had mandatory requirements for ballast water management since 2000 and has actively participated in the development of the Ballast Water Convention (BWM 2005) and its associated guidelines. Ballast water is managed under the Quarantine Act (1908) and all vessels must either undertake mid-ocean exchange or receive a low risk ranking after accessing the Australian Ballast Water Decision Support System. Australia has also introduced a protocol to address biofouling on small international vessels (<25 m) which requires them to be clean on arrival in Australia or to be slipped shortly thereafter for cleaning in an approved facility (i.e., where wastes are contained). This protocol will become mandatory under the Quarantine Act after a voluntary phase-in period.

Arrangements are under development to refine the international ballast water arrangements to also include addressing the risks of movements of organisms by vessels on domestic routes. This will be addressed through State/NT legislation in a manner consistent with the Ballast Water Convention. For biofouling management domestically a series of voluntary guidelines or codes of conduct are under development for all marine sectors (e.g., fishing, aquaculture, recreational boats). Also, Emergency Preparedness and Response arrangements are in place with an Emergency Marine Pest Plan that outlines roles and responsibilities for action. A consultative committee on marine pest emergencies provides advice on responses including recommending provision of funds to scope and/or fund an eradication attempt. Where eradication has failed, ongoing management and control of pests is underpinned by National Control Plans (NCP) currently under development for key species (an NCP for *Asterias amurensis* is already in place).

A variety of supporting arrangements underpin the National System. These include monitoring, which builds on a series of comprehensive marine pest baseline surveys to establish an ongoing marine pest monitoring program targeted at high risk species and high risk locations. Communication is seen as critical to the success of the National System, and will be targeted for all sectors to increase awareness of marine bioinvasion issues, roles and responsibilities and to encourage behavioural change. The National System also has a strategy which identifies research and development needs for the next 5–10 years, which will help ensure that information needs for the national system are met. Finally, clear objectives for all parts of the National System are being established to ensure that the success of the system in addressing marine bioinvasions can be measured and the System adaptively improved.

19.4.2 New Zealand

Much like Australia, New Zealand has experienced significant biodiversity and economic losses as a result of intentional and accidental bioinvasions (see also Chap. 24, Hayden et al.). New Zealand is one of the few countries in the world to have drafted specific pieces of legislation aimed at the comprehensive prevention and management of non-indigenous species: the Biosecurity Act (1993), and subsequent amendments, and the Hazardous Substances and New Organisms (HSNO) Act (1996). The Biosecurity Act is oriented towards the management of unintentional introductions of species and sets out the standards for creating pre-border quarantine systems as well as the post-border incursion response and continued management. The HSNO Act is oriented towards the intentional introductions of new species or genotypes and is managed by the Environmental Risk Management Authority (ERMA).

Biosecurity until recently was delivered through different sectors within government (e.g. Human Health, Animals, Plants, Forests, Marine, and Conservation). Following a number of evaluations and Parliamentary reports (PCE 2000), government established the Biosecurity Council with a mandate to develop a Biosecurity Strategy, released in 2003 (Biosecurity Council 2003). As a direct consequence, biosecurity delivery has been reorganised into a new agency, Biosecurity New Zealand (BNZ), established within the Ministry of Agriculture and Forestry (MAF) in November 2004. BNZ brings together core MAF biosecurity functions and transfers responsibilities from Department of Conservation (conservation biosecurity) and the Ministry of Fisheries (marine biosecurity).

A comprehensive system for marine biosecurity has been established within the Ministry of Fisheries (Hewitt et al. 2004), with new funding sought from government to increase the marine biosecurity budget 450% (up to 5% of the total biosecurity budget). This system is split into three functional groups: pre-clearance (prior to quarantine inspection), post-clearance (after quarantine inspection) and policy (including strategic science).

Current pre-clearance activities include the implementation of the BWM Convention and management of the current Ballast Water Import Health Standard (IHS) which requires mid-ocean ballast water exchange in compliance with IMO Standards. In addition, significant effort is underway to develop appropriate and timely risk assessments to aid both prevention (development of IHS) and eradication activities. These assessments include: identifying the risk profile of the next species likely to be transferred to New Zealand; the relative risks of the primary vectors (ballast water, biofouling, aquaculture) for domestic and international transfers; and an understanding of the spatial context in which impacts are likely to occur.

Post-clearance activities include surveillance and monitoring systems to detect new incursions or changes in distribution and abundance of existing species, incursion response and long term control. New Zealand has embarked on a series of central government-funded baseline evaluations of high risk entry points to determine the current scale of introductions (Inglis, personal communication; Wotton and Hewitt 2004) using internationally accepted protocols (Hewitt and Martin 2001). A surveillance program has been in place for the early detection of incursions of a suite of six notifiable organisms (Wotton and Hewitt 2004) with the link to explicit incursion response and management plans guided by a standard protocol. The management of domestic translocation pathways has been identified as a high priority to prevent the movements of invasive alien marine species already in New

Zealand and is tightly coupled with the risk analysis of international pathways mentioned above.

The activities embedded within the policy group include obtaining New Zealand's desired outcomes for marine biosecurity consistency at international negotiations (e.g. IMO). New Zealand has been and continues to be a strong proponent of the BWM Convention and has participated in the development of appropriate guidelines underpinning implementation of the Convention. In addition, New Zealand has indicated an interest in IMO considering biofouling management to reduce bioinvasion risks as a matter of importance. The policy group is also responsible for the development of strategic science relationships, both domestically and internationally, to guarantee that future operational needs are met.

The full integration of marine biosecurity delivery has had significant consequences. Marine outcomes are now recognised as having greater importance within the system, and risk analyses and the development of Import Health Standards are deemed a priority for the next several years. In addition, the ability to maintain consistency across the various ecosystems and sectors will allow for a more realistic and transparent expenditure of public monies.

19.4.3 USA

In the U.S.A., the prevention and management of alien species was initially aimed at weeds, pests, pathogens, and injurious wildlife, with a primary focus on protecting crops and livestock. As laws were amended and new laws were enacted, the federal regime became quite complex, to the extent that in a landmark 1993 overview of the status of nonindigenous species invasions and government responses the U.S. Congress Office of Technology Assessment (OTA 1993), concluded that "The federal framework is a largely uncoordinated patchwork of laws, regulations, policies, and programs. Some focus on narrowly drawn problems. Many others peripherally address NIS. In general, present federal efforts only partially match the problems at hand".

For the pre-border prevention of fish, wildlife, and plants, the primary laws were, and continue to be, the Lacy Act (1900) and the Federal Noxious Weed Act (1974), the latter largely replaced by the Plant Protection Act (2000). In practice, these acts and their implementing regulations are largely based on "dirty lists" of a relatively small number of prohibited organisms. Unfortunately, these laws allow the prohibition of injurious species only after they have been added to the blacklists by the promulgation of specific regulations. There are often strong political oppositions to listing species, particularly by the aquaculture and pet trades (OTA 1993).

In the absence of a strong federal regime, it is left to the respective states to regulate the importation and introduction of non-indigenous species into their jurisdictions in marine and freshwater ecosystems. Among the 50 states, the mix of protective regimes is varied, and includes the use of "clean" lists of organisms allowed to be introduced (only Hawaii has a complete clean list approach), dirty lists of prohibited taxa (among the states using this approach, the OTA found that in 1993 that the median number of banned species was eight), and no lists – but with a variety of approaches to regulating introductions through permits and approvals.

The first federal attempt to specifically address bioinvasions in marine and freshwaters of the U.S. was the National Aquatic Nuisance Prevention and Control Act of 1990 (NANPCA 1990), which was passed in response to the severe problems caused by the invasion of the Great Lakes by the zebra mussel *Dreissena polymorpha*. NANPCA (1990) directed the promulgation of regulations to reduce the risk of introductions of invasive aquatic species via the discharge of ships' ballast water in the Great Lakes, and also established the interagency Aquatic Nuisance Species Task Force (ANSTF) to develop and coordinate a national aquatic nuisance species program. Following continuing invasions in other aquatic ecosystems of the U.S., Congress reauthorized and amended NANPCA with the National Invasive Species Act (NISA) of 1996. NISA broadened the coverage of ballast water management requirements to cover the rest of the U.S. outside of the Great Lakes system.

As a result of the regulations and policies established to implement NANPCA and NISA, the U.S. has one of the most protective ballast water management regimes in the world (The U.S. ballast water management requirements are described in the Code of Federal Regulations as 33 CFR 151.). The current regulatory regime addresses both ballast water and biofouling. Ships entering U.S. waters from outside the 200 nautical mile exclusive economic zone (EEZ) are required to manage ballast water using a set of options including retention on-board, mid-ocean ballast water exchange, or the use of alternative management method approved by the Coast Guard. Although no alternative treatment methods have yet been approved, the Coast Guard is actively developing regulations that will establish a ballast water discharge standard and the procedures for type approving ballast water management systems designed to achieve those standards of treatment. The threat of bioinvasions due to biofouling is addressed by a requirement that vessels operating in U.S. water regularly clean their hulls to reduce the build-up of fouling assemblages. All vessels entering ports or places in the U.S. are required to report their ballast water discharges and management actions to the Coast Guard (http://www.uscg.mil/hq/g-m/mso/ans.htm).

Interagency coordination under the umbrella of the ANSTF has grown since the establishment of the group. Originally consisting of 6 federal and 4 ex-officio members, the Task Force has grown to comprise 10 federal and 12 ex-officio members to date. In addition to serving as a venue for the coordination of federal activities, the ANSTF was directed to encourage state activities on invasive aquatic species through the development of comprehensive state invasive aquatic species management plans, and to encourage interstate cooperation and coordination through the establishment of regional panels comprised of state and non-governmental organizations. Two regional panels were originally directed – the Western and Great Lakes Panels, respectively. Additional panels were also authorized to be formed as appropriate, and the ANSTF has subsequently established panels representing the Northeast, Gulf of Mexico, Mid-Atlantic, and Mississippi River basin regions. Regional panels serve as venues for interstate coordination, and provide a conduit for information exchange between the regional and national efforts.

The primary mission elements of the ANSTF are the prevention of introduction and dispersal of non-indigenous aquatic species; detection and control of established populations of such species; and the education of the U.S. public about the need for prevention and control of these species (http://www.anstaskforce.gov). The primary prevention (pre-border) emphases of NANPCA and NISA were initially on the problem of unintentional ballast water-mediated invasions. However, the ANSTF is also seeking ways to use existing authorities of its members to identify and address other pathways of introductions, including intentional introductions. The second goal, on (post-border) control and management of established populations, focuses on developing control plans for identified species of concern, including rapid response capabilities, survey and monitoring efforts, state management plans, and research and education on the various elements of monitoring and control. The ANSTF has also worked to expand its involvement in international efforts to identify and understand invasion pathways, and foster information transfer.

Additional coordination of federal efforts to prevent and control bioinvasions is occurring more broadly, beyond the aquatic focus of the ANSTF. In response to the lack of coordination identified in the 1993 OTA report, a Presidential Executive Order (No. 13112) in 1999 created the National Invasive Species Council (NISC). NISC is tasked with coordinating and enhancing efforts to prevent and control invasive species government wide. A key effort of the NISC has been developing the National Invasive Species Management Plan. The NISC Plan (NISC 2001) is a critical coordination tool, serving as a national "blueprint" for federal action on invasive species. The Plan consists of 57 action items comprised of 170 distinct identifiable actions, 70 of which are "discrete" (expected to require finite efforts), and 100 of which are "on-going" (expected to require a continuing commitment of resources). Under the Plan, NISC developed the first invasive species crosscut budget for federal agencies for the fiscal year 2004. The Crosscut Budget was designed to enhance interagency cooperation and coordination on invasive species issues; promote the use of interagency performance-based approaches; and provide a clear and comprehensive picture of invasive species efforts across the federal government.

The broad mix of authorities and jurisdictions over the introduction and management of invasive aquatic species continues to complicate the development of a comprehensive and effective U.S. national regime. However, the coordinating activities of the ANSTF and the NISC are resulting in significant improvements in coordination among agencies. Internationally, the U.S. has been a strong proponent of the IMO Ballast Water Management Convention, and continues to participate actively in the development of the numerous implementing guidelines.

19.5 Conclusions

Due to the open nature of marine ecosystems, and the frequency and magnitude of international shipping and commerce, international coordination and harmonized regulations are particularly important for the effective prevention of marine bioinvasions. Marine bioinvasions are a trans-boundary issue, with the inaction of one country causing the potential for significant harm to other coastal states. Consequently marine bioinvasion regulatory frameworks need to be established at all levels – global, regional, and national.

It is imperative that all stakeholder interests are engaged in this process, including transport, primary industries, and environmental interests. By doing so, preventative measures will more rapidly be developed and transferred to potential exacerbators (parties whose actions transport non-native species) and reduce the need for hard regulation as a solution to the marine bioinvasions problem.

Effective national regimes, while reflecting differing social, economic, and environmental philosophies, are based broadly on similar frameworks of pre-border and post-border prevention and management efforts. Increasingly, effective national efforts also require an outward-looking policy stance that recognizes the need to engage at the regional multi-lateral and global international scales to achieve desired reductions in the threat of marine bioinvasions.

In order to guarantee equity and prevent shifting problems to offshore (and frequently smaller) nations, there is a continuing need to develop capacity building and information sharing arrangements at regional and global levels, including sharing policy and legislative developments, implementation schemes and research outcomes. These activities are highlighted in numerous multi-lateral agreements, and provide focus for future actions within national regimes.

Acknowledgements We would like to thank Gil Rilov and Jeff Crooks for the invitation to contribute to this volume. We also thank Stephan Gollasch for providing input on HELCOM, and Marnie Campbell for comments during preparation of the manuscript. We would like to acknowledge the members of the Invasive Marine Species Program of the Australian Department of Agriculture, Fisheries and Forestry (NP), the Marine Biosecurity team in Biosecurity New Zealand (CLH), and the Environmental Standards Division at US Coast Guard (RAE).

References

AFS (2001) International Convention on the Control of Harmful Anti-Fouling Systems on Ships, International Maritime Organization, London, United Kingdom. <<www.imo.org>> accessed 30 January 2006

APEC (2002) First Asia-Pacific Economic Cooperation Ocean-related Ministerial Meeting, Seoul, Korea, 22–26 April, 2002. <<www.apec.org/apec/ministerial_statements/sectoral_ministerial/ocean-related/ocean-related.html>> accessed 30 January 2006

APEC (2005) APEC Senior Officials' Report on Economic and Technical Cooperation 2005. <<www.apec.org>> accessed 30 January 2006

Bax NJ (1999) Eradicating a dreissenid from Australia. Dreissena! 10:1–5

Biosecurity Council (2003) Tiakina Aotearoa – Protect New Zealand: the biosecurity strategy for New Zealand. Wellington, New Zealand, Ministry of Agriculture and Forestry. 63p. <<www.maf.govt.nz/biosecurity-strategy>> accessed 30 January 2006

Bolch CJS, de Salas MF (2007) A review of the molecular evidence for ballast water introduction of the toxic dinoflagellates *Gymnodinium catenatum* and the *Alexandrium* "*tamarense* complex" to Australia. Harmful Algae 6:465–485.

BWM (2005) International Convention on the Control and Management of Ship's Ballast Water and Sediments, International Maritime Organization, London, United Kingdom. <<www.imo.org>> accessed 30 January 2006

Byers JE, Reichard S, Randall JM, Parker IM, Smith CS, Lonsdale WM, Atkinson IAE, Seastedt TR, Williamson M, Chornesky E, Hayes F (2002) Directing research to reduce the impacts of non-indigenous species. Conserv Biol 16:630–640

Carlton JT (1996) Pattern, process, and prediction in marine invasion ecology. Biol Conserv 78:97–106

Carlton JT (2001) Introduced species in U.S. coastal waters: environmental impacts and management priorities. Arlington, Virginia, USA, Pew Oceans Commission, p 28

Carlton JT, Ruiz GM (2004) The magnitude and consequences of bioinvasions in marine ecosystems: implication for conservation biology. In: Norse EA, Crowder LB (eds) Marine conservation biology: the science of maintaining the sea's biodiversity. Island Press, Washington, USA, pp 123–148

CEP VIII (2005) Report of the Committee for Environmental Protection, CEP VIII, Stockholm, Sweden, June 2005. <www.sweden.gov.se/sb/d/5430>, accessed 30, January 2006

Cohen AN, Carlton JT (1998) Accelerating invasion rate in a highly invaded estuary. Science 279:555–558

CBD (1992) Convention on Biological Diversity, Rio de Janeiro, Brazil. <www.cbd.org> accessed 30 January 2006

Convention on Wetlands (1971) Ramsar Convention, Tehran, Iran. <<www.ramsar.org>> accessed 30 January 2006

Drake JA, Mooney HA, di Castri F, Groves RH, Kruger FJ, Rejmanek M, Williamson M (eds) (1989) SCOPE 37. Biological invasions: a global perspective. Wiley, London, United Kingdom

Elton CS (1958) The ecology of invasions by animals and plants. Methuen, London, United Kingdom

FAO (1995) Code of conduct for responsible fisheries. Food and Agriculture Organisation, Rome, Italy. <www.fao.org> accessed 30 January 2006

FAO (1996) Precautionary approach to capture fisheries and species introductions. Elaborated by the technical consultation on the precautionary approach to capture fisheries (including species introductions). FAO Technical Guidelines for Responsible Fisheries, No 2. Food and Agriculture Organisation, Rome, Italy. <www.fao.org> accessed 30 January 2006.

FAO (1997) Aquaculture development. FAO Technical Guidelines for Responsible Fisheries, No 5. Food and Agriculture Organisation, Rome, Italy. <www.fao.org> accessed 30 January 2006.

Gollasch S (2002) The importance of ship hull fouling as a vector of species introductions into the North Sea. Biofouling 18:105–121

Grosholz ED, Ruiz GM, Dean CA, Shirley KA, Maron JL, Connors PG (2000) The impacts of a nonindigenous marine predator in a California bay. Ecology 81:1206–1224

Hallegraeff GM (1993) Review of harmful algal blooms and their apparent global increase. Phycologia 32:79–99

Hedley J (2004) The International Plant Protection Convention and invasive species. In: Miller ML, Fabian RN (eds) Harmful invasive species: legal responses. Environmental Law Institute, Washington, DC, USA, pp 185–201

Hewitt CL (2003) Marine biosecurity issues in the world oceans: global activities and Australian directions. Ocean Yearbook 17:193–212

Hewitt CL, Martin RB (2001) Revised protocols for baseline port surveys for introduced marine species – design considerations, sampling protocols and taxonomic sufficiency. CRIMP Technical Report Number 22. CSIRO Marine Research, Hobart, 46 pp

Hewitt CL, Campbell ML, Thresher RE, Martin RB (eds) (1999) The introduced species of Port Phillip Bay, Victoria. Centre for Research on Introduced Marine Pests Technical Report No. 20. CSIRO Marine Research, Hobart

Hewitt CL, Willing J, Bauckham A, Cassidy AM, Cox CMS, Jones L, Wotton DM (2004) New Zealand Marine Biosecurity: delivering outcomes in a fluid environment. N Z J Mar Freshwater Res 38:429–438

Hines AH, Ruiz GM, Chapman JM, Hansen GI, Carlton JT, Foster N, Feder HM (2000) Biological invasions of cold-water coastal ecosystems: ballast-mediated biological introductions in Port Valdez/Prince William Sound, Alaska. Regional Citizen's Advisory Committee of Prince William Sound, Valdez

ICES (1994) ICES Code of Practice Concerning Introductions and Transfers of Marine Species. International Council for the Exploration of the Seas, Copenhagen, Denmark.

ICES (2005) ICES Code of Practice on the introductions and transfers of marine organisms. International Council for the Exploration of the Seas, Copenhagen, Denmark

IPPC (1951) International Plant Protection Convention. Food and Agriculture Organization, Rome, Italy. <www.fao.org/Legal/treaties/004s-e.htm> accessed 30 January 2006

Lewis PN, Hewitt CL, Riddle M, McMinn A (2003) Marine introductions in the Southern Ocean: an unrecognised hazard to biodiversity. Mar Pollut Bull 46:213–223

Lubchenco JA, Olson M, Brubaker LB, Carpenter SR, Holland MM, Hubbell SP, Levin SA, MacMahon JA, Matson PA, Melillo JM, Mooney HA, Peterson CH, Pulliam HR, Real LA, Regal PJ, Risser PG (1991) The sustainable biosphere initiative: an ecological research agenda. Ecology 72:371–412

Miller ML, Fabian RN (eds) (2004) Harmful invasive species: legal responses. Environmental Law Institute, Washington, D.C., USA

Mooney HA, Mack RN, McNeely JK, Neville LE, Schei PJ, Waage JK (eds) (2005) SCOPE 63. Invasive alien species – a new synthesis. Island Press, Washington, DC

NANPCA (1990) US National Aquatic Nuisance Prevention and Control Act of 1990. US Congress, Washington DC. <http://nis.gsmfc.org/pubs/NANPCA.pdf>

Niimi AJ (2004) Environmental and economic factors can increase the risk of exotic species introductions to the Arctic region through increased ballast water discharge. Environ Manag 33(5):712–718

NISC (2001) The national invasive species management plan: meeting the invasive species challenge. National Invasive Species Council, Washington, D.C. <<www.invasivespecies.gov>> accessed 30 January 2006

OTA (1993) Harmful non-indigenous species in the United States. OTA-F-565. U.S. Congress, Office of Technology Assessment, Washington, DC: U.S. Government Printing Office. 397pp. <<www.wws.princeton.edu/ota/ns20/year_f.html>> accessed 30 January 2006

PCE (2000) New Zealand under siege: a review of the management of biosecurity risks to the environment. Wellington, New Zealand, Office of the Parliamentary Commissioner for the Environment

Pimentel D (2002) Biological invasions: economic and environmental costs of alien plant, animal, and microbe species. CRC Press, Boca Raton, USA

Pimentel D, Lach L, Zuniga R, Morrison D (2000a) Environmental and economic costs of nonindigenous species in the United States. BioScience 50:53–65

Pimentel D, McNair S, Janecka J, Wightman J, Simmonds C, O'Connell C, Wong E, Russell L, Zern J, Aquino T, Tsomondo T (2000b) Economic and environmental threats of alien plant, animal, and microbe invasions. Agric Ecosyst Environ 84:1–20

Reiser A, Gray Hudson C, Roady SE (2004) The role of legal regimes in marine conservation. In: Norse EA, Crowder LB (eds) Marine conservation biology: the science of maintaining the sea's biodiversity. Island Press, Washington, USA, pp 362–374

Ruiz GM, Hewitt CL (2002) Toward understanding patterns of coastal marine invasions: a prospectus. In: Leppäkoski E, Gollasch S, Olenin S (eds.) Invasive aquatic species of Europe – distribution, impact and management. Kluwer Academic Publishers, Dordrecht, The Netherlands. pp 529–547

Ruiz GM, Fofonoff PW, Carlton JT, Wonham MJ, Hines AH (2000) Invasion of coastal marine communities in North America: apparent patterns, processes, and biases. Ann Rev Ecol Syst 31:481–531

Schaffelke B, Hewitt CL (2007) Impacts of introduced macroalgae. Bot Mar 50:397–417

UNCLOS (1982) United Nations Convention on the Law of the Sea. United Nations General Assembly, Montego Bay, Jamaica. <<www.un.org/Depts/los/index.htm >>, accessed 30 January 2006

UNEP/CBD/COP/8/3 (2005) Report of the subsidiary body on scientific, technical and technological advice on the work of its eleventh meeting. Convention on Biological Diversity, 8th Conference of Parties, Meeting Paper 3, Montreal, Canada. << www.biodiv.org/doc/meetings/cop/cop-08/official/cop-08-03-en.doc>>, accessed 30 January 2006

UNESCO (1972) Convention Concerning the Protection of the World Cultural and Natural Heritage, Paris, France. << unesdoc.unesco.org >>, accessed 30 January 2006

Vermeij GJ (1996) An agenda for invasion biology. Biol Conserv 78:3–9

Williamson AT, Bax NJ, Gonzalez E, Geeves W (eds) (2002). Development of a regional risk management framework for APEC economies for use in the control and prevention of introduced marine pests. APEC Marine Resource Conservation Working Group, Final Report. <<crimp.marine.csiro.au>> accessed 30 January 2006

Wotton DM, Hewitt CL (2004) Marine biosecurity post-border management: developing incursion response systems for New Zealand. N Z J Mar Freshwater Res 38:553–559

Wyatt ASJ, Hewitt CL, Walker DI, Ward TJ (2005) Marine introductions in the Shark Bay World Heritage Property, Western Australia: a preliminary assessment. Divers Distrib 11:33–44

Chapter 20
An Overview of Risk Assessment in a Marine Biosecurity Context

Marnie L. Campbell

20.1 Introduction

Our ability to manage the variety of human induced stresses in the marine environment is hampered by limited resources, a lack of fundamental knowledge and the absence of appropriate tools (Lubchenco et al. 1991; Norse 1993). This is particularly true when faced with introduced marine species. Structured and transparent evaluation techniques that both determine and justify management decisions are needed to effectively deal with introduced marine species in both an ecological and socio-political sense (as discussed by Hewitt et al., Chap. 33). Coupling this need with knowledge, resource and data limitations has led decision makers and management to use risk assessment as a means to direct their actions.

In simple terms, risk assessment is a method of evaluating the likelihood that an event may occur and the consequences of such an event. In general, ecological risk assessment proceeds by establishing the context (e.g., introduced species in a region; hazard analysis); identifying the risk, hazards and effects (e.g., impacts on core values); assessing those risks (analyse and evaluate the risks); and treating the risk(s) (e.g., incursion response activity, mitigation, Australian Risk Management Guidelines; Standards Australia 2000, 2004). A measure of risk is derived by multiplying likelihood by consequence. Hazard analysis (a technique often confused with risk assessment) determines the actions, events, substances, environmental conditions, or species that could result in an undesired event, but does not identify the likelihood or the level of consequence. Introduced species, vectors or transport pathways are all examples of hazards.

Likelihood is the probability that an event may occur. Typically, likelihood will range from rare occurrence to highly likely (or frequent). Consequence, on the other hand, measures the impact an event may have on the values being assessed and can be derived by measuring the change in value from a pre- and post impacted system. Although monetary units are often used to measure change in value (because they are easily understood and facilitate comparison) this does not have to be the unit of measure; semi-quantitative categorical ranking (e.g., low, medium, high value) is also possible.

G. Rilov, J.A. Crooks (eds.) *Biological Invasions in Marine Ecosystems.*
Ecological Studies 204,

In general, when management uses risk assessment to evaluate introduced marine species, they ask the question "what is the likelihood that a species will arrive in our region", or "what impacts (consequences) will that species have on our native biota".

20.1.1 Defining Endpoints

Before undertaking a risk assessment, the risk endpoint must be identified and agreed. Endpoint selection will determine the null hypothesis that is tested during the analysis. With introduced species risk assessments, the endpoint tends to be either: (a) quarantine related – where the species has arrived, and therefore barrier control has been breached resulting in a quarantine failure; or (b) impact driven – where the risk assessment examines the effect, impact, and/or harm the introduced species will have as the basis of decision making.

If a quarantine stance is taken, then all introduced species consequences are classified as 'significant' and the likelihood must be determined to derive risk. The International Convention for the Control and Management of Ship's Ballast Water and Sediments (BWM 2005) approaches introduced marine species from a quarantine stance (see also Chap. 19, Hewitt et al.), which tends to blanket all introduced species as causing significant consequences. In reality this may not be the case, as species can be assessed against environmental, economic, social and cultural values. The convention identifies "harmful aquatic organism or pathogens" as the management target, implying that some impact assessment is necessary (BWM 2005; Gollasch et al. 2007).

If the assessment is determined to be impact driven, then both the likelihood of arrival (and survival) and the impact of the arrival (consequence) must be determined to derive risk. An impact approach is typically followed when determining if an incursion should be eradicated or managed based on its likely spread and subsequent impact. Similarly, an impact driven assessment will occur to identify species that have not yet arrived but are of greatest concern, a method similar to identifying a blacklist. If a species is seen as causing negligible to low risk, then it is likely to be monitored and no further action taken due to the cost of eradication being greater than the benefit (M. Cassidy, Biosecurity New Zealand, personal communication 2005).

20.1.2 Core Values

To aid the prioritisation of management actions for an introduced species incursion or an import request, the real and perceived impacts the species may have must be examined against core values (environment, economic, social, and cultural) in the import/incursion region and other potential regions that may be capable of sustaining the species of concern. Using core values increases the transparency of decision making and places management action into a context of objectively assessing introduced species across environmental and socio-political issues.

Individual core values have typically been assessed separately using tools such as species impact assessment (SIS) (New South Wales Department of Urban Affairs and Planning 1996, 2000; Thomas and Elliott 2005), economic valuation analysis (e.g., Costanza et al. 1997; Toman 1998; Pagiola 2004; Kalof and Satterfield 2005), social impact assessment (SIA) (Lang and Armour 1981; Thomas and Elliott 2005), environmental impact assessment (EIA) (Thomas and Elliott 2005), and strategic environmental assessment (SEA) (Marsden and Dovers 2002; Dalal-Clayton and Sadler 2005). Within the introduced species context, the effort is now being made to assess all core values under the one method (combining risk analysis and risk assessment) and defines the core values as such:

- Environment – everything from the biological to physical characteristics of an ecosystem being assessed, excluding extractive (economic) use and aesthetic value. Examples include floral and faunal biodiversity, habitat, rare, endangered and protected species and marine protected areas.
- Economics – components within an ecosystem that provide a current or potential economic gain or loss. Examples include the infrastructure associated with ports, marinas and shipping channels, moorings and allocated mariculture and fisheries areas.
- Social – the values placed on a location in relation to human use for pleasure, aesthetic, generational values. This value may also include human health. Examples include tourism, recreation, education and aesthetics.
- Cultural – those aspects of the marine environment that represent an iconic or spiritual value, including those that create a sense of local, regional or national identity.

Each core value consists of a variety of different subcomponents (examples given above) that will differ both spatially and temporally. A risk assessment can occur at the level of the core value or at the level of core value subcomponent(s).

20.1.3 Uncertainty and the Precautionary Principle/Approach

Regardless of the method used, risk assessment will have uncertainty surrounding the outcomes. Uncertainty exists because of natural and stochastic variation in our environments that are difficult to capture, and incomplete understanding of the biological, physical and anthropogenic systems (Cooney and Dickson 2005; Peel 2005). As ecosystems are highly complex and interconnected, varying both spatially and temporally, it is often impossible to predict ecosystem dynamics (see Burgman et al. 1993; Harwood and Stokes 2003).

Although uncertainty exists, there is a fundamental need for environmental management to make decisions. To aid decision makers to overcome uncertainty, the precautionary principle/approach was developed and has been widely adopted in environmental management (Gullett 1997; Cooney 2005; Peel 2005). As stated by Gullett (1997), the precautionary principle/approach imposes an environmental duty of care meant to prevent spatial and temporal damage. The principle/approach acknowledges the intrinsic value of ecosystems, the 'economic utilitarianism' of these systems (i.e., bequest value; Handl 1990), and the 'moral right' to protect these systems (Cameron 1993; Gullett 1997).

As various definitions exist for the precautionary principle/approach, in this paper I use the definition from Cooney (2005): "Complete certainty regarding an environmental harm should not be a prerequisite for taking action to avert it." This is a preferred definition as it addresses precaution from a Convention on Biological Diversity perspective and hence is more environmentally conservative, placing the environment as a higher priority than trade. I have also used the terms 'principle' and 'approach' interchangeably, although acknowledging the problems associated with both terms.

A poignant example of this dilemma can be seen in the differing management strategies implied by the legislative requirements of the Sanitary and Phytosanitary Agreement (SPS Agreement 1994), requiring a risk assessment before any restrictions can be imposed, and the guidelines of the Convention on Biological Diversity (CBD), which requires risk assessment before any new species should be admitted (introduced) to a country (Cooney 2004). The Sanitary and Phytosanitary Agreement was developed by the World Trade Organisation (WTO) and addresses the issue of food safety, animal and plant health (typically via importation of products) and is applicable to all current 151 WTO members. The WTO has no specific agreement on the environment, although it acknowledges the concept of sustainable development and environmental protection. Alternatively, the CBD was developed by United Nations Environmental Programme (UNEP) to conserve biological diversity, to use nature's components sustainably and to share equitably benefits arising from the use of genetic resources. Currently, the CBD is signed by 150 countries.

20.1.4 *Quantitative vs Qualitative*

Quantitative and qualitative risk assessment procedures have been used for marine introduced species in a number of countries such as Australia (Kahn et al. 1999; Hayes and Hewitt 2001; Hewitt and Hayes 2001, 2002; Hayes 2003; Anon 2005); Chile (Campbell 2005b; Hewitt and Campbell 2005; Hewitt et al. 2006); Germany (Gollasch 1996); New Zealand (Campbell 2005a, c); and Nordic waters (Gollasch and Leppäkoski 1999). Similarly, countries in the Mediterranean (through the Regional Activities Center for Specially Protected Areas of the UNEP Mediterranean Action Plan [RAC/SPA]) are now moving towards development of a standard risk assessment process for introduced marine species (Campbell 2005d). Many countries have developed risk assessment processes but they do not specifically address marine bioinvasions, or marine biosecurity (management) and hence are omitted from this chapter. Typically, quantitative risk assessment (e.g., Hayes and Hewitt 2001; Hewitt and Hayes 2002) is highly robust, but requires significant levels of data and information, which requires considerable input of funds and time. Semi-quantitative and qualitative risk assessment methods are useful when low or variable levels of information are available and the lack of quarantine failure in countries using these methods suggests their effectiveness (Kahn et al. 1999; Anon 2005; Campbell 2005a, c).

No matter if quantitative or qualitative methods are used, a trustworthy risk assessment can only be produced if well defined procedures for determining appropriate consequence and likelihood measures exist. These procedures need to establish, in a clear, transparent and scientific manner, a consistent process that identifies and evaluates risk, providing adequate and robust response mechanisms for the risk assessment outcomes. Typically, this involves informed stakeholder input, taking into account all available information, and explicitly stating uncertainties, assumptions and trade-offs.

This chapter provides an overview of qualitative and semi-qualitative risk assessment methods that have been applied in the context of introduced marine species management (i.e., marine biosecurity). Few quantitative methods are used at the decision making level[1] due to their onerous data and information requirements. Examples of where qualitative and semi-qualitative types of risk assessment are being successfully applied on an international and regional basis are also provided. This chapter is marine and management focused and, hence, introduced risk analyses that are freshwater in focus (e.g., Kolar and Lodge 2002; Herborg et al. 2007) and/or are not currently used by management (e.g., Lodge et al. 2006; Leung and Dudgeon 2007) are beyond the scope of this chapter.

20.2 Types of Risk Assessment

Introduced marine species risk assessment tends to use three approaches:

- *Species level risk assessments* that may be applied to intentional and unintentional introductions or translocations to help identify high risk introduced species, generally prior to importation
- *Vector based risk assessments* that allow for the differentiation within a vector of high risk items (e.g., vessels, pieces of gear, farms) or activities to aid management outcomes
- *Pathway level risk assessments* that allow for a cross comparison between different vectors or between different "**nodes**" such as ports and marinas

[1] A notable exception is the AQIS Decision Support System; based on the risk assessment developed by CSIRO (Hayes and Hewitt 2001; Hewitt and Hayes 2002).

20.3 Species Level Risk Assessment

There are a variety of circumstances for which species level risk assessment are suitable, such as assessing intentional introductions prior to import certification, or post-hoc analyses after an incursion (unintentional) has been detected. To undertake a species risk assessment successfully, Hewitt and Hayes (2001) suggest the following information is needed:

- Propagule pressure: the amount of biological material arriving into a specific location (e.g., country, state, region, port)
- The number of sites of release for the species
- The number of introduction events
- To a lesser extent, the environmental tolerances of a species' native distribution compared to the region being assessed

20.3.1 Species Level Risk Assessment Examples

Three common methods employed to assess a species level risk in both intentional and unintentional situations are the development of next pest lists (Hewitt and Hayes 2001; Hayes and Sliwa 2003), Organism Impact Assessments (for post-hoc assessments of incursions; Campbell 2005a), and the development of Import Health Standards (for intentional importation of species; Kahn et al. 1999; Anon 2005). A fourth method, the ICES Code of Practice for the Introduction and Transfers of Marine Organisms (ICES 2005), provides an example of a procedural methodology that incorporates the risk assessment and decision making process for intentional introductions. The ICES Code (ICES 2005) evaluates on the basis of individual planned species movements, with the intent to identify whether the target species is likely to cause harm, and whether any associated species living in, on, or with the target are likely to cause harm, including parasites, disease agents, and human pathogens. It is a useful tool for intentional introductions.

20.3.1.1 Next Pest Lists

Identification of species of concern is a difficult and often controversial task. Nonetheless, several countries (e.g., Australia, New Zealand) have adopted a target species approach to marine biosecurity (i.e., management context). Scientific approaches have also been applied to assess potential new pests but these have not been adopted by management at this point in time (e.g., Herborg et al. 2007; Leung and Dudgeon 2007) and/or lack a marine focus (e.g., Kolar and Lodge 2002; Rixon et al. 2005). The next pest lists approach generates target species that are "blacklisted" or identified as "unwanted organisms" and hence are unable to be imported into a country (through import health standards) unless an exemption is granted. This assessment uses a quarantine endpoint.

Development of next pest species lists rely on evaluating species against set-criteria. The criteria provide a clear, explicit, transparent and non-discriminatory method for evaluating and identifying potential species hazards. One possible set of criteria (based on hull fouling and ballast water) are (Hayes and Sliwa 2003):

- The species has been reported in a shipping vector or has a ship-mediated history
- The vector still exists
- The species has been responsible for environmental and/or economic harms
- The species is introduced to *[country/region]* or present in *[country/region]* but subject to official control (i.e., listed, restricted or otherwise legislated by an authorised national authority)

20.3.1.2 Organism Impact Assessments

An organism impact assessment (OIA) evaluates species risk using an endpoint of impact: does or will the introduction of the species cause an impact on environmental, economic, social, and/or cultural values? OIAs are used to evaluate potential impacts from the unintentional incursion of an introduced species using heuristic knowledge drawn from the literature and from expert panels/technical advisory groups (e.g., Campbell 2005a). This method is similar to a 'relative risk assessment' as discussed by Roberts et al. (2002).

If there is a paucity of published, empirical scientific data on the impacts of a particular introduced species, a *delphi* approach is adopted. The delphic approach utilises a number of focus groups from different regions, with focus group membership drawn from a range of stakeholder interest, thus representing a wide range of community perceptions. This approach creates a statistical population of beliefs that captures a wide range of community opinions with the central tendency being the perceived risk. A focus group aims to assess perceived value of an area and then assess the perceived impact(s) to this value if an introduced species incursion occurs in that region. The data collected from these focus groups is then analysed and a risk assessment of the introduced species impact on environmental, economic, social and cultural values is determined.

An important aspect of the OIA is the use of valuation methods to determine "value". Numerous methods exist for determining value and in general they fall into the categories of revealed preference, such as replacement cost, travel cost, hedonic pricing, or stated preference methods, such as contingent valuation and choice modelling (Pagiola 2004; Farber et al. 2005). Each method has its benefits and limitations and should be selected carefully to represent the value accurately. OIAs also have the ability to display uncertainty by providing the range of likelihood, consequence and valuation data as determined by the focus groups.

To a certain extent an OIA is subjective and imprecise; however it does have strong inherent advantages such as the ability to produce a result when empirical data is insufficient or lacking, stakeholder input across a range of regions leading to high stakeholder understanding and buy-in, transparency and education (data on introduced species and effects is provided to stakeholders), and stakeholder participation by providing perceived risk.

20.3.1.3 Import Health Standards (IHS)

IHS are legislative procedural documents that are established to ensure that the internationally agreed standard of quarantine (typically SPS agreement) and scientific evaluation are met to reduce the unwarranted restrictions of trade when importing goods. In this context, an IHS is used to assess risk associated with intentional introductions of species (Anon 2005). Because the species are being intentionally imported the likelihood is considered as 'almost likely' in every assessment, with the consequences on human, animal and plant health being assessed. Rarely is the impact that an import species may have on the environment assessed by management. For example, an IHS for 'vannamei' prawns (e.g., *Litopenaeus vannamei*) would examine if a pathogen listed by the World Organisation of Animal Health (OIE) is associated with the imported species. Consequently the risk posed by the associated pathogen to human, animal and plant health is assessed within the importing country (e.g., Biosecurity Australia 2000; Inland Water Resources and Aquaculture Services [FAO] 2003). The impact that *L. vannamei* has on local prawn or other native species is not assessed within an IHS. This is a failure of the IHS system where the focus is on the import species pathogens, rather than the impact the imported species may have on native populations.

IHS are similar to the ICES Code of Practice (ICES 2005), combining both risk assessment and the decision making process for intentional introductions.

When a request for an importation of a species is received, it initiates a series of steps that lead to both risk analyses and risk assessment being undertaken. In this instance the endpoint is to assess what impact this species will have on the values of the recipient region (most often defined as the widest possible range a species may attain; its fundamental niche). Most IHS assessments are species-specific; assessing the individual species and its possible associated species, however some are vector based. For example, a request to import adult oysters for aquaculture purposes would involve a risk analysis of the oyster species itself, and risk analyses of all possible epi- and endo-biont associated species (species growing on and in the imported species) known from the donor region. This would then involve overlaying the risk analysis outcomes with social, economic and cultural imperatives to provide a risk assessment. Both positive and negative impacts are assessed in the IHS process. Typically, low to negligible risk species are granted approval for importation, with moderate to extreme risk species being rejected. However, moderate to extreme risk species can be granted importation approval (through exemption) if quarantine/containment standards are applied, met, monitored and reported upon.

The outcome of the IHS and its associated analyses is a list of species ('white' list) that is appended to the IHS document. The white list contains negligible to low risk species that have been assessed and approved for importation. Once added to the white list a species is granted future importation approval, which allows the rigour of the risk analysis, risk assessment and importation process to be bypassed. To be effective the IHS document and its associated white list of exempted species need to be regularly re-evaluated and updated, especially when new information becomes available. Two examples of efficient and active IHS documents are the *Australian Import Risk Analysis for Live Ornamental Finfish* (Kahn et al. 1999) and the *New Zealand Import Health Standard for the Importation Into New Zealand of Ornamental Fish and Marine Invertebrates from All Countries* (Anon 2005).

20.4 Vector Based Risk Assessment

Vector based risk assessments identify which shipments or potential incursions are more risky than others (e.g., ballast water risk assessment undertaken in Australia). There are a large number of vectors that are known to be responsible for the transfer of marine introduced species (Carlton 2001; Chap. 5, Minchin et al.). Typically, the examples of ballast water and associated sediments, hull fouling and mariculture (aquaculture) have been concentrated upon (see Chap. 6, Hewitt et al.).

The most widely established vector based risk assessments have been applied to the management of ballast water and sediments. These assessments have been performed by a number of countries and organisations, and have been based on two primary types of assessment: environmental matching where two environments are compared for similarity (or dissimilarity) across a range of environmental variables believed to have ecological significance; and species based assessments where a chain-of-events model is used to determine the likelihood of a species arriving and establishing in the receiving environment (Hewitt and Hayes 2002). Both types of vector based risk assessments can be applied at varying geographic scales, such as at the bioprovince, down to smaller regions (e.g., nation, state, marine protected area).

Environmental matching typically evaluates similarity in a statistical sense, with no biological determinant of the cut-off between similar and dissimilar. Similarly, the selection of environmental parameters for evaluation is rarely based on species' requirements for survival, but instead are readily accessible environmental characteristics of the donor and recipient regions. As a result, while environmental matching assessments have a reduced data requirement, they typically result in less conservative outcomes with greater likelihood of Type I error (finding a difference where none exists, resulting in an erroneous low risk).

In contrast, species based risk assessments rely on detailed knowledge of the species' distributions, reproductive periodicity, physiological constraints and environmental preferences. Species level risk assessments have a high data requirement, and typically result in overly conservative outcomes with greater likelihood of Type II error (finding no difference where one exists resulting in an erroneous high risk).

The International Convention on the Management and Control of Ships Ballast Water and Sediments has developed a Risk Assessment Guideline (G7) that underpins the ability of a State to grant exemptions from the obligations of the Convention. The current formulation of G7 (adopted in July 2007; Resolution MEPC.162(56)) develops a framework in which both environmental matching and species based assessments are used. This formulation suggests that environmental matching risk assessments should be used only in circumstances where the environments are at biological extremes, such as between purely freshwater and purely marine environments. In these circumstances, those species that can survive at both extremes should be individually assessed.

In contrast, species based assessments should only be used within a single bioprovince (such as the Mediterranean) where the assumption is that the majority of native species are shared. In these circumstances, the unknown species can be assumed to be native, reducing the number of species assessments required. For donor ports, introduced species known to cause harm should be assessed for the ability to establish and cause harm in the recipient port (and adjacent localities). Harm should be assessed according to specific impact on core values and resources. Species based assessments need to be reviewed regularly because newly available information may alter the risk outcomes.

20.4.1 Vector Based Risk Assessment Examples

The development of import health standards (IHS) such as the New Zealand Import Health Standard for the Importation of Ballast Water (Biosecurity New Zealand), the Chilean Aquaculture Species Import Process and the New Zealand microalgae import decision tree, are examples of risk analyses that evaluate vector risks.

20.4.1.1 Import Health Standards

As previously stated, IHS work by investigating the validity and risk posed by all requests to import a species or a vector (i.e., the emphasis of the analysis is placed on the vector itself). There are a number of specific IHS that apply to vectors. A current example of a vector IHS is the ballast water exchange at sea requirements. Other vectors include fishing equipment, marine rock (including live rock from the aquarium trade), imported recreational vessels, ropes and anchors. Vector based IHS are used for regulatory purposes and when consequence has been demonstrated. They provide action to mitigate the likelihood, by providing information such as where ballast water exchange can occur, quarantine, and cleaning and dumping standards.

20.4.1.2 Aquaculture Species Import Model

Mariculture and aquaculture are growing global industries that are attempting to address the problem of expanding populations and decreasing fish stocks.

A number of regions have decided that economic stability and food security can be improved by utilising introduced marine species to either aid in providing food to the region's population, or aid in providing an export product that is highly valued elsewhere and therefore marketable. Both these reasons have merit, with the ethical use of introduced marine species needing to be considered against the social and economic security that such a use may provide. Few models exist that specifically target introduced species importation for aquaculture/mariculture purposes. The following model is one that has been adopted in Chile, South America, and has operated with reasonable success (Campbell 2005b; Hewitt and Campbell 2005; Hewitt et al. 2006). It has a quarantine endpoint.

The model is initiated when a request to import a non-indigenous species or non-indigenous genome occurs. The request is made using standardised templates, thus allowing a transparent assessment process. At a minimum, the request should include information that allows the decision makers to determine:

- *Species*: the species and associated species involved in the request; known impacts the target species has had elsewhere, if any; what the species will be used for; can a local species be used instead; will the target species require the importation of a specific food source that is also introduced.
- *Export facilities*: where does the importation originate from (bioprovince, water temperature, salinity, disease information); certification and quarantine procedures followed by the exporting region; how the importation will occur (specify whether it is importation of larvae, eggs, juveniles, adults; what measures will be taken to reduce fouling of adults; what practices are used to detect disease); are the imported stock from wild stocks, mariculture/aquaculture facilities; and are the imported stock genetically modified or been fed with a genetically modified food source.
- *Import Facilities*: who is making the request (person, company, local, regional, national, international); the containment and quarantine procedures that will be followed (if these need to be established, how will they be peer reviewed); does the facility meet regional, national, and/or international certification; information about the recipient aquaculture facility (is it an open or closed facility; filtration systems used; does translocation of species between facilities occur); is there any likely release of material into the marine environment; what emergency containment procedures exist; what contingencies exist for disease outbreak containment within the facility; are there any requirements for the transfer of species between facilities within the country (e.g., establishing a brood stock facility); and the proximity of the facility to high value areas, specifically those protected by national or international obligations.
- *Monitoring*: what type of environmental health monitoring will be established; what type of environmental monitoring will occur; what is the frequency of monitoring; is the monitoring peer reviewed and provided to a statutory body for assessment; and what provisions (contingency measures) exist if an accidental release of the introduced species occurs.

It is the role of the decision makers to use this import risk model. To be efficacious the risk process needs to define what impacts are unacceptable, what methods will be used for the risk assessments, set an acceptable level of risk, establish a scientific overview and review committee and develop contingency and action plans or guidelines to deal with the accidental release of a non-indigenous species (Fig. 20.1). The values (and/or the subcomponents) that the decision makers are attempting to protect and manage must be identified a priori. This can occur through a simple evaluation of national and international obligations (e.g., CBD), or it can be as complex as evaluations of individual subcomponents of the environmental, economic, social and cultural values. In order to have a clear, transparent and consistent process it's ideal to identify the values a priori, instead of identifying values with each solicitation.

In some instances, it may be necessary to conduct experimental trials with a species to determine its ability to survive, grow and be controlled in certain conditions. These may occur in the donor (risk minimisation) or recipient country. Upon completion of the risk assessment a decision maker is able to determine whether to reject an application or move into a second phase: a cost-benefit analysis. A cost-benefit analysis determines the net benefits of an introduced species to the ecosystem, economy, socially and culturally, and assesses the costs associated with an introduced species incursion (e.g., destruction of infrastructure, loss of jobs, loss of industry, loss of marine resources, extinction of species, etc.). Consistency is maintained across all solicitations by ensuring the valuation methodology and its limitations are stated a priori to the analysis. Finally, based on the outcomes of the cost-benefit analysis a decision is made whether to reject or accept the import request.

20.4.1.3 Microalgae Import Decision Tree Model

A further model that can be used in conjunction with IHS procedures is a decision tree that leads the decision maker through a series of questions with "if/then" statements to direct actions regarding whether to approve an importation or not. In New Zealand, a risk assessment process that uses a decision tree model exists for the importation of microalgae (native and non-indigenous species; Campbell 2004). By answering a series of simple yes/no questions the decision tree progress through the process indicating where importation should be rejected, approved with stipulations or approved without stipulations. The model can be qualitative, semi-quantitative or quantitative and is driven by the data input. As with IHS procedures, likelihood is almost certain since the species is being imported. Each step is assessed against a risk mitigation context (such as a management procedure) with the endpoint derived by the questions asked at each step in the process. Decision tree models invariably consider specific national and international obligations. In New Zealand, an eight-step model was developed for the importation of microalgae typically used for laboratory purposes (colour standards) and aquaculture feed that is released directly into the marine environment (Campbell 2004; Fig. 20.2). Such models are

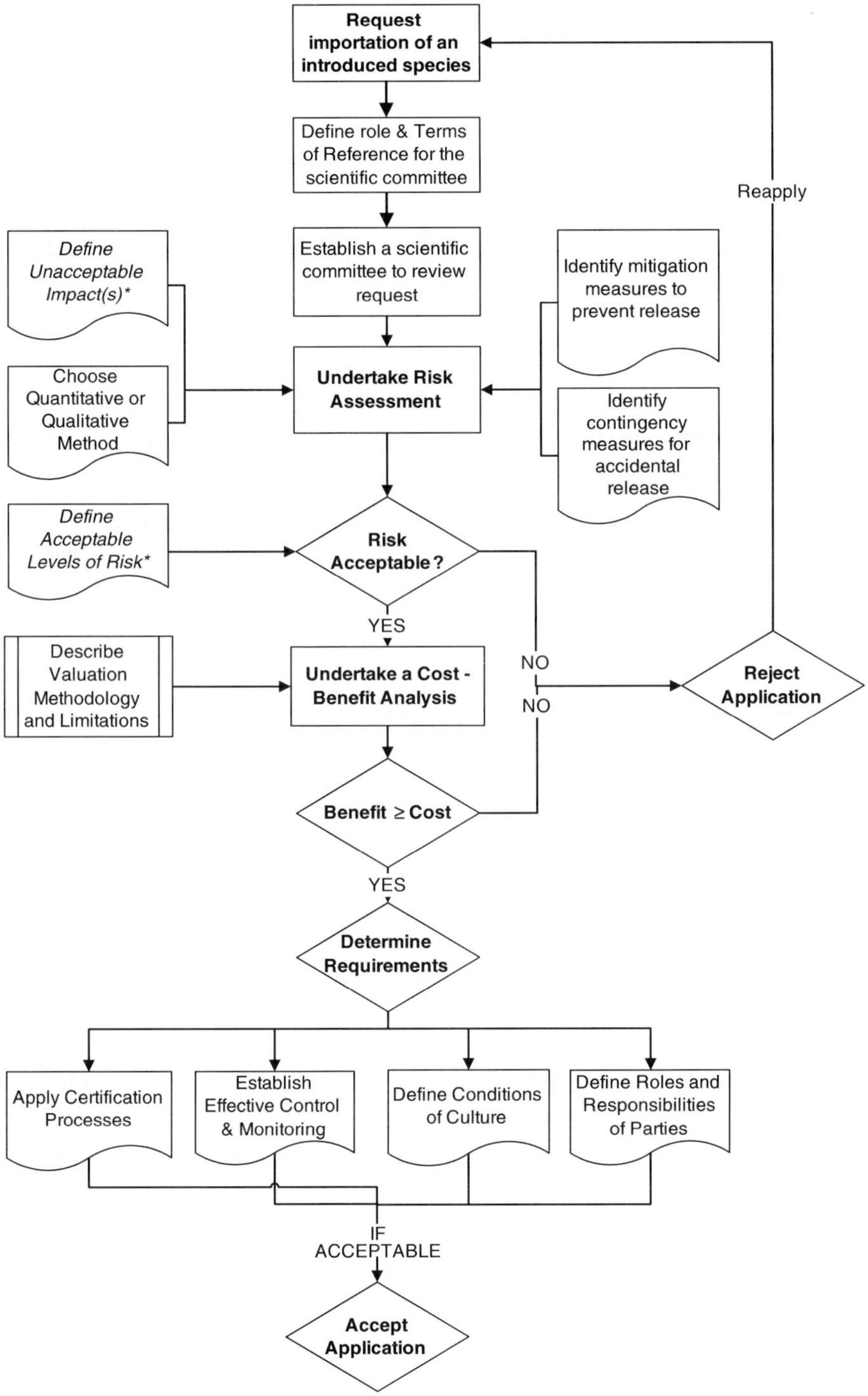

Fig. 20.1 Conceptual risk framework used in Chile for the importation of non-indigenous species for aquaculture purposes (modified from A. Brown, personal communication). * denotes actions that can occur a priori to the risk assessment

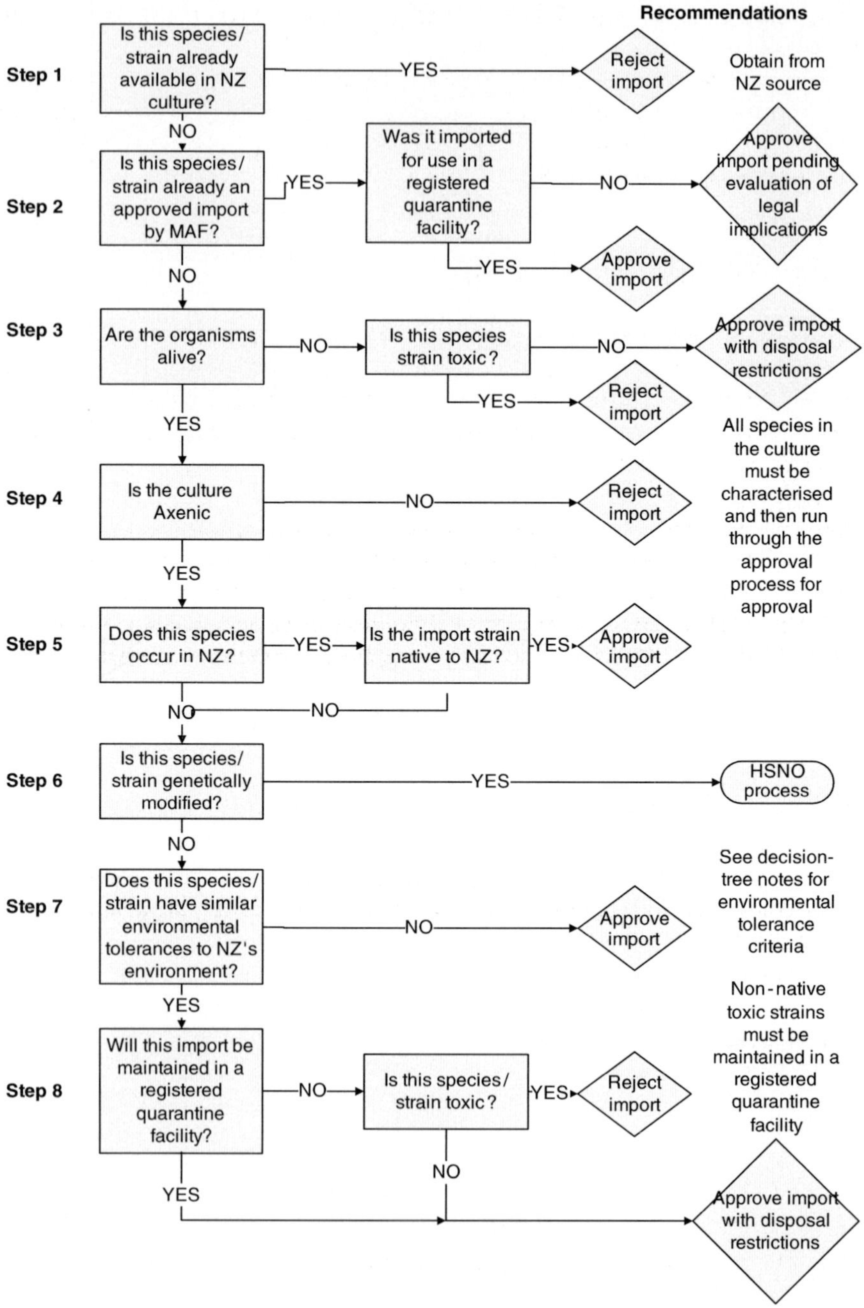

Fig. 20.2 Microalgae decision tree model used to assess the risk associated with the importation of microalgae in New Zealand. HSNO denotes the Hazardous Substances and New Organism Act (1996)

readily adaptable to other countries, regions and/or taxa (such as fish and invertebrates).

A strength of the decision tree model is the ability to incorporate multi-level analyses that deal with introduced species and genetically modified organisms. A further strength is that this model combines biological and social information, as well as legal obligations, into clear instructions for decision makers.

20.4.1.4 Fishing and Domestic Shipping

In both New Zealand and Australia, fisheries activities and the potential risk these activities pose via either entraining or translocating "pest" species are currently being investigated using vector risk assessments. These studies provide hazard information (e.g., vector movements by fishery, vessel type, time, origin, destination, etc.) that feed into the development of guidelines. It is envisaged that education about the problem of translocation via fishing and aquaculture methods, coupled with consultation will be required for the guidelines to be truly effective. Similarly, domestic or regional shipping is also being investigated in New Zealand and Australia, to develop vector risk assessments for differing shipping types. A general outcome of this research is the development of best practice guidelines that provide preventative advice for recreational vessel owners.

20.5 Pathway Risk Assessment

Pathway risk assessments combine aspects of species and vector assessments to identify their intersections and overlaps (e.g., Biosecurity New Zealand pathway analysis, GloBallast assessment). Typically, this method concentrates on nodes such as ports or marina's, the vector exposure strength and their transport corridors; examining which nodes are more likely to receive a new organisms. These analyses have a quarantine endpoint; attempting to assess risk before a species arrives within a region.

20.5.1 Pathway Risk Assessment Examples

Unlike the terrestrial environment (e.g., Andow 2003), few marine pathway analyses have been completed for the marine environment, although a number of research projects are currently attempting to elucidate the relationship between pathway strength, transport corridor and vector type. The following examples are of research that is currently underway in New Zealand and Australia and as yet remain unpublished.

20.5.1.1 Hull Fouling and Pathways

To capture fully the risk associated with hull fouling (or other vectors) requires robust, empirically derived data. The following example from New Zealand illustrates how a pathway analysis can be used to determine risk. Currently in New Zealand, a three-year study is underway to determine the realised risk associated with hull fouling (via vector and pathway analyses). This research examines the extent of fouling and fouling species identity on the hulls of arriving international vessels. Categories of vessels being examined are: fishing, passenger, merchant, slow-moving barges, oil platforms, and recreational vessels. The research investigates seasonal trends in vessel fouling for each vessel type, associated trade routes and target source/donor regions (IUCN bioregions) based on a priori analyses of previous shipping (merchant and recreational) history and customs data. This type of research is data and effort intensive but surprisingly inexpensive (NZ$<3 million) considering the detailed data that is generated and the multiplicity of this data's uses.

It is anticipated that the outcomes of this research will feed into 'realised' hazard and risk assessments associated with each port and marina dealing with international vessels within a country or region. This in turn greatly improves the ability of decision makers in the development of introduced marine species guidelines and standards.

20.5.1.2 Nodal Analysis

Nodal analyses aim to examine the strength of different vectors such as hull fouling (commercial and recreational), ballast water, and aquaculture into specific nodes (such as ports, marinas, protected areas). The nodal analysis investigates donor/recipient interactions and likely flow-on-effects. This type of analysis is currently being undertaken across Australia by the Department of Agriculture, Fisheries and Forestry.

20.5.1.3 Single Vector Pathway Analysis

One component of GloBallast risk assessment is a pathway analysis. In this instance, the GloBallast risk assessment concentrates on a single vector, examining the relative strength of ballast water movement (i.e., surrogate for propagule strength) between various source ports and receiving ports. These analyses were implemented for the six GloBallast ports in Brazil, China, India, Iran, South Africa and the Ukraine and are coupled with the GloBallast environmental matching exercise to aid in the recommendation of management strategies for ballast water management between ports (Awad et al. 2004; Alexandrov et al. 2004; Anil et al. 2004; Clarke et al. 2003, 2004a, b).

20.6 Conclusions

In conclusion, risk assessment for marine bioinvasions is not a single tool but a process that can be applied across a range of protected values (environmental, economic, social and cultural). As discussed above, there are a variety of different risk assessment methods that are currently used by decision makers to manage marine bioinvasions. Although assessment methods may differ, a number of commonalities exist: for example, risk assessments may evaluate species, vectors, or pathways; endpoints may be quarantine or impact driven; and the assessment may be quantitative, semi- quantitative, or qualitative.

Management has moved towards using risk assessment to seek a balance between complete environmental protection and the social use of public funds, while satisfying WTO obligations. It is insufficient for management to halt the importation of a species merely because it is non-indigenous, the burden of proof lies with the receiving nation to demonstrate that impact will deleteriously affect their economy. Current debate concerning national obligations to the CBD with respect to WTO limitations will continue into the foreseeable future. Similarly, incursion response is tempered by the realities of cost-benefit; a species must in all likelihood cause a high level of risk before management can stop its importation or before an incursion response can occur. This is driven by the fact that public funds are limited, managers require decisions to be made in a clear and transparent manner, with a minimum time delay; decisions must be scientifically robust under the WTO and CBD; and local, national and international obligations must be met.

As a consequence of the conflicting requirements placed upon managers and decision makers, risk assessment forms the basis for consistent and transparent decision making under a precautionary approach (as interpreted by national government). The current formulations of risk assessment are largely qualitative with limited data requirements in order to achieve functionality. Those few examples of quantitative marine bioinvasion risk assessments have been found too data onerous by decision makers and hence have been simplified or abandoned (e.g., international application of the Australian Decision Support System for Ballast Water Management). To achieve effective risk assessment based decision making in the future, both quantitative and qualitative methods will need to be developed that provide immediate results with robust information. To do this, significant research efforts must be undertaken that identify the linkages between species, vectors and pathways, as well an understanding of the likely degree of impact that individual species will have on specific values. This research agenda is significant and costly and will best be achieved through capacity building and information sharing between the World's nations.

Acknowledgements I thank the following people for valuable discussions on risk assessment and their current risk assessment practices: C. Reed, S. Sathyapala and M. Cassidy (Biosecurity New Zealand); N. Parker (formerly Department of Agriculture, Fisheries and Forestry, Australia), C. Hewitt (National Centre for the Marine Environment, Australia); J. Matheickal (International Maritime Organisation, GloBallast); I. Melianne (IUCN, global); Souha El Asmi (RAC/SPA, UNEP, Mediterranean); Alex Brown (SubPesca, Chile). Figure 20.2 is used with permission from

Biosecurity New Zealand. I'd also like to thank Gil Rilof and Jeff Crooks for their invitation to contribute to this volume.

References

Alexandrov B, Bashtannyy R, Clarke, Hayes T, Hilliard R, Polglaze J, Rabotnyov V, Raaymakers S (2004) Ballast water risk assessment, port of Odessa, Ukraine. GloBallast Monogr Ser No 10. IMO, London. <http://globallast.imo.org/index.asp?page=monograph.htm&menu=true> accessed 04/09/2007

Andow DA (2003) Pathways-based risk assessment of exotic species invasions. In: Ruiz GM, Carlton JT (eds) Bioinvasions: pathways, vectors, and management strategies. Island Press, Washington DC, pp 439–455

Anil AC, Clarke C, Hayes T, Hilliard R, Joshi G, Krishnamurthy V, Polglaze J, Sawant S, Raaymakers S (2004) Ballast water risk assessment, ports of Mumbai and Jawaharlal Nehru, India. GloBallast Monogr Ser No 11. IMO, London.<http://globallast.imo.org/index.asp?page=monograph.htm&menu=true> accessed 04/09/2007

Anon (2005) Import health standard for the importation into New Zealand of ornamental fish and marine invertebrates from all countries. Biosecurity New Zealand, Wellington. <http://www.biosecurity.govt.nz/files//imports/animals/standards/fisornic.all.pdf> accessed 28/01/2006

Awad A, Clarke C, Greyling L, Hilliard R, Polglaze J, Raaymakers S (2004) Ballast water risk assessment, port of Saldanha Bay, Republic of South Africa. GloBallast Monogr Ser No 13. IMO, London. <http://globallast.imo.org/index.asp?page=monograph.htm&menu=true> accessed 04/09/2007

Biosecurity Australia (2000) Draft import risk analysis report for prawns and prawn products. Biosecurity Australia, Canberra. <http://www.daff.gov.au/__data/assets/pdf_file/0012/42141/00-041a.pdf> accessed 02/09/2007

Burgman MA, Ferson A, Akcakaya HR (1993) Risk assessment in conservation biology. Chapman and Hall, London

BWM (2005) The International Convention for the Control and Management of Ship's Ballast Water and Sediments. Adopted 13 February 2004. The International Maritime Organisation, London

Cameron J (1993) The precautionary principle – core meaning, constitutional framework and procedures for implementation. Precautionary Principle Conference, Institute of Environmental Studies, University of New South Wales, September 1993

Campbell ML (2004) Microalgae importation decision-tree – user notes. Biosecurity New Zealand, Wellington

Campbell ML (2005a) Risk assessment (modified organism impact assessment) to update information on *Undaria pinnatifida*. All Oceans Ecology Client Report AOE2005-03

Campbell ML (2005b) Risk Analysis for introducing marine species for aquaculture purposes: practical examples. Chilean Aquaculture Risk Assessment Workshop, Valparaiso, Chile. 20–31 March 2005

Campbell ML (2005c) Organism impact assessment (OIA) for potential impacts of *Didymosphenia geminata*. All Ocean Ecology Client Report AOE2005-02 prepared for Biosecurity New Zealand, Wellington. <http://www.biosecurity.govt.nz/files/pests-diseases/plants/didymo/didymo-org-ia-oct-05.pdf> accessed 04/09/2007

Campbell ML (2005d) A draft guide to risk analysis/assessment. All Ocean Ecology Client Report AOE2005-04 prepared for the Regional Activities Center for Specially Protected Areas (RAC/SPA, UNEP), Tunisia

Carlton JT (2001) Introduced species in US coastal waters: environmental impacts and management priorities. Pew Oceans Commission, Arlington, Virginia, USA

Clarke C, Hayes T, Hilliard R, Kayvanrad N, Taymourtash H, Parhizi A, Yavari V, Raaymakers S (2003) Ballast water risk assessment, port of Khark Island, Islamic Republic of Iran. GloBallast Monogr Ser No 8. IMO, London. <http://globallast.imo.org/index.asp?page=monograph.htm&menu=true> accessed 04/09/2007

Clarke C, Hilliard R, Junqueira A de OR, Neto A, Polgaze J, Raaymakers S (2004a) Ballast water risk assessment, port of Sepetiba, Federal Republic of Brazil. GloBallast Monogr Ser No 14. IMO, London. <http://globallast.imo.org/index.asp?page=monograph.htm&menu=true> accessed 04/09/2007

Clarke C, Hilliard R, Liuy Y, Polglaze J, Zhao D, Xu X, Raaymakers S (2004b) Ballast water risk assessment, port of Dalian, People's Republic of China. GloBallast Monogr Ser No 12. IMO, London. <http://globallast.imo.org/index.asp?page=monograph.htm&menu=true> accessed 04/09/2007

Cooney R (2004) Precaution and invasive alien species: challenges at the interface of trade and environment regimes. Paper given at 'Biodiversity loss and species extinction: managing risk in a changing world.' World Conservation Forum, 18–20 November 2004, Bangkok, Thailand. <www.iucn.org/congress/wcforum/forum_th_biodiversity-proceed.htm> ; accessed 28/01/2006

Cooney R (2005) From promise to practicalities: the precautionary principle in biodiversity conservation and sustainable use. In: Cooney R, Dickson B (eds) Biodiversity and the precautionary principle. Risk and uncertainty in conservation and sustainable use. Earthscan, London, pp 3–17

Cooney R, Dickson B (2005) Precautionary principle, precautionary practice: lessons and insights. In: Cooney R, Dickson B (eds) Biodiversity and the precautionary principle. Risk and uncertainty in conservation and sustainable use. Earthscan, London, pp 288–298

Costanza R, d'Arge R, de Groot R, Farber S, Grasso M, Hannon B, Limburg K, Naeem S, O'Neill RV, Paruelo J, Raskin RG, Sutton P, Van der Belt M (1997) The value of the world's ecosystem services and natural capital. Nature 387:253–260

Dalal-Clayton B, Sadler B (2005) Strategic environmental assessment. A sourcebook and reference guide to international experience. Earthscan; London

Farber SC, Costanza R, Wilson MA (2005) Economic and ecological concepts for valuing ecosystem services In: Kalof L, Satterfield T (eds) Environmental values. Earthscan, London, pp 18–40

Gollasch S (1996) Untersuchungen des Arteintrages durch den internationalen Schiffsverkehr unter besonderer Berücksichtigung nichtheimischer Arten. Dr. Kovac; Hamburg (Diss.)

Gollasch S, Leppäkoski E (1999) Initial risk assessment of alien species in Nordic waters. Nordic Council of Ministers, Copenhagen

Gollasch S, David M, Voight M, Dragsund E, Hewitt C, Fukuyo Y (2007) Critical review of the IMO international convention on the management of ships' ballast water and sediments. Harmful Algae 6(4):585–600

Gullett W (1997) Environmental protection and the "precautionary principle": a response to scientific uncertainty in environmental management. Environ Plan Law J 14(1):52–69

Handl G (1990) Environmental security and global change: the challenge to international law. YBK Int'l Environmental L 3

Harwood J, Stokes K (2003) Coping with uncertainty in ecological advice: lessons from fisheries. Trends Ecol Evol 18:617–622

Hayes KR (2003) Biosecurity and the role of risk-assessment. In: Ruiz GM, Carlton JT (eds) Bioinvasions: pathways, vectors, and management strategies. Island Press, Washington DC, pp 382–414

Hayes KR, Hewitt CL (2001) Ballast-water risk assessment – vol II. CRIMP Tech Rep 21. CSIRO Division of Marine Research, Hobart, Australia

Hayes KR, Sliwa C (2003) Identifying potential marine pests – a deductive approach applied to Australia. Mar Poll Bull 46:91–98

Hazardous Substances and New Organism Act (1996) New Zealand Ministry for the Environment. < http://www.mfe.govt.nz/laws/hsno.html> accessed 04/09/2007

Herborg L-M, Jerde CL, Lodge DM, Ruiz GM, MacIsaac HJ (2007) Predicting invasion risk using measures of introduction effort and environmental niche models. Ecol Appl 17(3):663–674

Hewitt CL, Campbell ML (2005) Applying risk analysis to marine aquaculture species. Chilean Aquaculture Risk Assessment Workshop, Valparaiso, Chile. 20–31 March 2005

Hewitt CL, Hayes KR (2001) Marine biosecurity and risk assessment. Proceedings of the Quarantine and Market Access Conference, Canberra, act, Australia, October

Hewitt CL, Hayes KR (2002) Risk assessment of marine biological invasions. In: Leppäkoski E, Gollasch S, Olenin S (eds) Invasive aquatic species of Europe: distribution, impact and management. Kluwer Academic Publishers, Dordrecht, The Netherlands, pp 456–466

Hewitt CL, Campbell ML, Gollasch S (2006) Alien species in aquaculture. Considerations for responsible use. IUCN, Gland

ICES (2005) ICES code of practice on the introductions and transfers of marine organisms 2005 <http://www.ices.dk/reports/general/2004/ICES%20Code%20of%20Practice%202005.pdf.> ; accessed 28/01/2006.

Inland Water Resources and Aquaculture Services [FAO] 2003. Health management and biosecurity maintenance in white shrimp (*Penaeus vannamei*) hatcheries in Latin America. FAO Fisheries Tech Pap 450. FAO, Rome. <ftp://ftp.fao.org/docrep/fao/007/y5040e/y5040e00.pdf> accessed 04/08/2007.

Kahn SA, Wilson DW, Perera RP, Hayder H, Gerrity SE (1999) Import risk analysis on live ornamental finfish. AQIS, Canberra

Kalof L, Satterfield T (2005) Environmental values. Earthscan, London

Kolar CS, Lodge DM (2002) Ecological predictions and risk assessment for alien fishes in North America. Science 298:1233–1236

Lang R, Armour A (1981) The assessment and review of social impacts. Federal Environmental Assessment Office, Ottawa

Leung KMY, Dudgeon D (2007) Ecological (pest) risk assessment and management. FAO/NACA Workshop on Understanding and Applying Risk Analysis in Aquaculture, 8–11 June 2007. Rayong, Thailand

Lodge DM, Williams S, MacIsaac HJ, Hayes KR, Leung B, Reichard S, Mack RN, Moyle PB, Smith M, Andow DA, Carlton JY, McMichael A (2006) Biological invasions: recommendations for US policy and management. Ecol Appl 16(6):2035–2054

Lubchenco J, Olson AM, Brubaker LB, Carpenter SR, Holland MM, Hubbell SP, Levin SA, MacMahon JA, Matson PA, Melillo JM, Mooney HA, Peterson CH, Pulliam HR, Real LA, Regal PJ, Risser PG (1991) The sustainable biosphere initiative: an ecological research agenda. Ecology 72:371–412

Marsden S, Dovers S (2002) Strategic environmental assessment in Australasia. The Federation Press, Marrickville

New South Wales Department of Urban Affairs and Planning (1996) EIS guidelines: extractive industries, dredging and other extraction in riparian and coastal areas. Department of Urban Affairs and Planning; Sydney

New South Wales Department of Urban Affairs and Planning (2000) Coal mines, and associated infrastructure: EIS guideline. Department of Urban Affairs and Planning; Sydney

Norse EA (ed) (1993) Global marine biological diversity: a strategy for building conservation into decision making. Centre for Marine Conservation, IUCN/WWF/UNEP, The World Bank. Island Press, Washington, DC

Pagiola S (2004) How much is an ecosystem worth? Assessing the economic value of conservation. IUCN, The Nature Conservancy, The World Bank; Washington DC

Peel J (2005) The precautionary principle in practice. Environmental decision-making and scientific uncertainty. The Federation Press, Sydney

Rixon CAM, Duggan IC, Bergeron NMN, Ricciardi A, MacIssac HJ (2005) Invasion risks posed by the aquarium trade and live fish markets on the Laurentian Great Lakes. Biol Invas 14(6):1365–1381

Roberts MH, Newman MC, Hale RC (2002) Overview of ecological risk assessment in coastal and estuarine environments. In: Newman MC, Roberts MH, Hale RC (eds) Coastal and estuarine risk assessment. Lewis Publishers, Boca Raton, pp 1–13

SPS Agreement (1994) Sanitary and phytosanitary measures. World Trade Organisation <http://www.wto.org/english/tratop_e/sps_e/spsagr_e.htm>; accessed 28/01/2006

Standards Australia (2000) Environmental risk management, edn 3. Standards Australia, Homebush, NSW

Standards Australia (2004) Risk management guidelines: companion to AS/NZS4360:2004. Standards Australia; Homebush, NSW

Thomas I, Elliott M (2005) Environmental impact theory and practice assessment in Australia, edn 4. The Federation Press, Marrickville

Toman M (1998) Why not to calculate the value of the world's ecosystem services and natural capital. Ecol Econ 25:57–60

Chapter 21
Differing Consequences of Removing Ecosystem-Modifying Invaders: Significance of Impact and Community Context to Restoration Potential

Sally D. Hacker and Megan N. Dethier

21.1 Introduction

In recent years there has been a strong focus on the colonization of invasive species and their impacts on native and nonindigenous species (Parker et al. 1999; Ruiz et al. 1999; Simberloff and Von Holle 1999; Davis and Thompson 2000; Kolar and Lodge 2001; Grosholz 2002). Invasive species of particular concern, known as ecosystem modifiers or "engineers" (see Jones et al. 1994), are those that cause community and ecosystem level effects by creating, modifying, or maintaining physical habitat (Vitousek 1990; Crooks 2002; Grosholz 2002; Chap. 16, Crooks). These effects may alter processes such as nutrient cycling (e.g., Vitousek et al. 1987), disturbance regime (e.g., Mack and D'Antonio 1998), species interactions (e.g., Callaway and Aschehoug 2000), or structural and physical characteristics of the community itself (e.g., Crooks 1998). A recent review by Crooks (2002) illustrates the complex ways invasive ecosystem modifiers can transform communities, resulting in both positive and negative effects for other species. It is clear that these invaders can have extraordinary influence on community structure because they alter ecological processes in multiple ways and they often create positive feedbacks that benefit their continued expansion and impact (Cuddington and Hastings 2004).

Much less is known about the consequences of removing invasive species, especially ecosystem modifiers (Hobbs and Humphries 1995; Myers et al. 2000a; Zavalata et al. 2001). Invasive species removal can have positive effects for some communities, with restoration occurring after removal (Fig. 1a) (Myers et al. 2000a). However, in many cases, the results have been mixed with unexpected and widespread impacts on natural communities (Zavalata et al. 2001; D'Antonio and Meyerson 2002). For example, removal has been shown to increase invasion by other species (North et al. 1994), enhance predation of native species via hyperpredation (Courchamp et al. 1999), and depress population sizes of endangered species (Van Riel et al. 2000; Myers et al. 2000b). A few conceptual models have used idealized food webs to understand the consequences of removing

G. Rilov, J.A. Crooks (eds.) *Biological Invasions in Marine Ecosystems.*
Ecological Studies 204,

invasive predators, herbivores, or plants (Zavalata et al. 2001; Courchamp et al. 1999, 2000). Some models predict that removing certain top predators can have counterintuitive consequences, indirectly leading to the decline of endemic prey. In addition, especially in the case of biocontrol, the removal process may lead to indirect detrimental effects (e.g., Louda et al. 1997; Callaway et al. 1999; but see Gratton and Denno 2005).

In the case of invasive ecosystem modifiers, communities may be so changed by the invasion that removal results in conditions not conducive to the re-colonization and growth of native species. One marine example where this has been measured experimentally is the removal of the invasive alga, *Caulerpa racemosa*, in the Mediterranean Sea (Piazzi and Ceccherelli 2006). *Caulerpa* causes a significant drop in native macroalgal diversity, which consists of a rich array of encrusting, turf-like,

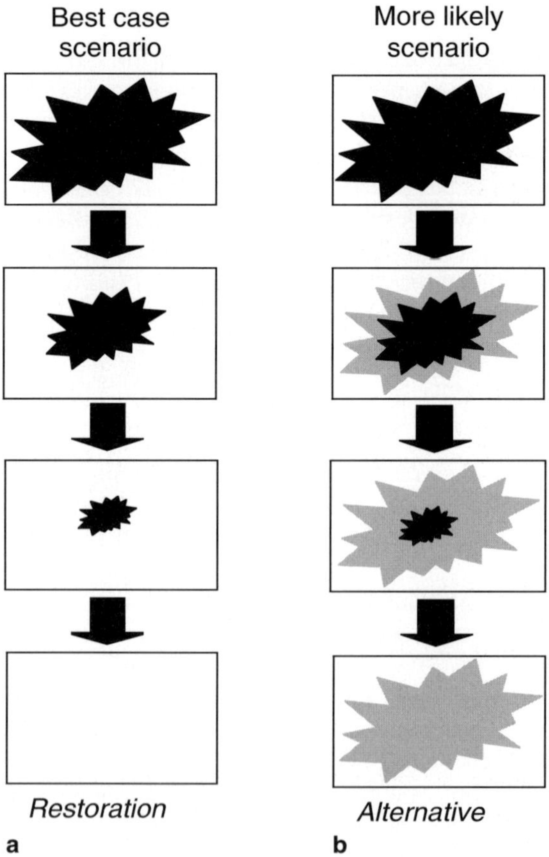

Fig. 21.1 Stages of invasive species removal (*arrow*) under: **a** the best case scenario for which the invader modification is lost (*black*) and the community is restored; **b** the more likely scenario for which the modification remains after the removal (*gray*) and the community follows an alternative trajectory

and erect species. Removal of *Caulerpa* did not result in the restoration of the original community but instead led to the domination by selected opportunistic turf species. Although it may be too early to tell, these species apparently excluded other native species that cannot recruit or compete with the new species assemblage.

Examples such as these suggest that, just as the invasion of species can create unforeseen changes to communities, their removal can have unpredictable consequences. Communities may not simply return to their former state in a straightforward reversal of the invasion process but instead be so modified by the invasion that they do not recover (Fig. 1b) (Hobbs and Humphries 1995; D'Antonio and Meyerson 2002). These modifications are likely to vary in degree, depending on characteristics of the invader, the invaded community, and the time since invasion, but could cause the community to never fully be restored after the invader is removed. Given the numerous removal programs underway, good information is critical to predict how communities will respond to management actions that are primarily designed to remove invasive species but not necessarily the "legacy" they may leave behind.

21.2 A Predictive Conceptual Model for Post-removal Restoration

One such theory that deals with how communities are influenced by the disturbance of strong ecosystem modifiers is the alternative stable state theory, originally proposed by Lewinton (1969) and field tested by Sutherland (1974). It was designed to explain the observation that different species assemblages can occur in the same general locality at different times (or different localities at the same time) because historical events or contingencies play an important role in creating community structure. Although clear evidence for alternative community stable states is clouded by (1) unresolved issues about the criteria needed to demonstrate their existence (Connell and Sousa 1984; Peterson 1984; Sutherland 1990; Knowlton 1992; Petraitis and Latham 1999; Morgan 2001; Beisner et al. 2003) and (2) the lack of good manipulative experiments (Petraitis and Latham 1999; but see Bertness et al. 2002), we suggest that it can provide a useful framework for identifying the processes important to post-removal community structure. In particular, it defines the processes that could lead a community away from or towards a restored state. Here we define restored and recovery of a system as a return to its original state after it has been degraded or destroyed. Alternative stable state theory provides a conceptual foundation that can be used to design experiments important for predicting the community consequences of invasive species removal and restoration potential (Beisner et al. 2003; Suding et al. 2004; Byers et al. 2006).

To illustrate the application of this theory to predictions of invasive species removal, we have adapted and expanded a conceptual model presented by Petraitis and Latham (1999) (Fig. 21.2). In the modified version, we have included three community states: (1) the invasion state in which the invader dominates the community through positive feedback mechanisms that continue its expansion and dominance (Fig. 21.2, middle), (2) a restored state defined as the replacement of the lost species assemblage with the original assemblage, and its function, after the invader is removed (Fig. 21.2, left), and (3) an alternative state defined as one in which a new species assemblage colonizes (either native or non-native) and persists; it could include reinvasion by the original invading species or other nonindigenous species (Fig. 21.2, right). In the interest of simplicity, the model shows that invaded communities will transition into the restored or alternative state when the invader is removed without proceeding through a series of successional stages. These stages are nonetheless important to the timing and tempo of the transition from one community to another.

In their model, Petraitis and Latham (1999) make an important distinction between transition events and the maintenance of species assemblages through positive feedback processes (Fig. 21.2), an issue that has been under-appreciated and is relevant to the establishment of different post-removal community outcomes. Processes important to transitions from the invaded state to the restored or alternative state are those that create the initial conditions for community development; they include processes that promote the loss of physical or biological effects of the invader as well as the recruitment of species that will shape the composition of the new community. Examples of processes that promote the loss of the physical or biological effects of invaders include water movement, wind, or fire. Many times the loss of the impact occurs simply by removing the invader and thus the maintenance of the effect (e.g., bioturbation will stop when the burrowing invader is removed).

One can apply these transitional processes by considering invasive species removal as the initiating event that results in either (1) a restored state in which there is a loss of the physical and biological effects of the invader and sufficient recruitment of the original species assemblage to recover the community (Fig. 21.2, left), or (2) an alternative state in which there is little or no loss of the effects of the invader coupled with

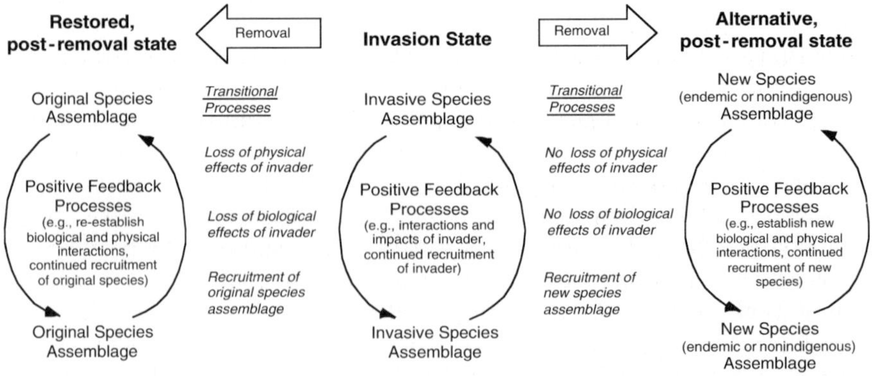

Fig. 21.2 Depiction of alternative community stable state theory as applied to community structure alternatives following invasive species removal (modified from Petraitis and Latham 1999)

recruitment by a new set of species that may be regionally endemic and/or nonindigenous (Fig. 21.2, right). Recruitment of the original species assemblage is critical to restoration of the original community but species identity alone may be only part of the necessary requirements; more complex factors such as propagule quantity (e.g., Law and Morton 1993; Seabloom et al. 2003; Chap. 7, Johnston et al.) and recruitment sequence (assembly rules; Diamond 1975; Robinson and Dickerson 1987; Drake 1991; Law and Morton 1993) are clearly important as well. As suggested in Weiher and Keddy (1999), assembly rules in particular play an important role in restoration and deserve more experimental attention. Possibly most important to our model, however, is the influence of the remaining invader modifications on the recruitment of the original species assemblage. If the habitat is modified in such a way as to discourage this recruitment, and there is no loss of this effect when the invader is removed, community restoration will be unlikely and an alternative community will develop. Moreover, due to chance events, it is possible that recruitment of new species will be favored over that of the original species, creating the initial conditions for an alternative post-removal community type (Fig. 21.2, right).

Once an assemblage is established, alternative community stable state theory predicts that communities must persist via positive feedback processes such as species interactions, physical conditions, and recruitment (Fig. 21.2). Understanding the processes that serve to maintain vs destabilize the original, pre-invasion assemblage or an alternative one will be key to predicting restoration success. For example, if the original species assemblage colonizes, but is unable to persist due to the legacy effects of the invader or a rare disturbance event (or a combination of both), then restoration of the community will be compromised. If new species assemblages are unable to persist for similar reasons, transitional processes could drive the community back to a restored state. Finally, if reinvasion is allowed to occur or new invaders are facilitated by the removal of the original invader (similar to the "invasional meltdown" described by Simberloff and Von Holle 1999), restoration will obviously be hindered and an alternative community (dominated by the invader) will be produced once again.

We can apply this model to the earlier example of the consequences of removing the invasive species, *Caulerpa racemosa*. In the invasion state, *Caulerpa* grows vegetatively via stolons, weaving its way through native macroalgal communities and competing for resources such as light and nutrients. The impact is particularly strong for encrusting and turf species that are essentially smothered by the invader. Over time, sediments accumulate around the stolons further reducing light and oxygen available for other species but enhancing the growth of *Caulerpa* (Piazzi et al. 2001). After a year of *Caulerpa* removal, restoration of the original macroalgal community did not occur. Instead selected native opportunistic turf species colonized and dominated the removal plots while encrusting and erect forms were essentially absent (Piazzi and Ceccherelli 2006). In this case, the removal process caused a loss of the physical and biological effects of the invader (Fig. 21.2, left) but only a few native species colonized the removal plots, thus shifting the community into an alternative state (Fig. 21.2, right). The mechanisms responsible for the persistence of this new state are in need of investigation but could include active exclusion by the turf species or the passive result of recruitment limitation by many of the original community members.

21.3 Context Dependent Restoration: Invasive English Cordgrass as an Example

The consequences of removing invasive species is likely to be highly context dependent with communities displaying restoration or some alternative depending on the context under which the invasion and removal is taking place. For example, in the *Caulerpa* study, restoration may be more successful if it takes place early in the invasion timeline when more native species are available to provide propagules to the restoration sites. Or it may be more likely in soft sediment habitats where native seagrasses are better able to respond to the sedimentation left behind by *Caulerpa*. Here we explore these ideas in more detail using an example from an estuarine system where a marine grass invader colonizes different habitats and produces varying impacts that potentially lead to alternative community outcomes after its removal.

English cordgrass, *Spartina anglica*, was first introduced into Puget Sound, Washington USA in 1961 to stabilize a dike system in a local estuary. It did not become a management priority until the late 1990s when it had spread to a total of 3300 ha of intertidal habitat at 77 sites (Hacker et al. 2001). Cordgrass species are strong ecosystem modifiers that accrete estuarine sediment around their dense root system and change sediment biogeochemistry, all of which can have important community-wide consequences (Thompson 1991; Chap. 17, Grosholz and Ruiz). *Spartina anglica* in Puget Sound grows in a range of communities that include mudflats and cobble beaches, which are normally devoid of vascular plants, and low and high salinity marshes, where native vascular plants are the main biological component (Hacker et al. 2001). Low salinity marsh and mudflat sites have much larger infestations than cobble beach and high salinity marsh sites driven mostly by variability in physical conditions across the four habitat types (Dethier and Hacker 2005).

Spartina anglica produces different structural and chemical changes to the sediment depending on the habitat it invades (Hacker and Dethier 2006). Sediment accretion, driven by *S. anglica*, is greatest in mudflats and salt marshes but least in cobble beaches. A likely explanation for this difference is wave energy; water movement and resuspension of fine sediments is higher in cobble beaches. Sediment differences lead to different chemical conditions including a decline in sediment water content in mudflats and high salinity marshes. In cobble beaches, the opposite was seen; root mat sediments had higher water content than the unmodified cobble sediments. Cordgrass generally increases oxygen content in all sediments likely due to oxygen leakage from root aerenchyma (Maricle and Lee 2002). It also shades the sediment surface, thus decreasing water evaporation and salt accumulation compared to unvegetated areas (Hacker and Bertness 1999).

Biological effects of the cordgrass invasion also differ with habitat type (Hacker and Dethier 2006). In mudflats, cobble beaches, and high salinity marshes, the invasion caused an increase in native vascular plant cover and decline in algal cover. By elevating sediments, increasing oxygen content, and decreasing salinity, cordgrass clearly provides a more suitable habitat for native vascular salt marsh plants but a less suitable habitat for algae, which require greater tidal inundation to avoid desiccation. In low salinity marshes, native vascular plant diversity and abundance declined precipitously presumably due to competitive dominance of cordgrass. We also compared marine invertebrates in mud and cobble sediments versus adjacent cordgrass patches with the same sediment. Uninvaded mudflats and cobble beaches often had abundant clams and a variety of polychaete worms, while those with cordgrass had more epifauna such as amphipods and insects (presumably due to the three-dimensional vegetated structure) (Hacker and Dethier, unpublished data).

Removal of English cordgrass, involving mowing and herbicide applications, began in 1997 (Hacker et al. 2001). Although local eradication has occurred at some sites, most have required repeated removal spanning multiple years. Preliminary study of restoration suggests that irrespective of habitat type, removing invasive *S. anglica* resulted in an increase in native vascular plants (Reeder and Hacker 2004). However, this increase does not constitute "restoration" for all habitat types; the recruiting plants are normally uncommon in mudflat and cobble beach communities, and thus do not represent return to a "natural" state. Instead, the legacy effects of cordgrass produce alternative short term outcomes, which may or may not continue for some time.

We can use the conceptual model presented earlier to predict the consequences of *S. anglica* removal. We predict that cobble beaches will assume a restored state due to the interaction of both transition and maintenance processes as outlined in Fig. 21.3. If we assume that vascular plant recruitment occurs in cobble beaches after *Spartina* has been removed, then an alternative state can only be produced if vascular plants can maintain elevated sediments and altered biogeochemical processes (Fig. 21.3a, right). Active water movement and the scouring action of shifting cobbles and gravel should hamper this process (Fig. 21.3a, left). Ultimately, we predict that an alternative state will not be maintained due to the transitional process of water movement increasing sediment erosion around the old root mat of *Spartina* and decreasing vascular plant and infaunal recruitment. As a result, negative feedback processes in the maintenance component of the alternative state will shift cobble beach community structure to a restored state (Fig. 21.3a, left). Algal recruitment will increase as sediment erosion occurs and cobbles re-emerge, pushing the community into a positive feedback loop that includes loss of sediment, decreased native vascular plant recruitment, and continued increases in algal and infaunal recruitment (Fig. 21.3a, left).

We predict that mudflat habitats, because they experience lower water movement, will have increased vascular plant recruitment, and allow for the maintenance of cordgrass sediment accretion and biogeochemical processes (Fig. 21.3a, right). As such, we predict that they will move into an alternative rather than restored state. We expect that a positive feedback loop generated by low water movement and the

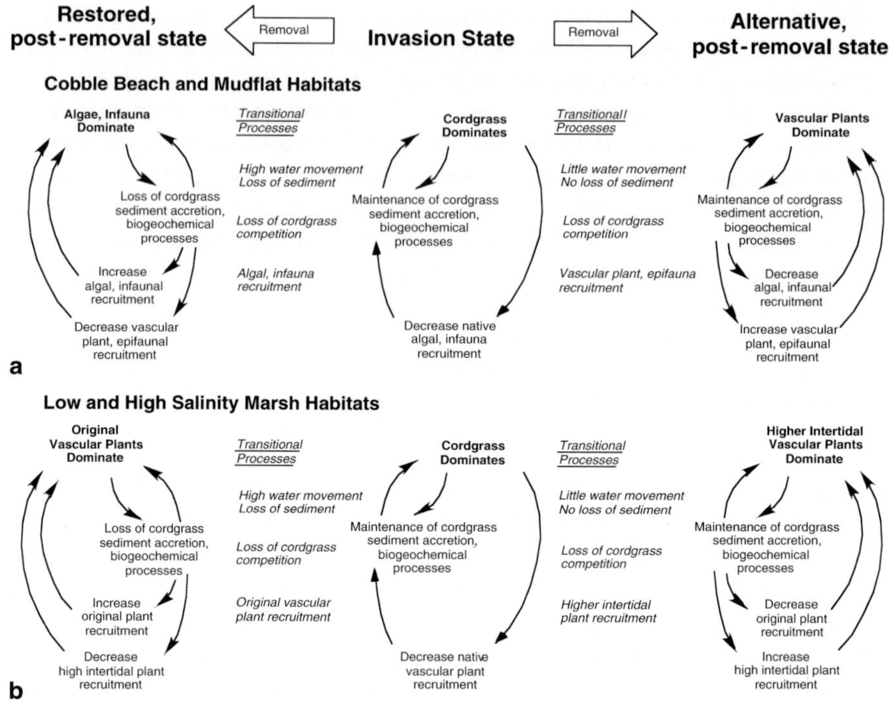

Fig. 21.3 The conceptual theory as applied to different post-removal outcomes (restored or alternative) for the removal of marine cordgrass, *Spartina anglica*, from four habitat types. Habitats include: **a** cobble beach and mudflat habitats (original communities have algae and infauna but lack native vascular plants); **b** high and low salinity marshes (original communities dominated by native vascular plants)

presence of vascular plants will maintain sediment characteristics created by cordgrass, decrease algal and infaunal recruitment, and continue to increase native vascular plant recruitment (Fig. 21.3a, right). This new vascular plant community could be a permanent feature of mudflat habitats unless it is additionally removed and sediments are allowed to erode.

Finally, we predict that high and low salinity marshes previously invaded by cordgrass will experience recruitment of higher intertidal vascular plant assemblages, rather than the original assemblage, due to the increased tidal elevation produced by cordgrass-accreted sediments (Fig. 21.3b, right). We expect that these plants will be good at maintaining sediment depth and biogeochemical processes originally created by cordgrass. Their presence will be maintained via a positive feedback loop that increases their own continued recruitment while decreasing that of the lower intertidal plant community present before the invasion (Fig. 21.3b, left). However, if water movement is sufficient to erode sediments, or if plant recruitment is low or delayed, marsh communities may shift into a restored state (Fig. 21.3b, left).

Testing these hypotheses will require both large and small-scale experiments in areas where cordgrass has been removed. It may be that active removal of native vascular plant assemblages will be necessary in habitats such as mudflats where restoration could be continually hindered by the recruitment and positive feedbacks produced by these plants. This example illustrates the importance of understanding the context dependent nature of the invasion because this context will likely dictate the outcome of removal efforts.

21.4 Conclusions

Removal and restoration are critical components in the management of many problematic invaders but the physical and biological legacy of invasion can hinder restoration and contribute to the formation of alternative communities. We suggest that by understanding and anticipating this hysteresis, it may be possible to better predict the consequences of removing certain invaders and include measures that will best promote restoration.

In the particular example given, we consider how removing a single species from habitats that vary in the degree of modification and physical disturbance can result in different community outcomes. Alternative community outcomes could also be produced when different species are removed from the same habitat. For example, if one species is a major ecosystem modifier, its removal might result in an alternative community while the removal of a less influential species would lead to restoration. Equally important might be the time since invasion. Not only could time determine the degree of modification produced by the invader but it could also decrease the availability of propagules of the original species for recruitment.

It is clear that we must view invasive species management not merely as the removal of a single species or population within the community, but as a community-wide change that will either result in restoration or an alternative community wholly different than those proceeding it. Understanding the processes that lead to one or the other result will be key to restoration success.

References

Beisner BE, Haydon DT, Cuddington K (2003) Alternative stable states in ecology. Front Ecol Environ 1:376–382

Bertness MD, Trussel GC, Ewanchuk PJ, Silliman BR (2002) Do alternative stable community states exist in the Gulf of Maine rocky intertidal zone? Ecology 83:3434–3448

Byers JE, Cuddington K, Jones CG, Talley TS, Hastings A, Lambrinos JG, Crooks JA, Wilson WG (2006) Using ecosystem engineers to restore ecological systems. Trends Ecol Evol 21:493–500

Callaway RM, Aschehoug ET (2000) Invasive plants versus their new and old neighbors: a mechanism for exotic invasion. Science 290:521–523

Callaway RM, DeLuca TH, Belliveau WM (1999) Biological-control herbivores may increase competitive ability of the noxious weed, *Centaurea maculosa*. Ecology 80:1196–1201

Connell JH, Sousa WP (1984) On the evidence needed to judge ecological stability or persistence. Am Nat 121:789–824

Courchamp F, Langlais M, Sugihara G (1999) Cats protecting birds: modeling the mesopredator release effect. J Anim Ecol 68:282–292

Courchamp F, Langlais M, Sugihara G (2000) Rabbits killing birds: modeling the hyperpredation process. J Anim Ecol 69:154–164

Crooks JA (1998) Habitat alternation and community-level effects of an exotic mussel, *Musculista senhousia*. Mar Ecol Prog Ser 162:137–152

Crooks J (2002) Characterizing ecosystem-level consequences of biological invasions: the role of ecosystem engineers. Oikos 97:153–166

Cuddington K, Hastings A (2004) Invasive engineers. Ecol Model 178:335–347

D'Antonio C, Meyerson LA (2002) Exotic plant species as problems and solutions in ecological restoration: a synthesis. Restorat Ecol 10:703–713

Davis MA, Thompson K (2000) Eight ways to be a colonizer; two ways to be an invader: a proposed nomenclature scheme for invasion biology. Bull Ecol Soc Am July:226–230

Dethier MN, Hacker SD (2005) Physical factors vs. biotic resistance in controlling the invasion of an estuarine marsh grass. Ecol Appl 15:1273–1283

Diamond JM (1975) Assembly of species communities. In: Cody ML, Diamond JM (eds) Ecology and evolution of communities. Belknap Press of Harvard University, Cambridge, MA, pp 342–444

Drake JA (1991) Community-assembly mechanics and the structure of an experimental species ensemble. Am Nat 137:1–26

Gratton C, Denno RF (2005) Restoration of arthropod assemblages in a *Spartina* salt marsh following removal of the invasive plant *Phragmites australis*. Restor Ecol 13:358–372

Grosholz E (2002) Ecological and evolutionary consequences of coastal invasions. Trends Ecol Evol 17:22–27

Hacker SD, Bertness MD (1999) Experimental evidence for factors maintaining plant species diversity in a New England salt marsh. Ecology 80:2064–2073

Hacker SD, Dethier MN (2006) Community modification by a grass invader has differing consequences for marine habitats. Oikos 113:279–286

Hacker SD, Heimer D, Hellquist CE, Reeder TG, Reeves B, Riordan T, Dethier MN (2001) A marine plant (*Spartina anglica*) invades widely varying habitats: potential mechanisms of invasion and control. Biol Invas 3:211–217

Hobbs RJ, Humphries SE (1995) An integrated approach to the ecology and management of plant invasions. Conserv Biol 9:761–770

Jones CG, Lawton JH, Shachak M (1994) Organisms as ecosystem engineers. Oikos 689:373–386

Knowlton N (1992) Thresholds and multiple stable states in coral reef community dynamics. Am Zool 32:674–682

Kolar CS, Lodge DM (2001) Progress in invasion biology: predicting invaders. Trends Ecol Evol 16:199–204

Law R, Morton RD (1993) Alternative permanent states of ecological communities. Ecology 74:1347–1361

Lewinton RC (1969) The meaning of stability. In: Diversity and stability in ecological systems. Brookhaven Symposia in Biology No. 22. Brookhaven National Laboratories, Brookhaven, NY, pp 13–24

Louda SM, Kendall D, Connor J, Simberloff D (1997) Ecological effects of an insect introduced for the biological control of weeds. Science 277:1088–1090

Mack MC, D'Antonio CM (1998) Impacts of biological invasions on disturbance regimes. Trends Ecol Evol 13:195–198

Maricle BR, Lee RW (2002) Aerenchyma development and oxygen transport in the estuarine cordgrasses *Spartina alterniflora* and *S. anglica*. Aquat Bot 74:109–120

Morgan SG (2001) The larval ecology of marine communities. In: Bertness MD, Gaines SD, Hay ME (eds) Marine community ecology. Sinauer Associates, Sunderland, MA, pp 159–181
Myers JH, Simberloff D, Kuris AM, Carey JR (2000a) Eradication revisited: dealing with exotic species. Trends Ecol Evol 15:316–320
Myers JH, Simberloff D, Kuris AM, Carey JR (2000b) Reply from J. Myers, D. Simberloff, A. Kuris, and J. Carey. Trends Ecol Evol 15:515–516
North SG, Bullock DJ, Dulloo ME (1994) Changes in the vegetation and reptile populations on Round Island, Mauritius, following eradication of rabbits. Biol Conserv 67:21–28
Parker IM, Simberloff D, Lonsdale WM, Goodell K, Wonham M, Kareiva PM, Williamson MH, Von Holle B, Moyle PB, Byers JE, Goldwasser L (1999) Impact: toward a framework for understanding the ecological effects of invaders. Biol Invas 1:3–19
Peterson CH (1984) Does a rigorous criterion for environmental identity preclude the existence of multiple stable points? Am Nat 124:127–133
Petratis PS, Latham RE (1999) The importance of scale in testing the origins of alternative community states. Ecology 80:429–442
Piazzi L, Ceccherelli G (2006) Persistence of biological invasion effects: recovery of macroalgal assemblages after removal of *Caulerpa racemosa* var. *cylindracea*. Estuarine Coast Shelf Sci 68:455–461
Piazzi L, Ceccherelli G, Cinelli F (2001) Threat to macroalgal diversity: effects of the introduced green alga, *Caulerpa racemosa* in the Mediterranean. Mar Ecol Prog Ser 210:161–165
Reeder TG, Hacker SD (2004) Factors contributing to the removal of a marine grass invader (*Spartina anglica*) and subsequent potential for habitat restoration. Estuaries 27:244–252
Robinson JV, Dickerson JE (1987) Does invasion sequence affect community structure? Ecology 68:587–595
Ruiz GM, Fofonoff P, Hines AH, Grosholz ED (1999) Nonindigenous species as stressors in estuarine and marine communities: assessing impacts and interactions. Limnol Oceanogr 44:950–972
Seabloom EW, Harpole WS, Reichman OJ, Tilman D (2003) Invasion, competitive dominance, and resource use by exotic and native California grassland species. Proc Natl Acad Sci 100:13384–13389
Simberloff D, Von Holle B (1999) Positive interactions of nonindigenous species: invasional meltdown? Biol Invas 1:21–32
Suding KN, Gross KL, Houseman GR (2004) Alternative states and positive feedbacks in restoration ecology. Trends Ecol Evol 19:46–53
Sutherland JP (1974) Multiple stable points in natural communities. Am Nat 108:859–873
Sutherland JP (1990) Perturbations, resistance, and alternative views of the existence of multiple stable points in nature. Am Nat 136:270–275
Thompson JD (1991) The biology of an invasive plant: what makes *Spartina anglica* so successful? BioScience 41:393–401
Van Riel P, Jordaens K, Frias Martins AM, Backeljau T (2000) Eradication of exotic species. Trends Ecol Evol 15:515
Vitousek PM (1990) Biological invasions and ecosystem process: towards an integration of population and ecosystem studies. Oikos 57:7–13
Vitousek PM, Walker LR, Whiteaker D, Mueller-Dombios D, Matson PA (1987) Biological invasion by *Myrica faya* alters ecosystem development in Hawaii. Science 238:802–804
Weiher E, Keddy PA (1999) Assembly rules in ecological communities: perspectives, advances, retreats. Cambridge University Press, Cambridge, UK
Zavaleta ES, Hobbs RJ, Mooney HA (2001) Viewing invasive species removal in a whole-ecosystem context. Trends Ecol Evol 16:454–459

Section VI
Geographic Perspectives

Chapter 22
Geographic Perspectives of Invasions in the Sea

Jeffrey A. Crooks and Gil Rilov

22.1 Introduction

In this section, geographic perspectives of invasions from coastal waters around the world are presented. These include invasion assessments from Africa (Chap. 23, Griffiths and Robinson), Australasia (Chap. 24, Hayden et al.; Chap. 25, Sliwa et al.), South America (Chap. 26, Castilla and Neill; Chap. 27, Ferreira et al.), the eastern United States (Chap. 28, Fofonoff et al.), Europe (Chap. 29, Gollasch et al.; Chap. 30, Leppäkoski et al.), the Mediterranean (Chap. 31, Rilov and Galil), and Korea/China (Chap. 32, Seo and Lee). The section concludes with a synthesis comparing estuarine versus coastal invasions across multiple regions (Chap. 33, Preisler et al.). These case histories represent a sampling of a growing set of invader assessments, including examples from the west coast of North America (Cohen and Carlton 1995), the Azores (Cardigos et al. 2006), Argentina and Uruguay (Orensanz et al. 2002), and Japan (Otani 2004, 2006).

22.2 The Emerging Global Picture of Marine Invasion – Possibilities and Precautions

The growing number of invasion assessments begins to provide a picture of the pervasiveness of marine invasions, and the one that is emerging has the world's coastal oceans, and particularly sheltered embayments and estuarine systems, filling with an amalgam of non-native marine species. Particularly striking is how ubiquitous some of these invaders are. Species like the green crab *Carcinus meanas*, the mussel *Musculista senhousia*, and the tunicate *Ciona intestinalis* are now globally distributed, aided by the rampant dispersal opportunities afforded by anthropogenic activities.

That the global picture of marine invasions is now coming into focus makes it tempting to draw broad conclusions about how invaded different parts of the world are, and what the factors influencing these patterns might be. The ability to paint this broad picture is important for theoretical reasons, such as understanding the

G. Rilov, J.A. Crooks (eds.) *Biological Invasions in Marine Ecosystems.*
Ecological Studies 204,

roles of propagule pressure (Chap. 7, Johnston et al.) and invasibility (Chap. 12, Olyarnik et al.), and is also valuable for more applied efforts, such as steering control efforts and assessing efficacy of invader management. However, a central theme emerging from this book clearly indicates that such comparisons must be done with extreme caution. The potential problems associated with such treatments are described in chapters throughout the book, and in particular by Preisler et al. (Chap. 33) in their more focused examination of coastal vs estuarine invaders. All stem in large part from the highly uneven nature of knowledge related to invasions in different parts of the world, and rapidly changing invasion patterns due to shifting environmental conditions, species relationships, and vector operation. As indicated in Chap. 2 (Carlton), and emphasized throughout this section, there are solutions to these issues. They include reconsidering the way in which we identify and assign biogeographic origins to marine species (e.g. Chap. 2, Carlton; Chap. 28, Fofonoff et al.), implementing standardized assessments across areas (e.g. Chap. 19, Hewitt et al.; Chap. 25, Sliwa et al.), and making commitments to monitoring and vigilance (e.g. Chap. 20, Campbell).

The chapters in this section, beyond providing assessments of invasion status (with differing degrees of acknowledged limitations), describe how many of the other topics described in this book are put into practice in real-world situations. Many chapters deal with the temporal and spatial aspects of invasion vectors, as well as ecological impacts of invaders. They address supply-side dynamics and forecasting (Chap. 24, Hayden et al.) and the importance of changes in paleo-ecological conditions (Chap. 30, Leppäkoski et al.; Chap. 31, Rilov and Galil). Differing legal frameworks for addressing invasions are also presented (Chap. 26, Castilla and Neill; Chap. 27, Ferreira), as are economic and management perspectives (Chap. 26, Castilla and Neill; Chap. 23, Griffiths et al.; Chap. 30, Leppäkoski et al.). The section concludes with a comparison of invasions in estuaries vs the open coast (Chap. 33, Preisler et al.), which highlights intriguing patterns, points to the relative paucity of extensive datasets across multiple habitat types within a region, and articulates approaches to address such issues in the future.

References

Cardigos FT, Tempera S, Avila J, Goncalves A, Colaco, Santos RS (2006) Non-indigenous species of the Azores. Helgol Mar Res 60:160–169

Cohen AN, Carlton JT (1995) Biological study. Nonindigenous aquatic species in a United States estuary: a case study of the biological invasions of the San Francisco bay and delta. A report for the United States Fish and Wildlife Service, Washington, D.C., and The National Sea Grant College Program, Connecticut Sea Grant, NTIS Report Number PB96-166525, p 246

Orensanz JM, Schwindt E, Pastorino G, Bortolus A, Casas G, Darrigran G, Elías R, López Gappa JJ, Obenat S, Pascual M, Penchaszadeh P, Piriz ML, Scarabino F, Spivak ED, Vallarino EA (2002) No longer the pristine confines of the world ocean: a survey of exotic marine species in the southwestern Atlantic Ocean. Biol Invas 4:115–143

Otani M (2004) Introduced marine organisms in Japanese coastal waters, and the processes involved in their entry. Jpn J Benthol 59:45–57 (English abstract, in Japanese; English translation available from author)

Otani M (2006) Important vectors for marine organisms unintentionally introduced to Japanese waters. In: Koike F, Clout MN, Kawamichi M, de Poorter M, Iwatsuki K (eds) Assessment and control of biological invasion risks. Shoukado Book Sellers, Kyoto, Japan and World Conservation Union (IUCN), Gland, Switzerland, pp 92–103

Chapter 23
The Status and Distribution of Marine Alien Species in South Africa

Charles L. Griffiths, Tamara B. Robinson, and Angela Mead

23.1 Introduction

No geographcally or taxonomically comprehensive census of marine alien species has ever been undertaken in South Africa and the state of knowledge of the taxonomy of many marine invertebrate groups remains poor in this region (Gibbons 1999), compromising the ability to detect introduced species. Given these constraints, 22 confirmed extant marine aliens, plus 18 cryptogenic species, have been recorded from South African waters to date, with one additional species recently found in on-land mariculture facilities (Simon and Booth, in press). The true number of introduced species may well exceed these estimates by several times (see also Chap. 2, Carlton).

All 22 of the marine alien species reported from the wild support well-established populations, but the majority of these remain restricted to the few sheltered bays, estuaries and harbours on this wave-exposed coastline. Interestingly, only three species are known to have become invasive (defined here as having spread significantly beyond their points of origin). These are the Mediterranean mussel *Mytilus galloprovincialis,* the European green crab *Carcinus maenas* (Griffiths et al. 1992; Robinson et al. 2005a) and the recently detected barnacle *Balanus glandula* (Laird and Griffiths, 2008). A list of species known to be introduced to the region, and which presently support populations is given in Table 23.1. Sites mentioned in this chapter are shown in Fig. 23.1. The various established species are discussed by taxonomic group below.

23.2 Cnidaria

In 1995, the anemone *Metridium senile,* which is native to both coasts of North America and the Atlantic coast of Europe, was reported from Table Bay harbour (Fig. 23.1), where it occurred in densities of up to ca. 10 individuals/m^2 (Griffiths et al. 1996). The ecological impacts of this invasion have not

G. Rilov, J.A. Crooks (eds.) *Biological Invasions in Marine Ecosystems.*
Ecological Studies 204,

Table 23.1 Alien species present along the South African coast

Taxonomic grouping	Species name	First record	Present distribution	Known impacts
Cnidaria				
Actiniaria	*Metridium senile*	1995	Table Bay Harbour	-
	Sagartia ornate	2002	Langebaan Lagoon	-
Arthropoda				
Cirripedia	*Balanus glandula*	2007	West coast from Cape Point to Elands Bay	-
Isopoda	*Paracerceis sculpta*	2006	Port Elizabeth	-
Amphipoda	*Corophium acherusicum*	1976	All major harbours west of Durban	-
	Jassa marmorata	1951	Table Bay and Durban	-
	Jassa morinoi	1972	False Bay, Port Elizabeth and Durban	-
	Jassa slatteryi	1972	Saldanha Bay, False Bay and Knysna	-
Decapoda	*Carcinus maenas*	1983	West coast between Table Bay Harbour and Hout Bay Harbour	Potential ecological and economic impacts
Bryozoa	*Membranipora membranacea*	1972	Saldanha Bay, Table Bay, Port Elizabeth, Durban	-
Mollusca				
Bivalvia	*Mytilus galloprovincialis*	1979	Entire west coast, south coast up to 20 km west of East London	Significant ecological and economic impacts
	Crassostrea gigas	2001	Breede, Goukou and Knysna Estuaries	-
	Ostrea edulis	2007	Alexander Bay	-
Gastropoda	*Littorina saxatilis*	1974	Langebaan Lagoon and Knysna Lagoon	-
Echinodermata				
Echinoidea	*Tetrapygus niger*	2007	Alexander Bay	-
Chordata				
Ascidiacea	*Botryllus schlosseri*	1955	Alexander Bay, Saldanha Bay, Hout Bay and Durban	-
	Ciona intestinalis	1955	Harbours along the whole South African coast	Significant economic impact
	Clavelina lapadi-formis	2001	Knysna Estuary and Port Elizabeth Harbour	-
	Diplosoma listerianum	2001	All harbours from Saldanha Bay to Port Elizabeth	-
	Microcosmus squamiger	1955	Alexander Bay, Table Bay and Knysna	-
Kingdom Plantae				
Rhodophyta	*Schimmelmannia elegans*	2002	Table Bay Harbour	-

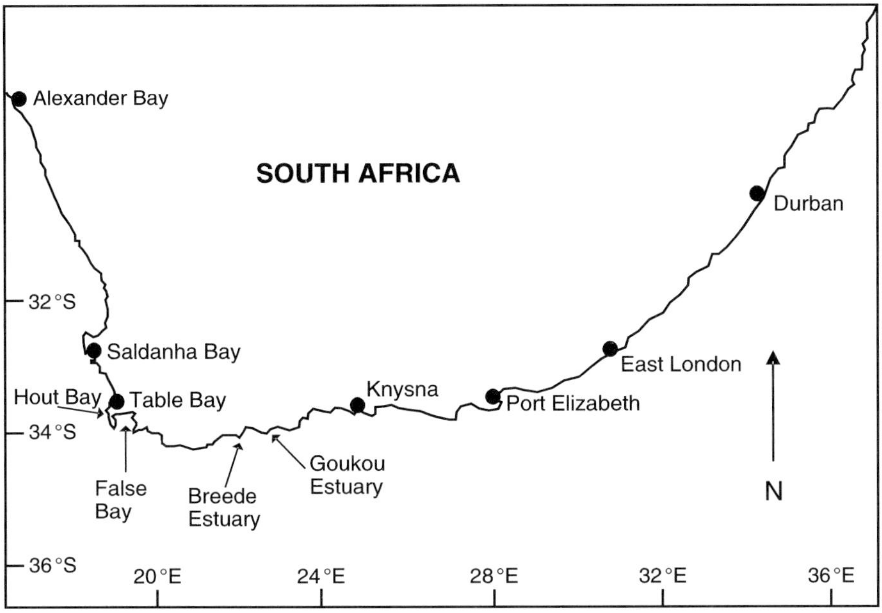

Fig. 23.1 Map of South Africa showing place names mentioned in the text

been investigated, but are unlikely to be significant as long as the anemone remains confined to this harbour, which supports a depauperate fauna dominated by other introduced species. A recent unconfirmed photographic observation that appears to represent a dense population of this species living in deep water off the south coast of South Africa could be of much greater concern, if validated.

A second anemone, *Sagartia ornata*, widely distributed throughout Western Europe, the United Kingdom and the Mediterranean (Manuel 1981), was first recorded in South Africa in 2002 (Acuna et al. 2004), although it is suspected to have been present long before this. At this time, this species is known only from the intertidal zone within Langebaan Lagoon, where it occurs in considerable densities of up to 426 ± 81 (SD) individuals/m^2, mostly in *Spartina maritima* beds and on rocks covered by sand (Robinson et al. 2004). This is in contrast to its habitat along British coasts, where it occurs in crevices on rocky shores and on kelp holdfasts (Gibson et al. 2001). Should it invade similar habitats in South Africa there is thus potential for this species to spread extensively along the coast, which offers cold waters and vast kelp beds typical of its home range. At present, the ecological influences of this invasion appear to be restricted to local effects on small invertebrate prey, but other impact could come into play if the population expands.

23.3 Polychaeta

To date the only introduced polychaete reported from South Africa is the polydorid *Boccardia proboscidea*, which has been reported by Simon and Booth (in press) infesting the shells of the abalone *Haliotis midae* in two on-shore abalone farms in Saldanha Bay and Hermanus (to the east of False Bay). This species has yet to be detected in the open ocean, although it appears likely that it occurs there, either in wild abalone of the same species or in the shells of introduced oysters, *Crassostrea gigas*, which are also farmed in Saldanha Bay.

23.4 Cirripedia

The North East Pacific barnacle *Balanus glandula* has recently been noted on the South African west coast (Simon-Blecher et al., 2008; Laird and Griffiths, 2008). Significant populations have been confirmed on intertidal shores along some 400 km of the west coast, between Cape Point and Lamberts Bay. The barnacle is concentrated within the balanoid zone, where it forms dense populations covering almost 90% of primary substratum at some sites in Saldanha Bay. Ecological effects of this newly-reported species have not yet been investigated, but are the subject of a planned research project to commence in 2008.

23.5 Isopoda

Paracerceis sculpta, a species introduced to many areas in the world (Hewitt and Campbell 2001), was first recorded in Port Elizabeth Harbour in 2006. As it currently supports a significant population in this busy South African port and intraregional movement of vessels is common, it is surprising that this isopod has not yet been recorded from any other South African harbours. This is, however, likely due to lack of appropriate sampling effort at these sites, few of which have been surveyed for small crustaceans.

23.6 Amphipoda

Amphipods such as *Jassa marmorata, J. morinoi, J. slatteryi* and *Corophium acherusicum*, which are all common members of fouling communities, have been widely distributed by shipping (Lewis et al. 2006) and are found in harbour areas around the South African coast (see Table 23.1). Although the ecological impacts of these species have not been considered in South Africa, they may compete with native organisms for food and space and in turn may act as easily accessible food items for a range of predatory fish and invertebrates.

23.7 Decapoda

South African populations of the European green crab *Carcinus maenas* were first detected in Table Bay Harbour in 1983 (Joska and Branch 1986). It has been proposed that these crabs reached the port via fouling of international oil exploration vessels, which have docked within the harbour since 1969 (Le Roux et al. 1990). By 1990, the species had been recorded at seven intertidal sites along the west coast of South Africa, six in the vicinity of Cape Town, and the seventh in Saldanha Bay, some 100 km to the north, where a single isolated record was noted by Le Roux et al. (1990). Fifteen years later no intertidal range extension was evident, although a dense but isolated population has established in Hout Bay Harbour, a small boat harbour about 20 km south of Table Bay docks (Robinson et al. 2005a). The lack of intertidal range expansion by *C. maenas* is probably a reflection of the wave-exposed nature of South African shores, and this crab's apparent inability to inhabit wave-exposed habitats due to its low tenacity, as experimentally demonstrated by Hampton and Griffiths (2007). Mark-recapture experiments conducted in 2002 suggested substantial subtidal populations of 133,568 individuals (95% confidence limits = 97,694–166,862) and 9,180 individuals (95% confidence limits = 5,870–12,003) in Table Bay Harbour and Hout Bay Harbour, respectively (Robinson et al. 2005a). As small rock lobster fishing vessels often move between these harbours, it is highly likely that adult crabs from Table Bay Harbour were inadvertently transported to Hout Bay by these boats. Despite extensive subtidal sampling within Saldanha Bay, no further specimens of this species have been recorded since the isolated record reported by Le Roux et al. (1990), suggesting that no viable population has yet established there (Robinson et al. 2004). Given the reputation of *C. maenas* as a highly successful invasive species, the lack of a well-established population within Saldanha Bay, 12 years after its initial discovery there, is curious. An extensive invasion of this important marine protected area in the future appears almost inevitable and could be potentially disastrous for the local biota, which is likely to be highly vulnerable to predation by *C. maenas* (Le Roux et al. 1990; see also Chap. 15, Rilov; Chap. 17, Grosholz and Ruiz).

23.8 Bryozoa

The encrusting bryozoan *Membranipora membranacea* has been reported coating the fronds of large algae, particularly the kelps *Laminaria pallida* and *Ecklonia maxima*, along much of the South African coast and has been present there for at least the past 35 years. The impacts of these encrustations have not been investigated in South Africa and might include reduced photosynthetic and hence growth rates. In other regions this bryozoan is also known to dramatically affect native Laminarian kelps by significantly reducing spore formation (Saier and Chapman 2004). Such an impact could be detrimental along the South African west coast, where dense kelp beds play a vital role in the near-shore environment (Bustamante et al. 1995).

23.9 Bivalvia

The most significant invasion along the South African coast is that of the Mediterranean mussel *Mytilus galloprovincialis*. Although first noted in Saldanha Bay in 1979 (Branch and Steffani 2004), genetic confirmation of this species identification was only gained in 1984 (Grant et al. 1984), by which time the species was already the dominant intertidal mussel along much of the west coast. *M. galloprovincialis* first appeared on the south coast of the country in 1989 (McQuaid and Phillips 2000) as an isolated population in Port Elizabeth Harbour, where it was deliberately introduced from the west coast for mariculture purposes. Subsequently this population was removed and the small populations it had spawned died out. Natural spread from the west to the south and east coasts began about the same time (Phillips 1994), and Robinson et al. (2005a) recorded this species along the entire west coast of South Africa, with populations extending eastwards around Cape Point and intermittently as far as 20 km west of East London on the southeast coast. *M. galloprovincialis* presently occupies 2050 km of the South African coast and supports a total standing stock estimated at 35,403.7 tons (±9099.6 SD), 88% of which occurs on the west coast (Robinson et al. 2005a).

The ecological effects of the *M. galloprovincialis* invasion are wide-ranging and have been most profound on the west coast (see also Chap. 16, Crooks). In comparison to the indigenous mussels *Choromytilus meridionalis* and *Aulacomya ater*, *M. galloprovincialis* exhibits a heightened growth rate, fecundity and tolerance to desiccation (Van Erkom Schurink and Griffiths 1990, 1991, 1992; Hockey and Van Erkom Schurink 1992). Consequently, there has been a marked upshore movement in the centre of gravity of intertidal mussel beds, as this species has become the dominant mussel along the west coast (Hockey and Van Erkom Schurink 1992). It is now only in sand-inundated areas that *C. meridionalis* remains the dominant mussel species. Coupled with the fact that *M. galloprovincialis* beds consist of multiple layers and support a higher biomass/m^2 than the single-layered beds of indigenous mussels, the increased vertical range of *M. galloprovincialis* beds has lead to a massive increase in overall mussel biomass along the South African west coast (Griffiths et al. 1992), and a simultaneous increase in the density and species richness of associated infauna (Hammond and Griffiths 2004; Robinson et al. 2007b). In contrast to the west coast situation, *M. galloprovincialis* on the more oligotrophic south coast forms mono-layered beds (Phillips 1994). *M. galloprovincialis* is immune to the trematode parasites that are common in indigenous mussels and that reduce both individual growth rates and population reproductive output by castrating females (Calvo-Ugarteburu and McQuaid 1998). On the south coast, this mussel has not yet completely replaced the indigenous mussel *Perna perna*. Instead, the two exhibit spatial segregation, with *P. perna* dominating the low shore, *M. galloprovincialis* the high shore and an overlap zone between the two (Robinson et al. 2005a).

The most important predator of *M. galloprovincialis* along the west coast is the whelk *Nucella cingulata* (Branch and Steffani 2004). However, due to the extremely high rate of recruitment of this mussel of up to 8,700,000 recruits/m^2 (Robinson et al. 2007a) and the relatively low numbers of *N. cingulata*, whelk predation is unable to control South African *M. galloprovincialis* populations (Branch and Steffani 2004). These high rates of recruitment have also allowed *M. galloprovincialis* to dominate primary rock surfaces at the expense of various competitively inferior limpet species. By excluding *Scutellastra granularis* from open rock, *M. galloprovincialis* has reduced the number of individuals occurring directly on rock, but at the same time has increased the overall density of this species by providing a favourable settlement and recruitment substratum for juveniles (Hockey and Van Erkom Schurink 1992). Associated with this increase in density, *S. granularis* has shown a decrease in mean size, as the maximum size of limpets occurring within the mussel beds is limited by the size of the host mussels (Griffiths et al. 1992). A second limpet species, *S. argenvillei*, has also been significantly affected by the invasion of *M. galloprovincialis*, although the strength of the interaction between these two species is mediated by wave action (Steffani and Branch 2003a, b). On exposed shores, *M. galloprovincialis* displaces *S. argenvillei* and dominates the primary substratum, while on semi-exposed shores the mussel becomes relatively scarce and *S. argenvillei* maintains dominance of open rock space (Steffani and Branch 2003a, b). Additional impacts on *S. argenvillei* include reductions in reproductive output and mean size of individuals, which now occur on mussels (Griffiths et al. 1992; Branch and Steffani 2004).

M. galloprovincialis has also affected some sandy shores, though to a lesser degree. In 1992, *M. galloprovincialis* invaded the centre banks of Langebaan Lagoon, an important marine conservation area adjacent to Saldanha Bay. Here it significantly altered the natural community composition by inducing a replacement of sandbank communities by those more typical of rocky shores (Robinson and Griffiths 2002). Interestingly, after supporting a biomass of 7.7 tons in 1998 (Robinson et al. 2004), the beds present on the centre banks decreased in size by 88% by 2001 (Hanekom and Nel 2002), and by 2003 only empty shells remained (Robinson et al. 2007c). The reason for this decline remains unclear.

Despite the many negative ecological impacts resulting from this invasion, one species, the near-threatened African Black Oystercatcher *Haematopus moquini*, has benefited from the presence of the introduced mussel. This endemic intertidal forager has shown a shift in diet since the arrival of *M. galloprovincialis*, and now feeds predominantly on this mussel (Hockey and Van Erkom Schurink 1992). Concurrent with this change in diet has come a dramatic increase in breeding success of *H. moquini* as a result of increased food supply (Hockey and Van Erkom Schurink 1992). From an economic perspective, the invasion of *M. galloprovincialis* has also had significantly positive impacts, as the entire mussel culture industry in South Africa is based on this alien species.

In line with global trends, the South African oyster industry is based on the Japanese oyster *Crassostrea gigas*, which was first introduced into Knysna Estuary in the early 1950s (De Moor and Bruton 1988). Due to difficulties in inducing predictable spawning and subsequent settlement under South African conditions, the industry is currently fuelled by spat imported from Chile, the United Kingdom and France. As *C. gigas* had appeared unable to complete its life cycle under local environmental conditions, this species was not previously considered likely to become invasive along South African shores (Griffiths et al. 1992). However, in 2001, oysters unlike any indigenous species were recorded in several estuaries along the south coast. The identification of these oysters as *C. gigas* was confirmed by Robinson et al. (2005b), who recorded populations of 184,206 ± 21,058.9 (SE), 876 ± 604.2 (SE) and 1228 ± 841.8 (SE) individuals in the Breede, Goukou and Knysna Estuaries, respectively. To date, however, this species has not been recorded on the open coast, and the invasion appears to be restricted to estuarine environments. The rate of spread and ecological impacts of this invasion remain undocumented, but similar invasions elsewhere have resulted in a variety of serious impacts, including the simultaneous introduction of associated fauna (Kaiser et al. 1998), the introduction of disease organisms (Ford 1992), genetic pollution of local oyster species (Gaffney and Allen 1993), and the reduction of indigenous oyster populations to threatened levels (Mann et al. 1991).

Ostrea edulis, the European flat oyster, was first introduced into Knysna Estuary between 1946 and 1951 for mariculture purposes (Korringa 1956). Despite the import of high numbers of spat, these populations died out and it was thought that no naturalized populations persisted (Griffiths et al. 1992; Robinson et al. 2005a). *O. edulis* has, however, recently re-emerged as an extant naturalized population, surviving and breeding in enclosed oyster dams in Alexander Bay, where the Japanese oyster *Crassostrea gigas* is now farmed. It is highly unlikely that this is the result of the spread of the distant original Knysna population, but rather represents a separate introduction. However, whether or not this species was originally introduced into the culture dams intentionally remains unclear.

23.10 Gastropoda

Littorina saxatilis, a small intertidal periwinkle, was first recorded in South Africa in 1974 (Day 1974). The only known populations occur in two discrete locations: Langebaan Lagoon and Knysna Estuary (Hughes 1979), and it has been proposed that these introductions may have resulted from early European shipping (McQuaid 1996). Despite occurring in crevices on rocky shores within its European home range (Gibson et al. 2001), along the South African coast *L. saxatilis* is restricted to sheltered salt marshes and lagoons, where it is found on the stems of the cord grass *Spartina maritima*. In 2002, densities of up to 433 ± 123 (SD) individuals/m^2 were recorded in Langebaan Lagoon (Robinson et al. 2004). The present status of the Knysna population is unknown. Despite its > 30 year presence along the South African coast, this species has remained geographically restricted. No ecological effects of the invasion are known, although these small gastropods could form an abundant food source for wading birds and crabs (Robinson et al. 2004).

23.11 Echinoidea

The urchin *Tetrapygus niger* originates from South America, where it occurs naturally in Peru and Chile (Rodriguez and Ojeda 1993). Along the South African coast it has recently been discovered in a single oyster dam in Alexander Bay where a self-sustaining population has apparently been present for several years. In its home range, *T. niger* is known to create 'urchin barrens' through destructive grazing. This is of great concern in the South African context, as such barrens could threaten the unique ecology of the near-shore environment along the kelp-dominated west coast (Bustamante et al. 1995). Importantly, an eradication programme may still be possible while the population remains confined within the oyster dam.

23.12 Ascidiacea

Botryllus schlosseri is an introduced compound ascidian that is confined to harbour and lagoonal areas in South Africa. First recorded in 1955 from Durban harbour on the east coast (Monniot et al. 2001), it has subsequently been recorded in Alexander Bay, Saldanha Bay, and Hout Bay on the west coast. In its natural range this species is a well recognised fouling organism that also occurs on eelgrass. This is of concern, as dense beds of the eelgrass *Spartina maritina* occurring within the West Coast National Park in Langebaan Lagoon may become invaded.

Ciona intestinalis is the earliest known accidental introduction to South African shores (Millar 1955). It occurs in harbours along the entire coast (Monniot et al. 2001), where it is a dominant fouling organism. This distribution pattern suggests that shipping has been the dispersal vector for this species. Despite *C. intestinalis* being well documented and of common occurrence, the ecological impacts of this invasion have not been quantified. Economic impacts have, however, been reported by mussel farmers in Saldanha Bay, who spend up to $15,000 per annum in an effort to maintain their mussel ropes free of this ascidian, which smothers the mussels, reducing growth and survival (Heasman 1996).

A review of South African ascidians by Monniot et al. (2001) documented three other introduced species, *Clavelina lapadiformis, Cnemidocarpa humilis* and

Diplosoma listerianum. C. lapadiformis appears to be limited to Knysna Estuary Hont Bay and Port Elizabeth Harbour and it seems likely that these populations represent a spread of the species rather than separate invasions. Such dispersal may have been aided by mariculture operations that regularly translocate oysters between these localities (Robinson et al. 2005b). *C. humilis* has only been recorded in Table Bay and Hout Bay harbours. *D. listerianum* is widely distributed in all harbours between Saldanha Bay and Port Elizabeth (Monniot et al. 2001). This may indicate numerous invasions, or intra-regional transport between harbours. The ecological and economic impacts of these ascidians in South African waters are unknown, but as both are relatively small encrusting species and appear to occur at relatively low densities, it is unlikely that they currently have significant ecological effects.

Originating from Australia, *Microcosmus squamiger* currently has a cosmopolitan distribution and it is highly likely that the South African populations are alien (Monniot et al. 2001). As a fouling organism, it has been recorded in Alexander Bay, Table Bay and Knysna Estuary, where it occurs on boats, buoys and rope lines.

23.13 Pisces

In contrast to the situation in South African freshwater systems, which have been invaded by numerous deliberately and accidentally introduced fish species (de Moor and Bruton 1988), with profound ecological impacts, no know marine fish introductions have been reported from South Africa to date.

23.14 Rhodophyta

Only a single alien algal species, *Schimmelmannia elegans*, is known from South Africa. First recorded in the Two Oceans Aquarium in Cape Town in 2002 (De Clerck et al. 2002), this species was also found growing below a water outlet where aquarium water enters Table Bay docks. Previously only known from the islands of Tristan da Cunha and Nightingale, this alga has no history as an invasive species. Its status as alien in South Africa is, however, well established, as it has not been detected in extensive surveys of the west coast conducted by Stegenga et al. (1997) and Bolton (1999). Due to its very limited distribution, it is unlikely that *S. elegans* presently exerts any significant ecological or economic impacts.

23.15 Patterns and Adequacy of Data

The relatively small number of alien species recorded in this region might suggest a significantly lower prevalence of non-indigenous marine species than has been reported for many other regions of the world. The apparently low number of alien

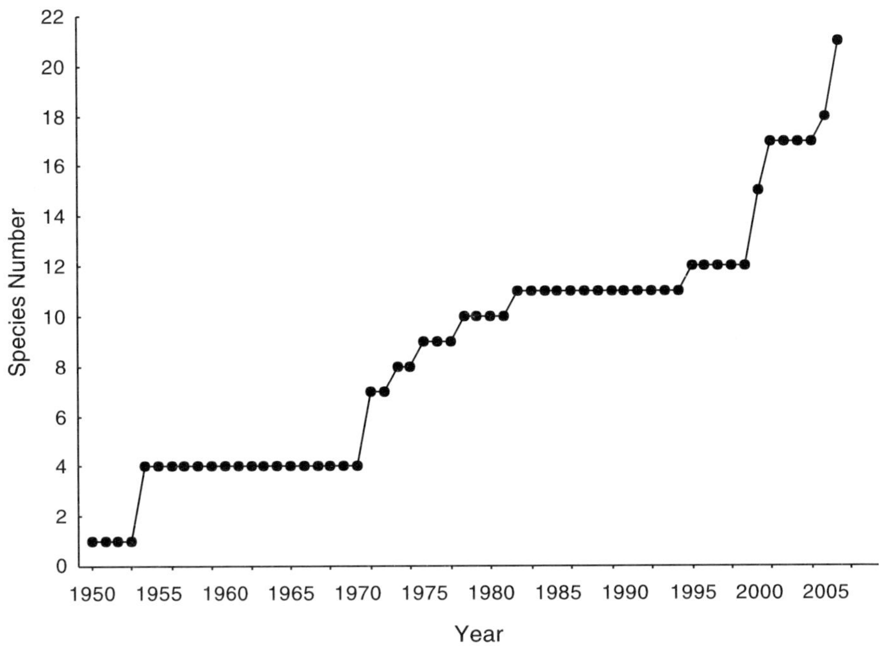

Fig. 23.2 Rate of discovery of marine invasive species in South Africa 1950–2007

species in southern Africa should, however, be treated with caution, as the true pervasiveness of invasions in the region may be obscured by several artefacts (see Chap. 2, Carlton). First, large areas of the South African coast remain poorly or un-surveyed with regards to non-indigenous species, with the Indian Ocean coast in particular having received little consideration. Second, the taxonomy of many marine groups is poorly developed within South Africa (Gibbons 1999; Griffiths 1999). At present, only four full-time professional marine taxonomists are working within the region, and their research is restricted to a few taxa (seaweeds and the phyla Porifera, Bryozoa and Mollusca). Awareness of, and research specifically directed towards, bioinvasions are relatively recent in this region, and this is demonstrated by the rapid recent rate of species discovery (Fig. 23.2). In our opinion, it is more likely that this represents an increased effort directed towards research into introduced species, rather than any actual spurt in the rates of introductions over recent years.

As is common in many parts of the world, most South African introductions remain confined to sheltered areas (see also Chap. 33 , Preisler et al.). The translocation of marine alien species has a well established link with shipping and ballast water tends to be loaded in sheltered harbours and unloaded at similarly sheltered destination ports (Awad et al. 2003). Thus the majority of alien species tend to originate from sheltered habitats in their native ranges and are thus most likely to survive if translocated to similar sheltered habitats. The fact that the majority of the South African coast is very wave exposed (Field and Griffiths 1991) may thus explain why only three alien species are known to have become invasive on the open coast in this region. A similar pattern is reported for marine alien species in Germany, where much of the coastline is also exposed (Gollasch and Nehring 2006).

Mariculture is commonly acknowledged as a major vector of marine introductions (Kaiser et al. 1998; Chap. 5, Minchin et al.; Chap. 20, Campbell), and the presence of oyster farms appears to be a significant factor contributing to the introduction and spread of marine alien species in South Africa. This is illustrated by the fact that the oyster farm at Alexander Bay, which is distant from any port, still supports one of the largest concentrations of marine invasive species on the South African coast (first tall bar in Fig. 23.2). These species seem likely to have been introduced along with oyster spat, or translocated with adult oysters introduced to the farm. In fact, the assemblages of introduced species found in Saldanha Bay, Table Bay and Port Elizabeth are very similar to those in mariculture sites at Alexander Bay and Knysna (Table 23.1, Fig. 23.3). This suggests that the common sheltered nature of these sites may be more important than the spatial variations in vector of introduction.

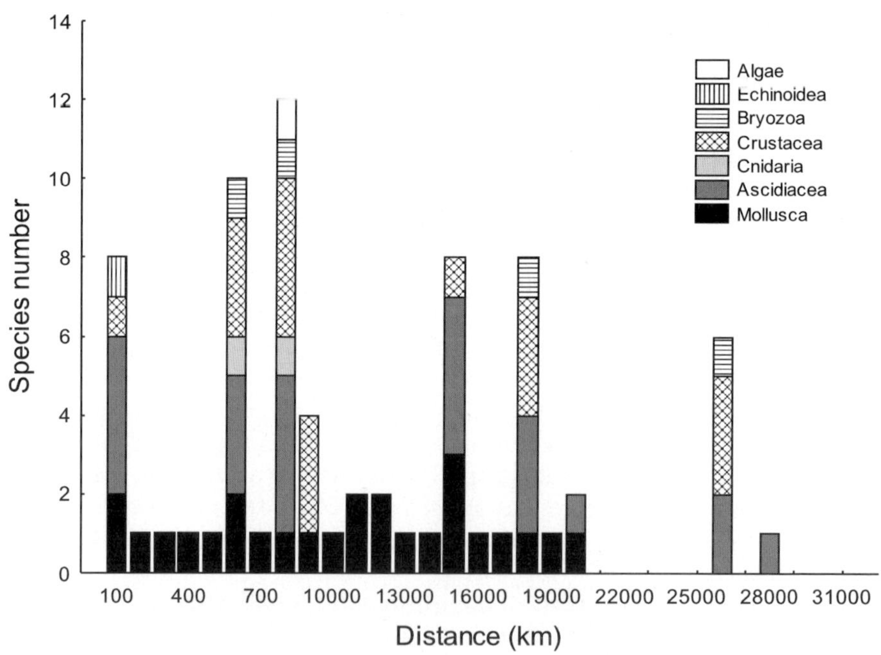

Fig. 23.3 Numbers of marine invasive species recorded in each 100-km stretch of coastline around South Africa from west to east (i.e. from the Namibian border at 0 km to the Mozambique border at 3100 km). The *six tallest bars* represent the oyster farm at Alexander Bay and sheltered harbour sites of Saldanha Bay, Table Bay, Knysna. Port Elizabeth and Durban (see Fig. 23.1 for locations of these sites)

It is interesting to note from Fig. 23.2 that marine alien species appear to be less common on the South African east coast than on the west and south coasts. This may be due to the predominant trade routes being between South Africa and the cooler temperate regions of Europe, from where most of our marine introductions originate. Alternatively, this may be at least partially an artefact of there being comparatively fewer studies on the east coast, although an extensive surveys by Sink et al. (2005) failed to record a single alien species on the open coastline of KwaZulu-Natal.

23.16 Additional Cryptogenic Species

Aside from the confirmed aliens discussed above, an additional 18 cryptogenic species are reported from South Africa. Such lists are very dependent on the current taxonomic knowledge of the different groups. This is strongly reflected in the dominance within Table 23.2 of amphipods, which are one of the best-studied

Table 23.2 Cryptogenic species along the South African coast

Taxonomic grouping	Species name
Porifera	*Cliona spp.*
Cnidaria	
Hydroida	*Obelia dichotoma*
	Obelia geniculata
Arthropoda	
Cirripedia	*Balanus amphitrite*
Isopoda	*Limnoria quadripunctata*
	Sphaeroma terebrans
Amphipoda	*Caprella equilibra*
	Caprella penantis
	Cerapus tubularis
	Chelura terebrans
	Cymadusa filosa
	Ericthonius brasiliensis
	Ischyrocerus anguipes
Bryozoa	*Bugula spp.*
Mollusca	
Bivalvia	*Bankia carinata*
Gastropoda	
Echinodermata	
Asteroidea	*Marthasterias glacialis*
Kingdom Plantae	
Rhodophyta	*Antithamnionella spirographidis*
	Antithamnionella ternifolia

marine taxa in South Africa. It is predicted that as the taxonomic knowledge base of South African marine organisms improves, many more cryptogenic species will be added to this list, or species currently regarded as cryptogenic will be moved to the confirmed list.

Based upon global distribution, five of the species currently regarded as cryptogenic have been highlighted as being very likely marine alien species. They are the isopod, *Limnoria quadripunctata*, the shipworm, *Bankia carinata*, the starfish, *Marthasterias glacialis* and the two algal species, *Anthithamnionella spirographidis* and *A. ternifolia*. However, in order to confirm the status of these five species within South Africa, comprehensive genetic analyses are required.

References

Acuna FH, Excoffon AC, Griffiths CL (2004) First record and redescription of the introduced sea anemone *Sagartia ornata* (Holdsworth, 1855) (Cnidaria: Actinaria: Sagartiidae) from South Africa. Afr J Zool 39:314–318

Awad AA, Clarke C, Greyling L, Hillard R, Polglaze J, Raaymakers S (2003) Ballast water risk assessment, Port of Saldanha Bay, Republic of South Africa, Final Report November 2003. Global Ballast Water Management Programme. Globallast Monograph Series 13. 64 pp

Bolton JJ (1999) Seaweed systematics and diversity in South Africa: an historical account. Trans R Soc S Afr 54:167–177

Branch GM, Steffani CN (2004) Can we predict the effects of alien species? A case-history of the invasion of South Africa by *Mytilus galloprovincialis* (Lamarck). J Exp Mar Biol Ecol 300:189–215

Bustamante RH, Branch GM, Eekhout S, Robertson B, Zoutendyk P, Schleyer M, Dye A, Hanekom N, Keats D, Jurd M, McQuaid CD (1995) Gradients of intertidal primary productivity around the coast of South Africa and their relationships with consumer biomass. Oecologia 102:189–201

Calvo-Ugarteburu G, McQuaid CD (1998) Parasitism and invasive species: effects of digenetic trematodes on mussels. Mar Ecol Prog Ser 169:149–163

Day JA (1974) A guide to marine life on South African shores. Balkema, Cape Town, 300 pp.

De Clerck O, Anderson RJ, Bolton JJ, Robertson-Anderson D (2002) *Schimmelmannia elegans* (Gloiosiphoniaceae, Rhodophyta): South Africa's first introduced seaweed? Phycologia 41: 184–190

De Moor IJ, Bruton MN (1988) Atlas of alien and translocated indigenous aquatic animals in southern Africa. S Afr Nat Sci Prog Rep 144:1–310

Field JG, Griffiths CL (1991) Littoral and sublittoral ecosystems of Southern Africa. In: Mathieson AC Nienhuis PH (eds) Intertidal and littoral ecosystems. Ecosyst World 24:323–346

Ford SE (1992) Avoiding the transmission of disease in commercial culture of mollusks, with special reference to *Perkinsus marinus* (Dermo) and *Haplosporidium nelsoni* (MSX). J Shellfish Res 11:539–546

Gaffney PM, Allen SK (1993) Hybridization among *Crassostrea* species: a review. Aquaculture 116:1–13

Gibbons MJ (1999) The taxonomic richness of South Africa's marine fauna: a crisis at hand. S Afr J Sci 96:8–12

Gibson R, Hextall B, Rogers A (2001) Photographic guide to the sea and shore life of Britain and north-west Europe. Oxford University Press, New York, 436 pp

Gollasch S, Nehring S (2006) National checklist for aquatic alien species in Germany. Aquat Invas 1:245–269

Grant WS, Cherry MI, Lombard AT (1984) A cryptic species of *Mytilus* (Mollusca: Bivalvia) on the west coast of South Africa. S Afr J Mar Sci 2:149–162
Griffiths CL (1999) Crustacean systematics in South Africa – status and historical overview. Trans R Soc S Afr 54:43–52
Griffiths CL, Hockey PAR, Van Erkom Schurink C, Le Roux PJ (1992) Marine invasive aliens on South African shores: implications for community structure and trophic functioning. S Afr J Mar Sci 12:713–722
Griffiths CL, Kruger LM, Smith EC (1996) First record of the sea anemone *Metridium senile* from South Africa. S Afr J Zool 31:157–158
Hammond W, Griffiths CL (2004) Influence of wave exposure on South African mussel beds and their associated infaunal communities. Mar Biol 144:547–552
Hampton SL, Griffiths CL (2007) Why *Carcinus maenas* cannot get a grip on South Africa's wave-exposed coastline. Afr J Mar Sci 29:123–126
Hanekom N, Nel P (2002) Invasion of sandflats in Langebaan Lagoon, South Africa, by the alien mussel *Mytilus galloprovincialis*: size, composition and decline of the populations. Afr Zool 37:197–208
Heasman KG (1996) The influence of oceanographic conditions and culture method on the dynamics of mussel farming in Saldanha Bay, South Africa. MSc thesis, Department of Zoology and Entomology, Rhodes University, 89 pp
Hewitt CL, Campbell ML (2001) The Australian distribution of the introduced sphaeromatid isopod, *Paracerceis sculpta*. Crustaceana 74:925–936
Hockey PAR, Van Erkom Schurink C (1992) The invasive biology of the mussel *Mytilus galloprovincialis* on the southern African coast. Trans R Soc S Afr 48:123–139
Hughes RN (1979) South African populations of *Littorina rudis*. Zool J Linn Soc 65:119–126
Joska MAP, Branch GM (1986) The European shore-crab – another alien invader? Afr Wildlife 40:63–65
Kaiser MJ, Laing I, Utting DS, Burnell GM (1998) Environmental impacts of bivalve mariculture. J Shellfish Res 17:59–66
Korringa P (1956) Oyster culture in South Africa. Hydrographical, biological and ostreological observations in the Knysna Lagoon, with notes on conditions in other South African waters. Invest Rep Div Sea Fish S Afr 20:84–102
Laird MC, Griffiths CL (2008) Present distribution and abundance of the introduced barnacle *Balanus glandula* Darwin in South Africa. Afr J Mar Sci 30:93–100
Le Roux PJ, Branch GM, Joska MAP (1990) On the distribution, diet and possible impact of the invasive European shore crab *Carcinus maenas* (L.) along the South African Coast. S Afr J Mar Sci 9:85–93
Lewis PN, Bergstrom DM, Whinam J (2006) Barging in: a temperate marine community travels to the subartic. Biol Invas 8:787–795
Mann R, Burreson E, Baker P (1991) The decline of the Virginia oyster fishery in Chesapeake Bay: considerations for introduction of a non-endemic species, *Crassostrea gigas* (Thunberg 1793). J Shellfish Res 10:379–388
Manuel RL (1981) British Anthozoa. Linnean Society, London, 241 pp
McQuaid CD (1996) Biology of the gastropod family Littorinidae. II. Role in the ecology of intertidal and shallow marine ecosystems. Oceanogr Mar Biol Annu Rev 42:263–302
McQuaid CD, Phillips TE (2000) Limited wind-driven dispersal of mussel larvae: *in situ* evidence from the plankton and the spread of the invasive species *Mytilus galloprovincialis* in South Africa. Mar Ecol Prog Ser 201:211–220
Millar RH (1955) On a collection of ascidians from South Africa. Proc R Soc Lond 125:169–221
Monniot C, Monniot F, Griffiths CL, Schleyer M (2001) South African ascidians. Ann S Afr Mus 108:1–141
Phillips TE (1994). Dispersal, settlement and recruitment: their influence on the dynamics of intertidal mussels. PhD thesis, Department of Zoology and Entomology, Rhodes University, 246 pp
Robinson TB, Griffiths CL (2002) Invasion of Langebaan Lagoon, South Africa, by *Mytilus galloprovincialis* – effects on natural communities. Afr Zool 37:151–158

Robinson TB, Griffiths CL, Kruger N (2004) Distribution and status of marine invasive species in and bordering the West Coast National Park. Koedoe 47:79–87

Robinson TB, Griffiths CL, McQuaid CD, Rius M (2005a) Marine alien species of South Africa – status and impacts. Afr J Mar Sci 27:297–306

Robinson TB, Griffiths CL, Tonin A, Bloomer P, Hare MP (2005b) Naturalized populations of *Crassostrea gigas* along the South African coast: distribution, abundance and population structure. J Shellfish Res 24:443–450

Robinson TB, Branch GM, Griffiths CL, Govender A (2007a) Effects of experimental harvesting on recruitment of an alien mussel *Mytilus galloprovincialis*. J Exp Mar Biol Ecol 345:1–11

Robinson TB, Griffiths CL, Branch GM, Govender A (2007b). The invasion and subsequent die-off of *Mytilus galloprovincialis* in Langebaan Lagoon, South Africa: effects on natural communities. Mar Biol 152:225–232

Robinson TB, Branch GM, Griffiths CL, Govender A, Hockey PAR (2007c) Changes in South African rocky intertidal invertebrate community structure associated with the invasion of the mussel *Mytilus galloprovincialis*. Mar Ecol Prog Ser 340:163–171

Rodriguez SR, Ojeda FP (1993) Distribution patterns of *Tetrapygus niger* (Echinodermata: Echinoidea) off the central Chilean coast. Mar Ecol Prog Ser 101:157–162

Saier B, Chapman AS (2004) Crusts of the alien bryozoan *Membranipora membranacea* can negatively impact spore output from native kelps (*Laminaria longicruris*) Bot Mar 47:265–271

Simon CA, Booth AJ (in press) Population structure and growth of polydorid polychaetes that infest the cultured abalone, *Haliotis midae*. Afr J Mar Sci

Simon-Blecher N, Granevitze Z, Achituv Y (2008) *Balanus glandula*: from North-West America to the west coast of South Africa. Afr J Mar Sci 30:85–92

Sink KJ, Branch GM, Harris JM (2005) Biogeographic patterns in rocky intertidal communities in KwaZulu-Natal, South Africa. Afr J Mar Sci 27:81–96

Steffani CN, Branch GM (2003a) Spatial comparisons of populations of an indigenous limpet *Scutellastra argenvillei* and an alien mussel *Mytilus galloprovincialis* along a gradient of wave energy. Afr J Mar Sci 25:195–212

Steffani CN, Branch GM (2003b) Temporal changes in an interaction between an indigenous limpet *Scutellastra argenvillei* and an alien mussel *Mytilus galloprovincialis*: effects of wave exposure. Afr J Mar Sci 25:213–229

Stegenga H, Bolton JJ, Anderson RJ (1997) Seaweeds of the South African west coast. Contr Bolus Herb 18:1–655

Van Erkom Schurink C, Griffiths CL (1990) Marine mussels of southern Africa – their distribution patterns, standing stocks, exploitation and culture. J Shellfish Res 9:75–85

Van Erkom Schurink C, Griffiths CL (1991) A comparison of reproductive cycles and reproductive output in four southern African mussel species. Mar Ecol Prog Ser 76:123–134

Van Erkom Schurink C, Griffiths CL (1992) Physiological energetics of four South African mussel species in relation to size, ration and temperature. Comp Biochem Physiol 101:779–789

Chapter 24
Marine Invasions in New Zealand: A History of Complex Supply-Side Dynamics

Barbara J. Hayden, Graeme J. Inglis, and David R. Schiel

24.1 Introduction

New Zealand's recent ecological history is often held up as a textbook example of the havoc that can be wrought by non-native species (Clout and Lowe 2000). The first human inhabitants of New Zealand arrived (by boat) just 800 years ago, and brought with them food crops and dogs. They arrived in a country where, already, many elements of the endemic fauna were in serious decline; an apparent legacy of the introduction of the Polynesian rat (*Rattus exulans: 'kiore'*) by transient human visitors, some 1000 years before (Holdaway et al. 2002). In the eighteenth and nineteenth centuries, British immigrants to New Zealand brought with them a wave of new predators, plant pests and grazing animals. When Charles Darwin stopped in NZ on his *Beagle* voyage in 1835, the settled European population in NZ numbered fewer than 2000 but Darwin lamented the rampant spread of "very troublesome" weeds which had already "overrun whole districts" and the loss of native flightless birds, "annihilated" sic by introduced Norway rats (*Rattus norvegicus*) (Darwin 1889). Now, 170 years later, there are more than 4 million human inhabitants and 25,000 introduced plant species in New Zealand, with established exotics outnumbering native species (Beston 2005; NZ Plant Conservation Network 2006).

Introduced species have significantly altered the natural landscape and ecological functioning of New Zealand's environments. Deliberate and accidental introductions of organisms continue to occur at an alarming rate. In this chapter we discuss the status of marine invasions in NZ, some of the impediments to accurately defining that status and the importance of taking account of "supply-side" dynamics when assessing the risks of new introductions. "Supply-side" ecology is the term introduced into marine ecology in the late 1980s to describe the study of the processes of arrival of new members of populations (see also Johnston et al. this volume). Its importance was to re-emphasise the consequences of variability in the supply of recruits to adult populations. While not new, the concept served to refocus attention on the dynamics of reproductive success, oceanographic influences on dispersal, larval behaviour, the process of settlement, and features of the receiving environment that cause variations in

G. Rilov, J.A. Crooks (eds.) *Biological Invasions in Marine Ecosystems.*
Ecological Studies 204,

numbers of recruits from place to place and time to time (Underwood and Fairweather 1989). Supply-side dynamics are also an important part of marine invasions because establishment of new populations of introduced species is contingent on variability in the same processes described above that influence the success of new recruits. However, because human activities spread marine invaders, the supply-side of the equation also encompasses variability in the transport pathways in which the species are carried to new environments. We use data on changing trade patterns in New Zealand to demonstrate the importance of including that aspect of supply-side dynamics into assessments of incursion risk. Other equally important processes that form part of supply-side ecology, such as dispersal of propagules and settlement success, are not discussed in this chapter.

24.2 Status of Marine Invasions in NZ

Non-native species have been introduced by humans into New Zealand's marine environment at least since the arrival of European ships in the mid-1860s, and almost certainly earlier, during the first wave of Maori settlement. Invasion of the Polynesian rat ca. 2000 years ago is thought to have had consequences for New Zealand's marine ecosystems even before settlement. Fossil evidence suggests the kiore was associated with the demise of vast numbers of sea birds, which nested in coastal burrows (Holdaway 1989).

Within a month of Captain James Cook setting foot on New Zealand shores in 1769, he ordered his crew to careen his vessel, the *Endeavour*, and "heel'd and scrubed both sides of the Ship". The rate of fouling was so great that, just two months later, Cook once again gave the order to careen the ship, where the "barnacles and seaweed" were scraped off. Later, in the early 1800s, whalers and sealers from America, England and other European nations operated in increasing numbers around the coast of New Zealand. They and the other sailing vessels of that era are likely to have brought with them on their hulls a range of fouling organisms that were introduced into a marine biota but had not yet been scientifically described (Cranfield et al. 1998).

Like many other countries, New Zealand has only recently begun to document the extent of invasions in its coastal environments. It is possible that some of the species that arrived on the hulls of early vessels are now so widespread and abundant that they cannot readily be distinguished as non-native. Species that are currently considered "cosmopolitan" or even "native" could well have been spread before any records were made (Cranfield et al. 1998; Ruiz et al. 2000). A few species introduced in the eighteenth and nineteenth centuries characteristically have very limited distributions and are more readily identifiable as adventives. For example, the brown alga *Chnoospora minima* is found in only one location in New Zealand, Port Underwood, where it grows abundantly unattached in about 8 m of water. In the early 1800s, Port Underwood was an important destination of the southern right whaling fleet arriving from the tropical Pacific where *C. minima* is widespread (Nelson and Duffy 1991). Similarly, the red alga *Chondria harveyana* is known only from Porirua Harbour, and from Tasmania, Australia. In the early nineteenth century, Porirua Harbour contained a whaling station that was regularly supplied from Tasmania (Nelson 1994).

Hayward (1997) and Cranfield et al. (1998) used published and unpublished biological records to compile inventories of the known and suspected introduced marine species that are present in New Zealand. The more comprehensive of these lists, that by Cranfield et al. (1998), tested adventism against nine biogeographic criteria developed by Chapman and Carlton (1991). They identified 139 species that satisfied at least 3 of the criteria and a further 20 species that have been found in New Zealand, but which did not become established. The organisms included macroalgae, estuarine grasses, "Protozoa", Porifera, Cnidaria, Annelida, Mollusca, Arthropoda, Entoprocta, Bryozoa, and Chordata. Most (96%) of these species arrived accidentally, as fouling organisms on vessel hulls (69%), in ballast (6%), or as either fouling or ballast (21%).

Since the inventory was compiled, a further 40 suspected introduced marine species and 27 cryptogenic species (see Chap. 2, Carlton) have been described from New Zealand waters (NIWA and MAF Biosecurity NZ unpublished data). The dramatic increase in the rate of discovery in the intervening nine years coincides with greater awareness and reporting of bioinvasions among the scientific community and general public, and the commencement, in 2001, of national baseline port surveys and targeted surveillance for unwanted marine organisms that have the expressed aim of identifying introduced marine species (Hewitt et al. 2004). Between 2001 and 2003, baseline port surveys were completed of all 13 major commercial ports in New Zealand and the 3 main first-marinas-of-entry for recreational vessels.

24.3 What is Non-native?

Inventories of the type prepared by Cranfield et al. (1998) are subject to several types of unsystematic bias that are difficult to estimate or control (Ruiz et al. 2000; Chap. 2, Carlton; but see Solow and Costello 2004). Perhaps the most significant of these is the variable quality of taxonomic and biogeographic information for many marine groups. "Cryptogenic" species – species which are not demonstrably native or non-native – can comprise up to 30% of some coastal marine assemblages (Ruiz et al. 2000). The patchy status of marine taxonomy, systematics and biogeography continues to provide a significant challenge for discriminating native from non-native species (Gordon 2001). New Zealand has the world's fourth-largest Exclusive Economic Zone (EEZ), which at more than 4 million km^2 is 15 times the land area (Blezard 1980). Although specimens have been collected from >9000 stations in this zone, sampling of the biota is estimated to cover less than 2 km^2 or one five-millionth of the EEZ area (Nelson and Gordon 1997). Since the early 1980s, the rate of discovery of new species and records of marine invertebrates has been approximately three per week, although this estimate is likely to be conservative because not all marine invertebrate groups are being studied in New Zealand. The paucity of specialist taxonomic expertise in New Zealand means that formal descriptions are unable to keep up with the rate of discovery of new species (Nelson and Gordon 1997). For instance, the origin of 160 species detected in New Zealand in the national port baseline surveys since 2000 cannot be determined decisively. Of these, 106 species are thought to be new to science and are yet to be described.

In north-eastern New Zealand, the problem of distinguishing native from non-native species is compounded by occasional natural immigration of subtropical species during warm summers (Dell 1968; Francis et al. 1999; see also Chap. 3, Lonhart). Some of these immigrants manage to establish viable breeding populations and would satisfy at least three of the criteria of Chapman and Carlton (1991) (i.e., sudden appearance, subsequent spread and restricted distribution). Natural arrivals of this type are also a prominent feature of the fossil records in New Zealand. Indeed, New Zealand's extant marine biota represents a blend of species with Indo-Pacific affinities and colder-water Antarctic affinity (Knox 1975). This mix is the product of a long series of incursions and subsequent extinctions of warmer water species that were associated with latitudinal shifts in the position of the subtropical and Antarctic convergences (sharp gradients in water temperature that intersect the New Zealand archipelago).

The prevalence of cryptogenic species (Chap. 2, Carlton) in the marine biota creates an uncertain regulatory environment for managers of marine pests (see also Chap. 19, Hewitt et al.). Eradication and control measures available under The Biosecurity Act, 1993, can be initiated only if the target organism is deemed to be a pest, and authorities are unwilling to take action against native species. Decisions to act, therefore, frequently revolve around the geographic origin of the species. This difficulty was highlighted in 2001, when the harbourmaster of a small harbour in north-eastern New Zealand reported an unusual growth blanketing wharf piles and some boat hulls in the harbour. Subsequent surveys revealed that the organism was a colonial ascidian in the genus *Didemnum*, *D. vexillum* sp. nov (Kott 2002). The species was subsequently recorded in the nearby port of Tauranga, and on a barge in Picton Harbour that had been relocated from Tauranga. Picton Harbour is located in Marlborough Sounds, where NZ's large industry of cultured greenshell mussels is located. Because of its habit of overgrowing other fouling species including mussels, discovery of the ascidian in Marlborough Sounds raised alarm bells in the mussel industry. While facing pressure from the industry to initiate an eradication programme, MAF Biosecurity NZ had to evaluate conflicting opinions from international taxonomists about the likely origin of the organism before they could act. Appearance of the New Zealand didemnid was followed closely by reports of nuisance species in this genus from the Atlantic coast of the USA, Mediterranean, North Sea and English Channel. Detailed morphological comparison suggests that these are, in fact, different species (Kott 2004). Because *D. vexillum* sp. nov has not been recorded elsewhere, it is currently assumed to be a native New Zealand species which underwent a sudden unusual bloom in abundance.

24.4 Evidence of the Leaky Border

The large increase in detected incursions in the last nine years is most likely to be a result of increased search effort rather than an increase in introductions. Nevertheless, exotic species continue to arrive, probably at an increasing rate. Visits from merchant vessels discharging ballast water have increased by 10% per annum since 2000 (Hewitt et al. 2004). Although increased vessel speed, faster port turnaround times, and more effective antifouling paints might be expected to make it more difficult for exotic species to reach New Zealand via hull fouling than in the past, Cranfield et al. (1998) observed that the number of introduced species that arrived on hulls and established in New Zealand between 1958 and 1998 was similar to the number that arrived on hulls in the 50 years previous to that period.

Evidence that New Zealand's marine border remains very leaky is illustrated by *Undaria pinnatifida* (Laminariales), the Asian brown alga that was first discovered in New Zealand in 1987 (Hay and Luckens 1988). Thought to have arrived in ballast water or attached to the hull of a ship, *Undaria* was progressively found in 11 other ports and coastal locations in the 10 years following its first discovery. Since 1997, *Undaria* has continued to spread around New Zealand with discoveries at a further six locations. Because it was also found on the hulls of vessels in some of these locations, the vector for the spread was presumed to be the coastal movement of vessels. Fouled vessels are indeed likely to be the main vector of spread from already established populations but a recent study of the genetic diversity of native and introduced populations of *Undaria* worldwide has shown that there have been multiple new introductions of the species to NZ since 1987 (Uwai et al. 2006).

24.5 Reasons Why Introductions Continue

New Zealand's biosecurity system, designed to protect both the terrestrial and aquatic environments from impacts of non-indigenous species, is based around (1) border control, (2) Import Health Standards, (3) post-entry quarantine, (4) surveillance, and (5) management of pests once in New Zealand (Hayden and Whyte 2003; Chap. 19, Hewitt et al.). In the marine realm, this has taken the form of national baseline surveys of biota in high risk ports, active surveillance for a range of unwanted marine organisms, assessments of fouling on vessel hulls and the efficacy of hull cleaning facilities as discussed in Hewitt et al. (2004). Border and pre-border control is the first line of defence against the introduction of exotic species and inspection by a range of techniques takes place at all entry pathways (passengers, aircraft, mail, personal effects, commercial cargo) except marine. The lack of effective marine border controls is one reason why organisms continue to arrive and establish themselves.

24.6 Most Common Marine Pathways

24.6.1 Ballast Water

Marine species are most commonly introduced in ships' ballast water and as fouling assemblages on the hulls of vessels (Carlton 1985; Ruiz et al. 2000; Hewitt et al. 2004; Chap. 6, Hewitt et al.) and submerged structures such as oil platforms (Foster and Willan 1979). Alternative pathways of transfer (Chap.5, Minchin et al.; Chap. 20, Campbell), such as the release of species for fisheries and aquaculture, or escape of aquarium species, account for a much smaller proportion of known establishments in New Zealand (Cranfield et al. 1998).

New Zealand has had voluntary guidelines for the management of ballast water discharges, based on the International Maritime Organisation's Guidelines, since 1992. The voluntary guidelines were transferred into an Import Health Standard (IHS) in 1998 (Ministry of Fisheries 1998) but the IHS lacked a legal framework until the enactment of The Biosecurity Act in 1993. Because it is "reasonable to suspect" that ballast water arriving into New Zealand poses a risk to the flora and fauna already in New Zealand, ballast water is classified as a "risk good" and the powers of the Biosecurity Act can be used to authorise controls. Under the Act, no "risk goods" can be imported into New Zealand unless they have complied with a relevant Import Health Standard (IHS).

The main features of the ballast water IHS are that ballast water should not be discharged within New Zealand if at all possible. If it must be discharged, then it should be ballast that has been exchanged or loaded in the open ocean. Other options include disinfection of the ballast water prior to discharge, discharge into an approved area or to an onshore facility, or to have the ballast tested to show it is not a risk. Effective treatment options are still in the development stage and there are currently no areas approved for ballast dumping nor any onshore discharge or ballast water treatment facilities. Thus, the only practical option available is to exchange the ballast water with oceanic water before the ship arrives in New Zealand's territorial waters, a practice known to be only partially effective at minimising the risk of exotic introductions (Rigby et al. 1993).

No sediment or mud from the cleaning of the holds, ballast tanks or equipment on the vessel can be discharged in New Zealand waters.

24.6.2 *Hull Fouling and Sea Chests*

Not all vessels that enter New Zealand waters are ballasted merchant vessels. Since the 1970s, many of New Zealand's major fishing companies have chartered foreign vessels on a seasonal basis and increasing numbers of private pleasure craft and passenger cruise liners are also entering New Zealand's waters. Emphasis in the past two decades on ballast water as a primary vector of marine introductions has temporarily diverted attention away from fouling assemblages on the hulls and other external structures of vessels. The shift in focus has in part been based on the assumption that modern antifoulant paints, the high speed of merchant vessels and their rapid port turn-around times would minimise the risk of introductions via fouling. However, there is evidence that hull fouling and sea chests remain significant vectors of marine introductions (James and Hayden 2000; Gollasch 2002; Coutts et al. 2003; Coutts and Taylor 2004; Floerl and Inglis 2005; Floerl et al. 2005a, b). An Import Health Standard (IHS) has yet to be established for fouled hulls or sea chests although MAF Biosecurity NZ is currently assessing the risks associated with hull fouling on all vessel types entering New Zealand. This is a comprehensive study that is using a standardized sampling approach to measure the biomass and identity of fouling organisms on the external structures of 450 merchant, fishing, recreational and passenger vessels and towed barges. Data on associated risk factors such as travel and maintenance history of the vessels have also been collected. The challenge to find management options for vessels deemed to be high risk is being addressed by additional MAF Biosecurity NZ projects to evaluate the efficacy of a range of hull cleaning operations.

24.7 The Dynamic Nature of Incursion Risk

Successful establishment of invasive species is likely to be a highly probabilistic outcome that depends on the coincidence between delivery of the species to the new location and suitable conditions for establishment, including the absence of enemies and the availability of resources (see also Chap. 7, Johnston et al.; Chap. 8, Miller and Ruiz; Chap. 12, Olyarnik et al.; Chap. 11, Torchin and Lafferty). Both the supply of colonizing stages of invasive organisms ("propagule supply") and the opportunity for their establishment ("niche opportunity") are likely to be highly variable in space and time. For a range of organisms, the most consistent correlate of invasion success tends to be propagule supply (Veltman et al. 1996; Forsyth and Duncan 2001; Mack et al. 2000; Kolar and Lodge 2001; Lester 2005; Verling et al. 2005; Wonham et al. 2005; Herborg et al. 2007). However, in marine invasive species research, propagule supply is rarely measured directly because of the difficulty in obtaining representative samples from ballast tanks and the cost of identifying all viable species present. Risk assessments tend to use the most cost-effective proxies available, which are related to the amount, type and frequency of shipping or ballast discharge from different locations (e.g., ACIL 1994; Hilliard and Raaymakers 1997; Clarke et al. 2004). In many cases these are not calibrated against actual propagule transport (i.e., abundance or frequency of delivery of species or species groups) and are static (i.e., cross-sectional) representations of the vector and transport pathway risk. However, historical vessel voyage data from New Zealand indicate that the vector risk is far from static, especially at the scale of individual ports, highlighting the need for more dynamic, quantitative predictors of risk.

24.8 Changes in the Source of Invaders

At a coarse level, the geographic origins of marine invaders tend to be correlated with the predominant shipping routes into and out of a country (e.g. Carlton 1996; Ruiz et al. 2000). This is also true of New Zealand invaders. In Fig. 24.1 we used the putative date of discovery of each introduced species that has established in New Zealand waters to construct four historical time periods with roughly equal numbers of discoveries: 1800 to 1925 ($n = 40$), 1926 to 1960 ($n = 37$), 1961 to 1990 ($n = 37$), 1991 to 2005 ($n = 35$). This allowed changes in the relative proportions of species coming from different regions to be compared with trade from those regions. Several patterns are immediately obvious in these data. First, as we noted earlier, the rate of discovery has quickened over time; the first 40 species were recorded over a 125 year period, the following 37 occurred within 34 years, the next 37 in 29 years, and the final group of 35 discoveries occurred in less than half that time, at a rate of 2.3 per year. As Costello and Solow (2003) point out, this does not necessarily imply an increasing rate of incursion, particularly because it does not account for increasing search effort as scientific knowledge of New Zealand's marine biota has increased and surveys have been specifically tasked with finding introduced species.

Nevertheless, when the presumed origin of these species is considered, the data do reveal a noticeable shift in the regions from which the invaders are arriving. Between 1800 and 1960, more than 90% of the introduced species recorded in New Zealand were thought to have derived from Europe, Australia or North America, or were considered "cosmopolitan". Often these cosmopolitan species were also first described from British and European specimens. Since 1960, the relative proportions of species from Europe and North America have declined and there has been a concomitant rise in the proportion of species

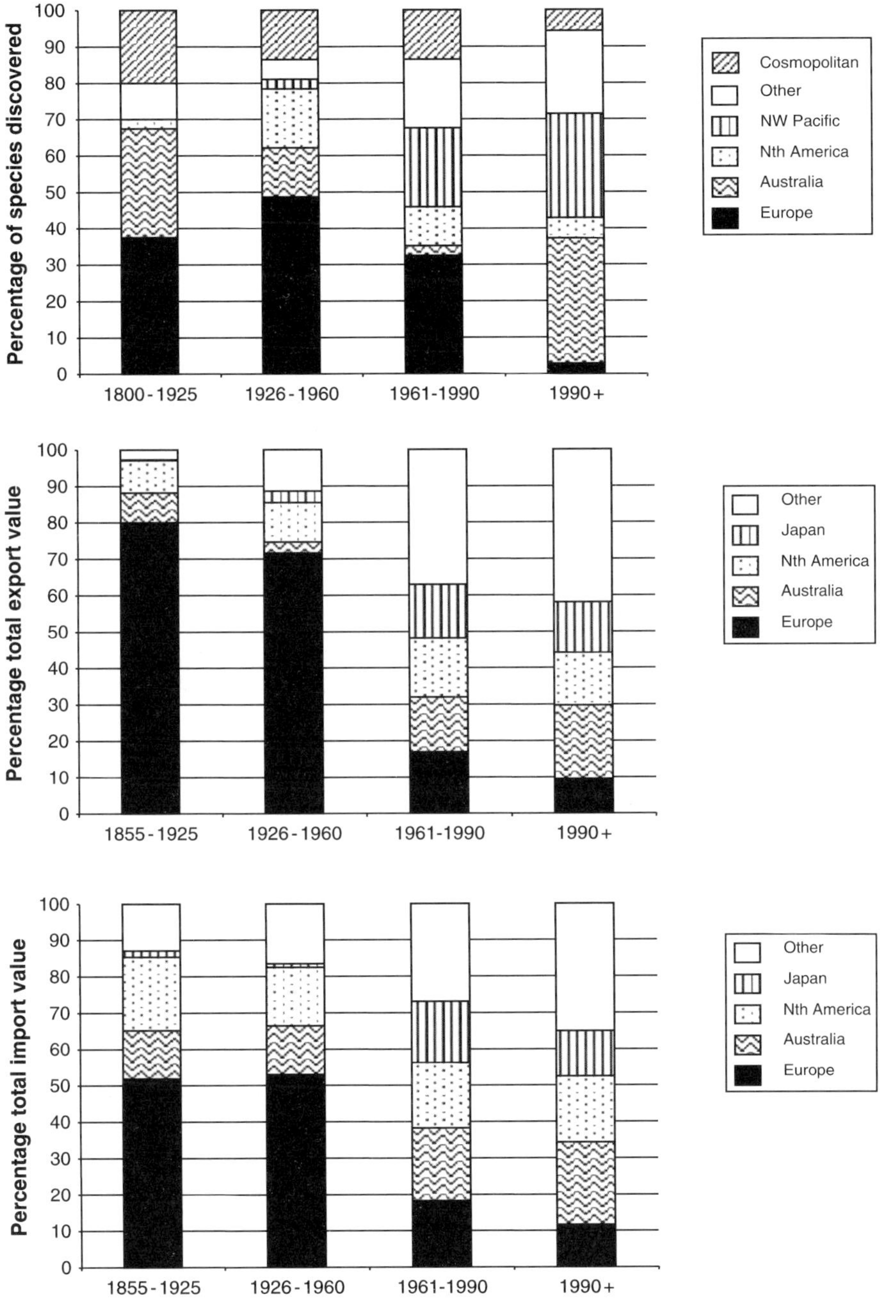

Fig. 24.1 Historical changes in the percentages of introduced marine species (**a**), exports (**b**), and imports (**c**) in New Zealand from different source regions. Roughly equal numbers of introduced species were discovered in each time period (1855–1925 n = 40, 1926–1960 n = 37, 1961–1990 n = 37, 1991–2005 n = 35). (Sources: Cranfield et al. 1998, NIWA & MAF Biosecurity NZ unpublished data, Statistics NZ. Overseas Trade Statistics)

whose biogeographic origins are in the North West Pacific and 'other' regions (such as South East Asia, South America, the Indian Ocean and Arabian Seas). This change mirrors long-term shifts in New Zealand's international trade patterns over the same period (Fig. 24.1).

During the early 1800s, most of New Zealand's shipping trade was with Australia and the United Kingdom, when New Zealand was administered as part of the New South Wales colony. Following the first refrigerated shipment of meat and dairy products to the United Kingdom in 1882, exports to more distant markets became possible and the economy became increasingly agriculturally based. Trade with North America increased in the mid-1800s with the gold-field discoveries that linked California, Australia, and New Zealand. Later, in the 1940s, the German blockade of allied shipping in the English Channel forced New Zealand shipping and goods to be diverted from the traditional British market to the USA and Canada. Nevertheless, the United Kingdom remained New Zealand's largest trading partner until the 1970s, when it joined the European Union. Since that time there has been significant diversification of the countries with which New Zealand trades and a decline in the relative importance of the United Kingdom. Japan emerged as a major market for New Zealand exports in the 1960s and 1970s following the first shipments of frozen meat in 1956. Although Australia remains New Zealand's single largest trading partner, the burgeoning economies of the Northwest Pacific - Japan, Korea, China and Taiwan –accounted for more than 64% of the gross tonnage of New Zealand exports in the period from 1996 to 2004 and, collectively, comprise the largest regional source of discharged ballast water (Hewitt et al. 2004).

24.9 The Complexity of Making Forecasts

Gross national trends such as these overlay more complex patterns of trade with individual ports. It is at this local scale that invasions begin. We are still a long way from developing predictive relationships at this scale between propagule supply and marine invasions. Patterns of global shipping are becoming increasingly complex as faster transit times, increased global demand, and lowering trade restrictions open access to new markets. At the same time, this is offset by globalization of production, so that competition to supply markets is intensifying. These two factors – increased market diversity and global competition – have a major impact on the trade patterns of ports, which can change rapidly over relatively short time periods.

To illustrate this point, in Figs. 24.2 and 24.3 we present data on the changing patterns of trade in forest products in New Zealand ports. Forest products are typically transported in dedicated bulk vessels that discharge large volumes of ballast water in New Zealand ports. Forestry related products have dominated recent bulk export growth in New Zealand and account for around 13% of total

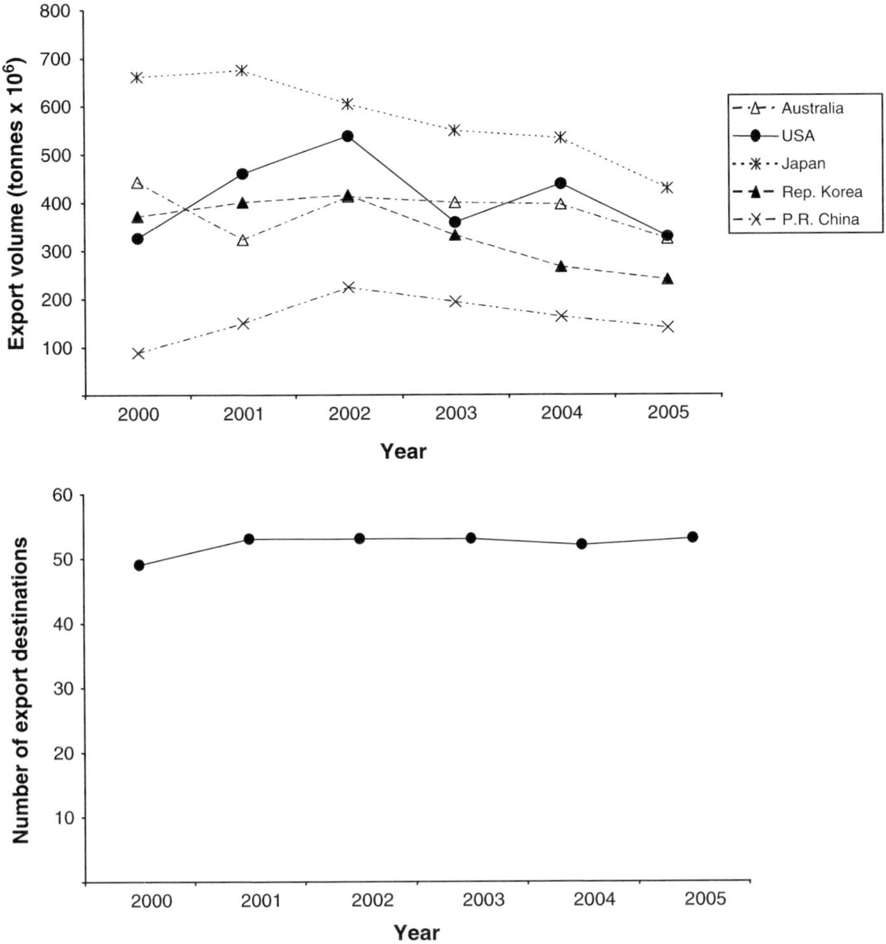

Fig. 24.2 Changes in the gross export volume of wood and wood articles from New Zealand to the top five destinations (**a**) and the total number of export destinations between 2000 and 2005 (b). (Source: Statistics NZ. Overseas Trade Statistics Merchandise Trade Exports)

exports. Since the early 1980s exports of forest products have more than doubled, with the volume of wood available expected to increase by 74% between 1996 and 2010. Over the past five years, total exports of wood and wood products to New Zealand's principal markets show (1) a doubling in export volumes to China, (2) steady decline in volumes leaving for Japan and Korea, (3) relatively consistent trade with Australia and the USA, and (4) no overall change in the diversity of export destinations (Fig. 24.2). Within individual ports, however, the trading patterns are much more dynamic and diverse (Fig. 24.3). At this level, trade in forest products is driven by local supply (i.e., the maturation of nearby forest plantations), global market demand and competition with other

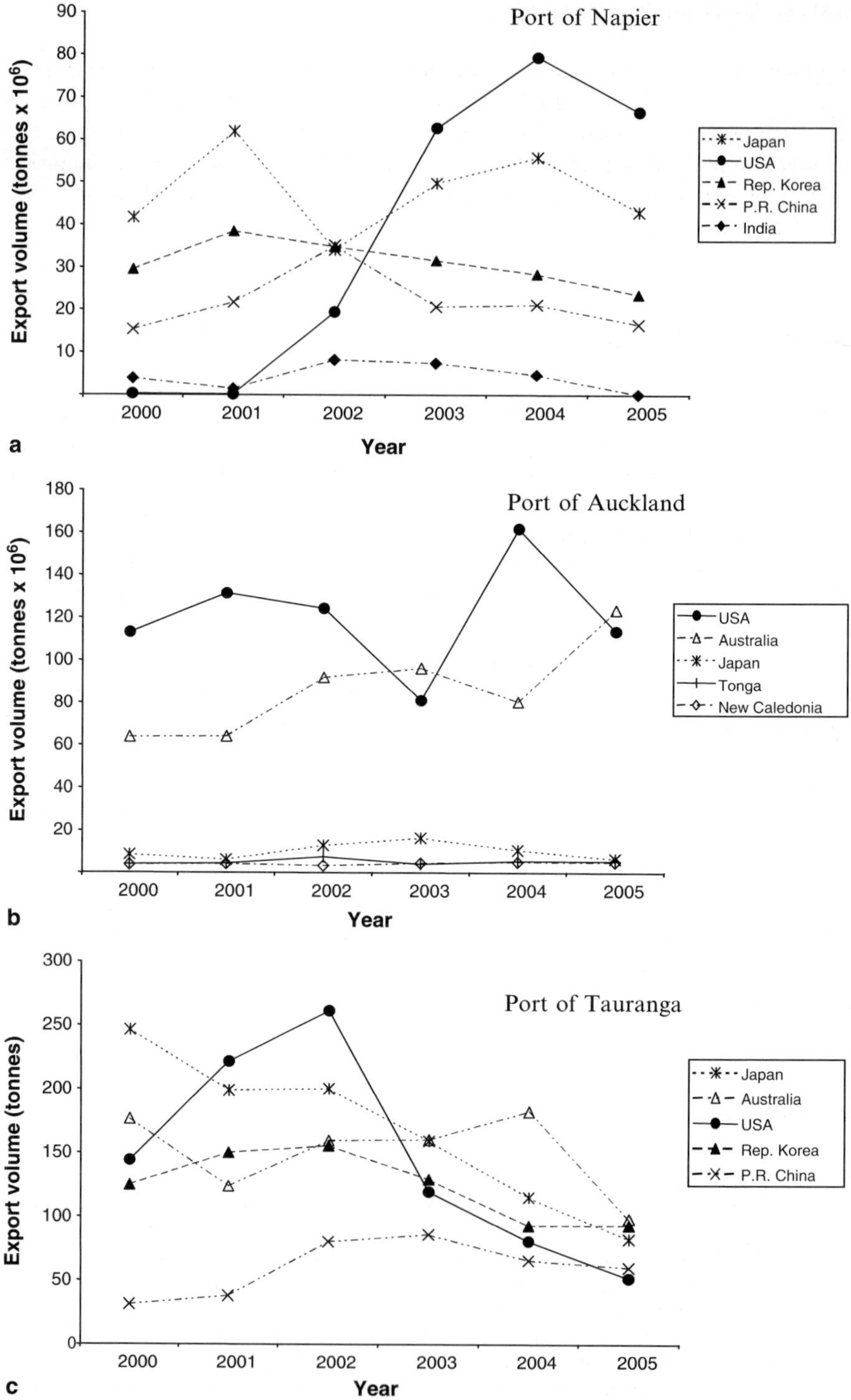

Fig. 24.3 Changes in export of wood and wood articles from three New Zealand ports: **a** Napier; **b** Auckland; **c** Tauranga. (Source: Statistics NZ. Overseas Trade Statistics Merchandise Trade Exports)

suppliers, both national and international. Currently, New Zealand exports bulk forest products to more than 50 countries worldwide, into 13 of the 18 bioregions in the IUCN's marine classification (Kelleher et al. 1995). Individual ports exhibit markedly different patterns of trade with different countries and these individual port dynamics will influence risk assessments over relatively short time-frames.

24.10 Conclusion

Despite the gaps in our taxonomic expertise and the difficulty in determining the origin of cryptogenic species, New Zealand has made major advances in describing the introduced marine biota in high risk entry points over the last decade. Our knowledge of marine invasions that may have occurred outside those entry points is less advanced. Multiple research projects to underpin the development of policy for managing marine invasive pests have also been initiated in the same time period (Hewitt et al. 2004; Chap. 20, Campbell) but positive outcomes from the research that will allow the leaks in the marine border to be plugged are still some way off. Propagule supply will remain a primary focus of incursion management. However, if we are to develop better predictive understanding of propagule supply (and thereby invasion dynamics), study of marine invasions needs to move from being a purely descriptive biological undertaking to an interdisciplinary science that incorporates economic and social drivers of risk, including those that drive changes in shipping patterns. Recent studies of risk factors associated with hull fouling of maritime vessels have revealed a similarly complex mix of social, economic and biological determinants of risk (Floerl et al. 2004, 2005a, b; Floerl and Inglis 2005). To paraphrase Underwood and Keough (2001), we contend that unless unpredictability caused by variations in the supply of introduced species is also included in conceptual understanding of marine invasions, our ability to effectively manage them will be severely limited.

References

ACIL Economics and Policy Pty Ltd (1994) Bio-economic risk assessment of the potential introduction of exotic organisms through ships' ballast water. Australian Quarantine and Inspection Service, Canberra, 47 pp

Beston A (2005) Native plants outnumbered. NZ Herald 2 December 2005:A7

Blezard RH (1980) Calculated sea area of the New Zealand 200 nautical mile Exclusive Economic Zone. N Z J Mar Freshwater Res 14(2):137–138

Carlton JT (1985) Transoceanic and interoceanic dispersal of coastal marine organisms: the biology of ballast water. Oceanogr Mar Biol Annu Rev 23:313–371

Carlton JT (1996) Pattern, process, and prediction in marine invasion ecology. Biol Conserv 78:97–106

Chapman JW, Carlton JT (1991) A test of criteria for introduced species: the global invasion by the isopod *Synidotea laevidorsalis* (Miers, 1881). J Crustacean Biol 11:386–400

Clarke C, Hilliard R, Junqueira Ade OR, Neto Ade CL, Polglaze J, Raaymakers S (2004) Ballast water risk assessment, Port of Sepetiba, Federal Republic of Brazil, December 2003: Final Report. GloBallast Monograph Series No 14, IMO, London

Clout MN, Lowe SJ (2000) Invasive species and environmental changes in New Zealand. In: Invasive species in a changing world. Mooney HA, Hobbs RJ (eds) Island Press, Washington, DC, pp 369–383

Costello CJ, Solow AR (2003) On the pattern of discovery of introduced species. Proc Natl Acad Sci 100:3321–3323

Coutts ADM, Taylor MD (2004) Variation in fouling patterns between different areas on the hulls of merchant vessels. N Z J Mar Freshwater Res 38:215–229

Coutts ADM, Moore KM, Hewitt CL (2003) Ships' sea-chests: an overlooked transfer mechanism for non-indigenous marine species. Mar Pollut Bull 46(11):1510–1512

Cranfield HJ, Gordon DP, Willan RC, Marshall BA, Battershill CN, Francis MP, Nelson WA, Glasby CJ, Read GB (1998) Adventive marine species in New Zealand. NIWA Tech Rep 34, 48 pp

Darwin C (1889) Journal of researches into the natural history and geology of the countries visited during the voyage of the H.M.S. "Beagle" round the world. Ward, Lock, London, 381 pp

Dell RK (1968) Composition and distribution of the New Zealand Brachyuran fauna. Trans R S N Z 10:225–240

Floerl O, Inglis GJ (2005) Starting the invasion pathway: the interaction between source populations and human transport vectors. Biol Invas 7:589–606

Floerl O, Pool TK, Inglis GJ (2004) Positive interactions between non-indigenous species facilitate transport by human vectors. Ecol Appl 14:1724–1736

Floerl O, Inglis GJ, Hayden BJ (2005a) A risk-based predictive tool to prevent accidental introductions of nonindigenous marine species. Environ Manag 35:765–778

Floerl O, Inglis GJ, Marsh H (2005b) Selectivity in vector management: an investigation of the effectiveness of measures used to prevent transport of non-indigenous species. Biol Invas 7:459–475

Forsyth DM, Duncan RP (2001) Propagule size and the relative success of exotic ungulate and bird introductions to New Zealand. Am Nat 157:583–595

Foster BA, Willan RC (1979) Foreign barnacles transported to New Zealand on an oil platform. N Z J Mar Freshwater Res 13:143–149

Francis MP, Worthington CJ, Saul PJ, Clements KD (1999) New and rare tropical and subtropical fishes from northern New Zealand. N Z J Mar Freshwater Res 33:571–586

Gollasch S (2002) The importance of ship hull fouling as a vector of species introductions into the North Sea. Biofouling 18(2):105–121

Gordon DP (2001) Marine biodiversity. Alpha 108. The Royal Society of New Zealand, Wellington, New Zealand

Hay CH, Luckens PA (1988) The Asian kelp *Undaria pinnatifida* (Phaeophyta: Laminariales) found in a New Zealand harbour. N Z J Bot 25(2):329–332

Hayden BJ, Whyte CF (2003) Invasive species management in New Zealand. In: Ruiz GM, Carlton JT (eds) Invasive species: vectors and management strategies. Island Press, Washington, DC, pp 270–291

Hayward BW (1997) Introduced marine organisms in New Zealand and their impact in the Waitemata Harbour, Auckland. Tane 36:197–223

Herborg L-M, Jerde CL, Lodge DM, Ruiz GM, MacIsaac HJ (2007) Predicting invasion risk using measures of introduction effort and environmental niche models. Ecol Appl 17(3):663–674

Hewitt CL, Willing J, Bauckham A, Cassidy AM, Cox CMS, Jones E, Wotton DM (2004) New Zealand marine biosecurity: delivering outcomes in a fluid environment. N Z J Mar Freshwater Res 38:429–438

Hilliard RW, Raaymakers S (1997) Ballast water risk assessment for twelve warm-water ports in Queensland, Australia. Proceedings of the Combined Australasian Coastal Engineering and Ports Conference, Christchurch, pp 717–722

Holdaway RN (1989) New Zealand's pre-human avifauna and its vulnerability. N Z J Ecol 12(Suppl):11–25

Holdaway RN, Roberts RG, Beavan-Athfield NR, Olley JM, Worthy TH (2002) Optical dating of quartz sediments and accelerator mass spectrometry ^{14}C dating of bone gelatin and moa eggshell: a comparison of age estimates for non-archaeological deposits in New Zealand. J R Soc N Z 32(3):463–505

James P, Hayden BJ (2000) The potential for the introduction of exotic species by vessel hull fouling: a preliminary study. NIWA Client Rep WGN00/51, 57 pp

Kelleher G, Bleakeley C, Wells S (1995) A global representative system of marine protected areas. Vols 1–4. Great Barrier Reef Marine Park Authority; The World Conservation Union (IUCN); The World Bank, Washington, DC

Knox GA (1975) The marine benthic ecology and biogeography. In: Kuschel, G (ed). Biogeography and ecology in New Zealand. Dr W. Junk Publishers, The Hague, pp 353–458

Kolar CS, Lodge DM (2001) Progress in invasion biology: predicting invaders. Trends Ecol Evol 16:199–204

Kott P (2002) A complex didemnid ascidian from Whangamata, New Zealand. J Mar Biol Assoc UK 82:625–628

Kott P (2004) A new species of *Didemnum* (Ascidiacea, Tunicata) from the Atlantic coast of North America. Zootaxa 732:1–10

Lester PJ (2005) Determinants for the successful establishment of exotic ants in New Zealand. Divers Distrib 11(4):279–288

Mack RN, Simberloff D, Lonsdale WM, Evans H, Clout M, Bazzaz FA (2000) Biotic invasions: causes, epidemiology, global consequences, and control. Ecol Appl 10:689–710

Ministry of Fisheries (1998) Import Health Standard (Biosecurity Act 1993) for ships' ballast water from all countries, 3pp. http://www.fish.govt.nz/sustainability/biosecurity/ballast_health.html

Nelson WA (1994) Marine invaders of New Zealand coasts: Auckland Botanical Society Lucy Cranwell Lecture 6 October 1993, University of Auckland. Auckland Bot Soc J 49(1):4–14

Nelson WA, Duffy CAJ (1991) *Chnoospora minima* (Phaeophyta) in Port Underwood, Marlborough – a curious new algal record for New Zealand. N Z J Bot 29:341–344

Nelson WA, Gordon DP (1997) Assessing New Zealand's marine biological diversity – a challenge for policy makers and systematists. N Z Sci Rev 54(3/4):58–66

NZ Plant Conservation Network (2006) http://www.nzpcn.org.nz/exotic_plant_life_and_weeds/index.asp

Rigby GR, Steverson IG, Bolch CJ, Hallegraeff GM (1993) The transfer and treatment of shipping ballast waters to reduce the dispersal of toxic marine dinoflagellates. In: Smayda TJ, Shimizu Y (eds) Toxic phytoplankton blooms in the sea. Elsevier, pp 169–176

Ruiz GM, Fofonoff PW, Carlton JT, Wonham MJ, Hines AH (2000) Invasion of coastal marine communities in North America: apparent patterns, processes, and biases. Annu Rev Ecol Syst 31:481–531

Solow AR, Costello CJ (2004). Estimating the rate of species introductions from the discovery record. Ecology 85:1822–1825

Underwood AJ, Fairweather PG (1989) Supply side ecology and benthic marine assemblages. TREE 4:16–20

Underwood AJ, Keough MJ (2001) Supply-side ecology: the nature and consequences of variations in recruitment of intertidal organisms. In: Bertness MD, Gaines SD, Hay ME (eds) Marine community ecology. Sinauer Associates, Sunderland, Maine, pp 183–200

Uwai S, Nelson W, Neill K, Wang WD, Aguilar-Rosas LE, Boo SM, Kitayama T, Kawai H (2006) Genetic diversity in *Undaria pinnatifida* (Laminariales, Phaeophyceae) deduced from mitochondria genes – origins and succession of introduced populations. Phycologia 45:687–695

Veltman CJ, Nee S, Crawley MJ (1996) Correlates of introduction success in exotic New Zealand birds. Am Nat 147:542–557

Verling E, Ruiz GM, Smith LD, Galil B, Miller AW, Murphy KR (2005) Supply-side invasion ecology: characterizing propagule pressure in coastal ecosystems. Proc R Soc B 272(1569):1249–1256

Wonham MJ, Lewis MA, MacIsaac HJ (2005) Minimizing invasion risk by reducing propagule pressure: a model for ballast-water exchange. Front Ecol Environ 3(9):473–478

Chapter 25
Marine Bioinvasions in Australia

Cathryn Sliwa, Sasha Migus, Felicity McEnnulty, and Keith R. Hayes

25.1 Introduction

Australia has been a regular port of call for ocean going vessels since the beginning of the nineteenth century, and the first records of non-native species followed soon after (Table 25.1). However, it has taken over one hundred years and several prominent invasions by demonstrably harmful species, for marine invasions to attract significant management attention (Thresher 1999; Ferguson 2000; Bax et al. 2001, 2002; Hewitt 2003).

The detection of the Japanese sea star, *Asterias amurensis* (Turner 1992; Morrice 1995; Goggin 1998), the European shore crab, *Carcinus maenas* (Gardner et al. 1994) and the population explosion of the European fan worm, *Sabella spallanzanii* (Clapin and Evans 1995; Thresher et al. 1999) in Australia led to the establishment of a Centre for Research on Introduced Marine Pests (CRIMP) at the marine division of the Commonwealth Scientific and Industrial Research Organization (CSIRO) in 1994 by the Federal Government. Initial research of the Centre focused on investigating the impacts and ecology of these highly visible non-native species which had potential impacts on aquaculture and shipping (Thresher 1999). More recent research has concentrated on risk assessment, management and control strategies and early detection/identification methods such as genetic identification probes for larval stages of marine pest species (e.g. Deagle et al. 2003).

In the early 2000s, a number of university laboratories in Australia also began researching marine bioinvasions. These research units have since provided large amounts of information on a number of introduced species such as *Undaria pinnatifida* (Sliwa et al. 2006; Valentine and Johnson 2003, 2005, 2004), *Asterias amurensis* (Ross et al. 2002, 2003a, b, 2004; Morris 2002; Ling 2000), *Sabella spallanzanii* (Holloway and Keough 2002a, b), *Maoricolpus roseus* (Reid 2003) and *Caulerpa taxifolia* (West 2004), as well as investigating fouling communities (Coutts 1999; Lewis 2001; Lewis et al. 2003), and a relationship between pollution and invasive species (Clark and Johnston 2005; Chap. 7, Johnston et al.)[1].

[1] Please note, this list is not intended to be exhaustive

G. Rilov, J.A. Crooks (eds.) *Biological Invasions in Marine Ecosystems.*
Ecological Studies 204,

Table 25.1 A history of marine bioinvasions in Australia 1862–2004

Year	Significant event
1862	First non-native species in Australia (*Electra pilosa*) recorded from Port Phillip Bay, Victoria (MacGillivray 1869; Thresher 1999)
1900	*Carcinus maenas* first identified in Port Phillip Bay (Fulton and Grant 1900)
1940s–1950s	Several studies of sub-tidal fouling communities identified a number of species had translocated to Australia (Allen and Wood 1949; Allen 1950, 1953)
1965	First record of *Sabella spallanzanii* in Australia (Clapin and Evans 1995)
1972	Toxic dinoflagellate *Gymnodinium catenatum* introduced to Tasmania (McMinn et al. 1997)
1972	Tributyltin (TBT) used as a paint additive to prevent fouling of ship hulls
1973	First documented ballast water sampling study undertaken on a vessel travelling from Japan to Australia (Medcof 1975)
1986	*Asterias amurensis* first collected in Tasmania, Australia (Turner 1992)
1989	Use of TBT-based paints on vessels <25 m in length prohibited and leaching rate regulations were imposed on vessels >25 m (IMO 2002)
Late 1980s	62 exotic species listed in Australian waters with one or more found in almost all states and territories (Pollard and Hutchings 1990a, b)
1990	National guidelines introduced for voluntary ballast water management by ships entering Australia from overseas ports (AQIS 1992; Oemcke 1998)
1994	Centre for Research on Introduced Marine Pests (CRIMP) established
1995	Australian Quarantine and Inspection Service (AQIS) established the National Ballast Water Management Strategy
1995	*A. amurensis* first recorded in Port Phillip Bay (Parry et al. 2000)
1999	*Mytilopsis* sp. discovered and eradicated from a marina in Darwin (Bax 1999; Bax et al. 2002)
2000	Total population size of *A. amurensis* in Port Phillip Bay estimated at 150 million (Parry and Cohen 2001)
2001	Mandatory ballast water management arrangements introduced for all international vessels arriving in Australian waters (during July 2001)
2003	*A. amurensis* found at Inverloch, Victoria – 150 km south-east of the only known populations in Port Phillip Bay (Inverloch has since been resurveyed and no *A. amurensis* were found)
2004	CSIRO Marine Research identifies 129 non-native and 214 cryptogenic species in Australia (Hayes et al. 2005)

At present in Australia, a coordinated approach to the management of non-native marine species is being undertaken through collaboration between Federal, State and Territory governments (see also Chap. 19, Hewitt et al.; Chap. 20, Campbell). These governments are working with industry, research and conservation groups to develop a National System for the Prevention and Management of Marine Pest Incursions. An important component of this system is a new Single National Interface for the management of domestic ballast water. It is proposed that this interface will operate species-specific risk assessment, managing ballast water discharge in relation to the translocation risk of designated target species between Australian ports.

Management of non-native marine species in Australia is both species- and vector-orientated. Ballast water management and monitoring activities are largely

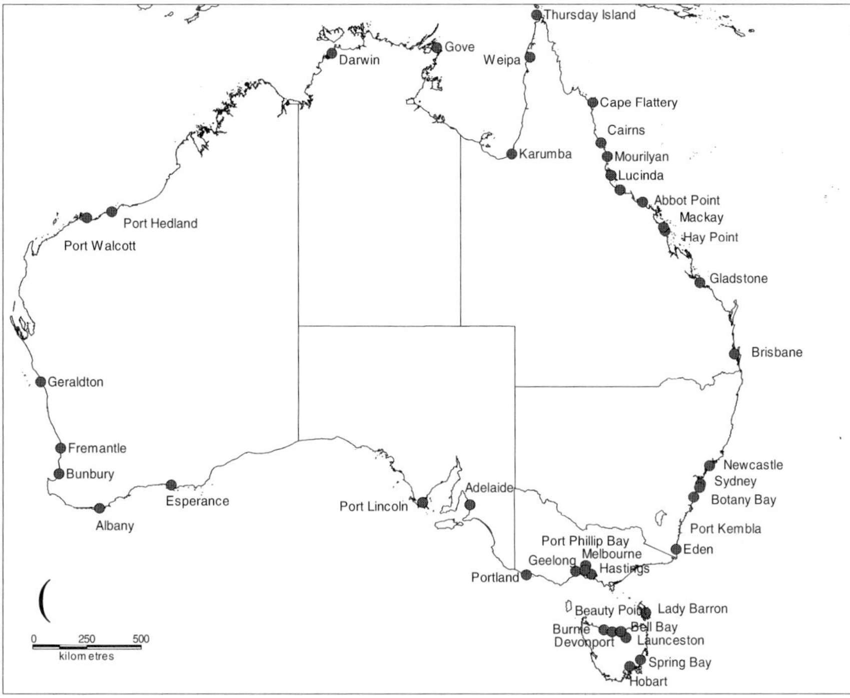

Fig. 25.1 Ports surveyed in Australia between 1995 and 2004

planned and implemented around lists of target species (see also Chap. 6, Hewitt et al.). Fishing vessels, recreational vessels and aquaculture equipment however, are managed without reference to particular species. During the last ten years, at least three marine target lists have been adopted and modified by various management authorities. The most notable of these are the Australian Ballast Water Management Advisory Council (ABWMAC), and the Australian Introduced Marine Pest Advisory Committee (AIMPAC) (see Hayes and Sliwa 2003). These lists were often compiled and modified in an arbitrary and inaccurate manner. In 2001 CRIMP began an extensive literature review (the "priority pest project") to identify and prioritise non-native and cryptogenic species in Australian waters using a rigorous, defensible process (Hayes and Sliwa 2003; Hayes et al. 2002, 2005). The priority pest project was restricted to non-native marine species that were transported via a shipping vector.

The results of Hayes et al. (2002, 2005) have recently been strengthened by the collation of port survey data. Between 1995 and 2004, 41 ports have been surveyed for non-native marine species in Australia (Fig. 25.1), the vast majority (95%) using standardised survey protocols developed by Hewitt and Martin (1996, 2001). In addition to this, a bay wide survey of Port Phillip Bay in Victoria was also conducted, including non-port regions (Fig. 25.1). All surveys sampled the benthos, hard structures such as pylons and rock breakwaters, plankton and nekton communities using quantitative and qualitative methods. The primary objective of the port surveys was to determine the presence/absence of 12 species listed by the ABWMAC[2]. Data from 39 of these surveys have been collated, verified and entered into a National Port Survey Database (NPSD). The NPSD currently holds over 11,000 records, however only a third of these are fully described to species level (i.e. they have both a genus and species name, e.g. *Asterias amurensis*). More than 99% of the genus and species names in the database have been taxonomically validated and allocated an invasion status.

This chapter provides an overview of the non-native and cryptogenic marine species in Australian waters, based on the results of the priority pest project and information currently held in the NPSD. We use the term "non-native" to refer to any species that has been transported by human activity to a new location beyond its known native range. We use the term "cryptogenic" to refer to those species that cannot be confirmed as either native or non-native (Carlton 1996; Chap. 2, Carlton). Hayes et al. (2002, 2005) include the additional term "establishment status" (yes, no or unknown), for non-native and cryptogenic species in Australian waters. Establishment status was determined by the presence of a first record in Australia, and additional records from other locations. Non-natives recorded as "not established" (or of "unknown" establishment status) are not included in these analyses. Species records in the NPSD are allocated an invasion status: introduced (non-native), cryptogenic, Australian native, not valid or non-Australian native (used for species that belong to a species complex, and therefore possibly representing misidentified Australian natives). Only introduced and cryptogenic NPSD records are used in this analysis.

25.2 Non-native and Cryptogenic Marine Species in Australia

Prior to 2005, invasion patterns in Australian waters were not well documented. However, there are a few notable exceptions. The works of Hutchings et al. (1987) and Pollard and Hutchings (1990a, b) listed 62 non-native species that were known to occur in Australian waters; and Hewitt et al. (2004) identified 99 non-native species and 61 cryptogenic species in Port Phillip Bay, Victoria using a mixture of literature studies, museum collections and field sampling.

Hayes et al. (2002, 2005) identified 129 non-native species in Australian waters; the data held in the NPSD from the port surveys identified 87 species.

[2] The ABWMAC listed species were: *Alexandrium catenella, A. minutum, A. tamarense, Gymnodinium catenatum, Asterias amurensis, Carcinus maenas, (Vari)corbula gibba, Crassostrea gigas, Musculista senhousia, Sabella spallanzanii, Undaria pinnatifida* and *Vibrio cholerae*. Two additional species *Mnemiopsis leidyi* and *Corbula amurensis*) were also listed but were not, and have not, been identified in Australian waters

The overlap between the two datasets is high, with 81 species in common; the total number of non-native species across the two data sets being 132 (Table 25.2). To the best of our knowledge, Table 25.2 includes all known non-native species in Australian waters. Almost 36% of the non-native marine species were identified only through the priority pest project (i.e. literature survey). Of that 36%, the largest proportion of species were the Heterokontophyta (brown algae; 5 out of 6 species), Rhodophycota (red algae; 6 out of 10 species), Chordata (fish and ascidians; 8 out of 14 species), Mollusca (11 out of 20 species) and Porifera (2 out of 4 species) (Fig. 25.2). Only one species was listed in the priority pest project for each of the following phyla – Chrysophyta, Entoprocta, Magnoliophyta and Platyhelminthes – none of which were detected in the port surveys.

Three of the non-native species detected in the port surveys – *Balanus improvisus, Grahamina gymnota* and *Mytilopsis sallei* – were identified in the priority pest project but were not listed as established non-native species. *Balanus improvisus* (listed as not established in the priority pest project) had only recently been detected in the Townsville port survey (November 2000). *G. gymnota* (listed as cryptogenic in the priority pest project) and has since been reviewed and classed as non-native in Australian waters (Clements et al. 2000; Hickey et al. 2004). *M. sallei* is no longer established in Australian waters after its eradication in 2001 (Bax 1999; Willan et al. 2000; Bax et al. 2002) and therefore was not listed as established in the priority pest project. These three examples provide support for the use of multiple techniques in developing lists of non-native and cryptogenic species, and the need for regular revision of species that are listed.

The Australian port surveys were conducted over a period of nine years (1995–2004), at a total cost of approximately AUS$6 million. The surveys underestimated the number of invasive species in Australia for many reasons: a large proportion of material in the port surveys was not identified to species level; the port surveys were designed to look for specific "target" species and therefore did not necessarily identify all species sampled; and the coverage of taxa collected was highly dependent on the survey methods adopted. Therefore, despite representing a large biodiversity dataset, the port survey list of non-native species, on its own, does not provide a comprehensive picture of marine bioinvasions in Australia.

The literature review conducted for the priority pest project was conducted over a period of three years (2000–2003) at a total cost of approximately AUS$180,000. The review was solely focused on species that were ship-vectored (i.e. those transported by ballast water or hull fouling). Other vectors were assessed but were not comprehensively investigated in the project, therefore any assessment of non-ship-vectored species as a result of this work is likely to be misrepresentative. As a result, Table 25.2 is potentially missing some non-native, non-ship-vectored species that are established in Australian waters.

Cryptogenic species are a more difficult group to assess than non-natives. Hayes et al. (2002, 2005) identified 214 cryptogenic species whereas the port surveys identified 217 species. The overlap between the two datasets is low with only 90 species common to both. The port surveys identified 127 cryptogenic species that

Table 25.2 Non-native marine species recorded through the port surveys and priority pest project

Phylum	Class	Species name	Recorded in port surveys	Recorded through literature surveys
Annelida	Polychaeta	*Alitta succinea*	1	1
Annelida	Polychaeta	*Boccardia proboscidea*	1	1
Annelida	Polychaeta	*Euchone limnicola*	1	1
Annelida	Polychaeta	*Hydroides diramphus*	1	1
Annelida	Polychaeta	*Hydroides ezoensis*	1	1
Annelida	Polychaeta	*Hydroides sanctaecrucis*	-	1
Annelida	Polychaeta	*Polydora cornuta*	1	1
Annelida	Polychaeta	*Polydora websteri*	-	1
Annelida	Polychaeta	*Pseudopolydora paucibranchiata*	1	1
Annelida	Polychaeta	*Sabella spallanzanii*	1	1
Arthropoda	Cirripedia	*Balanus improvisus*	1	-
Arthropoda	Cirripedia	*Balanus reticulatus*	1	1
Arthropoda	Cirripedia	*Megabalanus occator*	1	-
Arthropoda	Cirripedia	*Megabalanus rosa*	1	1
Arthropoda	Cirripedia	*Megabalanus tintinnabulum*	1	1
Arthropoda	Cirripedia	*Megabalanus zebra*	1	1
Arthropoda	Cirripedia	*Notomegabalanus algicola*	-	1
Arthropoda	Malacostraca	*Apocorophium acutum*	1	1
Arthropoda	Malacostraca	*Caprella californica*	1	1
Arthropoda	Malacostraca	*Carcinus maenas*	1	1
Arthropoda	Malacostraca	*Cirolana harfordi*	1	1
Arthropoda	Malacostraca	*Elasmopus rapax*	1	1
Arthropoda	Malacostraca	*Halicarcinus innominatus*	1	1
Arthropoda	Malacostraca	*Jassa marmorata*	1	1
Arthropoda	Malacostraca	*Metacarcinus novaezealandiae*	1	1
Arthropoda	Malacostraca	*Corophium acherusicum*	1	1
Arthropoda	Malacostraca	*Monocorophium insidiosum*	1	1
Arthropoda	Malacostraca	*Oratosquilla oratoria*	1	1
Arthropoda	Malacostraca	*Paracerceis sculpta*	1	1
Arthropoda	Malacostraca	*Paradella dianae*	1	1
Arthropoda	Malacostraca	*Paradexamine pacifica*	1	1
Arthropoda	Malacostraca	*Petrolisthes elongatus*	1	1
Arthropoda	Malacostraca	*Pyromaia tuberculata*	-	1
Arthropoda	Malacostraca	*Sphaeroma serratum*	1	1
Arthropoda	Malacostraca	*Sphaeroma walkeri*	1	1
Chlorophycota	Chlorophyceae	*Cladophora prolifera*	1	1
Chlorophycota	Chlorophyceae	*Codium fragile* ssp *tomentosoides*	1	1
Chordata	Ascidiacea	*Ciona intestinalis*	1	1
Chordata	Ascidiacea	*Styela clava*	1	1
Chordata	Osteichthyes	*Acanthogobius flavimanus*	1	1

(continued)

Table 25.2 (continued)

Phylum	Class	Species name	Recorded in port surveys	Recorded through literature surveys
Chordata	Osteichthyes	*Acentrogobius pflaumi*	1	1
Chordata	Osteichthyes	*Forsterygion varium*	-	1
Chordata	Osteichthyes	*Gambusia holbrooki*	-	1
Chordata	Osteichthyes	*Grahamina gymnota*	1	-
Chordata	Osteichthyes	*Oncorhynchus mykiss*	-	1
Chordata	Osteichthyes	*Oreochromis mossambicus*	-	1
Chordata	Osteichthyes	*Poecilia latipinna*	-	1
Chordata	Osteichthyes	*Salmo trutta trutta*	-	1
Chordata	Osteichthyes	*Salvelinus fontinalis*	-	1
Chordata	Osteichthyes	*Tilapia mariae*	-	1
Chordata	Osteichthyes	*Tridentiger trigonocephalus*	1	1
Chrysophyta	Xanthophyceae	*Vaucheria piloboloides*	-	1
Cnidaria	Hydrozoa	*Amphisbetia operculata*	-	1
Cnidaria	Hydrozoa	*Antennella secundaria*	1	1
Cnidaria	Hydrozoa	*Bougainvillia muscus*	1	1
Cnidaria	Hydrozoa	*Cordylophora caspia*	1	1
Cnidaria	Hydrozoa	*Ectopleura crocea*	1	1
Cnidaria	Hydrozoa	*Ectopleura dumortieri*	1	1
Cnidaria	Hydrozoa	*Eudendrium carneum*	1	1
Cnidaria	Hydrozoa	*Filellum serpens*	-	1
Cnidaria	Hydrozoa	*Halecium vasiforme*	1	1
Cnidaria	Hydrozoa	*Monotheca obliqua*	-	1
Echinodermata	Asteroidea	*Asterias amurensis*	1	1
Echinodermata	Stelleroidea	*Astrostole scaber*	-	1
Echinodermata	Stelleroidea	*Patiriella regularis*	1	1
Ectoprocta	Gymnolaemata	*Anguinella palmata*	-	1
Ectoprocta	Gymnolaemata	*Bowerbankia gracilis*	1	1
Ectoprocta	Gymnolaemata	*Bowerbankia imbricata*	1	1
Ectoprocta	Gymnolaemata	*Bugula flabellata*	1	1
Ectoprocta	Gymnolaemata	*Bugula neritina*	1	1
Ectoprocta	Gymnolaemata	*Bugula simplex*	-	1
Ectoprocta	Gymnolaemata	*Bugula stolonifera*	1	1
Ectoprocta	Gymnolaemata	*Celleporella hyalina*	-	1
Ectoprocta	Gymnolaemata	*Conopeum reticulum*	1	1
Ectoprocta	Gymnolaemata	*Conopeum tubigerum*	-	1
Ectoprocta	Gymnolaemata	*Cryptosula pallasiana*	1	1
Ectoprocta	Gymnolaemata	*Electra pilosa*	1	1
Ectoprocta	Gymnolaemata	*Fenestrulina malusii*	1	1
Ectoprocta	Gymnolaemata	*Schizoporella errata*	1	1
Ectoprocta	Gymnolaemata	*Schizoporella unicornis*	1	1
Ectoprocta	Gymnolaemata	*Scruparia ambigua*	1	1
Ectoprocta	Gymnolaemata	*Scrupocellaria bertholetti*	-	1
Ectoprocta	Gymnolaemata	*Scrupocellaria scruposa*	-	1
Ectoprocta	Gymnolaemata	*Tricellaria occidentalis*	1	1
Ectoprocta	Gymnolaemata	*Watersipora arcuata*	1	1
Ectoprocta	Gymnolaemata	*Watersipora subtorquata*	1	1

(continued)

Table 25.2 (continued)

Phylum	Class	Species name	Recorded in port surveys	Recorded through literature surveys
Ectoprocta	Gymnolaemata	*Zoobotryon verticillatum*	1	1
Entoprocta	Loxosomatida	*Barentsia benedeni*	-	1
Heterokontophyta	Phaeophyceae	*Asperococcus compressus*	-	1
Heterokontophyta	Phaeophyceae	*Elachista orbicularis*	-	1
Heterokontophyta	Phaeophyceae	*Sphacella subtilissima*	-	1
Heterokontophyta	Phaeophyceae	*Stictyosiphon soriferus*	-	1
Heterokontophyta	Phaeophyceae	*Striaria attenuata*	-	1
Heterokontophyta	Phaeophyceae	*Undaria pinnatifida*	1	1
Magnoliophyta	Liliopsida	*Spartina anglica*	-	1
Mollusca	Bivalvia	*Crassostrea gigas*	1	1
Mollusca	Bivalvia	*Musculista senhousia*	1	1
Mollusca	Bivalvia	*Mytilopsis sallei*	1	-
Mollusca	Bivalvia	*Neilo australis*	-	1
Mollusca	Bivalvia	*Ostrea edulis*	-	1
Mollusca	Bivalvia	*Paphies ventricosa*	-	1
Mollusca	Bivalvia	*Raeta pulchella*	1	1
Mollusca	Bivalvia	*Ruditapes largillierti*	1	1
Mollusca	Bivalvia	*Teredo navalis*	-	1
Mollusca	Bivalvia	*Theora lubrica*	1	1
Mollusca	Bivalvia	*Varicorbula gibba*	1	1
Mollusca	Gastropoda	*Aplysiopsis formosa*	-	1
Mollusca	Gastropoda	*Godiva quadricolor*	-	1
Mollusca	Gastropoda	*Maoricolpus roseus*	1	1
Mollusca	Gastropoda	*Phytia myosotis*	-	1
Mollusca	Gastropoda	*Polycera capensis*	-	1
Mollusca	Gastropoda	*Polycera hedgpethi*	-	1
Mollusca	Gastropoda	*Thecacera pennigera*	1	1
Mollusca	Gastropoda	*Zeacumantus subcarinatus*	-	1
Mollusca	Polyplacophora	*Chiton glaucus*	-	1
Platyhelminthes	Turbellaria	*Euplana gracilis*	-	1
Porifera	Demospongiae	*Aplysilla rosea*	-	1
Porifera	Demospongiae	*Dysidea avara*	-	1
Porifera	Demospongiae	*Dysidea fragilis*	1	1
Porifera	Demospongiae	*Halisarca dujardini*	1	1
Pyrrophycophyta	Dinophyceae	*Alexandrium minutum*	1	1
Pyrrophycophyta	Dinophyceae	*Gymnodinium catenatum*	1	1
Rhodophycota	Rhodophyceae	*Antithamnion cruciatum*	-	1
Rhodophycota	Rhodophyceae	*Antithamnionella spirographidis*	-	1
Rhodophycota	Rhodophyceae	*Chondria arcuata*	-	1
Rhodophycota	Rhodophyceae	*Deucalion levringii*	-	1
Rhodophycota	Rhodophyceae	*Gymnogongrus crenulatus*	-	1
Rhodophycota	Rhodophyceae	*Medeiothamnion lyallii*	-	1
Rhodophycota	Rhodophyceae	*Polysiphonia brodiei*	1	1
Rhodophycota	Rhodophyceae	*Polysiphonia senticulosa*	1	1
Rhodophycota	Rhodophyceae	*Schottera nicaeensis*	1	1
Rhodophycota	Rhodophyceae	*Solieria filiformis*	1	1

(continued)

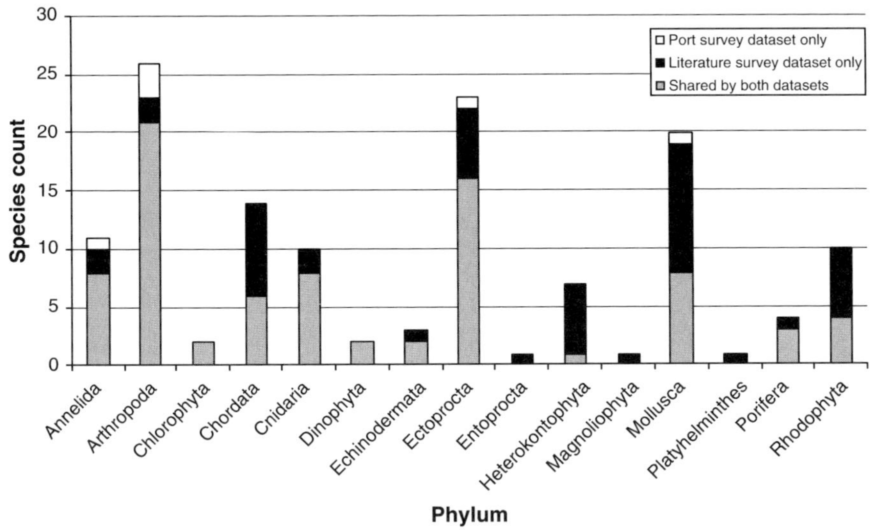

Fig. 25.2 Summary of the number of non-native marine species identified in the port survey dataset and the priority pest project dataset, and the number of taxa shared by both datasets

were not recorded in the literature review. Similarly, the literature review identified 124 cryptogenic species that were not found in any of the port surveys.

The cryptogenic species recorded through the port surveys but not through the literature review could represent new records; be misidentifications or synonyms; be present in museum collections, but yet to be reported in the literature; or were allocated an unknown establishment or invasion status during the priority pest project. The project identified over 1000 species with an unknown invasion or establishment status, nine of which were found in the port surveys. Similarly, the 121 cryptogenic species identified in the literature but not found in the port surveys could represent misidentifications or synonyms; may have been present for a long time and considered native by many biologists; or have been found in non-port environments. Furthermore some of the port survey material, particularly in genera not represented by the target species, was not fully sorted or identified.

25.3 Problems with Lists of Non-Native and Cryptogenic Species

The lists resulting from the two projects discussed within this chapter could both be expanded. In many instances, the results of the port surveys highlight the difficulty of systematically searching large ports with high sensitivity and a lack of knowledge regarding Australia's marine fauna (see also Chap. 2, Carlton). Many of the invertebrate specimens collected as part of the survey, for example, could not be identified to species level. These unidentified species may be native, but could also include some non-native introductions (Hutchings and Glasby 2004).

A number of the port surveys were completed without specialist taxonomic assistance, and therefore the accuracy of the identifications could be questioned. Indeed the port survey data, when collated, was assessed for its 'veracity' (i.e. our confidence that the identification was correct). This analysis suggests that the dinoflagellates and brown algae were the most reliably identified taxa (McEnnulty et al. 2005), but these two groups were also among those with the lowest proportion of samples identified to species level in the port surveys (Fig. 25.2) (McEnnulty et al. 2005). In comparison, the dinoflagellates and brown algae were among the highest numbers of taxa identified to species level through the literature review.

Creating lists of non-native and cryptogenic species from literature reviews is also potentially problematic. The available information resides solely in published literature, and the priority pest project concentrated on species primarily transported by either ballast water or hull fouling. Other vectors were included in the dataset, but were not given equal attention. If unpublished literature were used, and if all vector types were equally considered, then it is likely that the number of species on the list (Table 25.2) would increase.

In trying to define a standard approach for creating invasive species lists and databases, we have concluded that the following would be helpful:

- The geographic distribution of species throughout the world need to be recorded in a standardized format
- A detailed referencing system
- Standardized vocabularies for key concepts such as vectors, habitats, impacts, and establishment status
- A consistent taxonomic framework (i.e. using standardized taxonomic references, and a consistent application of taxonomic names across the database)
- A clear understanding of the scope of each database. Some regional databases could be restricted to holding original local survey data only. Other databases (such as our own) hold summarized data from many other sources and replication of this information would not be cost-effective
- A mechanism to update fields; especially important for taxonomic records that will change through time.

Non-native species lists inevitably end up within management frameworks, with some species being determined "pests" or "target species", around which control or eradication programs are developed. Once these lists are taken into the management framework, they are often used for long periods without revision. It is important that these types of lists, and the data that underpin them, are constantly reviewed and updated to incorporate new introductions and changes in species status based on further research. This is especially important when considering species that are not listed in any form because they are not as yet introduced or they are species which are considered "harmless" in their introduced range.

This chapter represents the first attempt to pull together information on marine bioinvasions from port surveys and the literature on an Australia-wide scale. The two techniques are both complementary (non-native species) and conflicting (cryptogenic species). Although there is little doubt that this assessment of non-native and cryptogenic species will be improved upon in the future, to the best of our knowledge it is currently the most accurate and up-to-date account of non-native and cryptogenic species in Australia.

References

Allen FE (1950) Investigations on underwater fouling 3. Notes on the fouling organisms attached to naval mines in Northern Queensland waters. Aust J Mar Freshwater Res 1:107–109

Allen FE (1953) Distribution of marine invertebrates by ships. Aust J Mar Freshwater Res 4:307–316

Allen FE, Wood EJ (1949) Investigations on underwater fouling 2. The biology of fouling in Australia: results of a years research. Aust J Mar Freshwater Res 1:92–105

AQIS (1992) Voluntary controls on the discharge of ballast water and sediment from entering Australia from overseas. Notice (General Quarantine) 90/01. Updated in July 1992 as Notice 92/2 <http://www.aph.gov.au/library/pubs/bd/1997-98/98bd057.htm>

Bax NJ (1999) Eradicating a dreissenid from Australia. Dreissena! 10:1–5

Bax N, Carlton JT, Mathews-Amos A, Haedrich RL, Howarth FG, Purcell JE, Rieser A, Gray A (2001) The control of biological invasions in the world's oceans. Conserv Biol 15(5):1234–1246

Bax N, Hayes K, Marshall A, Parry D, Thresher R (2002) Man-made marinas as sheltered islands for alien marine organisms: establishment and eradication of an alien invasive marine species. In: Veitch CR, Clout MN (eds) Turning the tide: the eradication of invasive species.Invasive Species Specialist Group of the World Conservation Union (IUCN) Auckland NZ, pp 26–39

Carlton JT (1996) Biological invasions and cryptogenic species. Ecology 77(6):1653–1655

Clapin G, Evans DR (1995) The status of the introduced marine fanworm *Sabella spallanzanii* in Western Australia: a preliminary investigation. CRIMP Technical Report Number 2, CSIRO Marine Research, Hobart, Tasmania, Australia

Clark GF, Johnston EL (2005) Manipulating larval supply in the field: a controlled study of marine invasibility. Mar Ecol Prog Ser 298:9–19

Clements KD, Jawad LA, Stewart AL (2000) The New Zealand triplefin *Grahamina signata* (Teleostei; Tripterygiidae): a junior synonym of *G. gymnota* from Tasmania. J R Soc N Z 30(4):373–384

Coutts ADM (1999) Hullfouling as a modern vector for marine biological invasions: investigation of merchant vessels visiting northern Tasmania. MSc. Thesis (unpublished), Australian Maritime College, Beauty Point, Tasmania, 283 pp

Deagle BE, Bax N, Hewitt CL, Patil JG (2003) Development and evaluation of a PCR-based test for detection of *Asterias* (Echinodermata: Asteroidea) larvae in Australian plankton samples from ballast water. Mar Freshwater Res 54:709–719

Ferguson R (2000) The effectiveness of Australia's response to the black striped mussel incursion in Darwin, Australia. Report of the Marine Pest Incursion Management Workshop. 27–28 August, 1999, Department of Environment and Heritage, Canberra, Australia

Fulton SW, Grant FE (1900) Note on the occurrence of the European crab, *Carcinus maenas*, Leach, in Port Phillip. Victorian Nat 17(8):147–148

Gardner NC, Kwa S, Paturusi A (1994) First recording of the European shore crab, *Carcinus maenas* in Tasmania. Tasmanian Nat 116:26–28

Goggin CL (1998) Proceedings of a meeting on the biology and management of the introduced seastar *Asterias amurensis* in Australian waters, 19 May 1998. CRIMP Tech Rep No 15, CSIRO Marine Research, Hobart, Tasmania, Australia

Hayes KR, Sliwa C (2003) Identifying potential marine pests – a deductive approach applied to Australia. Mar Pollut Bull 46:91–98

Hayes KR, McEnnulty FR, Sliwa C (2002) Identifying potential marine pests – an inductive approach. Final report for Environment Australia, National Priority Pests Project, CSIRO Marine Research, Hobart, Tasmania, Australia

Hayes K, Sliwa C, Migus S, McEnnulty F, Dunstan P (2005) National priority pests. Part II. Ranking of Australian marine pests. Independent report for Department of Environment and Heritage, Australian Government Department of Environment and Heritage, Canberra, Australia

Hewitt CL (2003) Marine biosecurity issues in the world oceans: global activities and Australian directions. Ocean Yearb 17:193–212

Hewitt C, Martin RB (1996) Port surveys for introduced marine pests. CRIMP Tech Rep No 4, CSIRO Marine Laboratories, Hobart, Tasmania, Australia

Hewitt CL, Martin RB (2001) Revised protocols for baseline port surveys for introduced marine species: survey design, sampling protocols and specimen handling. CRIMP Tech Rep No 22, CSIRO Marine Research, Hobart, Australia

Hewitt C, Campbell M, Thresher R, Martin R, Boyd S, Cohen B, Currie D, Gomon M, Keough M, Lewis J, Lockett M, Mays N, McArthur M, O'Hara T, Poore G, Ross D, Storey M, Watson J, Wilson R (2004) Introduced and cryptogenic species in Port Phillip Bay, Victoria, Australia. Mar Biol 144:183–202

Hickey AJR, Lavery SD, Eyton SR, Clements KD (2004) Verifying invasive marine fish species using molecular techniques: a model example using triplefin fishes. N Z J Mar Freshwater Res 38:439–446

Holloway MG, Keough MJ (2002a) An introduced polychaete affects recruitment and larval abundance of sessile invertebrates. Ecol Appl 12(6):1803–1823

Holloway MG, Keough MJ (2002b) Effects of an introduced polychaete, *Sabella spallanzanii*, on the development of epifaunal assemblages. Mar Ecol Prog Ser 236:137–154

Hutchings PA, Glasby C (2004) Port surveys for introduced marine species – and the fate of the material collected. Mar Poll Bull 48:1009–1011

Hutchings PA, van der Velde JT, Keable SJ (1987) Guidelines for the conduct of surveys for detecting introductions of non-indigenous marine species by ballast water and other vectors – and a review of marine introductions to Australia. Occasional Reports of the Australian Museum No 3. The Australian Museum, Sydney, Australia

IMO (2002) Focus on IMO: anti-fouling systems. http://www.imo.org/home.asp

Lewis PN (2001) In the wake of the *Aurora*: an investigation into the hazard of marine introductions in the Southern Ocean. B Antarctic Sc Honours Thesis, Institute of Antarctic and Southern Ocean Studies, University of Tasmania, Hobart, Australia, 133 pp

Lewis PN, Hewitt CL, Riddle M, McMinn A (2003) Marine introductions in the Southern Ocean: an unrecognised hazard to biodiversity. Mar Pollut Bull 46:213–223

Ling SD (2000) Anthropogenic structure and reproduction of *Asterias amurensis*. BSc Honours Thesis, University of Tasmania, Hobart, 66 p

MacGillivray PH (1869) Descriptions of some new genera and species of Australian polyzoa. Trans Proc R Soc Victoria 9:126–148

McEnnulty F, Hayes K, Cooper S, Sliwa C, Dunstan P, Migus S, Green M (2005) National port survey collection: consolidation, analysis and integration. Final Report. Submitted to: Australian Government Department of Agriculture, Fisheries and Forestry

McMinn A, Hallegraeff GM, Thomson P, Jenkinson AV, Heijnis H (1997) Cyst and radionucleotide evidence for the recent introduction of the toxic dinoflagellate *Gymnodinium catenatum* into Tasmanian waters. Mar Ecol Prog Ser 161:165–172

Medcof JC (1975) Living marine organisms in a ships ballast water. Proc Natl Shellfish Assoc 65:11–12

Morrice MG (1995) The distribution and ecology of the introduced northern Pacific seastar, *Asterias amurensis* (Lutken), in Tasmania. In: The introduced northern Pacific seastar, *Asterias amurensis,* in Tasmania. Australian Nature Conservation Agency 1996, Canberra, Australia, pp 1–70

Morris AE (2002) Early life history of the introduced seastar *Asterias amurensis* in the Derwent Estuary, Tasmania: the potential for ecology-based management. PhD Thesis, University of Tasmania, Hobart, 117 pp

Oemcke D (1998) The treatment of ships' ballast water to remove exotic marine pests: research in north Queensland. Reef Res 8(2):1–5

Parry G, Cohen B (2001) *Asterias amurensis* biomass, distribution and abundance in Port Phillip Bay following a survey during March-April 2001. Marine and Freshwater Resources Institute, Queenscliff Victoria, Australia

Parry GD, Cohen BF, McArthur MA, Hickman NJ (2000) Victorian incursion management report, No 2. *Asterias amurensis* incursion in Port Phillip Bay: Status at May 1999, Marine and Freshwater Resources Institute Report No. 19. Marine and Freshwater Resources Institute, Queenscliff, Victoria, Australia

Pollard DA, Hutchings PA (1990a) A review of exotic marine organisms introduced into the Australian region. II. Invertebrates and algae. Asian Fish Sci 3:223–250

Pollard DA, Hutchings PA (1990b) A review of exotic marine organisms introduced to the Australian region. I. Fishes. Asian Fish Sci 3(303):205–221

Reid AP (2003) Utilisation of the introduced New Zealand screwshell (*Maoricolpus roseus*) by native hermit crabs in eastern Tasmania: indications of potential impact. BSc Honours Thesis, University of Tasmania, Hobart, 74 pp

Ross DJ, Johnson CR, Hewitt CL (2002) Impact of introduced seastars *Asterias amurensis* on survivorship of juvenile commercial bivalves *Fulvia tenuicostata*. Mar Ecol Prog Ser 241:99–112

Ross DJ, Johnson CR, Hewitt CL (2003a) Assessing the ecological impacts of an introduced seastar: the importance of multiple methods. Biol Invas 5:3–21

Ross DJ, Johnson CR, Hewitt CL (2003b) Variability in the impact of an introduced predator (*Asterias amurensis*: Asteroidea) on soft-sediment assemblages. J Exp Mar Biol Ecol 288:257–278

Ross DJ, Johnson CR, Hewitt CL, Ruiz GM (2004) Interaction and impacts of two introduced species on a soft-sediment marine assemblage in SE Tasmania. Mar Biol 144:747–756

Sliwa C, Johnson CR, Hewitt CL (2006) Mesoscale dispersal of the introduced kelp *Undaria pinnatifida* attached to unstable substrata. Bot Mar 49:396–405

Thresher RE (1999) Diversity, impacts and options for managing invasive marine species in Australian waters. Aust J Environ Manag 6:137–148

Thresher RE, Hewitt CL, Campbell ML (1999) Synthesis: introduced and cryptogenic species in Port Phillip Bay. In: Hewitt CL, Campbell ML, Thresher RE, Martin RB (eds) Marine biological invasions of Port Phillip Bay, Victoria. Centre for Research on Introduced Marine Pests. Tech Rep No 20. CSIRO Marine Research, Hobart, pp 283–295

Turner E (1992) A northern Pacific seastar, *Asterias amurensis*, in Tasmania. AMSA Bull 120:18–19

Valentine JP, Johnson CR (2003) Establishment of the introduced kelp *Undaria pinnatifida* in Tasmania depends on disturbance to native algal assemblages. J Exp Mar Biol Ecol 295:63–90

Valentine JP, Johnson CR (2004) Establishment of the introduced kelp *Undaria pinnatifida* following dieback of the native macroalga *Phyllospora comosa* in Tasmania, Australia. Mar Freshwater Res 55:223–230

Valentine JP, Johnson CR (2005) Persistence of the exotic kelp *Undaria pinnatifida* does not depend on sea urchin grazing. Mar Ecol Prog Ser 285:43–55

West EJ (2004) The role of anthropogenic activity in the fragmentation and dispersal of the invasive alga, *Caulerpa taxifolia*. In: Virtue P, Hall N, Thompson P (eds) Program and abstracts for the 2004 meeting of the Australian Marine Sciences Association, 6–9 July 2004, Hobart, Tasmania, Australia Australian Marine Sciences Association, Toowong, Qld, p 134

Willan RC, Russel BC, Murfet NB, Moore KL, McEnnulty FR, Horner SK, Hewitt CL, Dally GM, Campbell ML, Bourke ST (2000) Outbreak of *Mytilopsis sallei* (Recluz, 1849) (Bivalvia: Dreissenidae) in Australia. Molluscan Res 20(2):25–30

Chapter 26
Marine Bioinvasions in the Southeastern Pacific: Status, Ecology, Economic Impacts, Conservation and Management

Juan C. Castilla and Paula E. Neill

26.1 Introduction

Knowledge of the status of marine non-indigenous species (NIS) in temperate southeastern Pacific countries (Chile and Peru) is incipient. Nevertheless, at least in Chile, the problem has recently been addressed by taxonomists and ecologists (e.g. Baez et al. 1998; Gajardo and Laikre 2003; Castilla et al. 2005; Camus 2005). While numerous NIS have been intentionally introduced for aquaculture purposes (e.g. *Crassostrea gigas*, *Haliotis rufescens*, in both Chile and Peru, and various salmonid species in Chile) there are few studies that have made an explicit examination of the marine environment for the presence of NIS. In Chile, there is a large work-force of researchers studying marine environments for other purposes, and from this assemblage of individuals only a few reports have been made regarding abundant or aggressive exotic species: currently only one thriving invader is known: *Codium fragile* ssp. *tomentosoides*. In this sense, it appears that Chile, and probably Peru, may be relatively "more pristine" than other coasts in terms of the presence of conspicuous, highly invasive NIS. Nevertheless, the status of non-invasive and cryptogenic NIS in these countries still needs to be fully evaluated. If these marine environments actually are less invaded than other parts of the world it is important to take action now to prevent future introductions, and to study the factors that may limit exotic species introductions and/or their spread. In this chapter our aims are threefold: (1) to discuss the state of knowledge of marine NIS in Chile and report on preliminary evidence from Peru; (2) to provide the most current listing of marine NIS in these countries (Chile and Peru), with information on exotic species' distributions in Chile, the most probable introduction pathway(s) and donor area(s), and known or potential ecological, economic and conservation effects; (3) to discuss the current legislative status of NIS in Chile and make recommendations for further studies in this country based on experiences from other parts of the world.

G. Rilov, J.A. Crooks (eds.) *Biological Invasions in Marine Ecosystems.*
Ecological Studies 204,

26.2 Background Information on the Chilean Marine System

The Chilean marine system spans approximately 37 degrees of latitude, with the southern tip bordering the southern Atlantic Ocean. The Chilean coastline is characterized as a temperate coast, and shares some physical similarities with other temperate coasts of the world (Castilla et al. 1993). Nevertheless, there are at least four notable biogeographical traits that may provide this region with a "natural" physical resistance to exotic species introductions and spread (Camus 2001; Castilla et al. 2005): (1) a high frequency of dissolved-oxygen deficits in coastal waters, particularly in northern Chile (Levin 2003; Ulloa and De Pol 2005), which generate severe anaerobic/hypoxic conditions that could limit available habitat for certain benthic or pelagic species; (2) the continental Chilean coast, north of Chiloé Island, has relatively few gulfs, bays and estuaries, which are known to be hot-spots for species introductions (Cohen and Carlton 1998; Castilla et al. 2005; Camus 2001; Grosholz 2005; Chap. 33, Preisler et al.); (3) a cold, north flowing current, north of 42°S, which could inhibit warm water species from spreading further south; (4) low temperatures and salinities in southern Chile (approximately south of 42°S), which could be a barrier to marine species with low tolerance to freshwater.

While these physical conditions may indeed provide some level of "natural" resistance to species introductions, and act as barriers to the subsequent spread of NIS, these same traits have also been considered to enhance NIS colonizing opportunities relative to other areas. For example, the wide range of water temperatures, especially associated with ENSO events, and biotopes may be seen as an opportunity for cold-adapted or tropical species to encounter the ideal habitat for their establishment (Ribera 2003). Furthermore, many anthropogenic activities are known to be able to surpass such biogeographical barriers (e.g. aquaculture activities, ship traffic) (Carlton 1999), making historically isolated regions just as susceptible to species introductions as other areas. For example, the southeastern Pacific coasts have been exposed to international shipping activities for over 500 years (Maino 1985; Larraín 2001). Chile possesses several ports of historical importance, which served as a gateway to the Pacific prior to the opening of the Panama Canal in 1914 (e.g. key ports in Punta Arenas, Talcahuano, San Antonio, Valparaíso, Antofagasta, and Iquique). Chilean ports have played a key role in trade, with ship traffic arriving from all over the globe, including North America, Europe, Asia, Africa, and Australia. During the nineteenth and twentieth centuries (1840–1940) international whaling, bird guano and nitrate fleets were based at ports in central and northern Chile (Maino 1985; Arce 1997). In addition, during the nineteenth century Chile entered into the aquaculture industry and commenced with various species' introduction programs (Basulto 2003). Today, Chile is one of the most important centers of exotic fish species aquaculture in the world (e.g., since 1990 Chile has been the world's second largest producer of salmon and trout, and in 2003 it contributed approximately 35% of total global production) (Gajardo and Laikre 2003; http://www.salmonchile.cl/). The historic and current connections between the Chilean coast and other coasts of the world, as a result of these diverse anthropogenic activities, raises concerns regarding the deliberate and fortuitous introduction of species, and the current status of biodiversity in Chile's marine communities. Furthermore, legislation regarding marine biosecurity in Chile is weak at best.

26.3 Marine NIS Introductions in Chile

An extensive examination of marine NIS in Chile was previously conducted by Castilla et al. (2005). Here we have updated this NIS database, including additions to the list of species, likely donor areas, introduction pathways and reported ecological and economic effects (Table 26.1). This list of marine NIS includes 51 species (Castilla et al. 2005 reported 47 species; note that we were unable to include strictly estuarine species due to lack of information, and we excluded species of fish or pelagic crustaceans showing ephemeral range extensions, mainly from Peru, during ENSO events): 1 higher marine plant, 15 algae species (1 Chlorophyta, 13 Rhodophyta, 1 Phaeophyta), 24 invertebrate species (3 Porifera, 3 Polychaeta, 4 Crustacea, 2 Bivalvia, 4 Gastropoda, 5 Ascidia, 2 Bryozoa, and 1 Cnidaria, the latter whose taxonomic status and NIS status is currently unresolved and has therefore been excluded from subsequent analyses), and 11 bony fish species of the genera *Salmo*, *Oncorhynchus*, *Paralichthys*, *Hippoglossus*, and *Scophthalmus*. This list includes marine species that have been introduced into natural environments, as well as species introduced for aquaculture; some of which are reported to have escaped from these facilities and are quite possibly established (Soto et al. 2006).

In Table 26.1 we have indicated the most probable vectors of introduction for the 51 marine NIS listed, together with the current NIS status of each species (i.e. introduced, reasonable candidate for introduction, deliberate introduction for aquaculture purposes, and species undergoing likely "natural" range expansions). The majority of the listed species are considered to be introduced to the Chilean coast (including introductions via various accidental vectors) or deliberately introduced for aquaculture purposes (20 and 16 species, respectively). Species in the introduced category mainly include algae (5 Rhodophyta and 1 Chlorophyta) and invertebrate species (13 species including Ascidiacea, Polychaeta, Crustacea, Bryozoa, Porifera, and Bivalvia), as well as 1 marine plant. In contrast, deliberately introduced species were principally bony fish (11 species), 2 gastropod species, 1 bivalve species, 1 crustacean and 1 red alga. A small proportion of the species were considered to be reasonable candidates for introduction (6 algae and 2 invertebrates) or species undergoing "natural" range expansions (2 algae, 4 invertebrate species).

Marine NIS have been reported along the entire coast of Chile (approximately 18–55°S). Deliberate species introductions have been made to controlled, ocean-based aquaculture facilities (e.g. caged salmon, oysters suspended in the water column, abalone in plastic boxes), which are concentrated principally in Southern Chile (39–48°S) and secondarily in northern Chile (26–32°S). The majority of the reported marine NIS (both introduced and reasonable candidates

Table 26.1 Marine non-indigenous species (NIS) introduced to Chilean waters with their most probable vectors of introduction based on: Direct Evidence (**DE**, published document or official record in museum collection), Indirect Evidence (**IE**, published evidence for transfer of this species or a congener via the indicated pathway), or Unclear (**?**, best guess by a researcher studying this species).

Taxa	Imported for aquaculture	Escaped from aquaculture facilities	Nontarget species associated with aquaculture	Shipping (on ship hulls, in ballast water)	Geographic range extensions	Unknown (including rafting)	Likely donor zone	Current status
Monocotyledoneae								
Heterozostera tasmanica						?	SPAC	INTR
Chlorophyta								
Codium fragile ssp. *tomentosoides*			IE	IE			NPAC	INTR
Rhodophyta								
Mastocarpus papillatus				IE			NPAC	INTR
Polysiphonia morrowii[a,b]			DE	DE			NPAC	INTR
Prionitis lyallii						?	NPAC	INTR
Schizymenia pacifica						?	NPAC	INTR
Schottera nicaensis						?	MANY	INTR
Porifera								
Stelleta clarella				IE			NPAC	INTR
Polychaeta								
Dispio uncinata				IE	?		MANY	INTR
Spiophanes bombyx				IE			MANY	INTR
Crustacea								
Acartia omori				IE			NPAC	INTR
Centropages abdominalis				IE			NPAC	INTR
Oithona davisae				IE			NPAC	INTR
Bivalvia								
Mytilus gallo-provincialis			DE				MED	INTR
Ascidiacea								
Asterocarpa humilis			DE				SPAC	INTR
Ciona intestinalis			DE	DE			NATL	INTR
Molgula ficus			DE				SPAC	INTR
Pyura praeputialis				?		?	SPAC	INTR
Bryozoa								
Bugula flabellata				DE			SPAC, SATL	INTR
Bugula neritina				DE			NPAC, SPAC	INTR
Phaeophyta								
Scytosiphon tenellus[a,c]						?	NPAC	RC

Table 26.1 (continued)

Taxa	Imported for aquaculture	Escaped from aquaculture facilities	Nontarget species associated with aquaculture	Shipping (on ship hulls, in ballast water)	Geographic range extensions	Unknown (including rafting)	Likely donor zone	Current status
Rhodophyta								
Porphyra linearis				IE			NPAC	RC
Porphyra pseudolinearis				IE			NPAC	RC
Porphyra torta				IE			NPAC	RC
Pugetia mexicana						?	NPAC	RC
Schimmelmannia plumosa						?	NPAC	RC
Porifera								
Mycale doellojuradoi				IE			SATL	RC
Ascidiacea								
Botryllus schlosseri			DE	DE			NATL, SPAC	RC
Rhodophyta								
Gracilaria spp.	DE	DE					SPAC	DEL
Gastropoda								
Haliotis discus hannai	DE						NPAC	DEL
Haliotis rufescens	DE						NPAC	DEL
Crustacea								
Litopenaeus vannamei	DE						SPAC	DEL
Bivalvia								
Crassostrea gigas	DE						NPAC	DEL
Osteichties								
Oncorhynchus gorbuscha	DE	DE[d]					NPAC	DEL
Oncorhynchus keta	DE	DE[d]					NPAC	DEL
Oncorhynchus kisutch	DE	DE[d]					NPAC	DEL
Oncorhynchus masou masou	DE	DE[d]					NPAC	DEL
Oncorhynchus mykiss[a]	DE	DE[d]					NSEA	DEL
Oncorhynchus nerka	DE	DE[d]					NPAC	DEL
Oncorhynchus tschawytscha	DE	DE[d]					NPAC	DEL
Salmo salar	DE	DE[d]					NSEA	DEL
Hippoglossus hippoglossus	DE						NATL	DEL
Paralichthys olivaceus	DE						NPAC	DEL
Scophthalmus maximus	DE						NATL	DEL

(continued)

Table 26.1 (continued)

Taxa	Imported for aquaculture	Escaped from aquaculture facilities	Nontarget species associated with aquaculture	Shipping (on ship hulls, in ballast water)	Geographic range extensions	Unknown (including rafting)	Likely donor zone	Current status
Rhodophyta								
Ahnfeltia gigartinoides						?	NPAC	RE
Rhodoglossum affine					IE		NPAC	RE
Porifera								
Timea authia					IE		NPAC	RE
Polychaeta								
Parandalia fauveli					DE		MANY, SPAC	RE
Gastropoda								
Aplysia (Aplysia) *juliana*					DE		SPAC	RE
Linatella wiegmanni					DE		NPAC, SPAC	RE
Cnidaria								
Anemonia alicemartinae n. sp.[e]					?		?	UN-CLEAR

[a]Species not included in Castilla et al. 2005

[b]Kim et al. (2004)

[c]Camus C et al. 2005

[d]Personal communication by Dr. Doris Soto

[e]The status of this anemone as an NIS in Chile is unclear, therefore this species was not considered in our analyses

Note that we have not included reported range extensions for fish or crustaceans linked with ENSO events for which there is no reported evidence of establishment of local populations in Chilean waters (e.g. Kong et al. 1985; Guzmán et al. 1999; Guzmán and Soto 2000; Sielfeld et al. 2003). We also indicate the likely donor zone from which the NIS was introduced to Chile. Abbreviations are: **NPAC**, North Pacific; **SPAC**, South Pacific; **NATL**, North Atlantic; **SATL**, South Atlantic; **NSEA**, North Sea; **MED**, Mediterranean Sea; **MANY**, species with numerous possible donor zones. Finally, we report the current NIS status of each species based on information provided in Castilla et al. 2005 and other sources. Abbreviations are: **INTR**, Introduced – species documented in the literature at least twice and/or repetitively collected; **DEL**, Deliberate – species deliberately introduced for aquaculture purposes; **RC,** Reasonable Candidate for introduction – species with a single documentation in the literature and showing a notorious biogeographically incongruous range of distribution, mainly along the Eastern Pacific coast; **RE,** Range Extension – species that are reasonable candidates for NIS invaders to Chilean waters via southern extensions of their geographical ranges in connection with El Niño (ENSO) events as introduced species) are observed in central and southern Chile, being found in and near shipping ports (frequently in bays and gulfs), as well as in natural environments (including rocky intertidal platforms, subtidal zones, fjords, and inlets). The greatest number of NIS in the country correspond to areas where the principal international shipping ports of the country are located (north and south-central Chile), which suggests that maritime traffic is an important pathway for marine NIS introductions. Nevertheless, the greater number of NIS found in these areas may be the result of observer bias (e.g. easier access to such sites, more observers working in these areas, etc.).

26.3.1 Likely Pathways of Marine NIS Introductions

For the majority of the marine NIS taxa contained on our list the precise introduction pathway is not known with certainty (notable exceptions are species intentionally introduced for aquaculture purposes, which bear some form of explicit documentation). Nevertheless, we have identified several potential pathways for NIS introductions in Chile based on indirect evidence (i.e. published reports of pathways used by congeneric or biologically similar species introduced to other marine systems around the world), or from direct evidence of the presence of marine NIS utilizing specific vectors in Chile (e.g. present on ship hulls, in ballast water, as fouling organisms on aquaculture equipment, including multiple pathways). Here we only discuss pathways related to the long-distance transfer of marine NIS or geographic range expansions. Numerous other pathways are likely to play important roles in the local dispersal of NIS (e.g. fishing gear, small crafts, rafting), but to date no studies are available.

As reported in other parts of the world (Naylor et al. 2001), aquaculture has played an important role as a pathway for species introductions in Chile, including the deliberate introduction of NIS for exploitation (3 mollusks, 1 crustacean, and 11 bony fish species), as well as the escape of NIS from controlled cultures to natural environments (reported cases for at least 8 salmon species) (Soto et al. 2006), and the introduction of non-target species associated with aquaculture (2 algal species, 4 ascidians, and 1 bivalve). In addition, the economically important red alga *Gracilaria* spp., considered native to southern Chile, has been transferred to several sites in northern and central Chile (Santelices 1989). As pointed out by Ribera (2003) the transfer of a species within a country also constitutes a species introduction, since NIS do not necessarily respect political boundaries.

Some of the escaped, deliberately introduced aquaculture species are known to have formed viable populations outside aquaculture facilities (e.g., salmonids; and *Gracilaria* spp., which grows vegetatively outside facilities, but does not complete its life-cycle) (Leonardi et al. 2006). In contrast, various non-target species associated with aquaculture have established viable populations at their presumed point of introduction (e.g. several ascidians, one mussel species). Furthermore, one introduced macroalga shows evidence of invasion of new sites (*Codium fragile* ssp. *tomentosoides*).

Globally, maritime transport is considered one of the greatest mechanisms of long-distance dispersal for marine NIS (Fofonoff et al. 2003; Ribera 2003; Chap. 6, Hewitt et al.). In Chile, vectors associated with shipping are attributed to the arrival of 30–38% of total marine NIS in Chile (15–19 species; Table 26.1). Shipping activities have connected Chile with every continent. Historical records indicate that the greatest amount of ship traffic (in terms of metric tons of cargo transported for export or import) occurs between Chile and Asia (e.g. Japan, Korea, China, Taiwan), followed by North America (e.g. Mexico, USA, Canada) and subsequently Europe (e.g. Holland, France, Italy) (www.directemar.cl; Boletín Estadístico Marítimo 1992–2004). Species transported as fouling on ship hulls frequently correspond to small, physically flexible species with high growth rates and a wide range of ecological tolerances (Fofonoff et al. 2003; Ribera 2003). Sessile organisms likely introduced to Chile via this pathway include several species of algae (six Rhodophyta, one Chlorophyta), and six species of sessile invertebrates (two Bryozoa, two Ascidiacea, and two Porifera). Species introduced in ballast water frequently include organisms with resistant propagules (e.g. cysts, zygotes, spores, fruits) (Ribera 2003). In Chile, species introduced in ballast water likely include three copepod species (probably arriving from Japan or the northeastern Pacific coast).

In our list of NIS we also include species that have likely experienced recent events of range expansion and have established populations in Chile (i.e. not ephemeral expansion events), although they may not be directly linked to anthropogenic activities (see Chap. 3, Lonhart). This category includes a variety of algal, invertebrate and vertebrate taxa (two Rhodophyta, one Porifera, one Polychaeta and two Gastropoda). Most cases refer to species from Peru that are now found in northern Chile, most of which are associated with ENSO events that bring warmer, northern waters southward towards Chile. In Chile, there have been no published reports of introductions of marine NIS due to scientific experiments (for Europe see Ribera 2003).

26.3.2 Likely Donor Areas of Marine NIS

In order to determine potential donor areas we conducted a search of the literature (including gray literature) for information on the current range of distribution of each species (including both native and introduced ranges), as well as reports of the importation or first records of each species in Chile. We utilized information from published journal papers and theses, as well as information freely available on the internet from reliable sources (i.e. http://www.algaebase.org/, http://www.ices.dk/reports/ACME/2003/WGITMO03.pdf, http://www.fao.org/figis/servlet/static?dom=root&xml=index.xml, http://nis.gsmfc.org/, http://nas.er.usgs.gov/).

An analysis of potential donor areas (excluding species for which many donor areas were possible, e.g. cosmopolitan species and deliberately introduced species) indicated that the majority of NIS enter Chile from the Northern Pacific (i.e. North American west coast, Asian east coast; 30 species). Other important donor areas are the South Pacific (i.e. Australia, Peru; two species), the North Atlantic (i.e. North American east coast; four species), the South Atlantic (i.e. Argentina, the west coast of South Africa, two species) and a few other sources (i.e. Mediterranean Sea, North Sea; nine species). The high number of NIS arriving from the North Pacific is consistent with the high amount of ship traffic entering from this area (www.directemar.cl; Boletín Estadístico Marítimo 1992–2004), further supporting ship traffic as an important pathway for marine species introductions to Chile.

26.4 Socio-economic Impacts of NIS

26.4.1 Positive Impacts

Several marine NIS constitute important marine resources for Chile, where the maritime economy relies significantly on the direct harvesting of numerous established populations of non-indigenous fish, invertebrate and algae species. Salmon aquaculture in Chile's southern interior seas, fjords and channels represents over 2.5 billion dollars per year in export products, and is presently within the five most important export items in the country. Three introduced mollusk species, the Pacific oyster (*Crassostrea gigas*) and two species of abalone (*Haliotis rufescens* and *H. discus hannai*) also constitute species that are cultivated for export (e.g. in 2003 the primary production of these mollusks comprised over 2200 tons) (SERNAPESCA 2004). Another intensively cultivated marine resource is the red alga, *Gracilaria* spp., which has been transplanted to northern Chile, and is amply cultivated to produce agar (in 2003 more than 71,000 tons of *Gracilaria* spp. were harvested) (SERNAPESCA 2004). All of these aquaculture activities are managed by the Chilean Fisheries Administration (SERNAP), which requires specific permits for the importation and farming of species (see details in Sect. 8: Marine NIS Regulations in Chile).

In addition to the extraction of species in controlled aquaculture facilities, Castilla et al. (2005) pointed out the importance of wild population extractions of the introduced red alga *Mastocarpus papillatus,* (locally known as luga-luga), which has been extracted in southern Chile for the past 15–20 years. Two other red algae, *Porphyra linearis* and *P. pseudolinearis* (González 1998), and the exotic tunicate, *Pyura praeputialis*, also represent items in the Chilean diet and constitute potential resources for exploitation by small-scale or subsistence fisheries (Bustamante and Castilla 1987; Castilla and Defeo 2001; Castilla et al. 2004).

Other indirect positive economic impacts of marine NIS in Chile have also been identified. For example, numerous escaped introduced fish species sustain important sport fishing activities for tourists (Gajardo and Laikre 2003; Soto et al. 2001; Castilla et al. 2005). These activities are managed by SERNAPESCA (Servicio National de Pesca), which requires fishing licenses and establishes closures.

26.4.2 Negative Impacts

The importance of aquaculture to Chile makes the country vulnerable to invasions and associated impacts by marine pests in several inter-connected ways: (1) non-target NIS may be introduced with target aquaculture species, (2) aquaculture facilities may provide adequate habitat conditions (e.g. nutrients, temperature, substrate) facilitating the arrival and recruitment of NIS, (3) many NIS represent pests or fouling organisms to aquaculturists, negatively affecting financial gain. For instance, the tunicate, *Ciona intestinalis* seriously affects rope-scallop aquaculture activities in northern Chile (mainly in Tongoy Bay; approximately 30°S latitude; J.C. Castilla, personal observation). Furthermore, the green alga, *Codium fragile* ssp. *tomentosoides* has been shown to negatively impact *Gracilaria* spp. farms by increasing processing time and producing additional costs for unwanted species removal and disposal (Neill et al. 2006). Nevertheless, the negative socio-economic impacts of NIS in Chile are limited in scope, and there have been no reports of marine NIS in Chile causing widespread socio-economic damage such as that reported in other countries (e.g. with the introduction of *Carcinus maenas* in South Africa or *Limnoperna fortunei* in Argentina) (Robinson et al. 2005; Orensanz et al. 2002). Recently, Gajardo and Laikre (2003) stated that the introduction and restocking of NIS for aquaculture purposes presents a conservation paradox for Chile given the great economic success that often results from such ventures contrasted with the potential ecological impacts of these activities (Camus 2005). Since preventive measures aimed at controlling the entry and spread of NIS are frequently more efficient and less costly than reacting after an introduction has occurred (Kolar and Lodge 2002), it is important for Chile to develop species-specific management plans for dealing with the negative impacts of marine NIS before they become significant problems.

26.5 Ecology and Conservation of Marine NIS in Chile

Unfortunately, in Chile there is a lack of published studies regarding the ecological and conservation impacts caused by NIS or imported aquaculture invertebrate or fish species (but see reports and comments by Soto and Mena 1999; Soto and Jara 1999; Gajardo and Laikre 2003; Soto and Norambuena 2004; Camus 2005; Castilla et al. 2005). In any case, it is notable that, to date, studies of marine species introductions along the Chilean coast (including coastal marine environments, as well as estuarine and brackish waters) have not revealed the presence of aggressive or highly competitive sedentary species, which often have profound effects on ecological systems and important consequences for the conservation of local communities. Such species have been reported for other southern hemisphere countries, including: the rocky intertidal mussel, *Mytilus galloprovincialis*, along Atlantic and Indian Ocean coasts of South Africa (Robinson et al. 2005); the barnacle, *Balanus glandula*, and the mussel, *Limnoperna fortunei*, in Rio de la Plata, Argentina, the barnacle *Balanus glandula,* along the Patagonian coast of Argentina (San Clemente to the south; E. Schwindt, personal communication), the seastar, *Asterias amurensis*, in Australia, and the laminarian kelp, *Undaria pinnatifida*, in Australia, New Zealand and Argentina. Following their introduction, these species have spread significantly during the past 30–50 years, invading new sites where they have had substantial impacts on local biota, modifying communities and altering ecological balances (Pastorino et al. 1993; Vallarino and Elias 1997; Elias and Vallarino 2001; Darrigan and Ezcurra de Drago 2000; Valentine and Johnson 2003; Hough and Dommissee 2004; Robinson et al. 2005). The Mediterranean mussel species introduced to South Africa, *M. galloprovincialis*, has also been reported in Chile (first reported in 1998 associated with mussel aquaculture activities; Toro 1998; Daguin and Borsa 2000), at two localities on the Chilean coast separated by approximately 700 km: Castro, Chiloe Island and Dichato Bay, Concepción (Table 26.1; see also Castilla et al. 2005), but there are no reports indicating that *M. galloprovincialis* has spread to other mussel cultures. This invasion pattern (e.g. rate of spread) of *M. galloprovincialis* in Chile is completely different from the one reported for the same species on South African rocky shores, where the mussel rapidly invaded new sites following its introduction during the 1970s, and now occupies the entire west coast of South Africa (Steffani 2001; Robinson et al. 2005). Furthermore, other marine NIS that have been present for decades in the southwestern Atlantic, causing ecological impacts and concerns regarding the conservation of local biota (e.g. the mussel *L. fortunei*—an invasive species in estuaries, the rocky intertidal barnacle *B. glandula* and the kelp *Undaria pinnatifida*; Orensanz et al. 2002) have not invaded Chilean waters. Nor has Chile experienced the invasion of top marine invertebrate predators, such as the green crab *Carcinus maenas,* which was recently reported in Argentina (Hidalgo et al. 2005) and in South Africa (Robinson et al. 2005). Perhaps the only invasive marine species causing ecological and conservation concerns in Chile is the introduction of the Asian green alga, *Codium fragile* ssp. *tomentosoides*, known as Broccoli weed or Dead man's fingers. This alga was first reported in 1998 as a weedy/pest species associated with aquaculture activities (i.e. plantations of the red alga, *Gracilaria* spp., Chilean scallop farms, Pacific oyster farms), and was likely introduced from Japan or the eastern coast of the USA (Provan et al. 2005). This species has expanded to rocky intertidal and subtidal zones in natural environments located near aquaculture facilities, and presents a patchy, discontinuous distribution along the Chilean coast (P. Neill, personal observation). Current studies are evaluating the magnitude of the invasion by *C. fragile*, and its effects on biodiversity, including effects on the recruitment of invertebrate and algal species.

In Chile, two introduced marine bioengineer species have had important consequences on local ecological systems and conservation issues (see detailed discussion in Castilla et al. 2005; Chap. 16, Crooks): the tunicate *Pyura praeputialis* (found in Chile exclusively inside the Bay of Antofagasta, approximately 24°S latitude, Table 26.1) and the sea grass *Heterozostera tasmanica* (the only sea grass along the southeastern Pacific, found in Chile mainly in Puerto Aldea, around 30°S latitude; González 1992; Table 26.1). In both cases these introduced species have been associated with increases in local biodiversity and productivity, as well as the enhancement of nursery areas for invertebrate species (Castilla et al. 2004; Ortiz and Wolff 2002).

26.6 Marine NIS in Peruvian Waters

We compiled information regarding marine NIS along the Peruvian coast from four sources: (1) Canepa et al. (1998), a report on the ecological effects of introduced freshwater and marine species in Peru; (2) Yépez and Cisneros (1999), a summary of information regarding the effects of aquaculture (including NIS) in the Peruvian marine environment; (3) Del Rio et al. (2002), a summary of NIS from all environments in Peru; and (4) personal interviews with 10 aquaculture experts and marine biologists from IMARPE, who have conducted research along the entire Peruvian coast. These experts recognize the deliberate introduction of *Scophthalmus maximus* (turbot), *Crassotrea gigas* (the Pacific oyster) and *Haliotis rufescens* (the red abalone) to Peruvian waters. All of these species were deliberately introduced from Chile for aquaculture purposes. The possible fortuitous introduction and spread of the green alga, *Codium fragile* is currently under study in northern Peru (Cherrepe Bay, 7°S latitude; Paquita Olivares, IMARPE, Peru, personal communication), but no detailed information is yet available. Finally, while future studies will likely uncover the presence of NIS or cryptogenic marine species, it is important to note that, as in Chile, no aggressive intertidal rocky shore competitors (i.e. sessile invertebrates or algae), nor introduced top invertebrate predators have been reported in Peruvian marine systems (C. Yamashiro, IMARPE, personal communication).

26.7 Comparative Analysis of NIS in Chile

We present a comparative analysis of the number of marine NIS in Chile compared with North America, and several Southern Hemisphere countries (i.e. Peru, Argentina, South Africa, Australia and New Zealand), based primarily on a previous analysis conducted by Castilla et al. (2005), together with new information compiled for this chapter. In this analysis we only consider species that are "introduced" to natural environments or are "reasonable candidates" for introduction. We have excluded species deliberately introduced for aquaculture purposes (e.g. abalone, bony fishes) and vascular plants in order to make comparisons valid between studies.

As noted by Castilla et al. (2005), NIS introductions in Chile appear to be up to one order of magnitude less than those reported for northern hemisphere studies in North America: 19 introduced and 8 reasonable candidates reported in Chile vs 298 NIS in North America (Ruiz et al. 2000). Clearly this comparison may be biased in terms of the geographical area surveyed (e.g. the study by Ruiz et al. 2000 includes both Pacific and Atlantic coasts), observer/researcher effort. Furthermore, since there is no information on introduced estuarine species on the Chilean and Peruvian coasts, it is difficult to make comparisons with studies from other regions. Within the Southern Hemisphere, Chile continues to present a comparatively low number of NIS. Studies from Australia, New Zealand, and Argentina report considerably greater numbers of NIS than in Chile: 305 NIS in New Zealand (Cranfield et al. 1998), 83 introduced and 71 reasonable candidates in Port Phillip Bay Australia (Hewitt et al. 2004), 30 introduced and 46 reasonable candidates in Argentina (Orensanz et al. 2002). Two other southern hemisphere countries, South Africa and Peru, report comparably low numbers of NIS: 10 introduced and 22 reasonable candidates in South Africa (Robinson et al. 2005), and in Peru there are only 3 deliberately introduced aquaculture species (Canepa et al. 1998; Yépez and Cisneros 1999; Del Río et al. 2002; Rita Orozco, personal communication and this chapter). Beyond pure numbers, it is interesting to note that in Chilean and Peruvian marine systems no NIS are known to competitively displace other species (although current studies on *Codium fragile* ssp. *tomentosoides* may show otherwise) nor are there introduced top invertebrate predators.

Castilla et al. (2005) proposed that the relatively low number of marine NIS in Chile may be due to several factors, such as the coastline physiography (i.e. the Chilean coastline is considered to be naturally less stressed than other coasts of the world due to its scarcity of sheltered bays, gulfs and estuaries in northern and central Chile, where the majority of the population lives), together with specific oceanographic characteristics (i.e. low salinity in the fjords of southern Chile, an extensive minimum oxygen layer, and the existence of intense, high frequency upwelling areas (see Pizarro and Montecinos 2005; Ulloa and De Pol 2005), and biotic resistance due to benthic predators (Dumont et al. 2006). As discussed earlier, these characteristics may function to prevent the establishment and spread of NIS in the Chilean marine environment.

The (preliminary) low number of NIS reported in Peru supports the hypothesis of abiotic factors restricting invasion success, given the similarities between the marine environments of south and central Peru with the north and south central Chilean coast (Ulloa and De Pol 2005). While the total number of NIS in Chile and Peru seem to be at least one order of magnitude lower than other Southern Hemisphere coasts (but see Robinson et al. 2005, for South Africa), the lack of intensive surveys and monitoring programs in and around ports and harbors in Chile and Peru may underestimate the real number of marine NIS present in these systems. Nevertheless, we highlight the fact that no aggressive invaders or notorious ecological unbalances due to NIS have been reported, in contrast to other Southern Hemisphere marine systems (e.g. Argentina, Australia, New Zealand), including other temperate coasts, such as South Africa (Robinson et al. 2005; Steffani 2001), which are also exposed to intensive upwelling events.

The presence of aggressive invaders and top invertebrate predators in other temperate systems (e.g. South Africa) does not necessarily refute the hypothesis proposed by Castilla et al. (2005) of a possible anti-bioinvasion scenario in the southeastern Pacific geographical area, due to relatively particular environmental conditions along the coast. Although the Oxygen Minimum Zone (OMZ; i.e. dissolved oxygen concentrations below 0.5 mL/L) is practically continuous from the Gulf of Alaska to central Chile (Helly and Levin 2004), one of the main differences between the oceanographic and biogeochemical characteristics of the Humboldt Upwelling System (Chile and Peru) is the depth of the OMZ. The Humboldt Upwelling System is characterized by an extremely coastal OMZ (Ulloa and De Pol 2005), reaching shallow depths of <50 m (Rivadeneira 2005), and in Antofagasta Bay occurs as shallow as 10–15 m (Castilla and Guiñez, unpublished data). We hypothesize that these hypoxic conditions may result in important oceanographic barriers to NIS invasions in the southeastern Pacific, potentially affecting nearshore, shallow water species (also see Castilla et al. 2005).

26.8 Marine NIS Regulations in Chile

In Chile there is national legislation specifically intended to prevent the introduction, establishment or spread of diseases (pests) that affect indigenous populations, animals, plants or ecosystems. The Fishery and Aquaculture Law, FAL (Editora Jurídica Manuel Montt SA 1995) specifically addresses marine and freshwater biota. The FAL, which falls under the jurisdiction of the Chilean Fisheries Administration, has general authority regarding the importation of marine and freshwater species from other countries, as well as the transport of both native and introduced species within the country. The law specifically addresses the management of sanitary risks (mainly the prevention of diseases) and indicates that the importation of species (1) requires a permit and (2) is the responsibility of the importer, who must carry out the necessary impact analyses regarding sanitary risks as well as ecosystem (environmental) impacts. According to the FAL, the Chilean Fisheries Administration may issue a provisional and limited species importation permit allowing for up to a one year testing period. Nevertheless, such legislation may be too general to effectively prevent the introduction of NIS to Chile (Camus 2005), especially species that arrive unintentionally (e.g. associated with aquaculture or through ship traffic). In addition, there are no specific monitoring programs that aim to evaluate the presence of NIS in susceptible coastal areas (e.g. in and near shipping ports and aquaculture facilities). In this sense, inadequacies in the national legislation leaves Chile open to potential problems of marine

NIS biosecurity. Furthermore, proposals for the introduction and cultivation of marine NIS such as *Mytilus galloprovincialis* (http://www.fondef.cl/bases/fondef/PROYECTO/03/I/D03I1095.HTML) and, recently *Laminaria japonica* (http://www.fondef.cl/bases/fondef/PROYECTO/99/I/D99I1101.HTML) have been approved for national funding. At an international level, Chile has voluntarily signed several agreements aimed at controlling the entry of NIS to coastal environments (see Camus 2005). However, these international agreements may also be insufficient to protect natural ecosystems while continuing to look after the socio-economic interests of the country.

26.9 Conclusions and Recommendations

In this chapter we review the state of knowledge of marine NIS in Chile, and provide preliminary evidence for Peru. Based on our analysis of marine NIS distributions, introduction pathways, donor areas, ecological, economic and conservation effects, together with national and international legislation we present seven major conclusions:

1. NIS introductions in Chile and Peru appear to be up to one order of magnitude less than those reported for northern hemisphere studies; even within the southern hemisphere, Chile and Peru, continue to present relatively low numbers of marine NIS.
2. In the southeastern Pacific there are marine NIS from a variety of taxa, however highly abundant, aggressive invaders are absent, indicating that these coasts may be “relatively pristine” in terms of invasive NIS.
3. Evidence regarding the ecological effects of NIS in Chile is scarce.
4. The most important introduction pathways for NIS are related to aquaculture and shipping, with the Northern Pacific Ocean being a principal donor area of NIS to Chile (e.g. Japan, USA Pacific Coast).
5. There is evidence of both positive and negative socio-economic impacts of NIS in Chile (e.g. aquaculture and sport fishing vs fouling pests).
6. The potential anti-bioinvasion scenario proposed for the southeastern Pacific may be related to the existence of less stressed coasts, specific oceanographic factors (e.g. low dissolved oxygen conditions of the Chilean and Peruvian coasts) and/or biotic resistance, which may prevent or limit the establishment and spread of marine NIS.
7. National legislation may be too general to effectively protect natural ecosystems from NIS introductions while balancing the socio-economic interests of these countries.

Given that exotic species introductions is a topic of global concern, and the fact that there are reports of marine NIS related disturbances in ecological communities and socio-economic systems for southern hemisphere countries (e.g. Argentina, Australia, South Africa), it is important for researchers and managers to explicitly evaluate the status of NIS in Chile and Peru, to develop a strategy to prevent future introductions and to create a plan to control or eradicate NIS that have already been introduced. In particular, the status of cryptogenic species needs urgent attention. Strategies for dealing with NIS are divided into two types: (1) preventive measures and (2) post-introduction, reactive measures (Kolar and Lodge 2002), with the former frequently being more efficient and less costly. To ensure marine biosecurity, it is essential for Chile to implement preventive measures, including monitoring programs to survey "NIS hotspots" (e.g. docks and pilings, aquaculture facilities, fishing equipment, areas within and near ports, bays and estuaries), as well as baseline evaluations in susceptible areas, to detect NIS introductions early on and determine propagule pressure. Molecular techniques now constitute important tools for identifying cryptogenic species and/or evaluating the taxonomic and population status of NIS vs native species. Proactive legislation to control the entrance of NIS to Chile should focus on primary introduction pathways and principal donor areas, as well as identify species that are likely to become pests. The deliberate introduction of species for aquaculture should be more rigorously regulated to prevent and monitor escapes from facilities, and to stop introductions of non-target species. In cases where marine NIS have already been introduced into the system, researchers and managers should conduct studies to evaluate ecological and socio-economic impacts. Legislation should focus on the implementation of precautionary principles (Simberloff 2005), eradicating or managing unwanted marine NIS in introduced areas, and controlling further spread.

Acknowledgements We are extremely grateful to many people for information regarding NIS in Peru, Chile and Argentina, especially Carmen Yamashiro and Paquita Olivares (IMARPE Peru), Rita Orozco (CONAM Peru), Evangelina Schwindt (Instituto de la Patagonia, Puerto Madryn, Argentina) and Martin Thiel (Universidad Católica del Norte). Some of these data were presented at the Marine Bioinvasions International Conference in New Zealand (August 2005). We are grateful to the Center for Advanced Studies in Ecology and Biodiversity for support (CASEB, FONDECYT Grant #1501-0001). PEN thanks CASEB Program 8 for a travel grant. We thank three reviewers for constructive comments on this chapter.

References

Arce I (1997) Narraciones Históricas de Antofagasta. Lama Industrial S.A., Iquique, Chile, pp 570

Baez PR, Meléndez M, Ramírez S, Letelier A, Brown M, Campos C, Alday C, Jelvez C, Labial A (1998). Efectos ecológicos de la introducción de especies exóticas en el medio marino y costero Chileno. In: Reunión de expertos para analizar los efectos ecológicos de la introducción de especies exóticas en el Pacífico sudeste, Viña del Mar, Chile, Comisión Permanente del Pacífico Sur, pp 1–83

Basulto S (2003) El largo viaje de los Salmones, una crónica olvidada. Impresión Maval Limitada, Santiago, Chile, pp 299

Boletín Estadístico Marítimo (1992–2004) Armada de Chile. 3 May 2005, www.directemar.cl

Bustamante R, Castilla JC (1987) The shellfisheries in Chile: an analysis of 26 year of landings. Biol Pesq Chile 16:79–97

Camus C, Meynard A, Faugeron S, Kogame K, Correa JA (2005) Differential life history phase expression in two coexisting species of *Scytosiphon* (Phaeophyceae) in northern Chile. J Phycol 41:931–941

Camus P (2001) Biogeografía marina de Chile continental. Rev Chil Hist Nat 74:587–617

Camus PA (2005) Introducción de especies en ambientes marinos chilenos: no solo exóticas, no siempre evidentes. Rev Chil Hist Nat 78:155–159

Canepa JR, Camella A, Mogollón V (1998) Efectos ecológicos de la introducción de especies exóticas en el Pacifico Sudeste: El caso del Perú. Comisión Permanente del Pacífico Sur – CPPS. Informe de la Reunión de Expertos (Viña del Mar – Chile), pp 127–139

Carlton JT (1999) The scale and ecological consequences of biological invasions in the World's oceans. In: Sandlund, OT (ed) Invasive species and biodiversity management. Kluwer Academic Publishers, The Netherlands, pp 195–212

Castilla JC, Defeo O (2001) Latin-American benthic shellfisheries: emphasis on co-management and experimental practices. Rev Fish Biol Fish 11:1–30

Castilla JC, Lubchenco J, Navarrete SA (1993) Southeastern Pacific coastal environments: main features, large-scale perturbations and global climatic change. In: Mooney HA, Fuentes E, Kronsberg B (eds) Earth system response to global change: contrasts between North and South America. Academic Press, pp 167–188

Castilla JC, Guiñez R, Caro A, Ortiz V (2004) Invasion of a rocky intertidal shore by the tunicate *Pyura praeputialis* in the Bay of Antofagasta, Chile. Proc Natl Acad Sci 101(23):8517–8524

Castilla JC, Uribe M, Bahamonde N, Clarke M, Desqueyroux-Faúndez R, Kong I, Moyano H, Rozbaczylo N, Santelices B, Valdovinos C, Zavala P (2005) Down under the Southeastern Pacific: marine non-indigenous species in Chile. Biol Invas 7:213–232

Cohen AN, Carlton JT (1998) Accelerating invasion rate in a highly invaded estuary. Science 279:555–558

Cranfield HJ, Gordon DP, William RC, Marshall BA, Battershill CN, Francis MP, Nelson WA, Glasby CJ, Read GB (1998) Adventive marine species in New Zealand. NIWA Tech Rep 34, pp 48

Daguin C, Borsa P (2000) Genetic relationships of *Mytilus galloprovincialis* Lamarck populations worldwide: evidence from nuclear-DNA markers. In: Harper EM, Taylor JD, Crame JA (eds) The evolutionary biology of the Bivalvia. Geol Soc London Spec Publ 177:389–397

Darrigan G, Ezcurra de Drago I (2000) Invasion of freshwater mussel, *Limnoperna fortunei* (Dunke, 1857) (Mytilidae) in the Americas. Nautilus 114:69–73

Del Río ML, Pariona D, Córdova JH, Salmón G (2002) Las especies exóticas invasoras en el Perú. Informe Nacional

Dumont CP, Harris LH, Gaymer CF (2006) Role of benthic predators in structuring piling communities. XXVI Congreso de Ciencias del Mar, Iquique, April, Abstract, p 78

Editora Jurídica Manuel Montt SA (1995) Ley General de Pesca y Acuicultura

Elías R, Vallarino E (2001) The introduced Barnacle *Balanus glandula* (Darwin) in the Mar del Plata port as a structuring species in the intertidal community. Invest Mar 29:37–46

Fofonoff PW, Ruiz GM, Steves B, Carlton JT (2003) In ships or on ships? Mechanisms of transfer and invasion for nonnative species to the coasts of North America. In: Ruiz GM, Carlton JT (eds) Invasive species. Vectors and management strategies. Island Press, pp 152–182

Gajardo G, Laikre L (2003) Chilean aquaculture boom is based on exotic salmon resources: a conservation paradox. Conserv Biol 17(4):1173–1174

González A (1998) Caracterización de especies y/o morfotipos de *Porphyra* (Rhodophyta, Bangiales), 'luche' de las bahías de Valparaíso y Laguna Verde, V Región, Chile. Thesis, Instituto de Oceanología, Universidad de Valaparaíso, pp 149

González S (1992) *Heterozostera tasmanica* (Martens en Aschers) den Hartog y comunidad asociada en el norte de Chile. Undergraduate thesis, Universidad Católica del Norte, Facultad de Ciencias del Mar, Chile

Grosholz ED (2005) Recent biological invasion may hasten invasional meltdown by accelerating historical introductions. Proc Natl Acad Sci 102:1088–1091

Guzmán G, Soto R (2000) Camarones peneidos (Decapoda, Dendobranchiata) frente al norte de Chile: evento "El Niño" 1997–1998. Not Mensual Museo Nacional Hist Nat 343:12–15

Guzmán G, Nicosia W, Peredo R (1999) Cangrejos Portunidae (Crustacea, Decapoda) asociados al evento "El Niño" 1997–1998 en la zona norte de Chile con el primer registro de *Areneus mexicanus* (G.). Not Mensual Museo Nacional Hist Nat 338:4–6

Helly JJ, Levin LA (2004) Global distribution of naturally occurring marine hypoxia on continental margins. Deep-Sea Res 51:1159–1168

Hewitt CL, Campbell ML, Thresher RE, Martin RB, Boyd S, Cohen BF, Currie DR, Keough MJ, Lewis JA, Lockett MM, Mays N, McArthur MA, O'Hara,TD, Poore GCB, Ross DJ, Storey MJ, Watson JE, Wilson RS (2004) Introduced and cryptogenic species in Port Phillip Bay, Victoria, Australia. Mar Biol 144:183–202

Hidalgo FJ, Barón PJ, Orensanz JM (2005) A prediction comes true: the green crab invades the Patagonian coast. Biol Invas 7(3):547–552

Hough D, Dommisse M (2004) Controlling the Northern Pacific seastar (*Asterias amurensis*) in Australia. Final Report for the Australian Government Department of the Environment and Heritage. Department of Sustainability and Environment, pp 52

Kim M, Yang EC, Mansilla A, Boo SM (2004) Recent introduction of *Polysiphonia morrowii* (Ceramiales, Rhodophyta) to Punta Arenas, Chile. Bot Mar 47(5):389–394

Kolar CS, Lodge DM (2002) Ecological predictions and risk assessment for alien fishes in North America. Science 298:1233–1236

Kong I, Tomicic J, Zegers J (1985) Ictiofauna asociada al fenómeno El Niño 1982–1983 en la zona norte del Chile. Invest Pesq (Chile) 32:215–224

Larraín G (2001) Diego de Almagro, Biografía. Editorial Luxemburgo, pp 408

Leonardi PI, Miravalles AB, Faugeron S, Flores V, Beltran J, Correa JA (2006) Diversity, phenomenology and epidemiology of epiphytism in farmed *Gracilaria chilensis* (Rhodophyta) in northern Chile. Eur J Phycol 41(2):247–257

Levin L (2003) Oxygen minimum zone benthos: adaptation and community response to hypoxia. Oceanogr Mar Biol Annu Rev 41:1–45

Maino V (1985) Islas oceánicas Chilenas y el desarrollo de las rutas de comunicación en el Pacífico Sur Oriental. Editorial Antártica S.A., pp 91

Naylor R, Williams S, Strong D (2001) Aquaculture – a gateway for exotic species. Science 294:1655–1656

Neill P, Alcalde O, Faugeron S, Navarrete SA, Correa JA (2006) Invasion of *Codium fragile* ssp. *tomentosoides* in northern Chile: a new threat for *Gracilaria* farming. Aquaculture 259:202–221

Orensanz JM, Schwindt E, Pastorino G, Bortolus A, Casas G, Darrigran G, Elías R, López-Grappa J, Obenat S, Pascual M, Penchaszadeh P, Piriz M, Scarabino F, Spivak E, Vallarino E (2002) No longer the pristine confines of the world ocean: a survey of exotic marine species in the Southwestern Atlantic. Biol Invas 4:115–143

Ortiz M, Wolff M (2002) Trophic models of four benthic communities in Tongoy Bay (Chile): comparative analysis and preliminary assessment of management strategies. J Exp Mar Biol Ecol 268:205–235

Pastorino G, Darrigan G, Martin S, Lunaschi L (1993) *Limnoperna fortunei* (Dunker, 1857) (Mytilidae) nuevo bivalvo invader del Río de la Plata. Neotropica 39

Pizarro O, Montecinos A (2005) El Niño y la Oscilación del Sur. In: Werlinger C (ed) Biología Marina y Oceanografía: Conceptos y Procesos. Trama Impresores SA, Concepción, Chile, pp 197–228

Provan JP, Murphy S, Maggs CA (2005) Tracking the invasive history of the green alga *Codium fragile* ssp. *tomentosoides*. Mol Ecol 14:189–194

Ribera MA (2003) Pathways of biological invasions of marine plants. In: Ruiz GM, Carlton JT (eds) Invasive species. Vectors and management strategies. Island Press, pp 183–226

Rivadeneira M (2005) Macroecologia evolutiva de los bivalvos marinos de la costa Pacifica de Sudamérica. Pontificia Universidad Católica de Chile, Santiago, Chile, PhD Thesis, p 76

Robinson TB, Griffiths CL, McQuaid CD, Rius M (2005) Marine alien species of South Africa — status and impacts. Afr J Mar Sci 27(1):297–306

Ruiz GM, Fofonoff PW, Carlton JT, Wonham MJ, Hines AH (2000) Invasion of coastal marine communities in North America: apparent patterns, processes, and biases. Annu Rev Ecol Syst 31:481–531

Santelices B (1989) Algas marinas de Chile: distribución, ecología, utilización, biodiversidad. Ediciones Universidad Católica de Chile, Santiago, Chile, pp 399

SERNAPESCA (2004) Anuario Estadístico de Pesca. Ministerio de Economía Fomento y Reconstrucción, Chile

Sielfeld W, Vargas M, Kong I (2003) Primer registro de *Etropus ectenes* Jordan, 1889, *Bothus constellatus* Jordan and Goss, 1889, *Achirus klunzingeri* (Steindachner, 1880) y *Symphurus elongatus* (Günther, 1868) (Pisces, Pleuronectiformes) en Chile, con comentarios sobre la distribución de los lenguados chilenos. Invest Mar Valparaíso 31(1):51–65

Simberloff D (2005) The politics of assessing risk for biological invasions: the USA as a case study. TREE 5:216–222

Soto D, Jara F (1999) Relevance of ecosystemic services provided by species assemblages: coupling salmon farming with biodiversity use and management. UN and Norway Conference on the Ecosystem Approach for Sustainable Use of Biological Diversity, Trondheim, Norway, pp 133–136

Soto D, Mena GP (1999) Filter feeding by the freshwater mussel *Diplodon chilensis* as a biocontrol of salmon farming eutrophication. Aquaculture 171:65–81

Soto D, Norambuena F (2004) Evaluation of salmon farming effects on marine systems in the inner seas of southern Chile: a large-scale mensurative experiment. J Appl Ichthyol 20:493–501

Soto D, Jara F, Moreno C (2001) Escaped salmon in the inner southern Chile: facing ecological and social conflicts. Ecol Appl 11:1750–1762

Soto D, Arismendi I, González J, Sanzana J, Jara F, Jara C, Guzmán E, Lara A (2006) Southern Chile, trout and salmon country: invasion patterns and threats for native species. Rev Chil Hist Nat 79(1):97–117

Steffani CN (2001) Interactions between an indigenous limpet, *Scutellastra argenvillei*, and an alien invasive mussel, *Mytilus galloprovincialis*: moderation by wave action. University of Cape Town, Cape Town, pp 190

Toro JE (1998) PCR-based nuclear and mtDNA markers and shell morphology as an approach to study the taxonomic status of the Chilean blue mussel, *Mytilus chilensis* (Bivalvia). Aquat Living Resour 11:347–353

Ulloa O, De Pol R (2005) Zonas de Mínima de Oxígeno. In: Werlinger C (ed) Biología Marina y Oceanografía: Conceptos y Procesos. Trama Impresores S.A. Concepción, Chile, pp 521–536

Valentine JP, Johnson CR (2003) Establishment of the introduced kelp *Undaria pinnatifida* in Tasmania depends on disturbance to native algal assemblages. J Exp Mar Biol Ecol 295:63–90

Vallarino E, Elías R (1997) The dynamics of an introduced *Balanus glandula* population in the Southwestern Atlantic rocky shores. The consequences on the intertidal community. P.S.Z.N.I. Mar Ecol 18:319–335

Yépez V, Cisneros C (1999) Efectos de la maricultura en los ambientes marinos y costeros adyacentes en el Pacífico Sudeste. Estado situacional en la Costa Peruana. Plan de Acción para la Protección del Medio Marino y Áreas Costeras del Pacífico Sudeste. Puerto Montt, Chile, October 5–8

Chapter 27
Marine Bioinvasions in the Brazilian Coast: Brief Report on History of Events, Vectors, Ecology, Impacts and Management of Non-indigenous Species

Carlos Eduardo Leite Ferreira, Andrea de Oliveira Ribeiro Junqueira, Maria Célia Villac, and Rubens Mendes Lopes

27.1 Introduction

The Brazilian coast extends for about 8000 km from Cape Orange (4°N) to Chui (34°S) (Fig. 27.1). This long coastline comprises a variety of ecosystems under the influence of oligotrophic waters transported by two western boundary currents, the Brazil and North Brazil currents (Stramma and England 1999), together with continental influences related to a wide spectrum of river inputs, the largest of which being the massive Amazon River plume in the north and the combination of the La Plata and Patos Lagoon outflows in the south (Castro Filho and Miranda 1998). Seasonal or intermittent intrusions of cold and nutrient-rich oceanic waters carried underneath the Brazil and North Brazil currents (the so-called South Atlantic Central Water) is another important physical forcing on regional shelf ecosystems of Brazil, particularly on the Southern Brazilian Bight and more southern areas (Lopes et al. 2006). Regionally important coastal ecosystems are (1) sandy beaches, occurring from north to south, with the largest ones in southernmost areas of the state of Rio Grande do Sul; (2) mangrove forests, which occur from the northern tip of the country to the state of Santa Catarina in the south, (3) coral reefs, ranging from Maranhão to Bahia including the largest coral reef system of the South Atlantic, the Abrolhos Reefs (Leão et al. 2003); (4) rocky shores, spread along the entire coast from the northeast to the south, but more extensive in the southeast; (5) coastal lagoons, such as the Cananéia-Iguape estuarine complex in the state of São Paulo and the Patos Lagoon in the state of Rio Grande do Sul; (6) saltmarshes, of regional importance only south of the state of São Paulo; and (7) sandy to muddy bottoms of the infralitoral realm down to the shelf break (Seeliger and Kjerfve 2000). Given such a large array of marine ecosystems and the extent of the Brazilian coastline, the country is undoubtedly a major receptor and donor of tropical and subtropical organisms in the world's oceans.

Research on marine bioinvasions is a relatively new topic in Brazil. The first comprehensive lists of introduced and invasive species are starting to be compiled and the understanding of patterns of invasion strategies is far from being accom-

G. Rilov, J.A. Crooks (eds.) *Biological Invasions in Marine Ecosystems.* Ecological Studies 204,

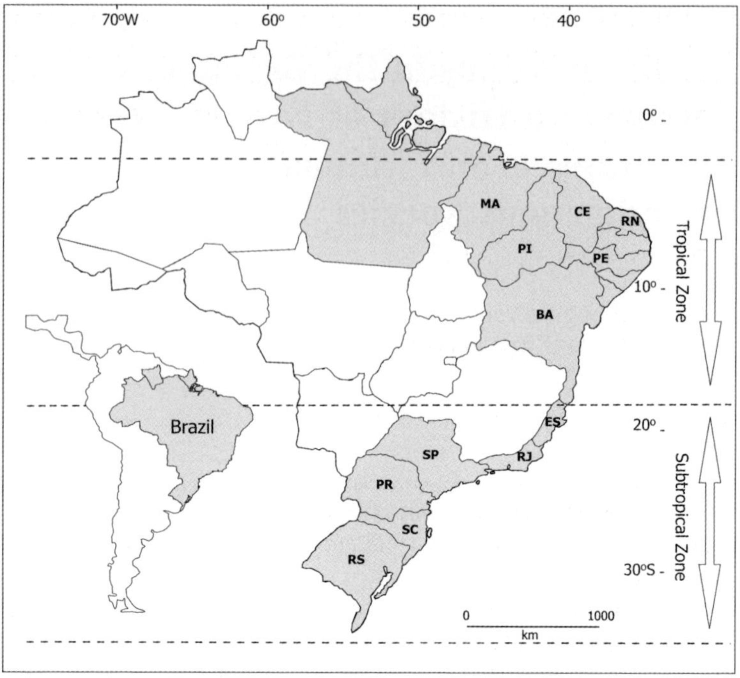

Fig. 27.1 Map of Brazil showing coastal states in *gray* with subdivision of tropical and subtropical zones. Acronyms for the names of the states mentioned in the text are, from north to south: Maranhão (MA), Piauí (PI), Ceará (CE), Rio Grande do Norte (RN), Pernambuco (PE), Bahia (BA), Espírito Santo (ES), Rio de Janeiro (RJ), São Paulo (SP), Paraná (PR), Santa Catarina (SC), and Rio Grande do Sul (RS)

plished. There is a clear trend of increasing bioinvasion events in regional coastal ecosystems, but whether invasion rates are actually increasing or are a result of more intensive research efforts in the recent past is still an open question. The data available is sparse and locally produced, that is, spatial trends might reflect specific research interests rather than actual introduction and dispersal patterns. Some temporal trends can be pointed out, however, as will be discussed below.

There is no doubt that commerce through shipping and offshore oil exploration have increased exponentially in Brazil in the last 20 years and, in fact, there is sound indication that ballast and fouling are the major vectors of some of the aliens detected (Souza and Silva 2004). Interestingly, if one goes back 10 or more years and checks the literature related to the taxonomy and distribution of Brazilian marine species, one notices that most publications considered unreported species as new occurrences of native species (see Chap.2, Carlton). Nowadays, most publications do not disregard the possibility of new occurrences as prospective alien species. Nevertheless, detailed information on possible transport vectors and their pathways are still needed to support introduction hypotheses.

Our goal in this chapter is to present our current understanding of possible trends of introduction of marine species in the Brazilian coast based on the best investigated case studies, including introduction vectors and further dispersal, information on the status of populations, actual and/or potential impacts, and management initiatives conducted so far. The locations mentioned in the text can be found in Fig. 27.1.

27.2 The Plankton Realm

27.2.1 Phytoplankton Species

Designating phytoplankton species as nonindigenous can be quite complex and controversial. Awareness of the role and risks of invasive species has led many researchers to immediately assign new reports in the cryptogenic category as a cautionary measure. For Brazilian waters, there are only three cases that hold strong evidence to support their status as introduced: the dinoflagellates *Alexandrium tamarense* (Lebour) Balech, 1992 and *Gymnodinium catenatum* Graham, 1943, and the diatom *Coscinodiscus wailesii* Gran and Angst, 1931. The sequence of events leading to their introduction, distribution and present/potential impacts on the Brazilian coast are discussed, respectively, in Persich et al. (2004), Proença et al. (2001), Fernandes et al. (2001), and summarized in Proença and Fernandes (2004).

The first record of *A. tamarense* in the western South Atlantic was in Argentina, as a coastal bloom in 1980. In 1991, the species bloomed along the Uruguayan coast, and the phenomenon was again detected in 1991, 1993, 1995 and 1996. In Brazil, it was first detected in August 1996 along Cassino Beach (state of Rio Grande do Sul). Cultures isolated from Brazilian waters showed two genetic signatures, from the eastern and western coasts of North America (Persich et al. 2004). The lack of more molecular data for populations isolated from the western South Atlantic makes difficult the interpretation of possible transport and introduction mechanisms to Brazilian waters. There are three possible, non-mutually exclusive explanations for the introduction of this species in Brazil: (1) transport by marine currents from their initial site of introduction, that is, from Argentina to Uruguay, and then to southern Brazil; (2) secondary introduction by port-hopping between these South American ports; and/or (3) direct introduction to a Brazilian port. Indeed, *A. tamarense* is a cyst-forming species and transport in ballast tanks is a likely vector. This bloom-forming species can produce toxins that may cause paralytic shellfish poisoning with impacts to mariculture activities and public health (Hallegraeff et al. 2003).

The biogeography of the early and more recent records of *Gymnodinium catenatum* is not very conclusive in tracing the history of its world-wide distribution. The earliest record was in the Gulf of California, eastern North Pacific in 1940; it was later detected in Argentina in 1962; since the 1970s, the number of records and locations increased and also include a citation for Uruguayan waters (Hallegraeff and Fraga 1998). In Brazil, it was first detected in the state of Santa Catarina in 1998, but it is also found at present in neighboring states (Paraná and São Paulo). One of the main arguments that favor the introduction hypothesis is the fact that it is a conspicuous species (large size and chain-forming) that would have hardly gone unnoticed in routine phytoplankton studies for so long. Moreover, like *A. tamarense*, it is a cyst-forming species that favor ballast tanks as a possible transport mechanism. It is also known to have caused blooms in different parts of the world, and it can produce toxins that may lead to paralytic shellfish poisoning.

Coscinodiscus wailesii is a centric diatom originally described from the Pacific coast of North America (state of Washington) in 1931. At that time, it was also recorded in several other locations along the coastline down to California, as well as in Japan. In the 1970s to 1980s, new records appeared in European and South American waters. In Brazil, its first published record was in the state of Paraná in 1983, but it is now known to be a regular component of phytoplankton populations in different locations between the states of Bahia and Rio Grande do Sul. This species is quite conspicuous in size and shape and would likely be noticed. Transport in ballast tanks is a likely vector, since *C. wailessii* can form resting cells. It has caused harmful blooms in Brazilian waters (Fernandes et al. 2001) and elsewhere (Boalch and Harbour 1977): high densities of this diatom can cause temporary exclusion of other phytoplankton species with detrimental effects to filter-feeders; it can lead to oxygen depletion that is deleterious to marine biota; and it can produce mucilage that may inhibit predation and also clog fishing nets with negative impacts on fishing activities.

27.2.2 Zooplankton Species

Six zooplankton species have been reported as non-native to Brazilian coastal waters, only three of which have widened their distribution range since initial reports (Lopes 2004). Hence, our brief account will focus on those well-established species.

The calanoid copepod *Temora turbinata* Dana, 1849 was first recorded in the Vasa-Barris estuary in the Sergipe coast in the late 1980s (Araújo and Montú 1993) and spread over southeastern (Lopes et al. 1999) and southern waters (Muxagata and Gloeden 1995), where it is currently one of the dominant pelagic copepods (Lopes et al. 2006). Its arrival at the Brazilian coast likely occurred through ballast water release, but the possibility of unintentional introduction with shrimp breed stocks brought from Southeast Asia to establish aquaculture initiatives in Northeastern Brazil in the late 1970s cannot be discounted. Before the establishment of the invasive species, *T. stylifera* (Dana, 1849) was the only representative of the genus on the Brazilian coast. However, there are no large pre-introduction datasets available to compare present and past distributions of *Temora* spp. in the region. The lack of robust historical information is a recurrent problem in building introduction and dispersion hypotheses.

Another exotic marine copepod thrives in Northeast Brazil, although at lower abundances compared to *T. turbinata*. In this case, there are more clear indications of how the introduction took place. *Pseudodiaptomus trihamatus* (Wright 1937), native to the Indo-Pacific, was first detected in shrimp ponds in the Potengí river estuary (Rio Grande do Norte) in 1977, and then subsequently found in the natural habitat under a relatively strict distribution range (Medeiros et al. 1991). The relationship of *P. trihamatus* introduction in Brazil to aquaculture is an obvious explanation because the establishment of shrimp farms at that time relied on breed stocks of *Penaeus monodon* (Fabricius, 1798) imported from Taiwan, where the copepod occurs as a natural component of the zooplankton community (Lo et al. 2004). Since the early records, *P. trihamatus* has been found in several estuarine ecosystems of the Northeastern coast from 6°19′60″S in Rio Grande do Norte to 3°02′12″S in Ceará (Medeiros et al. 2006). Additional sampling is urgently needed in estuaries of the relatively less-studied eastern coast of Brazil (state of Bahia) to determine whether the species is restricted to the presently reported locations, or is expanding its occurrence toward more southern areas. Like *Temora turbinata*, the existence of potential ecological impacts of *P. trihamatus* on native *Pseudodiaptomus* and other zooplankters from estuarine and coastal systems of Northeast Brazil remains unclear to date.

A small pelagic cladoceran, *Pleopis schmakeri* (Poppe, 1889), also native to the Indo-Pacific, was recorded by Rocha (1985) in a river estuary of southeastern Brazil, and in other inshore areas thereafter (Lopes 2004). Although not as common as the above-mentioned microcrustaceans, *P. schmakeri* has been reported at relatively high abundances in coastal and estuarine waters (Lopes 1997), suggesting a successful spread of the species in the nearshore environment, thanks to its parthenogenetic reproduction strategy and high growth rates (Onbé 1983).

Despite the fast and extensive geographical dispersion of exotic species of marine zooplankton in Brazil, which has been in effect for at least 30 years, we still know very little about their potential community and ecosystem level impacts, a picture that unfortunately mirrors the worldwide scenario (Bollens et al. 2002). Additional monitoring programs and experimental studies are needed to address the problem.

27.3 The Benthic Realm

27.3.1 Phytobenthos

Caulerpa scalpelliformis (R. Brown ex Turner) C. Agardh 1817, a pantropical green algae, has been reported along the tropical region of the Brazilian coast with a southern limit of distribution in the state of Espirito Santo (Mitchell et al. 1990). In 2001, *C. scalpelliformis* was found in Ilha Grande Bay, within the subtropical Southeastern Brazilian shelf (Falcão and Széchy 2005). Since then, the species spread and became dominant on the rocky and sand substrata, replacing the native species, particularly *Sargassum* spp. Significant changes in the macroalgae community structure were detected following the appearance of *C. scalpelliformis*, which reinforces its invasive character. Many possible transport vectors were suggested by the authors, including fouling, aquaculture and the aquarium trade.

27.3.2 Zoobenthos

Most reported introductions of zoobenthic species in the Brazilian coast are considered isolated occurrences. Some species can be ranked as well-established in the country. Possible impacts of these species have not been detected and/or assessed at present. Only five species have presented evident impacts and are expanding their range: *Isognomon bicolor* (Adams 1845), *Tubastraea coccinea* Lesson 1829, *Tubastraea tagusensis* Wells 1982, *Chromonephthea braziliensis* (Ofwegen 2005), and *Charybdis helleri* (Milne-Edwards 1867). These species are highlighted below.

27.3.2.1 Bivalves

One of the most important introductions to the benthic realm was the bivalve *Isognomon bicolor*. This species, originated from the Caribbean region, was first reported for the Brazilian coast in 1994 (Domaneschi and Martins 2002). Initially, it was identified as another species (*Isognomon alatus* (Gmelin, 1791), recorded in Atol das Rocas in 1970) but its expansion and great abundance caused concern and detailed studies were performed. Currently, *I. bicolor* distribution extends from the Northeast to the South coast of Brazil (Domaneschi and Martins 2002). Ballast water and fouling are the suggested introduction vectors (Rocha 2002).

I. bicolor prevails in steep rocky coasts (>70°) with moderate wave exposure (Whorff et al. 1995; Domaneschi and Martins 2002). In the intertidal zone of rocky shores of the Arraial do Cabo region, state of Rio de Janeiro, the species was first observed in 1996 (Rocha 2002). Nowadays, this species is dominant in the mid-intertidal of Arraial do Cabo rocky shores and is responsible for reducing 50% of the native barnacle population, *Tetraclita stalactifera* Lamarck, 1818 (López 2003). This pattern of *I. bicolor* dominance is present in other rocky shores along the southeastern coast, where densities between 200 and 800 individuals/100 cm^2 can be found (Magalhães 1999; Rapagnã 2004; Breves-Ramos 2004) (Fig. 27.2). Despite its high abundance, this invasive species has not been successful in colonizing and establishing itself on bare substrate, appearing only at late succession stages in the intertidal habitat (Rocha 2002) or recruiting inside holes, crevices or other complex substrates (Moyses 2005). Recent ecological studies in Arraial do Cabo show that this alien bivalve is now part of the diet of the major predator in this habitat, the gastropod *Stramonita haemastoma* (Linnaeus, 1767) (López 2003). Although *I. bicolor* is superior to other native preys (mussels and barnacles) in caloric terms, it is not yet preferentially selected (López and Coutinho 2005). The

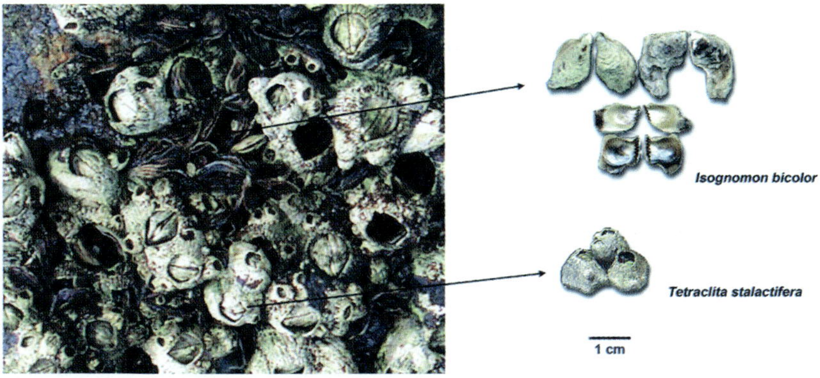

Fig. 27.2 The invasive bivalve *Isognomon bicolor* and the native barnacle *Tetraclita stalactifera* on the rocky coasts of Angra dos Reis, state of Rio de Janeiro. This invasive species is not successful in colonizing and establishing itself on bare substrate, appearing only at late succession stages in the intertidal habitat or recruiting inside holes, barnacle tests, crevices or other complex substrates. In the mesolittoral of Arraial do Cabo, state of Rio de Janeiro, it is responsible for reducing 50% of the population of *Tetraclita stalactifera*

authors suggested that 10 years of establishment and trophic interactions probably are not sufficient for *I. bicolor* to be recognized as a better caloric prey and that prey manipulation is not yet efficient to compensate its consumption. The impact of predation by *S. haemastoma* on *I. bicolor* populations has not yet been investigated but it was suggested to be an important mortality factor.

The ecological impacts of *I. bicolor* have not been assessed in detail to date. This is a difficult task because little quantitative data of the pre-invasion communities is available. Nevertheless, the high densities of this species in the Brazilian coast suggest it has caused profound changes in the native rocky coast communities. Besides the competition and displacement of barnacle populations, *I. bicolor* probably competes with a commercial species of Mytilidae, *Perna perna* (Linnaeus 1758), a very abundant mussel found in the region (Rapagnã 2004; Breves-Ramos 2004). This species, native to Africa, was probably introduced between the eighteenth and nineteenth centuries during the slave trade. This hypothesis is supported by the absence of *P. perna* in prehistoric deposits of shells that are close to present-day populations (Souza et al. 2004). Nowadays, *P. perna* is completely established. Actually, it plays an important role in the community structure of rocky coasts between the states of Espírito Santo and Rio Grande do Sul. In the past, *P. perna* probably diminished the density of *Pinctata imbricata* Rodin, 1867, a very abundant bivalve in the prehistoric deposits of shells and rare at present (Rapagnã 2004).

27.3.2.2 Corals

In the subtidal zone, three exotic coral species also deserve attention because of their invasive potential and ecological significance. The first is an Alcyonacean, recently described as *Chromonephthea braziliensis* (Ofwegen 2005) whose origin is, in spite of the name, the Indo-Pacific. The species was first found approximately 15 years ago inhabiting a sand bottom on the boundary of shallow (8–10 m) rocky shores of the southeastern coast (Arraial do Cabo, state of Rio de Janeiro) (Fig. 27.3). The first scientific observations of the establishment event started with precisely eight individuals ranging from 20 to 60 cm in height (Ferreira 2003; Ferreira et al. 2004). Monitoring dives to assess abundance and size showed that, after one year, the colony had increased to about 40 individuals near the sand-rock interface and also in a direction away from the rocky shore. Although information about the reproductive biology of the coral was not assessed, the high abundance of juveniles (5 cm) near "adults" and the restricted area of their distribution highly suggest that this soft coral has expanded only by asexual reproduction modes. The mortality of young brooded monitored in the field was about 70% and exposure to water temperature <17 °C was probably the main factor restricting the expansion of the species. Recent experiments show that this soft coral alien has chemical defenses against generalist fishes, and can also cause physical damage to other native coral species (Lages et al. 2005). Indeed, these experiments have shown that the species

Fig. 27.3 *Chromonephthea braziliensis* sharing space with *Sargassum furcatum* in unconsolidated bottom of Arraial do Cabo, state of Rio de Janeiro. This species finds better habitat in the sand interface near rocky shore, but in the past few years it began to colonize the hard substrate as well

uses chemical defense to slowly expand its distribution in the environment. Despite such invasion potential, the distribution of *C. braziliensis* is still restricted to about a few hundred meters even though more than a decade has passed after its first discovery in Arraial do Cabo. Most colonies found 200 m away from the original dispersal point could be related to transposition by unwarned scuba divers, as the species site became a famous diving point. Their sizes (maximum 90 cm high) and habitat (sand bottom near the interface of the rocky shore) make them easy to detect and mechanically remove. Eradication in this case is controversial because this is considered a protected area by the National Environmental Agency of Brazil.

The second alien coral, *Tubastraea coccinea* (Fig. 27.4) is an azooxanthellate species that was first recorded in the mid-1980s in association with oil drilling platforms located in the Campos Basin, offshore of the northern coast of the state of Rio de Janeiro (Castro and Pires 2001). Later, in the mid-1990s, this species and also *Tubastraea tagusensis* (the third one) were recorded at Ilha Grande Bay, state of Rio de Janeiro, southward from Campos Basin (Paula and Creed 2004, 2005). More recently, *T. coccinea* was confirmed recruiting in subtidal rocky shores of Arraial do Cabo (Ferreira 2003; Ferreira et al. 2004). Today, this species has been

Fig. 27.4 *Tubastraea coccinia* inhabiting overhangs of rocky shores of Arraial do Cabo, state of Rio de Janeiro. Succession and competition experiments suggested the species is expanding its percent cover rapidly. Despite high diversity of benthic organisms, no other species seems to be competitively effective to preclude *T. coccinea* expansion

recorded in the southeastern and southern coasts in natural rocky shores and artificial structures, and there are also some records in the northeastern coast associated with oil platforms. The suggested introduction vector is most likely fouling as the area of the first detection is a large oil exploration basin and Ilha Grande Bay is near an oil ship port. The fact that these organisms have short-lived larvae (between 3 and 14 days until final recruiting) (Reyes-Bonilla et al. 1997) does not favor ballast water as a very successful long-distance transport mechanism.

The literature describes *Tubastraea* as an asymbiotic, azooxanthellate coral, having no requirements for sunlight and consequently occurring on overhangs or drop-offs (Cairns 1994; Paula and Creed 2005). In Ilha Grande Bay, both species are abundant from very shallow (0.5–1.0 m) to deep parts of the rocky shores, prefer vertical substrata, but utilize all angles (Paula and Creed 2005). Reports from that region have shown the strong invasive potential of both *Tubastraea* species excluding other native benthic organisms, including other coral species (Creed 2006). In Arraial do Cabo, where only *T. coccinea* was recorded, the species is found recruiting only in overhangs (CELF, personal observation) and spreading over to more illuminated substrate by budding. Succession experiments have shown percent cover to increase by an average 20% per year, although three years of consecutive monitoring make evident an asymmetric growth per year (CELF, unpublished data). In spite of a high diversity of benthic organisms (algae, sponges, bryozoans, corals, hydrozoans, ascidians among others) on natural substrate where *Tubastraea* recruits, no species has apparently shown any competitive ability to preclude the alien expansion in subtidal rocky shores of Arraial do Cabo. Chemical defense is highly evident in the expansion phase of such species (Koh and Sweatman 2000). The expansion process of *T. coccinea* in Arraial do Cabo seems to be at least tenfold faster than that of *Chromonephthea braziliensis*. The former species is comparatively a better competitor, reproducing by either asexual or sexual modes. Its fast expansion along the coast associated with its cryptic habits (recruiting in overhangs) make an eradication program a difficult task. Like *C. braziliensis*, *Tubastraea* species are appreciated by divers, that is, they are not perceived as threats.

27.3.2.3 Crustaceans

Charybdis helleri, a portunid crab native to the Indo-West Pacific region, was first detected in the Brazilian coast in Guanabara Bay, state of Rio de Janeiro, in 1995 (Tavares and Mendonça Junior 1996). This species was probably introduced in one or more sites in the Caribbean in the mid-1980s, coming from the Mediterranean Sea where it was introduced through the Suez Canal. Therefore, it appears that the invasion into Brazilian waters was a result of secondary introduction and/or natural dispersal that followed its arrival to the American Continent. One of the possible vectors is ballast water. Nowadays, it is found from Maranhão (North) to Santa Catarina (South) (Carqueija and Gouvêa 1996; Calado 1996; Mantelatto and Dias 1999; Tavares and Mendonça Junior 2004).

C. helleri has many traits that favors the invasion of new areas: a relatively long larval life (44 days), rapid growth and maturation within about one year, ability to store sperm and produce multiple broods of high fecundity, generalized carnivorous diet, ability to explore different habitats and a cryptic behavior (protection from visual predators) (Dineen et al. 2001). On the southern Brazilian coast, populations with high densities of adults, young and ovigerous females confirm that *C. helleri* is well established (Mantelatto and Dias 1999). Ovigerous females were present throughout most of the year and the estimated size at sexual maturity was 35.0 mm of carapace width (Mantelatto and Garcia 2001). The fecundity estimated in laboratory in a female of 50.0 mm carapace width was of 47,000 zoeal stage I (Tavares and Mendonça Junior 2004).

C. helleri is of commercial interest in Southeast Asia (Lemaitre 1995) but in Brazil it has no value to fishermen. Possible impacts on native communities are still to be quantified. Although in some areas (state of São Paulo) this crab is less abundant than the native ones (Mantelatto and Fransozo 2000), in other areas (state of Bahia) it is already more abundant than the native species, *Callinectes larvatus* (Ordway, 1863) (Carqueija 2000). Furthermore, the ecological consequences of its introduction into reef habitats of the northeastern coast are yet unforeseen. It is also a potential host of the WSSV (White Spot Syndrome Virus) (Tavares and Mendonça Junior 2004). For these reasons, the expansion and relationships of this species with other brachyuran species should be closely monitored in the region (Mantelatto and Garcia 2001). The predation of *C. helleri* by a native *Octopus* species at Northeastern Brazilian reefs was recently documented. Unfortunately, the populations of this octopus species are declining due to overfishing; therefore, it probably cannot serve as a controlling agent (Sampaio and Rosa 2005).

Many species of barnacles are commonly associated with fouling and probably their ranges were expanded through marine-related human activities (Young 1995). Three species of barnacles are considered introduced in the Brazilian coast. The first is *Megabalanus coccopoma* (Darwin, 1854), recorded in the 1970s in Guanabara Bay, state of Rio de Janeiro (Young 1994). Its main dispersal mode is through fouling, including ship hulls and oil platforms (Apolinário 2003). This species dominates the same zone as *M. tintinnabulum* Pilsbry, 1916, a cryptogenic species found in the Brazilian coast since the beginning of the twentieth century (Apolinário 2003).

Amphibalanus reticulatus (Utinomi, 1967) is a recent introduction that is currently expanding its range in the Southwestern Atlantic. In 1990, it was first sighted in the state of Pernambuco; by 1992, it was found in the state of Bahia (Young 1995); and by 1996, in the state of Rio de Janeiro. Nowadays, it is also found in the state of Paraná (Neves 2006). The density found on port structures in Sepetiba Bay, state of Rio de Janeiro, was very high (4410 individuals/m^2). In Ilha Grande Bay, state of Rio de Janeiro, it is already the dominant species on artificial substrates (Mayer-Pinto and Junqueira 2003).

The third barnacle species is *Chirona (Striatobalanus) amaryllis* (Darwin, 1854). It was detected in 1982 in the state of Piauí (Young 1989), in 1990 in the state of Pernambuco (Farrapeira-Assunção 1990), and in 1992 in the state of Bahia

(Young 1995). Recently (2005), it was found in the state of Paraná (Neves 2006). Although this species occurs in the sub-littoral in its native area (Indo-Pacific) it was found in the intertidal area in Brazil (Young 1989).

27.4 Fishes

There are no reports on fish introductions in Brazilian coastal waters related to mariculture developments, in contrast to fresh water systems in the country (Gomiero and Braga 2004). Few records of non-native reef fish species were reported for the southeastern coast (Moura 2000), all of them based on sights of one or two individuals and considered as originated from sporadic ornamental aquarium releasing with no further possibilities of population establishment.

At least one species, *Acanthurus monrovie* Steindachner, 1876, originally only distributed in the Eastern Atlantic, was recently found (three individuals) to reach a coastal island off Santos, state of São Paulo (Luiz-Junior et al. 2004). In this case, however, the invasion mechanism was suggested as a natural dispersion, by means of planktonic larvae crossing the mid-Atlantic ridge barrier through a well known trans-Atlantic route. Actually, this route is said to function in both eastward-westward directions (Luiz-Junior et al. 2004). Recent works have genetically proven that fishes colonized both sides of the Atlantic using these routes (Bowen et al. 2001).

Two fishes that are associated with fouling in ships and oil platforms have recently been indicated as introduced species. One is the tessellated blenny, *Hypsoblennius invemar* Smith-Vaniz & Acero P., 1980, originally from Mexican Gulf to Venezuela including the lesser Antilles (Cervigon 1994). This species was reported to be associated with oil platforms in the Mexican Gulf and was first recorded in the Brazilian coast at oil platforms of the southern region (Hostim-Silva et al. 2002). Since then, the species has been found all along the south and southeastern region inhabiting empty barnacles in shallow hard substrate (Ferreira et al. 2004), the same substrate they utilize on ship and platform hulls. The other species is also a blenny, the muzzled blenny *Omobranchus punctatus* (Valenciennes, 1836), with wide-spread distribution including the Indo-Pacific, Mediterranean and some parts of the Atlantic. Recently, this species was recorded inhabiting mussel (*P. perna*) cultivation stands in the state of Santa Catarina (Gerhardinger et al. 2006) and also further north, in the states of Rio de Janeiro and Bahia (Gerhardinger et al. 2006). Cryptic fishes like blennies and gobies are predisposed to bioinvasion associated with fouling. However, their cryptic behavior makes their detection and identification a hard task. Because they seek refuge and lay eggs in small holes and empty barnacles, they are prone to inhabit artificial hard substrates like ship hulls, pilings in ports, etc. Additionally, their tolerance to salinity variations allows them to take advantage of ballast-intake holes on ship hulls and offshore oilrigs (Wonham et al. 2000). There is no doubt that these fishes are successful in "taking a ride" as fouling and travel great distances.

27.5 Intentional Introductions Related to Marine Farming

Shrimp farming remained rare in the Brazilian coastline until the mid-1980s, when substantial efforts were employed by private and public enterprises to establish a cost-effective strategy, again testing several shrimp species. Target native species including *Farfantepenaeus paulensis* (Pérez Farfante, 1967) were not productive enough and were discarded against exotic species such as *Marsupenaeus japonicus* (Bate, 1888) and *Litopenaeus vannamei* (Boone, 1931). The latter gradually proved to be the most viable species for cultivation because of its high adaptability to the varying hydrochemical conditions of the tropical estuaries of the Northeast. At the present stage, *L. vannamei* is virtually the only marine shrimp species cultivated in Brazil.

Marine shrimp farming is a growing economic activity, currently accounting for 50% of the total exportation of marine products by the country, with an average annual yield of approximately US$ 140 million for 2002–2005. However, such immediate socioeconomic benefits have obscured the environmental impacts associated with the implementation and operation of shrimp farms under non-sustainable practices. These include – but are not limited to – mangrove deforestation for the establishment of cultivation tanks, outflow of highly eutrophic effluents into natural water bodies, and the potential dissemination of shrimp virus diseases to native crustacean populations (but see Boeger et al. 2005). While an intense debate persists over environmental impacts caused by shrimp farming against actual or prospective socioeconomic benefits of the activity, an increasing number of reports on *L. vannamei* occurrence in estuarine and coastal waters of Brazil – often at high densities – has emerged in recent years (Santos and Coelho 2002; Barreto et al. 2000; Barbieri and Melo, in press). To what extent these fugitive specimens of poorly managed shrimp farms will be able to establish self-sustaining populations in the natural environment is still an open question.

A different strategy for the intentional introduction of a marine species occurred in the case of the Indo-Pacific red alga *Kappaphycus alvarezii* (Doty) Doty ex. P. Silva 1996, cultivated for production of linear sulphated polysaccharides, the so-called carrageenins. These are employed in food products as thickening and stabilizing agents. A controlled introduction plan for *K. alvarezii* has been carried out in the southeastern coast of Brazil (Ubatuba region, state of São Paulo) since 1995. The initial Philippine propagules were brought from Japan and kept under quarantine conditions in the laboratory for 10 months before their release in cultivation structures in the field (Paula et al. 1999). Careful monitoring of the potential spread of the alga toward natural habitats continues since the initial introduction, and results obtained have shown that the Ubatuba strands were unable to thrive outside aquaculture facilities (Paula et al. 2002; Oliveira and Paula 2003; Paula and Pereira 2003). However, parallel introductions of the same species have occurred in other parts of Brazil without taking into account the same careful procedures observed in Ubatuba. The outcome is similar to that of *L. vannamei*: records of the exotic species have been reported in the natural marine environment in several regions of Brazil with the most challenging scenarios occurring again in the northeastern coast. Studies about the true establishment, dispersal rates and possible ecological impacts of *K. alvarezii* populations in those areas are still needed.

27.6 Final Remarks

Brazil was one of the six developing countries engaged in the Global Ballast Water Management Program (GloBallast) implemented from 2000 to 2004 by the International Maritime Organization together with the Global Environmental Facility and the United Nations Development Program. Several Brazilian institutions were involved in a National Task Force whose goal was to develop activities to prepare the country for the implementation of the "International Convention for the Control and Management of Ships Ballast Water and Sediments" and, thus, help minimize the transfer of harmful marine organisms (Leal Neto and Jablonski 2004). One of the initiatives of this program, which is relevant in this context, was the pilot study conducted at a demonstration site, the Port of Sepetiba, state of Rio de Janeiro, that led to two concrete outcomes: (1) the development of a port-specific bioinvasion risk assessment that included studies of shipping patterns, an environmental similarity analysis between the receiving port and donor ports and also a risk species analysis (Clarke et al. 2004); and (2) a port survey following standardized protocols (Hewitt and Martin 2001) that made possible the detection of six introduced species.

The Brazilian maritime authority has established a legal instrument, the Regulation Norman 20, to coordinate the control and management of ballast water in Brazilian ports which is in accordance with the IMO Convention. It is expected that, in the near future, the inspection and control of ballast water will be a routine and widespread activity. Nevertheless, we still do not have any legal instrument to prevent/minimize invasions by hull fouling. This vector is no doubt critical to bioinvasion worldwide and deserves special attention due to the future ban of tributyl tin (TBT), an effective biocide in anti-fouling paints. Brazilian marine biodiversity, that is still being revealed, relies on the development of an effective management system of both vectors, ballast water and hull fouling.

In addition to the GloBallast program, since 2001 the Ministry of the Environment of Brazil has funded research initiatives related to bioinvasions, two of which applied to the marine environment: (1) the ALARME project has dealt with the assessment of exotic species and the establishment of a ballast water management plan for the Port of Paranaguá, state of Paraná, one of the busiest ports in southern Brazil; and (2) the INFORME/PROBIO project has provided a comprehensive list of exotic species in coastal areas of Brazil, accompanied by a series of ecological and socioeconomic data on the various species (including their actual or potential impacts). In addition, the project has provided information on prevention and control practices under development in the country to face the problem of marine bioinvasions (MMA 2006). Along the lines discussed by Colautti and MacIsaac (2004), the INFORME/PROBIO approach was to assign exotic species recorded in Brazil to three categories (detected, established and invasive) according to their populational status following initial introduction, and to their ecological, economical or sanitary impact. Apart from probable high number of cryptogenic species, a total of 53 marine and estuarine non-indigenous species has been confirmed by this national assessment, 8 of which are within the "invasive" category (FUNDESPA 2006): *Coscinodiscus wailesii, Alexandrium tamarense* (phytoplankton), *Temora turbinata* (zooplankton), *Caulerpa scalpelliformis* (macroalga), *Charybdis hellerii, Tubastraea coccinea, T. tagusensis and Isognomon bicolor* (zoobenthos). These species are dealt with in the present chapter. Large (>1 cm) benthic animals dominate the INFORME/PROBIO list, with 34 species (64% of total).

Awareness of the impacts imposed by invasive marine species has truly increased in Brazil in the past decade, not only within academia, but also in other sectors such as environmental and public health agencies, the maritime and port authorities and the shipping industry. Although funding agencies have initiated programs geared towards marine bioinvasions, the amount of money allocated does not fulfill the needs that a comprehensive survey of the extensive Brazilian coast requires. Similarly, although the shipping industry has been directly involved in R&D issues and has provided subsidies to test the efficiency of on-route ballast exchange methods (e.g., Villac et al. 2001), initiatives of the private sector are minor. There is no system of return of revenues from shipping activities to support research about bioinvasion issues. Considering that Brazil sustains an exuberant diversity in its coastal systems that provides habitat to both tropical and subtropical components, the experience and data that can be generated by studying its extensive coastline will certainly produce knowledge to be applied elsewhere.

References

Apolinário M (2003) Dinâmica populacional de duas espécies de *Megabalanus* Hoek, 1913 (Crustacea: Balanidae) no litoral do Rio de Janeiro, Brasil. PhD Dissertation, Universidade Federal do Rio de Janeiro, Brazil

Araujo HMP, Montú M (1993) Novo registro de *Temora turbinata* (Dana, 1849) (Copepoda, Crustacea) para águas atlânticas. Nauplius 1:89–90

Barbieri E, Melo G (in press) Ocorrência da espécie exótica *Litopenaeus vannamei* (Boone, 1931) no complexo estuarino-lagunar de Cananéia-Iguape-Ilha Comprida. Biota Neotropical (in press)

Barreto W, Mendonça JT, Barbieri E (2000) Ocorrência e possíveis influências de espécies exóticas na pesca artesanal em uma comunidade banhada pelo Rio Ribeira de Iguape. Proceedings of the V Simpósio de Ecossistemas Brasileiros: Conservação (Brazil), pp 310–317

Boalch GT, Harbour DS (1977) Unusual diatom off the coast of south-west England and its effect on fishing. Nature 269:687–688

Boeger WA, Pie M, Ostrensky A, Azambuja LP (2005) Identity of the putative agent of the lethargic crab disease (LCD), responsible by massive mortalities of the mangrove crab, *Ucides cordatus* (Brachyura, Ocypodidae), in Northeastern Brazil. Mem Inst Oswaldo Cruz 100:161–167

Bollens SM, Cordell JR, Avent S, Hooff R (2002) Zooplankton invasions: a brief review, plus two case studies from the northeast Pacific Ocean. Hydrobiologia 480:87–110

Bowen BW, Bass AL, Rocha LA, Grant WS, Robertson DR (2001) Phylogeography of the Trumpetfishes (*Aulostomus*): ring species complex on a global scale. Evolution 55:1029–1039

Breves-Ramos A (2004) Distribuição, abundância e estrutura populacional de *Isognomon bicolor* (Adams, 1845) no litoral do estado do Rio de Janeiro, Brasil. MSc. Thesis, Universidade Federal do Rio de Janeiro, Brazil

Cairns SD (1994) Scleractinia of the temperate North Pacific. Smith Contr Zool 557:1–150

Calado TCS (1996) Registro de *Charybdis hellerii* (Milne Edwards, 1867) em águas do litoral brasileiro (Decapoda: Portunidae). Bol Estudos Ciências Mar 9:175–180

Carqueija CRG (2000) Situação atual e impactos da introdução da espécie exótica *Charybdis hellerii* (Decapoda, Brachyura, Portunidae) na costa da Bahia. Abstracts of the I Congresso Brasileiro sobre Crustáceos (Brazil)

Carqueija CRG, Gouvêa EP (1996) A ocorrência, na costa brasileira, de um Portunidae (Crustacea, Decapoda), originário do Indo-Pacífico e Mediterrâneo. Nauplius 4:105–112

Castro CB, Pires DO (2001) Brazilian coral reefs: what we already know and what is still missing. Bull Mar Sci 69(2):357–371

Castro Filho BM, Miranda LB (1998) Physical oceanography of the western Atlantic continental shelf located between 4°N and 34°S. In: Robinson AR Brink K (eds) The sea, vol 11. New York, Wiley, pp 209–252

Cervigon F (1994) Los peces marinos de Venezuela, vol III. Editora Ex Libris, Caracas

Clarke C, Hilliard R, Junqueira AOR, Leal Neto AC, Polglaze J, Raaymakers S (2004) Ballast water risk assessment: Port of Sepetiba, Federal Republic of Brazil. GloBallast Monograph Series 14, 63 pp

Creed JC (2006) Two invasive alien azooxanthellate corals, *Tubastraea coccinea* and *Tubastraea tagusensis*, dominate the native zooxanthellate *Mussismilia hispida* in Brazil. Coral Reefs 25:350

Colautti RI, MacIsaac HJ (2004) A neutral terminology to define 'invasive' species. Diversity Distrib 10:135–141

Dineen JF, Clark PF, Hines AH, Reed S, Walton HP (2001) Life history, larval description, and natural history of *Charybdis hellerii* (Decapoda, Portunidae), an invasive crab in the Western Atlantic. J Crustacean Biol 21(3):774–805

Domaneschi O, Martins CM (2002) *Isognomon bicolor* (C.B. Adams) (Bivalvia: Isognomonidae): primeiro registro para o Brasil, redescrição da espécie e considerações sobre a ocorrência e distribuição de *Isognomon* na costa brasileira. Rev Bras Zoo 19(2):611–627

Falcão C, Széchy MTM (2005) Changes in shallow phytobentic assemblages in southeastern Brazil, following the replacement of *Sargassum vulgare* (Phaeophyta) by *Caulerpa scalpelliformis* (Chlorophyta). Bot Mar 48:208–217

Farrapeira-Assunção CM (1990) Ocorrência de *Chirona* (*Striatobalanus*) *amaryllis* Darwin, 1854 e de *Balanus reticulatus* Utinomi, 1967 (Cirripedia, Balanomorpha) no Estado de Pernambuco. Abstracts of the XVII Congresso Brasileiro de Zoologia (Brazil)

Fernandes FL, Zehnder-Alves L, Bassfeld JC (2001) The recently established diatom *Coscinodiscus wailesii* (Coscinodiscales, Bacillariophyta) in Brazilian waters. I: Remarks on morphology and distribution. Phycol Res 49:89–96

Ferreira CEL (2003) Non-indigenous corals at marginal sites. Coral Reefs 22:498

Ferreira CEL, Gonçalves JEA, Coutinho R (2004) Cascos de navios e plataformas como vetores na introdução de espécies exóticas. In: Silva JSV, Souza RCCL (org) Água de lastro e bioinvasão. Interciência, Rio de Janeiro, pp 273–274

FUNDESPA (2006) Informe sobre as espécies exóticas invasoras no Brasil: Organismos que afetam o ambiente marinho. Technical Report to the Brazilian Ministry of the Environment, PROBIO Program, 406 pp

Gerhardinger LC, Freitas MO, Bertoncini AA, Rangel CA (2006) *Omobranchus punctatus* (Teleostei: Blenniidae), an exotic blenny in the Southwestern Atlantic. Biol Invas 8(4):941–946

Gomiero LM, Braga FMS (2004) Cannibalism as the main feeding behaviour of Tucunarés introduced in southeast Brazil. Braz J Biol 64(3B):625–632

Hallegraeff GM, Fraga S (1998) Bloom dynamics of the toxic dinoflagellate *Gymnodinium catenatum*, with emphasys on Tasmanian and Spanish coastal waters. In: Anderson DM, Cembella AD, Hallegraeff GM (eds) Physiological ecology of harmful algal blooms. Springer, Berlin Heidelberg New York, pp 59–80

Hallegraeff GM, Anderson DM, Cembella AD (2003) Manual on harmful marine microalgae. Monographs on Oceanographic Methodologies 11. IOC-Unesco, Paris, 793 pp

Hewitt CL, Martin RB (2001) Revised protocols for baseline port surveys for introduced marine species – survey design, sampling protocols and specimen handling. Tech Rep 22, Center for Research on Introduced Marine Pests – CSIRO Marine Research, Tasmania, 49 pp

Hostim-Silva M, Fontes J, Afonso P, Serpa N, Sazima C, Barreiros JP, Sazima I (2002) Plataformas de petróleo: pontos de encontro de peixes em alto mar. Ciência Hoje 31(183):20–26

Koh EGL, Sweatman H (2000) Chemical warfare among scleractinians: bioactive natural products from *Tubastraea faulkneri* Wells kill larvae of potential competitors. J Exp Mar Biol Ecol 251:141–160

Lages BG, Fleury BG, Ferreira CEL, Pereira RC (2005) Chemical defense of an exotic coral as invasion strategy. J Exp Mar Biol Ecol 328:127–135

Leal Neto ACL, Jablonski S (2004) O Programa GloBallast no Brasil. In: Silva JSV, Souza RCCL (org) Água de lastro e bioinvasão. Interciência, Rio de Janeiro, pp 11–20

Leão ZMAN, Kikuchi RKP, Testa V (2003) Corals and coral reefs of Brazil. In: Cortés J (ed) Latin American coral reefs. Elsevier Science, Amsterdam, pp 9–52

Lemaitre R (1995) *Charybdis hellerii* (Milne Edwards, 1867), a nonindigenous portunid crab (Crustacea: Decapoda : Brachyura) discovered in the Indian River lagoon system of Florida. Proc Biol Soc Washington 108(4):643–648

Lo WT, Chung CL, Shih CT (2004) Seasonal distribution of copepods in Tapong Bay, Southwestern Taiwan. Zool Stud 43:464–474

Lopes RM (1997) Zooplâncton da Baía de Paranaguá: distribuição espacial, variação temporal e interações tróficas. PhD Dissertation, Universidade Federal do Paraná, Brazil

Lopes RM (2004) Bioinvasões aquáticas por organismos zooplanctônicos: uma breve revisão.. In: Silva JSV, Souza RCCL (org) Água de lastro e bioinvasão. Interciência, Rio de Janeiro, pp 113–131

Lopes RM, Brandini FP, Gaeta SA (1999) Distributional patterns of epipelagic copepods off Rio de Janeiro State, Southeastern Brazil. Hydrobiologia 411:161–174

Lopes RM, Katsuragawa M, Dias JF, Montú MA, Muelbert JH, Gorri C, Brandini FP (2006) Zooplankton and ichthyoplankton distribution on the southern Brazilian shelf: an overview. Sci Mar 70(2):189–202

López MS (2003) Efecto de la potencial presa exótica *Isognomon bicolor* (Adams, 1845) sobre la ecología trófica de *Stramonita haemastoma* (Kool, 1987) en el intermareal rocoso de Arraial do Cabo, RJ, Brasil. MSc Thesis, Universidad Internacional de Andalucía, Spain

López MS, Coutinho R (2005) Una nueva presa en el intermareal rocoso de Arraial do Cabo (RJ, Brasil): la invasión del bivalvo *Isognomon bicolor* (C.B.Adams, 1845) y la selección de presas del gasterópodo predador nativo *Stramonita haemastoma* (Kool, 1987). Abstracts of the IX Congreso Latinoamericano de Ciencias del Mar (Chile)

Luiz-Junior OJ, Floeter SR, Gasparini JL, Ferreira CEL, Wirtz P (2004) The occurrence of *Acanthurus monroviae* (Perciformes: Acanthuridae) in the south-western Atlantic, with comments on other eastern Atlantic reef fishes occurring in Brazil. J Fish Biol 65:1173–1179

Magalhães CA (1999) Partilha de recursos em guilda de gastrópodes predadores em costões de São Sebastião, SP. PhD Dissertation, Universidade Estadual de Campinas, Brazil

Mantelatto FLM, Dias LL (1999) Extension of the known distribution of *Charybdis hellerii* (Milne Edwards, 1867) (Decapoda: Portunidae) along the Western Tropical South Atlantic. Crustaceana 72(6):617–620

Mantelatto FLM, Franzoso A (2000) Brachyuran community in Ubatuba Bay, northern coast of São Paulo State, Brazil. J Shellfish Res 19(2):701–709

Mantelatto FLM, Garcia RB (2001) Biological aspects of the nonindigenous portunid crab *Charybdis hellerii* in the western tropical South Atlantic. Bull Mar Sci 68(3):469–477

Mayer-Pinto M, Junqueira AOR (2003) Effects of organic pollution on the initial development of fouling communities in a tropical bay, Brazil. Mar Pollut Bull 46:1495–1503

Medeiros GF, Rocha CEF, Silva ML (1991) A note on the occurrence of *Pseudodiaptomus trihamatus* Wright, 1937 (Crustacea Copepoda) in Natal, Brazil. Bol Dep Oceanogr Limnol Univ Fed Rio Grande do Norte 8:113

Medeiros GF, Medeiros LF, Henriques DMF, Carlos MT, Faustino GVBS, Lopes RM (2006) Current distribution of the exotic copepod *Pseudodiaptomus trihamatus* Wright, 1937 along the northeastern coast of Brazil. Braz J Oceanogr 54(4):241–245

Mitchell GJP, Nassar CAG, Maurat MCS, Falcão C (1990) Tipos de vegetação marinha da Baía do Espírito Santo sob influência da poluição. Espírito Santo, Brasil. Publ Acad Ciênc Estado de São Paulo 71:202–214

MMA (2006) Espécies exóticas invasoras: situação brasileira. Ministério do Meio Ambiente, Secretaria de Biodiversidade e Florestas [available online at http://www.mma.gov.br/invasoras/capa/publicacoes.html]

Moura RL (2000) Non-indigenous reef fishes in the southwestern Atlantic. Abstracts of the IX International Coral Reef Symposium, vol 1

Moyses DN (2005) Influência da heterogeneidade do substrato no recrutamento de invertebrados bentônicos e sucessão ecológica do médiolitoral do costão rochoso da Ilha do Brandão, Angra dos Reis, RJ. MSc Thesis, Universidade Federal do Rio de Janeiro, Brazil

Muxagata E, Gloeden IM (1995) Ocorrência de *Temora turbinata* Dana, 1849 (Crustácea: Copepoda) no estuário da Lagoa dos Patos, RS, Brasil. Náuplius 3:163–164

Neves CS (2006) Bioinvasão mediada por embarcações de recreio na Baía de Paranaguá, Pr e suas implicações para a conservação. MSc Thesis, Universidade Federal do Paraná, Brazil

Ofwegen LP van (2005) A new genus of nephtheid soft corals (Octocorallia: Alcyonacea: Nephtheidae) from the Indo-Pacific. Zool Meded 1:236

Oliveira EC, Paula EJ (2003) Exotic seaweeds: friends or foes? In: Chapman ARO, Anderson RJ, Vreedland VJ, Davison IR (eds) Proceedings of the 17th International Seaweed Symposium. Oxford University Press, Cape Town

Onbé T (1983) Preliminary observations on the biology of a marine cladoceran *Pleopis* ("*Podon*") *schmackeri* (Poppe). J Fac Appl Biol Sci Hiroshima 22:55–64

Paula AF, Creed JC (2004) Two species of the coral *Tubastraea* (Cnidaria, Sclerectinia) in Brazil: a case of accidental introduction. Bull Mar Sci 74(1):175–183

Paula AF, Creed JC (2005) Spatial distribution and abundance of nonindigenous coral genus *Tubastraea* (Cnidaria, Scleractinia) around Ilha Grande, Brazil. Braz J Biol 65(4): 661–673

Paula EJ, Pereira RTL (2003) Factors affecting growth rates of *Kappaphycus alvarezii* (Doty) Doty ex. P. Silva (Rhodophyta, Solieriaceae) in subtropical waters of São Paulo State, Brazil. In: Chapman ARO, Anderson RJ, Vreedland VJ, Davison IR (eds) Proceedings of the 17th International Seaweed Symposium. Oxford University Press, Cape Town

Paula EJ, Pereira RTL, Ohno M (1999) Strain selection in *Kappaphycus alvarezii* var. *alvarezii* (Solieriaceae, Rhodophyta) using tetraspore progeny. J Appl Phycol 11:111–121

Paula EJ, Pereira RTL, Ohno M (2002) Growth rate of the carrageenophyte *Kappaphycus alvarezii* (Rhodophyta, Gigartinales) introduced in subtropical waters of São Paulo State, Brazil. Phycol Res 50:1–9

Persich GR, Kulis DM, Lilly EL, Anderson DM, Garcia VMT (2004) Genetic variability and toxin profile of *Alexandrium tamarense* (Lebour) Balech from Southern Brazil. In: Steidinger KA, Landsberg JH, Tomas CR, Vargo GA (eds) Harmful algae 2002. Florida Fish and Wildlife Conservation Commission, Florida Institute of Oceanography, and Intergovermental Oceanographic Commission of UNESCO, pp 437–439

Proença LAO, Fernandes LF (2004) Introdução de microalgas no ambiente marinho: impactos negativos e fatores controladores. In: Silva JSV, Souza RCCL (org) Água de lastro e bioinvasão. Interciência, Rio de Janeiro, pp 77–97

Proença LAO, Tamanaha, MS, Souza NP (2001) The toxic dinoflagellate *Gymnodinium catenatum* Graham in southern Brazilian waters: occurrence, pigments and toxins. Atlântica 23:59–65

Rapagnã LC (2004) Estudo da estrutura das populações dos bivalves *Isognomon bicolor* (CB Adams, 1845), *Perna perna* (Linnaeus, 1758) e *Pinctada imbricata* (Roding, 1798) nos costões rochosos de Arraial do cabo, Rj, Brasil. MSc Thesis, Universidade Federal Fluminense, Brazil

Reyes-Bonilla H, Pérez-Vivar TL, Ketchum JT (1997) Nuevos registros del coral ahermatípico *Tubastraea coccinea* Lessen, 1829 (Scleractinia: Dendrophyllidae) en el Pacifico de Mexico. Rev Invest Científica UABCS 8:31–34

Rocha CEF (1985) The occurrence of *Pleopis schmackeri* (Poppe) in the southern Atlantic and other marine cladocerans on the Brazilian coast. Crustaceana 49:202–204

Rocha FM (2002) Recrutamento e sucessão de uma comunidade bentônica de mesolitoral dominada pela espécie invasora *Isognomon bicolor* (Bivalvia: Isognomidae) C.B. Adams, 1748 em dois costões rochosos submetidos a diferentes condições de batimento de ondas. MSc Thesis, Universidade Federal do Rio de Janeiro, Brazil

Sampaio CLS, Rosa IL (2005) Predation of an alien species of crab (*Charybdis hellerii* Milne Edwards) by a native *Octopus* species on NE Brazilian reefs. Coral Reefs 25:58

Santos MCF, Coelho PA (2002) Espécies exóticas de camarões peneídeos (*Penaeus monodon* Fabricius, 1798 e *Litopenaeus vannamei* Boone, 1931) nos ambientes estuarino e marinho do nordeste do Brasil. Bol Técnico-Científico CEPENE, Tamandaré, Brazil 10(1):65–77

Seeliger U, Kjerfve B (eds) (2000) Coastal marine ecosystems of Latin America. Springer, Berlin Heidelberg New York

Souza, RSCL, Silva JV (2004) *Água de Lastro e Bioinvasão*, Interciência, Rio de Janeiro, Brazil, 224 pp

Souza RCCL, Fernandes FC, Silva EP (2004) Distribuição atual do mexilhão *Perna perna* no mundo: um caso recente de bioinvasão. In: Silva JSV, Souza RCCL (org) Água de lastro e bioinvasão. Interciência, Rio de Janeiro, pp 157–172

Stramma L, England M (1999) On the water masses and mean circulation of the South Atlantic Ocean. J Geophys Res 104(C9):20863–20883

Tavares M, Mendonça Junior JB (1996) *Charybdis hellerii* (A. Milne Edwards, 1867) (Brachyura: Portunidae), eighth nonindigenous marine decapod recorded from Brazil. Crustacean Res 25:151–157

Tavares M, Mendonça Junior JB (2004) Introdução de Crustáceos Decápodes Exóticos no Brasil: Uma Roleta Ecológica. In: Silva JSV, Souza RCCL (org) Água de lastro e bioinvasão. Interciência, Rio de Janeiro, pp 59–76

Villac MC, Persich G, Fernandes L, Paranhos R, Dias C, Bonecker S, Garcia V, Odebrecht C, Tenenbaum D, Tristão ML, Andrade S, Fadel A (2001) Ballast water exchange: testing the dilution method (Petrobrás, Brazil). In: Hallegraeff GM, Blackburn SI, Bolch C, Lewis RJ (eds) Harmful algal blooms 2000. IOC, Paris, pp 470–473

Whorff JS, Whorff LL, Sweet MH (1995) Spatial variation in an algal turf community with respect to substratum slope and wave height. J Mar Biol Ass U 75:429–444

Wonham MJ, Carlton JT, Ruiz GM, Smith LD (2000) Fish and ships: relating dispersal frequency to success in biological invasions. Mar Biol 136:1111–1121

Young PS (1989). Establishment of an Indo-Pacific barnacle (Cirripedia, Thoracica) in Brazil. Crustaceana 56(2):212–214

Young PS (1994) The Balanoidea (Crustacea: Cirripedia) from the Brazilian coast. Bol Mus Nac Zool 356:1–36

Young PS (1995) New interpretations of South American patterns of barnacle distribution. In: Schram FR, Hoeg JT (eds) New frontiers in barnacle evolution. Crustacean issues. Balkema, Rotterdam, pp 229–253

Chapter 28
Four Centuries of Biological Invasions in Tidal Waters of the Chesapeake Bay Region

Paul W. Fofonoff, Gregory M. Ruiz, Anson H. Hines, Brian D. Steves, and James T. Carlton

28.1 Introduction

Biological invasions are prevalent in marine ecosystems throughout the world. Several studies demonstrate that the number and abundance of non-native species have increased dramatically in recent time (Cohen and Carlton 1998; Cranfield et al. 1998; Reise et al. 1999; Ruiz et al. 2000a; Hewitt et al. 2004). Although the impact of many non-native populations remains unexplored, it is also evident that some species have fundamentally altered the structure and function of marine systems (Ruiz et al. 1999; Crooks 2001; Carlton 2001).

Most marine invasions are known from protected waters of bays and estuaries, instead of exposed outer coasts (Chap. 33, Preisler et al.). This results at least partly from the concentration of human activities surrounding estuaries, creating many transfer mechanisms (vectors) for the human-aided movement of organisms from other global regions. Most of the world's trade occurs by shipping among ports, concentrated in bays and estuaries, creating opportunities for species transfers associated with ships' hulls and ballasted materials (Carlton 1985). In addition, bays are foci for many other activities known to transfer organisms, such as aquaculture, fishing, and outdoor recreation. Estuaries also represent an intersection between marine, freshwater, and terrestrial environments, and potentially can be invaded by organisms from each of these adjacent regions. Although estuaries include a diverse range of habitats and have undergone many anthropogenic changes, both potentially affecting colonization by non-native species, it appears certain that the propagule supply moved among bays is an important driver for the predominance of non-native species in more protected waters.

For North America, analyses and syntheses of marine invasions now exist for several estuaries along the Pacific coast (Carlton 1979; Cohen and Carlton 1995; Cohen et al. 1998, 2001; Wasson et al. 2001; Boyd et al. 2002; Wonham and Carlton 2005). European colonization and modern human activities are relatively recent here, with the major expansion in shipping in the 1800s. In contrast, extensive colonization and shipping to eastern North America began in the 1600s, and

G. Rilov, J.A. Crooks (eds.) *Biological Invasions in Marine Ecosystems.*
Ecological Studies 204,

a comprehensive analysis of invasions for such an Atlantic coast estuary has not been published.

In this chapter we provide an overview of invasion patterns for the Chesapeake Bay, a major estuary on the Atlantic coast and one of the earliest sites of continuous European settlement in North America. Following the first European settlement at the mouth of the Chesapeake in 1608, the region experienced rapid and sustained growth in human population size, shipping, fishing, and agriculture. Today, the Chesapeake remains a major hub of human activity, and the combined ports of Baltimore and Norfolk have the second largest number of ship arrivals in the U.S. (Smith et al. 1999). This long history of modern human activities suggests the Chesapeake Bay region has been exposed to non-native biota delivered by many vectors.

We compiled information on species in the Chesapeake Bay region from a variety of sources, including published literature, "gray literature", Internet datasets, and interviews with scientists. Records were included in the database when a museum specimen was reported, or other evidence was given to verify the identity and occurrence of a species in the study area. We also conducted intensive field surveys of sessile invertebrates in the lower Chesapeake Bay (for description see Ruiz and Hewitt 2002; NEMESIS 2005), providing additional information and several new species records for the region.

With these information sources, we classified species using several categories, which describe their invasion history and distribution in the Chesapeake Bay region, as follows:

- Invasion Status [Introduced, Cryptogenic, Native].
- Population Status [Established, Extinct, Failed, Unknown].
- Residency [Regular Resident, Boundary Resident, Unconfirmed]. Boundary resident species occur commonly in terrestrial or freshwater habitats, and less frequently in tidal or marine waters.
- Native Region [Western Atlantic, Eastern Atlantic, Pacific, Unknown Marine, North America, South America, Eurasia, East Africa, Africa]. Marine species are attributed to ocean basin and others (freshwater/terrestrial) to continental regions.
- Source Region [Categories as described for Native Region].
- Date of First Record – First documented date of sighting, collection or report.
- Vector(s) of Introduction – Plausible mechanism(s) of introduction.

These classifications and detailed histories were entered into our database (the National Estuarine and Marine Exotic Species Information System, NEMESIS) and used for analysis of invasion patterns by taxonomic group, time, transport mechanism (vector), and origin. We provide further description and detailed information in supplemental materials (http://www.serc.si.edu/labs/marine_invasions/publications/supplements.jsp). Additional information for each species is also available on-line at http://invasions.si.edu/nemesis/chesapeake.html.

28.2 Description of the Chesapeake Bay Region

We defined the Chesapeake Bay region as the tidal waters, including tidal wetlands of the Chesapeake Bay and its tributaries, the adjacent Atlantic waters of Virginia and Maryland, and the chain of coastal Atlantic bays north of the mouth of the Chesapeake and up to the Maryland-Delaware border (see Fig. 28.1). The landward boundary of our study area is the monthly-mean high-tide line of shores and wetlands, and the limit of tidal influence in tributaries.

28.2.1 Physical Features

Chesapeake Bay is the largest estuary in the United States, with a total surface area of 11,500 km^2. The watershed includes 163,170 km^2 across six states, with ten major tributary rivers (Fig. 28.1. The estuary is less than 10,000 years old, and the history geological, climatic and ecological change has received considerable attention (Schubel and Pritchard 1987; Brush 2001; Kutzbach and Webb 2001). The Bay's large freshwater inflow (~48% of which comes from the Susquehanna River, entering at the north) and shallow depth result in a gradual salinity gradient and extensive regions of tidal freshwater regions in the Upper Bay and the major tributaries, with large areas grading from oligo- and mesohaline to polyhaline waters in the Lower Bay (Schubel and Pritchard 1987). Fully marine salinities (euhaline, 30–35 PSU) occur only in the adjacent Atlantic waters, and in the coastal bays (from Assawoman and Chincoteague Bays, south to Hog Island Bay) along the Atlantic shoreline (Fig. 28.1). The low elevation of much of the surrounding Coastal Plain means that, even with a relatively small tidal range (0.3–0.9 m) (Schubel and Pritchard 1987), the Bay is surrounded by more than 79,000 hectares of freshwater, brackish, and marine tidal wetlands (Chesapeake Bay Program 2005).

The climate of the Chesapeake Bay region is marked by drastic seasonal changes in temperature, with typical mid-Bay ranges from 0 to 2 °C in winter and ≥30 °C in summer in water, and −5 to 40 °C in air (Schubel and Pritchard 1987). Spatially, the Chesapeake Bay region, which spans ~2° of latitude, has a noticeable North-South climate gradient, with mean air temperatures about 2 °C higher at the mouth of the Bay, and at least 30% fewer days below freezing, compared to the head of the Bay (Kutzbach and Webb 2001). Along the major tributaries, summer temperatures are coolest in tidal freshwater just below the Fall Line (the boundary between the Piedmont and Coastal Plain), and increase moving bay-ward across the low Coastal Plain. The Bay's mouth and the adjacent Atlantic waters, subject to a more marine climate, have a narrower seasonal temperature range, in both summer and winter (Kutzbach and Webb 2001).

The Chesapeake Bay region includes a great diversity of habitats. Among the major habitat types are unstructured sediments (including intertidal mudflats and beaches), oyster beds, freshwater to marine submerged aquatic vegetation (SAV), coarse woody debris, tidal marshes (fresh to salt), and freshwater tidal swamps.

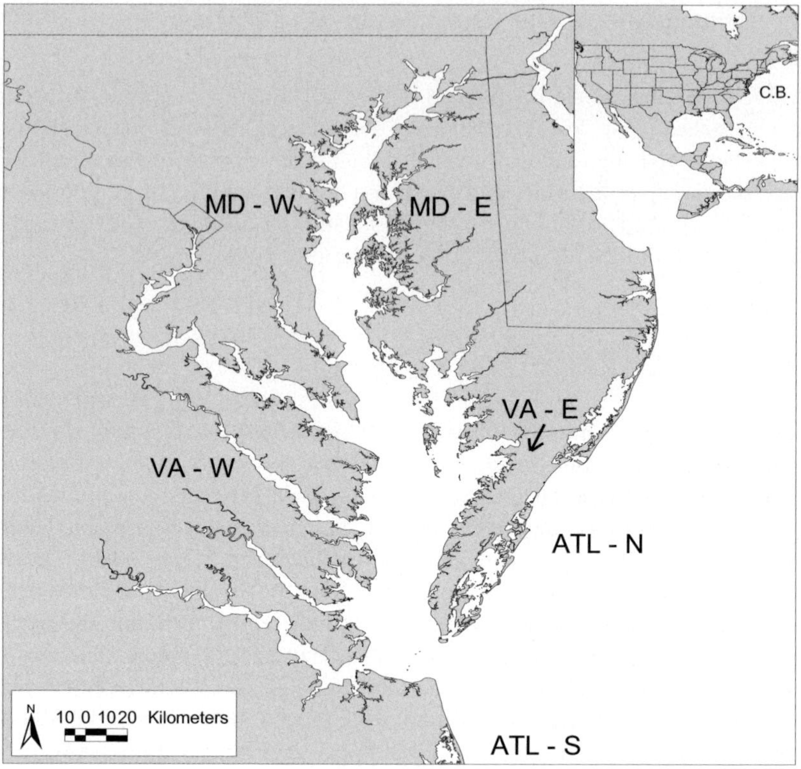

Fig. 28.1 Map of Chesapeake Bay region. The labeled sub-regions are: Maryland-West (MD-W), Maryland-East (MD-E), Virginia-West (VA-W), Virginia-East (VA-E), Atlantic-North (ATL-N) and Atlantic-S (ATL-S). Each region extends inland to the monthly mean high-tideline on shores, and to the head of tide of tributaries

However, natural rocky substrates are confined to small portions of the uppermost tidal fresh portions of tributaries, just below the edge of the Piedmont plateau (the Fall Line) (Jenkins and Burkhead 1993). The major natural hard substrates for attached organisms are logs and bivalve shells, especially oysters, which, until twentieth century over-harvesting, formed massive reefs (Kennedy 1995). As a major center of human activity, a large amount of anthropogenic hard substrate now exists as seawalls, rock rip-rap, docks, and piers.

28.2.2 *History of Biological Studies*

Our knowledge of the occurrence of biological invaders in a particular region, and especially the timing of their arrival, is dependent on the history (especially extent and timing) of biological studies in the region. In the Chesapeake Bay region, botanical collections began in the late seventeenth and early eighteenth centuries (Gronovius 1739; Reveal 1983; Brown et al. 1987), while regional species lists of such economically important animal groups as fishes, mollusks, and decapod crustaceans were not published until the late nineteenth century (fishes – Cope 1869; Uhler and Lugger 1876; mollusks – Dall 1889; decapod crustaceans – Stimpson 1859, 1871; Kingsley 1879).

A few studies of other groups, such as hydroids (Clark 1878, 1882) and polychaetes (Webster 1879), were published in the late nineteenth century, but many invertebrate groups were not extensively collected in the region until the twentieth century. Surveys of macroalgae in the Chesapeake Bay region were first published in the 1960s (Zaneveld 1966; Wulff et al. 1968; Zaneveld and Barnes 1965; Mathieson and Fuller 1969). In the years preceding and following World War I, an extensive biological survey of Chesapeake Bay was carried out (Cowles et al. 1930), and in the 1960s, the Virginia Institute of Marine Sciences compiled lists of the biota of the lower Chesapeake Bay (Wass 1963, 1972). However, the last publication to give extensive species lists of Chesapeake Bay benthic invertebrate fauna was published in 1984, and was based on field work completed in 1978 (Dauer et al. 1984). Similarly, the most recent publication on Chesapeake Bay seaweeds was in 1980 (Orris 1980). The results of more recent monitoring programs are available as “gray literature” and in computer databases, but tracing identifications of species in these records can be difficult (Fofonoff, personal observation). Our knowledge of the invertebrate and algal fauna of the Chesapeake Bay region is thus confined to a narrow temporal window. Many invaders arriving before the late nineteenth and earliest twentieth century have doubtless been overlooked, while it is likely that some recent arrivals have been undiscovered because of the lack of researchers or knowledgeable taxonomists.

28.3 Patterns of Invasion in the Chesapeake Bay Region

28.3.1 Taxonomic Composition and Residency

We have documented a total of 170 species introduced and established in the tidal waters and wetlands of the Chesapeake Bay region. This total includes 121 regular residents and 49 boundary residents (primarily terrestrial or non-tidal freshwater species, occasionally entering tidal wetlands or waters). Eleven of the species, all regular residents of North American origin (six fish, one reptile, two birds, and two vascular plants), are native/cryptogenic in parts of the region (most frequently in the tidal James River, near the southern edge), but are well documented as introduced elsewhere in the Chesapeake. [See supplemental material at http://www.serc.si.edu/labs/marine_invasions/publications/supplements.jsp and NEMESIS 2005, for complete list and species-level information.]

These 170 established species are distributed among 17 different phyla, which we have combined into 4 major groups for comparison. These groups include invertebrates plus algae (7 species of algae, 58 invertebrates), vascular plants (68 species), fishes (27 species), and air-breathing vertebrates (reptiles, birds, mammals, 10 species) (Fig. 28.2). Among the 65 invertebrate species, the three most numerous groups are insects (Hexapoda) (13 species, 20%), mollusks (12 species, 18%), and crustaceans (11 species, 17%) (Fig. 28.3).

The major groups of organisms vary greatly in the proportion of regular and boundary residents. Non-indigenous vascular plants in the Chesapeake Bay region are almost equally divided between regular (33 species) and boundary residents (35 species), whereby 33 of the latter are predominantly terrestrial in habitat preferences (exceptions are *Rorippa nasturtium-aquaticum*, Watercress; *Landoltia punctata* – Dotted Duckweed). By contrast, almost all the invertebrates/algae group (63 of 65 species), are considered regular residents of tidal waters or wetlands. All of the introduced fishes are freshwater species, of which 19 regularly occur in estuarine waters, while 8 are predominantly species of non-tidal freshwater streams, but occasionally are collected in upper reaches of tributaries. We consider six of the air-breathing vertebrates (one turtle, three waterfowl, two mammals) to be regular residents of the estuary, while four terrestrial species (one bird, three mammals) occasionally reside or feed in tidal wetlands.

In addition to the established species, we have recorded at least 36 introduced species as having "unknown" population status, most of which are known from single or scattered records. Most (21) of these species are invertebrate/algal species which are

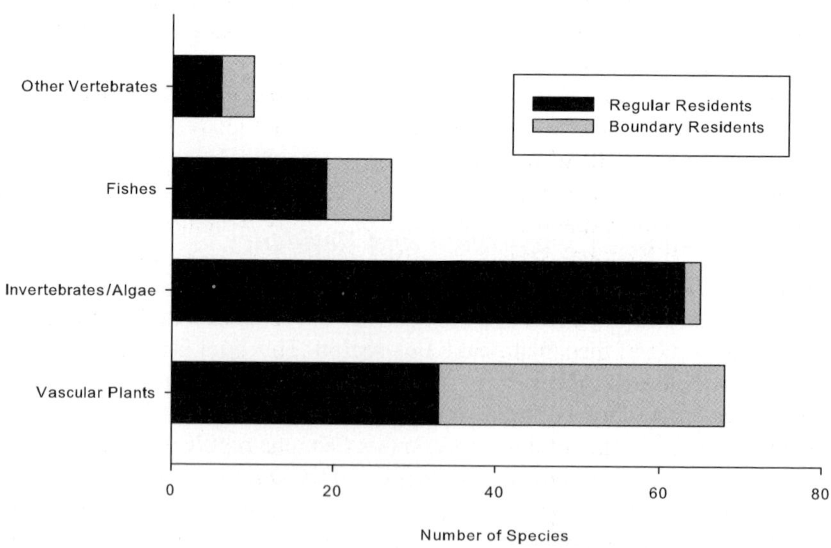

Fig. 28.2 Composition of Chesapeake Bay region non-indigenous species by broad taxonomic categories, showing each regular and boundary residents (n=170 species)

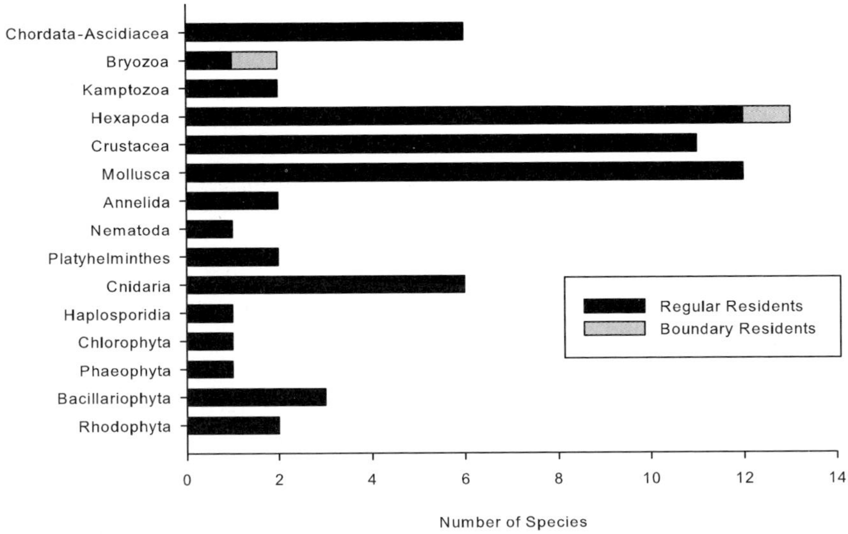

Fig. 28.3 Composition of Chesapeake Bay region non-indigenous invertebrates/algae by phyla, showing each regular and boundary residents (n=65 species)

likely to be overlooked in current sampling programs, because of small size or the scarcity of taxonomic expertise. Seven species, five fishes and the exotic oysters *Crassostrea ariakensis* (Suminoe Oyster, 1998) and *C. gigas* (Pacific Oyster, 1962), were all deliberately introduced and considered to have "unknown" status because of uncertain fertility. Most introductions of the oysters (all documented introductions of *C. ariakensis*) and the fish *Ctenopharyngodon idella* (Grass Carp, 1989) were of sterile triploid individuals, in order to reduce the likelihood of adverse ecological impacts from reproducing populations (NEMESIS 2005). However, reversion of triploids to diploid status, which occurs at a low probability, becomes more likely as the scale and time-span of stocking expands, as does the possibility of human error (Jacobson and Kartalia 1994; National Research Council 2003). Five "species" of fishes were artificially produced hybrids introduced for sport purposes, believed to be sterile or having reduced fertility, also with the intention of reducing unexpected impacts (Christmas et al. 1998). Establishment of some of these "unknown" species is likely to be confirmed in the future.

The establishment of 170 non-indigenous species in the Chesapeake Bay region implies that the flux of introduced species into the region must be many times larger, since most invasions fail (Williamson 1996). For most accidentally introduced invertebrates and algae, failed invasions are difficult to document, given the small size, scarcity, and difficulty of detection and identification of many species (and especially larval or immature forms). We have documented at least 22 failed invasions of tidal waters and wetlands, including freshwater and anadromous fishes (12 species), wetland plants found in piles of dry ballast (4 species), and reptiles (3 species). Most of the fish introductions were attempted as part of state and federal fish acclimatization programs between 1874 and 1916. Seven species of cold-water salmonid fishes were introduced in large numbers (e.g. ~8 million *Oncorhynchus tschawytsha*, Chinook Salmon, 1876–1899; NEMESIS 2005), with no evidence of prolonged survival or reproduction. More recent failed introductions have been single or scattered captures of released pet fishes (two tropical species) and reptiles (three species). Failures of most of the fish and reptile introductions can be attributed to mismatches in climate. In addition, discarded pets are usually released as single individuals or in small numbers, making reproduction unlikely.

28.3.2 Changing Patterns of Invasion Over Time

Reports of introduced species in tidal waters and wetlands of the Chesapeake Bay region have varied over time, and the number of newly discovered species has increased sharply in the last 50 years. Taxonomic composition, native and source regions, and vectors of introduction of introduced species have all exhibited strong shifts through time, as outlined below.

28.3.2.1 Changing Taxonomic Composition and Residency Status

Until the late nineteenth century, vascular plants were the predominant group of introduced organisms first reported in tidal waters and wetlands of the Chesapeake Bay region, comprising 79% (33 of 42) of species reported before 1880 (Fig. 28.4). The majority of these early introduced plants (22 of 33 species) were terrestrial boundary resident species, primarily weedy species (e.g. *Rumex crispus* – Curly Dock; *Chenopodium ambrosioides* – Mexican Tea; *Echinochloa crusgalli* – Barnyard Grass) (Fig. 28.5). Between 1880 and 1955, more typically aquatic and regular resident species outnumbered the boundary residents, comprising 63% (18 of 29) species introduced during that period. Significant introductions during this period included *Lythrum salicaria* (Purple loosestrife), *Trapa natans* (Water Chestnut, first record 1923), *Myriophyllum spicatum* (Eurasian Watermilfoil, 1942), and the invasive form of *Phragmites australis* (1881). After the 1930s, the numbers of newly reported plant species in tidal waters decreased sharply, with only six new introductions reported after 1955 (Fig. 28.4). The most prominent recent introduction has been the submerged plant *Hydrilla verticillata* (Hydrilla, 1982) (NEMESIS 2005). The decline in reported vascular plant invasions has been somewhat puzzling, since possible vectors for plant transport, including shipping, agriculture, and water-gardening are still active (see Sect. 28.3.2.4)

The first documented introduction of a fish to tidal waters in the Chesapeake Bay region occurred by 1864, when *Micropterus dolomieu* (Smallmouth Bass),

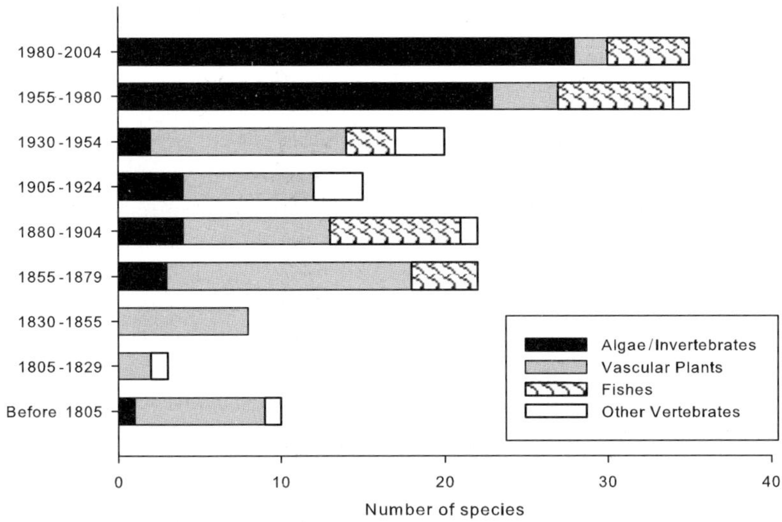

Fig. 28.4 Changes in taxonomic composition of newly reported non-indigenous species in the Chesapeake Bay region over time, by 25-year periods, using dates of first record (n=170 species)

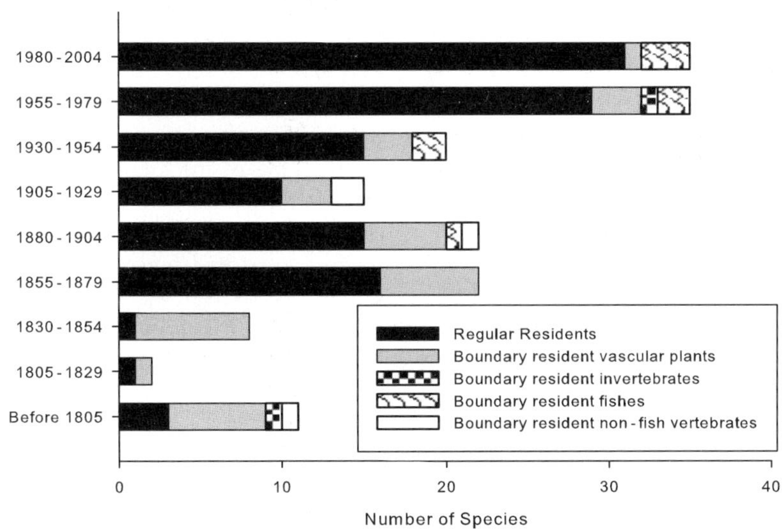

Fig. 28.5 Changes in residency composition of newly reported non-indigenous species in the Chesapeake Bay region over time, by 25-year periods, using dates of first record (n=170 species)

released in the Potomac near Harper's Ferry, West Virginia, in 1854, reached the tidal river. Subsequently, from 1869 to 1900, 11 additional species of freshwater fishes were newly reported from Chesapeake Bay tidal waters (Uhler and Lugger 1876; Smith and Bean 1898; Smith 1907; Jenkins and Burkhead 1993). Prominent species include *Cyprinus carpio* (Common Carp, 1882), *Micropterus salmoides* (Largemouth Bass, 1869) *Ictalurus punctatus* (Channel Catfish, 1889), and *Lepomis macrochirus* (Bluegill Sunfish, 1900). A second wave of fish introductions occurred in the post-World War II period, when nine more species became established in Chesapeake Bay tidal waters, from 1949 to 1979, including *Ictalurus furcatus* (Blue Catfish, 1974), *Pylodictis olivaris* (Flathead Catfish, 1965) and *Dorosoma petenense* (Threadfin Shad, 1953) (Jenkins and Burkhead 1993). The rate of fish introductions appears to have decreased somewhat since 1980, but five new species have been reported since then (Fig. 28.4). The best-known recent fish invader is *Channa argus* (Northern Snakehead), adults and juveniles of which were discovered in the tidal fresh Potomac in 2004 (Orrell and Weigt 2005; NEMESIS 2005). Fluctuations in the number of fish invasions have been driven largely by private and government interest in deliberate stocking for fisheries purposes (see Sect. 28.3.2.4).

Air-breathing vertebrates have been introduced to the region sporadically, beginning with *Rattus norvegicus* (Norway Rat), introduced around 1775, which we consider a regular resident, because of its frequent use of aquatic habitats (Paradiso 1969). Prominent regular resident invaders include *Trachemys scripta* (Red-Eared Slider Turtle, 1941), *Myocastor coypus* (Nutria, 1943), *Cygnus olor* (Mute Swan, 1962), and breeding populations of two waterfowl species (*Anas platyrhynchos* – Mallard Duck, 1913; *Branta canadensis maxima/moffati*, resident Canada Geese, 1935) (NEMESIS 2005).

The apparent dramatic increase in invasions in the Chesapeake Bay region in the last 50 years is due to the discovery of 44 species of invertebrates and 7 species of algae since 1955 (Fig. 28.4). This represents 78% of the total number (65) of non-indigenous invertebrates and algae known from tidal waters and wetlands. Among the invaders reported early are the boundary resident insect *Stomoxys calcitrans* (Stable Fly, before 1800; *Stomoxys calcitrans* breeds in washed-up vegetation in strandlines on shores and in marshes, as well as barnyard manure (Simmons and Dove 1941; Bickley and Seek 1975), *Carcinus maenas* (Green Crab, 1874), *Cordylophora caspia,* (Freshwater Hydroid, 1877) and *Teredo navalis* (Naval Shipworm, 1878). In the last 50 years, many ecologically or economically significant invertebrate/algal invaders have been reported as established, including *Haplosporidium nelsoni* (MSX disease of oyster, 1958), *Rangia cuneata* (Gulf Wedge Clam, 1960), *Corbicula fluminea* (Asian Freshwater Clam), *Codium fragile* ssp. *tomentosoides* (Green Fleece), *Hemigrapsus sanguineus* (Asian Shore Crab, 1994), *Anguillicola crassus* (Eel Swimbladder Nematode, 1997), and *Rapana venosa* (Veined Rapa Whelk, 1998) (Ruiz et al. 2000a; NEMESIS 2005). The observed increase in invertebrate invasions appears to be due largely to shipping (see Sect. 28.3.2.4).

28.3.2.2 Changing native regions

As shown in Fig. 28.6, we have documented 44 non-indigenous species of marine origin (7 algae, 37 invertebrates, all regular residents) and 126 species of continental origin (freshwater to largely terrestrial). The latter category includes 21 invertebrates, and all of the species of vascular plants, fishes, and other vertebrates, in the Chesapeake Bay region.

No marine fishes or truly marine vascular plants (e.g. mangroves, seagrasses, *Spartina* spp.; excluding salt-tolerant species of the upper intertidal, here considered continental) have been introduced to the Chesapeake Bay region. One marine fish, *Pterois volitans*, Red Lionfish, an Indo-Pacific native, is established in waters south of Chesapeake Bay, and can be expected to occur in Atlantic coastal waters in the Chesapeake Bay region (Whitfield et al. 2002).

The total number of continental invaders to the Chesapeake is increased by our decision to include in our totals 49 boundary resident species, 40 of which are predominantly terrestrial, but nonetheless frequently occur in tidal wetlands. However, the majority (65%, 82 of 126) of our aquatic, regular resident species, are also of continental, freshwater origin. The importance of continental invaders, in part, reflects the extensive areas of tidal fresh and oligohaline waters in the region, as well as the Bay's huge watershed, which can collect a large number of introduced species and concentrate them in tidal waters.

The biogeographical origins of newly reported introduced species have changed over time, with the first marine species being reported in the region in 1874

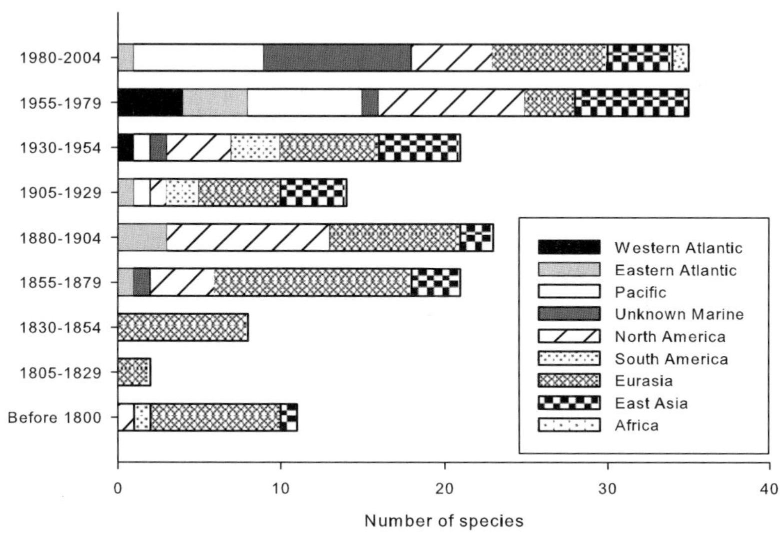

Fig. 28.6 Changes in native regions of non-indigenous species in the Chesapeake Bay region over time, by 25-year periods, using dates of first record (n=170 species). Marine regions are indicated by *solid colors*, and continental regions by *patterns*

(*Carcinus maenas,* Green Crab), with the numbers and proportion of newly discovered marine invaders (relative to continental species) increasing sharply in the second half of the twentieth century, from 22% in 1855–1954 (10 of 79 species) to 49% in 1955–2004 (34 of 70 species). Among marine species themselves, there also is an apparent shift in native regions, with Eastern Atlantic species making up 50% (5 of 10) marine species in 1855–1954, but only 15% of the marine forms in 1955–2004 (5 of 34 species). In the latter period, species of Pacific (50%, 17 of 34 species) and "unknown-marine" (29%, 11 of 34 species) origin comprised the bulk of newly reported marine invertebrates and algae (Fig. 28.6), representing a complementary shift through time. These shifts in the native regions of marine invaders likely reflect the globalization of trade, combining increases in inter-oceanic shipping and delivery of species of Pacific and cosmopolitan species of "unknown-marine" origin.

Six of the 10 species of Eastern Atlantic origin are species that occur on the open coast of the Northeast Atlantic (*C. maenas* – a crab; *Littorina littorea* – a periwinkle; *Striaria attenuata* – a brown alga; *Mysosotella mysotis* – a pulmonate snail; *Anisolabis maritima* – an earwig; *Gyrodactylus anguillae* – an eel gill trematode). The other four Eastern Atlantic species are all hydrozoan (Cnidaria) with Ponto-Caspian affinities (*Cordylophora caspia*; *Blackfordia virginica*; *Maeotias marginata*; *Moerisia lyonsi*), which have been collected in brackish waters of the Bay and its tributaries (NEMESIS 2005).

All but two of the 17 Pacific species are native to the Northwest or Indo-Western Pacific. Two diatoms (*Coscinodiscus wailesii*, *Thalassiosira punctigera*), introduced to the Chesapeake Bay region have broad amphi-Pacific ranges (NEMESIS 2005). Ten species, including many prominent invaders in the Chesapeake Bay region, are of northwest Pacific origin, native to the coasts of Japan, Korea, and China. Examples include: *Diadumene lineata* (Striped Sea Anemone, 1928), *Haplosporidium nelsoni* (MSX disease of oyster, 1958), *Codium fragile* ssp. *tomentosoides* (Green Fleece), *Hemigrapsus sanguineus* (Asian Shore Crab, 1994), *Anguillicola crassus* (Eel Swimbladder Nematode, 1997), and *Rapana venosa* (Veined Rapa Whelk, 1998) (Ruiz et al. 2000a; NEMESIS 2005). The five species with Indo-West Pacific ranges including the Indian Ocean and tropical West Pacific include several common-to-abundant species in the Chesapeake Bay region – *Ligia exotica* (Sea Roach, 1924), *Odontella sinensis* (a diatom, 1961), and *Loxosomatoides laevis* (an entoproct, 1994) (Wasson et al. 2000; NEMESIS 2005).

The 10 introduced species of "unknown marine" origin now have cosmopolitan ranges, but they have been so widely dispersed by shipping and other vectors that their original native regions are a source of speculation. Examples include *Teredo navalis* (Naval Shipworm, 1878); *Garveia franciscana* (Rope Grass Hydroid, 1946); and *Ficopomatus enigmaticus* (a serpulid tubeworm, 1994) (NEMESIS 2005).

The five Western Atlantic species, considered introduced to the Chesapeake Bay region, are all native to the North American coast south of Cape Hatteras, but are presumed to have been transported northward by human activities, and were discovered between 1953 and 1966. These species were: *Cyrenoida floridana* (Florida

Marsh Clam, 1953); *Stramonita haemastoma* (Southern Oyster Drill, 1955); *Rangia cuneata* (Gulf Wedge Clam, 1960); *Loxothylacus panopei* (mud crab parasitic barnacle, 1964); and *Ecteinascidia turbinata* (Mangrove Tunicate, 1966) (NEMESIS 2005).

Among continental invaders, the native regions of species have also shifted, with Eurasian species dominating newly reported invaders before 1855 (86%, 18 of 21 species), but decreasing in successive periods, to 39% in 1855–1954 (31 of 79 species) and to 14% in 1955–2004 (10 of 70 species). Most Eurasian invaders are vascular plants (71%, 42 of 59 species), so that the generally decreasing dominance of newly discovered Eurasian species primarily reflects the trends in plant invasions. However, seven insects associated with Eurasian wetland plants (*Typha angustifolia* – Narrowleaf Cattail, 1806; *Lythrum salicaria* – Purple Loosetrife, 1896; *Phragmites australis* – Common Reed, 1881, invasive genotype) have been released or discovered in the Chesapeake Bay region since 1955, contributing to an increase in Eurasian invaders in the last 25 years (Fig. 28.6). The insects associated with *Typha* and *Phragmites* could have been introduced with the plants and discovered long after, while the *Lythrum* herbivores were deliberately introduced for biocontrol (NEMESIS 2005).

Continental introductions of North American species have also fluctuated greatly over time, peaking in 1880–1904 and in 1955–1980 (Fig. 28.6). This largely reflects the temporal pattern of fish introductions, which comprise 67% of the species of North American origin. East Asian species constituted only 6% (1 of 18) of introductions before 1855, but 18% (14 of 79) in 1880–1954 and 16% (11 of 70) in 1955–2004. Most of these (61%, 16 of 26 species) were vascular plants, but East Asian invertebrates (5 species), fishes (3 species) and mammals (2 species) have also been introduced (Fig. 28.6). The increasing number of East Asian species, many of them ornamental, also likely reflects the globalization of trade and growing Asian economy.

28.3.2.3 Changing Source Regions

Species can spread from previously invaded regions, which serve as "stepping stones" for secondary introduction. Source regions may correspond more closely to patterns of transport than native regions. Since the Chesapeake Bay region was an early center of European settlement, most (81%, 31 of 42 species) of its early introductions (before 1880), primarily vascular plants, are presumed to have come from, or by way of Europe, including four East Asian and one South American native. However, in later periods, the relative importance of Europe as a source region declined, to 34% (19 of 58 species) in 1880–1954 and 13% (9 of 70 species) in 1955–2005 (Fig. 28.7). As continental trade developed within North America, the spread of introduced species along the coasts and within the continent also increased, and as regional natural history collecting developed, became easier to document. Thus, from the late nineteenth century, many Eurasian species, as well as those from other continents introduced to other parts of North America, can be documented as spreading secondarily, by anthropogenic or natural means, into the Chesapeake Bay region. This is manifest by the total number of species with a North American source region (Fig. 28.7) exceeding the number of North American natives (Fig. 28.6). In 1880–1954, 15 North American continental natives were recorded, but at least 9 additional species, 4 native to Eurasia, 3 from East Asia, 2 from South America, spread into the Chesapeake Bay region from other parts of North America. In 1955–2004, in addition to 14 North American natives, 9 species (1 Eurasian; 7 East Asian; 1 African) dispersed into the region.

Examples of species with a well-documented spread through North America are: *Sturnus vulgaris* (Common Starling, North America-NA 1890; Chesapeake Bay-CB 1896; Kessel 1953); *Bithynia tentaculata* (Faucet Snail, NA 1871, CB 1927; Mills et al. 1993, 1997; NEMESIS 2005); (East Asian) *Corbicula fluminea* (Asian Freshwater Clam. NA 1924, CB 1971; Counts 1986); *Cipangopaludina chinensis* (Chinese Mystery Snail, NA 1892, CB 1960; Jokinen 1982); *Murdannia keisak* (Asian Dewflower, NA 1927, CB 1939; Dunn and Sharitz 1990).We have probably underestimated the number of invaders spreading from elsewhere on the continent, since we have not included species which were cultivated in the Chesapeake Bay region, and could have dispersed either from local garden or captive populations or else from wild North American populations.

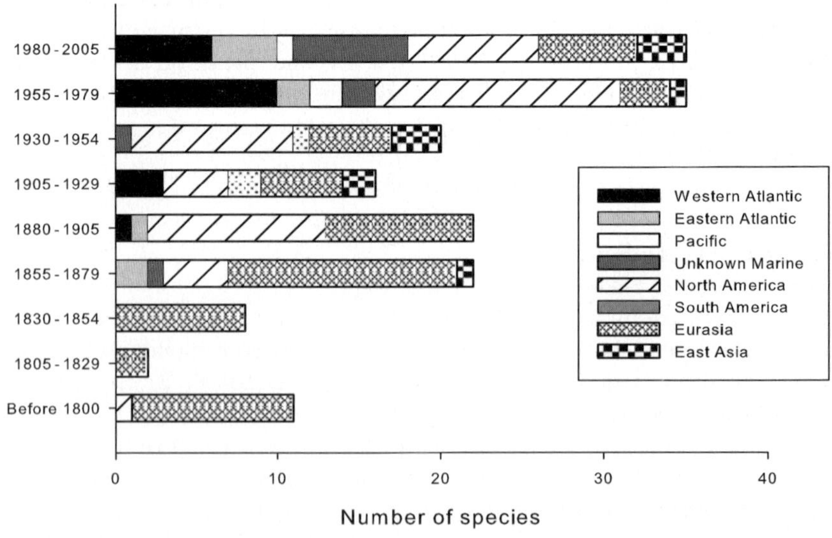

Fig. 28.7 Changes in probable population source regions of non-indigenous species in the Chesapeake Bay region over time, by 25-year periods, using dates of first record (n=170 species). The source region is the area from which the individuals invading the region resided prior to their introduction. Species can be either native or introduced to their source region. Marine regions are indicated by *solid colors* and continental regions by *patterns*

Although few marine invaders were documented for Chesapeake prior to 1900, a similar mismatch occurs overall between native and source regions, indicating secondary spread to Chesapeake from previous sites of invasion. While only 5 Western Atlantic species are documented as invaders to the Chesapeake Bay region, the Western Atlantic is a source region for 12 additional species, 4 species native to the Eastern Atlantic and 8 Pacific natives, which invaded the Atlantic coast to the north or south, and then spread, either by anthropogenic, or natural means, into the Chesapeake Bay region. Similarly, the Eastern Atlantic has been a source region for three Pacific species, as well as six Eastern Atlantic natives introduced to Chesapeake Bay.

Because of its early settlement and long history of trade, Chesapeake Bay could be expected to be a center for invasions into the rest of North America. Thirty species (2 algae, 9 invertebrates, 1 fish, 1 mammal and 17 vascular plants) had their first North American records in the region. However, the two algal species (planktonic diatoms *Coscinodiscus wailesii*, and *Odontella sinensis*, both first reported in 1961) and most of the invertebrates (e.g. the entoproct *Loxosomatoides laevis*, 1994; the hydrozoan *Moersia lyonsi* 1965), are small, obscure organisms which could have easily been introduced elsewhere, but overlooked. Six of the vascular plant species were collected in the seventeenth and eighteenth century botanical surveys, among the first made in North America (Gronovius 1739; Reveal 1983; Brown et al. 1987). The early history and spread of these plants, mostly widespread agricultural weeds and boundary residents of tidal wetlands (e.g. *Plantago major*, Common Plantain, 1739), is obscure, and most of these species were probably introduced independently at many of the sites of early European agriculture. The importance of Chesapeake Bay as an invasions center is thus difficult to assess, given historical gaps in biological knowledge. However, several recent and ecologically important invasions had their first North American records in Chesapeake Bay, including *Rapana venosa* (Veined Rapa Whelk, 1998), *Channa argus* (Northern Snakehead, 2003), *Myriophyllum spicatum* (Eurasian Watermilfoil, 1942), *and Typha angustifolia* (Narrow-leaved Cattail, 1806) (NEMESIS 2005).

Only four Chesapeake Bay invaders are apparently confined to the Chesapeake Bay region, based on reported records. Three are obscure invertebrate species (*Ilyocryptus agilis,* cladoceran, 1974; *Gitanopsis* sp., amphipod, 1994; *Loxosomatoides laevis*, entoproct, 1994) which could be overlooked in other locations. *Rapana venosa* (Veined Rapa Whelk) is a large marine gastropod, which so far has only been collected in Chesapeake Bay, but is expected to greatly extend its range on the Atlantic Coast (Mann and Harding 2000).

28.3.2.4 Changing Vectors of Transport

For 108 species introduced to the Chesapeake Bay region, we assigned a single broad category for the vector of introduction (e.g., shipping, fisheries, ornamental escape, agriculture, etc.), whereas two or more (multiple) vectors seemed plausible for the other 62 species (Fig. 28.8). Of those species attributed to a sole vector,

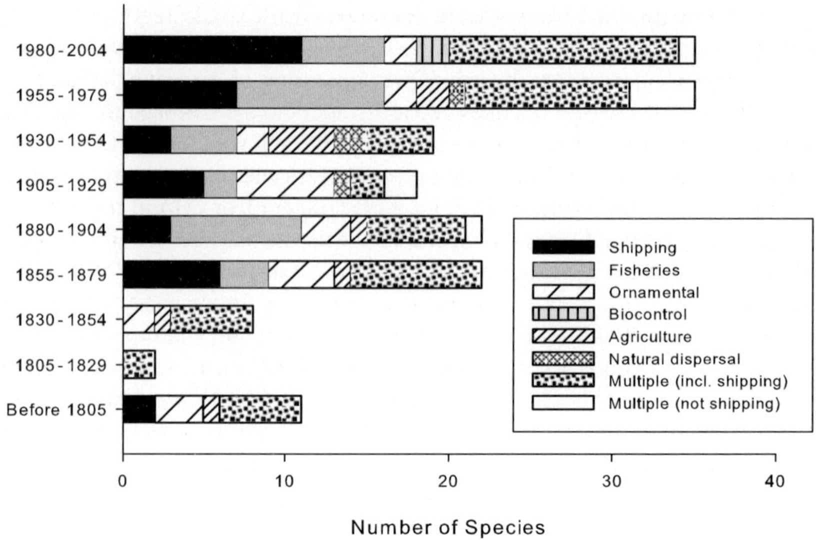

Fig. 28.8 Changes in vectors for newly detected non-indigenous species into the Chesapeake Bay region over time, by 25-year periods, using dates of first record (n=170 species)

shipping was the most frequent (37 species), followed by fisheries (31), and ornamental activities (24). Agriculture (14 species), biological control (2) and natural dispersal (4) dispersing secondarily from other invaded North American or Northwest Atlantic regions) were less important as sole vectors.

When considering species assigned to the multiple vector category, the potential importance of shipping is further increased. For such polyvectic species (see Cohen 1997), shipping was considered a possible vector for 56 species, compared to 27 for ornamental activities, 29 for agriculture, 13 for fisheries, and 16 for natural dispersal (Fig. 28.8).

28.3.2.4.1 Shipping

The number of species attributed solely to shipping introductions has fluctuated through time, with the highest number detected in the last two 25-year time intervals (Fig. 28.8). However, the number of newly recorded polyvectic species for which shipping is one of several possible vectors has risen from 12 in 1880–1954 (0.16/year, averaged) to 24 (0.48/year) in 1955–2004 (Fig. 28.8). Over this time, the mechanisms of shipping transport have changed drastically, as dry ballast has been replaced by ballast water, while fouling and other forms of transport have been affected by the increased speed and size of modern ships, and by the use of metal hulls and antifouling paint (Carlton 1985; Ruiz and Carlton 2003; Chap. 6, Hewitt et al.).

The earliest reported marine species introduced to the Chesapeake Bay region are the hydroid *Cordylophora caspia* (1877), the shipworm *Teredo navalis* (1878), and the Green Crab *Carcinus maenas* (1874). *Carcinus maenas* may have been collected in the region before 1817, as it was included on a list of Crustacea of the United States by Thomas Say, based on his collections on the Atlantic coasts of New Jersey, Maryland and Florida, but the location was given only as 'bays and inlets of the United States' (Say 1817). The first two species were probably introduced by fouling, while *C. maenas* could have been introduced either by fouling or solid ballast. The Chesapeake Bay region's earliest possible ballast-water introduction was the Black Sea hydromedusan *Blackfordia virginica,* first collected and described from the region in 1904.

A more detailed analysis of the shipping vector further indicates a large increase in the role of ballast water and hull fouling in the past 50 years, and this is driven primarily by an increase of invertebrates and algae for this time period (Fofonoff et al. 2003; see also Fig. 28.4).

28.3.2.4.2 Fisheries/Hunting

Introductions of aquatic species, either intentional introductions for improvement of fisheries and hunting, or accidental ones arising from fisheries/hunting activities, have been important in the Chesapeake Bay region since the late 1870s (Fig. 28.8). Probable fisheries introductions have been dominated by fishes (81%, 25 of 31 species), all of freshwater origin. The most frequent mode of introduction was direct stocking by federal or state agencies (15 species, all freshwater fishes). In the late nineteenth and early twentieth centuries, a federal "fish acclimatization" program, centered in Washington DC (Smith 1907), was responsible for many fish introductions to the Chesapeake Bay region. However, some species (e.g., *Micropterus dolomieu*, Smallmouth Bass, 1853 in upper Potomac, reaching the Bay by 1863) (NEMESIS 2005) were first introduced deliberately or accidentally by private individuals, but later stocked by government agencies. Major modes of accidental introductions include (1) escape from hatcheries or other holding facilities -*Pylodictis olivaris* and *Myocastor coypus* (Flathead Catfish and Nutria) (2) introduction with transported oysters (*Haplosporidium nelsoni* and *Loxothylacus panopei* (MSX disease and mud crab parasitic barnacle), both transported with planted oysters), and (3) introduction with discarded bait (*Orconectes virilis* and *Etheostoma zonale*, Virile Crayfish and Banded Darter) (NEMESIS 2005). Fisheries activities were a possible vector for at least 13 additional species, including 6 marine species for which transport with oysters was possible.

The frequency of fisheries introductions (Fig. 28.8) largely corresponds to federal and state interest in stocking of non-native fishes, peaking in the late nineteenth century and in the post World War II period. The last major intentional governmental introductions of fertile non-native fishes to the Chesapeake Bay watershed were those of *Ictalurus furcatus* – Blue Catfish, in 1974 and

Micropterus punctulatus – Spotted Bass in 1976, in Virginia (Jenkins and Burkhead 1993). Concerns about impacts of stocked predatory fishes have been one factor limiting recent state-government introductions of new game fishes (Christmas et al. 1998). In the last 25 years, accidental introductions of fishes and other organisms with discarded bait, contaminated hatchery stock, and illegal introductions by private individuals appear to be major sources of new introductions.

28.3.2.4.3 Ornamental

The cultivation of plants and animals for ornamental purposes has been responsible for the introduction of at least 24 aquatic/wetland species (21 of them vascular plants, escaped from terrestrial and water gardens) to the Chesapeake Bay region. Other ornamental introductions include released pets such as *Carassius auratus* – Goldfish, *Trachemys scripta* – Slider Turtle, and *Cygnus olor* – Mute Swan. Ornamental activities have been one of several possible vectors for 27 species, including 11 plants, 15 freshwater invertebrates, and 1 fish. Most of the invertebrates could have been transported accidentally in shipments of aquatic plants, but two freshwater snails (*Cipangopaludina chinensis*; *Viviparus georgianus*) were sold as aquarium pets and scavengers. Species, for which ornamental activities were the probable sole vector, were most numerous in 1905–1929, but temporal trends are not clear, given the large number of species for which ornamental activities are one of several multiple vectors (Fig. 28.8). The rearing of ornamental plants and fishes is the focus of several major commercial operations in the Chesapeake watershed, valued at ~$3 million in Maryland (Maryland State Archives 2006). In addition, the region's growing population suggests the likelihood of further pet and ornamental plant escapes.

28.3.2.4.4 Agriculture

Agriculture activities were considered a sole probable vector for 9 species (8 plants and one mammal), but a possible vector for 29 other species (20 plants, 8 invertebrates and 1 mammal). The most frequent type of agricultural invaders were weedy vascular plants, transported with contaminated seed, farm implements, farm animals, etc. ("agricultural weed", 5 probable, 17 species possible). The other major agricultural subvector was the use of plant material (rice straw or European marsh grasses) as packing material, a probable vector for three species and a possible mechanism for at least nine other species (two plants, seven invertebrates). Most (55%, 21 of 38 species) probable/possible agricultural introductions were terrestrial boundary resident species, invading primarily the upper edges of tidal wetlands. Agriculture was a probable or possible vector for 42% (18 of 43) species first reported before 1880, but only 20% (11 of 54) of species reported in 1880–1954, and 11% (8 of 70) in 1955–2004.

28.3.2.4.5 Biocontrol

Two species of beetles (*Galerucella calmariensis; G. pusilla*) were first introduced to wetlands in the Chesapeake Bay region in 1992 for the control of the invasive plant *Lythrum salicaria* (Purple Loosestrife).

28.3.2.4.6 Natural Dispersal

Four species were considered to have well-documented natural dispersal into the Chesapeake Bay region from other invaded parts of North America: *Littorina littorea* – Common Periwinkle; *Bassia hirsuta* – Hairy Seablite; *Murdannia keisak* – Asian Dewflower; *Sturnus vulgaris* – Common Starling). Natural dispersal was considered a possible vector for least 19 other species. Modes of natural dispersal include ocean currents (e.g. larvae of *L. littorina*, seeds of *B. hirsuta*), bird dispersal (*M. keisak*), and flight (e.g. *S. vulgaris*).

28.4 Conclusions

Non-indigenous species are a conspicuous component of the Chesapeake Bay biota in terms of species richness, abundance, and function. We know of 170 non-indigenous species with established, self-sustaining populations, and this must be viewed as a minimum estimate. Some of the established populations are relatively large and are known to have significant impacts as predators, competitors, pathogens, and physical structure (Carter and Rybicki 1994; Phelps 1994; Burreson et al. 2000; see also review by Ruiz et al. 1999). Although the direct and indirect impacts of most non-indigenous species in the Chesapeake remain unexplored, it is evident that invasions play a significant role in the ecology of Chesapeake Bay.

The number of newly detected invasions exhibits a strong increase in the last 50 years, climbing from a rate of 15–22 species per 25-year interval (1855–1954) to 35 species in each of the last 25-year intervals (1955–2004). This increase is driven by a sharp rise in the number of invertebrate and algal species reported, even as the reported number of plant invasions (previously the dominant component) has declined (Fig. 28.4).

We urge some caution in interpreting these temporal patterns of invasion. Although these are indeed the patterns from reported invasions, there are inherent biases in the data. As we have discussed earlier (Ruiz et al. 2000a), these records derive from historical sampling efforts that are unevenly distributed among time intervals, taxonomic groups, and habitats. The available data are essentially by-catch from a broad mix of prior studies, instead of a routine monitoring program designed to rigorously evaluate changes in species composition and abundance. Importantly, sampling effort was sparse in the first few centuries and episodic through time for many taxonomic groups (see Sect. 28.2.2

– History of Biological Studies), placing obvious constraints on the detection of new invasions in particular intervals and possibly inflating estimates of the overall rate increase.

More broadly, a lagtime in detection of new invaders may result from sampling effort operating in combination with population dynamics and species-level attributes (Crooks and Soulé 1999; Crooks 2005). Given a fixed level of sampling effort (field surveys), the likelihood of detecting a species will depend upon its abundance and the observer's ability to recognize it as unique from native (or previously described) residents. Clearly, if an organism occurs in very low abundance in only one very small area, the likelihood of detection is relatively low compared to an organism that is common over a large area. Likewise, a non-indigenous species that is small in body size or not easily identified may avoid detection, and this may explain the relative paucity of microorganisms among marine invasions (Ruiz et al. 2000a, b). At the present time, it remains a significant challenge to predict the population dynamics of invasions (Carlton 1996; Kolar and Lodge 2002; Drake 2004), making estimates of actual date of colonization uncertain.

In Chesapeake Bay, these issues of detection are illustrated by our recent surveys of the sessile invertebrate community. Using substrate deployed as passive collectors in the lower Chesapeake Bay, we have detected 15 non-indigenous species since 1994 that were previously undescribed for the bay (NEMESIS 2005), representing a significant fraction of the 35 species newly reported in the past 25-year interval. Although many of these species appear to be recent arrivals, surveys of the Chesapeake's fouling community have been very limited in the past few decades (Calder 1971; Wass 1972; Thompson 1993; Wasson et al. 2000), creating uncertainty about the actual date of colonization.

Despite the lack of precision, we have considerable confidence that the overall rate of invasions by marine invertebrates and algae have increased in the Chesapeake in the past 50 years. Many of the newly detected species are conspicuous such that they are unlikely to avoid detection for long (e.g., the whelk *Rapana venosa*, the rhizocephalan barnacle *Loxothylacus harrisii*, the serpulid polychaete *Ficopomatus enigmaticus*, the clam *Rangia cuneata*, the tunicate *Styela plicata*), or they have well documented patterns of spread (e.g., the shorecrab *Hemigrapsus sanguineus*, the clam *Corbicula fluminea*, the alga *Codum fragile*) (see NEMESIS 2005 for details). In previous 25-year intervals, the number of newly reported invertebrates and algae never exceeded 4 species (Fig. 28.4). Thus, given that the number of conspicuous or well-documented arrivals exceeds this number, we surmise a recent increase in invasion rate has indeed occurred.

Our analysis suggests that the shipping vector contributes strongly to the observed increase in newly detected invasions. Chesapeake Bay is one of the largest port systems in the U.S., both in terms of number of ship arrivals and ballast water discharge (Carlton et al. 1995; Smith et al. 1999). For 1991, the Chesapeake received an estimated 12 million metric tons of ballast water from foreign arrivals, the second largest in the country, and Smith et al. (1999) have characterized the diverse taxa present in this ballast. The number and size of ships arriving to the Chesapeake has certainly increased greatly over the past century, likely resulting in an increasing transfer of organisms in ships' ballast water and outer surfaces (e.g., hull, rudder, propeller, etc.) to the region, but the magnitude of this change has not been quantified.

While shipping appears to be delivering an increasing number of marine species to the Chesapeake Bay region, invasions are also continuing in low salinity (including freshwater) and terrestrial habitats of the watershed. At least 67 aquatic and wetland species (18 plants, 26 freshwater invertebrates; 23 fishes) have been successfully introduced into the Chesapeake Bay watershed, but have not yet reached tidal waters (Fofonoff, unpublished data). Some of these species are unlikely to colonize the estuary because of habitat preferences, but others are probable future invaders. Examples include *Dreissena polymorpha* (Zebra Mussel), which was discovered to be established in the headwaters of the Susquehanna River in 2001, *Scardinius erythrophthalmus* (Rudd, a minnow), first collected in the watershed in 1991, and also established in the Susquehanna headwaters (NEMESIS 2005), and *Marsilea mutica* (Water-Clover, an aquatic fern), collected in 2001 near tidal wetlands near the city of Chesapeake, Virginia (Knepper et al. 2002). Vectors such as the cultivation of ornamental animals and plants, transfer of organisms with trailered boats, bait, and fishing gear, and release of live food organisms continue to be active in the region.

Changes in local conditions of the Chesapeake may also play a role in the observed invasion patterns, interacting with propagule supply. As an urbanized estuary with a large and growing human population in the surrounding watershed, the bay has been subjected to many changes in hydrology, eutrophication, sediment loading, fishing pressure, and habitat alteration (Brush et al. 2001; Kennedy and Mountford 2001). Major declines have occurred in the area occupied by submerged aquatic vegetation and native oyster reefs, the abundance of commercial shellfish and finfish, and the frequency of hypoxia events (Davison et al. 1997; Dauer et al. 2000; Paul 2001; Wennersten 2001). These changes represent major disturbance agents that may operate alone or in combination to affect susceptibility to invasion (Elton 1959; Cohen and Carlton 1995; Occhipinti-Ambrogi and Savini 2003; Jewett et al. 2003). To date, the relationship between these disturbances and invasion susceptibility is not well understood in estuaries (see Ruiz et al. 1999 and references therein).

It is interesting to compare magnitude of invasions in Chesapeake Bay to other marine bays and estuaries along the Pacific coast of North America that have been well studied. Studies exist for several Pacific coast estuaries in the continental U.S., including San Francisco Bay (Carlton 1979; Cohen and Carlton 1995), Elkhorn Slough (Wasson et al. 2001), Coos Bay (Wonham and Carlton 2005; Carlton unpubl. data), Willapa Bay (Cohen et al. 2001), and Puget Sound (Cohen et al. 1998). Four general features stand out:

1. There are more non-indigenous marine species known from San Francisco Bay than Chesapeake Bay and other Pacific coast estuaries. Cohen and Carlton (1995) reported 212 (150 marine, 62 continental) species and several dozen more have been reported in the last ten years. Their analysis focused primarily

on regular residents, of which we documented 121 (44 marine, 77 continental) species in the Chesapeake.

2. There is a higher proportion of non-indigenous plants in the Chesapeake (27%) compared to those reported for Pacific coast estuaries (~10% for San Francisco Bay, Cohen and Carlton 1995; probably fewer for other estuaries, Wonham and Carlton 2005), even when comparing only regular residents. This may be due in part to search effort, or to differing definitions of what should be considered "aquatic plants" (e.g., see recent survey for San Francisco Bay Delta by Light et al. 2005).
3. The number of introduced invertebrates and algae in Chesapeake Bay (65 species, 44 of them marine) are similar to those in west coast estuaries (43–56), with the exception of San Francisco Bay (~160 species). The numbers of non-indigenous species for these taxa have been estimated in all of many Pacific coast estuaries, providing some basis for these comparisons (but see discussion below).
4. There exists considerable overlap in introduced species between the Chesapeake and Pacific coast estuaries. For example, Chesapeake Bay and San Francisco Bay have 43 introduced regular resident species in common (8 vascular plants, 1 alga, 16 marine invertebrates, 5 freshwater invertebrates, 12 fishes, and 1 reptile). Many of these shared species have been introduced to other west coast estuaries, as well as many coastal regions around the world (Ruiz et al. 2000a). Moreover, this does not include the species that are native to the Chesapeake but introduced to San Francisco Bay and other Pacific coast estuaries (see references above).

We might expect to see far more non-indigenous species in Chesapeake Bay than Pacific coast estuaries, which did not experience rapid population growth and modern human activities until the nineteenth century, lagging roughly two centuries behind that in the Chesapeake Bay and Atlantic coast more broadly. The unusually high number of marine invasions in San Francisco Bay is attributable partly to a massive influx of Atlantic and Pacific oysters into this bay, transporting large numbers of associated species (Cohen and Carlton 1995; Miller et al. 2007). Some of these oyster-mediated introductions spread to other Pacific coast estuaries. Oysters were transported regionally to Chesapeake Bay but not across ocean basins or continents, limiting such oyster-mediating introductions relative to San Francisco Bay.

Nonetheless, it is still surprising that the extent of reported invasions in the Chesapeake is not greater than that of Pacific coast estuaries. Not only was there a relatively long duration of modern human activities (i.e., transport mechanisms) to this estuary, but the strength of shipping and ballast delivery to the Chesapeake has been relatively high, far exceeding that for San Francisco Bay and the other estuaries (Carlton et al. 1995; Smith et al. 1999; Ruiz et al. 2000a).

Several different mechanisms, operating alone or in concert, may explain why Chesapeake does not appear to be more heavily invaded than Pacific coast estuaries. These fall into three general categories, associated with regional differences in the historical record, trade patterns (source regions), and recipient regions.

It is clear that the historical baseline knowledge of biota for Chesapeake Bay and Atlantic coast estuaries was poor, developing long after European colonization, and many early invasions may have gone undetected. There is a high degree of species overlap between the eastern and western North Atlantic coasts. An initial survey of 780 marine species from Chesapeake Bay found that 34% also occurred in Europe (Fofonoff, unpublished data). Roughly 5% of these species are considered non-indigenous to the Chesapeake, but the invasion status of most has not been evaluated. There is also some overlap in wetland flora and freshwater biota, with many species usually considered "Holarctic", but with unexplored or disputed native/introduced status. Many of these taxa were first recorded in the Chesapeake decades to centuries after extensive European trade became established and may have been transported prior to early species inventories to the region.

In contrast, biotic inventories for Pacific coast estuaries did not lag far behind the rapid increase in human population and transport mechanisms. Intensive human activity in San Francisco Bay and Pacific coast estuaries commenced in the mid-nineteenth century and major species inventories commenced within 50–60 years, compared to a lag-time of centuries for the Chesapeake. As a result, many more invaders may in fact be undetected as such in the Chesapeake that Pacific coast estuaries. To our knowledge, a formal comparison of the proportion of cryptogenic species between Atlantic and Pacific coast estuaries has not yet been conducted.

Potential differences in propagule supply may explain some observed invasion patterns among estuaries, and these have not been adequately evaluated to date. It is interesting that the recent level of propagule supply to San Francisco Bay does not appear greater than that for the Chesapeake Bay. Historically, the shipment of oysters resulted in a large flux of species to San Francisco Bay that did not occur in similar fashion in the Chesapeake, but this activity ceased by the mid-twentieth century. In recent times, the number of ship arrivals and amount of ballast water discharged to the Chesapeake exceed that to San Francisco Bay (Carlton et al. 1995; Smith et al. 1999; see also http://invasions.si.edu/nbic/). It seems likely that propagule supply from ships, a leading vector in both bays, parallels this pattern, suggesting that quality of propagules may be much more important.

There are many conspicuous differences in the trade patterns between the Chesapeake and Pacific coast estuaries that may have affected the source and quality of propagules. For example, most recent overseas shipping traffic to the Chesapeake arrives from the northeastern Atlantic, whereas that to the Pacific coast arrives from the northwestern Pacific (Carlton et al. 1995). This results in a different species pool arriving to the two coasts in ships' ballast materials and on hulls. The effect of these different trade patterns and source regions on species richness, or physiological condition of propagules, delivered to the respective coasts has not been examined to date but may explain considerable variation in observed invasion patterns.

Recipient regions also differ dramatically in environmental and biotic conditions that can affect colonization. The continental climate in Chesapeake regions clearly differs from the Mediterranean climate of San Francisco Bay and the Pacific coast estuaries, and many differences exist in the biotic composition and disturbance regimes (Chapman 2000). Although there are likely considerable differences in susceptibility to invasion between coasts (see Ruiz et al. 2000a and references therein), which also interact with different species assemblages being introduced to each coast from the respective source regions, the magnitude and direction of any such differences in susceptibility remain to be measured.

Our study provides the first comprehensive analysis of non-indigenous species for tidal waters of the Chesapeake Bay region, but there is still clearly much to learn about the number, abundance, and effects of non-indigenous species in the Chesapeake, as well as the mechanisms that underlie the patterns described in our analyses. While advances in these areas require both descriptive and experimental research in the Chesapeake Bay system, comparative analyses among estuaries are also especially critical to explain observed spatial and temporal variation in invasions. Only by measuring responses to different vectors, trade patterns, and source/recipient environments can we gain a robust understanding of invasion ecology and better guide management and policy in this area.

Acknowledgements We thank Whitman Miller, Richard Everett, Jeff Crooks, and Chad Hewitt for discussion of Chesapeake invasion history. Tami Huber, Kristin Larson, Linda McCann, Natasha Hitchcock Gray, and Esther Collinetti assisted in field surveys for Chesapeake Bay and subsequent sample analysis. We benefited greatly from taxonomic analyses and discussions with Gretchen and Charlie Lambert, Judy Winston, LeAnn Henry, Dale Calder, Francis Kerckhof, and Pam Fuller. Sue Cheek assisted with references in NEMESIS. Rob Aguilar assisted with figures. This project was supported by funding from the U.S. Fish and Wildlife Service and Smithsonian Institution. Recent field surveys and analyses were funded by the Maryland and National Sea Grant Programs, U.S. Department of Defense Legacy Program, and Smithsonian Institution.

References

Bickley WE, Seek TR (1975) Insects in four Maryland marshes. Agric Exp Sta Univ Maryl College Park, MD: Misc Publs 870:1–27

Boyd MJ, Mulligan TJ, Shaughnessy FJ (2002) Non-indigenous marine species of Humboldt Bay, California. Calif Dep Fish Game. State Water Resource Control Bd, Natl Fish Wildlife Foundation, Sacramento

Brown ML, Reveal JL, Broome CR, Frick GF (1987) Comments on the vegetation of colonial Maryland. Huntia 7:247–283

Brush GS (2001) Forests along the colonial Chesapeake. In: Curtin PD, Brush GS, Fisher GW (eds) Discovering the Chesapeake: the history of an ecosystem. Johns Hopkins University Press, Baltimore, pp 40–59

Burreson EM, Stokes NA, Friedman CS (2000) Increased virulence in an introduced pathogen: *Haplosporidium nelsoni* (MSX) in the Eastern Oyster *Crassostrea virginica*. J Aquat Anim Health 12:1–8

Calder DR (1971) Hydroids and hydromedusae of southern Chesapeake Bay. Spec Sci Rep Virginia Inst Mar Sci 1:1–125

Carlton JT (1979) History, biogeography, and ecology of the introduced marine and estuarine invertebrates of the Pacific Coast of North America. PhD dissertation, University of California, Davis

Carlton JT (1985) Transoceanic and interoceanic dispersal of coastal marine organisms: the biology of ballast water. Oceanogr Mar Biol Annu Rev 23:313–371

Carlton JT (1996) Patterns, process, and prediction in marine invasion ecology. Biol Conserv 78:97–106
Carlton JT (2001) Introduced species in coastal waters: environmental impacts and management priorities. Pew Oceans Commission, Arlington
Carlton JT, Reid D, van Leeuwen, H (1995) The role of shipping in the introduction of nonindigenous aquatic organisms to the coastal waters of the United States (other than the Great Lakes) and an analysis of control options. Report to US Coast Guard, Washington DC
Carter V, Rybicki, NB (1994) Invasions and declines of submersed macrophytes in the tidal Potomac River and estuary, the Currituck Sound-Back Bay system, and the Pamlico River estuary. Lake Reserv Manag 10:39–48
Chapman JW (2000) Climate effects on the geography of nonindigenous peracaridan crustacean introductions in estuaries. In: Pederson J (ed) Marine bioinvasions, Proceeding of a conference, January 24–27. MIT Sea Grant College Program, pp 66–80
Chesapeake Bay Program (2000) Chesapeake Bay: introduction to an ecosystem- Wetlands. Retrieved from http://www.chesapeakebay.net/info/ecoint5a.cfm on 30 January 2006
Christmas J, Eades R, Cincotta D, Shiels A, Miller R, Siemien J, Sinnot T, Fuller P (1998) History, management, and status of introduced fishes in the Chesapeake Bay basin. In: Therres GD (ed) Conservation of biological diversity: a key to restoration of the Chesapeake Bay and beyond. Maryland Department of Natural Resources, Annapolis
Clark SF (1878) A new locality for *Cordylophora*. Am Nat 12:231–234
Clarke SF (1882) New and interesting hydroids from Chesapeake Bay. Mem Boston Soc Nat Hist 3:135–141
Cohen AN (1997) Have claw, will travel. Aquat Nuisance Species Digest 2:16–17
Cohen AN, Carlton JT (1995) Nonindigenous aquatic species in a United States estuary: a case study of the biological invasions of the San Francisco Bay and Delta. U.S. Fish and Wildlife Service and National Sea Grant College Program (Connecticut Sea Grant). Washington DC, Silver Spring
Cohen AN, Carlton JT (1998) Accelerating invasion rate in a highly invaded estuary. Science 279:555–558
Cohen AN, Mills CE, Berry H, Wonham MJ, Bingham B, Bookheim B, Carlton J, Chapman J, Cordell J, Harris L, Klinger T, Kohn A, Lambert C, Lambert G, Li K, Secord D, Toft J (1998) Puget Sound expedition: a rapid assessment survey of non-indigenous species in the shallow waters of Puget Sound. Washington State Department of Natural Resources, Olympia, WA
Cohen AN, Berry HD, Mills CE, Milne D, Britton-Simmons K, Wonham MJ, Secord DL, Barkas JA, Bingham B, Bookheim BE, Byers JE, Chapman JW, Cordell JR, Dumbauld B, Fukuyama A, Harris LH, Kohn AJ, Li K, Mumford TF Jr, Radashevsky V, Sewell AT, Welch K (2001) Washington State Exotics Expedition 2000: a rapid survey of exotic species in the shallow waters of Elliott Bay, Totten and Eld Inlets, and Willapa Bay. The Nearshore Habitat Program, Washington State Department of Natural Resources, Olympia, WA
Cope ED (1869) On the distribution of fishes in the Allegheny region of southwestern Virginia. J Acad Nat Sci Phil 6:207–249
Counts CL III (1986) The zoogeography and history of the invasion of the United States by *Corbicula fluminea* (Bivalvia: Corbiculidae). Am Malacol Bull, Spec Ed 2:7–39
Cowles RP (1930) A biological study of the offshore waters of Chesapeake Bay. Fish Bull 46:277–381
Cranfield HJ, Gordon DP, Willan RC, Marshall BA, Battershill CN, Francis MP, Nelson WA, Glasby CJ, Read GB (1998) Adventive marine species in New Zealand. NIWA Tech Rep 34:1–48
Crooks JA (2001) Characterizing ecosystem-level consequences of biological invasions. Oikos 97:153–166
Crooks JA (2005) Lag times and exotic species: the ecology and management of biological invasions in slow motion. Ecoscience 12:316–329
Crooks JA, Soulé ME (1999) Lag times in population explosions of invasive species: causes and implications. In: Sandlund OT, Schei PJ, Viken Å (eds) Invasive species and biodiversity management. Kluwer, Dordrecht. pp 103–125

Dall WH (1889) A preliminary catalogue of the shell-bearing marine mollusks and brachiopods of the south-eastern coast of the United States. Bull US Natl Mus 37:1–221

Dauer DM, Stokes TL, Barker HR, Ewing MR, Sourbreer JW (1984) Macrobenthic communities of the lower Chesapeake Bay. IV. Bay-wide transects and the inner continental shelf. Int Rev Gesamten Hydrobiol 69:1–22

Dauer DM, Weisberg SB, Ranasinghe (2000) Relationships between benthic community condition, water quality, sediment quality, nutrient loads, and land use patterns in Chesapeake Bay. Estuaries 23:80–96

Davison SG, Merwin JG, Capper J, Power G, Shivers FR Jr (1997) Chesapeake waters: four centuries of controversy, concern, and legislation. Tidewater Publishers, Centerville MD

Drake JM (2004) Risk analysis for invasive species and emerging infectious diseases: concepts and applications. Am Midl Nat 153:4–11

Dunn CP, Sharitz R (1990) The history of *Murdannia keisak* (Commelinaceae) in the southeastern United States. Castanea 66:122–129

Elton CS (1959) The ecology of invasions by animals and plants. Methuen, London

Fofonoff PW, Ruiz GM, Steves B, Carlton JT (2003) In ships or on ships? Mechanisms of transfer and invasion for nonnative species to the coasts of North America. In: Ruiz GM, Carlton JT (eds) Invasive species: vector and management strategies. Island Press, Washington, pp 152–182

Gronovius JF (1739) Flora Virginica exhibens plantas quas V. C. Johannes Clayton in Virginia observavit atque collegit. Cornelius Haak, London

Hewitt CL, Campbell ML, Thresher RE, Martin RB, Boyd S, Cohen BF, Currie DR, Gomon MF, Keough MJ, Lewis JA, Lockett MM, Mays N, McArthur MA, O'Hara TD, Poore GC, Ross DJ, Storey MJ, Watson JE, Wilson RS (2004) Introduced and cryptogenic and species in Port Phillip Bay, Victoria, Australia. Mar Biol 144:183–202

Jacobson PT, Kartalia SD (1994) Ecological risk assessment of the effects of grass carp on submerged aquatic vegetation in Chesapeake Bay. In: Hill P, Nelson S (eds) Toward a sustainable coastal watershed: the Chesapeake experiment. Proceedings of a Conference. Chesapeake Research Consortium, Baltimore

Jenkins RE, Burkhead NM (1993) Freshwater fishes of Virginia. American Fisheries Society, Bethesda MD

Jewett EB, Hines AH, Ruiz GM (2003) Epifaunal disturbance by periodic low levels of dissolved oxygen: native vs. invasive species response. Mar Ecol Prog Ser 304:31–44

Jokinen EH (1982) *Cipangopaludina chinensis* (Gastropoda: Viviparidae) in North America, review and update. Nautilus 96:89–95

Kennedy VS (1995) Ecological role of the Eastern Oyster, *Crassostrea virginica* with remarks on disease. J Shellfish Res 15:177–183

Kennedy VS Mountford K (2001) Human influences on aquatic resources in the Chesapeake Bay watershed. In: Curtin PD, Brush GS, Fisher GW (eds) Discovering the Chesapeake: the history of an ecosystem. Johns Hopkins Baltimore, pp 191–219

Kessel B (1953) Distribution and migration of the European starling in North America. Condor 55:49–67

Kingsley JS (1879) On a collection of Crustacea from Virginia, North Carolina, and Florida, with a revision of the genera of Crangonidae and Palaemonidae. Proc Acad Nat Sci Phila 31:383–427

Knepper DA, Johnson DM, Musselman LJ (2002) *Marsilea mutica* in Virginia. Am Fern J 92:243–244

Kolar CS, Lodge DM (2002) Ecological predictions and risk assessment for alien fishes in North America. Science 298:1233–1236

Kutzbach JE, Webb T III (2001) Climate and climate history in the Chesapeake Bay region. In: Curtin PD, Brush GS, Fisher GW (eds) Discovering the Chesapeake: the history of an ecosystem. Johns Hopkins, Baltimore, pp 15–39

Light T, Grosholz E, Moyle P (2005) Delta ecological survey (phase1): nonindigenous aquatic species in the Sacramento-San Joaquin Delta, a literature review. US Fish and Wildlife Service Stockton, CA

Mann R, Harding JM (2000) Invasion of the North American Atlantic Coast by a large predatory Asian mollusc. Biol Invas 2:7–22

Maryland State Archives (2006) Maryland at a glance – aquaculture. Retrieved from http://www.msa.md.gov/msa/mdmanual/01glance/html/aqua.html

Mathieson AC, Fuller SW (1969) A preliminary investigation of the benthonic marine algae of the Chesapeake Bay region. Rhodora 71:524–534

Miller AW, Ruiz GM, Minton MS, Ambrose RF (2007) Differentiating successful and failed molluscan invaders in estuarine ecosystems. Mar Ecol Prog Ser 332:41–51

Mills EL, Leach JH, Carlton JT, Secor CL (1993) Exotic species in the Great Lakes: a history of biotic crises and anthropogenic introductions. J Great Lakes Res 19:1–54

Mills EL, Scheuerell MD, Carlton JT, Strayer D (1997) Biological invasions in the Hudson River: an inventory and historical analysis. New York State Mus Circ 57:1–51

National Research Council (2003) Non-native oysters in the Chesapeake Bay. National Academies Press, Washington DC

NEMESIS (2005) National Estuarine and Marine Exotic Species Information System. Smithsonian Environmental Research Center, Marine Invasions Laboratory. Retrieved from http://invasions.si.edu/nemesis/chesapeake.html in 2005

Occhipinti-Ambrogi A, Savini D (2003) Biological invasions as a component of global change in stressed marine ecosystems. Mar Pollut Bull 46:542–551

Orrell TM, Weigt L (2005) The Northern Snakehead *Channa argus*, a non-indigenous fish in the Potomac River, U.S.A. Proc Biol Soc Wash 118:407–415

Orris PK (1980) A revised species list and commentary on the macroalgae of the Chesapeake Bay in Maryland. Estuaries 3:200–206

Paul RW (2001) Geographical signatures of Middle Atlantic estuaries: historical layers. Estuaries 24:151–166

Paradiso JL (1969) Mammals of Maryland. U.S. Bureau of Sport Fisheries and Wildlife, Washington

Phelps HL (1994) The Asiatic clam (*Corbicula fluminea*) invasion and system-level ecological change in the Potomac. Estuaries 17:614–621

Reise K, Gollasch S, Wolff WJ (1999) Introduced marine species of the North Sea coasts. Helgol Meeresunters 52:219–234

Reveal JL (1983) Significance of pre-1753 botanical explorations in temperate North America on Linnaeus' first edition of *Species Plantarum*. Phytologia 53:1–96

Ruiz GM, Carlton JT (2003) Invasion vectors: a conceptual framework for management. In: Ruiz GM, Carlton JT (eds) Invasive species: vectors and management strategies. Island Press, Washington, pp 459–504

Ruiz GM, Hewitt CL (2002) Toward understanding patterns of coastal marine invasions: a prospectus. In: Leppäkoski E, Gollasch S, Olenin S (eds) Invasive aquatic species of Europe: distribution, impacts and management. Kluwer, Dordrecht

Ruiz GM, Fofonoff P, Hines AH (1999) Non-indigenous species as stressors in estuarine and marine communities: assessing invasion impacts and interactions. Limnol Oceanogr 44:950–972

Ruiz GM, Fofonoff PW, Carlton JT, Wonham MJ, Hines AH (2000a) Invasion of coastal marine communities in North America: apparent patterns, processes, and biases. Annu Rev Ecol Syst 31:481–531

Ruiz GM, Rawlings TK, Dobbs FC, Drake LA, Mullady T, Huq A, Colwell RR (2000b) Global spread of microorganisms by ships. Nature 408:49–50

Say T (1817) An account of the Crustacea of the United States. J Acad Nat Sci Phila 1:57–63, 65–80, 97–101

Schubel JR, Pritchard DW (1987) A brief physical description of the Chesapeake Bay. In: Majumdar SK, Hall LW Jr, Austin, HM (eds) Contaminant problems and management of living Chesapeake Bay resources. Pennsylvania Academy of Sciences, Easton, pp 1–32

Simmons SW, Dove WE (1941) Breeding places of the Stablefly or 'Dog Fly' *Stomoxys calcitrans* (L) in northwestern Florida. J Econ Entomol 34:457–462

Smith HM (1907) Our fish immigrants. Natl Geogr 18:385–400
Smith HM, Bean BA (1898) List of fishes known to inhabit the waters of the District of Columbia and vicinity. Bull US Fish Commiss 18:179–187
Smith LD, Wonham MJ, McCann LD, Ruiz GM, Hines AH, Carlton JT (1999) Invasion pressure to a ballast-flooded estuary and an assessment of inoculant survival. Biol Invas 1:67–87
Stimpson W (1859) Notes on North American Crustacea, in the Museum of the Smithsonian Institution. Ann Lyceum Nat Hist New York 7:49–93
Stimpson W (1871) Notes on North American Crustacea, in the Museum of the Smithsonian Institution, No.III. Ann Lyceum Nat Hist New York 7:92–136
Thompson ML (1993) Dynamics of an oligohaline, macrofaunal, fouling community. MS. Thesis, College of William and Mary, Williamsburg
Uhler PR, Lugger O (1876) List of the fishes of Maryland. In: Report of Commissioners of Fisheries of Maryland, pp 81–208
Wass ML (1963) Check list of the marine invertebrates of Virginia. Spec Sci Rep Virginia Inst Mar Sci 24:1–36
Wass ML (1972) A checklist of the biota of lower Chesapeake Bay. Spec Sci Rep Virginia Inst Mar Sci 65:1–20
Wasson K, Toft J, Von Holle B, Ruiz GM (2000) Detecting invasions of marine organisms: kamptozoan case histories. Biol Invas 2:59–74
Wasson K, Zabin CJ, Bedinger L, Diaz MC, Pearse JS (2001) Ecological invasions of estuaries without international shipping: the importance of intraregional transport. Biol Conserv 102:143–153
Webster HE (1879) Annelida Chaetopoda of the Virginia coast. Trans Albany Inst 9:202–269
Wennersten JR (2001) The Chesapeake: an environmental biography. Maryland Historical Society, Baltimore
Whitfield P, Gardner T, Vives SP, Gilligan MR, Courtenay WR Jr, Ray GC, Hare J (2002) Biological invasion of the Indo-Pacific lionfish *Pterois volitans*, along the Atlantic coast of North America. Mar Ecol Prog Ser 234:289–297
Williamson M (1996) Biological invasions. Chapman and Hall, London
Wonham MJ, Carlton JT (2005) Trends in marine biological invasions at local and regional scales: the Northeast Pacific Ocean as a model system. Biol Invas 7:369–392
Wulff BL, Wulff EM, Robison BH, Lowry JK, Humm HJ (1968) Summer marine algae of the jetty at Ocean City, Maryland. Chesapeake Sci 6:56–60
Zaneveld J (1966) The marine algae of the American coast between Cape May NJ and Cape Hatteras, NC. Bot Mar 9:101–128
Zaneveld J, Barnes WD (1965) Reproductive periodicities of some benthic algae in lower Chesapeake Bay. Chesapeake Sci 6:17–32

Chapter 29
Introduced Aquatic Species of the North Sea Coasts and Adjacent Brackish Waters

Stephan Gollasch, Deniz Haydar, Dan Minchin, Wim J. Wolff, and Karsten Reise

29.1 Introduction

Introduced aquatic species have received more attention in north-western Europe following the summaries from the German North Sea coast (Gollasch 1996; Nehring and Leuchs 1999), Britain (and Ireland) (Eno et al. 1997; Minchin and Eno 2002), Norway (Hopkins 2002) and a more general account for the North Sea (Reise et al. 1999). Since then, several inventories have appeared: for the German coast (Nehring 2005), the Dutch coast (Wolff 2005) and the Danish coast (Jensen and Knudsen, 2005). In this account we review, summarise and update all those previous accounts. We have also included NIS (=non-indigenous introduced species) which were known from the North Sea but most probably are extinct in this area today, and species that have been recorded, but for which we have no proof of self-sustaining populations.

For the purpose of this account:

- The North Sea is defined from a line between Dover and the Belgian border in the south-west to a parallel line from the Shetland Islands to Norway in the north, and also includes the Skagerrak in the east (modified after North Sea Task Force, 1993). The boundary between the North and Baltic Seas, as defined by the Helsinki Commission (www.helcom.fi), is the parallel of the Skaw in the Skagerrak at 57°44.43′N (Fig. 29.1).
- We define marine and brackish-water species as those aquatic species which do not complete their entire life cycle in freshwater (modified after ICES 2005). Marine species are those having their main distribution in salinities higher than 18 psu; brackish-water species have their main distribution in salinities between 1 and 18 psu.
- Introduced species (= non-indigenous, exotic or alien species, NIS) are species transported intentionally or accidentally by a human-mediated vector into habitats outside their native range. Note that secondary introductions may be transported by human-mediated vectors or by natural means (ICES 2005).
- A vector is any living or non-living carrier that transports living organisms intentionally or unintentionally (ICES 2005).

G. Rilov, J.A. Crooks (eds.) *Biological Invasions in Marine Ecosystems.*
Ecological Studies 204,

29.2 Non-indigenous Aquatic Species in the North Sea Region

In total, 167 NIS and cryptogenic species were reported in the North Sea. There appear to be more records from The Netherlands than from other parts (Fig. 29.1) which may be explained by the most intensive shipping (Port of Rotterdam) and aquaculture (Oosterschelde) in the North Sea region (Wolff 2005). The lower number of records for the British North Sea coast is more difficult to explain. With respect to red algae, Maggs and Stegenga (1999) suggest that the prevailing longshore currents from the north are less likely to spread NIS compared to the eastward currents from Norfolk and the Channel which pass the continental shores of the North Sea.

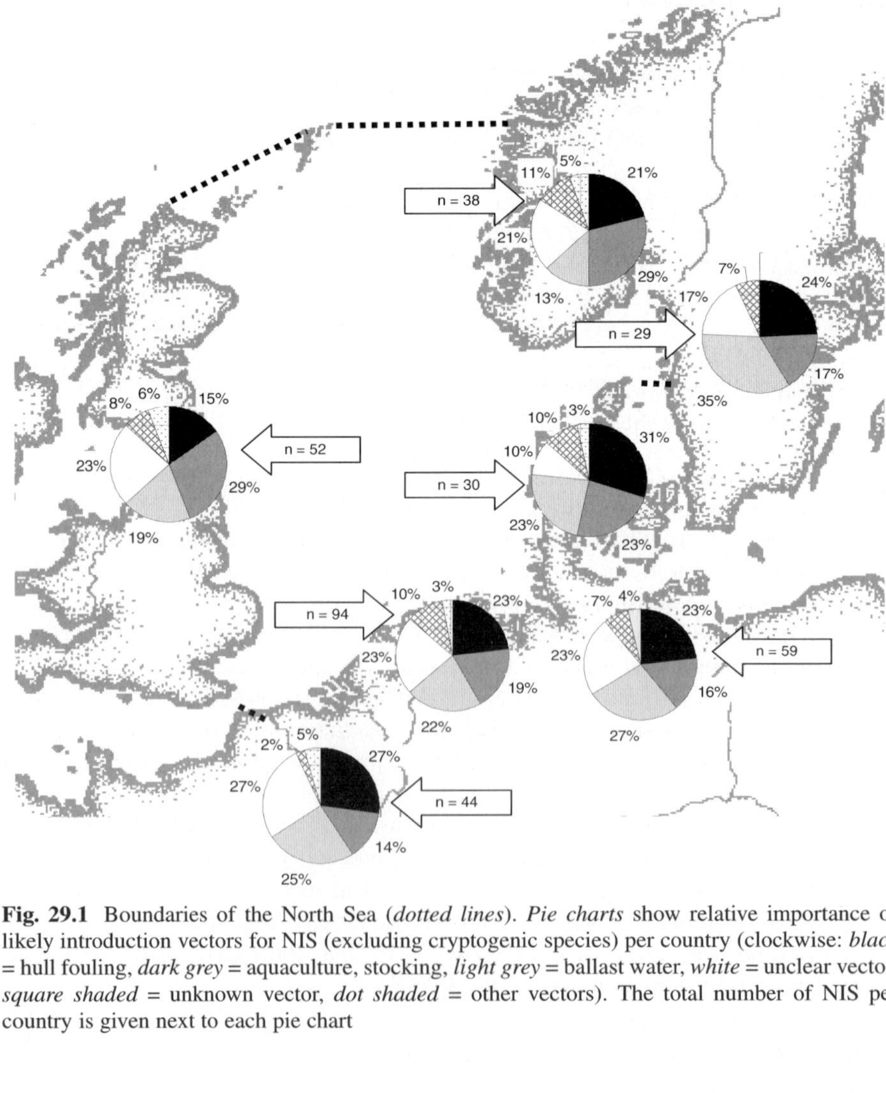

Fig. 29.1 Boundaries of the North Sea (*dotted lines*). *Pie charts* show relative importance of likely introduction vectors for NIS (excluding cryptogenic species) per country (clockwise: *black* = hull fouling, *dark grey* = aquaculture, stocking, *light grey* = ballast water, *white* = unclear vector, *square shaded* = unknown vector, *dot shaded* = other vectors). The total number of NIS per country is given next to each pie chart

The dominant introduction vectors are shipping and intentional introductions for stocking or aquaculture purposes (Table 29.1, Fig. 29.3). The most recently recorded NIS are *Rapana venosa* and *Neogobius melanostomus*, which were both recorded for the first time in the North Sea and adjacent waters in 2005 (Kerckhof et al. 2006; van Beek 2006). Shortly after the first version of this manuscript was submitted a new nonindigenous species of great concern was found in the North Sea (<2006): Mnemiopsis leidyi. This comb jelly was also introduced in other European Seas and contributed to the decline of fisheries.

Most introduced species in the North Sea are benthic species comprised mainly of animal taxa (Table 29.2). More than two thirds of the known NIS have established self-sustaining populations. For others the population status is unknown. For some species there are only single specimen records or occurrences in small numbers and some populations may have been present over varying time periods for which there are no recent records (Fig. 29.2).

The majority of NIS have local distributions (Table 29.3), although 18 taxa were found in six of the seven North Sea countries (i.e. Belgium, Denmark, Germany, The Netherlands, Norway, Sweden and United Kingdom). Many native species are widely distributed on the coasts of North Sea countries and this pattern is generally found with many species that were introduced at an early time and had an ability to become dispersed. Table 29.3 shows that many recent introductions, as well as cryptogenic species, were recorded in one or two North Sea countries, which may indicate a comparatively recent arrival.

Marine taxa formed 136 NIS (81.9%). However, the proportion of marine vs brackish water invaders varied by country, and marine species dominated. Investigations on alien species will have different levels of effort according to the degree of nuisance a species causes, its size, the available taxonomic expertise and diligence of monitoring surveys in each country. There will almost certainly be other species that exist that have as yet not been recognised. The absence of a species in neighbouring countries may reflect some of these issues.

For several species, the invasion vector cannot be easily determined, for example, Pacific oysters may be introduced either as adults attached to ship hulls, as larvae carried in ballast water of ships, with imports of stock for aquaculture purposes, or for direct human consumption but released to the wild. We have selected the most likely vector which in this case we believe to be stock movements of Pacific oysters because the evidence for this is strongest. For species most frequently associated with hull fouling, this form of arrival was assumed to be the responsible vector. For planktonic taxa and microscopic resting stages we have deemed ballast water to be the most likely vector since any such species associated with the hull might be expected to become flushed away during ships journeys at sea (Table 29.1). The human activities near to the site of the first records generally are assumed to be responsible for an arrival. However, such deductions are not always secure and for this reason we have indicated where the likely vector remains unclear (Table 29.1, Fig. 29.3).

In summary, the dominant vectors of introductions are the shipping-associated vectors (i.e. hull fouling of ships and small craft and ships' ballast water and its sediments) and aquaculture including their associated biota (Fig. 29.3: see also Chap. 5, Minchin et al.; Chap. 6, Hewitt et al.).

Table 29.1 Introduced marine and brackish and cryptogenic species (all of which will be established and for these no vector is provided) from the North Sea and records per country. Note added in proof: This table reflects the situation up to march 2006 and only a few remarkable species such as *Mnemiopsis leidyi* and new records of *Urosalpinx cinerea* were added.

Species	Taxonomic group	Country							Status	Habitat	Ballast water	Hull fouling	Aquacult. or stock.	Other	Unknown	Unclear
Coscinodiscus wailesii	Bacillariophyceae	NO	SE		GE	NL	BE	UK	Establ	Marine	x		x			x
Chaetoceros calcitrans f. *pumulis*	Bacillariophyceae							UK	Uncertain	Marine			x			
Odontella sinensis	Bacillariophyceae	NO	SE	DK	GE	NL	BE	UK	Establ	Marine	x					
Pleurosigma planctonicum (= *simonsenii*)	Bacillariophyceae					NL		UK	Establ (extinct in NL)	Marine	x					
Thalassiosira punctigera	Bacillariophyceae	NO	SE		GE	NL	BE	UK	Establ	Marine	x		x			x
Thalassiosira tealata	Bacillariophyceae	NO					BE	UK	Establ	Marine	x		x			x
Alexandrium angustita bulatum	Dinophyceae		SE						Uncertain	Marine	x					
Alexandrium leei	Dinophyceae					NL			Establ	Marine	x		x			x
Alexandrium minutum	Dinophyceae	NO	SE	DK					Crypto	Marine						
Alexandrium tamarense	Dinophyceae	NO	SE			NL		UK	Crypto	Marine						
Dicroerisma psilonereiella	Dinophyceae		SE						Establ	Marine	x					
Gyrodinium corallinum	Dinophyceae		SE						Establ	Marine	x					
Karenia (= *Gymnodinium*) *mikimotoi* (= *aureolum*) (= *Gymnodinium* cf. *nagasakiense*)	Dinophyceae	NO	SE	DK	GE	NL	BE	UK	Establ	Marine	x					
Prorocentrum minimum	Dinophyceae		SE			NL			Crypto	Marine						
Prorocentrum redfieldii	Dinophyceae				GE	NL			Crypto	Marine						
Thecadinium yashimaense (= *mucosum*)	Dinophyceae				GE				Uncertain	Marine	x					
Verrucophora cf. *fascima*	Dinophyceae		SE						Establ	Marine	x					
Chattonella antiqua	Rhaphidophyceae				GE	NL			Establ	Marine	x					
Chattonella marina	Rhaphidophyceae				GE	NL			Establ	Marine	x					
Fibrocapsa japonica	Rhaphidophyceae				GE	NL			Establ	Marine	x					
Heterosigma akashiwo (= *carterae*) (= *Olisthodiscus luteus*)	Rhaphidophyceae	NO				NL		UK	Crypto	Marine						
Verruca (= *Chattonella*) cf. *verruculosa*	Rhaphidophyceae	NO	SE	DK					Crypto	Marine						

Species	Taxonomic group	Country							Status	Habitat	Ballast water	Hull fouling	Aquacult. or stock.	Other	Unknown	Unclear
Isochrysis sp. (Tahitian strain)	Haptophyte							UK	Uncertain	Marine			x			
Codium fragile ssp. *atlanticum*	Chlorophyta	NO						UK	Establ	Marine		x				
Codium fragile ssp. *scandinavicum*	Chlorophyta	NO	SE	DK					Establ	Marine		x				
Codium fragile ssp. *tomentosoides*	Chlorophyta	NO	SE	DK	GE	NL	BE	UK	Establ	Marine		x	x			x
Ulva pertusa	Chlorophyta					NL			Establ	Marine					x	
Acrochaetium densum	Rhodophyta	NO				NL			Crypto	Marine						
Agardhiella subulata	Rhodophyta					NL			Establ	Marine		x				
Aglaothamnion halliae	Rhodophyta	NO	SE		GE				Establ	Marine		x				
Anotrichium furcellatum	Rhodophyta					NL			Extinct	Marine		x				
Antithamnionella spirographidis	Rhodophyta					NL	BE		Establ	Marine		x				
Antithamnionella ternifolia	Rhodophyta				GE	NL	BE		Establ	Marine		x				
Asparagopsis armata	Rhodophyta					NL		UK	Establ	Marine		x				
Bonnemaisonia hamifera	Rhodophyta	NO	SE	DK	GE			UK	Establ	Marine		x				
Colaconema dasyae	Rhodophyta					NL			Establ	Marine		x				
Dasya baillouviana	Rhodophyta	NO	SE	DK	GE	NL			Establ	Marine		x				
Gracilaria gracilis	Rhodophyta	NO							Crypto	Marine						
Gracilaria vermiculophylla	Rhodophyta		SE	DK	GE	NL			Establ	Marine		x				
Grateloupia turuturu (= *doryphora*)	Rhodophyta					NL			Establ	Marine		x	x			x
Heterosiphonia japonica	Rhodophyta	NO	SE			NL		UK	Establ	Marine					x	
Mastocarpus stellatus	Rhodophyta			DK	GE				Crypto	Marine						
Neosiphonia (= *Polysiphonia*) *harveyi*	Rhodophyta	NO	SE	DK	GE	NL	BE	UK	Establ	Marine		x				
Polysiphonia senticulosa	Rhodophyta					NL	BE		Establ	Marine			x			
Porphyra miniata	Rhodophyta			DK					Establ	Marine		x				
Sphaerococcus coronopifolius	Rhodophyta	NO	SE						Crypto	Marine						
Asperococcus scaber	Phaeophyta					NL			Establ	Marine		x	x			x
Colpomenia peregrina	Phaeophyta	NO	SE	DK	GE	NL		UK	Establ	Marine					x	
Elachista sp.	Phaeophyta					NL			Establ	Marine				x		
Fucus evanescens	Phaeophyta	NO	SE	DK					Establ	Marine		x				
Laminaria ochotensis	Phaeophyta				GE				Uncertain	Marine		x				
Leathesia verruculiformis	Phaeophyta					NL			Establ	Marine		x				

(continued)

Table 29.1 (continued)

Species	Taxonomic group	Country							Status	Habitat	Ballast water	Hull fouling	Aquacult. or stock.	Other	Unknown	Unclear
Laminaria ochotensis	Phaeophyta				GE				Uncertain	Marine		x				
Leathesia verruculiformis	Phaeophyta					NL			Establ	Marine		x				
Sargassum muticum	Phaeophyta	NO	SE	DK	GE	NL	BE	UK	Establ	Marine		x	x			x
Scytosiphon dotyi	Phaeophyta							UK	Establ	Marine			x			
Undaria pinnatifida	Phaeophyta					NL	BE		Establ	Marine		x	x			x
Cotula coronopifolia	Anthophyta				GE	NL			Establ	Brackish					x	
Spartina anglica	Anthophyta				GE	NL	BE	UK	Establ	Marine				x		
Spartina x townsendii	Anthophyta			DK	GE	NL	BE	UK	Establ	Marine				x		
Bonamia ostreae	Protozoa					NL			Establ	Marine			x			
Haplosporidium armoricanum	Protozoa					NL			Extinct	Marine			x			
Marteilia refringens	Protozoa					NL			Extinct	Marine			x			
Haliclona (= Acervochalina) loosanoffi	Porifera				GE	NL			Crypto	Marine						
Haliclona cf. *simplex*	Porifera					NL			Crypto	Marine						
Haliclona rosea	Porifera					NL			Crypto	Marine						
Hymeniacidon perlevis	Porifera					NL			Crypto	Marine						
Mycale micracanthoxea	Porifera					NL			Crypto	Marine						
Scypha scaldiensis	Porifera					NL			Crypto	Marine						
Bougainvillia macloviana	Cnidaria				GE				Extinct	Marine		x				
Clavopsella (= Thieliana) navis	Cnidaria				GE	NL			Uncertain	Brackish		x				
Cordylophora caspia	Cnidaria	NO	SE	DK	GE	NL	BE		Crypto	Brackish						
Diadumene cincta	Cnidaria				GE	NL			Crypto	Marine						
Diadumene lineata	Cnidaria				GE	NL	BE	UK	Establ	Marine		x	x			x
Garveia (= Bimeria) franciscana	Cnidaria				GE	NL	BE		Establ	Brackish		x				
Gonionemus vertens	Cnidaria	NO	SE		GE	NL	BE	UK	Establ	Marine	x	x	x			x
Nemopsis bachei	Cnidaria	NO			GE	NL	BE		Establ	Brackish	x	x				x
Ostroumovia inkermanica	Cnidaria					NL			Extinct	Brackish	x	x				x
Rhizogeton nudum	Cnidaria	NO							Crypto	Marine						
Cercaria sensifera	Trematoda							UK	Uncertain	Marine	x		x			
Mnemiopsis leidyi	Ctenophora			DK	GE	NL	BE		Establ	Brackish	x					
Gyrodactylus salaris	Trematoda	NO							Establ	Brackish			x			
Pseudobacciger harengulae	Trematoda		SE						Establ	Marine			x			

Species	Taxonomic group	Country							Status	Habitat	Ballast water	Hull fouling	Aquacult. or stock.	Other	Unknown	Unclear
Pseudodactylogyrus anguillae	Trematoda	NO		DK					Establ	Brackish			x			
Pseudodactylogyrus bini	Trematoda	NO							Establ	Brackish			x			
Anguillicola crassus	Nematoda	NO	SE	DK	GE	NL	BE	UK	Establ	Marine			x			
Euplana gracilis	Turbellaria					NL			Uncertain	Brackish		x				
Imogine necopinata	Turbellaria					NL			Uncertain	Brackish			x			
Stylochus flevensis	Turbellaria					NL			Crypto	Brackish						
Alkmaria romijni	Polychaeta	NO		DK		NL			Crypto	Brackish						
Aphelochaeta (= Tharyx) marioni	Polychaeta				GE	NL			Crypto	Marine						
Boccardia ligerica (= Polydora redeki)	Polychaeta		SE		GE	NL	BE		Crypto	Brackish						
Clymenella torquata	Polychaeta							UK	Uncertain	Marine	x	x	x			x
Ficopomatus enigmaticus	Polychaeta				GE	NL	BE	UK	Establ	Brackish	x	x	x			x
Hydroides elegans	Polychaeta				GE	NL			Establ	Brackish	x	x				x
Janua brasiliensis	Polychaeta					NL			Uncertain	Marine		x	x			x
Marenzelleria neglecta	Polychaeta	NO		DK	GE	NL	BE	UK	Establ	Brackish	x					
Marenzelleria viridis	Polychaeta		SE	DK	GE	NL		UK	Establ	Brackish	x					
Microphthalmus similis	Polychaeta				GE	NL			Crypto	Marine						
Proceraea cornuta	Polychaeta					NL			Crypto	Marine						
Scolelepis cf. *bonnieri*	Polychaeta	NO				NL			Establ	Marine					x	
Tharyx killariensis	Polychaeta				GE				Establ	Marine	x	x				x
Tubificoides heterochaetus	Oligochaeta				GE	NL	BE		Crypto	Brackish						
Corambe obscura (= batava)	Gastropoda					NL			Extinct	Brackish					x	
Crepidula fornicata	Gastropoda	NO	SE	DK	GE	NL	BE	UK	Establ	Marine			x			
Rapana venosa	Gastropoda					NL		UK	Uncertain	Marine					x	
Urosalpinx cinerea	Gastropoda					NL		UK	Uncertain	Marine			x			
Aulacomya ater	Bivalvia							UK	Unestabl	Marine		x				
Crassostrea gigas	Bivalvia	NO		DK	GE	NL	BE	UK	Establ	Marine			x			
Crassostrea virginica	Bivalvia				GE	NL		UK	Unestabl	Marine			x			
Ensis directus (= americanus)	Bivalvia	NO	SE	DK	GE	NL	BE	UK	Establ	Marine	x					
Mercenaria mercenaria	Bivalvia					NL	BE	UK	Establ	Marine			x	x		x
Mya arenaria	Bivalvia	NO	SE	DK	GE	NL	BE	UK	Crypto	Marine						
Mytilopsis (= Congeria) leucophaeta	Bivalvia				GE	NL	BE		Establ	Brackish	x					

(continued)

Table 29.1 (continued)

Species	Taxonomic group	Country							Status	Habitat	Ballast water	Hull fouling	Aquacult. or stock.	Other	Unknown	Unclear
Petricola pholadiformis	Bivalvia	NO	SE	DK	GE	NL	BE	UK	Establ	Marine			x			
Ruditapes (= *Tapes*) *philippinarum*	Bivalvia	NO						UK	Establ	Marine			x			
Teredo navalis	Bivalvia	NO	SE	DK	GE	NL	BE	UK	Crypto	Marine						
Limulus polyphemus	Xiphosura			DK	GE	NL			Unestabl	Marine					x	
Artemia salina	Branchipoda							UK	Unestabl	Marine			x			
Eusarsiella zostericola	Ostracoda							UK	Establ	Brackish			x			
Acartia tonsa	Copepoda		SE	DK	GE	NL	BE	UK	Establ	Marine	x					
Eurytemora americana	Copepoda					NL	BE	UK	Establ	Brackish	x					
Myicola ostreae	Copepoda					NL			Uncertain	Marine			x			
Mytilicola intestinalis	Copepoda			DK	GE	NL	BE	UK	Establ	Marine		x	x			x
Mytilicola orientalis	Copepoda					NL			Establ	Marine			x			
Nucellicola holmanae	Copepoda							UK	Crypto	Marine						
Balanus amphitrite amphitrite	Cirripedia					NL	BE	UK	Establ	Marine		x				
Balanus eburneus	Cirripedia					NL			Uncertain	Marine		x				
Balanus improvisus	Cirripedia	NO	SE	DK	GE	NL	BE	UK	Crypto	Brackish						
Elminius modestus	Cirripedia			DK	GE	NL	BE	UK	Establ	Marine		x				
Megabalanus coccopoma	Cirripedia					NL	BE		Establ	Marine		x				
Megabalanus tintinnabulum	Cirripedia						BE		Establ	Marine		x				
Solidobalanus fallax	Cirripedia	NO					BE	UK	Establ	Marine		x				
Caprella mutica	Amphipoda	NO			GE	NL	BE	UK	Establ	Marine	x	x		x		x
Chelicorophium curvispinum	Amphipoda				GE	NL			Establ	Brackish	x					
Corophium (= *Monocorophium*) *sextonae*	Amphipoda	NO			GE	NL	BE	UK	Crypto	Marine						
Gammarus tigrinus	Amphipoda		SE		GE	NL	BE		Establ	Brackish			x			
Incisocalliope aestuarius	Amphipoda					NL	BE		Establ	Brackish	x					
Melita nitida	Amphipoda					NL			Establ	Marine	x	x				x
Orchestia cavimana	Amphipoda				GE	NL	BE		Establ	Brackish	x					
Platorchestia platensis	Amphipoda				GE	NL			Crypto	Marine						
Callinectes sapidus	Decapoda				GE	NL	BE		Establ	Marine	x					
Eriocheir sinensis	Decapoda	NO	SE	DK	GE	NL	BE	UK	Establ	Marine	x					

Species	Taxonomic group	Country							Status	Habitat	Ballast water	Hull fouling	Aquacult. or stock	Other	Unknown	Unclear
Hemigrapsus penicillatus	Decapoda				GE	NL	BE		Establ	Marine	x	x				x
Hemigrapsus sanguineus	Decapoda				GE	NL	BE		Establ	Marine	x	x				x
Homarus americanus	Decapoda	NO						UK	Unestabl	Marine				x		
Marsupenaeus (= *Penaeus*) *japonicus*	Decapoda	NO							Unestabl	Marine				x		
Palaemon macrodactylus	Decapoda				GE	NL	BE	UK	Establ	Marine	x					
Rhithropanopeus harrisii	Decapoda				GE	NL	BE		Establ	Brackish		x				
Telmatogeton japonicus	Insecta	NO		DK	GE	NL	BE	UK	Establ	Marine					x	
Bugula simplex	Bryozoa					NL	BE		Crypto	Marine						
Bugula stolonifera	Bryozoa					NL	BE	UK	Crypto	Marine						
Smittoidea prolifica	Bryozoa					NL			Establ	Marine			x			
Tricellaria inopinata	Bryozoa					NL	BE		Establ	Marine		x				
Victorella pavida	Bryozoa		SE		GE	NL			Crypto	Marine						
Botrylloides violaceus	Tunicata					NL	BE		Establ	Marine		x				
Botryllus schlosseri	Tunicata	NO	SE	DK	GE	NL	BE	UK	Crypto	Marine						
Didemnum sp.	Tunicata					NL			Establ	Marine					x	
Diplosoma listerianum	Tunicata					NL			Crypto	Marine						
Molgula manhattensis	Tunicata	NO			GE	NL	BE	UK	Crypto	Marine						
Styela clava	Tunicata			DK	GE	NL		UK	Establ	Marine		x	x			x
Lebistes reticulatus	Pisces				GE	NL			Establ	Brackish			x			
Micropogonias undulatus	Pisces					NL	BE		Uncertain	Marine	x					
Neogobius melanostomus	Pisces					NL			Establ	Brackish	x					
Oncorhynchus gorbuscha	Pisces	NO						UK	Extinct	Marine			x			
Oncorhynchus keta	Pisces	NO						UK	Extinct	Marine			x			
Oncorhynchus kisutch	Pisces					NL			Extinct	Marine			x			
Oncorhynchus mykiss (= *Salmo gairdneri*)	Pisces	NO		DK	GE				Establ	Brackish			x			
Trinectes maculatus (= *Achirus fasciatus*)	Pisces					NL			Unestabl	Marine	x					

BE = Belgium, DK = Denmark, GE = Germany, NL = The Netherlands, NO = Norway, SE = Sweden and UK = United Kingdom. “Aquaculture” includes species imports for stocking; “hull fouling” includes ships, vessels and pleasure craft; “other” includes ornamental species, species imported for direct human consumption, but released into the wild; “unclear” refers to species were no single vector was assumed responsible (in these cases possible vectors are indicated). Key references: Gollasch (1996); Eno et al. (1997); Maggs and Stegenga (1999); Reise et al. (1999); Wolff (1999); Weidema (2000); Minchin and Eno (2002); Jensen and Knudsen (2005); Nehring (2005); Wolff (2005); Kerckhof et al. (2007).

29.3 Cryptogenic Species

Some species (n=38) whose origin remains unknown or undecided may be identified as NIS at some future time with, for example, the use of genetic markers. In the meantime these species are deemed to be species of uncertain origin, i.e. they can not be fully ascribed as being native or introduced, and were termed cryptogenic species by Carlton (1996; Chap. 2). These species may have been introduced during the time of early sea voyages by several European countries from where species may have either deliberately or inadvertently become imported to Europe on return. Some of these almost certainly became established and will have spread under a wide range of circumstances within Europe and may today be considered native. The arrival of the soft-shelled clam is one such species thought to have been

Table 29.2 Number of non-indigenous species in the North Sea per functional group

Group	Number
Zoobenthos	83
Phytobenthos	36
Phytoplankton	22
Parasite/pathogen	12
Nekton	8
Zooplankton	6
Total	167

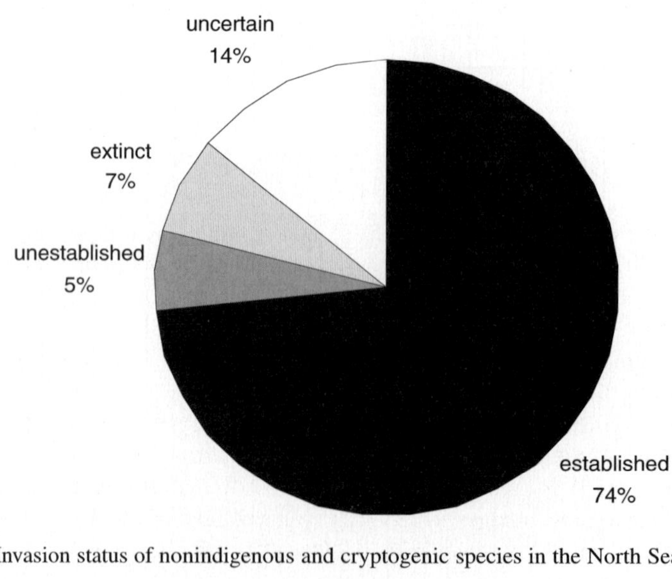

Fig. 29.2 Invasion status of nonindigenous and cryptogenic species in the North Sea

Table 29.3 Occurrence of all nonindigenous and cryptogenic species per number of North Sea countries

Number of Countries	All nonindigenous species	Cryptogenic species
1	48	11
2	25	11
3	23	8
4	6	3
5	9	
6	8	1
7	10	4
Total	129	38

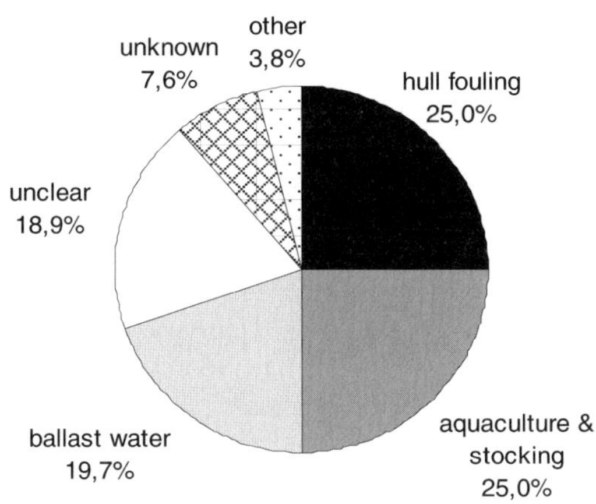

Fig. 29.3 Vectors of first introduction for non-indigenous species found in the North Sea

introduced to Europe by returning Viking expeditions in the 1200s (Petersen et al. 1992; Strasser 1999) and an introduction of the Pacific oyster, previously known as the Portuguese oyster *Crassostrea angulata,* may have been carried with returning sailing ships from Taiwan in the 1500s. It is because the study of taxonomy and ecology developed at a later time, from the eighteenth century, that the changes in distribution have been more carefully recorded. During these years, ships will have had wooden hulls, which may have been subject to intensive fouling, holes created by boring organisms, travelled at low speeds and remained immersed in the water over long periods.

There are potentially many overlooked introductions, often belonging to the less conspicuous, and less studied groups, such as interstitial fauna, polychaetes, microalgae, protozoans, hydroids, and bryozoans (Carlton 2003). Estimating the total number of cryptogenic species in the North Sea is impossible, although some indication may be obtained by examining each taxon and its ability to foul or bore in ship hulls or to survive voyages associated with solid ballast. Indications of an alien origin may be provided from identifying species with disjunct distributions, low dispersal potential, high fouling capacity and the likelihood of interacting with a human mediated vector and route that will have occurred at a specific time.

29.4 Nonindigenous Species Recorded in the North Sea as a Result of Natural Dispersal

In the introduction, an overview was given of the published accounts of NIS in the North Sea region. These accounts tend to list only those species that have known impacts or have been commonly encountered. Species recorded as non-indigenous in these country reports may actually be native to another North Sea country, or to the biogeographic region encompassing the North Sea and then spread by human activities. These include species not previously recorded but may have been introduced with for example, oyster stock movements. Natural events such as exceptional water inflow due to rare hydrodynamic events or storms can result in (mostly) temporary occurrence of species outside their normal ranges (e.g. Berge et al. 2005). Vagrant species such as fishes (i.e. *Mola mola* and *Carcharinus longimanus*), neustonic species (i.e. *Lepas anatifera*) and plankton occasionally appear in the North Sea under such circumstances naturally. Wiltshire (personal communication) and Franke and Gutow (2004) have indicated that many species newly found in the North Sea had previously eastern distribution limits in the British Channel but these have been extended into the North Sea in recent decades most probably on account of climate change (Stachowicz et al. 2002; Beare et al. 2004; Perry et al. 2005).

Some species native to warmer climate regimes have colonised lagoons or docks that have generally higher temperatures or areas in summer or areas where there are thermal plumes. For example, the polychaete *Ficopomatus enigmaticus* was first recorded at the London Docks in the United Kingdom in 1922 (Eno et al. 1997), in the port of Vlissingen (The Netherlands) in 1967 near a power plant (Wolff 2005) and also in the German port of Emden in close proximity to a power plant (Kühl 1977). Today, the species is widespread in the south-western North Sea and is established in four countries (Belgium, Germany, the Netherlands and United Kingdom).

29.5 Case Histories

Three NIS that pose a significant impact in the North Sea and are found in all seven countries are selected as case studies. These include: the slipper limpet *Crepidula fornicata*, the Chinese mitten crab *Eriocheir sinensis* and the 'ship-worm' *Teredo navalis*. These species have varying social, economic and ecological effects within the North Sea region.

29.5.1 *Crepidula fornicata* – the Slipper Limpet

This snail has a thin white shelf inside the shell aperture that protects the visceral mass, giving it a slipper-like appearance and it can attain a size of 5 cm. Individuals are most often found in a 'chain' with the oldest, female individual at the base. Following its planktonic phase the crawling male seeks to attach to the last member of a chain where it will remain confined. Over time the male gradually transforms to a female to which further wandering males may become attached to extend the chain to as many as twelve individuals. Those that do not find chains may self-fertilise (Cole 1952). In temperate waters they can produce more than one brood a year and survive up to ten years. This species is a successful invader because of its persistent recruitment success and ability to colonise a wide range of habitats. Its first known occurrence in Europe was in 1872 in Liverpool Bay, England, but it did not form an established population at this time but did so at a later time on the south-east coast of Britain following introductions of half-grown American oysters *Crassostrea virginica* laid on estuarine shores.

Once introduced, its population can develop to nuisance levels within ten years. It is tolerant of a wide range of conditions within its native range where it occurs from the Gulf of St Lawrence to northern Mexico. It occurs in shallow bays, estuaries and lagoons where temperatures range from –6 °C when exposed to frosts to >25 °C and salinities 25–35 psu, but can endure short periods of lower salinity (Walne 1956). Should mortalities arise from extreme weather events, recruitment from planktonic larvae can take place from deeper water. There is evidence that slipper limpet populations declined during cold winter periods (Thieltges et al. 2004). However, the current trend of warmer winters may have aided in its continued northward expansion. It now occurs as far north as 59°N on the Norwegian coast but has also extended its range southwards to the Spanish rias. For some reason they have not become abundant in the shallow Baie de Arcachon in France (Montaudouin et al. 2001). *C. fornicata* is also known to occur in Sicily in the Mediterranean Sea.

The routes and modes of spread of the slipper limpet are varied. It reached Europe tucked with American oysters inside wooden barrels dispatched as deck cargo on steam-ships from Long Island Sound (Minchin et al. 1995). These oysters were laid on shores and the limpets among them colonised an estuary on the south-east coast of Britain, first found there in 1893. It then spread, partly aided by its planktonic larval stage, to become dispersed along the south British coast. It has also been spread with flotsam. Specimens were stranded on Belgian shores in 1911 and soon after became established there. Korringa (1942) found many attached to stranded wreckage on a Dutch shore in 1926. A few years later the species was found in the Oosterschelde. In 1930 they had become common. During the Second World War in 1944 on 'D-day', large numbers were carried to Normandy, France, as hull fouling on the undersides of Mulberry Harbours used to deliver military equipment ashore. These floating units had acquired sufficient limpet fouling while awaiting deployment in sheltered British estuaries (Blanchard 1997).

Much of the slipper limpet expansion along North Sea coasts has involved the movement of oysters between estuaries and lagoons such as the Wadden Sea

(Thieltges et al. 2003) and the Limfjord. It has even spread to isolated islands such as Helgoland. Scallops often bear the slipper limpet and stocking with sowing sized scallops may also result in its spread.

Off the coast of Brittany, this limpet has become associated with mäerl deposits which are important for conservation. In some areas their populations are culled by dredging to reduce their competition with oysters. This has taken place in France in Marennes-Oleron (Deslous-Paoli 1985). Abundant slipper limpets resulted in changes to sediment structure from the accumulations of vast numbers of their vacant shells and fine particulates from fecal and pseudofecal 'rain' that accumulated within the drifts of shell. Its biomass in Europe in the 1980s probably exceeded one million tonnes (Quiniou and Blanchard 1987). Although during the Second World War 4000 tonnes of *C. fornicata* were processed for human consumption, the species has not been marketed since.

Soon after its arrival in Europe it was declared an 'oyster pest', although the evidence is somewhat equivocal. In field experiments, Montaudouin et al. (1999) could not find any effect on the growth of the Pacific oyster *Crassostrea gigas*, and by use of carbon and nitrogen isotopes, Riera et al. (2002) found differences in food sources; however, competition was shown between the slipper limpet and the mussel *Mytilus edulis*. In a separate study, Thieltges (2005a) found adverse effects of the slipper limpet on mussel growth and survival. Yet mussels with attached slipper limpets had a higher survival whereas those not fouled by them succumbed to higher levels of predation from sea-stars (Thieltges 2005b). Chauvaud et al. (2000) have suggested that the impact from harmful algal blooms can be lessened where the slipper limpet is abundant. It may be seen that there are a complex series of interactions within an ecosystem that show both negative and positive effects of an invader on other components of the ecosystem (Thieltges et al., 2006).

Outside of Europe, slipper limpets occur on the North American Pacific coast, Japan and Uruguay. The species has an ability to colonise other temperate estuaries and inlets of the world, such as on the southern coastline of Australia, Tasmania and New Zealand, South Africa and South America. It is possible that they may be distributed to these regions by movements of oyster stock or as hull fouling on ships. Vigilance in the monitoring of oyster consignments should aid in preventing their establishment in these regions. Areas where *C. fornicata* have become abundant have often been preceded by accounts of individuals or of some small chains. Early reporting, if soon acted on, may lead to their elimination. Following the 1993 European Trade agreement, the Pacific oyster, subject to some conditions, may be distributed within European waters. This is likely to lead to the further spread of the slipper limpet and of other species unless consignments are carefully monitored.

Despite management measures, the high dispersal ability of the slipper limpet has ensured that it would spread within Europe following its establishment over a century ago. This spread has been due to the varied human modes of distributing them but also as a result of the natural spread of their larvae and settled stages. A further species, the American tingle *Urosalpinx cinerea*, was introduced along with the slipper limpet to the south-east coast of Britain at about the same time.

This predatory snail has no pelagic life history stage and so this reduces its natural ability to be spread. The close regulation on the movement of oysters in Britain from the areas where it occurs has shown, even after a hundred years, that some control measures do work and is an example of successful management.

29.5.2 Eriocheir sinensis – the Chinese Mitten Crab

The crab's life-cycle is characterised by migrations between waters of different salinities. Larvae develop in marine waters and juveniles as well as young adults actively migrate upstream into freshwater habitats. Two-year-old adults migrate downstream to marine conditions, which may take several months and during this they become reproductively mature. There is no native crab in Europe with a similar catadromous mode of life. Its area of origin are waters in temperate and tropical regions between Vladivostock (Russia) and southern China (Peters 1933; Panning 1938). The centre of occurrence is the Yellow Sea, a temperate region off northern China (Panning 1952). The mitten crab was first recorded outside its native range in 1912 in the German River Aller. It was suggested that the crab was introduced to Germany with ballast water releases. The greatest abundance in Europe is in estuaries adjacent to the North Sea. The first mass development was documented during the 1930s – and was followed by other mass occurrences in 1940s, 1950s, 1980s and 1990s (Schnakenbeck 1924; Boettger 1933; Sukopp and Brande 1984; Anger 1990; Reise 1991; Michaelis and Reise 1994; Clark et al. 1998; Fladung, personal communication). After the last mass occurrence the crab population declined in Germany (Strauch, personal communication).

Soon after it was first found, the species spread to the Baltic Sea coast of Germany (1926) and Poland (1928), probably via the Kiel Canal. Today it is frequently found along southern and eastern Baltic coasts up to the eastern Gulf of Finland. This is >1500 km from the German Bight, its main centre of abundance (Ojaveer et al. 2007). While it seems unlikely that self-sustaining populations occur in the central and eastern Baltic due to the low salinity, unsuitable for larval development, an egg-carrying female was found recently in Lithuanian waters at very low salinities (Olenin, personal communication). Other records of the crab in Europe were known from the White Sea, Norway, Ireland, Portugal, Black and Caspian Seas, and even the French Mediterranean coast without any indication of establishment.

Mitten crabs also invaded other regions of the world. They were first found in San Francisco Bay in 1992 and have since spread up and down the coast (Cohen and Carlton 1995; Rudnick et al. 2000). Individuals were collected in the Great Lakes from 1965 to 1994 (Nepszy and Leach 1973) and from Quebec, on the St Lawrence River (de Lafontaine 2005). A single Chinese mitten crab was collected in the Mississippi River delta in 1987 (Felder, personal communication).

When abundant, the impact of this invader becomes obvious, with predation of native species, on fish caught in traps and nets and on pond cultured fishes as well as habitat structure effects, mainly burrowing in river embankments causing erosion and damages to dikes. Crabs also aggregate on water intake filters of industrial cooling water supplies and drinking water plants.

In its native range in Asia, the Chinese mitten crab is the second intermediate host for the human lung fluke parasite. The oriental lung fluke is a parasite which uses a snail as its primary host, freshwater crayfish and crabs as intermediate hosts, and a variety of mammals (including humans) as the final hosts. The fluke settles in the lungs and other parts of the body, and can cause severe bronchial illness (Ichiki et al. 1989). The disease is not known in Europe, but conditions may enable it to become established at some future time.

Since its first occurrence in 1912, the crab's economic impact in Germany is estimated at 80 million Euro (based on modified calculations of Fladung, personal communication). These costs include catchment gear installation and maintenance, impact on bank erosion and loss in commercial fisheries and pond-aquaculture (estuaries and in-land). However, they can be marketed at 1–3 € per kg for industrial use and for direct human consumption (Asian markets). During 1994–2004, crabs to the total value of approximately 3–4.5 million € were sold in Germany (Gollasch and Rosenthal 2006). This is still way below the cost of mitigation.

29.5.3 Teredo navalis – the Shipworm

The description of *Teredo navalis* by Linnaeus in 1758 was based on material collected by Sellius in The Netherlands in 1730–1732. Its massive occurrence during these years (Vrolik et al. 1860; Van Benthem Jutting 1943) suggests a nonindigenous origin. Mass occurrences have often been observed for many nonindigenous species some years after their introduction and in several cases this resulted in their discovery (e.g. Ostenfeld 1908).

The classical authors Aristotle, Ovid, and Pliny (Vrolik et al. 1860), living by the Mediterranean Sea, knew of shipworms, but the species involved are not known. Almost a thousand years later, from 1516, shipworms were reported from the West Indies and Atlantic Europe (Moll 1914). Vrolik et al. (1860) record fossil finds from NW Europe, but it is unclear whether these are of Holocene age and belong to this species. Moll (1914) lists only fossil finds belonging to other species. There seem to be no records of damage to Viking vessels in northern Europe (Hoppe 2002). However, in the historical museum of Haithabu (Germany), wood with boreholes from the stem of a Viking ship is on display (Minchin, personal observation). Since this vessel was found in a freshwater environment, later colonization by marine borers can be excluded. It is unclear, however, which species created these boreholes. The first confirmed accounts of *T. navalis* in Atlantic Europe are known from The Netherlands.

Van Benthem Jutting (1943) states that, before 1730, *T. navalis* occurred sporadically along the Dutch coast. She refers to Hooft (1580) who recorded damage to seawalls in Zeeland, however, without identifying the cause (Moll 1914). Vrolik et al. (1860) cite the 'Journal des Savants de l'an 1665' and state that vessels in the IJ estuary at Amsterdam were virtually destroyed by the shipworm (however, this may be due to a further species, e.g. *Psiloteredo megotara* or *Teredo norvegica*, and the "worms" may have colonized the ships elsewhere). Also Martinet (1778) records heavy damage to herring fishing vessels in 1714 and 1727. As a result, any records before 1730 concern either unspecified damage or the occurrence of shipworms in vessels. Hence, it seems that until the eighteenth century we have no clear indication that *T. navalis* occurred in wooden structures in The Netherlands.

In 1730 considerable damage to wooden constructions along seawalls was recorded from Zeeland and West-Friesland in the Netherlands (van Benthem Jutting 1943). Vrolik et al. (1860) record damage to seawalls in 1730, 1731, 1732, 1770, 1827, 1858 and 1859. Vrolik et al. (1860) found a relationship between the outbreaks of *Teredo* and dry, warm summers and periods of higher salinities. In the eighteenth century, however, its occurrence in the wood constructions protecting Dutch seawalls was considered a disaster which enforced a radical and costly switch to new dike protection methods. The former wooden poles at the seaward side of the dike had to be replaced by stones imported from abroad. In the eighteenth and nineteenth century, damage to the wooden tide gates and locks was also widespread in The Netherlands and Germany. In The Netherlands even a special governmental "shipworm committee" was installed to study causes of the problem and suggest solutions (Vrolik et al. 1860). The construction of the German naval base at Wilhelmshaven was seriously delayed when a protective dam constructed out of parallel pilings with earth in between them was damaged by a shipworm infestation and collapsed during a storm in January 1860 (Blackbourn 2006). Thereafter the occurrence of *Teredo* gradually declined because wood was no longer used for commercial ship building and dike construction whereas more resistant tropical hardwoods were being used for the doors of locks.

Recently, *T. navalis* showed up for the first time in the brackish waters of Bremerhaven in the Weser estuary, where it was most abundant in fir floating fenders (>10,000 m^{-2}) but less abundant in fir and oak pier posts (Tuente et al. 2002). It is also common in Dutch coastal waters today (Wolff 2005) and is apparently increasing in wooden coastal defense structures in the northern Wadden Sea (Reise, personal observation). Elsewhere in the North Sea *T. navalis* still causes minor economic damage occurring in driftwood, wrecks, and wooden poles.

Van Benthem Jutting (1943) considers *T. navalis* to be a cosmopolitan species probably originating from the North Sea area. Whereas eighteenth-century authors believed that ships returning from the East Indies were responsible for their introduction (see, e.g., Martinet 1778). However, during this period, North Sea states were trading worldwide from where *T. navalis* may have been introduced. It is for these reasons that this species is considered to be cryptogenic.

29.6 Conclusions

We presented a checklist of 167 nonindigenous and cryptogenic species in the North Sea. Shipping associated and aquaculture vectors we consider to be the dominant vectors. More than two thirds of the recorded nonindigenous species have established self-sustaining populations. The majority of nonindigenous species have localized distributions; only ten of these are known from all of the seven countries bordering the North Sea.

Crepidula fornicata, *Eriocheir sinensis* and *Teredo navalis* are examples of nonindigenous and cryptogenic species that have a significant impact on coastal systems of the North Sea. However, Reise et al. (1999) concluded that in the North Sea introduced species in most cases are more "additive" without causing major unwanted economic or ecological impacts. However, nowadays the introduced Pacific oyster *Crassostrea gigas* is spreading in the coastal waters of the North Sea (Reise et al. 2005) and is replacing the native blue mussel *Mytilus edulis*. This rapid spread is probably promoted by the recent warm summers which support the recruitment of the Pacific oyster (Diederich et al. 2005) and also due to the lack of cold winters which are required for good recruitment of *M. edulis*. It is assumed that the current abundance of *C. gigas* may become reduced should water temperatures decline (Nehls et al. 2006). However, this is unlikely because of a continued trend of rising seawater temperatures in the region.

In the North Sea, region paleoenvironmental history as well as strongly transformed modern coastal environments have contributed to a relatively low species richness. Many of the species that happened to become introduced and tolerate the physical regime became established, increased local diversity and together considerably modified ecosystem functioning in the nearshore zone (Reise et al. 2006). Plants like the introduced cordgrass *Spartina anglica* or the Japanese seaweed *Sargassum muticum* altered structural complexity, while abundant benthic filter feeders like the molluscs *Ensis directus*, *Crassostrea gigas* and *Crepidula fornicata* can be assumed to impact regional plankton dynamics in the coastal waters.

Some NIS have the capability of re-organising trophic relationships (see Chap. 17, Grosholz and Ruiz; Chap. 31, Rilov and Galil) within an ecosystem and influence economies both negatively and to advantage. Though potentially enormous, the impacts of introduced species are highly unpredictable. Those with noted impacts in other temperate regions are likely to have impacts in the North Sea. Others may develop unexpectedly high levels of abundance or cause disease and harm that could not be predicted. Since ballast water can carry millions of propagules that are being discharged into North Sea harbours each day, and because vectors may now also distribute these species there needs to be an improved understanding of the vector mechanisms involved and how this may be used to reduce unwanted species in the future.

The rate of invasions has increased in the North Sea (Reise et al. 1999), as it has increased worldwide, and it will probably continue to increase as a consequence of climate change and globalization. For each individual species, the potential number of transport vectors has also increased, for example the European shore crab *Carcinus maenas* is potentially dispersed by ten different vectors today, whereas

200 years ago there were two possible modes of transport and dispersal (Carlton and Cohen 2003).

Knowledge of the invasion process is essential in designing management plans to cope with the potential detrimental effects of invasive species, and to attempt to prevent their large-scale spread. The checklist of NIS in the North Sea provided here can serve as a basis for future studies of introduced species and design of management plans in this region, but as the list will inevitably continue to grow longer, and will need to be periodically updated.

Acknowledgements The authors express their grateful thanks to a large number of colleagues who contributed to this inventory, the list being too numerous to be mentioned here. S. Gollasch and D. Minchin were supported by the EC specific targeted project Delivering Alien Invasive Species Inventories for Europe (DAISIE, see also www.daisie.se) (SSPI-CT-2003-511202).

References

Anger K (1990) Der Lebenszyklus der Chinesischen Wollhandkrabbe (*Eriocheir sinensis*) in Norddeutschland: Gegenwärtiger Stand des Wissens und neue Untersuchungen. Seevögel 11(2):32–37

Beare DJ, Burns F, Greig A, Jones EG, Peach K, Kienzle M, Mckenzie E, Reid DG (2004) Long-term increases in prevalence of North Sea fishes having southern biogeographic affinities. Mar Ecol Prog Ser 284:269–278

Berge J, Johnsen G, Nilsen F, Gulliksen B, Slagstad D (2005) Ocean temperature oscillations enable reappearance of blue mussels in Svalbard after a 1000 year absence. Mar Ecol Prog Ser 303:167–175

Blackbourn D (2006) The conquest of nature. WW Norton, Company, New York, London, 466 pp

Blanchard M (1997) Spread of the slipper limpet *Crepidula fornicata* (L. 1758) in Europe. Current state and consequences. Sci Mar 61(Suppl 2):109–118

Boettger CR (1933) Die Ausbreitung der Wollhandkrabbe in Europa. Sitzungberichte Gesellschaft naturforschender Freunde, Berlin 1933, pp 399–415

Carlton JT (1996) Biological invasions and cryptogenic species. Ecology 77:1653–1655

Carlton JT (2003) Community assembly and historical biogeography in the North Atlantic Ocean: the potential role of human-mediated dispersal vectors. Hydrobiologia 503:1–8

Carlton JT, Cohen AN (2003) Episodic global dispersal in shallow water marine organisms; the case history of the European shore crabs *Carcinus maenas* and *C. aestuarii*. J Biogeog 30:1809–1820

Chavaud L, Jean F, Ragueneau O, Thozeau G (2000) Long-term variation of the Bay of Brest ecosystem: benthic-pelagic coupling revisited. Mar Ecol Prog Ser 200:35–48

Clark PFC, Rainbow PS, Robbins RS, Smith B, Yeomans WE, Thomas M, Dobson G (1998) The alien Chinese mitten-crab, *Eriocheir sinensis* (Crustacea: Decapoda: Brachyura), in the Thames catchment. J Mar Biol Assoc UK 78:1215–1221

Cohen AN, Carlton JT (1995) Biological study: non-indigenous aquatic species in a United States estuary: a case study of the biological invasions of the San Francisco Bay and Delta. US Fish and Wildlife and National Sea Grant College Program Report PB96-166525, Springfield, Virginia, USA, pp 273

Cole HA (1952) The American slipper limpet on Cornish oyster beds. Fisheries Investigation Ser 2 17(7):1–13

de Lafontaine Y (2005) First record of the Chinese mitten crab (i) in the St. Lawrence River, Canada. J Great Lakes Res 31:367–370

Deslous-Paoli J-M (1985) *Crepidula fornicata* L. (Gastropode) dans la bassin de Marennes-Oleron: structure, dynamique et production d'une population. Oceanol Acta 8:453–460

Diederich S, Nehls G, van Beusekom JEE, Reise K (2005) Introduced Pacific oysters (*Crassostrea gigas*) in the northern Wadden Sea: invasion accelerated by warm summers? Helgol Mar Res 59:97–106

Eno NC, Clark RA, Sanderson WG (1997) Non-native marine species in British waters: a review and directory. JNCC, Peterborough

Franke H-D, Gutow L (2004) Long-term changes in the macrozoobenthos around the rocky island of Helgoland (German Bight, North Sea). Helgol Mar Res 58:303–310

Gollasch S (1996) Untersuchungen des Arteintrages durch den internationalen Schiffsverkehr unter besonderer Berücksichtigung nichtheimischer Arten. Diss., Univ Hamburg; Verlag Dr. Kovac, Hamburg, pp 314

Gollasch S, Rosenthal H (2006) The Kiel Canal. In: Gollasch S, Galil BS, Cohen A (eds) Bridging divides. Maritime canals as invasion corridors. Springer, Dordrecht, The Netherlands, pp 5–90

Hooft PC (1580) Nederlands Historien

Hopkins CCE (2002). Introduced marine organisms in Norwegian waters, including Svalbard. In: Leppäkoski E, Gollasch S, Olenin S (eds) Invasive aquatic species of Europe. Distribution, impacts and management. Kluwer, Dordrecht, pp 240–252

Hoppe KN (2002) *Teredo navalis* – the cryptogenic shipworm. In: Leppäkoski E, Gollasch S, Olenin S (eds) Invasive aquatic species of Europe. Distribution, impacts and management. Kluwer, Dordrecht, pp 116–119

ICES (2005) ICES Code of practice on the introductions and transfers of marine organisms 2005, pp 30

Ichiki M, Suzumiya H, Hayakawa K, Imai JI, Nawa Y (1989) Two cases of *Paragonimiasis westermani* with pleural effusion in young girls living in the southern part of Miyazaki Prefecture, Japan. Jpn J Parasitol 38(6):392–395

Jensen KR, Knudsen J (2005) A summary of alien marine benthic invertebrates in Danish waters. Oceanol Hydrobiol Stud 34(Suppl 1):137–162

Kerckhof F, Vink RJ, Nieweg DC, Post JJN 2006. The veined whelk *Rapana venosa* has reached the North Sea. Aquat Invas 1:35–37

Kerckhof F, Haelters J & Gollasch S (2007) Alien species in the marine and brackish ecosystem: the situation in Belgian waters. Aquatic Invasions 2(3), 243–257

Korringa P (1942) *Crepidula fornicata*'s invasion in Europe. Basteria 7:12–23

Kühl H (1977) *Mercierella enigmatica* (Polychaeta: Serpulidae) an der deutschen Nordseeküste. Veröffentlichungen Institut für Meeresforschung Bremerhaven, 16:99–104

Maggs CA, Stegenga H (1999) Red algal exotics on North Sea coasts. Helgol Meeresunters 52:243–258

Martinet JF (1778) Katechismus der Natuur. Derde deel. Johannes Allart, Amsterdam, 387 pp

Michaelis H, Reise K (1994) Langfristige Veränderungen des Zoobenthos im Wattenmeer. In: Lozán JL, Rachor E, Reise K, Westernhagen, H v, Lenz W (eds), Warnsignale aus dem Wattenmeer. Bd. 2, Blackwell Wissenschafts-Verlag, Berlin, 106–116

Minchin D, Eno C (2002) Exotics of coastal and inland waters of Ireland and Britain. In: Leppäkoski E, Gollasch S, Olenin S (eds) Invasive aquatic species of Europe: distribution, impact and management. Kluwer, pp 267–275

Minchin D, McGrath D, Duggan CB (1995) The slipper limpet *Crepidula fornicata* (L.) in Irish waters, with a review of its occurrence in the north-east Atlantic. J Conchol London 35:247–254

Moll F (1914) Die Bohrmuschel (Genus *Teredo* Linné). Naturwissenschaftliche Zeitschrift für Forst- und Landwirtschaft 12:505–564

Montaudouin X de, Audemard C, Labourg P-J (1999) Does the slipper limpet (*Crepidula fornicata*, L.) impair oyster growth and zoobenthos biodiversity? A revisited hypothesis. J Exp Mar Biol Ecol 135:105–124

Montaudouin X de, Labarraque D, Giraud K, Banchelet G (2001) Why does the introduced gastropod *Crepidula fornicata* fail to invade Arcachon Bay (France)? J Mar Biol Assoc U K 81:97–104

Nehls G, Diederich S, Thieltges DW, Strasser M (2006) Wadden Sea mussel beds invaded by oysters and slipper limpets: competition or climate control? Helgol Mar Biol 60:135–143

Nehring S (2005) International shipping – a risk for aquatic biodiversity in Germany. In: Nentwig W, Bacher S, Cock MJW, Dietz H, Gigon A, Wittenberg R (eds) Biological invasions – from ecology to control. NEOBIOTA 6:125–143

Nehring S, Leuchs H (1999) Neozoa (Makrozoobenthos) an der deutschen Nordseeküste – Eine Übersicht. Bundesanstalt für Gewässerkunde Koblenz, Bericht BfG-1200, pp 131

Nepszy SJ, Leach JH (1973) First records of the Chinese mitten crab, *Eriocheir sinensis*, (Crustacea: Brachyura) from North America. J Fish Res Board Can 30(12):1909–1910

Ojaveer H, Gollasch S, Jaanus A, Kotta J, Laine AO, Minde A, Normant M, Panov V (2007). Chinese mitten crab *Eriocheir sinensis* (H. Milne-Edwards, 1853) (Crustacea, Decapoda, Varunidae) population in the Baltic Sea – a supply-side invader? Biol Invas 9:409–418

Ostenfeld CH (1908) On the immigration of *Biddulphia sinensis* Grev. and its occurrence in the North Sea during 1903–1907. Medd Komm Havunders Ser Plankton 1(6):1–46

Panning A (1938) The Chinese mitten crab. Smithsonian Rep 361–375

Panning A (1952) Die chinesische Wollhandkrabbe. Die neue Brehm-Bücherei 70:1–46

Perry AL, Low PJ, Reynolds JD (2005) Climate change and distribution shifts in marine fishes. Science 308:1912–1915

Peters N (1933) B. Lebenskundlicher Teil. In: Peters N, Panning A (eds) Die chinesische Wollhandkrabbe (*Eriocheir sinensis* H. MILNE-EDWARDS) in Deutschland. Akademische Verlagsgesellschaft mbH, Leipzig, pp 59–156

Petersen KS, Rasmussen JB, Heinemeier J, Rud N (1992) Clams before Columbus? Nature 359:679

Quiniou F, Blanchard M (1987) Etat de la proliferation de la crepidule (*Crepidula fornicata* L.) dans le secteur de Granville (Golfe Normano-Breton – 1985). Haliotus 16:513–529

Reise K (1991) Ökologische Erforschung des Wattenmeeres. Biologie der Meere, Spektrum Akademischer Verlag, Heidelberg, 68–79

Reise K, Gollasch S, Wolff WJ (1999) Introduced marine species of the North Sea coasts. Helgol Meeresunters 52:219–234

Reise K, Dankers N, Essink K (2005) Introduced species. In: Essink K, Dettmann C, Farke H, Laursen K, Lüerßen G, Marencic H, Wiersinga W (eds) Wadden Sea Quality Status Report 2004. Wadden Sea Ecosystem No.19. Common Wadden Sea Secretariat, Wilhelmshaven, Germany, pp 155–161

Reise K, Olenin S, Thieltges DW (2006) Are aliens threatening European aquatic coastal ecosystems? Helgol Mar Res 60:77–83

Riera P, Stal L, Nieuwenhuize J (2002) $\delta^{13}C$ versus $\delta^{13}N$ of co-occurring molluscs within a community dominated by *Crassostrea gigas* and *Crepidula fornicata* (Oosterschelde, The Netherlands). Mar Ecol Prog Ser 240:291–295

Rudnick DA, Halat KM, Resh VH (2000) Distribution, ecology and potential impacts of the Chinese mitten crab (*Eriocheir sinensis*) in San Francisco Bay. University of California, Berkeley, Water Resources Center, Contribution 26, pp 74

Schnakenbeck W (1924) Ueber das Auftreten chinesischer Krabben in der Unterelbe. Schriften für Süßwasser- und Meereskunde, 5

Stachowicz JJ, Terwin JR, Whitlach RB, Osman RW (2002) Linking climate change and biological invasions: ocean warming facilitates nonindigenous species invasions. Proc Natl Acad Sci 99:15497–15500

Strasser M (1999) *Mya arenaria* – an ancient invader of the North Sea coast. Helgol Meeresunters 52:309–324

Sukopp H, Brande A (1984) Beiträge zur Landschaftsgeschichte des Gebietes um den Tegeler See. Sitzungsber Ges Naturforsch Freunde Berlin 24:198–214

Thieltges DW (2005a) Impact of an invader: epizootic American slipper limpet *Crepidula fornicata* reduces survival and growth in European mussels. Mar Ecol Prog Ser 286:13–19

Thieltges DW (2005b) Benefit from an invader: American slipper limpet *Crepidula fornicata* reduces star fish predation on basibiont European mussels. Hydrobiologia 541:241–244

Thieltges DW, Strasser M, Reise K (2003) The American slipper limpet *Crepidula fornicata* (L.) in the northern Wadden Sea 70 years after its introduction. Helgol Mar Res 57:27–33

Thieltges DW, Strasser M, van Beusekom J, Reise K (2004) Too cold to prosper – winter mortality prevents population increase of the introduced slipper limpet *Crepidula fornicata* in northern Europe. J Exp Mar Biol Ecol 311:375–391

Thieltges DW, Strasser M, Reise K (2006) How bad are invaders in coastal waters? The case of the American slipper limpet Crepidula fornicata in western Europe. Biol Invas 8:1673–1680

Tuente U, Piepenburg D, Spindler M (2002) Occurrence and settlement of the common shipworm *Teredo navalis* (Bivalvia: Teredinidae) in Bremerhaven harbours, northern Germany. Hegol Mar Res 56:87–94

van Beek GCW (2006) The Round goby *Neogobius melanostomus* first recorded in the Netherlands. Aquat Invas 1:42–43

Van Benthem Jutting T (1943) Mollusca. C. Lamellibranchia. Fauna van Nederland 12:1–477

Vrolik W, Harting P, Storm Buysing DJ, van Oordt JWL, von Baumhauer EH (1860) Verslag over den Paalworm. Natuurkundige Afdeeling der Koninklijke Nederlandsche Akademie van Wetenschappen, Amsterdam, pp 153

Walne PR (1956) The biology and distribution of the slipper limpet *Crepidula fornicata* in Essex Rivers. Fish Invest Ser II 20(6):50

Weidema IR (2000) Introduced species in the Nordic countries. Nordic Council of Ministers, Copenhagen. Nord Environment 2000, 13, pp 242

Wolff WJ (1999) Exotic invaders of the meso-oligohaline zone of estuaries in the Netherlands: why are there so many? Helgol Meeresunters 52:393–400

Wolff WJ (2005) Non-indigenous marine and estuarine species in The Netherlands. Zool Meded Leiden 79:1–116

Chapter 30
European Enclosed and Semi-enclosed Seas

Erkki Leppäkoski, Tamara Shiganova, and Boris Alexandrov

The brackish-water seas of Europe, i.e. the Black (including the Sea of Azov), Caspian and Baltic Seas, can be regarded as "brackish-water islands", locked in by land masses and isolated from other major brackish-water bodies by physical (ocean and land) barriers. During the last two centuries, more than 300 alien species have been recorded in the four seas. Introduced species have contributed to species diversity and community structure, introduced novel functions and created new interspecific relationships in these seas. Their within-the-sea dispersal has been rapid and effective in all the seas compared, as demonstrated, e.g., by the recent dispersal history of the most successful invaders of American origin.

In all four seas, alien species have become a permanent and exponentially growing problem: they spread with unpredictable consequences, prey on native species or compete with them for food and space, degrade habitats, and alter food webs.

30.1 Introduction

The enclosed or semi-enclosed, brackish-water seas of Europe, i.e. the Black (including the Sea of Azov), Caspian and Baltic Seas, are of special interest to invasion biology. Their long history of research into flora and fauna and ongoing monitoring programmes make it possible to detect many newcomers with only a reasonable delay. The exchange of scientific data is well organized. Cooperation is a key issue, as six, five, and nine riparian countries surround the Black, Caspian, and Baltic Seas respectively.

Inland seas are extreme marginal marine ecosystems (Table 30.1). They can be regarded as "brackish-water islands", locked in by an "ocean of land" (thus presenting many semi-continental features), and isolated from other major brackish-water bodies by physical (ocean and land) barriers. The Baltic and Black Seas are very young seas. Practically all marine and brackish-water biota have invaded during the last 10,000 years, and it is obvious that this immigration continues. After the latest glaciation period, both were freshwater lakes; at present, they have permanently stratified conditions and are thus characterized by steep physical and chemical

G. Rilov, J.A. Crooks (eds.) *Biological Invasions in Marine Ecosystems.* Ecological Studies 204,

Table 30.1 Main characteristics of the European brackish seas

	Black Sea	Sea of Azov	Caspian Sea[a])	Baltic Sea[b])
Latitude °N	40.6–46.3	45.2–47.2	36–47	54–66
Surface area km^2	423,000	37, 860	422,000(1929; level −26 m) 365,000 (1978; level −29 m) 390,000 (1994; level −27 m)	377,400
Water volume km^3	547,000	324	78,600 (level −27 m)	21,200
Mean depth m	1,282	8.5	208	56
Max. depth m	2,212	14	1,025	459
Catchment area 10^6 km^2	2.3	0.57	3.5	1.7
Temperature range °C				
- winter	0–8	−0.8 to +1.2	0–11	0–2
- summer	24–26	24–30	24–28	14–18
Salinity range, surface psu	<0.5–22	<0.5–14	<0.5–13	<0.5–10

[a]The water level of the Caspian Sea is principally controlled by variations of the evaporation rate and the inflow from Volga River, which provides more than 80% of the total inflow
[b]Excludes the Kattegat but includes the three Danish straits (Great Belt, Little Belt and The Sound) that connect the Baltic Sea with the Kattegat

gradients, both horizontally and vertically; see Sorokin (2002) for the Black Sea, Leppäkoski and Bonsdorff (1989) for the Baltic, and for great differences in different regions of the Caspian Sea, Kosarev and Yablonskaya (1994).

Each of the four seas is connected to the ocean only by narrow inlets. The Black and Azov Seas are linked to the Mediterranean through the Bosporus Strait, the Baltic connects with the Atlantic via the Danish Straits, and the Caspian through a man-made canal, opened in 1952 (Grinevetsky et al. 2006), connecting it with the Black Sea through the Volga and Don rivers. Being both recipient, transit, and donor areas for nonindigenous species (NIS), e.g., predatory cladocerans (Cristescu et al. 2001) and zebra mussels (Mills et al. 1993) to the North American Great Lakes, these seas play an important role in global NIS transfers, and are involved in the process of homogenisation of the aquatic fauna and flora occurring mainly in the Northern hemisphere. NIS native to the Southern hemisphere are extremely rare in these seas.

During the last two centuries, more than 300 alien species have been recorded in the four seas. Introductions of NIS have taken place through intentional introductions, moving along rivers from adjacent freshwater bodies, as well as accidental introductions of non-target species. Today, the transfer of NIS with ships' ballast water, tank sediments, and hull fouling increasingly exceeds the importance of other vectors (Paavola et al. 2005; Shiganova et al. 2005). The multitude of invasion corridors opening into each of the seas became clear when both marine and freshwater NIS where recorded in increasing numbers from the 1970s. However, the rapidly increased scientific and public awareness of NIS and their impact on the ecosystem dates back to the early 1990s when several surprisingly successful invaders established themselves (e.g. the comb jellies *Mnemiopsis leidyi* and *Beroe ovata* in the Black Sea and, in the late 1990s, *M. leidyi* also in the Caspian[1]), the polychaetes *Marenzelleria* spp., and the predatory water flea *Cercopagis pengoi* in the Baltic).

In these seas, salinity is a key factor that governs species success in a new area. In a recent study (Paavola et al. 2005), the salinity range of established NIS in European brackish water seas was investigated and related to the classical species/salinity curve of Remane (1934), according to which indigenous species seem to reach a minimum species richness at intermediate salinities. It became evident that the number of alien species is the highest within the salinity intervals of lowest native species. It was also shown that established NIS have adapted to matching salinities already in their native areas of origin. The success of these euryhaline species in their new areas seems to be highly dependent on their salinity tolerance and adaptive plasticity (see also Chap. 10, Smith). For example, it is known from coastal lagoons and inlets of the Baltic Sea that salinity zones with the lowest native species richness are also the zones which support low functional diversity; here the established non-native fauna has considerably increased the functional diversity, particularly in the β-oligohaline (0.5–3 psu) zone (Olenin and Leppäkoski 1999). These authors found several novel functions introduced into the studied lagoon ecosystems with non-native animals, for example (1) the mud snail *Potamopyrgus antipodarum*, native to New Zealand: surface deposit feeding on extremely soft bottoms where the native *Hydrobia* spp. do not occur; (2) the Ponto-Caspian zebra mussel *Dreissena polymorpha*: filter feeding in oligohaline and freshwater parts of the lagoons where the blue mussel *Mytilus* is absent; (3) the bay barnacle *Balanus improvisus*: suspension filter feeding in the uppermost hydrolittoral zone; (4) the North American polychaetes *Marenzelleria* spp.: deep bioturbation of the sediment. Besides brackish water tolerance, the availability of empty niches might thus explain the success of NIS in brackish-water seas.

The European history of aquatic NIS can be divided into three eras (Leppäkoski and Olenin 2000): (1) early accidental introductions; (2) a period of experimentation with potentially beneficial species of economic interest; (3) modern time introductions (intentional ones more or less banned, but unintentional additions increasing due mainly to ship travellers). Species transfer and biological invasions have a long history. Ancient tribes, when moving across the European continent, probably carried not only terrestrial plants and animals but also stocked fish and crayfish from one lake to another. Such events in prehistory remain unknown and nameless. Similarly, early maritime explorers may have transported both terrestrial and aquatic NIS intentionally as well as unintentionally (see also Chap. 2, Carlton). Between the ninth and eleventh centuries, the Vikings sailed from Scandinavia not only to the remotest corners of the European continent (White Sea, British Isles, Iceland, Black Sea down to present-day Istanbul) but also crossed the Northern Atlantic to Greenland and, finally, discovered North America in ca. A.D. 1000. The Black and Caspian Seas were parts

[1] *Mnemiopsis leidyi* was first recorded in the western Baltic Sea (Kiel Bight) in 2006 (Javidpour et al. 2006).

of the Persian Empire (500 years B.C.); 200 years later Alexander the Great conquered the coasts of Asia Minor. Two thirds of the Black Sea coastline became annexed to the Roman Empire in the second century A.D.

Since the eighteenth century, waterways and canals have interconnected the Black and Baltic Seas. Canal construction created a heavily trafficked network of routes all over Central and Eastern Europe and further to the coastal European Atlantic. These canal systems have largely facilitated both the active and passive dispersal of NIS mainly from the southern seas to the Baltic. The most important waterway in this context is the Volga-Baltic connection.

The horizontal and vertical gradients (salinity, temperature, substrate, oxygen, primary production, presence/absence of competitors and predators, etc.) allow species of different origin to find favourable habitats and establish themselves. It could be expected that brackish seas are well protected against marine introductions, owing to their low salinity and wide annual temperature variations. However, many of the major harbours in the world are located at river mouths. The salinity gradient of these estuarine habitats covers the oligo- and mesohaline conditions prevailing in the European inland seas. Consequently, the role of increasing international shipping, faster ships, new trade connections, and thus better survival of stowaways in the ballast tanks, become increasingly important and the brackish waters offer favourable conditions for a number of mostly estuarine species.

The first sea-wide reviews of NIS in the four seas appeared in the mid-1980s. For the Black Sea, see Cvetkov and Marinov (1985) and for the Baltic Sea, Leppäkoski (1984). For the Caspian Sea there was no special review on invaders but reviews of Mordukhai-Boltovskoi (1960), Zenkevich (1963) and Karpevich (1975) provide historical information on the Caspian fauna, its origin, time of penetration to the sea and intentionally introduced species. For recent regional overviews, see Zaitsev and Öztürk (2001), Gomoiu et al. (2002), Leppäkoski et al. (2002), and Aladin et al. (2002), respectively.

Much of the seas' present structural and functional diversity is now of foreign origin. This human-mediated addition of NIS to the native biodiversity has been defined as xenodiversity (Gr. *xenos* – strange; Leppäkoski and Olenin 2000). In the most heavily invaded parts of the seas, xenodiversity tends to exceed even native biodiversity in terms of the number of species and life forms, and strongly affects the ecosystem functions. Moreover, even if the recipient ecosystems may become more diverse they also become more similar to each other as well as to water bodies with matching temperature and salinity all over the world.

30.2 The Black Sea

The Black Sea includes the Sea of Azov. Its only link with the Mediterranean Sea is through the 30 km long and 0.7–3.5 km wide Bosporus Strait. The water of the nontidal Black Sea is brackish and the transition zone between the Mediterranean and

Black Seas comprises a barrier, a corridor, or an acclimatisation zone for different organisms. The salinity gradient extends from >38 psu in the Dardanelles to 18.4 psu in the surface layer of the Black Sea decreasing to 11–14 psu in the Sea of Azov. Nearly 87% of the water volume of the Black Sea is anoxic and contains high concentrations of hydrogen sulphide (Bronfman 1995; Zaitsev and Mamaev 1997). Therefore, living space suitable for the majority of aquatic organisms makes up 10% of the total volume of the Black Sea. Since the early 1970s, noticeable changes have taken place in the Black Sea ecosystem because of anthropogenic eutrophication (e.g. Kideys 2002). The total amount of phosphates entering the Black Sea has increased 3.5-fold between the 1950s and 1980s, with a 2.2-fold increase in nitrates. After 1970, nutrient changes stimulated the development of numerous summer phytoplankton blooms. Since the 1920s, water transparency has steadily decreased in the open sea from 20–21 m (Secchi value); it reached 14–16 m in the early 1980s, and reduced to only 5–9 m in the 1990s (Zaitsev 1998).

30.2.1 Origin of Native Biota

The aquatic inhabitants of the Black Sea can be divided by their origin into four groups:

1. *Pontian relicts*, these average 10% of the total number of free-living metazoans (about 20 species are endemic).
2. *Atlantic-boreal relicts* comprised mainly of marine coldwater species.
3. *Mediterranean settlers* are the most numerous element of the Black Sea biota, which together with the Atlantic-Boreal relicts make up about 80% of the fauna.
4. *Freshwater species* represent some 10% of the metazoan fauna.

Salinity is one of the main factors that determine the distribution of aquatic biota in the Black Sea. The salinity interval at 5–8 psu acts as a barrier that prevents mixing between marine and freshwater fauna (Khlebovich 1974). In the Black Sea and the Sea of Azov, over 1700 metazoan species inhabit the < 3-psu zone, while about 1500 species (98% of which are of Mediterranean origin) have been recorded at 17–18 psu. Less than 30% of Mediterranean fauna penetrate into the Black Sea, and only < 3% into the Sea of Azov (Mordukhai-Boltovskoi 1972).

30.2.2 Nonindigenous Flora and Fauna

Non-native biota in the Black Sea has been documented by Zolotarev (1996), Gomoiu and Skolka (1998), Shadrin (2000), Zaitsev and Öztürk (2001), and Gomoiu et al. (2002).

One of the earliest anthropogenic invaders is the shipworm (*Teredo navalis*), which may have invaded as early as 750–500 B.C. during the Attic period of Greek conquests (Gomoiu and Skolka 1998). The first scientifically documented NIS are the North American barnacle species *Balanus improvisus* (first recorded in 1844) and *B. eburneus* in Sevastopol Bay (1892; Zaitsev and Öztürk 2001).

The history of accidental introductions of NIS into the Black Sea can be divided into three periods. The first (1920–1950) was a period of intensive shipping development. There are ten NIS known from that time, most of them being hull fouling species that penetrated the sea with an average speed of one species per three years. The second period (1951–1980) was characterized by a slight increase in the number of NIS (one species per two years). During this period, increasing salinity in the Sea of Azov (resulting from the infilling of the Tsemliansk reservoir in 1952 and regulated run-off of Don and Kuban rivers) facilitated the penetration of some new species from the Black Sea. During the third period (1981–2000), the Black Sea ecosystem was drastically disturbed due to large-scale eutrophication. In these years the intensity of alien species appearance increased to about two species per year. The principal vector of alien species during this period was ballast water. At the time of writing, the recognized non-native biota recorded in the Black Sea consists of 2 species of marine fungi, 81 species of aquatic flora, and 82 animal species (Alexandrov et al. 2004 and later observations). About 40 NIS developed mass occurrences, among them 9 mollusc species, 9 aquatic plants, and 6 species of phytoplankton. Two non-native brown algae (*Desmarestia viridis* and *Ectocarpus caspicus)* occur in high densities in the low-salinity northwestern part of the Black Sea and in the Sea of Azov. The bay barnacle *Balanus improvisus,* and the bivalves *Mya arenaria* and *Anadara inaequivalvis* are key species in the benthic communities. The predacious gastropod *Rapana venosa,* the comb jellies *Mnemiopsis leidyi* and *Beroe ovata,* and the copepod *Acartia tonsa* are the dominant species that control the state of the ecosystem. The latter three pelagic species are recent invaders and underpin the significance of ballast water as one of the main sources of biological pollution. In addition, there are 13 intentionally introduced species that have established permanent populations in the Black Sea (10 fish, 2 shrimp and 1 oyster species). Of these introductions, four fish species (*Gambusia holbrooki*, *Hypophthalmichthys molitrix*, *Liza haematochila* (*Mugil soiuy*) and *Oryzias latipes*) and the oyster *Crassostrea gigas* have very successfully acclimatized to the Black Sea conditions.

30.2.3 Case Histories

The comb jelly (*Mnemiopsis leidyi*), native to the North American Atlantic coast, is the best-known example of negative impact of NIS on the Black Sea ecosystem. First it was recorded in the Black Sea in 1982, and six years later (1988) a population explosion started in the Sea of Azov. By 1994, the total biomass of *Mnemiopsis* was estimated at 100 million tons and 17–18 million tons in the Sea of Azov in 1996–1997 (maximum biomass at 30 million tons was reached in 1989). The first outbreak of the *M. leidyi* population in the Black Sea occurred in 1989, with an increase in offshore biomass from 225 g wet wt m^{-2} in early 1988 to 2000 g m^{-2} in 1989. A second biomass peak was recorded in 1995 that was >50% the level of the first peak (2700 g.m^{-2} in 1995 compared to 4600 g·m^{-2} in 1989) in Russian offshore waters (Shiganova et al. 2001; Bilio and Niermann 2004 and citations therein). Since 1995, the abundance of *Mnemiopsis* began to decrease and stabilized at 300–800 g·m^{-2} in the Black Sea and at 500–600 g·m^{-2} in the Sea of Azov (Zaitsev and Öztürk 2001). After the invasion of another predacious comb jelly, *Beroe ovata* in 1997, the abundance of *M. leidyi* declined sharply and was maintained at a level more than four times lower than during the late 1980s (Kamburska et al. 2000). One possible explanation of the quick expansion of *Mnemiopsis* in the 1980s may be due to its wide range of salinity and temperature tolerance. It can live and reproduce between 1.3 and 32 °C and between 3.4 and 75 psu (Kremer 1994; Shiganova et al. 2001). After its invasion, the structure of the planktonic communities in coastal waters and the open sea changed significantly. The abundance of mesozooplankton declined by 2–2.5 times or more. A pronounced decrease (approximately 2–10 times) of meroplankton in summer also occurred, showing the grazing impact of *Mnemiopsis* upon the larvae of benthic animals and thus upon the benthos.

Three main impacts of *Mnemiopsis* on the fisheries were identified: (1) predation on fish eggs and larvae; in shelf waters *Mnemiopsis* was estimated to graze up to 70% of total ichthyoplankton stock (Tsikhon-Lukanina et al. 1993); (2) feeding on the food of larvae and adult fish, thus causing starvation (Bilio and Niermann 2004); (3) further accelerating of ongoing ecological change due to eutrophication. There is also believe that overfishing, pollution and eutrophication were major causatives of the fish decrease (Caddy and Griffiths 1990). These events resulted in a drastic decrease in fish production, e.g. a four- to fivefold decrease of kilka (*Clupeonella* spp.) and more than a tenfold decline of anchovy. The annual losses attributed to the *Mnemiopsis* plague were calculated to be ca. 200 million $US in the Black Sea and 30–40 million $US in the Sea of Azov (Zaitsev and Öztürk 2001).

The North Atlantic soft-shelled clam (*Mya arenaria*) was first found in the Gulf of Odessa in 1966. During the first decade, *Mya* biomass was as high as 17 kg·m^{-2} (Zaitsev and Mamaev 1997). In the 1980s, mean biomass had declined to 240 g·m^{-2}. At present, *Mya* is a key species of a biocoenosis covering about 1000 km^2 of the northwestern Black Sea shelf. It has replaced native dominant species such as the small bivalve *Lentidium mediterraneum* and caused a noticeable impact on the benthic community structure and its biodiversity (total number of invertebrate species in the *Mya* biocoenosis is 2.5 times lower than in the original *Lentidium* community) (Zaitsev and Öztürk 2001). The food base for bottom-feeding fish also changed. For example, the great sturgeon *Huso huso ponticus*, the starry sturgeon *Acipenser stellatus*, and the turbot *Psetta maxima maeotica* prefer *Lentidium* over *Mya* as food item because of the invader's large adult size (Kiseleva 1981).

The large predacious gastropod *Rapana venosa,* native to the Sea of Japan, was first found in Novorossiysk Bay in 1946. It became widespread mainly on rocky bottoms of the Crimean, Caucasian, Bulgarian, and Turkish coasts (Gomoiu et al. 2002). *Rapana* is a notorious predator that feeds on bivalves. In the 1950s, it depleted oyster banks on the Caucasus shelf, including economically important bivalves such as the oyster *Ostrea edulis*, scallop *Pecten ponticus*, and mussel *Mytilus galloprovincialis* (Zaitsev and Öztürk 2001). Both *Rapana* and the mullet haarder *Liza haematochila* (*Mugil soiuy*) are examples of NIS that have commercial importance in the Black Sea. *Rapana* is caught in Turkey, Bulgaria, and Russia and exported as frozen meat mainly to Japan and Korea. Shells are also sold to tourists. Along the Turkish Black Sea coast, there are several factories that process *Rapana* meat for export, although only the Turkish export of *Rapana* meat was over 1000 tonnes per year in the 1990s (Zaitsev and Öztürk 2001).

30.2.4 Vectors, Modes of Transmission

Data of pilot ship service in the Bosporus document the ship traffic into the Black Sea. Over the period 1995–2000, the total number of ships passing through the Bosporus reached 47,000–51,000 per year, including 2000–7000 ships longer than 200 m. In addition, the risk of introduction of NIS also can be evaluated based on the volume of transported ballast water. In 2001, in the ports of Ukraine alone, 11 million tonnes of ballast water were discharged. Oil or oil products comprise approximately 80% of cargo transported (Alexandrov 2004).

The Black Sea is a recipient area for NIS native to very different geographical donor areas, i.e. North Atlantic (35%), East Atlantic-Mediterranean (23%), West Pacific (13%), South-East Asia (8%), South-West Pacific (1%), Indo-Pacific (6%), and cosmopolitan species (15%) (Zaitsev et al. 2004).

The high probability of introducing new species is tied not only with the loss of "biological immunity" of the Black Sea due to eutrophication, declining biodiversity and as a result, the appearance of ecological niches filled up by introduced species. The greatest risk sources for exchange of NIS to and from the Black Sea are the marginal seas of the world with low salinity (2–20 psu), as well as deltas and estuaries that form some of the "hot spots" in the Black Sea for introduction of exotic species. When a ship enters a river from seawards, it inevitably discharges some of its ballast waters to facilitate flotation in freshwater. As a result, there is a large number of exotic species discovered on the "sea-river" boundary.

30.3 The Caspian Sea

The Caspian Sea, the largest inland water body on our planet, is situated at the southeastern boundary of Europe. Its waters wash the shores of the Russian Federation, Azerbaijan, Kazakhstan, Turkmenistan, and the Islamic Republic of Iran. The shelf zone (<100 m depth) occupies 62% of the surface area. According to its physical geography and bottom topography, the Caspian is divided into the Northern, Middle and Southern regions.

In the Northern Caspian, salinity ranges from 0.1 psu near the mouths of the Volga and Ural rivers, increasing abruptly to 10–11 psu near the Middle Caspian boundary. In the Middle and Southern Caspian areas, salinity varies from 12.6 to 13 psu and increases only slightly with depth (by 0.1–0.2 psu).

30.3.1 Origin of Native Biota

The inhabitants of the Caspian Sea belong to four groups in accordance to their origin: (1) the most ancient and most abundant autochthonous species (89% of all species) are descended from the Akchagyl Lake–Sea (2.5–2 mya; 5–12 psu), freshwater and Chauda Lake–Sea (0.9–0.4 mya; 5–8 psu; Reid and Orlova 2002) biota; (2) Arctic species (14 species) that arrived during the last glaciation; (3) Atlantic-Mediterranean fauna (7 species) that penetrated about 13,000 years ago. They have become full members of the Caspian communities, have evolved considerably, and generated new species and subspecies; (4) freshwater species that entered the Caspian Sea on several occasions (their numbers are changeable depending on freshwater input) (Mordukhai-Boltovskoi 1960).

30.3.2 Nonindigenous Flora and Fauna

In the twentieth century, a number of alien species of algae, invertebrates, and fish penetrated the Caspian Sea due to human activity, either accidentally through ship fouling or in ballast water tanks, associated with intentional introduction of other species, or were intentionally introduced. The accurate number of such organisms is hard to determine. Some species manifest themselves immediately, others distribute widely only after several years' incubation period, while some NIS appear for only a short term, sometimes increasing dramatically in abundance before becoming extinct.

The mussel *Mytilaster lineatus* was the first known NIS established in the Caspian and was transferred with fouling on boats transported by train from the Black Sea in 1919. Most of the NIS were introduced during two main phases. The first started with intentional acclimatization undertaken since the 1930s, as part of the acclimatization programme of the former USSR. Attempts to acclimate many commercial fish or edible benthic species into the Caspian were performed to increase its commercial or food resources. Most of them were unsuccessful while some species became very abundant, among them the polychaete *Nereis diversicolor* and the bivalve *Abra ovata*, which were introduced intentionally in order to create a new food base for fish (Aladin et al. 2002). Both species developed mass occurrences on soft bottoms. Two species of 30 grey mullets (*Liza saliens* and *L. aurata*) introduced from the Black Sea became species commercial in the Middle and Southern Caspian. The Black Sea flounder (*Platichthys flesus luscus*) was established at first, but has probably since become extinct. A total of 14 introduced fish species were not established. By comparison, the freshwater mosquitofish *Gambusia holbrooki*, intentionally introduced into Georgian Black Sea wetlands, spread widely and reached the Lenkoran area with Kura river runoff (Zenkevich 1963; Karpevich 1975).

Several non-target species were accidentally transferred together with introduced species, among them two species of shrimps, *Palaemon adspersus* and *P. elegans* from the Black Sea during the intentional acclimatization of grey mullets in the 1930s. These shrimps became valuable food resources for bottom-feeding fish. The diatom *Pseudosolenia* (*Rhizosolenia*) *calcar-avis* was also introduced accidentally with the grey mullet in 1934. The unusually rapid population growth and distribution of this euryhaline marine species led to the replacement of previously dominating species, e.g. the diatom *Pseudosolenia fragilissima* and the pyrrophyte *Prorocentrum cordatum* (=*Exuviaella cordata)* over a major part of the sea. The turbellarian *Pentacoelum caspium* was quite probably introduced at the same time.

The second phase of introductions started with the opening of the Volga-Don Canal in 1952. Thereafter, new vectors were added and a new group of species invaded the Caspian Sea. Most of them were carried from the Black Sea and the Sea of Azov by ships in ballast water or fouling assemblages (Karpevich 1975; Aladin et al. 2002). Two species of acorn barnacles, *Balanus improvisus* and *B. eburneus,* recorded in 1955 and 1956, were probably the first invaders to arrive through the Volga-Don Canal with hull fouling. Invasions of hydrozoans (*Bougainvillia megas*), bryozoans (*Conopeum seurati* and *Lophopodella carteri*), and camptozoans (*Barentsia benedeni*) followed. The polychaete *Ficopomatus* (*Mercierella*) *enigmaticus* was introduced between 1958 and 1961 and spread very rapidly in the Krasnovodsk Bay. Its biomass (including the tubes) reached 30 kg m^2, but it then became extinct and has not been found since, as with *B. eburneus* (Atlas 1968). Later, the benthic bivalve *Hypanis colorata* and, probably, the amphipod *Corophium volutator* penetrated to the Caspian Sea. The gastropod *Lithoglyphus naticoides*, the intermediate host of several parasitic trematodes that appeared with it, arrived and rapidly colonized the Volga delta in 1971.The alien crab *Rhithropanopeus harrisii* now inhabits and forms high densities in the southern part of the Northern Caspian Sea (Karpinsky 2002). During the last decades, the nudibranch *Tenellia adspersa* was recorded in the Southern Caspian, with the mussel *Dreissena bugensis* in the northern part (Orlova et al. 1999).

It is impossible to determine accurately the number of non-native fouling algae, as the species composition was first studied in detail only in the 1960s, and numerous epiphytic green, brown and red algae had already appeared in the 1950s, e.g. *Arochaete parasitica, Ectochaete* (*Entocladia*) *leptochaete, Enteromorpha flexuosa* (*E. tubulosa*), *E. maeotica, Ectocarpus confervoides* f. *fluviatilis* (*E. siliculosus*), *Entonema* (*Streblonema*) *oligosporum, Acrochaetium daviesti, Ceramium diaphanum, Polysiphonia variegata* (*P. denudata),*and *Monostroma latissimum*). Soon after its introduction, *C. diaphanum* became a dominant species in the Northern Caspian (Atlas 1968).

The arrival of phytoplankton species likely with ballast water from the Black Sea has greatly increased during recent years (diatoms *Pseudo-nitzchia* (*Nitzchia*) *seriata, Cerataulina pelagica* and occasionally *Tropidoneis lepidoptera*, dinoflagellates *Gymnodinium sanguineum* (*G. splendens*) and *Protoperidinium crassipes*; Shiganova et al. 2005).

After the opening of the Volga-Don Canal, the first zooplankton invaders from the Black Sea to arrive were the hydrozoans, *Blackfordia virginica*, native to the estuaries of North America, and the Black Sea species *Moerisia maeotica*. Other alien species from the Black Sea include the cladocerans *Pleopis polyphemoides* and *Podon intermedius,* and the copepods *Acartia clausi* and *A. tonsa* (Kurasheva and Abdulaeva 1984; Aladin et al. 2002). In two years, *A. tonsa* became one of the dominant species in the Middle and Southern Caspian, particularly in the coastal zone, and plays an important role as a food item for kilka in the eastern part of the Northern Caspian. *A. tonsa* became the only copepod species able to survive after the *Mnemiopsis leidyi* bloom, and it comprised 75–99% of zooplankton samples (Shiganova et al. 2004). In spring 2004, the copepods *Oithona similis* and *Calanus euxinus* and the chaetognath *Sagitta setosa* were found in the Middle Caspian, all of them probably carried from the Black Sea with ballast waters.

In 1999, the jellyfish *Aurelia aurita* and the ctenophore *M. leidyi* were recorded in the Middle Caspian (Ivanov et al. 2000). In 2000, *M. leidyi* spread across all areas of the Caspian Sea even to areas of low salinity (4.3 psu). In 2001, it greatly increased in population size, particularly in the Southern Caspian in August, where its abundance was twice as high as the maximum values recorded in the Black Sea in 1989. By 2002, *M. leidyi* attained its peak abundance in the Caspian Sea at densities of 1000 ind m^{-3} (Shiganova et al. 2004). The consequences of the introduction of *M. leidyi* into the Caspian are severe, even disastrous. In 2000, when it colonized the whole Caspian Sea, zooplankton biomass decreased by 5–20 times in all groups when compared to previous years, and this trend continued. Three species of kilka feed on zooplankton, and such a sharp decrease of their food reserve resulted in a rapid decline of their population, resulting in the near collapse of the formerly intensive kilka fishery (Shiganova et al., 2004). Kilka serves as an important food reserve for beluga sturgeon and other sturgeon species, as well as the Caspian seal. A decrease in the kilka abundance caused nutritional deficiency, which may lead to seal diseases and fecundity decrease. Zooplankton decrease also brought an increase of phytoplankton biomass, due to a decline in zooplankton grazing pressure (Kideys 2002; Shiganova et al. 2004), giving an example of the trophic cascade effect. The introduction of *M. leidyi* also affected benthic organisms, specifically molluscs, polychaetes, and crustaceans, which have pelagic larvae that are grazed by comb jellies.

30.3.3 Origin of Nonindigenous Species

Most NIS established in the Caspian Sea were introduced from the Black Sea (see above), others from the Sea of Azov. The largest group of the Black Sea NIS that invaded the Caspian Sea accidentally had an Atlantic origin from the inshore regions of North America (*Balanus improvisus, B. eburneus, Acartia tonsa,*

Rhithropanopeus harrisii, and *Mnemiopsis leidyi*). All these NIS are eurythermal although rather thermophilic, and, most importantly, highly euryhaline. All are widespread in the coastal areas of the ocean. As a result, they became dominant and suppressed native species in the Caspian (Aladin et al. 2002). Another group of Atlantic species first introduced into the Black Sea and later brought to the Caspian Sea includes inhabitants of brackish bays and estuaries of the North American Atlantic coast, e.g. the hydromedusae *Blackfordia virginica* and *Bougainvillia megas*). A few Black Sea NIS are of European Atlantic or Mediterranean origin.

30.3.4 Vectors, Modes of Transmission

Of the accidentally introduced benthic fauna, the majority (19 species) joined the ship-fouling community, while another 4 (*Mytilaster lineatus, Rhithropanopeus harrisii, Tenellia adpersa, Corophium volutator*) are related to it. Recently, however, a change has taken place. Now plankton organisms (17 species) are common among NIS. In 2004, three species of zooplankton and nine species of phytoplankton were found in the Caspian Sea, which most probably arrived with ballast water from the Black Sea. At present, the plankton community and its functioning determine the major processes occurring in the Caspian Sea. Altogether, 58 NIS species are established in the Caspian, and as recently as 2004, 5 new invaders were recorded.

30.4 The Baltic Sea

The Baltic Sea is a brackish-water, non-tidal small (area 0.1% of the world's seas) land-locked sea, isolated from the North Sea by both geographical and ecological (e.g. low temperature and salinity) barriers. Seasonal variations are distinct, increasing in proportion with increasing distance from the entrance area. The northernmost parts are covered by ice for 140–170 days/year.

Occasional inflows of salty and denser seawater from the North Sea into the deeps create strong stratification. In the bottom water, total oxygen deficit and the formation of hydrogen sulphide periodically kill all bottom-living animals. These lifeless bottoms, "Europe's largest deserts", comprise up to 100,000 km^2 (25% of the total area). Eutrophication is a major environmental concern in the Baltic. The drainage basin (population 85 million) is shared by 14 industrialised countries. Since the early 1900s, nitrogen and phosphorus inputs have increased four- and eightfold, respectively. During the twentieth century, water transparency in the coastal and open Baltic has decreased by 2–5 m. The macroalgal community has suffered from eutrophication, such that annual filamentous green and brown algae replaced perennial brown and red algae. In shallower areas, the biomass of soft-bottom fauna has increased three- to fivefold, which is related to an almost twofold increase in the deposition of organic matter in sediments since the 1920s (Leppäkoski and

Bonsdorff 1989 and references therein). Massive phytoplankton blooms of cyanobacteria are common in the Baltic proper, particularly in late summer, affecting up to one sixth of the total sea area.

30.4.1 Origin of Native Biota

The Baltic is a sea of invaders. Practically all animal and plant species are immigrants that arrived after the latest deglaciation. Few species survive in the brackish water of the Baltic (surface water salinity decreases from ca. 10 psu in the southwest to < 2 psu in the innermost parts). There are few endemic species and the biota of the Baltic consist of species of varied ecological origin. These include euryhaline invaders from the Boreal North Atlantic, Arctic relicts from previous periods, brackish water species of North Sea and Sarmatian (southeastern) origin, freshwater species, and alien species recently introduced by man. Marine diversity is low, especially in the northern Baltic Sea. For example, only five marine bivalves and seven native polychaete species live on the south coast of Finland.

30.4.2 Nonindigenous Flora and Fauna

The present Baltic Sea exists as an ecological continuum, being a result of large-scale natural alterations in its hydrography, from a freshwater lake to more saline conditions than those prevailing today, during the past 10,000 years. The NIS invasions represent a new contribution to this continuum. Today, the Baltic is exposed to other brackish- and freshwater biota of the world, the natural geographical barriers being weakened by ships' traffic and, during the twentieth century, numerous intentional introductions of both North American and Eurasian fish and crustacean species into adjacent freshwaters. These introductions largely increased the number of species moving from the inland waters along rivers into the coastal inlets of the Baltic.

Non-native species in the Baltic Sea have been documented by, e.g. Nikolaev (1951), Leppäkoski (1984), Jansson (1994, 2000), and Gollasch and Mecke (1996). For other references, see Baltic Sea Alien Species Database (2006). Until the establishment of permanent and expanding populations of the North American polychaetes *Marenzelleria* spp. and the Ponto-Caspian predatory cladoceran *Cercopagis pengoi*, scientific interest in NIS was mostly descriptive in the Baltic Sea area (e.g., invasion history, distribution, and abundance assessment). The arrival of *Marenzelleria* and *Cercopagis* contributed widely to the scientific awareness of aquatic bioinvasions. The first risk assessment study for selected ports along the salinity gradient from St. Petersburg (Russia) to Bergen (Norway) included risk profiles for five northwest European harbours (Gollasch and Leppäkoski 1999). The first shipping study in the Baltic Sea was undertaken during 1992–1996 in Germany (Gollasch 1996; Lenz et al. 2000). Another study followed in 1999 to quantify the survival of organisms in ballast tanks during ship voyages (Olenin et al. 2000).

By the year 2006, more than 120 species of non-native animals and plants have been recorded in the Baltic Sea (the Kattegat included; Baltic Sea Alien Species Database 2006), among them 10 phytoplankton, 7 zooplankton, 10 phytobenthos, 39 zoobenthos, 14 nektobenthos and 30 fish species. Almost 80 NIS species have established reproducing populations in the Baltic or at least in some parts of it. Counting the established NIS only, 40% are of freshwater origin. The number of NIS is lowest in the northernmost parts and highest in the coastal lagoons in the south as well as in the Kattegat (Swedish west coast). NIS are common members of the benthic and nektobenthic community in shallow waters, especially at river mouths and in coastal inlets. The deep bottoms were practically free from NIS until the mid-1980s when the North American spionid polychaetes *Marenzelleria* spp. started their expansion. These very successful invaders occupied major parts of the Baltic in less than ten years. Furthermore, *Marenzelleria* is a giant compared to previously dominating native burrowing organisms in the same habitats (chironomid larvae and oligochaetes), dwelling in muddy bottoms of the Baltic coastal lagoons down to a depth of 40 cm in the sediment (Olenin and Leppäkoski 1999). First recorded in the Southern Baltic in 1985, this polychaete successfully colonised most of the Baltic Sea by secondary spread before 1996 and developed into a major faunal element. In some Polish and German estuaries, it developed high densities (up to 5,000–30,000 ind m^{-2}) and biomasses (up to 400–800 g wet wt m^{-2}) and could comprise 95% of the total biomass of bottom fauna during the years of its most abundant occurrence (Zettler et al. 1995; Zmudzinski 1996).

There have been few properly quantified ecological or economic problems with NIS established in the Baltic Sea (see Leppäkoski 2002 for a review). Most known effects are relatively benign and, in some areas, the invaders have served to initially increase both species and functional diversity (Olenin and Leppäkoski 1999; Ojaveer et al. 2002). Today, some 20 species are, from man's point of view, considered harmful. Of the NIS occurring in the coastal waters, four fouling species (the hydrozoan *Cordylophora caspia* (first recorded in the early 1800s, the bay barnacle *Balanus improvisus* (1844), the fishhook water flea *Cercopagis pengoi* (1992; fouling gill nets), and the zebra mussel *Dreissena polymorpha* (early 1800s)) cause economic damage to fisheries, shipping, boating or fish farming, and industry (Leppäkoski 2002).

30.4.3 Origin of the NIS

The presence of both horizontal and vertical gradients makes the Baltic an interesting field laboratory for the study of species spread and bioinvasions. These gradients provide NIS of different origins an extended repertoire of hospitable abiotic conditions within a salinity range of up to >20 psu. For example, in the 400 km long Gulf of Finland, salinity increases gradually from almost 0 psu at the surface in the easternmost basin to >11 psu at the bottom of the entrance area. Consequently all highly euryhaline and eurythermal species are potential invaders. Since most originate from warmer areas, global warming can be expected to increase future invasions. The ability of these species to live and reproduce at the low salinity is a key factor to determine their invasion success (e.g. Paavola et al. 2005).

There is a pool of species, native to the Ponto-Caspian seas and their catchments to be kept on a next-to-arrive list. Of the 29 species examined on the basis of an environmental matching approach (Pienimäki and Leppäkoski 2004), six invertebrate species turned out to have a capacity for introduction via the innermost parts of the Baltic and establishment in the Finnish Lake District in the near future.

NIS in the Baltic Sea originate from all continents but South America and Antarctica. The most important donor area is the east coast of North America, starting with the soft-shell clam *Mya arenaria*, which most probably appeared already in the thirteenth century in Danish waters (Petersen et al. 1992). Today, Neo-Europeans of American origin constitute approximately 25% of all known introductions into the Baltic, including the semi-aquatic bird (the Canada goose *Branta canadensis)* and mammal species (the muskrat *Ondatra zibethicus* and the American mink *Mustela vison*). Ongoing Americanisation appears to be one of the most important processes that contribute to the xenodiversity of all semi-enclosed European seas, including the Baltic (Leppäkoski and Olenin 2000, 2001). Species of Ponto-Caspian origin constitute the second largest part of xenodiversity; of these, 23 species have been able to establish self-reproducing populations. Their proportion is greatest in the sheltered, low-salinity coastal lagoons along the southern and southeastem coast of the Baltic and in the eastern Gulf of Finland, but diminishes westward while the share of North American invaders increases.

Other than North American, NIS of transoceanic origin are rare in the Baltic; the exceptions are, for example, the New Zealand mud snail *Potamopyrgus antipodarum*, the Chinese mitten crab *Eriocheir sinensis* (not reproducing in the Baltic but adult specimens commonly recorded since the 1930s), and the Indo-Pacific diatom *Odontella sinensis*.

30.4.4 Vectors, Modes of Transmission

There is clear evidence of an accelerating invasion rate in recent decades, especially in the 1990s. This trend can be explained in part, however, by the increased public awareness and scientific efforts.

Species native to the Ponto-Caspian basins (the Black and Caspian Seas and their catchments) have spread and become established in inland Europe, the Baltic Sea, and the North American Great Lakes. More than 40 Ponto-Caspian species have expanded their ranges into Central and Western Europe (Jazdzewski 1980; Kinzelbach 1995). This active or passive within-continent dispersal was facilitated by the construction of numerous canals (opened between 1775 and 1952) and reservoirs along Ponto-Caspian rivers, which allowed species to disperse by active migration, attachment to barge hulls or by transport in ballast tanks. Several Ponto-Caspian crustaceans were also transplanted between the 1950s and 1980s to stimulate fish production in western lakes and reservoirs of the former USSR, with more than 30 species of amphipods and opossum shrimps from the "Caspian complex" used for these acclimatisation experiments (Gasiunas 1964). Several were able to move along rivers to the low-salinity coastal waters of the Baltic and spread further from their first bridgeheads.

Vectors for intentional introduction of NIS into the Baltic Sea include those for stocking and aquaculture (43) and ornamental purposes (3), while unintentional introductions have been associated with aquaculture (14) or carried by ships (55); for 8 species the vector remains unknown (Baltic Sea Alien Species Database 2006).

30.5 Conclusions

The European semi-enclosed (the Baltic and Black Seas and the Sea of Azov) or enclosed (the Caspian Sea) brackish seas are subject to worldwide biological homogenisation of the brackish-water fauna and flora. Because of their ecological and evolutionary history, they seem to be predominantly receiver areas of introduced species. It is apparent that several introduced species have contributed to species diversity and community structure, introduced novel functions and created new interspecific relationships in these seas. Their within-the-sea dispersal has appeared to be rapid and effective in all the seas compared, as demonstrated, e.g., by the recent dispersal history of the most successful invaders of American origin. Once transported with man's aid over physical and ecological barriers into an ecosystem, dispersal may be easier for aquatic than terrestrial species as water movements facilitate dispersal and there are fewer dispersal barriers in water (Lodge et al. 1998).

In all four seas, alien species have become a permanent and exponentially growing problem: they spread with unpredictable consequences, prey on native species or compete with them for food and space, degrade habitats, alter food webs and impair water quality, cause diseases or spread as parasites.

The seas included in this contribution are undoubtedly well studied, but there are some groups that are poorly known (e.g. NIS among phytoplankton, meiofauna, microorganisms). What is still more important is that the role of non-native species in ecosystem functions and services (e.g., nutrient cycling) and their ability to displace native species is still badly understood.

References

Aladin NV, Plotnikov IS, Filippov AA (2002) Invaders in the Caspian Sea. In: Leppäkoski E, Gollasch S, Olenin S (eds) Invasive aquatic species of Europe. Distribution, impacts and management. Kluwer Academic Publishers, Dordrecht, pp 351–359

Alexandrov BG (2004) Problem of aquatic organisms transportation by ships and some approaches for risk assessment of the new invasions. Mar Ecol J Sevastopol 3(1):5–17 (in Russian)

Alexandrov B, Bashtannyy R, Clarke C, Hayes T, Hilliard R, Polglaze J, Rabotnyov V, Raaymakers S (2004) Ballast water risk assessment, Port of Odessa, Ukraine, October 2003: Final Report. GloBallast Monogr Ser 10. IMO, London

Atlas (1968) Atlas of invertebrates of the Caspian Sea . Pizhevaya promyshlennostj, Moscow, 415 pp (in Russian)

Baltic Sea Alien Species Database (2006) http://www.corpi.ku.lt/nemo/

Bilio M, Niermann U (2004) Is the comb jelly really to blame for it all? *Mnemiopsis leidyi* and the ecological concerns about the Caspian Sea. Mar Ecol Prog Ser 269:173–183

Bronfman AM (1995) The Sea of Azov. In: Mandych A (ed) Enclosed seas and large lakes of Eastern Europe and Middle Asia. SPB Academic Publishing, Amsterdam, pp 1–32

Caddy JF, Griffiths RC (1990) A perspective on recent fishery-related events in the Black Sea. GFCM Stud Rev 63:43–71

Cristescu MEA, Hebert PDN, Witt JDS, MacIsaac HJ, Grigorovich IA (2001) An invasion history for *Cercopagis pengoi* based on mitochondrial gene sequences. Limnol Oceanogr 46:224–229

Cvetkov LP, Marinov TM (1985) Faunistic enrichment of the Black Sea and changes in its benthic ecosystems. Hidrobiologiya 27:3–21 (in Russian, with English summary)

Gasiunas I (1964) Acclimatisation of forage crustaceans into the Kaunas Waterpower Plant reservoir and possibility of their migration into other waters of Lithuania. Trudy Akad Nauk Lit SSR, Ser B 1(30) (in Russian)

Gollasch S (1996) Untersuchungen des Arteintrages durch den internationalen Schiffsverkehr unter besonderer Berücksichtigung nichtheimischer Arten. Dissertation, University of Hamburg; Verlag Dr. Kovac, Hamburg, 314 pp

Gollasch S, Leppäkoski E (eds) (1999) Initial risk assessment of alien species in Nordic coastal waters. Nord 1999:8. Nordic Council of Ministers, Copenhagen, 244 pp

Gollasch S, Mecke R (1996) Eingeschleppte Organismen. In: Lozan JL, Lampe R, Matthäus W, Rachor E, Rumohr H, Westernhagen H (eds) Warnsignale aus der Ostsee. Parey Buchverlag, Berlin, pp 146–150

Gomoiu MT, Skolka M (1998) Increasing of biodiversity by immigration of new species for the Romanian fauna, Ann Univ "Ovidius" Constanta, Ser Biol-Ecol 2:181–202 (in Romanian)

Gomoiu MT, Alexandrov B, Shadrin N, Zaitsev Y (2002) The Black Sea – a recipient, donor and transit area for alien species. In: Leppäkoski E, Gollasch S, Olenin S (eds) Invasive aquatic species of Europe. Distribution, impacts and management. Kluwer Academic Publishers, Dordrecht, pp 341–350

Grinevetsky SR, Zonn IS, Zhiltsov SS (2006) Black Sea encyclopaedia. Moscow: Mezhdunarodnie otnosheniya Publ, 664 pp (in Russian)

Ivanov PI, Kamakin AM, Ushivtzev VB, Shiganova T, Zhukova O, Aladin N, Wilson SI, Harbison GR, Dumont HJ (2000) Invasion of the Caspian Sea by the comb jellyfish *Mnemiopsis leidyi* (Ctenophora). Biol Invas 2:255–258

Jansson K (1994) Alien species in the marine environment. Swedish EPA, Rep 4357, 67 pp

Jansson K (2000) The marine environment. In: Weidema IR (ed) Introduced species in the Nordic countries. Nord 2000:13. Nordic Council of Ministers, Copenhagen, pp 43–86

Javidpour J, Sommer U, Shiganova S (2006) First record of *Mnemiopsis leidyi* A. Agassiz 1865 in the Baltic Sea. Aquat Invas 1:299–302

Jazdzewski K (1980) Range extensions of some gammaridean species in European inland waters caused by human activity. Crustaceana Suppl 6:84–107

Kamburska L, Moncheva S, Konsulov A, Krastev A, Prodanov K (2000) The invasion of *Beroe ovata* in the Black Sea – why a warning for ecosystem concern? In: Mediterranean marine science, vol 2. Internat Conference "Oceanography of the Eastern Mediterranean and Black Sea. Similarities and Differences of two Interconnected Basins", Athens, 23–26 February 1999, 15 pp

Karpevich AF (1975) Theory and practice of aquatic organisms acclimatization. Pischevaya promyshlennostj, Moscow, 432 pp (in Russian)

Karpinsky MG (2002) Ecology of the benthos of the middle and southern Caspian. VNIRO, Moscow, 283 pp (in Russian)

Khlebovich VV (1974) The critical salinity of biological processes. Nauka Publ., Leningrad, 236 pp (in Russian)
Kideys AE (2002) Fall and rise of the Black Sea ecosystem. Science 297:1482–1484
Kinzelbach R (1995) Neozoans in European waters – exemplifying the worldwide process of invasion and species mixing. Experientia 51:526–538
Kiseleva MI (1981) Benthos of the light sediments of the Black Sea. Naukova dumka Publ, Kiev, 168 pp (in Russian)
Kosarev AN, Yablonskaya EA (1994) The Caspian Sea. SPB Academic Publishing, The Hague, 260 pp
Kremer P (1994) Patterns of abundance for *Mnemiopsis* in US coastal waters: a comparative overview. ICES J Mar Sci 51:347–354
Kurasheva EK, Abdulaeva NM (1984) *Acartia clausi* (Calanoida, Acartiidae) in the Caspian Sea. Zool J 63:931–933 (in Russian)
Lenz J, Andres H-G, Gollasch S, Dammer M (2000) Einschleppung fremder Organismen in Nord- und Ostsee: Untersuchungen zum ökologischen Gefahrenpotential durch den Schiffsverkehr. UBA Project Water: 102 04 250, Umweltbundesamt, Berlin, Texte 5, 273 pp
Leppäkoski E (1984) Introduced species in the Baltic Sea and its coastal ecosystems. Ophelia Suppl 3:123–135
Leppäkoski E (2002) Harmful non-native species in the Baltic Sea – an ignored problem. In: Schernewski G, Schiewer U (eds) Baltic coastal ecosystems: structure, function and coastal zone management. Central and Eastern European Development Studies, Springer, Berlin Heidelberg New York, pp 253–275
Leppäkoski E, Bonsdorff E (1989) Ecosystem variability and gradients. Examples from the Baltic Sea as a background for hazard assessment. In: Landner L (ed) Chemicals in the aquatic environment. Springer, Berlin Heidelberg New York, pp 6–58
Leppäkoski E, Olenin S (2000) Xenodiversity of the European brackish water seas: the North American contribution. In: Pederson J (ed) Marine bioinvasions. Proceedings of the First National Conference, January 24–27, 1999, MIT, Cambridge, MA, pp 107–119
Leppäkoski E, Olenin S (2001) The meltdown of biogeographical peculiarities of the Baltic Sea: the interaction of natural and man-made processes. Ambio 30:202–209
Leppäkoski E, Olenin S, Gollasch S (2002) The Baltic Sea – a field laboratory for invasion biology. In: Leppäkoski E, Gollasch S, Olenin S (eds) Invasive aquatic species of Europe. Distribution, impacts and management. Kluwer Academic Publishers, Dordrecht, pp 253–259
Lodge DM, Stein RA, Brown KM, Covich AP, Brönmark C, Garvey JE, Klosiewski SP (1998) Predicting impact of freshwater exotic species on native biodiversity: challenges in spatial scaling. Austr J Ecol 23:53–67
Mills EL, Leach JH, Carlton JT, Seacor CL (1993) Exotic species in the Great Lakes: a history of biotic crises and anthropogenic introductions. J Great Lakes Res 19:1–54
Mordukhai-Boltovskoi FD (1960) Caspian fauna in the Azov and Black Sea basins. Acad Sci USSR, Moscow, 288 pp (in Russian)
Mordukhai-Boltovskoi FD (1972) General characteristics of the Black Sea and the Sea of Azov fauna. In: Descriptive catalogue of the fauna of the Black Sea and the Sea of Azov 3, Naukova dumka Publ, Kiev, pp 316–324 (in Russian)
Nikolajev II (1951) On new introductions in fauna and flora of the North and the Baltic Seas from distant areas. Zool J 30:556–561 (in Russian)
Ojaveer H, Leppäkoski E, Olenin S, Ricciardi A (2002) Ecological impact of Ponto-Caspian invaders in the Baltic Sea, European inland waters and the Great Lakes: an inter-ecosystem comparison. In: Leppäkoski E, Gollasch S, Olenin S (eds) Invasive aquatic species of Europe. Distribution, impacts and management. Kluwer Academic Publishers, Dordrecht, pp 412–425
Olenin S, Leppäkoski E (1999) Non-native animals in the Baltic Sea: alteration of benthic habitats in coastal inlets and lagoons. Hydrobiologia 393:233–243
Olenin S, Gollasch S, Jonusas S, Rimkute I (2000) En-route investigations of plankton in ballast water on a ship's voyage from the Baltic Sea to the open Atlantic coast of Europe. Int Rev Hydrobiol 85:577–596

Orlova MI, Arakelova ES, Komendantov AY (1999) On overlap of the habitat of *Dreissena bugensis* (Andr.) and *Dreissena polymorpha* (Pall.) in the Volga delta and in shallow waters of the Northern Caspian. Abstract of Conference, Astrahan 23–28 August, 1999, pp 67–69 (in Russian)

Paavola M, Olenin S, Leppäkoski E (2005) Are invasive species most successful in habitats of low native species richness across European brackish water seas? Estuarine Coastal Shelf Sci 64:738–750

Petersen KS, Rasmussen KL, Heinemeier J, Rud N (1992) Clams before Columbus? Nature 359:679

Pienimäki M, Leppäkoski E (2004) Invasion pressure to the Finnish Lake District: invasion corridors and barriers. Biol Invas 6:331–346

Reid DF, Orlova MI (2002) Geological and evolutionary underpinnings for the success of Ponto-Caspian species invasions in the Baltic Sea and North American Great Lakes. Can J Fish Aquat Sci 59:1144–1158

Remane A (1934) The fauna of brackish waters. Verh Dtsch Zool Ges 36:34–74

Shadrin NV (2000) New settlers in the Azov and Black Seas: reasons and consequences? In: New settlers in Russian European seas. Apatyty, pp 76–90 (in Russian)

Shiganova TA, Mirzoyan ZA, Studenikina EA, Volovik SP, Siokou-Frangou I, Zervoudaki S, Christou ED, Skirta AY, Dumont H (2001) Population development of the invader ctenophore *Mnemiopsis leidyi* in the Black Sea and other seas of the Mediterranean basin. Mar Biol 139:431–445

Shiganova TA, Dumont HJ, Sokolsky AF, Kamakin AM, Tinenkova DK, Kurashova EK (2004) Population dynamics of *Mnemiopsis leidyi* in the Caspian Sea, and effects on the Caspian ecosystem. In: Dumont H, Shiganova T, Niermann U (eds) Aquatic invasions in the Black, Caspian and Mediterranean Seas. Kluwer Academic Publishers, pp 71–111

Shiganova TA, Musaeva EI, Pautova LA, Bulgakova YV (2005) The problem of invaders in the Caspian Sea in the context of the findings of new zoo- and phytoplankton species from the Black Sea. Izvestiya RAN Ser Biol 1:78–87 (in Russian)

Sorokin YI (2002) The Black Sea. Ecology and oceanography. Backhuys Publ, 875 pp

Tsikhon-Lukanina, EA, Reznichenko OG, Lukasheva TA (1993) Predation rates on fish larvae by the ctenophore *Mnemiopsis* in the Black Sea inshore waters. Okeanologia 33:895–899 (in Russian)

Zaitsev Y (1998) Eutrophication of the Black Sea and its major consequences. In: Mee L, Topping G (eds) Black Sea pollution assessment. Black Sea Environmental Series 10. United Nations Publ, New York, pp 57–67

Zaitsev Y, Mamaev V (1997) Biological diversity in the Black Sea: a study of change and decline. Black Sea Environmental Series 3. United Nations Publ, New York, 208 pp

Zaitsev Y, Öztürk B (eds) (2001) Exotic species in the Aegean, Marmara, Black, Azov and Caspian Seas. Turkish Marine Research Foundation, Istanbul, 267 pp

Zaitsev Y, Alexsandrov B, Berlinsky N et al. (2004) Basic biological investigations of Odessa maritime port (August–December, 2001). GloBallast Programme Monographs Series 7, 171 pp (in Russian)

Zenkevich LA (1963) Biology of the seas of the USSR. Nauka, Moscow, 739 pp (in Russian)

Zettler ML, Bick A, Bochert R (1995) Distribution and population dynamics of *Marenzelleria viridis* (Polychaeta, Spionidae) in a coastal water of the southern Baltic. Arch Fish Mar Res 42:209–224

Zmudzinski L (1996) The effect of the introduction of the American species *Marenzelleria viridis* (Polychaeta: Spionidae) on the benthic ecosystem of Vistula Lagoon. Mar Ecol 17:221–226

Zolotarev V (1996) The Black Sea ecosystem changes related to the introduction of new mollusc species. Mar Ecol 17:227–236

Chapter 31
Marine Bioinvasions in the Mediterranean Sea – History, Distribution and Ecology

Gil Rilov and Bella Galil

31.1 Introduction

The Mediterranean Sea is in many ways a unique body of water. It is small, but deep compared to other bodies of water of its size, and for its size (0.82% in surface area of the world oceans and 0.32% in volume) it encompasses an impressive variety of ecosystems. From a biodiversity perspective, it can be considered relatively rich in species. Bianchi and Morri (2000) estimate that more than 8500 macroscopic marine species should live in the Mediterranean Sea, which is 4–18% of the world's marine species (depending on different estimates of global diversity). This means that it has high species density for its size (Bianchi and Morri 2000). The body of water that is now the Mediterranean Sea went through dramatic changes in its biota through most of its existence. It is a vestige of the Tethys Ocean, meaning that in prehistoric times it was inhabited by tropical biota. After it was squeezed between Eurasia and Africa and cut off from the rest of the Indo-Pacific at the end of the Miocene (ca. 10 million years ago) it slowly lost its tropical characteristics. It was also cut off from the Atlantic Ocean several times throughout its history, eventually becoming a warm-temperate to subtropical body of water once the Straits of Gibraltar opened at the late Pleistocene (ca. 5 million years ago). These changes in its environmental conditions, that were followed by changes in its biota (as evident from its fossil record; Ruggieri 1967; Sorbini 1988; Zaccaria 1968), naturally occurred over timescales of thousands to millions of years. But lately the rate of biotic change has been increasing dramatically. The biodiversity in the Mediterranean Sea has been altering at an alarmingly high rate for the past two centuries due to human-mediated arrival of new species, with an apparent acceleration in the rate of recorded invasions in the last four decades of the twentieth century. In this chapter we review the current status of the invasion process in the Mediterranean, examine spatio-temporal patterns of species from three major taxonomic groups of invaders, and explore the ecological and conservation implications of some of the most infamous invasions. Special emphasis is given to the major vector of invasion into the Mediterranean Sea – the Suez Canal, and to lagoons as important hotspots of invasion in the western Mediterranean.

G. Rilov, J.A. Crooks (eds.) *Biological Invasions in Marine Ecosystems.*
Ecological Studies 204,

In addition to a detailed review of the most recent literature, we analyzed the temporal, spatial and ecological patterns of fish, decapod crustaceans (plus one stomatopod) and molluscs, using a dataset accumulated and organized by CIESM (International Commission for the Scientific Exploration of the Mediterranean Sea) in the "Atlas of Exotic Species in the Mediterranean Sea" (http://www.ciesm.org/atlas/), with some additional, more updated, information that we have gathered. The CIESM website supplies a description for each introduced species, with additional data on its biology and ecology, where available. From this dataset we used the following information for our analysis: (1) the date of first publication, and in most cases, the date of first record; (2) the species establishment status; "established" is a species having self-maintaining populations, as evidenced by a minimum of two (three for fishes) published records from either different localities or in different periods; (3) species area of origin (categories are somewhat different for each taxonomic group because data were complied by different groups of people); and (4) maps of the current distribution for each species. We divided the sea into three regions: western, central and eastern Mediterranean (see map in Fig. 31.1), and categorized the distribution of each species accordingly. For fish we also analyzed available data on the biology and ecology of the species (mode of reproduction,

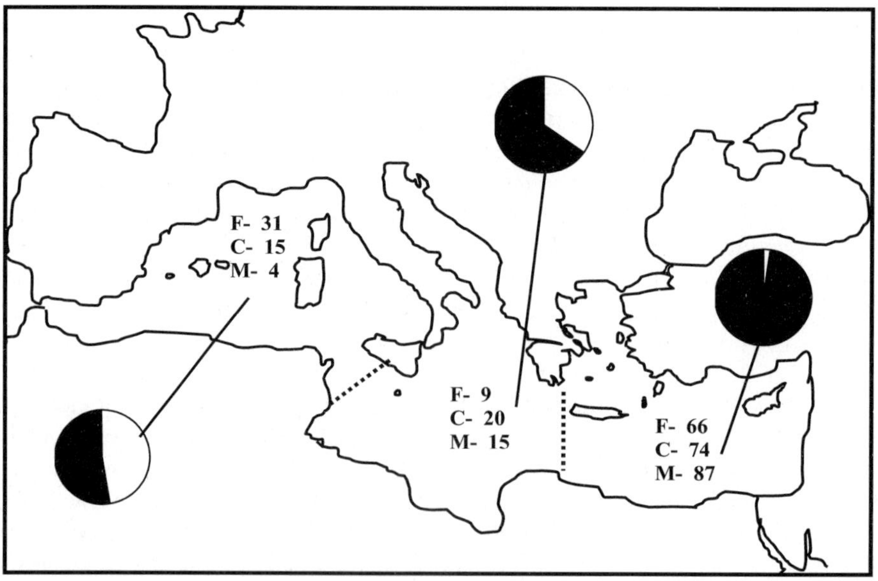

Fig. 31.1 Spatial patterns of invasive species in the Mediterranean. Pie charts indicate the percentage of species (all taxonomic groups pooled) with either a Pacific (*black*) or Atlantic (*white*) origin in the three basins. Basins are denoted by *broken lines* that mark the shallow and narrow "straits" that divide the Mediterranean. *Numbers* represent the percentage of alien fish (F), crustaceans (C), and molluscs (M) that occur in each basin from the total number of aliens in the Mediterranean Sea from each taxonomic group. The percentages add up to more than 100% per taxonomic group because some species occur in more than one basin

trophic level, habitat, depth). For the other two taxonomic groups, too many species had too little available data on the biology or ecology to allow proper analysis of this kind. It is important to note that the Black Sea with its multitude of invasive species is not covered in this chapter but in that on European enclosed and semi-enclosed seas (Chap. 30, Leppäkoski et al.).

Direct transport of species from the Red Sea through the Suez Canal (in the canal's water) is the main vector/corridor for introductions into the Mediterranean Sea, shipping (ballast water and fouling) being second in importance (Galil and Zenetos 2002; Galil 2000). The latter authors replaced the misleading term for this process, "Lessepsian migration" (Por 1978), with the original term, 'Erythrean' invasion. In ecology, "migration" is the cyclical and predictable movement between two geographic areas that is related to the spatio-temporal distribution of resources or the reproductive cycle. This certainly is not the case for the Red Sea species that entered the Mediterranean and that is why invasion here is more appropriate. Aquaculture is third, but the intentionally introduced species are by far outnumbered by the unintentionally introduced species that accompany them. Although introductions through the aquarium trade represent only a small fraction of the total number of invasive species, one of the most notorious of invaders in the Mediterranean, the tropical seaweed *Caulerpa taxifolia*, was accidentally introduced through this means, as we discuss below.

31.2 The Mediterranean Sea: A "Hotspot" of Marine Bioinvasions

In both absolute and relative (to its size) terms, the Mediterranean Sea can be considered one of the hottest hotspots of marine bioinvasions on earth. For example, by 2001, 98 exotic marine algae were recorded in the Mediterranean, compared to 49 on the European Atlantic coast, 26 on the Australian coast, 20 on the coast of New Zealand, 20 on the North-American Atlantic coast, and 19 on the North-American Pacific coast (Ribera Siguan 2002). The total number of invasive molluscs in the Mediterranean at the end of the millennia was almost twice as many as on the coasts of North America (135 compared to around 80; Fig. 31.1; see Ruiz et al. 2000), but the coastline of North America is almost an order of magnitude longer than that of the Mediterranean (398,835 vs 46,267 km respectively). This means that per unit length of coastline, invasion density in the Mediterranean outnumbers any other region. Since then, 25 more mollusc species have been added to the Mediterranean list.

The high number of invasions is not surprising given the criteria for an area highly susceptible to biological invasions (see Chap. 7, Johnston et al.; Chap. 12, Olyarnik et al.). Maritime traffic is high and increasing steadily, aquaculture is extensive, and anthropogenic disturbance is high, destabilizing ecosystems and opening opportunities for new species to settle and establish. Although the overall species richness in the Mediterranean is relatively high, in the eastern basin it is low compared to many other subtropical bodies of water, potentially leaving plenty of niches open for invasion. Furthermore, physical conditions in this small sea vary greatly on the geographical scale. The western basin is a warm-temperate sea (temperatures fluctuate between 12 and 22 °C) and with salinity similar to the Atlantic Ocean (~35 PSU), while the eastern basin is subtropical (15–31 °C) and saltier (38–39 PSU) due to its more arid nature. This should make the western basin more susceptible to temperate Atlantic/Pacific invasions and the eastern basin more susceptible to subtropical and tropical invasions, allowing for a wide range of potential invaders in the Mediterranean Sea.

31.2.1 Spatial Patterns

The distribution of introduced fish, decapods and molluscs in the Mediterranean reveals several interesting patterns. Most species (65–95%) originate from tropical areas, mainly from the Indo-Pacific (Figs. 31.1 and 31.2). Although we would expect that most invaders in the western basin would originate from temperate waters, a high number of aliens there (65%) are of tropical origin (including those originating from the Atlantic Ocean; Figs. 31.1 and 31.2). As expected, though, the species origin changes when going from west to east. In the western basin, invasive species almost equally originate from the Pacific and the Atlantic Oceans, while the proportion of Pacific species increases to 65% in the central region and reaches almost a 100% in the eastern basin (Fig. 31.1). In all taxonomic groups examined, most introduced species occur in the eastern basin; however, there are still differences among the three groups. Invasive molluscs were recorded almost solely in the eastern and central regions (87%, and 15%) and only a few (4% of 160 species) in the western basin (for a comprehensive review see Gofas and Zenetos 2003). Decapod crustaceans have a few more representatives in the western basin than molluscs (15%), and fish are more evenly distributed; 31% of the invasive species occur in the western basin and the rest (66%) are found in the eastern basin, with only a few in the central region. Of the invasive fish occurring in the western basin, 24 have a tropical Atlantic origin, 4 are Indo-Pacific (2 of them not recorded in the eastern basin) and 3 originate in Boreal Atlantic. Why most of the alien fish in the western Mediterranean originate from tropical rather than temperate Atlantic waters is a puzzling question that deserves attention in future studies.

The greater occurrence of introduced species in the eastern basin results from its proximity to the Suez Canal that connects the Red Sea to the Mediterranean, and the more favorable conditions (warmer, saltier waters) in the eastern basin for tropical species. The Suez Canal is the most probable vector for most invasions in the Mediterranean, especially in the eastern basin (Gofas and Zenetos 2003), and we discuss this specific invasion vector below. Another reason for the high number of invaders in the eastern basin compared to the western basin is the extreme faunal impoverishment of the former (less than half the number of the benthic species found in the entire Mediterranean Sea). This impoverishment was attributed to its erratic

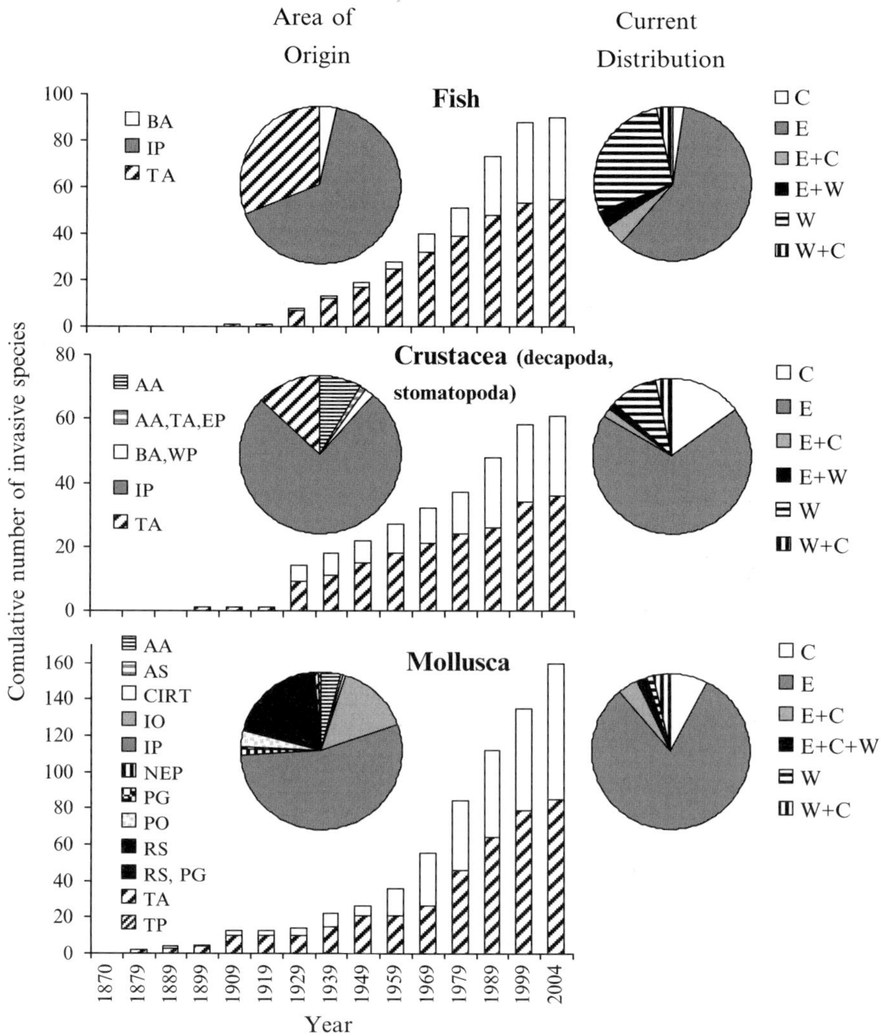

Fig. 31.2 Temporal and spatial patterns of invasions in the Mediterranean. The cumulative number of established (*hatched bars*) and non-established (*empty bars*) introduced species from three taxonomic groups. Pie charts represent the region of origin and the current distribution of the species in the Mediterranean. Origin: AA = American Atlantic, AS = Arabian Sea, BA = Boreal Atlantic, CIRT = Circumtropical, EP = Eastern Pacific, IO = Indian Ocean, IP = Indo-Pacific, NEP = North Eastern Pacific Ocean, PG = Persian Gulf, PO = Pacific Ocean, RS = Red Sea, SC = Suez Canal, TP = Tropical Pacific Ocean, TA = Tropical Atlantic, WP = Western Pacific Ocean. Distribution: C = Central, E = East, W = West

geographical, hydrological and climatic past, its late re-colonization with Atlantic species 5 millions years ago (after a long period of being a shallow hypersaline basin), and its ultra-oligotrophic nature (Fredj 1974; Galil and Zenetos 2002; Yacobi et al.

1995). Unfortunately, it is impossible to determine which of the factors, proximity to a major vector or biotic impoverishment, is more important for the high susceptibility of the eastern basin to invasion. There is also increasing evidence that water temperatures in the western Mediterranean are rising (Béthoux et al. 1990, 1998), and some reports exist on range expansion of native Mediterranean species that were once found only in more southeastern areas and now started to establish populations in the northwestern Mediterranean, while cold water species decrease in numbers (Chevaldonné and Lejeusne 2003; Gomez and Claustre 2003). This warming can also further facilitate the establishment of warm water species that arrive either from the Indo-Pacific or from the tropical Atlantic in the western Mediterranean.

31.2.2 Temporal Patterns

In all three taxonomic groups, there has been a steady increase in the number of invasive species that has accelerated since the 1960s (Fig. 31.2). The comparison between the average number of invaders per decade in the first half of the twentieth century and the second half is striking. For fish, the average number of new invaders per decade was 3.6 times higher in the second half of the century, in crustaceans 1.7, and in molluscs 5.2. With respect to molluscs, 0–10 species were added per decade until 1960, while in the past 40 years, 19–28 species were added each decade. The ratio between non-established and established species (as evident from the size of the bars in Fig. 31.2.) has been increasing over time, and most species that were introduced by the 1950s are already considered established. These trends suggest that many of the current non-established species might become established in the near future. We cannot exclude the possibility that some of the increase in non-established species may result from more intense sampling in more locations and systems in the last few decades, which increased the chance of finding rare invaders. However, as the trend is so strong it is hard to believe this is just an artifact.

31.2.3 Ecological Characteristics of Mediterranean Invaders (Fish)

As stated above, at present, reliable information on ecological characteristics such as mode of reproduction, habitat and depth is available for most introduced fish species but not for decapods and molluscs. The majority of fish invaders in the Mediterranean are spawning, shallow water, benthic carnivores (Fig. 31.3). These characteristics make sense ecologically. Spawners usually have longer dispersal capabilities than species with other modes of reproduction and can also spread faster from a given point of introduction. Whether by natural dispersal (larvae in the plankton, active migration by adults) via the shallow Suez Canal, or by passive transport in ballast water or in mariculture products, most invaders using these vectors are by definition shallow water species. The fact that the majority of the invasive

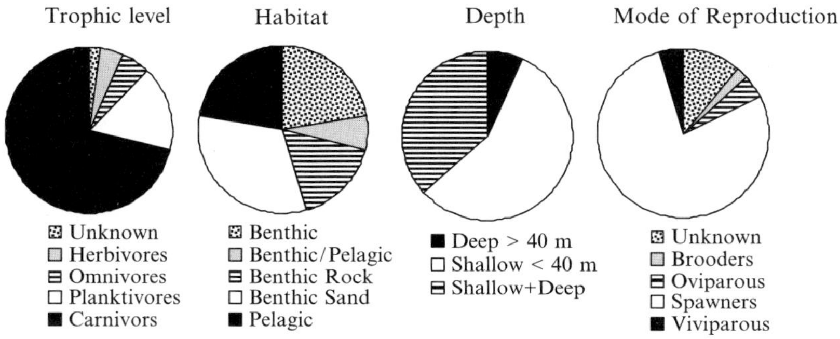

Fig. 31.3 Invasive fish in the Mediterranean Sea. Ecological classifications

fish species are benthic (80%) carnivores (64%, piscivores and/or invertebrate feeders) is not surprising given the fact that the majority of fish species that occur in the Red Sea are benthic (80%) and/or carnivores (79%; Goren 1993).

31.3 Erythrean Invasion Through the Suez Canal

The sharp increase in the number of invaders appearing in the Mediterranean since the mid-1960s may be, at least partially, associated with major changes related to the chief vector of invasions into this sea — the Suez Canal. The 163 km long canal was opened in 1869; it was then 7 m deep and 22 m wide. Since then, successive dredging campaigns have greatly enlarged the canal. By the mid-1960s the canal was deepened to handle laden tankers of 70,000 tons with a maximum draft of 14.6 m, and in the late 1970s it was enlarged again to accommodate fully laden 150,000 ton tankers. In the mid-1960s the construction of the Aswan high dam greatly reduced the Nile annual flood; consequently, the seawater salinity in the area adjacent to the Canal opening, as well as the silt-adsorbed nutrients in Lake Nasser, were sequestered, thus greatly altering the biota at the SE Levantine Basin (the eastern Mediterranean basin). Also, in the aftermath of the 'Six-Day' war, the traffic through the Suez Canal ceased for nearly a decade, reducing turbidity and allowing establishment of benthic communities. Today, 20,000 ships pass through the canal annually from the Red Sea to the Mediterranean Sea, alongside a multitude of Indo-Pacific species, making it the busiest canal in the world in both human and biotic terms.

Contrary to earlier expectations that the Canal serve as a bilateral conduit for exchange of fauna between the Red Sea and the Mediterranean (Steinitz 1919), by the time the results of the 1924 'Cambridge Expedition to the Suez Canal' were published in 1927, it was quite clear that the movement is nearly unidirectional: 138 Erythrean mollusc species are now known to be in the Mediterranean, but there is no evidence of a single Mediterranean mollusc established in the Red Sea. Only a few Mediterranean species established populations within the Canal and near its southern terminal, whereas despite impediments such as the Canal's length, shallowness, turbidity, temperature and salinity extremes, hundreds of Erythrean species traversed the Canal and settled in the Mediterranean, forming thriving populations along the Levantine coasts, with some invaders extending their range as far west as Tunis, Malta and Sicily (Galil 2000).

As soon as it was found that there is "great preponderance of Red Sea over Mediterranean species" (Fox 1927), it was hypothesized that, though some of the Erythrean species may be more adaptable to the hydrographic conditions within the canal, especially the much higher salinity and temperature variations that have formerly characterized it, the chief reason for the "preponderantly Erythrean, as far as concerns animals distributed by currents", is that the currents in the canal flow mostly northwards. In addition, the faunal impoverishment of the Levantine Sea is attributed to its comparatively late recolonization after the Messinian crisis, the pleistocenic climatic fluctuations, and the Basin's extreme oligotrophy, which places it at disadvantage compared with the Red Sea. Since the tropical Atlantic species are mostly barred and their niche is only partially occupied, the native Levantine biota is mostly composed of biota better adapted to colder, less haline water, with many of the taxa present presumably at the limit of their ecological tolerance, leaving this part of the Mediterranean vulnerable to invasion. The chances for Erythrean biota to pass the Canal and settle in the Mediterranean rose because the "Nasser plan" to deepen and widen the Canal increased the speed of tidal currents through the Canal and dissolved the bottom layer of salt in the Great Bitter Lake. The cessation of the Nile outflow with the completion of the Aswan high dam also caused the stabilization of salinity in the SE Levantine Basin.

It has been assumed that Erythrean aliens progress through the Suez Canal and along the coasts of the Levant as a result of "natural" dispersal, by active or passive larval or adult movements, unaided further either directly or indirectly by human activity. Indeed, a temporal succession of directional ("stepping stones") records from the Red Sea, the Suez Canal, and along the coasts of the Levant confirms a species status as a naturally dispersing Erythrean alien. However, dispersal could also result from anthropogenic translocation; already in 1927 Fox (1927) wrote "It is, of course, well known that ships have in more than one instance dispersed marine organisms from one part of the world to another. This must apply equally to transport through the Suez Canal. Possibly tugs and barges permanently employed in the Canal may take a larger share than other vessels in this transport from one end of the Canal to the other. There are coal barges, for instance, which remain for some months at one end of the Canal and then are towed through to stay for some months more at the other end. The time spent at either end would permit on the one hand the settling of larvae on the bottom of the barges, and on the other hand the liberation of eggs or larvae from mature individuals". Shipping is considered the largest single vector for the movement of alien marine species across the globe (Ruiz et al. 1997; Chap. 5, Minchin et al.; Chap. 6, Hewitt et al.). So it is entirely conceivable that shipping may serve to transport Erythrean aliens further on. Even where records are consistent with long-shore autochthonous dispersal, there might be a degree of uncertainty when fouling organisms (such as serpulid polychaetes or mussels) are considered, as they are more susceptible to shipping-mediated transfer. In some cases we suspect simultaneous mechanisms of transport. The small Erythrean mussel,

Brachidontes pharaonis, common in the Levantine basin, where it settles in dense clusters on midlittoral and infralittoral rocks, piers and debris (Barash and Danin 1992; Rilov et al. 2004) has spread as far west as Sicily (Di Geronimo 1971), probably in ship fouling. The pearl oyster, *Pinctada radiata*, was one of the first Erythrean molluscs recorded in the Mediterranean (Monterosato di 1878, as Meleagrina sp.) and has spread as far west as Tunisia, Malta, Sicily, and France (Di Natale 1982; Pallary 1912; Zibrowius 1979). Its rapid dispersal is attributed to ship-borne individuals (Zibrowius 1992), or marine turtles – it was recorded as an epibiont on a loggerhead turtle off Lampedusa Island (Oliverio et al. 1992). Other species may have spread with ballast water, or entangled in fishing gear. Anthropogenic dispersal may take place either from the source populations, or from established Erythrean alien populations within the Mediterranean (Zibrowius 1979).

31.3.1 A History of Dramatic Changes

The greatest change in the size of the Suez Canal was during the late 1970s when the canal was deepened from 7 to 19 m, and widened to 300–365 m, more than ten times its original width. This increase in size reduced salinity and temperature fluctuations, which allowed the survival of additional alien species in the canal waters. Depth is a very important physical impediment to invasion, as indicated by the scarcity of Red Sea invaders found in the Mediterranean waters deeper than 40 m (Fig. 31.3). Thus, the plans by the Egyptian government to widen and deepen the Suez Canal even further to permit passage of super tankers are of grave importance. The planned 5-m increase in canal depth will allow invasion of more species that until now were "banned", as larvae or adults, from the canal due to its current depth.

Salinity along the path of the Suez Canal has also changed greatly over the years. Two dry salt valleys became salty lakes (the Great and the Small Bitter Lakes) when the Suez Canal was cut through them. After the opening of the canal, the salty bottom of the lakes (estimated to have been 13 m thick) gradually dissolved over a period of 60 years. Right after the opening of the Canal, the salinity of the lakes was near 160 PSU (Vadiya and Shenuda 1985), but it dropped sharply to 70 PSU shortly afterwards, and a century later the levels were approaching the ones measured in the northern Gulf of Suez, with an average of 49 PSU in summer and 44 PSU in winter. The high levels of salinity in this area are expected to remain high due to the intense evaporation. It is quite probable that the initial hypersaline conditions in the southern end of the canal served as a barrier for most Indo-Pacific species, and especially to their larvae, particularly in the deeper, denser layers. The deepening of the canal, and the reduced salinity that equalized vertical density differences, have created conditions more favorable for exchange of water and biota along the canal. Salinity also changed at the northern (Mediterranean) end of the canal. Before the completion of the Aswan dam in 1964, Nile floodwater during autumn (September to November), with the help of the prevailing eastward current, spread as a surface layer of fresh water all the way to the Port Said entrance of the canal. The canal was still very shallow (8 m) and this freshwater layer may have acted as a barrier to larvae of species more sensitive to low water salinity, and reduced their chances of entering the Mediterranean Sea. With the damming of the Nile, the river floods ceased (the river's freshwater flow was cut to <10%), thus lifting that autumn barrier to invasion. Moreover, the mean salinity along the coast of Israel to the north increased from 38.8 to 39.1 PSU, and the salinity drops during the autumn months that used to be as low as 34 PSU have disappeared.

The damming of the Nile caused other dramatic changes in the Levant basin that may have also facilitated invasion. For example, phytoplankton blooms associated with the regular Nile floods ceased. This plummet in primary productivity was linked to an order of magnitude drop of sardine catches off the coast of Egypt in 1966 (see article by Pierre Madl, http://www.sbg.ac.at/ipk/avstudio/pierofun/lm/lesseps.htm). It may have, on the other hand, increased the survivorship of larvae of tropical species that are sensitive to highly productive waters. The general increase in salinity in the Levant basin could potentially raise productivity again in this region because heavier salty water would sink below 150 m during cold winters resulting in an intense mixing of cold, nutrient rich bottom waters with the upper photic zone. The sediment transport regime along the Levant shores has also changed. Millions of tons of silty sediments that once flowed to the Nile Delta together with nutrients are now withheld by the dam, affecting the structure of the delta and disrupting the dynamic equilibrium between coastal erosion and sedimentation. Sandy shores along the coast of Israel are getting narrower every year, and the possible change in the deposition of sand on rocky shores may be altering conditions in the receiving environment that may aid in the establishment of invasive species (see Rilov et al. 2004).

A probable scenario for the not so distant future is that global warming may facilitate Erythrean invasion even further. At present, the low (14–15 °C) winter water temperatures in the eastern Mediterranean is probably the main obstacle for many tropical species (such as reef building corals, which do not grow in temperatures below 18 °C) from establishing themselves in the Levant basin (assuming that some manage to pass through the canal today). An increase of a few degrees in winter water temperatures may lift this barrier as well.

31.3.2 The Warm Winter of 1954–55

The Erythrean aliens are thermophilic, originating in tropical waters, and to establish populations they are thought to require "temperatures high enough for the reproductive processes and development of eggs, and minimum winter temperatures above their lethal limits" (Ben Tuvia 1966). Thus, it stands to reason that should Mediterranean sea water temperature rise, the Erythrean aliens would gain a distinct advantage over the native biota. In a few cases there is clear concurrence between rising water temperatures and changes in the composition of resident biota that lead us to believe that higher temperatures facilitate invasions by favoring alien over native biota (see also Chap. 14, Byers for a discussion of "selection regime modification").

For some of the most successful Erythrean invasive species, the initiation of explosive population growth correlated with a rise in winter water temperatures. The "abrupt rise in catch of the lizard fish *Saurida undosquamis*, taken by otter trawlers with the usual gear on the regular fishing grounds" (Oren 1957) was attributed to a rise of 1–1.5 °C in sea temperature during the winter months of 1955 (Ben Yami 1955; Chervinsky 1959). Few individuals had been caught before, and the fish had been only "of taxonomic and zoogeographic interest" (Oren 1957), yet, following that warm winter it became commercially important, constituting for a few years up to one fifth of the total annual trawl catch along the Mediterranean coast of Israel, and over half of the total catch on the shallow shelf opposite El-Arish (northern Sinai).

In the late 1940s the Erythrean goldband goatfish, *Upeneus moluccensis*, made up 10–15% of the total mullid catches off the Israeli coast (Wirszubski 1953). Following the exceptionally warm winter of 1954–1955, goldband goatfish percentages increased to 83% of the catch, replacing the native red mullet, *Mullus barbatus*. Both native and Erythrean mullids have a similar diet, and occupy muddy bottoms shallower than 75 m, but whereas the red mullet spawns from April to June with a peak in May, the goldband goatfish spawns from June to September (Wirszubski 1953). The considerably higher water temperatures at depth of 75 m in May of 1955 and 1956 may have resulted in poor survival of the red mullet spawn; that year the goldband goatfish has had the same temperatures during its spawning period as in previous years, yet an unusually large year class survived. In previous years, the young red mullets would settle to the bottom during July through September, where they have had a distinct size advantage over the later-spawned goldband goatfish. The failure of the 1955 red mullet year class may have left their niche only partly occupied, to the advantage of the Erythrean species, and the unusually warm waters enhanced the latter species' survival rate (Oren 1957). This may suggest that increasing the minimum temperature favors thermophilic aliens by increasing their recruitment relative to native biota.

31.3.3 The Eastern Mediterranean Climatic Transient

The marine biota of the southwestern Anatolian coast, Turkey, and the nearby Dodecanese Islands had been well studied since the early twentieth century. In particular, the aim of the 1970 joint American-Israeli expedition was to investigate whether the Indo-West Pacific (IWP) biota invading the Mediterranean through the Suez Canal had reached Rhodes (Barash and Danin 1988; Lewinsohn 1976). Lewinsohn (1976) recorded no Erythrean alien decapods among the species collected there; neither did Barash and Danin (1988), who considered Rhodes to belong biogeographically to the Aegean Sea, so it was deemed that "Further work in Rhodes can be discontinued, unless different conclusions should result from the processing of material. It seems that the Indo-Pacific influence is very restricted". A study of benthic macrofauna off NW Rhodes conducted a decade later revealed no exotic crustaceans either, and the authors attributed the presence of two exotic decapods reported from Rhodes (Kevrekidis and Kevrekidis 1997) to ephemeral incursion or mariculture (Pancucci-Papadopoulou et al. 1999).

Then, two independent surveys, one carried out along the Levantine and Aegean shores of Rhodes between 1995 and 1999, and the other along the southwestern Anatolian coast between 1996 and 2000, revealed a considerable increase in the number of Erythrean fish, decapods and molluscs along the Turkish and in the southern Aegean Sea (Bilecenoglu et al. 2002; Corsini et al. 2002, Galil and Kevrekidis 2002; Katagan et al. 2004; Kumulu et al. 2002; Yokes and Galil 2004; Yokes and Rudman 2004). The sudden influx of Erythrean aliens in the 1990s cannot be assigned solely to the increased research effort.

A persistent drought in the period 1988–1992 and changes in the water mass pathways initiated a one to four times increase in salt transport from the Levantine into the Aegean in the upper 200 m layer between 1987 and 1994 (Theocharis et al. 1999). In 1991, the source of the Eastern Mediterranean Deep Water shifted from the Adriatic to the southern Aegean Sea (Nittis and Lascaratos 1999; Theocharis et al. 1992), though the process might have started as early as 1987. The increased outflow of the newly formed, denser water through the Cretan Arc Straits into the eastern Mediterranean has been compensated for by inflowing Levantine surface and intermediate depth water (Wu et al. 2000). The significant changes in the South Aegean water mass characteristics, which have considerably influenced the thermohaline circulation of the eastern Mediterranean, have been termed the Eastern Mediterranean Transient (Lascaratos et al. 1999; Theocharis and Lascaratos 2000).

The sudden influx of Erythrean aliens westwards along the Turkish Mediterranean coast and into the southeastern Aegean in the 1990s is attributed to the augmented salinity and to the more extensive inflow of the Asia Minor Current that runs along the Anatolian coastline carrying westwards warm, salty water and the biota from the Levantine Sea and passing northward through the eastern Cretan Arc Straits, mainly the Rhodes and Karapathos Straits. The hydrographic changes have far-reaching implications not only in terms of the formation and spreading of water masses, but for the invasion dynamics of the Erythrean aliens.

31.3.4 Ecological Interactions of Erythrean Invaders in the Receiving Environment

Not enough information exists on the ecological effects of Erythrean invaders in the Mediterranean Sea to allow drawing general conclusions at this stage. However, for some species there are different levels of evidence suggesting that their ecological effects are strong. Documented drops in the abundance of indigenous species that coincided with the increase in the populations of Erythrean invaders suggest for example competitive exclusion, but the experimental work needed to test this hypothesis is lacking. One such case is that of small sea stars along the Israeli coast; as the small Erythrean *Asterina burtoni* increased rapidly in numbers along the

Israeli coast, its indigenous congener *A. gibbosa* was nearly eliminated there (Achituv 1973). Another case is that of prawns in the same region; a native penaeid, *Penaeus* (*Melicertus*) *kerathurus*, was a common catch of trawlers along the Israeli coastal shelf on sandy and muddy bottoms, and supported a commercial fishery until the 1950s (Holthuis and Gottlieb 1958). This species has nearly disappeared since then, and its habitat is now swarming with Erythrean penaeid prawns, mainly *M. japonicus* (d'Udekem d'Acoz 1999).

One of the few species that has been studied quite extensively from an ecological perspective is the Red Sea mussel *Brachidontes pharaonis*. It is one of the earliest Erythrean invaders to the Mediterranean, already present in 1876 near the northern entrance of the Suez Canal (Pallary 1912). Since then, it spread along the Israeli coast and as far north-west as Sicily (Sara et al. 2000). Studies conducted in the late 1970s, when *B. pharaonis* was still relatively rare, predicted that it would not establish dense populations along the Israeli coast and would not outcompete the smaller indigenous mussel *Mytilaster minimus*, although it had shown strong negative effects on survival and growth of the native species (Safriel et al. 1980; Safriel and Sasson-Frosting 1988). The relative rarity of *B. pharaonis* was previously attributed to the invader's low intrinsic rate of population growth relative to that of the native species, and to high density-independent mortality caused by exposure to either high wave action or sedimentation. Contrary to these predictions, massive formations of *B. pharaonis* beds were found in the mid-1990s, about 120 years after initial introduction (Rilov et al. 2004). There is no south-north gradient in its abundance but a strong habitat-dependent colonization pattern. Dense *B. pharaonis* mussel beds are prominent on rocky platforms where beds of any mussel species were absent in the past. These platforms now lack the rim made of sedentary vermetid gastropods that is typical of this formation. On platforms protected by a biogenic rim, sediment accumulation is high and perennial algae flourish. None of the mussel species forms beds in such habitats. Rilov et al. (2004) suggested that the delayed formation of *B. pharaonis* beds along the Israeli coast is a consequence of a recent shift in habitat conditions on some platforms. It is possible that receding of the biogenic rim at the edge of these platforms (due to die-offs that are not well understood) allowed more effective washing, reduced sediment accumulation and reduced perennial algae cover, making platforms more suitable for mussels. Intense grazing by two invasive siganid fish on some platforms (see below) may have also aided in reducing the competition between algae and mussels. Now that sedimentation is lower, decreased density-independent mortality may allow *B. pharaonis* to dominate the indigenous species on such platforms. On beachrock, a habitat previously dominated by *M. minimus*, a rapid shift in numerical domination to *B. pharaonis* (1:7 to 1.4:1 *Brachidontes*/*Mytilaster* individuals) was recorded over a period of four years (1995 to 1999). Rilov et al. (2004) suggested that this is probably a result of saturation of this habitat by *B. pharaonis* recruits originating from the already established populations on nearby rocky platforms. Salinity changes and a potential genetic shift, or the introduction of new genotypes from other regions may also have contributed to the invasive mussel outbreak. Shefer et al. (2004) showed that about 30% of the Mediterranean haplotypes are unique and may have arrived from sources other than the Gulf of Suez and the Red Sea via ballast water in ships. The arrival of more larvae, with potentially more suitable characteristics, may have facilitated the invasion. (Shefer 2003) also investigated the reproduction cycle and recruitment dynamics of *B. pharaonis* and *M. minimus* and showed that *B. pharaonis* has ecological characteristics that make it a successful invader, including a long reproductive season compared to *M. minimus*, high competitive ability, and ecological plasticity. The establishment of dense populations of *B. pharaonis* in the eastern Mediterranean must have dramatic effects on the ecology of the vermetid platforms, a habitat that is unique to the Levant basin (Safriel 1974). Being a habitat-forming species (see Chap. 16, Crooks), its massive beds must have changed the identity and diversity of the species in the community on these platforms by excluding some species and facilitating others. One ecological effect was on the feeding habits of a large common predatory whelk, *Stramonita haemastoma*. When offered different types of prey the whelk favors this invader over indigenous species such as *M. minimus* and barnacles, and in the field it now feeds primarily on *B. pharaonis* (Rilov et al. 2002).

The next two examples also illustrate the effects of invaders on food-web interactions. Two of the most abundant Erythrean fish along the Israeli coast are the siganids, *Siganus rivulatus* and *S. luridus* (see Chap. 10, Smith). The absence of genetic differentiation between Mediterranean and Red Sea populations in both species indicates that many individuals participated in the colonization of the Mediterranean, excluding any bottleneck event (Hassan et al. 2003). According to Hassan et al. (2003), this successful invasion may be a result of eco-physiological plasticity in both *S. rivulatus* and *S. luridus*. A study on their behavior and diet on a vermetid reef south of Haifa Bay, indicated that both species feed on algae species on the platform, especially in spring (Lundberg et al. 2004). The fish exhibited a strong dietary niche-partitioning during spring (overlap of 36%, with *S. rivulatus* feeding mainly on green algae and *S. luridus* on brown algae), but this partitioning is weakened in autumn when the preferred algae are scarce (80% diet overlap). As these herbivores are now highly abundant along the coast and foraging on seaweeds on the rocky shore, we would expect that they affect food-web structure and potentially also community structure in this environment. However, a change in food web or community structure is yet to be experimentally demonstrated.

Leptochela pugnax, a small Erythrean pasiphaeid shrimp, is by far the most important prey of two Erythrean goatfish, *Upeneus asymmetricus* and *U. moluccensis*, and two indigenous goatfish, *Mullus barbatus* and *M. surmuletus*, in the eastern Mediterranean, which indicates a considerable trophic overlap between all four mullid species (Golani and Galil 1991). Because trophic separation does not play an important role in resource partitioning of those fish, Golani and Galil (1991) suggest that coexistence between colonizing and indigenous mullids is achieved by spatial segregation. An earlier study (Oren 1957) already suggested that *M. barbatus* is displaced to deeper, cooler waters by *U. moluccensis*.

With the vast number of invaders arriving from the Red Sea it is expected that related species will compete for resources also among themselves, and that species that are ecologically related in their native environment will facilitate establishment of each other (invasional meltdown; see for example Simberloff and Von Holle 1999; Grosholz 2005). In a few cases there is circumstantial evidence for competition among invaders. For example, the prawn *Trachysalambria palaestinensis*, that was first recorded in 1924 along the Mediterranean coast of Israel was the most common penaeid on sandy mud bottoms until 1987, when another Erythrean prawn, *Metapenaeopsis aegyptia*, was detected in that same habitat. By 1993, *M. aegyptia* outnumbered *T. palaestinensis* 3 to 1, and 3 years later the ratio was 25 to 1 (for earlier citations see Galil and Zenetos 2002). Another Erythrean prawn, *Metapenaeopsis mogiensis consobrina*, appeared in the same habitat in 1996 (Galil 1997). At present (2005), *M. aegyptia* is the most abundant of the three coexisting Erythrean penaeids.

A clear example of facilitation between Erythrean invaders is that of the swimming crab *Charybdis (Goniohellenus) longicollis* and its parasitic castrator, the sacculinid rhizocephalan barnacle, *Heterosaccus dollfusi*. *C. longicollis* was first recorded in the Mediterranean in 1959 off the Turkish coast (Holthuis 1961), and later occurred from Egypt to Cyprus (Lewinsohn and Holthuis 1986). At that time, up to 70% of the benthic biomass on sandy-silt bottoms off the Israeli coast was composed of this invader (Galil 1986). Of the thousands of specimens collected over three decades, none showed evidence of parasitism until 1992, when a few bore the sacculinid parasite (Galil and Lützen 1995). Within three years, *H. dollfusi* had spread as far as the eastern Anatolian coast (Øksnebjerg et al. 1997). The parasite affects the growth, morphology and behavior of the host, and castrates both male and female crabs (Innocenti et al. 1998, 2003). The infection rate in Haifa Bay rose to 77% in 1995 (Galil and Innocenti 1999). Yet a decade later, in spite of high levels of infestation, there is no noticeable reduction in the host population. This example raises doubts concerning the effectiveness of pathogens or parasites, such as sacculinid castrators, for the biocontrol of invasive species (as suggested in recent years; Kuris 1997; Thresher et al. 2000).

31.3.5 Impacts of the Erythrean Invasion on Human Activity

Some invasive Erythrean species are considered pests or cause nuisance, whereas others are of commercial value, collectively they have significant economic impacts in the Levantine Basin. Here are a few examples.

Each summer since the mid-1980s, swarms of the invasive Erythrean jellyfish, *Rhopilema nomadica*, have appeared along the Levantine coast (Galil et al. 1990). These swarms, some stretching for over 100 km, frequently draw nearer to the shore and adversely affect tourism, fisheries and coastal installations. Local municipalities report a decrease in vacationers frequenting the beaches because of the public's concern over the painful stings inflicted by the jellyfish. Coastal trawling and purse-seine fishing are disrupted for the duration of the swarming due to net clogging and inability to sort yield. Jellyfish-blocked water intake pipes pose a threat to engine cooling systems of ships and coastal power plants: in the summer of 2001

Israel Electric removed tons of jellyfish from its seawater intake pipes at its two largest power plants, at estimated costs of 50,000 US$.

Yet, other Erythrean invaders are considered a boon to the Levantine fisheries. Erythrean penaeid prawns make up most of the shrimp catches along the SE Levantine coasts. The Erythrean prawns are highly prized and a small fleet of Israeli coastal "mini" trawlers has specialized, since the mid-1980s, in shrimping, bringing in a quarter of the total trawl catch volume and a third of the trawl gross income (Snovsky and Shapiro 1999). An early Erythrean invader, the swimming crab *Portunus pelagicus*, was recorded from Port Said in 1898, where it soon became abundant, and by the early 1900s it was offered in the markets of Port Said, Alexandria and Haifa (Calman 1927; Fox 1924). It is now offered in many restaurants catering to the thriving tourist industry along the southern Turkish coast. Erythrean fish constitute nearly half of the trawl catches along the Israeli coast (Golani and Ben-Tuvia 1995). In the 1990s two invading mullids, *Upeneus moluccensis* and *U. pori*, formed 87% of the mullet catch off the coast of Israel at depths of 20 m, and the Red Sea obtuse barracuda, *Sphyraena chrysotaenia*, has outnumbered the native sphyraenids in inshore trawl and purse-seine catches (Grofit 1987). The rabbitfish, *Siganus rivulatus*, and lately, the jack, *Alepes djedaba*, are the mainstay of the artisan fisheries in the SE Levant, where they are commonly caught in trammel nets and purse seines. *Sillago sihama* and the Erythrean mackerel, *Scomberomorus commersoni*, two species that underwent explosive population growth in the early 1980s, are common in purse-seine landings. In addition, two Erythrean clupeids – *Dussummieria acuta* and *Herklotsichthys punctatus* – are of importance to the inshore-pelagic fishery. The Erythrean conch, *Strombus persicus*, and on occasion the Erythrean spiny oyster, *Spondylus spinosus*, are served in seafood restaurants in Israel and Lebanon. Population growth of these invasive species to the point that they are harvested commercially is an excellent index of how prevalent they have become. This pattern is not limited to the south-eastern Levant; in the Gulf of Iskendrun, Turkey, it is estimated that invaders constitute 62% of the demersal fish biomass (Gücü and Bingel 1994), and an increasingly important part of the commercially valuable landings.

31.4 Lagoons as Invasion Hotspots

The shallow, frequently brackish coastal lagoons, formed by recent changes in sea level and in the movement of sediment-laden coastal currents especially near large river mouths, are transitory and unstable environments. Those lagoonal environments differ from adjacent coastal regions in their hydrography and less diversified biota (Sacchi et al. 1989). Additionally, Mediterranean lagoons have long been exploited for culture of alien species. The coupling of unstable, polluted, eutrophic or physically-degraded lagoonal environments, of low biodiversity, with enhanced alien propagule pressure (see Chap. 33, Preisler et al.) resulting from mariculture introductions, have produced hot spots of both intentional and nonintentional introductions

(Verlaque 2001; Zibrowius 1992). Large-scale transplantations of shellfish in the twentieth century have served to inoculate Mediterranean coastal waters with many alien camp-followers (Ribera and Boudouresque 1995).

Crassostrea angulata, actually a strain of *C. gigas* from Taiwan, was imported to restock the oyster beds of Atlantic coast of France and subsequently introduced along the French Mediterranean coast (Raimbault 1964), as well as the Tyrrhenian, Ionian, and Adriatic coasts of Italy (Matta 1969). *Crassostrea gigas* was reintroduced to the Mediterranean coast of France and to the Adriatic lagoons of Italy by the late 1960s from the Atlantic French stock originating from brood stock imported from Canada and Japan, and has since been introduced to many Mediterranean localities from Cyprus to Tunisia. The Manila clam, *Venerupis philippinarum*, a native of the western Pacific, was introduced into the Mediterranean lagoons of Languedoc (France) in the late 1970s, in 1983 to the Venice Lagoon and in 1985 to other parts of the Italian coast, including Sardinia (Cesari and Pelizzato 1985).

A comprehensive compilation of marine macrophytes introduced into Thau lagoon by way of oyster farming includes 43 species, one fifth of the algal flora (Verlaque 2001). *Sargassum muticum*, successfully introduced into the coastal lagoons of Languedoc and northern Spain with *C. gigas*, has rapidly covered artificial substrates and negatively affected native algae. At Thau lagoon, *S. muticum* has locally displaced the native *Cystoseira barbata* by blocking light and thus inhibiting the recruitment of the native species (Verlaque 1994). The fronds of *S. muticum* may entangle propellers or block water intake pipes of small vessels (Ribera and Boudouresque 1995). The slipper limpet, *Crepidula fornicata*, native to the Atlantic coast of North America, was first recorded in the Mediterranean in 1957 from mussel beds near Toulon (Zibrowius 1992); it has since arrived in the lagoons of Languedoc (Bertrand 1993). In high densities it may compete with commercial shellfish crop for space and food and may enhance silting. A small mytilid mussel native to East Asia, *Musculista senhousia*, is known from the shell-farming lagoons of Languedoc (Clanzig 1989) and the northern Adriatic, and from Sardinia where it was likely spread with mussel or clam seed. *M. senhousia* occasionally attains densities as high as 1000 ind/m^2, forming byssal mats on the surface of soft sediments, altering native benthic assemblages (Crooks 1998) and competing with cultivated shellfish (Mistri 2004a). Two parasitic copepods, an intestinal parasite, *Mytilicola orientalis* and a branchial parasite, *Mytilicola ostreae*, have been accidentally introduced with *C. gigas* to the shell-farming lagoons of Languedoc (Clanzig 1989).

Occhipinti-Ambrogi (2000) lists 30 species of alien benthic algae and invertebrates (out of 111 that occur in Italian waters; Occhipinti Ambrogi 2002) that occur in Venice lagoon; most were recorded from the early 1980s onwards. Some of the aliens have established large populations in the past 20 years. These include species intentionally introduced for mariculture, such as *C. gigas* (that had quickly replaced the native oyster *Ostrea adriatica*), and *R. philippinarum*, and unintentionally introduced aliens, such as the Indo-Pacific mud clam *Scapharca inaequivalvis*, that has been replacing the native clam *Cerastoderma glaucum* on soft bottoms. It appears that *R. philippinarum* and *S. inaequivalvis* may be competing for space in some locations in the Po river delta, a highly invaded estuary only a short distance south of the Venice lagoon (Occhipinti Ambrogi 2002). The xanthid crab *Dyspanopeus sayi,* originally from the North Atlantic coast, has occurred in the Lagoon since the 1980s and is currently the most common crab, exceeding in abundance the native species, *Carcinus mediterraneus* and *Pilumnus hirtellus* (Mizzan 1995). The two Japanese brown algae, *Undaria pinnatifida* and *S. muticum*, that arrived in the Lagoon in the early 1990s have also established large populations there (Curiel et al. 1998; Occhipinti Ambrogi 2000, 2002). The Japanese red alga, *Antithamnium pectinatum*, an epiphyte on other algae and on mussels, is widespread in the lagoon (Curiel et al. 1998; Occhipinti Ambrogi 2002). The bryozoan *Tricellaria inopinata,* occurring in Japan, Australia, New Zealand and the USA west coast (although it is possible that this wide distribution represents earlier introduction events), appeared in the Venice lagoon in 1982 (Occhipinti Ambrogi 2000), and later spread in the northern Adriatic and in the southern coast of England (Dyrynda et al. 2000). Its expansion in the lagoon during the late 1980s corresponded with a decline in the distribution of many of the native bryozoans and with major changes in hydrodynamic conditions that may have facilitated this invasion (Occhipinti Ambrogi 2000). In the late 1990s all bryozoans, native as well as alien, declined, possibly due to increased turbidity and deteriorating environmental conditions (Occhipinti Ambrogi 2000). In summary, the combination of high intentional and unintentional species introductions with high rates of disturbance in Mediterranean lagoons resulted in severe hotspots of bioinvasions.

31.5 Infamous Invaders

Here we describe in more detail the invasion of two species of plants and one animal. The species were selected based on the availability of information on their invasion and on the intensity of their impacts on the ecology of the marine system and/or on humans. This is by no means a complete account of infamous invaders but it provides an example of the tremendous ecological and economical effects of invasions in this region.

31.5.1 Macrophytes

Out of the 124 introduced macrophytes in the Mediterranean Sea, 9 are currently known to have harmful ecological and/or economical impacts (Boudouresque and Verlaque 2002). Boudouresque and Verlaque (2002) indicate that most of these nine species share some common life history traits including large size, perennial existence, efficient vegetative reproduction, toxic metabolites and the lack of conspicuous grazers. Solely or in combination they can dramatically change the underwater seascape. The co-occurrence of four macrophytes in a test station in Italy, for example, has resulted in lower total cover and diversity of the macroalgae assemblage (Piazzi and Cinelli 2003). Here we discuss two of the most notorious and well-known invaders, both of which from the genus *Caulerpa*. The others, according to Boudouresque and Verlaque (2002), are *Acrothamnion preissii, Asparagopsis armata, Lophocladia lallemandii, Womersleyella setacea* (Rhodophyta), *Sargassumm muticum, Stypopodiumschim peri* (Fucophyceae).

31.5.1.1 *Caulerpa taxifolia*

Caulerpa taxifolia introductions are now a global problem with populations in the Mediterranean, Australia and California. The aquarium trade is thought to be partly responsible for spreading this notorious seaweed. *C. taxifolia* was first recorded in the Mediterranean in 1984 in the waters under the building of the Oceanographic Museum in Monaco (Meinesz and Hesse 1991). It had arrived at the museum two years earlier after several years of intensive genetic selection under laboratory conditions in the Wilhelmina Zoo in Stuttgart, which resulted in a variety that does not shrivel, grows with incredible speed, and resists cool water temperatures, unlike its original tropical parent populations. Such qualities make it a desirable aquarium plant but also an extremely harmful invader. It also possesses highly potent toxins that make it almost unpalatable by consumers outside its natural range increasing its resistance even further. The original 1 m^2 patch outside the museum in Monaco has since extended and the species dispersed to other places, now creating huge meadows that cover extensive coastal areas in six countries in the western Mediterranean (Monaco, France, Italy, Spain, Croatia, and recently Tunisia). The species multiplies vegetatively via fragmentation, and travels great distances attached or caught in anchors and fishing nets, or via ballast water, starting new populations mostly around harbors and marinas where boats anchor (Boudouresque et al. 1995). Fragments can also spread directly by regional currents (Chisholm et al. 1997). The ecological effects of *C. taxifolia* are devastating. It outcompetes local floral species, mostly those that are low-laying, (i.e., encrusting or turf algal species), while erect species are less vulnerable (Ceccherelli et al. 2002). Seagrass areas of *Cymodocea nodosa* and *Posidonia oceanica*, that were already denuded due to other factors, were readily overtaken by *C. taxifolia*, which further facilitated decay and regression of the seagrasses (Ceccherelli and Cinelli 1999). However, it appears that healthy seagrass meadows are less prone to this invasion, again suggesting that destabilized environments are more invasible.

31.5.1.2 *Caulerpa racemosa*

The story of this *Caulerpa* species is more complex than that of *C. taxifolia*. It was first identified in the eastern Mediterranean as an Erythrean invader in the 1920. In the early 1990s a different variety appeared in the western Mediterranean and since then has become rapidly widespread throughout this basin. A detailed morphological and molecular study by Verlaque et al. (2003) suggested that there are three varieties of *C. racemosa* in the Mediterranean, one of which is the recent western Mediterranean invader (*C. racemosa* var. *cylindracea* (Sonder)) that is endemic to south-west Australia. Piazzi et al. (2001a) demonstrated its devastating effects on macroalgal diversity on a local scale. The invasive alga grows rapidly (up to 2 cm of horizontal stolon elongation per day). Consequently, it completely covered the study area six months after the initiation of the invasion, overgrowing the native macroalgal species and reducing their diversity considerably. Mostly affected again were turf and encrusting species and the effects were long-lasting; even in the winter months, when the invader population diminishes, the macroalgal community did not bounce back to its original structure. The two highly invasive *Caulerpa* species, *C. racemosa* and *C. taxifolia*, finally met in 1996 along the Tuscany coast near Leghorn, Italy. This meeting naturally facilitated comparative experimental studies on their ecological effects and on their inter-specific interactions (Balata et al. 2004; Ceccherelli et al. 2002; Piazzi et al. 2001a,b, 2003; Piazzi and Ceccherelli 2002). These studies showed that *C. racemosa* has a stronger effect on the local algal community and is a stronger competitor than *C. taxifolia* (it has negative effects on the growth and survival of the latter). Both species grow faster in higher densities, meaning that this invasion facilitates itself.

31.5.2 Animals

Although hundreds of animal species invaded the Mediterranean Sea in the past two centuries, little is known about their overall ecological impacts or even on specific interactions with native species. It is clear that benthic ecosystem engineers (see Chap. 16, Crooks), such as corals, mussels, oysters or bryozoans, are changing their new environment dramatically, just like the *Caulerpa* species do, by having negative (competitive) and/or positive (facilitation) effects, but these effects have rarely been studied in relation to introduced species. Filter feeding species can also have food-web effects as has been demonstrated in the San Francisco Bay (Carlton 1999), but they have not been studied in the Mediterranean. The dramatic effect of the Asian whelk *Rapana venosa* on the Black Sea fauna (Zolotarev 1996), and of the ctenophore *Mnemiopsis leidyi* on the Black Sea plankton, have not been reported in the Mediterranean sites where these two species have been recorded (Shiganova et al. 2001; see Chap. 30, Leppäkoski et al.). Studies on the trophic effects of predators are also rare, despite the fact that there are many new predatory fish and crabs in the Mediterranean (see, for instance, Mistri 2004b), perhaps because none poses a direct economic threat. Among pelagic predators, the jellyfish *Rhopilema nomadica* is considered a highly conspicuous and hazardous species, but its ecological effects have not been studied yet. We discuss this notorious invader here.

31.5.2.1 *Rhopilema nomadica*

As mentioned above, since the mid-1980s, vast annual blooms of the Indo-Pacific scyphomedusa, *Rhopilema nomadica*, have occurred in the eastern Mediterranean near the Israeli coast and northward towards Lebanon and Syria during the summer months (Lotan et al. 1994). By the mid-1990s, specimens appeared off the southeastern coast of Turkey (Kideys and Gucu 1995). Comparison of the temperature regime in the Levant Basin and the dependence of polyp strobilation on temperature suggest that the spring rise in water temperature causes a rapid strobilation, whereas winter and summer water temperatures inhibit or diminish reproduction (Lotan et al. 1994). Lotan et al. (1994) predicted that the sensitivity of the polyps to low water temperatures will prevent the future dispersal of *R. nomadica* to the western Mediterranean. Interestingly, to date, the millions of tiny strobila of *R. nomadica* that must exist on the bottom of the sea somewhere have not been found. The summer swarms of millions of jellyfish, sometimes stretching for 100km, are usually found several kilometers offshore, but when they get closer to shore they can adversely affect humans as mentioned earlier. Apart from the impacts on human activities, this planktivorus jellyfish must have major effects on the food-web structure of this ultra-oligotrophic basin (Galil et al. 1991). To date, no studies have been conducted to evaluate these effects.

31.6 Conclusions

The Mediterranean Sea is a hotspot of marine bioinvasions, with a rate of introductions that has been dramatically accelerating in the last decades of the twentieth century. Boudouresque and Verlaque (2002) suggested that 5–10 alien macrophyte species that are currently benign are expected to become highly invasive and affect humans and the environment in the Mediterranean in the next 20 years. This is probably true for other taxonomic groups as well. The eastern Mediterranean is the major recipient region for invasive species; we have shown that most of the species from the three groups analyzed here that occur in the Mediterranean can be found in the Levant basin. This is most likely related (1) to its proximity to a major pathway, the Suez Canal, (2) to favorable conditions for tropical species and (3) to the impoverishment of species in this region. There is a growing need for studies on the ecological and human impacts of the current invasions and on the characteristics of successful vs unsuccessful invasions. Such studies could increase our ability to predict future invasions and their effects and help develop better management tools to reduce invasions and their impacts. The Mediterranean is a small, almost closed, body of water with a vast and growing human population around it. It is exposed to tremendous anthropogenic stresses – among them bioinvasions – that destabilize its environment, making it even more prone to invasions. With the expected increase in human population, in trade, and in global warming, and with the expected expansion of its prime invasion vector, the Suez Canal, the future of the Mediterranean seems grim. Like in other places, measures need to be taken now to allow for a better future for this unique sea.

References

Achituv Y (1973) On the distribution and variability of the Indo-Pacific sea star *Asterina wega* (Echinodermata: Asteroidea) in the Mediterranean Sea. Mar Biol 18:333–336

Balata D, Piazzi L, Cinelli F (2004) A comparison among assemblages in areas invaded by *Caulerpa taxifolia* and *C. racemosa* on a subtidal Mediterranean rocky bottom. Mar Ecol – Pubblicazioni Della Stazione Zoologica Di Napoli I 25:1–13

Barash A, Danin Z (1988) Marine Mollusca at Rhodes. Isr J Zool 35:1–74

Barash A, Danin Z (1992) Fauna Palaestina: Mollusca I. Annotated list of Mediterranean molluscs of Israel and Sinai. The Israel Academy of Sciences and Humanities, Jerusalem

Ben Tuvia A (1966) Red Sea fishes recently found in the Mediterranean. Copeia 2:254–275

Ben Yami M (1955) 1954–1955: over-fishing or bad season? Fishermen Bull Haifa 6:10–14 [Hebrew]

Bertrand M (1993) Sur la présence de crépidules en Méditerrannée (*Crepidula fornicata L.*). Ann Soc HorticHist Nat Hérault 133:13–14

Béthoux JP, Gentili B, Raunet J, Tailliez D (1990) Warming trend in the western Mediterranean deep water. Nature 347:660–662

Béthoux JP, Gentili B, Tailliez D (1998) Warming and freshwater budget change in the Mediterranean since the 1940s, their possible relation to the greenhouse effect. Geophys Res Lett 1023–1026

Bianchi CN, Morri C (2000) Marine biodiversity of the Mediterranean Sea: situation, problems and prospects for future research. Mar Pollut Bull 40:367–376

Bilecenoglu M, Taskavak E, Bogac Kunt K (2002) Range extension of three lessepsian migrant fish (*Fistularia commersoni*, *Sphyraena flavicauda, Lagocephalus suezensis*) in the Mediterranean Sea. J Mar Biol Ass U K 82:525–526

Boudouresque CF, Verlaque M (2002) Biological pollution in the Mediterranean Sea: invasive versus introduced macrophytes. Mar Pollut Bull 44:32–38

Boudouresque CF, Meinesz A, Ribera MA, Ballesteros E (1995) Spread of the green alga *Caulerpa taxifolia* (Caulerpales, Chlorophyta) in the Mediterranean: possible consequences of a major ecological event. Sci Mar (Barcelona) 59 (Suppl 1)

Calman WT (1927) Appendix to the report on the Crustacea Decapoda (Brachyura). Zoological results of the Cambridge expedition to the Suez Canal (1924). XIII. Trans Zool Soc London 217

Carlton JT (1999) Molluscan invasions in marine and estuarine communities. Malacologia 41:439–454

Ceccherelli G, Cinelli F (1999) Effects of *Posidonia oceanica* canopy on *Caulerpa taxifolia* size in a north-western Mediterranean bay. J Exp Mar Biol Ecol 240:19–36

Ceccherelli G, Piazzi L, Balata D (2002) Spread of introduced *Caulerpa* species in macroalgal habitats. J Exp Mar Biol Ecol 280:1–11

Cesari P, Pelizzato M (1985) Molluschi pervenuti in Laguna di Venezia per apporti volontari o casuali. Acclimazione di Saccostrea commercialis (Iredale & Roughely, 1933) e di Tapes philippinarum (Adams & Reeve, 1850). Boll Malacol 21:237–274

Chervinsky J (1959) A systematic and biological comparison between the lizardfish (Saurida grandisquamis) from the Mediterranean and the Red Sea. Fishermen Bull Haifa 19:10–14 [English Abstract]

Chevaldonné P, Lejeusne C (2003) Regional warming-induced species shift in north-west Mediterranean marine caves. Ecol Lett 6:371

Chisholm JRM, Fernex FE, Mathieu D, Jaubert JM (1997) Wastewater, seagrass decline and algal proliferation on the Côte d'Azur. Mar Pollut Bull 34:78–84

Clanzig S (1989) Invertebrates d'introduction récente dans les lagunes méditerranéennes du Languedoc-Roussillon (France). Bull Soc Zool Fr 114:151–152

Corsini M, Kondilatos G, Economidis PS (2002) Lessepsian migrant *Fistularia commersonii* from the Rhodes marine area. J Fish Biol 61:1061–1062

Crooks JA (1998) Habitat alteration and community-level effects of an exotic mussel, *Musculista senhousia*. Mar Ecol Prog Ser 162:137–152

Curiel D, Bellemo G, Marzocchi M, Scattolin M, Parisi G (1998) Distribution of introduced Japanese macroalgae *Undaria pinnatifida, Sargassum muticum* (Phaeophyta) and *Antithamnion pectinatum* (Rhodophyta) in the Lagoon of Venice. Hydrobiologia 385:17–22

Di Geronimo I (1971) Prima segnalazione sulle coste italiane di *Brachiodontes variabiles* Krauss. Bollettino delle Sedute dell Accademia Gioenia di Scienze Naturali in Catania 10:847–852

Di Natale A (1982) Extra-Mediterranean species of mollusca along the southern Italian coasts. Malacologia 22:571–580

Dyrynda PEJ, Fairall VR, d'Hondt JL, Occhipinti Ambrogi A (2000) The distribution, origins and taxonomy of *Tricellaria inopinata* d'Hondt and Occhipinti Ambrogi, 1985, an invasive bryozoan new to the Atlantic. J Nat Hist 34:1993–2006

Fox HM (1924) The migration of a Red Sea crab through the Suez Canal. Nature 93:714

Fox HM (1927) Introduction. Zoological results of the Cambridge Expedition to the Suez Canal (1924). Trans Zool Soc London 22:1–88

Fredj G (1974) Stockage et exploitation des données en écologie marine. C. Considérations biogéographiques sur le peuplement benthique de la Méditerrannée. Mém Inst Océanogr (Monaco) 7:1–88

Galil BS (1986) Red Sea decapods along the Mediterranean coast of Israel: ecology and distribution. In: Dubinsky Z, Steinberger Y (eds) Environmental quality and ecosystem stability. Bar-Ilan University Press, Ramat Gan, pp 179–183

Galil BS (1997) Two Lessepsian migrant decapods new to the coast of Israel. Crustaceana 70

Galil BS (2000) A sea under siege – alien species in the Mediterranean. Biol Invas 2:77–186

Galil BS, Innocenti G (1999) Notes on the population structure of the portunid crab *Charybdis longicollis* Leene, parasitized by the rhizocephalan *Heterosaccus dollfusi* Boschma, off the Mediterranean coast of Israel. Bull Mar Sci 64:233–241

Galil BS, Kevrekidis K (2002) Exotic decapods and a stomatopod off Rhodes Island (Greece) and the eastern Mediterranean transient. Crustaceana 75:925–930

Galil BS, Lützen J (1995) Biological observations on *Heterosaccus dollfusi Boschma* (Cirripedia: Rhizocephala), a parasite of *Charybdis longicollis* Leene (Decapoda: Brachyura), a lessepsian migrant to the Mediterranean. J Crustacean Biol 15:659–670

Galil B, Zenetos A (2002) A sea change – exotics in the eastern Mediterranean. In: Leppäkoski E, Olenin S, Gollasch S (eds) Invasive aquatic species of Europe : distributions, impacts and management. Kluwer Academic Publishers, Dordrecht, The Netherlands, pp 325–336

Galil BS, Spanier E, Ferguson WW (1990) The Scyphomedusae off the Mediterranean coast of Israel, including two lessepsian migrants new to the Mediterranean. Zool Meded 64:95–105

Gofas S, Zenetos A (2003) Exotic molluscs in the Mediterranean basin: current status and perspectives. Oceanogr Mar Biol 41:237–277

Golani D, Ben-Tuvia A (1995) Lessepsian migration and the Mediterranean fisheries of Israel. In: Armantrout NB (ed) Conditions of the world's aquatic habits. Proceedings of the World Fishery Congress Theme 1. Oxford and IBH Publ Co Pvt, New Delhi, pp 279–289

Golani D, Galil B (1991) Trophic relationships of colonizing and indigenous goatfishes (Mullidae) in the eastern Mediterranean with special emphasis on decapod crustaceans. Hydrobiologia 218:27–33

Gomez F, Claustre H (2003) The genus *Asterodinium* (Dinophyceae) as a possible biological indicator of warming in the western Mediterranean Sea. J Mar Biol Assoc U K 83:173–174

Goren M (1993) Statistical aspects of the Red Sea ichthyofauna. Isr J Zool 39:293–298

Grofit E (1987) The trawl fishery in 1986. Fish Fishbreeding Isr 20:6–42 [English abstract]

Grosholz ED (2005) Recent biological invasion may hasten invasional meltdown by accelerating historical introductions. Proc Natl Acad Sci U S A 102:1088–1091

Gücü AC, Bingel F (1994) Trawlable species assemblages on the continental shelf on the Northeastern Levant Sea (Mediterranean) with an emphasis on Lesseptian migration. Acta Adriat 35:83–100

Hassan M, Harmelin-Vivien M, Bonhomme F (2003) Lessepsian invasion without bottleneck: example of two rabbitfish species (*Siganus rivulatus* and *Siganus luridus*). J Exp Mar Biol Ecol 291:219–232

Holthuis LB (1961) Report on a collection of Crustacea, Decapoda and Stomatopoda from Turkey and the Balkans. Zool Verh (Leiden) 1–67

Holthuis LB, Gottlieb E (1958) An annotated list of the decapod crustacea of the Mediterranean coast of Israel, with an appendix listing the Decapoda of the Eastern Mediterranean. Bull Res Counc Isr 7:1–126

Innocenti G, Vannini M, Galil BS (1998) Notes on the behaviour of the portunid crab *Charybdis longicollis* Leene parasitized by the rhizocephalan *Heterosaccus dollfusi* Boschma. J Nat Hist 32:1577–1585

Innocenti G, Pinter N, Galil BS (2003) Observations on the agonistic behavior of the swimming crab *Charybdis longicollis* Leene infected by the rhizocephalan barnacle *Heterosaccus dollfusi* Boschma. Can J Zool 81:173–176

Katagan T, Kocatas A, Zengin M, Ates AS (2004) An indo-pasific Stomatopod from the sea of Marmara: *Erugosquilla massavensis* (Kossmann, 1880). Crustaceana 77:381–383

Kevrekidis K, Kevrekidis T (1997) First record of the lessepsian migrant *Trachypenaeus curvirostris* (Stimpson) (Crustacea, Penaeidae) in Rodos Island. In: Proccedings of the *5th Hellenic Symposium on Oceanography and Fisheries Kavala, Greece,* NCMR Athens, Greece, pp. 115–116. [In Greek]

Kideys AE, Gucu AC (1995) *Rhopilema nomadica*: A poisonous Indo-Pacific Scyphomedusan new to the Mediterranean coast of Turkey. Isr J Zool 41:615–617

Kumulu M, Eroldogan OT, Aktas M, Gocer M (2002) A new shrimp record for the Turkish Seas: *Melicertus Hathor* (Burkenroad, 1959) (Penaeidae: Crustacea). Isr J Zool 48:246–247

Kuris AM (1997) Conceptual framework for biocontrol of introduced marine pests. In: Thresher RE (ed) Proceedings of the First International Workshop on the Demography, Impacts and Management of Introduced Populations of the European Crab, Carcinus maenas. Centre for Research on Introduced Marine Pests Technical Report, pp 66–68

Lascaratos A, Roether W, Nittis K, Klein B (1999) Recent changes in deep water formation and spreading in the eastern Mediterranean Sea: a review. Prog Oceanogr 44:1–36

Lewinsohn CH (1976) Crustacea Decapoda von der Insel Rhodos, Griechenland. Zool Meded Leiden 49:237–254

Lewinsohn CH, Holthuis LB (1986) The Crustacea Decapoda of Cyprus. Zool Verhand Leiden 230

Lotan A, Fine M, Ben-Hillel R (1994) Synchronization of the life cycle and dispersal pattern of the tropical invader scyphomedusan *Rhopilema nomadica* is temperature dependent. Mar Ecol Prog Ser 109:59–64

Lundberg B, Ogorek R, Galil B, Goren M (2004) Dietary choices of siganid fish at Shiqmona reef, Israel. Isr J Zool 50:39–53

Matta F (1969) Sull'acclimatazione e l'accrescimiento di Gryphea angulata Lmk. In alcune regione Italiane. Boll Pesca Piscic Idrobiol 23:171–258

Meinesz A, Hesse B (1991) Introduction of the tropical alga *Caulerpa taxifolia* and Its Invasion of the Northwestern Mediterranean. Oceanol Acta 14:415–426

Mistri M (2004a) Effects of *Musculista senhousia* mats on clam growth and survival: much ado about nothing? Aquaculture 241:207–218

Mistri M (2004b) Prey preference of *Carcinus aestuarii*: possible implications with the control of an invasive mytilid and Manila clam culture in a northern Adriatic lagoon. Aquaculture 230:261–272

Mizzan L (1995) Notes on presence and diffusion of *Dyspanopeus sayi* (Smith, 1869) (Crustacea, Decapoda, Xanthidae) in the Venitian Lagoon. Boll Museo Civico Storia Nat Venezia 44:121–129

di Monterosato TA (1878) Enumerazione e sinonima delle conchiglie mediterranee. G Sci Nat Econ Palermo 13:61–115

Nittis K, Lascaratos A (1999) Intermediate water formation in the Levantine Sea: the response to interannual variability of atmospheric forcing. In: Malanotte-Rizzoli P, Eremeev VN (eds) The eastern Mediterranean as a laboratory basin for the assessment of contrasting ecosystems. Kluwer Academic Publishers, Dordrecht, pp 441–446

Occhipinti Ambrogi A (2000) Biotic invasions in a Mediterranean lagoon. Biol Invas 2:165–176

Occhipinti Ambrogi A (2002) Current status of aquatic introductions in Italy. In: Leppäkoski E, Olenin S, Gollasch S (eds) Invasive aquatic species of Europe : distributions, impacts and management. Kluwer Academic Publishers, Dordrecht, The Netherlands, pp 311–324

Øksnebjerg B, Enzenross R, Enzenross L (1997) First record of Rhizocephala (Crustacea; Cirripedia) from Turkish waters, with notes on Lessepsian migration. Stuttg Beitr Naturkd Biol 557

Oliverio M, Gerosa G, Cocco M (1992) First record of *Pinctada radiata* (bivalvia, Pteriidae) epibiont on the loggerhead sea turtle *Caretta caretta* (Chelonia, Cheloniidae). Boll Malacol 28:149–152

Oren OH (1957) Changes in the temperature of the Eastern Mediterranean Sea in relation to the catch of the Israel trawl fishery during the years 1954/55 and 1955/56. Bull Inst Océanogr Monaco 1102:1–12

Pallary P (1912) Catalogue des mollusques du littoral méditerranéen de l'Egypte. Mém Inst Egypte 7:69–207

Pancucci-Papadopoulou MA, Simboura N, Zenetos A, Thessalou-Legaki M, Nicolaidou A (1999) Benthic invertebrate communities of Rodos (Rhodes) Island (SE Aegean Sea) as related to hydrological regime and geographical location. Isr J Zool 45:371–393

Piazzi L, Ceccherelli G (2002) Effects of competition between two introduced Caulerpa. Mar Ecol Prog Ser 225:189–195

Piazzi L, Cinelli F (2003) Evaluation of benthic macroalgal invasion in a harbour area of the western Mediterranean Sea. Eur J Phycol 38:223–231

Piazzi L, Balata D, Ceccherelli G, Cinelli F (2001a) Comparative study of the growth of the two co-occurring introduced green algae *Caulerpa taxifolia* and *Caulerpa racemosa* along the Tuscan coast (Italy, Western Mediterranean). Cryptogamie Algologie 22:459–466

Piazzi L, Ceccherelli G, Cinelli F (2001b) Threat to macroalgal diversity: effects of the introduced green alga *Caulerpa racemosa* in the Mediterranean. Mar Ecol Prog Ser 210:149–159

Piazzi L, Balata D, Cecchi E, Cinelli F (2003) Co-occurrence of *Caulerpa taxifolia* and *C. racemosa* in the Mediterranean Sea: interspecific interactions and influence on native macroalgal assemblages. Cryptogamie Algologie 24:233–243

Por FD (1978) Lessepsian migration: the influx of Red Sea biota into the Mediterranean by way of the Suez Canal. Springer, Berlin Heidelberg New York

Raimbault R (1964) Croissance des huitres atlantiques élevées dans les eaux méditerranéennes francaises. Sci Pêche 126:1–10

Ribera MA, Boudouresque CHF (1995) Introduced marine plants, with special reference to macroalgae: mechanism and impact. In: Round FE, Chapman DJ (eds) Progress in phycological research. Biopress Ltd, Bristol, p 188

Ribera Siguan MA (2002) Review of the non-native marine plants in the Mediterranean Sea. In: Leppäkoski E, Gollasch S, Olenin S (eds) Invasive aquatic species of Europe. Distribution, impacts and management. Kluwer Academic Publishers, Dordrecth Boston London, pp 291–310

Rilov G, Gasith A, Benayahu Y (2002) Effect of an exotic prey on the feeding pattern of a predatory snail. Mar Environ Res 54:85–98

Rilov G, Benayahu Y, Gasith A (2004) Prolonged lag in population outbreak of an invasive mussel: a shifting-habitat model. Biol Invas 6:347–364

Ruggieri G (1967) The Miocene and later evolution of the Mediterranean Sea. In: Adams CG, Ager DV (eds) Aspects of tethyan biogeography. Systematics Association Publishing, London

Ruiz GM, Carlton JT, Grosholz ED, Hines AH (1997) Global invasions of marine and estuarine habitats by non-indigenous species: mechanisms, extent, and consequences. Am Zool 37:621–632

Ruiz GM, Fofonoff PW, Carlton JT, Wonham MJ, Hines AH (2000) Invasion of coastal marine communities in North America: apparent patterns, processes, and biases. Annu Rev Ecol System 31:481–531

Sacchi CF, Occhipinti Ambrogi A, Sconfietti R (1989) Les lagunes nord-adriatiques: un environement conservateur ouvert aux nouveautés. Bull Soc Zool Fr 144:47–60

Safriel UN (1974) Vermetid gastropods and intertidal reefs in Israel and Bermuda. Science 186:1113–1115

Safriel UN, Gilboa A, Felsenburg T (1980) Distribution of rocky intertidal mussels in the Red Sea coasts of Sinai, the Suez Canal, and the Mediterranean coast of Israel, with special reference to recent colonizer. J Biogeog 7:39–62

Safriel UN, Sasson-Frosting Z (1988) Can colonizing mussel outcompete indigenous mussel? J Exp Mar Biol Ecol 117:211–226

Sara G, Romano M, Mazzola A (2000) The new Lessepsian entry *Brachidontes pharaonis* (Fischer P., 1870) (Bivalvia, Mytilidae) in the western Mediterranean: a physiological analysis under varying natural conditions. J Shellfish Res 19:967–977

Shefer S (2003) Factors and processes facilitating the invasion of *Brachidontes pharaonis* into the Levant basin. PhD dissertation Tel-Aviv University, Tel-Aviv, Israel

Shefer S, Abelson A, Mokady O, Geffen E (2004) Red to Mediterranean Sea bioinvasion: natural drift through the Suez Canal, or anthropogenic transport? Mol Ecol 13:2333–2343

Shiganova TA, Mirzoyan ZA, Studenikina EA, Volovik SP, Siokou-Frangou I, Zervoudaki S, Christou ED, Skirta AY, Dumont HJ (2001) Population development of the invader ctenophore Mnemiopsis leidyi, in the Black Sea and in other seas of the Mediterranean basin. Mar Biol 139:431–445

Simberloff D, Von Holle B (1999) Positive interactions of nonindigenous species: invasional meltdown? Biol Invas 1:21–32

Snovsky Z, Shapiro J (1999) The fisheries and Aquaculture of Israel 1997 in figures. Ministry of Agriculture, Department of Fisheries, xii+42 pp

Sorbini L (1988) Biogeography and climatology of Pliocene and Messinian fossil fish of Eastern Central Italy. Boll Museo Civico Storia Nat Verona 14:1–85

Steinitz W (1919) Memorial on the founding of a sea-station on the coast of Palestine for zoological investigations. Manuscript. 16 pp

Theocharis A, Lascaratos A (2000) Possible causes, origin, evolution and some consequences of the Eastern Mediterranean Transient during the period 1987–1999. In: Briand F (ed) The Eastern Mediterranean climatic transient, its origin, evolution and impact on the ecosystem. CIESM Workshop Series, Monaco, pp 1–86

Theocharis A, Georgopoulos D, Karagevrekis P, Iona A, Perivoliotis L, Charalambidis N (1992) Aegean influence in the deep layers of the Eastern Ionian Sea (October 1991). In: Rapports et Procès-verbaux des Réunions, Commission Internationale pour l'Exploration Scientifique de la Mer Méditerranée, 3, p 235

Theocharis A, Nittis K, Kontoyiannis H, Papageorgiou E, Balopoulos E (1999) Climatic changes in the Aegean Sea influence the eastern Mediterranean thermohaline circulation (1986–1997). Geophys Res Lett 26:1617–1620

Thresher RE, Werner M, Høeg JT, Svane I, Glenner H, Murphy NE, Wittwer C (2000) Developing the options for managing marine pests: specificity trials on the parasitic castrator, *Sacculina carcini*, against the European crab, *Carcinus maenas*, and related species. J Exp Mar Biol Ecol 254:37–51

d'Udekem d'Acoz C (1999) Inventaire et distribution des crustacés décapodes de l'Atlantique nord-oriental, de la Méditerranée et des eaux continentales adjacentes au nord de 25°N. Patrimoines naturels (M.N.H.N./S.P.N.) 40:1–383

Vadiya V, Shenuda S (1985) Role of the Suez Canal and flow from the Nile river in changing the salinity and fauna of the Mediterranean Sea. J Ichthyol 25:134–135

Verlaque M (1994) Inventaire des plantes introduites en Mediterranee: origines et repercussions sur l'environnement et les activites humaines. Oceanol Acta 17:1–23

Verlaque M (2001) Checklist of the macroalgae of Thau Lagoon (Herault, France), a hot spot of marine species introduction in Europe. Oceanol Acta 24:29–49

Verlaque M, Durand C, Huisman JM, Boudouresque CF, Le Parco Y (2003) On the identity and origin of the Mediterranean invasive *Caulerpla racemosa* (Caulerpales, Chlorophyta). Eur J Phycol 38:325–339

Wirszubski A (1953) On the biology and biotope of the red mullet, Mullus barbatus L. Bull Sea Fish Res St Caesarea 7:1–20

Wu P, Haines K, Pinardi N (2000) Toward an understanding of deep-water renewal in the eastern Mediterranean. J Phys Oceanogr 30:443–458

Yacobi YZ, Zohary T, Kress N, Hecht A, Robarts RD, Waiser M, Wood AM, Li WKW (1995) Chlorophyll distribution throughout the southeastern Mediterranean in relation to the physical structure of the water mass. J Mar Syst 6:179–190

Yokes B, Galil BS (2004) New record of alien decapods from the southwestern coast of Turkey. Rapp p-v Reun Comm Int Explor Sci mer Mediterr 37:556

Yokes B, Rudman WB (2004) Lessepsian opisthobranchs from southwestern coast of Turkey; five new records for Mediterranean. Rapp p-v Reun Comm Int Explor Sci mer Mediterr 37:557

Zaccaria M (1968) Studi geologici sulle isole del Dodecaneso (Mare Egeo) VIII. Fauna Calabriana dell'Isola di Rodi. Rivista italiana di Paleontologia e stratigrafia 74:275–306

Zibrowius H (1979) Serpulidae (Annelida Polychaeta) de l'Océand Indien arrives sur des coques de bateaux à Toulon (France, Méditerranée). Rapp p-v Reun Comm Int Explor Sci mer Mediterr 25/26 (4): 133–134.

Zibrowius H (1992) Ongoing modification of the Mediterranean marine fauna and flora by the establishment of exotic species. Mésogée 51:83–107

Zolotarev V (1996) The Black Sea ecosystem changes related to the introduction of new mollusc species. Mar Ecol (Pubblicazioni Della Stazione Zoologica Di Napoli I) 17:227–236

Chapter 32
A First Assessment of Invasive Marine Species on Chinese and Korean Coasts

Kyung Suk Seo and Yoon Lee

32.1 Introduction

Although the Asian coast of the Pacific Ocean is known to be a source of invaders to many other parts of the world, such as Australia and North America, there has been relatively little work on invasions into this region. However, a picture of invasions to Asia is beginning to emerge. For example, over 40 invasive marine species have been recognized in Japan (Otani 2004, 2006) – a number which certainly represents an underestimate (see Chap. 2, Carlton). Among these invaders are species native to regions often receiving species translocated from Japan, such as the west coast of North American. Examples of American invaders to Japan include the crab *Pyromaia tuberculata* and the barnacle *Balanus glandula*. Japan and North America also share species which are not native to either area, such as the mussel *Mytilus galloprovincialis* and the barnacle *Balanus amphitrite* (Crooks 2006). There are few published reports on the extent of invasions in other Asian countries, such as Korea and China, although it is assumed that many non-indigenous species have invaded here.

This chapter aims to summarize briefly the current knowledge on invasive marine species in these two countries, and will highlight species believed to be new in Chinese and Korean waters due to both anthropogenic transport as well as range expansions likely due to global warming and shifting temperature patterns (Park 2000, 2002; Kim 2002; Chap. 3, Lonhart).

32.2 Marine Invaders in Korea and China

Overall, there are 136 species suspected to be invaders in Chinese and Korean coastal waters (see Tables 32.1 and 32.2), resulting from either northward shifts in geographic ranges or anthropogenic introductions. These include a wide variety of taxa, including fish, invertebrates, macroalgae, phytoplankton, and pathogens (e.g., Kim IH 1992; Gong and Seo 2004; Lee and Kim 2006). Below we will highlight some of the more prominent.

G. Rilov, J.A. Crooks (eds.) *Biological Invasions in Marine Ecosystems.*
Ecological Studies 204,

Table 32.1 A list of marine animals considered invasive (either through introduction or range expansion) or cryptogenic in China and/or Korea

Classification	Species name
Annelida	*Balanus amphitrite*
Arthropoda	*Balanus eburneus*
	Balanus improvisus
	Balanus perfortatus
	Carcinus maenas
	Cercopagis pengoi
	Hydroides elegans
	Metapenaeus affinis
	Penaeus merguiensis
	Penaeus penicillatus
Bryozoans	*Amathia distans*
	Bugula californica
	Bugula neritina
	Celleporaria aperta
	Celleporina geminate
	Electra tenella
	Eurystomella bilabiata
	Escharoides excavate
	Fenestrulina malusii
	Mucronella perforate
	Schizoporella unicornis
	Tricellaria occidentilis
	Watersipora sutorquata
Cordata	*Cellepora lineaia*
	Ciona intestinalis
	Molgula manhattensis
	Styela clava
	Styela plicata
Cnidaria	*Mnemiopsis leidyi*
	Nemopilema nomurai
Echinodermata	*Asterias amurensis*
	Strongylocentrotus intermedius
Fishes	*Acipenser nudiventris*
	Amphirian akallopisos
	Amphirion ephippium
	Amphiprion sandaracinos
	Amphirion sebae
	Chelmon marginalis
	Chelmon muelleri
	Chelmon rostratus
	Chelmonops curiosus
	Chelmonops truncatus
	Calamus bajonado
	Clarias batrachus
	Ctenochaetus binotatus
	Ctenochaetus hawaiiensis
	Ctenochaetus marginatus
	Ctenochaetus truncatus
	Fugu rubripes
	Genicanthus bellus

(continued)

Table 32.1 (continued)

Classification	Species name
	Genicanthus watanabei
	Heniochus acuminatus
	Heniochus chrysostomus
	Heniochus intermedius
	Heniochus pleurotaenis
	Heniochus singuarius
	Heinochus varius
	Hemitaurichthys multispinosus
	Hemitaurichthys thompsoni
	Hemitaurichthys zoster
	Holacanthus tricolor
	Johnrandallia nigrirostris
	Lateolabrax maculates
	Lates calcarifer
	Lucioperca lucioperca
	Morone saxatilis
	Nemipterus japonicus
	Oncorhynchus kisutch
	Parachaetodon ocellatus
	Platichthys stellatus
	Pomacanthus maculosus
	Prognathodes acueatus
	Prognathodes aya
	Prognathodes brasiliensis
	Prognathodes dichrous
	Prognathodes falcifer
	Prognathodes guyotensis
	Pseudocaranx dentex
	Salmo salar
	Sciaenops ocellatus
	Scophthalmus maximus
	Trichoglaea lubricum
	Tubulipora flabellaris
	Verasper moseri
	Zanclus cornutus
Mollusca	*Chlamys senatoris nobilis*
	Crepidula onyx
	Dreissena polymorpha
	Haliotis laevigata
	Haliotis discus discus
	Haliotis gigantea
	Haliotis diversicolor eeue
	Haliotis diversicolor supertexta
	Haliotis iris
	Halocynthia roretzi
	Loligo japonica
	Mytilus galloprovincialis (M. edulis)
	Patinopecten yessoensis
	Pecten maxima
	Pomacea canaliculata

Table 32.2 A list of marine plants and pathogens considered invasive (either through introduction or range expansion) or cryptogenic in China and/or Korea

Classification	Species name
Pathogens	Human Papilloma Virus (shrimp)
	Irido Virus (fish)
	Nervous Necrosis Virus (fish)
	White-Spot Syndrom Virus (shrimp)
Phytoplankton	*Alexandrium catenella*
	Alexandrium tamarense
	Biddulphia sinensis
	Ceratium geniculatum
	Ceratium lamellicorne
	Ceratium lamellicorne
	Ceratium praelongum
	Chaetoceros concavicornis
	Cochlodinium polykrikoides
	Cylindrotheca closterium
	Dinophysis diegens
	Dinophysis shuttii
	Heterosigma akashiwo
	Histioneis highlei
	Nitzschia delicatissima
	Noctiluca scintillans
	Ornithocerus calolinae
	Ornithocerus serratus
	Oxytoxum reticulatum
	Peridinium peradiforme
	Phalacroma cuneus
	Pinnularia viridis
	Podolampas palmipes
	Prorocentrum micans
	Prorocentrum minimum
	Pyrocystis hamulus
	Pyrocystis lunula
	Skeletonema costatum
	Trichodesmium erythraeum
Seaweeds	*Desmarestia ligulata*
	Palmaria palmata
	Porphyra yezoensis

32.2.1 Pathogens in the Mariculture Industry

As described above, mariculture is of growing importance in the region, and this can serve as a vector for transfer of alien pathogens. For example, since the outbreak of White Spot Disease in China in 1993, which is caused by the White-Spot Syndrome Virus (WSSV), indigenous shrimp culture farming has been seriously damaged; 28.4% of shrimp farms in Korea were impacted by the virus during 2004. Without a strategic approach to eradicate such imported marine pests as soon as they are found, damage to the fishery industry could potentially spread along the entire coast (Subasinghe et al. 1998). Similarly, human papillomavirus (HPV) was recently found in 76% of shrimp farms, and is spreading gradually around the world (FAO 1993, 2004).

Other agents of disease have affected other fishery species. Indigenous turbot (*Scophthalmus maximus*), a typical mariculture species in Korea (NFRDI 2007), has been suffered from mortality of 15% owing to invasive virus diseases. Other fishes, such as *Pagrus major* and *Sebastes inermis*, have had populations impacted by the infections of iridovirus and viral nervous necrosis (VNN). It is assumed that the financial damage to mariculture by viruses or bacteria reaches about 250 million (US) dollars per year in Korea (NFRDI 2007). Infected fishes from one mariculture industry may transfer alien viruses to other mariculture industries, causing potential global disaster (Subasinghe et al. 1998).

32.2.2 Mussels

The Mediterranean mussel *Mytilus galloprovincialis* was introduced to Korean shores by an ocean-going ship during the 1950s, and has since spread along the shores of Korea and China and became dominant over the endemic species *M. coruscus* (Park 2004). As is common for this morphologically plastic taxa (Seed 1992; Gardner 2006), there has been some confusion regarding this invasion. In the 1990s, it was suspected that a new *Mytilus* species was suspected to be an invader from Japan (APEC 2004). However, based on allometric analysis, it was confirmed that both the Korean and Japanese blue mussels were identical to the Mediterranean species, *M. galloprovincialis* (Yoo 1992). It has also been suggested that mitochondrial DNA polymorphism of the blue mussel species complex indicates a possible evolutionary process, introgression from *M. edulis* to *M. galloprovincialis*, in the mussels in Korea (Kim et al. 1999a, b).

32.2.3 Sea Stars

The sea star, *Asterias amurensis*, known as a dominant invader in other regions such as Australia, has also invaded the south coast of Korea. As much as 4 tons of sea stars were harvested from two local surveys during 2005. This sea star spawns 20 million eggs once a year. The sea star is a generalist consumer, feeding on the most marine animals such as shellfishes, octopus, and fishes and is negatively impacting local mariculture and fishery industries (Jeon 2006). Due to its current and potential impacts on natural communities and mariculture there is an urgent need to eradicate this invader.

32.2.4 Tunicates

In the late 1990s, the invader *Ciona intestinalis* was first found on the east coast of Korea. Since then, it has become abundant on buoys and lines in mariculture areas, inhibiting the growth of fishery animals, and contaminating seawater by its dead debris, thereby leading to a mass destruction of fishery animals in the region (Chung 2000; Kim 2006). This species has heavily invaded other regions of the world, similarly causing massing damages to mariculture facilities. Four other non-indigenous tunicate species are recorded from Asian waters (see Table 32.1), but not much is known on their distribution and impacts.

32.2.5 Bryozoans

A total of 31 bryozoan species are known in Korea, and *Bugula neritina, Tricellaria occidentalis* and *Waterspipora subtorquata* are the most abundant species (Gong and Seo 2004; Hur et al. 1986). *Codonellia parviavicularia* is thought to be endemic, and *B. neritina* and *Schizoporella unicornis* are thought to be cosmopolitan (Gong and Seo 2004; Rho and Seo 1988; but see comments in Chap. 2, Carlton). Overall, 13 invasive bryozoan species are reported in Asia (Table 32.1).

32.2.6 Phytoplankton

There are currently 29 suspected phytoplankton non-native species in Chinese and Korean waters (Table 32.2), including some with a tropical origin (Kim DS 2006). Although the occurrence of tropical species of fishes, invertebrates, and macroalgae have been commonly observed in southern coastal area of Northeast Asia, the occurrence of non-native tropical phytoplankton was not recorded in the region until recent scientific surveys. The presence of tropical phytoplankton species might be attributed to recent changes in strength and direction the Kuroshio Current moving warmer southern waters close to shore of Northeast Asia, global climatic change, and/or introduction by ballast water discharges (Lee and Kim 2006). The establishment of marine phytoplankton species in new geographic areas faces many challenges including new regimes of salinity and nutrients. However, the intrinsic ability of phytoplankton to flourish in new environments (for example, the outbreaks of red tides), indicate that their risks could be extreme (North 1993).

32.3 Vectors of Invasions into Korean and Chinese Waters

The vectors of invasion which have brought species into Korea and China represent many of the vectors highlighted elsewhere throughout this book, including climate change and poleward expansion of populations. Of these many vectors, a potentially important one relates to the growth of mariculture and fisheries in the region. The fishing and mariculture industry has gradually expanded along Asian shores in the last several decades and continues to grow. In Korea, for example, it is expected that the catch (of either native or invasive species) through sport fishing in 2030 will be twice as large as in 1995 (Kim et al. 2004). With the expansion of the mariculture industry and commercial trade of fishery products, the possibility for invasion of non-indigenous species to Asian shores will no doubt increase (Kim DH et al. 2004). Careless treatment of imported non-indigenous fish can lead to the release of individuals and/or their pests. Such unintentional releases can lead to competition for habitat and/or food with endemic species, or other types of species interactions such as predation (Bang 2004).

In addition, the recent increase in international trade increases the potential for non-indigenous introductions in Asian waters through ballast water and hull fouling (OTA 1993; Hallegraeff and Bolch 1992; Chap. 6, Hewitt et al.). The first national project to study ballast water in Korea initiated in 2003 (MOMAF 2004, 2005). Various marine planktonic species were observed in the ballast water originating from the coastal waters of New Zealand, Taiwan, Singapore, Japan, Hong Kong and Pakistan, and entering the harbors of Inchen and Busan in Korea. Among the plankton observed in the ballast water, several harmful algae and non-indigenous aquatic species were identified (Yoo et al. 2006).

32.4 Conclusion

The number of known and suspected non-native marine species in Asian waters is relatively low compared to numbers in North America, Europe and the Mediterranean. As Asia has thousands of kilometers of coastline, many ports, and a very active international maritime trade, the numbers present are most probably hugely underestimated. The reasons for this are the paucity of surveys and studies on this topic and the lack of publications in international journals. It is clear that this information gap needs to be filled (Park et al. 1998). To do this, it will first be important to find the sources and the pathway of invasions of non-native species (APEC 2004). This would be possible with cooperation and international sharing of information between Asian countries and with other countries worldwide. Second, the development of a system to survey and identify non-indigenous species is essential. Adoption of successful concepts and techniques used in other regions for this purpose should be considered.

References

APEC (2004) Development of a regional risk management framework for APEC economies for use in the control and prevention of introduced marine pests. APEC MRC-WG Final Report, APEC.190 pp.

Bang SW (2004) Total control system on ecosystem risk foreign species. KEI Report (RE-02). pp 1–8 (Korean)

Chung SY (2000) Foreign species in East coast of Korea. Internet Hankyoreh, Oct. 2, 2000 (http://www.hani.co.kr/section-005100007/2000/005100007200010022055005.html) (Korean)

Crooks JA (2006) The arrival, establishment, and integration of an invasive marine mussel into foreign ecosystems. In: Koike F, Clout MN, Kawamichi M, De Poorter M, Iwatsuki K (eds) Proceedings of the International Conference on Assessment and Control of Invasion Risks. Shoukadoh Kook Sellers, Kyoto, Japan, and IUCN Press, Gland, Switzerland, pp 113–115

FAO (1993) Marine fisheries and law of the sea; a decade of change. Special chapter (revised) of The State of Food and Agriculture 1992, FAO Fisheries Circular No. 853. Rome, FAO. 66 pp.

FAO (2004) Preventive medicine. FAO Newsroom. Sept 2004

Gardner JPA (2006) A historical perspective of the genus *Mytilus* (Bivalvia: Mollusca) in New Zealand: multivariate morphometric analyses of fossil, midden, and contemporary blue mussels. Biol J Linn Soc 82:329–344

Gong YH, Seo J (2004) A taxonomic study on fouling Bryozoans from Korea. Underwater Sci Technol 5(1):11–16 (Korean)

Hallegraeff GM, Bolch CJ (1992) Transport of diatom and dinoflagellate resting spores in ships' ballast water: implications for plankton biogeography and aquaculture. J Plankton Res 14:1067–1084

Hur SB, Chun SC, Kim HS (1986) Chemical treatment of fouling organisms on fishing nets. Bull Korean Fish Soc 19(3):241–248. (Korean)

Jeon HH (2006) Story of ship(5); ballasted water. Digital Kookje, Sept. 14, 2006 (http://www.pusannews.co.kr/news2006/asp/news_print.asp?key=20060914.22025210340) (Korean)

Kim DH, Choi JS, Joung MS (2004) Control and counterplan for introduced marine pests. KMI Report (Korean)

Kim DS (2006) Found 12 foreign species on the coasts of Jeju Island. Digital Kookje, Aug. 1, 2006 (http://www.pusannews.co.kr/news2006/asp/center.asp?gbn=sr&code=1200&key=20060801.22014214653&sword1=제주&sword2=) (Korean)

Kim IH (1992) Invasion of foreign barnacles into Korean waters. Korean J Syst Zool 8:163–176 (Korean)

Kim IH (2006) Some non-indigenous invertebrates on Korean coasts. The 1st International Symposium on the Bio-invasion of Non-indigenous Species. Bukyung National University, July 31, 2006 (Korean)

Kim IS, Min BY, Yoon MH, Yoo MS, Kim DH (1999a) Unusual mitochondrial DNA polymorphism of the blue mussel (*Mytilus edulis*) species complex on the southern coast of Korea. Korean J Biol Sci 3:79–87 (Korean)

Kim IS, Min BY, Yoon MH, Kim DH (1999b) Mitochondrial DNA polymorphism of the blue mussel (*Mytilus edulis*) species complex on the east coast of Korea. Korean J Life Sci 9:262–267 (Korean)

Kim S (2002) Consideration on new research direction in marine environmental science in relation to climate change. Environ Pol Res 1:1–26

Lee JB, Kim HS (2006) Occurrence of tropic marine flagellates introduced from Kuroshio Current in the adjacent waters of Jeju Island. The 1st International Symposium on the Bio-invasion of Non-indigenous Species. Bukyung National University, July 31, 2006. (Korean)

MOMAF (2004) Research of response technique development discharge regulation ship ballast water (I). Ministry of Maritime Affairs and Fisheries (Korean)

MOMAF (2005) Research of response technique development discharge regulation ship ballast water (II). Ministry of Maritime Affairs and Fisheries (Korean)

NFRDI (2007) Website (http://portal.nfrdi.re.kr/bbs?id=fishecophoto). National Fisheries Research and Development Institute (Korean)

North EA (1993) Global marine biological diversity: a strategy for building conservation into decision making. Island Press, Washington D.C.

OTA (1993) Harmful non-indigenous species in the United States. OTA-F-565. OTA, U.S. Congress/Office of Technology Assessment. U.S. Government Printing Office

Otani M (2004) Introduced marine organisms in Japanese coastal waters, and the processes involved in their entry. Jpn J Benthol 59:45–57 (English abstract, in Japanese; English translation available from author)

Otani M (2006) Important vectors for marine organisms unintentionally introduced to Japanese waters. In: Koike F, Clout MN, Kawamichi M, de Poorter M, Iwatsuki K (eds) Assessment and control of biological invasion risks. Shoukado Book Sellers, Kyoto, Japan and World Conservation Union (IUCN), Gland, Switzerland, pp 92–103

Park YH (2000) Political suggestions on the foreign species control. Conserv Nat 110:49–53 (Korean)

Park YH (2002) International trends and policy recommendations related to non-indigenous species. Environ Pol Res 1:27–48 (Korean)

Park SH (2004) Fish Eye (44); mussels. Digital Kookje, Oct. 14, 2004 (http://www.pusannews.co.kr/news2006/asp/center.asp?gbn=sr&code=2500&key=20041014.22018201746&sword1=진주담치&sword2=) (Korean)

Park YH, Lee SD, Kim JW (1998) Suggestions on legal sanctions and programs related to non-indigenous species. KEI report (WO-05) (Korean)

Rho BJ, Seo JE (1988) One new species, *Codonellina parviavicularia*. N. sp., an Ascophoran Bryozoa from the Yellow Sea in Korea. Korean J Syst Zool 2:37–42 (Korean)

Seed R (1992) Systematics, evolution, and distribution of mussels belonging to the genus *Mytilus*: an overview. Am Malacol Bull 9:123–137

Subasinghe R, Arthur JR, Kumar D, Phillips MJ, Bernoth EM (1998) FAO's assistance for the responsible movement of live aquatic animals in Asia. FAO Aquaculture Newsletter 1020–3443, 19: 19–22

Yoo JK, Song TY, Hong HP, Jeong KM, Myung CS (2006) Marine plankton in ballast water of ship entering Korea. Ocean Polar Res 28:57–65 (Korean)

Yoo MS (1992) A taxonomical study of the shell morphology of blue mussel, *Mytilus edulis galloprovincialis* Lamarck in Korea and Japan. Bull Korean Fish Soc 25:165–170 (Korean)

Chapter 33
Invasions of Estuaries vs the Adjacent Open Coast: A Global Perspective

Rikke K. Preisler, Kerstin Wasson, Wim J. Wolff, and Megan C. Tyrrell

33.1 Habitat Differences in Marine Invasion Rates

Invasions by alien species have been reported from every marine habitat where surveys have been conducted for them. Conspicuous examples from around the globe include the brown alga *Sargassum mangarevense* in tropical coral reef systems (Andréfouët et al. 2004), the bivalve *Mytilus galloprovincialis* along temperate rocky shores (Steffani and Branch 2003), and the reef-building polychaete, *Ficopomatus enigmaticus* in estuaries (Schwindt et al. 2004). Despite numerous examples of marine invaders from a variety of habitats, little is known about how invasion rates of entire assemblages of organisms compare between different marine habitat types. And indeed most marine habitats have not been thoroughly surveyed – the majority of our understanding of marine invasions comes from shallow near-shore environments.

Some studies have attempted to quantify habitat differences in marine invasions, examining assemblages (both natives and aliens) at different scales. Within estuarine ecosystems, focus has been on comparisons between different salinities and substrates. (In this chapter an estuary is considered to be a ‘partly enclosed body of water by the coast in which sea water and fresh water mix‘ (Little 2000).) Wolff (1973) examined the benthic macroinvertebrates of four major estuaries in the Netherlands. He found that in the high salinity parts of these estuaries about 2% of the species were alien, in the brackish part about 20%, and in the tidal freshwater part about 8%. In non-tidal brackish waters the share of alien species was about 28%. Wolff (1999) re-analyzed these data and included three more estuaries in the northern Netherlands and Germany. He found that tidal and stagnant low salinity habitats of seven Dutch and German estuaries harbored a higher proportion of alien species (about 20%) than estuarine high salinity habitats (about 6%). This pattern was not clearly related to propagule pressure (harbors and aquaculture were not focused in the middle salinity). Lee et al. (2003) found that patterns of invasion varied along an estuarine gradient in San Francisco Bay; soft-bottom benthic communities at estuarine salinities were more invaded than communities at either brackish or marine salinities. Wasson et al. (2005) found hard substrates to be more invaded than soft substrates, and a site near the mouth of an estuary to be less

G. Rilov, J.A. Crooks (eds.) *Biological Invasions in Marine Ecosystems.*
Ecological Studies 204,

invaded than a site nearer the head of an estuary in Central California, despite the harbors in this estuary being closer to the mouth.

Few studies have made comparisons between habitats within versus outside of estuaries. Reise et al. (1999) examined benthic macroinvertebrates on the North Sea coast, and found that 6% of species were alien in coastal habitats as well as in high salinity estuarine habitats, while 20% of species in the brackish parts of estuaries were alien species. Wasson et al. (2005) found that while about 11% of the invertebrate species within a Californian estuary were alien, only 1.5% of those on the adjacent open coast were.

Understanding habitat differences in invasion rates is critical for development and implementation of management strategies (see also Chap. 7, Johnston et al.; Chap. 8, Miller and Ruiz; Chap. 12, Olyarnik et al.). As a first step, identifying those habitats that appear to be most vulnerable to establishment of alien species, and those most susceptible to major impacts from invasions is valuable for focusing prevention efforts. Second, unraveling the mechanisms behind invasion success vs failure in different habitats is critical for developing control strategies. Here we attempt to provide a global perspective on differences in invasion rates between estuaries vs adjacent open coast habitats in temperate zones, as one of many possible explorations of habitat differences in invasion rates at a broad geographic scale.

33.2 An Estuarine Emphasis to Marine Invasion Research

Traditionally, marine invasion research has been focused especially intensely on estuarine habitats. Pioneering studies were conducted in estuarine habitats, for instance by Dolgikh (1969) in the Tiligul Estuary, Ukraine and Carlton (1979) in San Francisco Bay, California. Most comprehensive recent syntheses of alien marine diversity (e.g., Cohen and Carlton 1995; Hewitt et al. 1999; Ruiz et al. 2000) focus on estuaries.

The focus on estuaries presumably stems from qualitative observations of their comparatively high invasion rates. Estuaries are often the sites of intensive human activities such as shipping and aquaculture, especially of oysters. These two human activities are considered responsible for the majority of marine introductions (Carlton 1989; Cohen et al. 1995; Carlton 1996; Ruiz et al. 1997; Reise et al. 1999; Emmett et al. 2000; Wolff 2005). Additionally, alien species may have a particularly good opportunity to become established in estuaries because low native species richness may render competition with natives weak (Elton 1958; Lodge 1993; Wolff 1973, 1999), because human alterations such as pollution and diking result in natives not being well adapted to current conditions (e.g., Byers 2002; Kennish 2002; Chap. 14, Byers), or because the semi-closed circulation of larvae allows for retention of small populations (Wasson et al. 2005). Despite these multiple explanations for high estuarine invasion rates, there has been no broad scale assessment of whether estuaries are in fact more invaded than adjacent coastal habitats. We therefore provide such an evaluation as a means to explore the above hypotheses in greater depth.

33.3 A Synthesis of Global Temperate Invertebrate Data on Invasions

33.3.1 Estuarine vs Open Coast Invasions

In order to broadly characterize estuarine vs coastal invasion rates globally, we solicited invertebrate datasets from researchers around the world. We requested data from surveys with comparable search effort in adjacent estuarine vs open coast habitats. We obtained four comprehensive invertebrate datasets (native, alien, and in some cases, cryptogenic species) and four additional datasets for only alien species, all for temperate zones (Fig. 33.1, Table 33.1). The datasets were collected for various purposes, with different sampling effort, taxonomic focus, and substrate types between regions. It would have been more desirable, of course, to have a single team consistently collect and identify specimens from all sites, using an identical protocol everywhere; such an investigation would yield sound, comprehensive results. Until such an investigation is undertaken, analysis and synthesis of data sets collected by different teams is useful for suggesting interesting preliminary trends in global invasion rates of estuaries vs coasts.

While search effort differed between coasts, search effort between estuaries vs adjacent coasts was broadly similar for each regional pairing. For the most part, similar substrate types (e.g., hard vs soft; vegetated vs unvegetated areas) and depths were searched in each regional pairing (Table 33.2). Hence, the comparisons of invasions between estuaries and open coasts appear robust. Comparisons

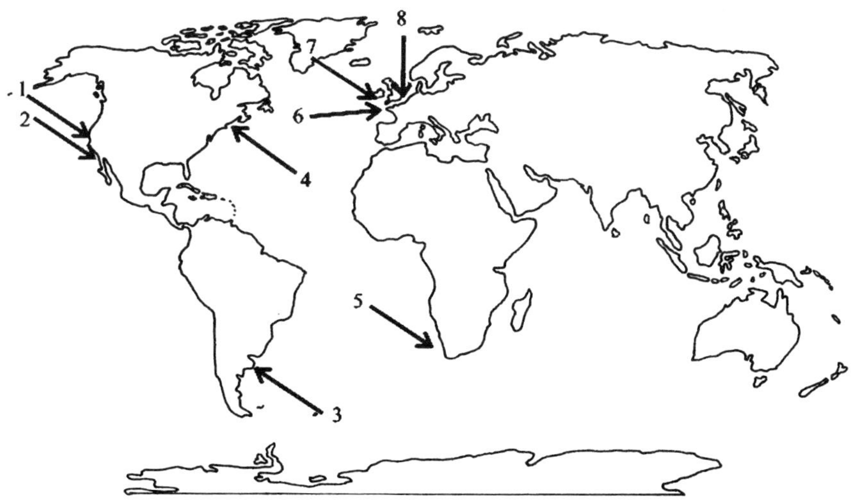

Fig. 33.1 Eight regions used for global comparison of estuarine vs coastal invertebrate invasion rates. Numbering corresponds to Table 33.1

Table 33.1 Sources of invertebrate data sets for global comparisons of habitat differences in invasion[a]

Bioregion	Location	Map	Estuarine habitat	Coastal habitat	Info on natives	Published source	Unpublished source
Northeast Pacific	Central California, USA	1	Elkhorn Slough	Rocky intertidal Carmel-Pigeon Point	yes	Wasson et al. 2001, Wasson et al. 2005	K. Wasson; J. Pearse
Northeast Pacific	Southern California, USA	2	Mission Bay	San Diego County	no	-	J. Crooks
Southwest Atlantic	Argentina	3	Mar Chiquita Coastal Lagoon	Mar del Plata	yes	-	O. Iribarne
Northwest Atlantic	New Hampshire, USA	4	Hilton Park	Odiorne State Park	yes	Tyrrell 2002	M. Tyrrell
Southeast Atlantic	South Africa	5	Langebaan Lagoon, False Bay, Knysna Estuary	Entire coast of South Africa	no	-	T. Robinson
Northeast Atlantic	France	6	Bay des Veys, Prevost Lagoon, Berre Lagoon, Bay of Morlaix, Charente Estuary	Calvados coast, Arcachon Basin, French Atlantic Coast	no	-	P. Garcia-Meunier
Northeast Atlantic	Ireland	7	Bantry Bay	Irish coast	no	-	D. Minchin
Northeast Atlantic	Netherlands	8	4 large estuaries in Delta area	Adjacent North Sea	yes	Wolff 1973	-
Northeast Atlantic	Netherlands, Germany	8	7 large estuaries in the Netherlands and Germany	High-salinity part of same estuaries	no	Wolff 1999	-

[a]The four surveys that included data for natives as well as alien species were used for the analyses summarized in Table 33.3; all studies were used for Tables 33.4 and 33.5

between regions in proportions of alien/native species are also likely sound, but comparisons of absolute numbers of alien species between regions are weaker, since search methods and effort differed. We report on trends between regions but emphasize that further studies, with consistent methods across regions, are needed for more rigorous assessments. We omitted species that are known to occur only in harbors in either habitat, thus our analysis pertains only to alien species that have established outside harbors. Finally, only established aliens were included in all databases; species that were only ever seen once, or were anecdotally reported as alien species were not included in the data sets.

For the four comprehensive datasets, we found that the percent of all invertebrates that were alien was higher, but not significantly (paired t-test, df=3, $p<0.203$), for estuaries (11.9%) vs coasts (7.9%) (Table 33.3). However, for three out of four of these regions, estuaries had a noticeably higher percentage of alien species than coasts (Fig. 33.2).

For all eight datasets of alien species, we found that on average 86.9% of the total number of alien species in a region occurred in estuaries whereas as only 33.2% occurred on the open coast. This result was highly significant (paired t-test, df=7, $p<0.006$). In six of the eight regions, there were clearly more alien species in the estuaries than on the open coasts (Fig. 33.3). For most regions, the majority of recorded alien invertebrates occurred only in estuaries, with only a few present on the coast or in both estuary and coast (Table 33.4).

Overall, our global dataset includes 198 alien invertebrates (Table 33.5). Of these, nine are found only in coastal habitats around the world, while the rest occur in estuaries (solely, or both in estuaries and on the open coast). This result provides a striking contrast. Patterns for the four major taxa included in most surveys are similar to the combined data for all taxa, but a few patterns are noteworthy. It appears that

Table 33.2 Surveyed habitats for four regions[a]

Habitat	Depth	Substrate	Northeast Pacific	Northwest Atlantic	Southwest Atlantic	Northeast Atlantic
Coast	Intertidal	Soft	No	No	Yes	Yes
		Hard	Yes	Yes	Yes	Yes
		Vegetation	No	Yes	Yes	Yes
	Subtidal	Soft	No	No	Yes	Yes
		Hard	No	No	Yes	Yes
		Vegetation	No	No	Yes	Yes
Estuary	Intertidal	Soft	Yes	No	Yes	Yes
		Hard	Yes	Yes	Yes	Yes
		Vegetation	No	Yes	Yes	Yes
	Subtidal	Soft	No	No	Yes	Yes
		Hard	No	No	Yes	Yes
		Vegetation	No	No	Yes	Yes

[a]List of habitat types, depth ranges (intertidal, subtidal), and substrate types surveyed in regions for which both alien and native species information was available

Table 33.3 Comparison of invasion rates across four regions[a]

	Estuary			Coast		
	Native	Cryptogenic	Alien	Native	Cryptogenic	Alien
North East Pacific (Central California)						
Annelids	90% (122)	6.5% (9)	3.5% (5)	99.5% (185)	0.5% (1)	0% (0)
Molluscs	91% (141)	0.5% (1)	8.5% (13)	99% (198)	0.5% (1)	0.5% (1)
Crustaceans	87% (134)	1.5% (2)	11.5% (18)	100% (79)	0% (0)	0% (0)
Bryozoans	35% (6)	24% (4)	41% (7)	87.5% (49)	7% (4)	5.5% (3)
All Taxa	84% (455)	5% (29)	11% (60)	96% (567)	2.5% (13)	1.5% (8)
North West Atlantic (New Hampshire)						
Annelids	100% (2)	0% (0)	0% (0)	100% (4)	0% (0)	0% (0)
Molluscs	91% (10)	0% (0)	9% (1)	89% (8)	0% (0)	11% (1)
Crustaceans	78% (7)	0% (0)	22% (2)	60% (3)	0% (0)	40% (2)
Bryozoans	33.3% (1)	33.3% (1)	33.3% (1)	100% (1)	0% (0)	0% (0)
All Taxa	78% (25)	3% (1)	19% (6)	79% (23)	0% (0)	21% (6)
South West Atlantic (Argentina)						
Annelids	83% (5)	0% (0)	17% (1)	100% (10)	0% (0)	0% (0)
Molluscs	100% (5)	0% (0)	0% (0)	100% (11)	0% (0)	0% (0)
Crustaceans	100% (6)	0% (0)	0% (0)	86% (12)	0% (0)	14% (2)
Bryozoans	0% (0)	0% (0)	0% (0)	100% (1)	0% (0)	0% (0)
All Taxa	89% (16)	0% (0)	11% (2)	95% (59)	0% (0)	5% (3)
North East Atlantic (Netherlands) – 1973						
Annelids	99% (79)	–	1% (1)	98% (51)	–	2% (1)
Molluscs	89% (56)	–	11% (7)	96% (25)	–	4% (1)
Crustaceans	91% (60)	–	9% (6)	91% (31)	–	9% (3)
–	–	–	–	–	–	–
All Taxa	93% (195)	–	7% (14)	96% (107)	–	4% (5)

[a]Percent (and absolute number in parentheses) of total invertebrate species that were native, cryptogenic and alien are shown for adjacent estuarine and open coast habitats, for four taxa surveyed in most studies and for all taxa combined

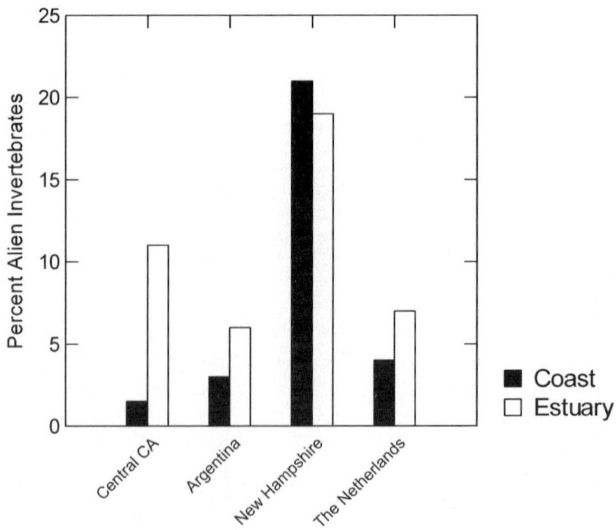

Fig. 33.2 Percent of total documented invertebrate species that were alien, in coastal vs estuarine habitats in four regions

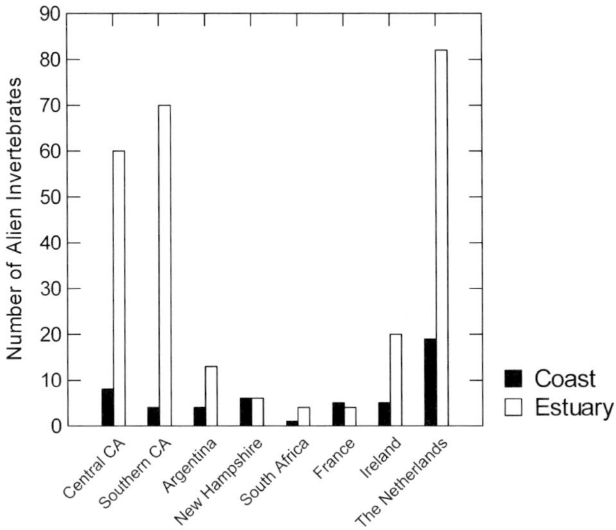

Fig. 33.3 Number of alien invertebrate species that were found in coastal vs estuarine habitats in eight regions

Table 33.4 Comparison of habitat associations of alien species across eight regions[a]

Taxon	Estuary only	Estuary & Coast	Coast only
Northeast Pacific (Central California)			
Annelids	100% (5)	0% (0)	0% (0)
Molluscs	93% (14)	7% (1)	0% (0)
Crustaceans	100% (17)	0% (0)	0% (0)
Bryozoans	63% (5)	37% (3)	0% (0)
All taxa	87% (52)	13% (8)	0% (0)
Northeast Pacific (Southern California, San Diego)			
Annelids	100% (11)	0% (0)	0% (0)
Molluscs	56% (5)	33% (3)	11% (1)
Crustaceans	100% (24)	0% (0)	0% (0)
Bryozoans	100% (6)	0% (0)	0% (0)
All Taxa	94% (66)	4% (3)	2% (1)
Southwest Atlantic (South America, Argentina)			
Annelids	0% (0)	100% (2)	0% (0)
Molluscs	100% (3)	0% (0)	0% (0)
Crustaceans	67% (4)	0% (0)	33% (2)
Bryozoans	0% (0)	0% (0)	0% (0)
All Taxa	73.3% (11)	13.3% (2)	13.3% (2)
Northwest Atlantic (New Hampshire)			
Annelids	0% (0)	0% (0)	0% (0)
Molluscs	0% (0)	100% (1)	0% (0)
Crustaceans	0% (0)	100% (2)	0% (0)
Bryozoans	0% (0)	100% (1)	0% (0)
All Taxa	0% (0)	100% (6)	0% (0)

(continued)

Table 33.4 (continued)

Taxon	Estuary only	Estuary & Coast	Coast only
Northeast Atlantic (Ireland)			
Annelids	100% (1)	0% (0)	0% (0)
Molluscs	86% (6)	14% (1)	0% (0)
Crustaceans	80% (8)	10% (1)	10% (1)
Bryozoans	0% (0)	0% (0)	0% (0)
All taxa	77% (17)	14% (3)	9% (2)
Northeast Atlantic (Netherlands)			
Annelids	86% (12)	7% (1)	7% (1)
Molluscs	70% (14)	25% (5)	5% (1)
Crustaceans	62% (16)	27% (7)	11% (3)
Bryozoans	100% (7)	0% (0)	0% (0)
All Taxa	78% (68)	16% (14)	6% (5)
Northeast Atlantic (France)			
Annelids	25% (1)	0% (0)	75% (3)
Molluscs	33% (1)	0% (0)	67% (2)
Crustaceans	100% (2)	0% (0)	0% (0)
Bryozoans	0% (0)	0% (0)	0% (0)
All Taxa	44% (4)	0% (0)	56% (5)
Southeast Atlantic (South Africa)			
Annelids	0% (0)	0% (0)	0% (0)
Molluscs	67% (2)	0% (0)	33% (1)
Crustaceans	100% (1)	0% (0)	0% (0)
Bryozoans	0% (0)	0% (0)	0% (0)
All Taxa	80% (4)	0% (0)	20% (1)

[a]Percent (and absolute number in parentheses) of alien species that were found only in estuaries, in both estuary and coast, or only on the open coast are shown for four taxa surveyed in most studies and for all taxa combined

bryozoans are almost exclusively invaders of estuarine, not coastal habitats. Of the four data sets that provided data on all types of species, a minority of the estuarine bryozoan species present are native (Table 33.3), and almost all reported alien bryozoans occur in estuaries (Tables 33.4 and 33.5). In contrast, crustaceans are the most common invaders of coastal habitats (Tables 33.4 and 33.5). Although we only synthesized invertebrate data in this study, greater numbers of estuarine vs coastal aliens have also been noted for algae (Wolff 2005, for the Netherlands) and fish (Crooks, personal communication, for southern California).

33.3.2 Regional Differences in Estuarine vs Coastal Invasions

Examination of our global comparison of estuarine vs coastal invertebrate invasions reveals striking regional differences (Table 33.3). Of course, these differences may in part be due to differences in search effort or sampling method, because these

Table 33.5 Alien species reported from eight regions[a–e]

Region / Species	Northeast Pacific		South West Atlantic	North West Atlantic	Northeast Atlantic			South East Atlantic			
	Central California	Southern California	Argentina	New Hampshire	Ireland	Netherlands	France	South Africa	# of regions that species is found in	Estuarine records	Coastal records
PORIFERA	-	-	-	-	-	-	-	-	-	-	-
Cliona celata var. lobata	C, E	-	-	-	-	-	-	-	1	1	1
Halichondria bowerbanki	C, E	-	-	-	-	-	-	-	1	1	1
Haliclona loosanoffi	E	-	-	-	-	E	-	-	2	2	0
Haliclona rosea	-	-	-	-	-	E	-	-	1	1	0
Haliclona cf. simplex	-	-	-	-	-	E	-	-	1	1	0
Hymeniacidon perlevis	-	-	-	-	-	E	-	-	1	1	0
Hymeniacidon sinapium	C, E	-	-	-	-	-	-	-	1	1	1
Mycale micracanthoxea	-	-	-	-	-	E	-	-	1	1	0
Scypha scaldiensis	-	-	-	-	-	E	-	-	1	1	0
CNIDARIA	-	-	-	-	-	-	-	-	-	-	-
HYDROZOA	-	-	-	-	-	-	-	-	-	-	-
Clavopsella navis	-	-	-	-	-	E	-	-	1	1	0
Cordylophora caspia	E	-	-	-	-	E	-	-	2	2	0
Ectopleura crocea	E	E	-	-	-	-	-	-	2	2	0
Garveia franciscana	-	-	-	-	-	E	-	-	1	1	0
Gonionemus vertens	-	-	-	-	-	E	-	-	1	1	0
Nemopsis bachei	-	-	-	-	-	E	-	-	1	1	0
Sarsia tubulosa	E	-	-	-	-	-	-	-	1	1	0
Ostroumovia inkermanica	-	-	-	-	-	E	-	-	1	1	0
Thieliana navis	-	-	-	-	-	E	-	-	1	1	0
ANTHOZOA	-	-	-	-	-	-	-	-	-	-	-
Bunodeopsis sp.	-	E	-	-	-	-	-	-	1	1	0

(continued)

Table 33.5 (continued)

Species / Region	Northeast Pacific		South West Atlantic	North West Atlantic	Northeast Atlantic			South East Atlantic			
	Central California	Southern California	Argentina	New Hampshire	Ireland	Netherlands	France	South Africa	# of regions that species is found in	Estuarine records	Coastal records
Diadumene cincta	-	-	-	-	-	C, E	-	-	1	1	1
Diadumene franciscana	E	E	-	-	-	-	-	-	2	2	0
Diadumene leucolena	E	-	-	-	-	-	-	-	1	1	0
Diadumene lineata	E	E	-	C, E	E	E	-	-	5	5	1
NEMATODA	-	-	-	-	-	-	-	-	-	-	-
Anguillicola crassus	-	-	-	-	C, E	E	-	-	2	2	1
PLATYHELMINTHES	-	-	-	-	-	-	-	-	-	-	-
TREMATODA	-	-	-	-	-	-	-	-	-	-	-
Cercaria batillariae	E	-	-	-	-	-	-	-	1	1	0
TURBELLARIA	-	-	-	-	-	-	-	-	-	-	-
Euplana gracilis	-	-	-	-	-	E	-	-	1	1	0
Stylochus flevensis	-	-	-	-	-	E	-	-	1	1	0
ANNELIDS	-	-	-	-	-	-	-	-	-	-	-
OLIGOCHAETES	-	-	-	-	-	-	-	-	-	-	-
Tubificoides brownea	-	E	-	-	-	-	-	-	1	1	0
Tubificoides heterochaetus	-	-	-	-	-	E	-	-	1	1	0
POLYCHAETA	-	-	-	-	-	-	-	-	-	-	-
Amblyosyllis speciosa	-	E	-	-	-	-	-	-	1	1	0
Aphelochaeta marioni	-	-	-	-	-	E	-	-	1	1	0
Boccardia ligerica	-	-	C, E	-	-	-	-	-	1	1	1
Boccardia polybranchia	-	-	-	-	-	-	C	-	1	0	1
Boccardia semibranchiata	-	-	-	-	-	-	E	-	1	1	0
Clymenella torquata	-	-	-	-	-	E	-	-	1	1	0
Demonax sp.	-	E	-	-	-	-	-	-	1	1	0

Ficopomatus enigmaticus	E	-	C, E	-	E	E	-	-	4	4	1
Heteromastus filiformis	E	-	-	-	-	-	-	-	1	1	0
Hydroides dirampa	-	E	-	-	-	-	-	-	1	1	0
Hydroides elegans	-	E	-	-	-	E	-	-	2	2	0
Janua brasiliensis	-	-	-	-	-	E	-	-	1	1	0
Lycatopsis pontica	-	E	-	-	-	-	-	-	1	1	0
Marenzelleria cf. wireni	-	-	-	-	-	E	-	-	1	1	0
Microphthalmus similis	-	-	-	-	-	C	-	-	1	0	1
Myrianida pachycera	-	E	-	-	-	-	-	-	1	1	0
Neanthes acuminata	-	E	-	-	-	-	-	-	1	1	0
Nereis virens	-	-	-	-	-	E	-	-	1	1	0
Nicolea sp. A	-	E	-	-	-	-	-	-	1	1	0
Polydora ciliata	-	-	-	-	-	-	C	-	1	0	1
Polydora hoplura	-	-	-	-	-	E	C	-	2	1	1
Polydora cornuta	E	-	-	-	-	-	-	-	1	1	0
Proceraea cornuta	-	-	-	-	-	E	-	-	1	1	0
Pseudopolydora paucibranchiata	E	E	-	-	-	-	-	-	2	2	0
Sabellaria spinulosa	-	-	-	-	-	C, E	-	-	1	1	1
Streblospio benedicti	E	E	-	-	-	-	-	-	2	2	0

MOLLUSCA	-	-	-	-	-	-	-	-	-	-	-
POLYPLACOPHORA	-	-	-	-	-	-	-	-	-	-	-
Lepidopleurus cancellatus	-	-	-	-	-	C, E	-	-	1	1	1
GASTROPODS	-	-	-	-	-	-	-	-	-	-	-
Aoelidella takanosimensis	-	C, E	-	-	-	-	-	-	1	1	1
Babakina festiva	-	E	-	-	-	-	-	-	1	1	0
Batillaria attramentaria	E	-	-	-	-	-	-	-	1	1	0
Callistoma zizyphinum	-	-	-	-	-	E	-	-	1	1	0
Calyptraea chinensis	-	-	-	-	E	E	-	-	2	2	0
Catriona rickettsi	-	C, E	-	-	-	-	-	-	1	1	1
Corambe batava	-	-	-	-	-	E	-	-	1	1	0

(continued)

Table 33.5 (continued)

Region / Species	Northeast Pacific		South West Atlantic	North West Atlantic	Northeast Atlantic			South East Atlantic			
	Central California	Southern California	Argentina	New Hampshire	Ireland	Netherlands	France	South Africa	# of regions that species is found in	Estuarine records	Coastal records
Crepidula fornicata	-	-	-	-	E	C, E	C	-	3	2	2
Cyclope neritea	-	-	-	-	-	-	E	-	1	1	0
Gibbula cineraria	-	-	-	-	-	E	-	-	1	1	0
Littorina littorea	-	-	-	C, E	-	-	-	-	1	1	1
Littorina saxatilis	-	-	-	-	-	-	-	E	1	1	0
Myosotella myosotis	E	-	-	-	-	-	-	-	1	1	0
Ocenebra erinacea	-	-	-	-	-	E	-	-	1	1	0
Okenia plana	E	-	-	-	-	-	-	-	1	1	0
Philine auriformis	E	C	-	-	-	-	-	-	2	1	1
Potamopyrgus antipodarum	-	-	-	-	E	E	-	-	2	2	0
Rapana venosa	-	-	E	-	-	-	C	-	2	1	1
Tenellia adspersa	E	-	-	-	-	-	-	-	1	1	0
Urosalpinx cinerea	E	-	-	-	-	E	-	-	2	2	0
BIVALVIA	-	-	-	-	-	-	-	-	-	-	-
Corbicula fluminalis	-	-	-	-	-	E	-	-	1	1	0
Crassostrea gigas	E	E	E	-	E	E	-	E	6	6	0
Crassostrea virginica	E	-	-	-	-	-	-	-	1	1	0
Dreissena polymorpha	-	-	-	-	-	E	-	-	1	1	0
Ensis directus	-	-	-	-	-	C, E	-	-	1	1	1
Gemma gemma	E	-	-	-	-	-	-	-	1	1	0
Limnoperna fortunei	-	-	E	-	-	-	-	-	1	1	0
Mercenaria mercenaria	-	?	-	-	-	E	-	-	2	1	0
Musculista senhousia	E	E	-	-	-	-	-	-	2	2	0
Mytilopsis leucophaeata	-	-	-	-	-	E	-	-	1	1	0

Mytilus galloprovincialis	C, E	C, E	-	-	-	-	-	C	3	2	3
Mya arenaria	E	-	-	-	E	E	-	-	3	3	0
Nutallia obscurata	E	-	-	-	-	-	-	-	1	1	0
Lyrodus pedicellatus	E	E	-	-	-	-	-	-	2	2	0
Petricola pholadiformis	-	-	-	-	-	C, E	-	-	1	1	1
Psiloteredo megotara	-	-	-	-	-	C	-	-	1	0	1
Ruditapes semidecussata	-	-	-	-	E	-	-	-	1	1	0
Spisula solidissima	-	-	-	-	-	E	-	-	1	1	0
Teredo navalis	-	-	-	-	C, E	C, E	-	-	2	2	2
Theora fragilis	-	E	-	-	-	-	-	-	1	1	0
Venerupis philippinarum	E	-	-	-	-	-	?	-	2	1	0
BRYOZOA	-	-	-	-	-	-	-	-	-	-	-
Alcyonidium polyoum	-	-	-	C, E	-	-	-	-	1	1	1
Amathia vidovici	E	-	-	-	-	-	-	-	1	1	0
Bowerbankia gracilis	E	E	-	-	-	E	-	-	3	3	0
Bowerbankia imbricata	-	-	-	-	-	E	-	-	1	1	0
Bugula neritina	C, E	E	-	-	-	?	-	-	3	2	1
Bugula simplex	-	-	-	-	-	E	-	-	1	1	0
Bugula stolonifera	E	-	-	-	-	E	-	-	2	2	0
Conopeum tenuissimum	E	-	-	-	-	-	-	-	1	1	0
Cryptosula pallasiana	C, E	E	-	-	-	-	-	-	2	2	1
Schizoporella unicornis	C, E	-	-	-	-	-	-	-	1	1	1
Tricellaria inopinata	-	-	-	-	-	E	-	-	1	1	0
Victoriella pavida	-	-	-	-	-	E	-	-	1	1	0
Walkeria uva	-	-	-	-	-	E	-	-	1	1	0
Watersipora arcuata	-	E	-	-	-	-	-	-	1	1	0
Watersipora subtorquata	E	E	-	-	-	-	-	-	2	2	0
Zoobotryon verticillatum	-	E	-	-	-	-	-	-	1	1	0
KAMPTOZOA	-	-	-	-	-	-	-	-	-	-	-
Barentsia benedeni	E	-	-	-	-	-	-	-	1	1	0

(continued)

Table 33.5 (continued)

	Northeast Pacific		South West Atlantic	North West Atlantic	Northeast Atlantic			South East Atlantic			
Species / Region	Central California	Southern California	Argentina	New Hampshire	Ireland	Netherlands	France	South Africa	# of regions that species is found in	Estuarine records	Coastal records
CRUSTACEA	-	-	-	-	-	-	-	-	-	-	-
AMPHIPODA	-	-	-	-	-	-	-	-	-	-	-
Ampithoe valida	E	-	-	-	-	-	-	-	1	1	0
Aoroides secunda	-	E	-	-	-	-	-	-	1	1	0
Caprella mutica	E	-	-	-	C	E	-	-	3	2	1
Chelicorophium curvispinum	-	-	-	-	-	E	-	-	1	1	0
Chelura terebrans	-	-	-	-	-	-	-	E	1	1	0
Corophium acherusicum	E	E	-	-	-	-	-	-	2	2	0
Corophium heteroceratum	-	E	-	-	-	-	-	-	1	1	0
Elasmopus rapax	-	E	-	-	-	-	-	-	1	1	0
Ericthonius brasiliensis	-	E	-	-	-	-	-	-	1	1	0
Gammarus tigrinus	-	-	-	-	-	E	-	-	1	1	0
Grandidierella japonica	E	E	-	-	-	-	-	-	2	2	0
Incisocalliope aestuarius	-	-	-	-	-	E	-	-	1	1	0
Jassa marmorata	E	E	-	-	-	-	-	-	2	2	0
Leucothoe alata	-	E	-	-	-	-	-	-	1	1	0
Liljeborgia sp.	-	E	-	-	-	-	-	-	1	1	0
Melita nitida	E	-	-	-	-	E	-	-	2	2	0
Monocorophium insidiosum	E	E	E	-	-	-	-	-	3	3	0
Monocorophium uenoi	E	E	-	-	-	-	-	-	2	2	0
Orchestia cavimana	-	-	-	-	-	E	-	-	1	1	0
Parapleustes derzhavini	E	-	-	-	-	-	-	-	1	1	0
Platorchestia platensis	-	-	-	-	-	E	-	-	1	1	0
Stenothoe valida	-	E	-	-	-	-	-	-	1	1	0

CIRRIPEDIA	-	-	-	-	-	-	-	-	-	-	-
Balanus amphitrite	-	E	E	-	-	C, E	-	-	3	3	1
Balanus balanus	-	-	-	-	-	C, E	-	-	1	1	1
Balanus eburneus	-	-	-	-	-	E	-	-	1	1	0
Balanus glandula	-	-	C	-	-	-	-	-	1	0	1
Balanus improvisus	E	-	-	-	E	C, E	-	-	3	3	1
Elminius modestus	-	-	-	-	C, E	C, E	-	-	2	2	2
Megabalanus coccopoma	-	-	-	-	-	C	-	-	1	0	1
Megabalanus tintinnabulum	-	-	-	-	-	C	-	-	1	0	1
COPEPODA	-	-	-	-	-	-	-	-	-	-	-
Acartia tonsa	-	-	-	-	-	E	-	-	1	1	0
Eurytemora americana	-	-	-	-	-	E	-	-	1	1	0
Herrmannella duggani	-	-	-	-	E	-	-	-	1	1	0
Myicola ostreae	-	-	-	-	E	E	-	-	2	2	0
Mytilicola intestinalis	-	-	-	-	E	E	-	-	2	2	0
Mytilicola orientalis	E	-	-	-	E	E	-	-	3	3	0
Pseudodiaptomus marinus	-	E	-	-	-	-	-	-	1	1	0
DECAPODA	-	-	-	-	-	-	-	-	-	-	-
Eriocheir sinensis	-	-	-	-	-	C, E	E	-	2	2	1
Callinectes sapidus	-	-	-	-	-	C, E	E	-	2	2	1
Carcinus maenas	E	-	-	C, E	-	-	-	?	3	2	1
Hemigrapsus penicillatus	-	-	-	-	-	E	-	-	1	1	0
Hemigrapsus sanguineus	-	-	-	C, E	-	C, E	-	-	2	2	2
Palaemon macrodactylus	E	E	-	-	-	-	-	-	2	2	0
Pyromaia tuberculata	-	-	?	-	-	-	-	-	1	0	0
Rhithropanopeus harrisii	-	-	-	-	-	E	-	-	1	1	0
ISOPODA	-	-	-	-	-	-	-	-	-	-	-
Halophiloscia couchii	-	-	?	-	-	-	-	-	1	0	0
Iais californica	E	E	-	-	-	-	-	-	2	2	0
Idotea baltica	-	-	C	-	-	-	-	-	1	0	1
Idotea metallica	-	-	E	-	-	-	-	-	1	1	0
Ligia exotica	-	-	E	-	-	-	-	-	1	1	0

(continued)

Table 33.5 (continued)

	Northeast Pacific		South West Atlantic	North West Atlantic	Northeast Atlantic			South East Atlantic			
Species / **Region**	Central California	Southern California	Argentina	New Hampshire	Ireland	Netherlands	France	South Africa	# of regions that species is found in	Estuarine records	Coastal records
Limnoria quadripunctata	E	E	-	-	-	C	-	-	3	2	1
Limnoria tripunctata	-	E	-	-	E	-	-	-	2	2	0
Paranthura japonica	-	E	-	-	-	-	-	-	1	1	0
Porcellanum ovatum	-	-	-	-	E	-	-	-	1	1	0
Sphaeroma quoyanum	E	E	-	-	-	-	-	-	2	2	0
Sphaeroma serratum	-	-	?	-	-	-	-	-	1	0	0
Sphaeroma walkeri	-	E	-	-	-	-	-	-	1	1	0
Synidotea laevidorsalis	-	-	?	-	-	-	-	-	1	0	0
MYSIDACEA	-	-	-	-	-	-	-	-	-	-	-
Hemimysis anomala	-	-	-	-	-	E	-	-	1	1	0
OSTRACODA	-	-	-	-	-	-	-	-	-	-	-
Aspidoconcha limnoriae	-	E	-	-	-	-	-	-	1	1	0
Redekea californica	-	E	-	-	-	-	-	-	1	1	0
TANAICIDAE	-	-	-	-	-	-	-	-	-	-	-
Sinelobus	E	-	-	-	-	-	-	-	1	1	0
Sinelobus? stanfordi	-	E	-	-	-	-	-	-	1	1	0
UROCHORDATA	-	-	-	-	-	-	-	-	-	-	-
Aplidium glabrum	-	-	-	-	-	E	-	-	1	1	0
Ascidia sp.	-	E	-	-	-	-	-	-	1	1	0
Ascidia zara	-	E	-	-	-	-	-	-	1	1	0
Bostrichobranchus pilularis	-	E	-	-	-	-	-	-	1	1	0
Botryllus schlosseri	-	E	E	-	-	-	-	-	2	2	0
Botryllus firmus	-	E	-	-	-	-	-	-	1	1	0
Botryllus sp. A	-	E	-	-	-	-	-	-	1	1	0

Botrylloides violaceus	E	E	-	C, E	-	E	-	-	4	4	1
Ciona intestinalis	-	E	E	-	-	-	-	-	2	2	0
Ciona robusta	-	-	E	-	-	-	-	-	1	1	0
Ciona savigny	-	E	-	-	-	-	-	-	1	1	0
Clavelina lapadiformis	-	-	-	-	-	-	-	E	1	1	0
Diplosoma listerianusm	-	-	-	-	-	E	-	-	1	1	0
Microcosmus squamiger	-	E	-	-	-	-	-	-	1	1	0
Molgula manhattensis	C, E	-	E	-	-	E	-	-	3	3	1
Phallusia mammilata	-	-	-	-	C	-	-	-	1	0	1
Polyandrocarpa zorritensis	-	E	-	-	-	-	-	-	1	1	0
Styela canopus	-	E	-	-	-	-	-	-	1	1	0
Styela clava	E	E	-	-	E	E	-	-	4	4	0
Styela plicata	-	E	-	-	-	-	-	-	1	1	0
Symplegma brakenhelmi	-	E	-	-	-	-	-	-	1	1	0
Symplegma reptans	-	E	-	-	-	-	-	-	1	1	0

[a]Species found in three or more localities listed in dark gray shading
[b]Species found only in coastal habitats listed in light gray shading
[c]Species whose habitat association is unknown marked with "?"
[d]'C' means that a species is found on an open coast
[e]'E' means that a species is found in an estuary

were not consistent across regions. We suspect, however, that at least the most pronounced differences are real.

For the four comprehensive datasets, we found that the difference between coast and estuary was most dramatic in California, where 1.5% of documented coastal invertebrates vs 11% of estuarine invertebrates were alien (Table 33.3). The same trend of lower percentage of aliens in coastal vs estuarine habitats was present in Argentina (5% vs 11%, for a very low sample size) and the Netherlands (4% vs 7%), but much less pronounced (mostly due to higher coastal invasion rates). In New Hampshire, coastal and estuarine habitats had similar percentages of aliens (21% vs 19%), much higher in both habitats than in the other regions. Survey methods differed between regions, with those in California and the Netherlands representing extensive searches over many years, while those for Argentina and New Hampshire consisted of more focused, shorter efforts. (A more focused, short-term survey at the same estuary in California detected only 47 total species, with aliens representing 21% of species in soft substrates, 52% on hard substrates – much higher than the 11% overall level calculated when every native species ever reported for the estuary is included in the analysis; Wasson et al. 2005). Nevertheless, we suspect that the general pattern – no difference in estuarine vs coastal invasion rates in New Hampshire, vs marked differences in the other three regions (Fig. 33.2) – is real and will be borne out by more consistently designed future comparisons.

In the comparison of habitat associations of alien species from eight regions, strong differences also emerged. In six regions (Central and Southern California, Argentina, South Africa, Ireland, and the Netherlands), there was a markedly higher number of alien species documented from estuarine vs coastal habitats (Fig. 33.3) – more than 70% of aliens from these regions were reported exclusively in estuaries (Table 33.4). In two regions (New Hampshire and France), numbers of alien species in estuaries vs coasts were similar (and in both cases low). Detailed patterns for these latter two differed: in New Hampshire, the six documented aliens all occurred in *both* estuarine and coastal habitats, while in France, none of the species occurred in both habitats (four aliens were found only in the estuary, and five only on the coast).

Not only did habitat associations of aliens differ among the eight regions, but also the absolute number of aliens recorded from each region (Table 33.4 and Fig. 33.3). California and the Netherlands have many more documented aliens (60 or more) than the other regions (which have 4–22 alien species). This result may simply reflect greater search effort, and should be compared to findings of other biogeographic syntheses (see also Chap. 2, Carlton).

Our compilation of alien species (Table 33.5) also reveals that 18 species are common invaders, shared across at least 3 regions. These are well-known aliens reported in other major syntheses of global invasions (Ruiz et al. 2000). None of these common invaders occurs only in coastal habitats – this cosmopolitan suite of aliens is mostly comprised of estuarine specialists, and some broadly tolerant species found in estuaries and on the coast.

33.4 Hypotheses for Higher Invasion Rates of Estuaries vs Coasts

What mechanisms are behind the higher number of aliens in estuaries vs adjacent open coast settings in most regions? We present five broad hypotheses for this phenomenon, and consider whether the patterns that emerge from our global analysis support them. We also draw on selected examples from other studies that support each hypothesis, and formulate examples of predictions that could be rigorously tested in future studies to determine the relative importance of each hypothesis for particular taxa or regions.

Differences in number of aliens in coastal vs estuarine habitats could stem both from differences in numbers of individuals introduced (propagule pressure) and from differences in establishment rates following introduction (Chap. 7, Johnston et al.; Chap.12 , Olyarnik et al.). The first hypothesis explores the former, and the remaining four hypotheses involve the latter. For establishment success, the mechanism could involve physical conditions (our second and third hypotheses) or biological interactions (our fourth and fifth ones). For a particular alien species' distribution in one region, a single mechanism may explain most of the pattern.

33.4.1 Greater Alien Propagule Pressure in Estuaries

33.4.1.1 Hypothesis

Estuaries are more invaded than adjacent coastal habitats because estuaries are subject to greater alien propagule pressure.

33.4.1.2 General Explanation

Because the two vectors considered responsible for the greatest number of marine introductions, shipping and oyster culture, are generally located in estuaries, it seems likely that many more alien species are transported among estuaries than open coast environments. This hypothesis has long been raised in the literature, and continues to be frequently invoked (e.g., Carlton 1979, 1985; Carlton and Geller 1993; Campbell and Hewitt 1999; Reise et al. 1999; Ruiz et al. 2000; Minchin and Eno 2002). Not only does it seem likely that more species are transported from estuarine than from coastal environments, it is also plausible that a greater proportion of estuarine species survive transport than do coastal ones. Conditions in ballast water tanks or in oyster transport from one aquaculture hub to another may involve fluctuating temperature and salinity conditions that broadly tolerant estuarine species are better adapted to than their coastal counterparts (Wolff 1999). Both higher transport rate and higher survival during transport would contribute to greater numbers of aliens in estuaries than coasts, given similar establishment rates following introduction.

33.4.1.3 Support from the Global Analysis

Overall, only nine alien species, a tiny fraction of the total, occurred exclusively on the coast (Table 33.5). In contrast, 37 species were found in both coastal and estuarine settings. There is no reason to expect that species that can tolerate estuarine conditions should be better at establishing on the coast than purely coastal species, so the contrast (37 mixed occurrence vs 9 purely coastal occurrences) seems likely due to greater propagule pressure in estuaries (and subsequent spread of these species from estuaries to coasts).

The contrast we report between US coasts, where similar numbers of aliens are found on the open coast but far more are found in the Pacific than Atlantic estuary, could also in part be due to the above explanation. For instance, while Elkhorn Slough (California) and Great Bay (New Hampshire) both have regular small boat traffic, only the former has a long history of alien oyster culturing, which is considered responsible for many of the introductions in Elkhorn Slough (Wasson et al. 2001). Thus, the low number of aliens on both the New Hampshire coast and estuary relative to other regions may be due, at least in part, to lower numbers of introductions.

33.4.1.4 Support from Other Observations

The highest number of alien species recorded for any one place is from an estuary, San Francisco Bay, no doubt in part due to high propagule pressure in this major shipping and former oyster culturing center (Cohen and Carlton 1998). One species-level example of propagule pressure being a key predictor of invasion patterns is the Asian mud snail *Batillaria attramentaria,* found on the US Pacific coast only in bays and estuaries where Asian oysters were cultured (Byers 1999). In this case, there appears to be a perfect correspondence of human introduction and invasion patterns (although if this mudflat species had been introduced to the open coast, it would probably not have survived, so the hypothesis in the following section would also be applicable). However, there are contrasting examples from this region. The European green crab *Carcinus maenas*, a coastal and estuarine species in its home and some introduced (e.g., US Atlantic coast) ranges is found only in estuaries along the US Pacific coast (Cohen et al. 1995). Its recent spread in this region was most likely due to larval transport on oceanic currents (Grosholz and Ruiz 1995), so propagule pressure should have been comparable in estuarine and coastal habitats along the hundreds of miles of coast where larvae have spread. Its absence from the open coast is very likely to be attributed to one of the other hypotheses described below, rather than due to introduction mechanisms.

33.4.1.5 Testable Predictions

If propagule pressure alone explains much of the variation in observed numbers of established aliens, then areas with comparable propagule pressure should have comparable numbers of established aliens. One fruitful comparison might be of small shipping harbors with comparable boat traffic along the open coast vs those in small estuaries, in a region not exposed to other vectors (no ballast water or oyster culturing). Given these constraints, the introduction mechanisms (small vessels transporting mostly hull-fouling organisms) and rates should be the similar between habitats. If numbers of established aliens (collected on settlement plates for example) are comparable in these paired coastal and estuarine harbors, it would provide support for the hypothesis that propagule pressure alone explains most of the variation in alien species richness. If, however, numbers of alien species are higher in the estuary, one of the subsequent hypotheses regarding post-introduction success must be invoked.

Another tactic is to examine the identity and number of established alien species in as opposed to near harbors from a survey as described above. If propagule pressure were the key predictor of invasion success, then we would expect an equal proportion of the alien species present to be established in natural habitats a short distance (e.g., 0.5 km) outside of harbors in both estuarine and coastal habitats. Anecdotally (Wasson, unpublished data), this does not seem to be the case for Central California – almost none of the few dozen alien invertebrates common in open coast harbors have established on the open coast. Data of this sort rigorously collected for multiple regions would refute the hypothesis that propagule pressure alone explains high estuarine invasion rates.

33.4.2 Estuarine Species are Better Suited to Estuarine Conditions

33.4.2.1 Hypothesis

Estuaries are more invaded than adjacent coastal habitats because the species that are transported between regions are typically adapted to estuarine conditions, and more likely to spread and establish within the new region in estuarine vs open coast habitats.

33.4.2.2 General Explanation

If, as postulated in the first hypothesis, more species are moved from estuary to estuary (by shipping and oyster culture in particular) than from open coast to open coast, then the identity of the species that are transported should also differ. Species that were transported from an estuary are likely to have adaptations for the physical and biological conditions of estuaries rather than open coast environments, and the transported species are therefore more likely to successfully establish in estuaries than on the open coast. This is a rather obvious observation that is simply a correlate of the first explanation. However, we raise it as a separate hypothesis because the predictions are somewhat different. The first explanation invokes high propagule pressure as the cause of high estuarine invasion rates, while this hypothesis attributes the difference in alien species numbers to higher establishment rates in estuarine vs open coast habitats following arrival, due to better matching between physical conditions in a species' native and invaded ranges. This applies both to initial establishment at the first site in a new region, and moreover to subsequent spread within a region.

33.4.2.3 Support from Global Analysis

The above logic would suggest that in regions where estuarine and coastal conditions are more similar, more alien species (largely transported from estuary to estuary) would establish on the open coast, while in regions where conditions contrast more sharply, aliens would fail to establish on the open coast. This might explain some of the difference we observed in coastal invasion rates between the Pacific and Atlantic coasts of the US. On the Pacific coast of California, high wave action may be intolerable for most estuarine species, which could account for the absence of all but 8 of the 60 species found in the estuary. On the Atlantic coast in the Gulf of Maine, wave action is much lower; thereby potentially lowering this barrier to invasion by estuarine species – and the six aliens reported from the estuary also occur on the open coast.

33.4.2.4 Support from Other Observations

Invasion rates vary along an estuarine gradient. Near the mouth, where conditions are most similar to those in adjacent marine habitats, the percentage of established aliens appears to be lower than near the head, where conditions are truly estuarine, fluctuating greatly in salinity and temperature, both daily and seasonally (Carlton 1979; Wolff 1999; Cohen and Carlton 1995; Wasson et al. 2005). In addition to salinity, water movement may also be a predictor of invasion success. In Central California (Wasson, personal observation) and New Hampshire (Tyrrell, personal observation), adjacent estuarine sites with high current speeds are far less invaded than those with slow currents. These observations of higher numbers of established aliens in more typically estuarine conditions (i.e., less water exchange, highly variable salinities) support this explanation, but are also consistent with hypotheses 33.4.3.1 and 33.4.5.1 below.

33.4.2.5 Testable Predictions

Many species arrive to a new region by one mechanism (e.g., ballast water transport), and then are secondarily transported to sites within this new region by other mechanisms (e.g., small boat traffic or natural larval transport) (Wasson et al. 2001). For such a species, secondary introduction rates to estuarine and open coast harbors should be similar following initial introduction to a region. However, if this hypothesis is supported, establishment rates should be lower in open coast harbors, because of poor matching of physical conditions under which the alien evolved. (Failure at establishment could also be due to interactions with natives or other previously introduced species, explanations 33.4.4.2 and 33.4.5.2 below, but this seems less likely in artificial harbor habitats than in natural settings.)

33.4.3 Establishment is Facilitated by the Limited Circulation in Estuaries

33.4.3.1 Hypothesis

Estuaries are more invaded than adjacent coastal habitats because they represent relatively closed systems in which small numbers of introduced individuals can more readily establish breeding populations than on the open coast.

33.4.3.2 General Explanation

Since it is probably common for only a few individuals of an alien species to be introduced in one event, establishment of an alien is unlikely if larvae or adults disperse too widely to allow for subsequent successful mating. Estuarine tidal circulation is more limited than that on the open coast, and reduced risk of gamete and larval dilution might facilitate establishment of a population from a small number of individuals in estuaries (J. Byers, personal communication; Wasson et al. 2005).

33.4.3.3 Support

None that we know of, from our data set or other observations.

33.4.3.4 Testable Predictions

This explanation should only apply to species with a long-distance dispersal stage. Therefore, a strong test of this hypothesis would be to compare estuarine vs coastal invasion rates by species with a mobile dispersal stage as a larva (e.g., benthic polychaete larva) or adult (e.g., jellyfish) to those with very limited dispersal capacity (e.g., a snail with crawl-away larvae). If the dispersive species are more likely to invade estuaries than coasts, but the non-dispersing ones are not, then this hypothesis is supported. Additionally, if species that rely on external fertilization (such as many bivalves) have higher establishment success in estuaries than on the open coast, then the circulation hypothesis is also supported. This type of analysis could readily be carried out on existing datasets of alien species, such as the one included here or that provided by Ruiz et al. (2000), with coding added for dispersal ability. Another approach would be to examine propagule densities of a single alien species with long-distance dispersal. For instance, *Carcinus maenas* larvae spreading on oceanic currents that enter estuaries may be retained and result in higher densities of recruiting juveniles than on the open coast, simply as a function of circulation patterns.

33.4.4 Estuaries Have Undergone More Human Alterations

33.4.4.1 Hypothesis

Estuaries are more invaded than adjacent coastal habitats because their communities offer less biotic resistance as a result of the dramatic anthropogenic alterations they have undergone. Thus native species may be less well adapted to the new conditions than are, by chance, some alien species.

33.4.4.2 General Explanation

Habitats that have been substantially altered by human activities may be particularly vulnerable to invasions by aliens because the environment has been so drastically altered that native species no longer enjoy a home court advantage in context-dependent interactions such as competition (Chap. 14, Byers). Estuaries are arguably the most altered aquatic ecosystems in the world (Edgar et al. 2000). In particular, excessive nutrient inputs have altered biogeochemical cycles, hydrological manipulations (diking, dredging, river diversion, etc.) have changed salinity levels and sedimentation rates, and addition of hard substrates (armored banks, harbor pilings, etc.) has caused a formerly rare habitat type to become quite common (Kennish 2002). Additionally, presence of previously established alien species can be considered an anthropogenically induced alteration, and these may facilitate the establishment and spread of other alien species (Simberloff and Von Holle 1999; Simberloff 2006). In general, open coasts have been far less altered than estuaries. Thus it is plausible that native estuarine invertebrates are less well adapted to current conditions than are their coastal counterparts, and therefore provide less biotic resistance to invasions (Wasson et al. 2005).

33.4.4.3 Support from Global Analysis

A few observations from our dataset support this hypothesis, although without replication the inference is weak. Humans have altered all the estuaries in the study, but those with lowest numbers of aliens (New Hampshire, South Africa) are those that have been less substantially altered, while those with highest numbers of aliens (California, the Netherlands) have been dramatically altered in terms of hydrology, pollution, and addition of hard substrates. In the Netherlands, open coast habitats have also been substantially altered (through hydrological manipulations and addition of hard substrates), and the number of aliens reported from them is higher than in the other seven regions in our study. These observations provide support for this hypothesis, though other explanations must still be invoked to explain why estuaries in the Netherlands are far more invaded than the open coast, since both have been highly altered.

33.4.4.4 Support from Other Observations

Previous studies have suggested that more disturbed marine habitats may be more invaded (e.g., Carlton 1979; Cohen and Carlton 1998; Byers 2002; Kennish 2002), but we do not know of studies with empirical data that compare estuarine and coastal invasions from this perspective.

33.4.4.5 Testable Predications

The most altered estuaries are typically also the ones that have the highest propagule pressure resulting from shipping and other vectors, which confounds testing of this hypothesis. It would be interesting to examine smaller estuaries without major shipping ports, to test the prediction that those with more alterations are more invaded than those with fewer ones. The same comparison could be carried out systematically for open coastal habitats with more vs fewer alterations.

To test the assumptions underlying this explanation, it would be important to determine whether native species indeed provide weaker biotic resistance (e.g., competition or predation) under altered conditions. Experiments to test consumption of resources under polluted vs more pristine conditions, or natural vs altered salinity regimes, would begin to address this question. One such example is Byers (2002) who examined whether alien species performed better under the altered vs natural conditions in the introduced range. He found that under low oxygen conditions (typical of eutrophic conditions) the survival rate of an alien snail (*Batillaria attramentaria*) was significantly higher than that of a native competitor, (*Cerithidea californica*). Such experiments could be complemented by an assessment of physical conditions in the home ranges of successful alien species. One would predict that home conditions would resemble those in the altered invaded range, e.g., a species from naturally nutrient-enriched estuaries should be likely to invade newly eutrophic estuaries elsewhere.

33.4.5 Estuaries Have More "Empty Niches"

33.4.5.1 Hypothesis

Estuaries are more invaded than adjacent coastal habitats because their communities offer less biotic resistance as a result of their lower native species richness related to their recent evolution or harsh conditions.

33.4.5.2 General Explanation

Alien species invading estuaries may face weaker negative biological interactions (competition, predation, disease) with native species than in adjacent open coast habitats, a concept often colloquially couched in terms of estuaries having more "empty niches" than the adjacent open coast (Wolff 1973, 1999; Carlton 1979; Cohen and Carlton 1998). This concept is particularly relevant to upper reaches of estuaries, where there is little representation by native marine species of the open coast, and where distinctive estuarine species – and often not many natives – are found. Conditions in these areas are physiologically challenging, with daily and seasonal fluctuation in salinity and temperature. Species richness is often low in harsh environments (Menge and Sutherland 1976; Nehring 2006), such as the upper reaches of estuaries. Species richness is also a function of habitat size, and estuarine habitats are much more rare than open coast habitats. Low species richness itself is not considered a predictor of low invasion rates at a regional scale – in terrestrial habitats, native and alien species richness are often positively correlated (Levine and D'Antonio 1999; Stohlgren et al. 2003; Chap. 12, Olyarnik et al.). In "mature" or "equilibrium" communities, it appears that the same factors that foster native richness (heterogeneity, moderate environmental conditions, etc.) also support alien species richness (Levine and D'Antonio 1999). However, it has been suggested that estuaries in some regions may not harbor as many species as they could accommodate, perhaps due to recent extinctions or due to insufficient time since their relatively recent geologic formation to allow for speciation or colonization by new species (Jones 1940; Hedgpeth 1968; Wolff 1971; Vermeij 1991). These "empty niches" would thus be available for alien species to fill.

33.4.5.3 Support from Global Analysis

This explanation may be supported by the contrast between invasion rates of the US Pacific and Atlantic coasts. Geologically, Pacific estuaries date back only to end of the last glaciation, while open coast habitats and their associated fauna have been continually present for much longer periods; in contrast, on the Atlantic coast, open coast habitats are geologically younger than most estuaries (Jones 1940; Hedgpeth 1968; Carlton 1979; Cohen and Carlton 1998; Ruiz et al. 2000; Emmett et al. 2000). The geologic youth of Pacific estuaries could explain their high numbers of aliens while the presence of a more ancient estuarine fauna on the Atlantic coast might explain why fewer alien species are established there. According to this hypothesis, one would also expect higher invasion rates on the open coast of the Atlantic than the Pacific. However, the absolute numbers of aliens reported from the coast of New Hampshire vs California are comparable – and very low – though the percentage of the fauna that is alien is markedly higher in New Hampshire (21%) vs California (1.5%).

33.4.5.4 Support from Other Observations

Data from San Francisco Bay suggest that alien species continue to accumulate, at a startling rate of one successful establishment every 14 weeks (Cohen and Carlton 1998); new invasions are apparently successful without displacing natives or earlier established aliens. This suggests that there were "empty niches" to be filled – that biotic interactions such as competition or predation were not intense enough to hamper establishment. This is in contrast to results from terrestrial habitats, where comparably high numbers of established aliens are typical in areas that are native biodiversity "hotspots" (e.g., California grasslands); there is a positive correlation between alien and native diversity at a broad regional scale (Stohlgren et al. 2003). Estuaries thus appear to provide an interesting example where the highest numbers of aliens are found in a habitat with the fewest native species (Wolff 1973, 1999). In contrast, the extremely low invasion rates (<2%; Wasson et al. 2005) of the Pacific coast in the same region as San Francisco Bay occur in one of the most species rich marine habitat types in temperate zones (though this pattern is also likely related to differing propagule pressure; see first hypothesis). Therefore, at least superficially in marine systems, invasion rates may be negatively correlated with native diversity, even if the underlying mechanisms relate more to "empty niches" and relatively depauperate communities than to low species numbers per se. However, along the US Pacific coast, numbers of both alien and native invertebrates decline from south to north (Ruiz et al. 2000; Townsend et al. 2000), refuting a simple inverse relationship between native and alien species numbers, and supporting instead the concept of "empty niches" as predictors of invasions. For terrestrial systems, several studies have examined the scale-dependence of relationships between species richness, diversity, and invasibility (Levine and D'Antonio 1999; Shea and Chesson 2002; Davies et al. 2005). However, for marine systems, examination of the scale-dependence of relationships between species richness, diversity, and invasions appears a fruitful area for further research.

33.4.5.5 Testable Predictions

In order to test the underlying assumption of this hypothesis, one could compare survival or growth of individuals of a broadly tolerant alien species in treatments excluding competitors or predators vs controls, in estuarine vs coastal habitats. If this hypothesis is supported, the alien species' fitness should be much more similar between caged treatment and control in the estuary than on the open coast.

To assess rigorously whether geologic age and evolutionary history indeed affect estuarine vs coastal invasion rates, one could perform more thorough, replicated analyses of the sorts carried out here, comparing numbers and proportions of aliens in regions where estuaries are geologically younger vs older than adjacent open coasts.

33.5 Directions for Future Research

As we prepared this global assessment of estuarine vs coastal invasion rates, it became clear how few consistently collected data are available for robust bioregional comparisons. Of the more than 70 invasion experts we contacted around the world, only 7 were able to provide data on both the target habitat types, and only 4 of those had data for both native and alien species. By far the most frequent response to our query was that data were only available for estuarine habitats, and only for aliens. This highlights the need for investigations that include multiple habitat types, and both native and alien species. An ideal sampling regime would be one that examines the same taxa, with the same methods and substrates (e.g., infaunal polychaetes sampled with benthic cores in low intertidal soft sediments; bryozoans on subtidal settlement plates; decapod crustaceans in traps), across habitat types and regions. Only with this sort of consistency can strong conclusions about habitat and regional differences be drawn. By sampling natives and aliens, invasion rates, not simply counts of aliens, can be analyzed. Finally, consistently sampling both within and outside of harbors in estuaries vs open coasts would allow for testing of hypotheses invoking differential introduction vs establishment success in these areas. Establishment of alien species outside harbors is also of much greater conservation concern than invasion of harbors; hence, focus on non-harbor habitats would be welcome from this perspective. To test among the five hypotheses outlined above, care must be taken to design correlative or manipulative experiments that avoid confounding of different causal factors. In particular, it is critical to separate the effects of differential propagule pressure (the first hypothesis) and habitat-matching (the second hypothesis) from other factors affecting establishment and spread (the remaining three hypotheses).

33.6 Conclusions

The global dataset compiled from eight temperate regions revealed that overall, estuaries harbor more alien invertebrate species than adjacent open coasts, but the magnitude of this difference varies regionally. The majority of the 198 alien species reported from all sites occur in at least some regions in estuaries; only 9 of them occur solely in coastal habitats. Four datasets that documented both native and alien species revealed that the proportion of the invertebrate fauna surveyed that is alien is generally higher in estuaries than on coasts. This trend was pronounced in Central California, weaker in Argentina and the Netherlands, and absent in New Hampshire.

Multiple mechanisms may account for the differences in invasion rates between estuaries and adjacent coasts. Propagule pressure is likely higher in estuaries, and estuarine species that are transported are more likely to establish in estuarine than coastal regions in the recipient regions. Establishment may also be higher in estuaries due to higher retention of dispersive stages and lower levels of negative biotic interaction, due to natives being poorly adapted to anthropogenically altered conditions in estuaries or due to the presence of "empty niches" in relatively depauperate and geologically young estuarine assemblages. Rigorous experimental studies and consistent biogeographic comparisons should be carried out to test these and other hypotheses about habitat differences in marine invasion rates. The answers derived from such studies would inform management practices and control strategies for alien species, which are one of the biggest threats to the integrity of marine and estuarine ecosystems.

Acknowledgements We are deeply indebted to Jeff Crooks, Pascale Garcia-Meunier, Oscar Iribarne, Dan Minchin, John Pearse, and Tammy Robinson who provided key datasets of estuarine vs coastal fauna. In particular, we thank Dan Minchin for providing many highly valued comments and suggestions. We are also grateful to Gonzalo Castillo, John Chapman, Eric Danner, Stephan Gollasch, Charlie Griffiths, Hiram Li, Vadim Panov, Maria Røjbek, and Thierry Vincent for supplying additional data and perspectives on estuarine vs coastal invasions. Finally, we thank the editors and two anonymous reviewers who also provided thoughtful comments that improved the chapter. Most importantly, we thank Gil Rilov and Jeff Crooks for the invitation to write this chapter and for suggesting we attempt a global synthesis of coastal vs estuarine differences in marine invasions.

References

Andréfouët S, Zubia M, Payri C (2004) Mapping and biomass estimation of the invasive brown algae *Turbinaria ornate* (Turner) J. Agardh and *Sargassum mangarevense* (Grunow) Setchell on heterogeneous Tahitian coral reefs using 4-meter resolution IKONOS satellite data. Coral Reefs 23(1):26–38

Byers JE (1999) The distribution of an introduced mollusk and its role in the long-term demise of a native confamilial species. Biol Invas 1(4):339–353

Byers JE (2002) Impact of non-indigenous species on natives enhanced by anthropogenic alteration of selection regimes. Oikos 97(3):449–458

Campbell ML, Hewitt CL (1999) Marine biological invasions of Port Phillip Bay, Victoria. CRIMP Technical Report Number 20. CSIRO Marine Research, Hobart, Australia, p 45

Carlton JT (1979) History, biogeography, and ecology of the introduced marine and estuarine invertebrates of the Pacific coast of North America. Ph.D. Dissertation, Ecology, University of California, Davis

Carlton JT (1985) Transoceanic and interoceanic dispersal of coastal marine organisms: the biology of ballast water. Oceanogr Mar Biol Annu Rev 23:313–371

Carlton JT (1989) Man's role in changing the face of the ocean: biological invasions and implications for conservation of near-shore environments. Conserv Biol 3(3):265–273

Carlton JT (1996) Pattern, process and prediction in marine invasion ecology. Biol Conserv 78:97–106

Carlton JT, Geller JB (1993) Ecological roulette: the global transport of nonindigenous marine organisms. Science 261:78–82

Cohen AN, Carlton JT (1995) Nonindigenous aquatic species in a United States estuary: a case study of the biological invasions of the San Francisco bay and delta. A report for the United States Fish and Wildlife Service, Washington D.C. and The National Sea Grant College Program Connecticut Sea Grant (NOAA Grant Number NA36RG0467)

Cohen AN, Carlton JT (1998) Accelerating invasion rate in a highly invaded estuary. Science 279:555–558

Cohen AN, Carlton JT, Fountain MC (1995) Introduction, dispersal and potential impacts of the green crab, *Carcinus maenas* in San Francisco Bay, California. Mar Biol 122:225–237

Davies KF, Chesson P, Harrison S, Inouye BD, Melbourne BA, Rice KJ (2005) Spatial heterogeneity explains the scale dependence of the native-exotic diversity relationship. Ecology 86:1602–1610

Dolgikh AV (1969) Invasion of Hydrobia-ventrosa of the Tiligul estuary by microphallidae larvae. Vestn Zool 3(4):32–35

Edgar GJ, Barrett NS, Graddon DJ, Last PR (2000) The conservation significance of estuaries: a classification of Tasmanian estuaries using ecological, physical and demographic attributes as a case study. Biol Conserv 92:383–397

Elton CS (1958) The ecology of invasions by animals and plants. Methuen, London, UK

Emmett R, Llansó R, Newton J, Thom R, Hornberger M, Morgan C, Levings C, Copping A, Fishman P (2000) Geographic signatures of North American West coast estuaries. Estuaries 23(6):765–792

Grosholz ED, Ruiz GM (1995) Spread and potential impact of the recently introduced European green crab, *Carcinus maenas*, in central California. Mar Biol 122:239–247

Hedgpeth JW (1968) Newcomers to the Pacific coast: the estuarine itinerants. In: Ricketts EF, Calvin J, Hedgpeth JW (eds) Between Pacific tides, 4th edn. Stanford University Press, Stanford, pp 376–380

Hewitt CL, Campbell ML, Thresher RE, Martin RB (1999) Marine biological invasions of Port Phillip Bay, Victoria. Crimp technical report number 20. December 1999

Jones LL (1940) An introduction of the Atlantic crab into San Francisco Bay. Proceedings of the 6th Pacific Scientific Congress, vol 3, pp 485–486

Kennish MJ (2002) Environmental threats and environmental future of estuaries. Environ Conserv 29(1):78–107

Lee H II, Thompson B, Lowe S (2003) Estuarine and scalar patterns of invasion in the soft-bottom benthic communities of the San Francisco Estuary. Biol Invas 5:85–102

Levine JM, D'Antonio CM (1999) Elton revisited: a review of evidence linking diversity and invasibility. Oikos 87:15–26

Little C (2000) The biology of soft shores and estuaries. Oxford University Press, New York, NY, USA, p 133

Lodge DM (1993) Biological invasions: lessons for ecology. Trend Ecol Evol 8:133–137

Menge BA, Sutherland JP (1976) Species diversity gradients synthesis of the roles of predation competition and temporal heterogeneity. Am Nat 110(973):351–369

Minchin D, Eno C (2002) Exotics of coastal and inland waters of Ireland and Britain. In: Leppäkoski E, Gollasch S, Olenin S (eds) Invasive aquatic species of Europe: distribution, impact and management. Kluwer, Dordrecht, pp 267–275
Nehring S (2006) Four arguments why so many alien species settle into estuaries, with special reference to the German river Elbe. Helgo Mar Res 60:127–134
Reise K, Gollasch S, Wolff WJ (1999) Introduced marine species of the North Sea coasts. Helgo Meeresunters 52:219–234
Ruiz GM, Carlton JT, Grosholz ED, Hines AH (1997) Global invasions of marine and estuarine habitats by non-indigenous species: mechanisms, extent, and consequences. Am Zool 37:621–632
Ruiz GM, Fofonoff PW, Carlton JT, Wonham MJ, Hines AN (2000) Invasion of coastal marine communities in North America: apparent patterns, processes, and biases. Annu Rev Ecol Evol Syst 31:481–531
Schwindt E, Francesco CGD, Iribarne OO (2004) Individual and reef growth of the invasive reef-building polychaete *Ficopomatus enigmaticus* in a south-western Atlantic coastal lagoon. J Mar Biol Assoc U K 84(5):987–993
Shea K, Chesson P (2002) Community ecology theory as a framework for biological invasions. Trend Ecol Evol 17:170–176
Simberloff D (2006) Invasional meltdown 6 years later: important phenomenon, unfortunate metaphor, or both? Ecol Lett 9:912–919
Simberloff D, Von Holle B (1999) Positive interactions of nonindigenous species: invasional meltdown? Biol Invas 1:21–32
Steffani CN, Branch GM (2003) Growth rate, condition, and shell shape of *Mytilus galloprovincialis*: responses to wave exposure. Mar Ecol Prog Ser 246:197–209
Stohlgren TJ, Barnett DT, Kartesz JT (2003) The rich get richer: patterns of plant invasions in the United States. Front Ecol Environ 1(1):11–14
Townsend CR, Harper JL, Begon M (2000) Essentials of ecology. Blackwell Science, Malden, MA, USA, pp 552
Tyrrell MC (2002) Impacts of the introduced crab species, *Carcinus maenas* and *Hemigrapsus sanguineus,* in northern New England. Ph.D. Dissertation, University of New Hampshire
Vermeij GJ (1991) When Biotas meet: understanding biotic interchange. Science 253:1099–1104
Wasson K, Zabin CJ, Bedinger L, Diaz MC, Pearse JS (2001) Biological invasions of estuaries without international shipping: the importance of intraregional transport. Biol Conserv 102:143–153
Wasson K, Fenn K, Pearse J (2005) Habitat differences in marine invasions of central California. Biol Invas 7:935–948
Wolff WJ (1971) Origin and history of the brackish-water fauna of NW Europe. Proc 5th European Mar Biol Symposium. Piccin Editore, Padova, pp 11–18
Wolff WJ (1973) The estuary as a habitat. An analysis of data on the soft-bottom macrofauna of the estuarine area of the rivers Rhine, Meuse, and Scheldt. Zool Verh Leiden 126:1–242
Wolff WJ (1999) Exotic invaders of the meso-oligohaline zone of estuaries in the Netherlands: why are there so many? Helgo Meeresunters 52:393–400
Wolff WJ (2005) Non-indigenous marine and estuarine species in the Netherlands. Zool Meded Leiden 79:1–116

Section VII
Concluding Thoughts

Chapter 34
Future Directions For Marine Invasions Research

Jeffrey A. Crooks and Gil Rilov

34.1 Introduction

The chapters in this book have synthesized some of the significant advances made related to patterns and processes of marine invasions. In so doing, they have also highlighted areas in which we need to learn more about the causes, consequences, and management of invasions. Below, we emphasize some specific points to consider as the field of marine invasion biology matures. For this, we draw upon specific points and general themes developed in the book, as well as suggestions presented in the rapidly expanding invasion science literature (e.g., Mooney and Hobbs 2000; Carlton 2001; Meliane and Hewitt 2005; Mooney et al. 2005; Simberloff et al. 2005). The suggestions made below are broken down by discipline, but, because of the very nature of invasion biology, many are in fact cross-cutting.

34.2 Scientific Considerations

A most important goal for marine invasion ecology should be to increase predictability of both invasions and their impacts on the abiotic environment, ecological communities, and humans. Such predictability is not only the hallmark of sound science; it will improve our ability to manage invasions. Although progress is being made in this regard (e.g., Hayes and Barry 2008), the science of invasions (and even more so the science of marine invasions) is still in its infancy. This stems from the relevantly recent coalescence of ideas central to biological invasions, but is also due in large part to the fact that most invasions themselves are also young. Although the rates at which different invasion processes can occur are highly variable, we know that many ecological and evolutionary effects can take decades, centuries, or millennia to manifest themselves. It will likely be many years before the consequences of the recent onslaught of invasion can be fully appreciated. The relative youth of most invasions also indicates that we are in a period of rapid change, and

G. Rilov, J.A. Crooks (eds.) *Biological Invasions in Marine Ecosystems.*
Ecological Studies 204,

there is thus a need for more explicit consideration of the temporal aspects of invasions (Crooks 2005; Strayer et al. 2006).

Another key goal of marine invasion science should be to more completely describe global patterns of marine invasion (Ruiz and Crooks 2001; Meyerson and Mooney 2007). As highlighted in this book, regional pictures of marine invasions are starting to emerge, but the depth and breadth of these studies is highly variable. We also still have huge gaps in knowledge. For example, there is relatively little information in international publications on the extent of marine bioinvasions along much of the Asian coast, and the majority of information on invasions in Africa comes from South Africa. The goal of obtaining a better picture of invasions will be aided by emerging technologies such as molecular tools (Holland 2000; Darling and Blum 2007), but old-fashioned approaches, such as taxonomy and natural history, must not fall out of favor (Chap. 2, Carlton). Part of the focus on assessing invasion patterns should also be quantifying impact, as indicated time and again in this book. These impact assessments will offer a better understanding of invader roles and, where appropriate, provide more persuasive tools for reaching out to the public and decision-makers regarding potential problems associated with invaders. We also urge investigators to examine invader interactions at broad, biogeographic scales, such as initiating experimental-comparative approaches with global invaders to examine processes when the same species encounters different physical and biological conditions.

Although understanding invasions in the sea is important in its own right, it is also of interest to place marine invasions into broader contexts so that unifying principles can be advanced. Therefore, marine invasions should be placed in the context of invasions in general, and still more broadly, the exchange of ideas between invasion biology as a whole and other ecological, biogeographical, and evolutionary disciplines will be fruitful (Sax et al. 2005; Cadotte et al. 2006). There is also a pressing need to study marine invasions as a cause and a consequence of anthropogenic global change (Occhipinti-Ambrogi 2007; Chap. 3, Lonhart), which can be accomplished by working closely with climatologists and physical oceanographers. In addition, it will be useful for invasion biologists to work with professionals in other applied fields. For example, we need to investigate further the role of invaders in Marine Protected Areas (Byers 2005), quantify the ecosystem services impacted (or provided) by marine invaders, and better couple economic principles with invader success and management (Levine and D'Antonio 2003; Batabyal 2007; Chap. 4, Wonham and Lewis).

34.3 Management Considerations

A common axiom of invader management, and one that we will repeat here, is that it is typically far easier to prevent an invader from getting in than controlling it once it has arrived (Bax et al. 2001). This is especially true in marine ecosystems, where the nature of the medium makes control and eradication efforts all the more difficult. In order to achieve effective vector control, we must continue to develop and implement approaches for dealing with the many vectors of marine invasions, such as new techniques for treating ballast water and hull fouling (Chaps. 6 and 18, Hewitt et al.). These efforts must be backed by effective policy for addressing invasions in the sea (Chap. 19, Hewitt et al.). In particular, international participation will be critical, given the global nature of the problem (e.g., Gollasch 2007).

Although a variety of potential tools are available for managing established (or establishing) invaders (Bax et al. 2001; Crombie et al. 2007; Chap. 18, Hewitt et al.), the implementation of such efforts has been much more limited in marine systems than on land due in large part to the aforementioned difficulty of working in the sea. However, progress can be made, and we urge for continued efforts to eradicate or control marine species, coupled with clear and thorough reporting of both successes and failures in these efforts. For example, adopting ecosystem approaches for marine invasion management seems especially fruitful (Meliane and Hewitt 2005), such as by improving the ability of ecosystems to resist invasion intrinsically by bolstering native species diversity and/or improving environmental quality. However, we need more real-world implementation and documentation to understand how effective this tool might actually be. Another important part of this effort will be to determine what further restoration actions will be needed if and when invaders are actually removed, which will be especially important for invaders with strong legacy effects such as habitat modification (Crooks 2002; Chap. 21, Hacker and Dethier). More broadly, a fuller consideration of the roles of non-native marine species (including wetland invaders) in restoration is needed, including cases where exotics are major problems as well as when they might actually be utilized effectively to achieve desired ecosystem goals (Ewel and Putz 2004).

Underlying all these efforts is the need for information on what is happening in the environment and how our actions are affecting desired outcomes. These must be based on robust monitoring programs (Ruiz and Crooks 2001; Wasson et al. 2002; Campbell et al. 2007). For example, in the case of ballast water management, the ultimate goal is ultimately not how much ballast water is treated or exchanged; it is decreasing the rate of invasion into receiving waters (Ruiz and Carlton 2003). This can only be assessed by examining species in the environment. Such monitoring is also central to much-needed efforts to detect and rapidly respond to invader incursion events, preferably soon after invader arrival and before any lag times in populations increase end. Monitoring will also provide valuable information for science-based forecasting, risk-assessment, and decision-making (Bax et al. 2001; Campbell et al. 2007; Chap. 20, Campbell).

34.4 Social Considerations

Although the invasion of species into ecosystems is inherently complex, the role of the invasion scientist can be equally complicated. Those working with invasions often straddle different realms, from objective purveyors of facts and figures, to environmental advocates with decidedly value-based positions, to ecosystem managers that often balance human and natural needs. Invasion biology needs continued discourse on the benefits and pitfalls of acting in any, some, or all of these arenas (Coates 2006; Larson 2007).

It is clear that, whatever the role of the invasion biologist, communication is key. We are well-served by garnering the inherent public interest in the topic and translating this into science-based education and outreach. This could include engaging citizen scientists in detection and management efforts (Delaney et al., in press), teaching invasion science in formal classroom settings, and training decision-makers (Meliane and Hewitt 2005). In so doing, we must address some potentially thorny issues head on. For example, what makes some invaders tolerable or even desirable (consider the great concern currently associated with the collapse of alien honeybee populations in North America), while others merit extermination for the "greater good?" Also, why, when there is such a public discourse on the value of biodiversity, do we worry about diversity increases wrought by the presence of invaders (e.g., Galil 2007)? Discussions centered around topics such as these can also serve to highlight broader issues that exist at the interface between science and action, such as ecosystem function (including the provision of ecosystem services), the importance of considering spatial and temporal scales, and the precautionary principle.

34.5 Conclusions

Fifty years ago, in his seminal volume, Charles Elton (Elton 1958) described an ecological explosion in the world around us. Therein he began to paint a picture of how biological invasions were reshaping a seemingly "inviolate" sea. We continue to paint that picture today, and it reveals a largely-unchecked explosion that has left the sea far different from than that found when Elton's book was first published. Hopefully, through continued work on the ecology and management of marine invaders, the seas 50 years hence will not seem so different from those of the past.

References

Batabyal AA (2007) International aspects of invasive species management: a research agenda. Stoch Environ Res Risk Assessment 21:717–727

Bax N, Carlton JT, Mathews-Amos A, Haedrich RL, Howarth FG, Purcell JE, Rieser A, Gray A (2001) The control of biological invasions in the world's oceans. Conserv Biol 15:1234–1246

Byers JE (2005) Marine reserves enhance abundance but not competitive impacts of a harvested nonindigenous species. Ecology 86:487–500

Cadotte MW, McMahon SM, Fukami T (2006) Conceptual ecology and invasion biology: reciprocal approaches to nature. Springer, Dordrecht, The Netherlands

Campbell ML, Gould B, Hewitt CL (2007) Survey evaluations to assess marine bioinvasions. Mar Pollut Bull 55:360–378

Carlton JT (2001) Introduced species in US coastal waters: environmental impacts and management priorities. Pew Oceans Commission, Arlington, Virginia

Coates P (2006) American perceptions of immigrant and invasive species: strangers on the land. University of California Press, Berkeley, California
Crombie J, Knight E, Barry S (2007) Marine pest incursions – a tool to predict the cost of eradication based on expert assessments. Commonwealth of Australia–Bureau of Rural Sciences, Canberra, Australia
Crooks JA (2002) Characterizing the consequences of invasions: the role of introduced ecosystem engineers. Oikos 97:153–166
Crooks JA (2005) Lag times and exotic species: the ecology and management of biological invasions in slow motion. Ecoscience 12:316–329
Darling JA, Blum MJ (2007) DNA-based methods for monitoring invasive species: a review and prospectus. Biol Invas 9:751–765
Delaney DG, Sperling CD, Adams CS, Leung B (in press) Marine invasive species: validation of citizen science and implications for national monitoring networks. Biol Invas
Elton CS (1958) The ecology of invasions by plants and animals. Chapman and Hall, New York
Ewel JJ, Putz FE (2004) A place for alien species in ecosystem restoration. Front Ecol Environ 2:354–360
Galil B (2007) Loss or gain? Invasive aliens and biodiversity in the Mediterranean Sea. Mar Pollut Bull 55:314–322
Gollasch S (2007) International collaboration on marine bioinvasions – the ICES response. Mar Pollut Bull 55:353–359
Hayes KR, Barry SC (2008) Are there consistent predictors of invasion success? Biol Invas 10:483–506
Holland BD (2000) Genetics of marine bioinvasions. Hydrobiologia 420:63–71
Larson BMH (2007) An alien approach to invasive species: objectivity and society in invasion biology. Biol Invas 947–956
Levine JM, D'Antonio CM (2003) Forecasting biological invasions with increasing international trade. Conserv Biol 17:322–326
Meliane I, Hewitt CL (2005) Gaps and priorities in addressing marine invasive species. IUCN Global Marine Programme, Gland, Switzerland
Meyerson LA, Mooney HA (2007) Invasive alien species in an era of globalization. Front Ecol Environ 5:199–208
Mooney HA, Hobbs RJ (2000) Invasive species in a changing world. Island Press, Washington
Mooney HA, Mack RN, McNeeley JA, Neville LE, Schei PJ, Waage JK (2005) Invasive alien species. A new synthesis. Island Press, Washington
Occhipinti-Ambrogi A (2007) Global change and marine communities: alien species and climate change. Mar Pollut Bull 55:342–352
Ruiz GM, Carlton JT (2003) Invasion vectors: a conceptual framework for managment. In: Ruiz GM, Carlton JT (eds) Invasive species. Vectors and management strategies. Island Press, Washington, pp 459–504
Ruiz GM, Crooks JA (2001) Biological invasions of marine ecosystems: patterns, effects, and management. In: Bendell-Young L, Gallaugher P (eds) Waters in Perl. Kluwer Academic Publishers, Norwell, Massachusetts, pp 3–18
Sax DF, Stachowicz JJ, Gaines SD (2005) Species invasions: insights into ecology, evolution, and biogeography. Sinauer Associates, Sunderland MA, USA
Simberloff D, Parker IM, Windle PN (2005) Introduced species policy, management, and future research needs. Front Ecol Environ 3:12–20
Strayer DL, Eviner VT, Jeschke JM, Pace ML (2006) Understanding the long-term effects of species invasions. Trend Ecol Evol 21:645–651
Wasson K, Lohrer D, Crawford M, Rumrill S (2002) Non-native species in our nation's estuaries: a framework for an invasion monitoring program. Nat Estuarine Res Reserve Tech Rep Ser 2002:1

Subject Index

Taxonomic Index

N

O

R

S

T

Printing: Krips bv, Meppel, The Netherlands
Binding: Stürtz, Würzburg, Germany